STUDENT'S SOLUTIONS MANUAL

DAVID ATWOOD
Rochester Community and Technical College

TERRY KRIEGER
Rochester Community and Technical College

with
THOMAS KRIEGER

ALGEBRA AND TRIGONOMETRY WITH MODELING & VISUALIZATION

FOURTH EDITION

PRECALCULUS WITH MODELING & VISUALIZATION

FOURTH EDITION

Gary Rockswold
Minnesota State University, Mankato

Addison-Wesley
is an imprint of

Reproduced by Pearson Addison-Wesley from electronic files supplied by the author.

ISBN-13: 978-0-321-57709-2
ISBN-10: 0-321-57709-4

1 2 3 4 5 6 EB 12 11 10 09

Addison-Wesley
is an imprint of

www.pearsonhighered.com

Preface

This solutions manual is written as an aid for students studying the text *Algebra and Trigonometry with Modeling and Visualization, 4/e,* by Gary Rockswold. It contains answers and solutions to the odd-numbered exercises.

This manual is consistent with the text and gives solutions that are accessible to college algebra students. The solutions are written in a clear format with hundreds of graphs and tables to help with the learning process. Solutions frequently include not only symbolic solutions but also graphical and numerical solutions to help students understand the problem completely. It is recommended that you make a genuine attempt to solve a problem before looking in this solutions manual for help. Regular class attendance is also recommended. Following these suggestions will promote insight and understanding of college algebra.

It is the author's hope that you will find this manual helpful when studying the text *Algebra and Trigonometry with Modeling and Visualization, 4/e*. Please feel free to send comments to either of the email addresses below. Your opinion is important. Best wishes for an enjoyable and successful college algebra and trigonometry course.

Terry Krieger

terry.krieger@roch.edu

David Atwood

david.atwood@roch.edu

Table of Contents

Chapter 1: Introduction to Functions and Graphs

1.1: Numbers, Data, and Problem Solving

1. $\frac{21}{24}$ is a real and rational number.

3. 7.5 is a real and rational number.

5. $90\sqrt{2}$ is a real number.

7. Natural number: $\sqrt{9} = 3$; integers: -3, $\sqrt{9}$; rational numbers: $-3, \frac{2}{9}, \sqrt{9}, 1.\overline{3}$; irrational numbers: $\pi, -\sqrt{2}$

9. Natural number: None; integer: $-\sqrt{4} = -2$; rational numbers: $\frac{1}{3}, 5.1 \times 10^{-6}, -2.33, 0.\overline{7}, -\sqrt{4}$; irrational number: $\sqrt{13}$

11. Shoe sizes are normally measured to within half sizes. Rational numbers are most appropriate.

13. Gasoline is usually measured to a fraction of a gallon using rational numbers.

15. Temperature is typically measured to the nearest degree in a weather forecast. Since temperature can include negative numbers, the integers would be most appropriate.

17. $|5 - 8 \cdot 7| = |5 - 56| = |-51| = 51$

19. $-6^2 - 3(2 - 4)^4 = -6^2 - 3(-2)^4 = -36 - 3(16) = -36 - 48 = -84$

21. $\sqrt{9 - 5} - \frac{8 - 4}{4 - 2} = \sqrt{4} - \frac{4}{2} = 2 - 2 = 0$

23. $\sqrt{13^2 - 12^2} = \sqrt{169 - 144} = \sqrt{25} = 5$

25. $\frac{4 + 9}{2 + 3} - \frac{-3^2 \cdot 3}{5} = \frac{13}{5} - \frac{-27}{5} = \frac{40}{5} = 8$

27. $-5^2 - 20 \div 4 - 2 = -25 - 5 - 2 = -32$

29. $184{,}800 = 1.848 \times 10^5$

31. $0.04361 = 4.361 \times 10^{-2}$

33. $2450 = 2.45 \times 10^3$

35. $0.56 = 5.6 \times 10^{-1}$

37. $-0.0087 = -8.7 \times 10^{-3}$

39. $206.8 = 2.068 \times 10^2$

41. $10^{-6} = 0.000001$

43. $2 \times 10^8 = 200{,}000{,}000$

45. $1.567 \times 10^2 = 156.7$

47. $5 \times 10^5 = 500{,}000$

49. $0.045 \times 10^5 = 4500$

51. $67 \times 10^3 = 67{,}000$

53. $(4 \times 10^3)(2 \times 10^5) = 4 \cdot 2 \times 10^{3+5} = 8 \times 10^8$; $800{,}000{,}000$

55. $(5 \times 10^2)(7 \times 10^{-4}) = 5 \cdot 7 \times 10^{2-4} = 35 \times 10^{-2} = 3.5 \times 10^{-1}$; 0.35

57. $\dfrac{6.3 \times 10^{-2}}{3 \times 10^{1}} = \dfrac{6.3}{3} \times 10^{-2-1} = 2.1 \times 10^{-3}; \ 0.0021$

59. $\dfrac{4 \times 10^{-3}}{8 \times 10^{-1}} = \dfrac{4}{8} \times 10^{-3-(-1)} = 0.5 \times 10^{-2} = 5 \times 10^{-3}; \ 0.005$

61. $\dfrac{8.947 \times 10^{7}}{0.00095}(4.5 \times 10^{8}) \approx 42381 \times 10^{15} = 4.2381 \times 10^{19} \approx 4.24 \times 10^{19}$

63. $\left(\dfrac{101 + 23}{0.42}\right)^{2} + \sqrt{3.4 \times 10^{-2}} \approx 87166 + 0.2 \approx 87166.2 \approx 8.72 \times 10^{4}$

65. $(8.5 \times 10^{-5})(-9.5 \times 10^{7})^{2} = (8.5 \times 10^{-5})(9.025 \times 10^{15}) \approx 76.7 \times 10^{10} = 7.67 \times 10^{11}$

67. $\sqrt[3]{192} \approx 5.769$

69. $|\pi - 3.2| \approx 0.058$

71. $\dfrac{0.3 + 1.5}{5.5 - 1.2} \approx 0.419$

73. $\dfrac{1.5^{3}}{\sqrt{2} + \pi - 5} \approx \dfrac{3.375}{-2.732} \approx -1.235$

75. $15 + \dfrac{4 + \sqrt{3}}{7} \approx 15.819$

77. $\dfrac{13 - 8}{8} \times 100 = \dfrac{5}{8} \times 100 = 62.5; \ 62.5\%$

79. $\dfrac{0.85 - 1.4}{1.4} \times 100 = \dfrac{-0.55}{1.4} \times 100 \approx -39.3; \ -39.3\%$

81. $\dfrac{B - 100}{100} \times 100 = 6 \ \Rightarrow \ B - 100 = 6 \Rightarrow B = \$106;$

$\dfrac{B - 106}{106} \times 100 = -6 \Rightarrow (B - 106)\left(\dfrac{50}{53}\right) = -6 \Rightarrow B - 106 = -6.36 \Rightarrow B = \$99.64$

83. $0.14 = 1.4 \times 10^{-1}$ watt

85. The distance Mars travels around the sun is $2\pi r = 2\pi(141{,}000{,}000) \approx 885{,}929{,}128$ miles.

The number of hours in 1.88 years is $365 \times 1.88 \times 24 \approx 16{,}469$ hours. So Mars' speed is

$\dfrac{885{,}929{,}128}{16{,}469} \approx 53{,}794$ miles per hour.

87. (a) $\dfrac{3.7 \times 10^{11}}{2.03 \times 10^{8}} \approx \$1{,}820$ per person

(b) $\dfrac{5.54 \times 10^{12}}{2.81 \times 10^{8}} \approx \$19{,}715$ per person

89. The area of the film is πr^{2} or $\pi(11.5)^{2} \approx 415.5$ cm^{2} and the volume of the drop is 0.12 cm^{3}. The thickness of

the film is equal to the volume divided by the area. $\dfrac{0.12}{415.5} \approx 2.9 \times 10^{-4}$ cm.

91. (a) It would take $\dfrac{5.54 \times 10^{12}}{100} \approx 5.54 \times 10^{10}$ or 55.4 billion \$100-dollar bills to equal the federal debt. The

height of the stacked bills would be $\dfrac{5.54 \times 10^{10}}{250} \approx 2.216 \times 10^{8}$ inches or $\dfrac{2.216 \times 10^{8}}{12} \approx 18{,}466{,}667$ feet.

(b) There are 5280 feet in one mile, so the stacked bills would span $\dfrac{18{,}466{,}667}{5280} \approx 3497$ miles. It would reach

farther than the distance between Los Angles and New York.

93. (a) $V = \pi r^2 h \Rightarrow V = \pi(1.3)^2(4.4) \Rightarrow V = 7.436\pi \approx 23.4 \text{ in}^3$

(b) $1 \text{ in}^3 = 0.55$ fluid ounce $\Rightarrow 23.4 \cdot 0.55 = 12.87$ fluid ounces; Yes, it can hold 12 fluid ounces.

95. $V = l \cdot w \cdot h \Rightarrow 125 = (100)(5)h \Rightarrow h = 0.25$ ft or 3 in.

1.2: Visualizing and Graphing Data

1. (a)

(b) Maximum: 6; minimum: -2

(c) $\dfrac{3 + (-2) + 5 + 0 + 6 + (-1)}{6} = \dfrac{11}{6} = 1.8\overline{3}$

3. (a)

(b) Maximum: 30; minimum: -20

(c) $\dfrac{-10 + 20 + 30 + (-20) + 0 + 10}{6} = 5$

5.

-30	-30	-10	5	15	25	45	55	61

(a) The maximum is 61 and the minimum is -30.

(b) The mean is $\dfrac{-30 - 30 - 10 + 5 + 15 + 25 + 45 + 55 + 61}{9} \approx 15.11$ and the median is 15.

7. $\sqrt{15} \approx 3.87, 2^{2.3} \approx 4.92, \sqrt[3]{69} \approx 4.102, \pi^2 \approx 9.87, 2^{\pi} \approx 8.82, 4.1$

$\sqrt{15}$	4.1	$\sqrt[3]{69}$	$2^{2.3}$	2^{π}	π^2

(a) The maximum is π^2 and the minimum is $\sqrt{15}$.

(b) The mean is $\dfrac{\sqrt{15} + 4.1 + \sqrt[3]{69} + 2^{2.3} + 2^{\pi} + \pi^2}{6} \approx 5.95$ and the median is $\dfrac{\sqrt[3]{69} + 2^{2.3}}{2} \approx 4.51$.

9. (a)

(b) Mean $= \dfrac{31.7 + 22.3 + 12.3 + 26.8 + 24.9 + 23.0}{6} = 23.5$; Median $= \dfrac{23.0 + 24.9}{2} = 23.95$. The

average area of the six largest freshwater lakes is 23,500 square miles. Half of the lakes have areas larger

than 23,950 square miles and half have less. The largest difference in area between any two lakes is 19,400

square miles.

(c) The freshwater lake with the largest area is Lake Superior.

11. *Answers may vary.* 16, 18, 26; No

13. $d = \sqrt{(5-2)^2 + (2-(-2))^2} = \sqrt{3^2 + 4^2} = \sqrt{25} = 5$

15. $d = \sqrt{(9-7)^2 + (1-(-4))^2} = \sqrt{2^2 + 5^2} = \sqrt{29} \approx 5.39$

17. $d = \sqrt{(-2.1 - 3.6)^2 + (8.7 - 5.7)^2} = \sqrt{(-5.7)^2 + 3^2} = \sqrt{41.49} \approx 6.44$

19. $d = \sqrt{(-3 - (-3))^2 + (10 - 2)^2} = \sqrt{0^2 + 8^2} = \sqrt{64} = 8$

21. $d = \sqrt{\left(\frac{3}{4} - \frac{1}{2}\right)^2 + \left(\frac{1}{2} - \left(-\frac{1}{2}\right)\right)^2} = \sqrt{\left(\frac{1}{4}\right)^2 + 1^2} = \sqrt{\frac{1}{16} + 1} = \sqrt{\frac{17}{16}} = \frac{\sqrt{17}}{4} \approx 1.03$

23. $d = \sqrt{\left(-\frac{1}{10} - \frac{2}{5}\right)^2 + \left(\frac{4}{5} - \frac{3}{10}\right)^2} = \sqrt{\left(-\frac{1}{2}\right)^2 + \left(\frac{1}{2}\right)^2} = \sqrt{\frac{1}{4} + \frac{1}{4}} = \sqrt{\frac{1}{2}} = \frac{\sqrt{2}}{2} = 0.71$

25. $d = \sqrt{(-30 - 20)^2 + (-90 - 30)^2} = \sqrt{(-50)^2 + (-120)^2} = \sqrt{2500 + 14{,}400} = \sqrt{16{,}900} = 130$

27. $d = \sqrt{(0 - a)^2 + (-b - 0)^2} = \sqrt{(-a)^2 + (-b)^2} = \sqrt{a^2 + b^2}$

29. $d = \sqrt{(3 - 0)^2 + (4 - 0)^2} = \sqrt{3^2 + 4^2} = \sqrt{9 + 16} = \sqrt{25} = 5$

    $d = \sqrt{(7 - 3)^2 + (1 - 4)^2} = \sqrt{4^2 + (-3)^2} = \sqrt{16 + 9} = \sqrt{25} = 5$

    The side between (0,0) and (3,4) and the side between (3,4) and (7,1) have equal length, so the triangle is isosceles.

31. (a) See Figure 31.

    (b) $d = \sqrt{(0 - (-40))^2 + (50 - 0)^2} = \sqrt{40^2 + 50^2} = \sqrt{1600 + 2500} = \sqrt{4100} \approx 64.0$ miles.

Figure 31

33. Use the midpoint formula: $M = \left(\dfrac{1980 + 2000}{2}, \dfrac{77.4 + 79.5}{2}\right) = (1990, 78.45)$. According to the midpoint estimate the life expectancy was 78.45 years. This compares favorably to the actual life expectancy of 78.8 years.

35. Assuming the distance of 0 meters requires 0 seconds to run, the midpoint between the data points (0, 0) and (200, 20) is equal to $M = \left(\dfrac{0 + 200}{2}, \dfrac{0 + 20}{2}\right) = (100, 10)$. According to the midpoint estimate the time required to run 100 meters is 10 seconds or half the time required to run 200 meters.

37. $M = \left(\dfrac{1 + 5}{2}, \dfrac{2 + (-3)}{2}\right) = (3, -0.5)$

39. $M = \left(\dfrac{-30 + 50}{2}, \dfrac{50 + (-30)}{2}\right) = (10, 10)$

41. $M = \left(\dfrac{1.5 + (-5.7)}{2}, \dfrac{2.9 + (-3.6)}{2}\right) = (-2.1, -0.35)$

43. $M = \left(\dfrac{\sqrt{2} + \sqrt{2}}{2}, \dfrac{\sqrt{5} + (-\sqrt{5})}{2}\right) = (\sqrt{2}, 0)$

45. $M = \left(\dfrac{a + (-a)}{2}, \dfrac{b + 3b}{2}\right) = (0, 2b)$

47. (a) $S = \{(-1, 5), (2, 2), (3, -1), (5, -4), (9, -5)\}$

(b) $D = \{-1, 2, 3, 5, 9\}$

$R = \{-5, -4, -1, 2, 5\}$

49. (a) $S = \{(1, 5), (4, 5), (5, 6), (4, 6), (1, 5)\}$

(b) $D = \{1, 4, 5\}$

$R = \{5, 6\}$

51. (a) The domain is $D = \{0, -3, -2, 7\}$ and the range is $R = \{5, 4, -5, -3, 0\}$.

(b) The minimum x-value is –3, and the maximum x-value is 7. The minimum y-value is –5, and the maximum y-value is 5.

(c) The axes must include at least $-3 \le x \le 7$ and $-5 \le y \le 5$. It would be appropriate to extend the axes slightly beyond these intervals and let each tick mark represent 2 units. Plot the points $(0, 5)$, $(-3, 4)$, $(-2, -5)$, $(7, -3)$, and $(0, 0)$.

(d) See Figure 51.

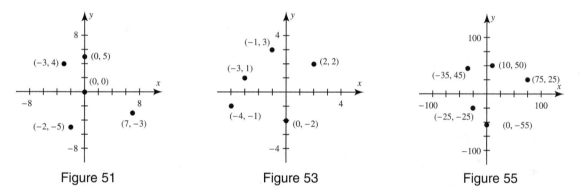

Figure 51   Figure 53   Figure 55

53. (a) The domain is $D = \{2, -3, -4, -1, 0\}$ and the range is $R = \{2, 1, -1, 3, -2\}$.

(b) The minimum x-value is -4, and the maximum x-value is 2. The minimum y-value is -2, and the maximum y-value is 3.

(c) The axes must include at least $-4 \le x \le 2$ and $-2 \le y \le 3$. Extend each axis slightly beyond these intervals and let each tick mark represent 1 unit. Plot the points $(2, 2)$, $(-3, 1)$, $(-4, -1)$, $(-1, 3)$, and $(0, -2)$.

(d) See Figure 53.

55. (a) The domain is $D = \{10, -35, 0, 75, -25\}$ and the range is $R = \{50, 45, -55, 25, -25\}$

(b) The minimum x-value is -35, and the maximum x-value is 75. The minimum y-value is -55, and the maximum y-value is 50.

(c) The axes must include at least $-35 \le x \le 75$ and $-55 \le y \le 50$, and let each tick mark represent 25 units. It would be appropriate to extend the axes slightly beyond these intervals. Plot the points $(10, 50)$, $(-35, 45)$, $(0, -55)$, $(75, 25)$, and $(-25, -25)$.

(d) See Figure 55.

57. $(1990, 5)$, $(1995, 34)$, $(2000, 109)$, $(2005, 208)$. See Figures 57a & 57b.

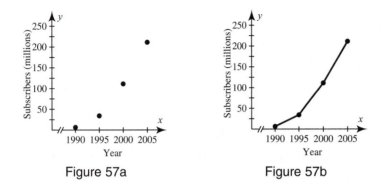

Figure 57a   Figure 57b

59. $x^2 + y^2 = 25 \Rightarrow (x - 0)^2 + (y - 0)^2 = 5^2 \Rightarrow$ Center: $(0, 0)$; Radius: 5

61. $x^2 + y^2 = 7 \Rightarrow (x - 0)^2 + (y - 0)^2 = (\sqrt{7}\,)^2 \Rightarrow$ Center: $(0, 0)$; Radius: $\sqrt{7}$

63. $(x - 2)^2 + (y + 3)^2 = 9 \Rightarrow (x - 2)^2 + (y - (-3))^2 = 3^2 \Rightarrow$ Center: $(2, -3)$; Radius: 3

65. $x^2 + (y + 1)^2 = 100 \Rightarrow (x - 0)^2 + (y - (-1))^2 = 10^2 \Rightarrow$ Center: $(0, -1)$; Radius: 10

67. Since the center is $(1, -2)$ and the radius is 1, the equation is $(x - 1)^2 + (y + 2)^2 = 1$.

69. Since the center is $(-2, 1)$ and the radius is 2, the equation is $(x + 2)^2 + (y - 1)^2 = 4$.

71. $(x - 3)^2 + (y - (-5))^2 = 8^2 \Rightarrow (x - 3)^2 + (y + 5)^2 = 64$

73. $(x - 3)^2 + (y - 0)^2 = 7^2 \Rightarrow (x - 3)^2 + y^2 = 49$

75. First find the radius using the distance formula: $r = \sqrt{(-3 - 0)^2 + (-1 - 0)^2} = \sqrt{10}$.

    $(x - 0)^2 + (y - 0)^2 = (\sqrt{10}\,)^2 \Rightarrow x^2 + y^2 = 10$

77. First find the center using the midpoint formula: $C = \left(\dfrac{-5 + 1}{2}, \dfrac{-7 + 1}{2} \right) = (-2, -3)$.

    Then find the radius using the distance formula: $r = \sqrt{(-2 - 1)^2 + (-3 - 1)^2} = \sqrt{25} = 5$.

    $(x - (-2))^2 + (y - (-3))^2 = 5^2 \Rightarrow (x + 2)^2 + (y + 3)^2 = 25$

79. x-axis: 10 tick marks; y-axis: 10 tick marks. See Figure 79.

81. x-axis: 10 tick marks; y-axis: 5 tick marks. See Figure 81.

[−10, 10, 1] by [−10, 10, 1]   [0, 100, 10] by [−50, 50, 10]   [1980, 1995, 1] by [12000, 16000, 1000]

Figure 79   Figure 81   Figure 83

83. x-axis: 16 tick marks; y-axis: 5 tick marks. See Figure 83..

85. Graph b

87. Graph a

89. Plot the points $(1, 3)$, $(-2, 2)$, $(-4, 1)$, $(-2, -4)$ and $(0, 2)$ in $[-5, 5, 1]$ by $[-5, 5, 1]$. See Figure 89.

91. Plot the points $(10, -20)$, $(-40, 50)$, $(30, 60)$, $(-50, -80)$, and $(70, 0)$

    in $[-100, 100, 10]$ by $[-100, 100, 10]$. See Figure 91.

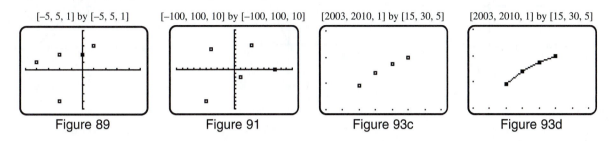

[–5, 5, 1] by [–5, 5, 1]   [–100, 100, 10] by [–100, 100, 10]   [2003, 2010, 1] by [15, 30, 5]   [2003, 2010, 1] by [15, 30, 5]

Figure 89   Figure 91   Figure 93c   Figure 93d

93. (a) x-min: 2005; x-max: 2008; y-min: 19.7; y-max: 25.0

    (b) $[2003, 2010, 1]$ by $[15, 30, 5]$. *Answers may vary.*

    (c) See Figure 93c.

    (d) See Figure 93d.

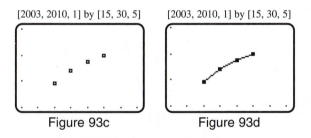

[2003, 2010, 1] by [15, 30, 5]   [2003, 2010, 1] by [15, 30, 5]

Figure 93c   Figure 93d

95. (a) x-min: 1950; x-max: 2000; y-min: 1.7; y-max: 5.5

    (b) $[1940, 2010, 10]$ by $[0, 7, 1]$ *Answers may vary.*

    (c) See Figure 95c.

    (d) See Figure 95d.

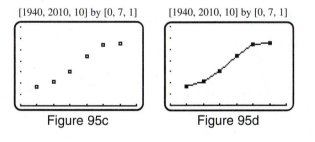

[1940, 2010, 10] by [0, 7, 1]   [1940, 2010, 10] by [0, 7, 1]

Figure 95c   Figure 95d

1.3: Functions and Their Representations

1. If $f(-2) = 3$, then the point $(-2, 3)$ is on the graph of f.

3. If $(7, 8)$ is on the graph of f, then $f(7) = 8$.

5. See Figure 5.

7. See Figure 7.

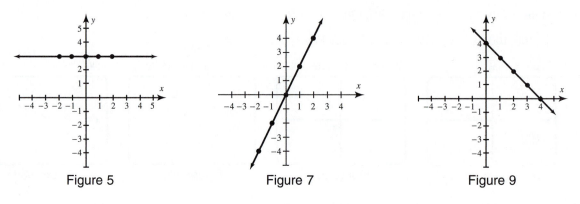

Figure 5   Figure 7   Figure 9

9. See Figure 9.

11. See Figure 11.

13. See Figure 13.

15. See Figure 15.

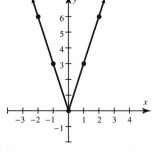

Figure 11   Figure 13   Figure 15

17. See Figure 17.

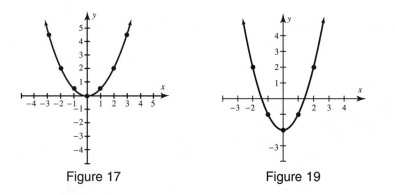

Figure 17   Figure 19

19. See Figure 19.

21. (a) $f(x) = x^3 \Rightarrow f(-2) = (-2)^3 = -8$ and $f(5) = 5^3 = 125$.

    (b) All real numbers.

23. (a) $f(x) = \sqrt{x} \Rightarrow f(-1) = \sqrt{-1}$ which is not a real number, and $f(a + 1) = \sqrt{a + 1}$.

    (b) All non-negative real numbers.

25. (a) $f(x) = 6 - 3x \Rightarrow f(-1) = 6 - 3(-1) = 6 + 3 = 9$ and

    $f(a + 1) = 6 - 3(a + 1) = 6 - 3a - 3 = 3 - 3a$

    (b) All real numbers.

27. (a) $f(x) = -7 \Rightarrow f(6) = -7$ and $f(a - 1) = -7$. (f is a constant function.)

    (b) All real numbers.

29. (a) $f(x) = \dfrac{1}{x^2} \Rightarrow f(4) = \dfrac{1}{4^2} = \dfrac{1}{16}$ and $f(-7) = \dfrac{1}{(-7)^2} = \dfrac{1}{49}$.

    (b) All real numbers not equal to 0 $(x \neq 0)$.

31. (a) $f(x) = \dfrac{1}{x^2 - 9} \Rightarrow f(4) = \dfrac{1}{4^2 - 9} = \dfrac{1}{16 - 9} = \dfrac{1}{7}$ and

    $f(a - 5) = \dfrac{1}{(a - 5)^2 - 9} = \dfrac{1}{a^2 - 10a + 25 - 9} = \dfrac{1}{a^2 - 10a + 16}$.

    (b) All real numbers not equal to 3 or -3 $(x \neq 3, x \neq -3)$.

33. (a) $f(x) = \dfrac{1}{\sqrt{2 - x}} \Rightarrow f(1) = \dfrac{1}{\sqrt{2 - 1}} = \dfrac{1}{\sqrt{1}} = \dfrac{1}{1} = 1$ and $f(a + 2) = \dfrac{1}{\sqrt{2 - (a + 2)}} = \dfrac{1}{\sqrt{-a}}$

    (b) $2 - x > 0 \Rightarrow -x > -2 \Rightarrow x < 2$

35. (a) All real numbers.

    (b) $g(x) = 2x - 1 \Rightarrow g(-1) = 2(-1) - 1 = -3$ and $g(2) = 2(2) - 1 = 3$

    (c) $g(-1) = -3$ and $g(2) = 3$

37. (a) All real numbers.

    (b) $g(x) = 2 - x^2 \Rightarrow g(-1) = 2 - (-1)^2 = 2 - 1 = 1$ and $g(2) = 2 - (2)^2 = 2 - 4 = -2$

    (c) $g(-1) = 1$ and $g(2) = -2$

39. (a) $-2 \leq x \leq 2$

    (b) $g(x) = x^2 - 3 \Rightarrow g(-1) = (-1)^2 - 3 = 1 - 3 = -2$ and $g(2) = (2)^2 - 3 = 4 - 3 = 1$

    (c) $g(-1) = -2$ and $g(2) = 1$

41. $D = \{x \mid -3 \leq x \leq 3\}$

    $R = \{y \mid 0 \leq y \leq 3\}$

    $f(0) = 3$.

43. $D = \{\text{all real numbers}\}$

    $R = \{y \mid y \leq 2\}$

    $f(0) = 2$

45. $D = \{x \mid x \geq -1\}$

    $R = \{y \mid y \leq 2\}$

    $f(0) = 0$

47. (a) $f(2) = 7$

(b) $f = \{(1, 7), (2, 7), (3, 8)\}$

(c) $D = \{1, 2, 3\}; \ R = \{7, 8\}$

49. Graph $f(x) = 0.25x^2$ in $[-4.7, 4.7, 1]$ by $[-3.1, 3.1, 1]$ by letting $Y_1 = 0.25X\hat{\ }2$. See Figure 49.

(a) From the graph, it appears that $f(2) = 1$.

(b) $f(2) = 0.25(2)^2 = 0.25(4) = 1$

(c) See Figure 49c.

$[-4.7, 4.7, 1]$ by $[-3.1, 3.1, 1]$ $[-4.7, 4.7, 1]$ by $[-3.1, 3.1, 1]$

Figure 49   Figure 49c   Figure 51   Figure 51c

51. Graph $f(x) = \sqrt{x + 2}$ in $[-4.7, 4.7, 1]$ by $[-3.1, 3.1, 1]$ by letting $Y_1 = \sqrt{(X + 2)}$. See Figure 51.

(a) From the graph, it appears that $f(2) = 2$.

(b) $f(2) = \sqrt{2 + 2} = \sqrt{4} = 2$

(c) See Figure 51c.

53. Verbal: Square the input x.

Graphical: Graph $Y_1 = X\hat{\ }2$. See Figure 53.

Numerical:

x	-2	-1	0	1	2
y	4	1	0	1	4

$f(2) = 4$

55. Verbal: Multiply the input x by 2, add 1, and then take the absolute value.

Graphical: Graph $Y_1 = \text{abs}(2X + 1)$. See Figure 55.

Numerical:

x	-2	-1	0	1	2
y	3	1	1	3	5

$f(2) = 5$

$[-10, 10, 1]$ by $[-10, 10, 1]$ $[-6, 6, 1]$ by $[-4, 4, 1]$ $[-10, 10, 1]$ by $[-10, 10, 1]$ $[-6, 6, 1]$ by $[-4, 4, 1]$

Figure 53   Figure 55   Figure 57   Figure 59

57. Verbal: Subtract the input x from 5.

Graphical: $Y_1 = 5 - X$. See Figure 57.

Numerical:

x	-2	-1	0	1	2
y	7	6	5	4	3

$f(2) = 3$

59. Verbal: Add 1 to the input x and then take the square root of the result.

Graphical: Graph $Y_1 = \sqrt{(X + 1)}$ See Figure 59.

Numerical:

x	-2	-1	0	1	2
y	—	0	1	$\sqrt{2}$	$\sqrt{3}$

$f(2) = \sqrt{3}$

61. (a) $g = \{(-1, 2), (0, 4), (1, -3), (2, 2)\}$

(b) $D = \{-1, 0, 1, 2\};\ R = \{-3, 2, 4\}$

63. It costs about $\$0.50$ per mile.

Symbolic: $f(x) = 0.50x$.

Graphical: See Figure 63a.

Numerical: See Figure 63b.

Miles	1	2	3	4	5	6
Cost	0.50	1.00	1.50	2.00	2.50	3.00

Figure 63a   Figure 63b

65. This is a graph of a function because every vertical line intersects the graph at most once. Both the domain and the range are all real numbers.

67. This is not a graph of a function because some vertical lines can intersect the graph twice. Because a vertical line can intersect the graph twice, two functions are necessary to create this graph.

69. This is a graph of a function because every vertical line intersects the graph at most once.

The domain is $\{x \mid -4 \le x \le 4\}$. The range is $\{y \mid 0 \le y \le 4\}$.

71. Yes. The calculation of the cube root requires the input of a real number. It produces a single output. A number has only one cube root.

73. No. On most English exams, more than one person passes. Therefore, for a given English exam as input, there is more than one output.

75. Yes, because the IDs are unique.

77. No. The ordered pairs (1, 2) and (1, 3) belong to the set S. The domain element 1 has more than one range element associated with it.

79. Yes. Each element in its domain is associated with exactly one range element.

81. No. The ordered pairs (1, 10.5) and (1, –0.5) belong to the set S. The domain element 1 has more than one range element associated with it.

83. No, for example, the ordered pairs (1, –1) and (1, 1) belong to the relation. The domain element 1 has more than one range element associated with it.

85. Yes. Each element in the domain of f is associated with exactly one range element.

87. No, for example, the ordered pairs $(0, \sqrt{70})$ and $(0, -\sqrt{70})$ belong to the relation. The domain element 0 has more than one range element associated with it.

89. Yes. Each element in the domain of f is associated with exactly one range element.

91. $g(x) = 12x \Rightarrow g(10) = 12(10) = 120$; there are 120 inches in 10 feet.

93. $g(x) = 0.25x \Rightarrow g(10) = 0.25(10) = 2.50$; there are 2.5 dollars in 10 quarters.

95. $g(x) = 60 \cdot 60 \cdot 24 \cdot x \Rightarrow g(x) = 86{,}400x \Rightarrow g(10) = 86{,}400(10) = 864{,}000$; there are 864,000 seconds in 10 days.

97. (a) $I = \{(N, 19162), (H, 26029), (B, 41681), (M, 51316)\}$

    (b) $D = \{N, H, B, M\}$; $R = \{19162, 26029, 41681, 51316\}$

99. $N(x) = 2200x$; $N(3) = 2200(3) = 6600$; in 3 years the average person uses 6600 napkins.

101. Verbal: Multiply the input x by –5.8 to obtain the change in temperature.

    Symbolic: $f(x) = -5.8x$.

    Graphical: $Y_1 = -5.8X$. See Figure 101a.

    Numerical: Table $Y_1 = -5.8X$. See Figure 101b.

    [0, 3, 1] by [–20, 20, 5]

Figure 101a   Figure 101b

103. Let x = time in seconds that elapsed for the observer to hear the thunder from the lightning bolt. The distance is found by multiplying the speed by the time. Since 5280 ft. = 1 mi., the distance in miles is given by $f(x) = \dfrac{1150}{5280}x = \dfrac{115}{528}x$; $f(15) = \dfrac{115}{528}(15) \approx 3.3$; with a 15-second delay, the lightning bolt was about 3.3 miles away.

1.4: Types of Functions

1. $f(x) = 5 - 2x \Rightarrow f(x) = -2x + 5;\ a = -2, b = 5$

3. $f(x) = -8x \Rightarrow f(x) = -8x + 0;\ a = -8, b = 0$

5. $f(x) = 7 \Rightarrow f(x) = 0x + 7;\ a = 0, b = 7$

7. $m = \dfrac{5 - 6}{2 - 4} = \dfrac{-1}{-2} = \dfrac{1}{2} = 0.5$

9. $m = \dfrac{-2 - 4}{5 - (-1)} = \dfrac{-6}{6} = -1$

11. $m = \dfrac{-8 - (-8)}{7 - 12} = \dfrac{0}{-5} = 0$

13. $m = \dfrac{0.4 - (-0.1)}{-0.3 - 0.2} = \dfrac{0.5}{-0.5} = -1$

15. $m = \dfrac{7.6 - 9.2}{-0.3 - (-0.5)} = \dfrac{-1.6}{0.2} = -8$

17. $m = \dfrac{7.9 - 5.6}{1994 - 1997} = -\dfrac{23}{30} \approx -0.7667$

19. $m = \dfrac{8 - 6}{-5 - (-5)} = \dfrac{2}{0} = \text{undefined}$

21. $m = \dfrac{\frac{7}{10} - \left(-\frac{3}{5}\right)}{-\frac{5}{6} - \frac{1}{3}} = \dfrac{\frac{13}{10}}{-\frac{7}{6}} = \dfrac{13}{10} \cdot \left(-\dfrac{6}{7}\right) = -\dfrac{39}{35} \approx -1.143$

23. Slope $= 2$; the graph rises 2 units for every unit increase in x.

25. Slope $= -\dfrac{3}{4}$; the graph falls $\dfrac{3}{4}$ unit for every unit increase in x, or equivalently, the graph falls 3 units for

    every 4-unit increase in x.

27. Slope $= 0$; the graph neither falls nor rises since the y-value is always –5.

29. Slope $= -1$; the graph falls 1 unit for every unit increase in x.

31. (a) Buying no carpet should and does cost $0.

    (b) Slope $= \dfrac{100}{5} = 20$

    (c) The carpet costs $20 per square yard.

33. (a) $P(x) = 19.4x \Rightarrow P(20) = 19.4(20) = 388$. Burning 20 gallons of gas produces 388 lbs. of CO_2.

    (b) Slope equals 19.4; 19.4 lbs. of CO_2 are produced for every gallon of gas that is burned.

35. (a) $D(2) = 75(2) = 150$ miles

    (b) Slope $= 75$; the car is traveling away from the rest stop at 75 miles per hour.

37. If the distance is constant, the car is not moving. The car's velocity is zero.

39. $f(x) = -2x + 5$ is a linear function, but not a constant function, with a slope of $m = -2$. See Figure 39.

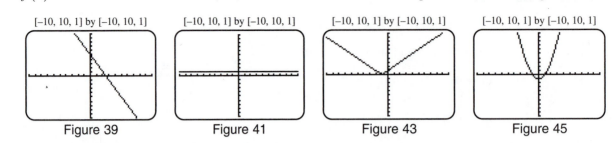

[–10, 10, 1] by [–10, 10, 1]   [–10, 10, 1] by [–10, 10, 1]   [–10, 10, 1] by [–10, 10, 1]   [–10, 10, 1] by [–10, 10, 1]

Figure 39   Figure 41   Figure 43   Figure 45

41. $f(x) = 1$ is a constant (and linear) function. See Figure 41.

43. From its graph, we see that $f(x) = |x + 1|$ represents a nonlinear function. See Figure 43.

45. From its graph, we see that $f(x) = x^2 - 1$ represents a nonlinear function. See Figure 45.

47. From its graph, we see that $f(x) = 2\sqrt{x}$ is a nonlinear function. See Figure 47.

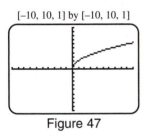

[–10, 10, 1] by [–10, 10, 1]

Figure 47

49. Between each pair of points, the y-values increase 4 units for each unit increase in x. Therefore, the data is linear. The slope of the line passing through the data points is 4.

51. The y-values do not increase by a constant amount for each 2-unit increase in x. The data is nonlinear.

53. The y-values decrease 0.5 unit for each unit increase in x, therefore, the data is linear. The slope of the line passing through the data points is –0.5.

55. (a) See Figure 55.

    (b) The function is linear since the slopes of the line segments are all equal.

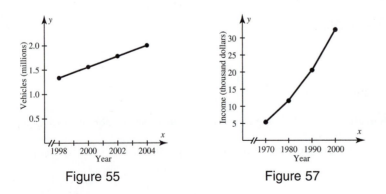

Figure 55   Figure 57

57. (a) See Figure 57.

    (b) From the line graph it can be seen that the data is not linear. The function f is nonlinear.

59. (a) Since the wind speed is not constant, it cannot be represented exactly by a constant function.

   (b) The average of the twelve wind speeds is 7 miles per hour. It would be reasonable to approximate the wind speed at Myrtle Beach by $f(x) = 7$.

   (c) A scatterplot of the data and the function $f(x) = 7$ are shown in Figure 59.

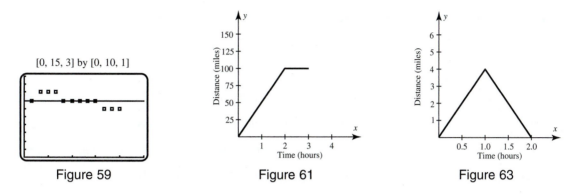

Figure 59   Figure 61   Figure 63

61. See Figure 61.

63. See Figure 63.

1.5: Functions and Their Rates of Change

1. $[5, \infty)$

3. $[4, 19)$

5. $(-\infty, -37]$

7. $[-1, \infty)$

9. $(-\infty, 1)\cup[3, \infty)$

11. $(-3, 5]$

13. $(-\infty, -2)$

15. $(-\infty, -2)\cup[1, \infty)$

17. f is decreasing on $(-\infty, \infty)$. In set-builder notation, the interval is $\{x \mid -\infty < x < \infty\}$.

19. f is increasing on $[2, \infty)$ and decreasing on $(-\infty, 2]$. In set-builder notation, these intervals are $\{x \mid x \geq 2\}$ and $\{x \mid x \leq 2\}$ respectively.

21. f is increasing on $(-\infty, -2]\cup[1, \infty)$ and decreasing on $[-2, 1]$. In set-builder notation, these intervals are $\{x \mid x \leq 2 \text{ or } x \geq 1\}$ and $\{x \mid -2 \leq x \leq 1\}$ respectively.

23. f is increasing on $[-8, 0]\cup[8, \infty)$ and decreasing on $(-\infty, -8]\cup[0, 8]$. In set-builder notation, these intervals are $\{x \mid -8 \leq x \leq 0 \text{ or } x \geq 8\}$ and $\{x \mid x \leq -8 \text{ or } 0 \leq x \leq 8\}$ respectively.

25. When x increases, $f(x)$ remains constant at $(-3) \implies$ the function is neither increasing or decreasing.

27. The graph of this equation is linear with a slope of 2 $\Rightarrow$ it is increasing: $(-\infty, \infty)$ and decreasing: never. In set builder notation, the interval is $\{x \mid -\infty < x < \infty\}$.

29. The graph of this equation is a parabola with a vertex $(0, -2)$ $\Rightarrow$ it is increasing: $[0, \infty)$ and decreasing: $(-\infty, 0)$. In set builder notation, these intervals are $\{x \mid x \geq 0\}$ and $\{x \mid x \leq 0\}$ respectively.

31. The graph of this equation is a parabola with a vertex $(1, 1)$, because of the negative coefficient before x^2 it opens downward $\Rightarrow$ it is increasing: $(-\infty, 1]$ and decreasing: $[1, \infty)$. In set builder notation, these intervals are $\{x \mid x \leq 1\}$ and $\{x \mid x \geq 1\}$ respectively.

33. The square root equation graph has a starting point $(1, 0)$, all x-values less than 1 are undefined $\Rightarrow$ it is increasing: $[1, \infty)$ and decreasing: never. In set builder notation, the interval is $\{x \mid x \geq 1\}$.

35. The absolute value graph has a vertex $(-3, 0)$ $\Rightarrow$ it is increasing: $[-3, \infty)$ and decreasing: $(-\infty, -3]$. In set builder notation, these intervals are $\{x \mid x \geq -3\}$ and $\{x \mid x \leq -3\}$ respectively.

37. The basic x^3 function is always increasing $\Rightarrow$ it is increaing: $(-\infty, \infty)$ and decreasing: never. In set builder notation, the interval is $\{x \mid -\infty < x < \infty\}$.

39. The graph of this cubic equation has turning points $\left(-2, \dfrac{16}{3}\right)$ and $\left(2, \dfrac{-16}{3}\right)$ $\Rightarrow$ it is increasing: $(-\infty, -2]$ or $[2, \infty)$; and decreasing: $[-2, 2]$. In set builder notation, these intervals are $\{x \mid x \leq -2 \text{ or } x \geq 2\}$ and $\{x \mid -2 \leq x \leq 2\}$ respectively.

41. The graph of this cubic equation has turning points $(-2, 20)$ and $(1, -7)$ $\Rightarrow$ it is increasing: $(-\infty, -2]$ or $[1, \infty)$; and decreasing: $[-2, 1]$. In set builder notation, these intervals are $\{x \mid x \leq -2 \text{ or } x \geq 1\}$ and $\{x \mid -2 \leq x \leq 1\}$ respectively.

43. The graph of this x^4 graph has a negative lead coefficient therefore is reflected through the x-axis has turning points $\left(-1, \dfrac{5}{12}\right)$, $(0, 0)$ and $\left(2, \dfrac{8}{3}\right)$ $\Rightarrow$ it is increasing: $(-\infty, -1]$ or $[0, 2]$; and decreasing: $[-1, 0]$ or $[2, \infty)$. In set builder notation, these intervals are $\{x \mid x \leq -1 \text{ or } 0 \leq x \leq 2\}$ and $\{x \mid -1 \leq x \leq 0 \text{ or } x \geq 2\}$ respectively.

45. According to the graph the water levels are increasing on the time intervals $[0, 2.4], [8.7, 14.7]$ and $[21, 27]$. In set builder notation, these intervals are $\{x \mid 0 \leq x \leq 2.4 \text{ or } 8.7 \leq x \leq 14.7 \text{ or } 21 \leq x \leq 27\}$.

47. The average rate of change from -3 to -1 is $\dfrac{f(-1) - f(-3)}{-1 - (-3)} = \dfrac{4 - 4}{2} = 0$.

   The average rate of change from 1 to 3 is $\dfrac{f(3) - f(1)}{3 - 1} = \dfrac{4 - 4}{2} = 0$.

49. The average rate of change from -3 to -1 is $\dfrac{f(-1) - f(-3)}{-1 - (-3)} = \dfrac{3.7 - 1.3}{2} = \dfrac{2.4}{2} = 1.2$.

   The average rate of change from 1 to 3 is $\dfrac{f(3) - f(1)}{3 - 1} = \dfrac{1.3 - 3.7}{2} = -\dfrac{2.4}{2} = -1.2$.

51. (a) $f(x) = x^2 \Rightarrow f(1) = 1^2 = 1$ and $f(2) = 2^2 = 4 \Rightarrow (1,1), (2,4)$; using the slope formula for rate of

change we get $\dfrac{4-1}{2-1} = \dfrac{3}{1} = 3.$

(b) See Figure 51.

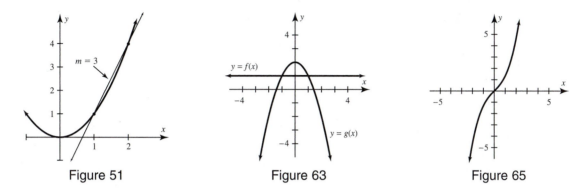

Figure 51   Figure 63   Figure 65

53. If $f(x) = 7x - 2$ then $\dfrac{f(4) - f(1)}{4 - 1} = 7$. The slope of the graph is 7.

55. If $f(x) = x^3 - 2x$ then $\dfrac{f(4) - f(2)}{4 - 2} = \dfrac{56 - 4}{2} = 26$. The slope of the line passing through the points

$(2, f(2))$ and $(4, f(4))$ is 26.

57. If $f(x) = \sqrt{2x - 1}$ then $\dfrac{f(3) - f(1)}{3 - 1} \approx 0.62$. The slope of the line passing through the points

$(1, f(1))$ and $(3, f(3))$ is approximately 0.62.

59. (a) The average rate of change from 1900 to 1940 is calculated as $\dfrac{182 - 3}{1940 - 1900} = \dfrac{179}{40} = 4.475$; from 1940 to

1980: $\dfrac{632 - 182}{1980 - 1940} = \dfrac{450}{40} = 11.25$; from 1980 to 2006: $\dfrac{371 - 632}{2006 - 1980} = \dfrac{-261}{26} \approx -10.04.$

(b) The average rates of change in cigarette consumption in the time periods 1900 to 1940, 1940 to 1980, and

1980 to 2006 were 4.475, 11.25, and –10.04, respectively.

61. (a) $D(t) = 16t^2 \Rightarrow D(2) = (16)(2)^2 = (16)(4) = 64$ and

$D(t) = 16t^2 \Rightarrow D(4) = (16)(4)^2 = (16)(16) = 256.$

(b) Average rate of change from 2 to 4 is $\dfrac{256 - 64}{4 - 2} = \dfrac{192}{2} = 96$. The object's average speed from 2 to 4

seconds is 96 ft/sec.

63. See Figure 63. *Answers may vary.*

65. See Figure 65. *Answers may vary.*

67. See Figure 67. *Answers may vary.*

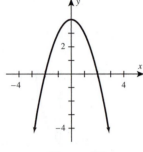

Figure 67

69. (a) $f(x + h) = 3$

    (b) $\dfrac{f(x + h) - f(x)}{h} = \dfrac{3 - 3}{h} = \dfrac{0}{h} = 0$

71. (a) $f(x + h) = -2(x + h) = -2x - 2h$

    (b) $\dfrac{f(x + h) - f(x)}{h} = \dfrac{-2x - 2h - (-2x)}{h} = \dfrac{-2h}{h} = -2$

73. (a) $f(x + h) = 2(x + h) + 1 = 2x + 2h + 1$

    (b) $\dfrac{f(x + h) - f(x)}{h} = \dfrac{2x + 2h + 1 - (2x + 1)}{h} = \dfrac{2h}{h} = 2$

75. (a) $f(x + h) = 3(x + h)^2 + 1 = 3(x^2 + 2xh + h^2) + 1 = 3x^2 + 6xh + 3h^2 + 1$

    (b) $\dfrac{f(x + h) - f(x)}{h} = \dfrac{3x^2 + 6xh + 3h^2 + 1 - (3x^2 + 1)}{h} = \dfrac{6xh + 3h^2}{h} = 6x + 3h$

77. (a) $f(x + h) = -(x + h)^2 + 2(x + h) = -(x^2 + 2xh + h^2) + 2x + 2h = -x^2 - 2xh - h^2 + 2x + 2h$

    (b) $\dfrac{f(x + h) - f(x)}{h} = \dfrac{-x^2 - 2xh - h^2 + 2x + 2h - (-x^2 + 2x)}{h} = \dfrac{-2xh - h^2 + 2h}{h} =$

    $-2x - h + 2$

79. (a) $f(x + h) = 2(x + h)^2 - (x + h) + 1 = 2(x^2 + 2xh + h^2) - (x + h) + 1 =$

    $2x^2 + 4xh + 2h^2 - x - h + 1$

    (b) $\dfrac{f(x + h) - f(x)}{h} = \dfrac{2x^2 + 4xh + 2h^2 - x - h + 1 - (2x^2 - x + 1)}{h} = \dfrac{4xh + 2h^2 - h}{h} =$

    $4x + 2h - 1$

81. (a) $f(x + h) = (x + h)^3 = x^3 + 3hx^2 + 3h^2x + h^3$

    (b) $\dfrac{f(x + h) - f(x)}{h} = \dfrac{x^3 + 3hx^2 + 3h^2x + h^3 - (x^3)}{h} = \dfrac{3hx^2 + 3h^2x + h^3}{h} = 3x^2 + 3hx + h^2$

83. (a) $d(t + h) = 8(t + h)^2 = 8(t^2 + 2th + h^2) = 8t^2 + 16th + 8h^2$

    (b) $\dfrac{d(t + h) - d(t)}{h} = \dfrac{8t^2 + 16th + 8h^2 - (8t^2)}{h} = \dfrac{16th + 8h^2}{h} = 16t + 8h$

    (c) If $t = 4$ and $h = 0.05$, the difference quotient becomes $16t + 8h = 16(4) + 8(0.05) =$

    $64 + 0.4 = 64.4$. The average speed of the car during the time interval from 4 to 4.05 seconds is 64.4 feet

    per second.

Chapter 1 Review Exercises

1. -2 is an integer, rational number, and real number. $\frac{1}{2}$ is both a rational and a real number. 0 is an integer, rational number, and real number. 1.23 is both a rational and a real number. $\sqrt{7}$ is a real number. $\sqrt{16} = 4$ is a natural number, integer, rational number, and real number.

3. $1{,}891{,}000 = 1.891 \times 10^6$

5. $1.52 \times 10^4 = 15{,}200$

7. (a) $\sqrt[3]{1.2} + \pi^3 \approx 32.07$

    (b) $\dfrac{3.2 + 5.7}{7.9 - 4.5} \approx 2.62$

    (c) $\sqrt{5^2 + 2.1} \approx 5.21$

    (d) $1.2(6.3)^2 + \dfrac{3.2}{\pi - 1} \approx 49.12$

9. $4 - 3^2 \cdot 5 = 4 - 9 \cdot 5 = 4 - 45 = -41$

11.

-23	-5	8	19	24

    (a) Maximum $= 24$; Minimum $= -23$

    (b) Mean $= \dfrac{-23 + (-5) + 8 + 19 + 24}{5} = 4.6$; Median $= 8$

13. (a) $S = \{(-15, -3), (-10, -1), (0, 1), (5, 3), (20, 5)\}$

    (b) $D = \{-15, -10, 0, 5, 20\}$ and $R = \{-3, -1, 1, 3, 5\}$

15. The relation $\{(10, 13), (-12, 40), (-30, -23), (25, -22), (10, 20)\}$ is plotted in Figure 15. It is not a function since both $(10, 13)$ and $(10, 20)$ are contained in the set. Notice that these points are lined up vertically.

[−50, 50, 10] by [−50, 50, 10]

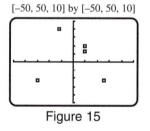

Figure 15

17. $d = \sqrt{(2 - (-4))^2 + (-3 - 5)^2} = \sqrt{6^2 + (-8)^2} = \sqrt{36 + 64} = \sqrt{100} = 10$

19. $M = \left(\dfrac{24 + (-20)}{2}, \dfrac{-16 + 13}{2} \right) = \left(\dfrac{4}{2}, \dfrac{-3}{2} \right) = \left(2, \dfrac{-3}{2} \right)$

21. Use the distance formula to find the side lengths:

    $d_1 = \sqrt{(1 - (-3))^2 + (2 - 5)^2} = \sqrt{4^2 + (-3)^2} = \sqrt{16 + 9} = \sqrt{25} = 5$

    $d_2 = \sqrt{((-3) - 0)^2 + (5 - 9)^2} = \sqrt{(-3)^2 + (-4)^2} = \sqrt{9 + 16} = \sqrt{25} = 5$

    $d_3 = \sqrt{(1 - 0)^2 + (2 - 9)^2} = \sqrt{1^2 + (-7)^2} = \sqrt{1 + 49} = \sqrt{50} = 7.07$

    Two sides are 5 units long and the other side is 7.07 units long, therefore, the triangle is isosceles.

23. First calculate the midpoint of the diameter: $x = \dfrac{6 + (-2)}{2} = \dfrac{4}{2} = 2$ and $y = \dfrac{6 + 4}{2} = \dfrac{10}{2} = 5 \Rightarrow$

the center is $(2, 5)$. Then calculate the length of the radius using points $(2, 5)$ and $(6, 6)$:

$d = \sqrt{(6 - 2)^2 + (6 - 5)^2} = \sqrt{4^2 + 1^2} = \sqrt{17}$. The equation is $(x - 2)^2 + (y - 5)^2 = 17$.

25. See Figure 25.

27. See Figure 27.

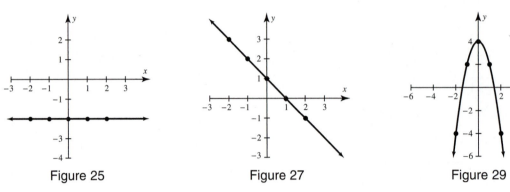

Figure 25   Figure 27   Figure 29

29. See Figure 29.

31. See Figure 31.

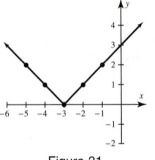

Figure 31

33. Symbolic: $f(x) = 16x$.

    Numerical: Table f starting at $x = 0$, incrementing by 25. See Figure 33a.

    Graphical: Graph $Y_1 = 16X$ in $[0, 100, 10]$ by $[0, 1800, 300]$. See Figure 33b.

$[0, 100, 10]$ by $[0, 1800, 300]$

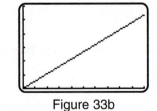

x	0	25	50	75	100
$f(x)$	0	400	800	1200	1600

Figure 33a   Figure 33b

35. (a) $f(x) = \sqrt[3]{x} \Rightarrow f(-8) = -2$ and $f(1) = 1$

    (b) All real numbers

37. (a) $f(x) = 5 \Rightarrow f(-3) = 5$ and $f(1.5) = 5$

(b) All real numbers

39. (a) $f(x) = x^2 - 3 \Rightarrow f(-10) = (-10)^2 - 3 = 97$ and $f(a + 2) = (a + 2)^2 - 3 =$

$a^2 + 4a + 4 - 3 = a^2 + 4a + 1$

(b) All real numbers

41. (a) $f(x) = \dfrac{1}{x^2 - 4} \Rightarrow f(-3) = \dfrac{1}{(-3)^2 - 4} = \dfrac{1}{5}$ and $f(a + 1) = \dfrac{1}{(a + 1)^2 - 4} =$

$\dfrac{1}{(a^2 + 2a + 1) - 4} = \dfrac{1}{a^2 + 2a - 3}$

(b) $D = \{x \mid x \neq \pm 2\}$

43. No, for example, an input $x = 6$ produces outputs of $y = \pm 1$.

45. Since any vertical line intersects the graph of f at most once, it is a function.

47. Yes, it is a function. Each input produces a single output.

49. $a = 0$, so the slope $= 0$.

51. $m = \dfrac{4 - 7}{3 - (-1)} = -\dfrac{3}{4}$

53. $m = \dfrac{4 - 4}{-2 - 8} = \dfrac{0}{-10} = 0$

55. $f(x) = 8 - 3x$ represents a linear function.

57. $f(x) = |x + 2|$ represents a nonlinear function.

59. See Figure 59.

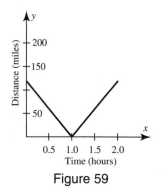

Figure 59

61. The best model is linear, but not constant, since the y-values decrease 8 units for every 2-unit increase in x.

63. $f(x + h) = 5(x + h) + 1 = 5x + 5h + 1$

$\dfrac{f(x + h) - f(x)}{h} = \dfrac{5x + 5h + 1 - (5x + 1)}{h} = \dfrac{5h}{h} = 5$

65. $\dfrac{2.28 \times 10^8}{3 \times 10^5} = 760$ seconds $= 12\dfrac{2}{3}$ minutes

67. (a) Sketching a diagram of the pool and sidewalk (Not Shown) gives the following dimensions:

    $l = 62$ ft. and $w = 37$ ft. Thus, $P = 2(62) + 2(37) = 198$ ft.

    (b) The area of the sidewalk would consist of the area of four 6×6 squares, two 50×6 rectangles, and two

    25×6 rectangles. $A = 4(6 \cdot 6) + 2(50 \cdot 6) + 2(25 \cdot 6) = 4(36) + 2(300) + 2(150) = 1044$ ft^2

69. (a) Plot the points (0, 100), (1, 10), (2, 6), (3, 3), and (4, 2) and make a line graph. See Figure 69. The survival

    rates decrease rapidly at first. This means that a large number of eggs never develop into mature adults.

    (b) Since any vertical line could intersect the graph at most once, this graph could represent a function.

    (c) From 0 to 1, $\dfrac{10 - 100}{1 - 0} = -90$; from 1 to 2, $\dfrac{6 - 10}{2 - 1} = -4$; from 2 to 3, $\dfrac{3 - 6}{3 - 2} = -3$;

    from 3 to 4, $\dfrac{2 - 3}{4 - 3} = -1$; during the first year, the population of sparrows decreased, on average, by 90

    birds. The other average rates of change can be interpreted similarly.

[-1, 5, 1] by [0, 110, 10]   [1, 5, 1] by [40, 70, 5]   [1, 5, 1] by [50, 63, 1]

Figure 69   Figure 71a   Figure 71c

71. (a) See Figure 71a. f is nonlinear.

    (b) $f(x) = 0.5x^2 + 50 \Rightarrow \dfrac{f(4) - f(1)}{4 - 1} = \dfrac{58 - 50.5}{3} = 2.5$

    (c) The average rate of change in outside temperature from 1 P.M. to 4 P.M. was 2.5° F per hour. The slope of

    the line segment from (1, 50.5) to (4, 58) is 2.5. See Figure 71c.

Chapter 2: Linear Functions and Equations

2.1: Linear Functions and Models

1. (a) $V(t) = -0.2t + 12.8 \Rightarrow V(0) = -0.2(0) + 12.8 = 12.8$, $V(4) = -0.2(4) + 12.8 = 12.0$

   (b) $V(t) = -0.2t + 12.8 \Rightarrow V(2) = -0.2(2) + 12.8 = 12.4$ million vehicles, estimate involves interpolation;

   $V(6) = -0.2(6) + 12.8 = 11.6$, estimate involves extrapolation.

   (c) The interpolation is more accurate.

3. Evaluating f for $x = 1, 2, 3, 4$ we get: $f(1) = 5(1) - 2 = 3$, value agrees with table;

   $f(2) = 5(2) - 2 = 8$, value agrees with table; $f(3) = 5(3) - 2 = 13$, value agrees with table;

   $f(4) = 5(4) - 2 = 18$, value agrees with table. Since all agree, f models the data exactly.

5. Evaluating f for $x = -6, 0, 1$ we get: $f(-6) = 3.7 - 1.5(-6) = 12.7$, value agrees with table;

   $f(0) = 3.7 - 1.5(0) = 3.7$, value agrees with table; $f(1) = 3.7 - 1.5(1) = 2.2$, value does not agree with

   table. Since $f(1)$ does not agree (but is close) and the others agree, f models the data approximately.

7. slope $= \dfrac{\text{rise}}{\text{run}} = \dfrac{-1}{2} = -\dfrac{1}{2}$; y-intercept: $3 \Rightarrow y = mx + b \Rightarrow y = -\dfrac{1}{2}x + 3 \Rightarrow f(x) = -\dfrac{1}{2}x + 3$

9. slope $= \dfrac{\text{rise}}{\text{run}} = \dfrac{2}{1} = 2$; $y = mx + b \Rightarrow y = 2x + b$

   Use $(1, 7)$ to find b: $7 = 2(1) + b \Rightarrow b = 5 \Rightarrow y = 2x + 5 \Rightarrow f(x) = 2x + 5$

11. (a) $f(x) = \dfrac{x}{16}$

    (b) $f(x) = 10x$

    (c) $f(x) = 0.06x + 6.50$

    (d) $f(x) = 500$

13. (a) Slope $= \dfrac{\text{rise}}{\text{run}} = \dfrac{2}{1} = 2$; y-intercept: -1; x-intercept: 0.5

    (b) $f(x) = ax + b \Rightarrow f(x) = 2x - 1$

    (c) 0.5

    (d) increasing

15. (a) Slope $= \dfrac{\text{rise}}{\text{run}} = \dfrac{-1}{3} = -\dfrac{1}{3}$; y-intercept: 2; x-intercept: 6

    (b) $f(x) = ax + b \Rightarrow f(x) = -\dfrac{1}{3}x + 2$

    (c) 6

    (d) decreasing

17. (a) Slope $= \dfrac{\text{rise}}{\text{run}} = \dfrac{100}{5} = 20$; y-intercept: -50; x-intercept: 2.5

    (b) $f(x) = ax + b \Rightarrow f(x) = 20x - 50$

    (c) 2.5

    (d) increasing

19. $f(x) = 3x + 2$; $m = 3$; y-intercept $= 2$. See Figure 19.

21. $f(x) = \dfrac{1}{2}x - 2$; $m = \dfrac{1}{2}$; y-intercept $= -2$. See Figure 21.

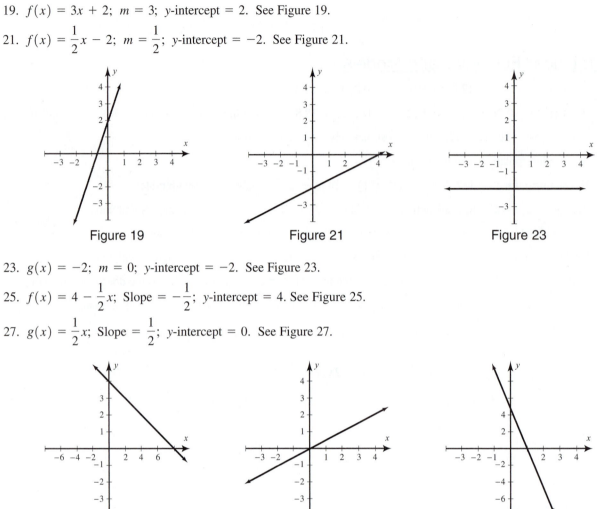

Figure 19   Figure 21   Figure 23

23. $g(x) = -2$; $m = 0$; y-intercept $= -2$. See Figure 23.

25. $f(x) = 4 - \dfrac{1}{2}x$; Slope $= -\dfrac{1}{2}$; y-intercept $= 4$. See Figure 25.

27. $g(x) = \dfrac{1}{2}x$; Slope $= \dfrac{1}{2}$; y-intercept $= 0$. See Figure 27.

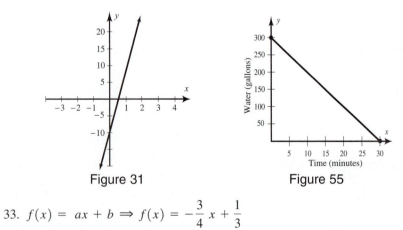

Figure 25   Figure 27   Figure 29

29. $g(x) = 5 - 5x$; Slope $= -5$; y-intercept $= 5$. See Figiure 29.

31. $f(x) = 20x - 10$; Slope $= 20$; y-intercept $= -10$. See Figure 31.

Figure 31   Figure 55

33. $f(x) = ax + b \Rightarrow f(x) = -\dfrac{3}{4}x + \dfrac{1}{3}$

35. $f(x) = ax + b \Rightarrow f(x) = 15x + 0$, or $f(x) = 15x$

37. Since the slope $= 0.5 = \dfrac{0.5}{1} = \dfrac{-0.5}{-1}$, to go from $(1, 4.5)$ to another point on the line, you can move 0.5 unit

    down and 1 unit to the left. This gives the point $(1 - 1, 4.5 - 0.5) = (0, 4)$, so the y-intercept is 4.

    $f(x) = mx + b \Rightarrow f(x) = 0.5x + 4$

39. $f(-2) = 10$ and $f(2) = 10 \Rightarrow$ we have points $(-2, 10)$ and $(2, 10)$; the slope of this line is $m =$

    $\dfrac{10 - 10}{2 - (-2)} = \dfrac{0}{4} = 0$. The average rate of change is 0. For two distinct real numbers a and b the points are

    $(a, 10)$ and $(b, 10)$; the slope of this line is $m = \dfrac{10 - 10}{b - a} = \dfrac{0}{b - a} = 0$. The average rate of change is 0.

41. $f(-2) = -\dfrac{1}{4}(-2) = \dfrac{1}{2}$ and $f(2) = -\dfrac{1}{4}(2) = -\dfrac{1}{2} \Rightarrow$ we have points $\left(-2, \dfrac{1}{2}\right)$ and $\left(2, -\dfrac{1}{2}\right)$; the slope of

    this line is $m = \dfrac{-\frac{1}{2} - \frac{1}{2}}{2 - (-2)} = \dfrac{-1}{4} = -\dfrac{1}{4}$. The average rate of change is $-\dfrac{1}{4}$. For two distinct real numbers a

    and b the points are given by $f(b) = -\dfrac{1}{4}b$ or $\left(a, -\dfrac{1}{4}a\right)$ and $\left(b, -\dfrac{1}{4}b\right)$; the slope of this line is

    $m = \dfrac{-\frac{1}{4}b - (-\frac{1}{4}a)}{b - a} = \dfrac{-\frac{1}{4}(b - a)}{b - a} = -\dfrac{1}{4}$. The average rate of change is $-\dfrac{1}{4}$.

43. $f(-2) = 4 - 3(-2) = 4 + 6 = 10$ and $f(2) = 4 - 3(2) = 4 - 6 = -2 \Rightarrow$ we have points $(-2, 10)$ and

    $(2, -2)$; the slope of this line is $m = \dfrac{-2 - 10}{2 - (-2)} = \dfrac{-12}{4} = -3$. The average rate of change is -3. For two

    distinct real numbers a and b the points are given by $f(a) = 4 - 3a$ and $f(b) = 4 - 3b$ or $(a, 4 - 3a)$ and

    $(b, 4 - 3b)$; the slope of this line is $m = \dfrac{(4 - 3b) - (4 - 3a)}{b - a} = \dfrac{-3b + 3a}{b - a} =$

    $\dfrac{-3(b - a)}{b - a} = \dfrac{-3}{1} = -3$. The average rate of change is -3.

45. The height of the Empire State Building is constant; the graph that has no rate of change is d.

47. As time increases the distance to the finish line decreases; the graph that shows this decline in distance as time

    increases is c.

49. $I(t) = 1.5t + 68$; t represents years after 2006;

51. $V(t) = 32t$; t represents time in seconds; $D = \{t \mid 0 \leq t \leq 3\}$

53. $P(t) = 21.5 + 0.581t$; t represents years after 1900; $D = \{t \mid 0 \leq t \leq 100\}$

55. (a) $W(t) = -10t + 300$

    (b) $W(7) = -10(7) + 300 = 230$ gallons

    (c) See Figure 55. x-intercept: 30, after 30 minutes the tank is empty; y-intercept: 300, the tank initially

    contains 300 gallons of water.

    (d) $D = \{t \mid 0 \leq t \leq 30\}$

57. (a) $f(x) = 4.3x + 40$

(b) Since 2006 corresponds to $x = 0$, 2012 corresponds to $x = 6$; $f(6) = 4.3(6) + 40 = 65.8$, which means that about 65,800,000 may be infected by 2012.

59. (a) $f(x) = 0.25x + 0.5$

(b) $f(2.5) = 0.25(2.5) + 0.5 = 1.125$ inches

61. (a) $(5, 84), (10, 169) \Rightarrow$ slope $= \dfrac{169 - 84}{10 - 5} = 17$

$(10, 169), (15, 255) \Rightarrow$ slope $= \dfrac{255 - 169}{15 - 10} = 17.2$

$(15, 255), (20, 338) \Rightarrow$ slope $= \dfrac{338 - 255}{20 - 15} = 16.6$

(b) $f(x) = 17x$

(c) See Figure 61. The slope indicates that the number of miles traveled per gallon is 17.

(d) $f(30) = 17(30) = 510$ miles. This indicates that the vehicle traveled 510 miles on 30 gallons of gasoline.

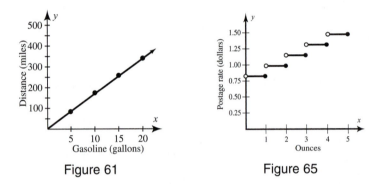

Figure 61   Figure 65

63. (a) The maximum speed limit is 55 mph and the minimum is 30 mph.

(b) The speed limit is 55 for $0 \le x < 4, 8 \le x < 12$, and $16 \le x < 20$. This is $4 + 4 + 4 = 12$ miles.

(c) $f(4) = 40, f(12) = 30$, and $f(18) = 55$.

(d) The graph is discontinuous when $x = 4, 6, 8, 12$, and 16. The speed limit changes at each discontinuity.

65. (a) $P(1.5) = 0.97$; it costs \$0.97 to mail 1.5 ounces. $P(3) = 1.14$; it costs \$1.14 to mail 3 ounces.

(b) See Figure 65. $D = \{x \,|\, 0 < x \le 5\}$

(c) $x = 1, 2, 3, 4$

67. (a) $f(1.5) = 30$; $f(4) = 10$

(b) $m_1 = 20$ indicates that the car is moving away from home at 20 mph; $m_2 = -30$ indicates that the car is moving toward home at 30 mph; $m_3 = 0$ indicates that the car is not moving; $m_4 = -10$ indicates that the car is moving toward home at 10 mph.

(c) The driver starts at home and drives away from home at 20 mph for 2 hours. The driver then travels toward home at 30 mph for 1 hour. Then the car does not move for 1 hour. Finally, the driver returns home in 1 hour at 10 mph.

(d) Increasing: $0 \le x \le 2$; Decreasing: $2 \le x \le 3$ or $4 \le x \le 5$; Constant: $3 \le x \le 4$

69. (a) $D = \{x \mid -5 \le x \le 5\}$

   (b) $f(-2) = 2, f(0) = 0 + 3 = 3, f(3) = 3 + 3 = 6$

   (c) See Figure 69.

   (d) f is continuous.

71. (a) $D = \{x \mid -1 \le x \le 2\}$

   (b) $f(-2)$ is undefined, $f(0) = 3(0) = 0, f(3)$ is undefined

   (c) See Figure 71.

   (d) f is not continuous.

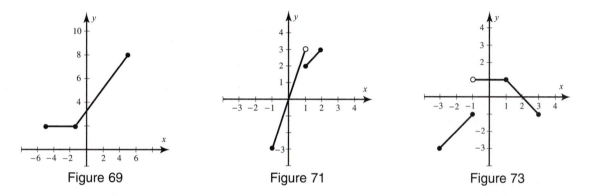

Figure 69   Figure 71   Figure 73

73. (a) $D = \{x \mid -3 \le x \le 3\}$

   (b) $f(-2) = -2, f(0) = 1, f(3) = 2 - 3 = -1$

   (c) See Figure 73.

   (d) f is not continuous.

75. $f(-4) = -\dfrac{1}{2}(-4) + 1 = 3, (-4, 3); f(-2) = -\dfrac{1}{2}(-2) + 1 = 2, (-2, 2);$ graph a segment from

   $(-4, 3)$ to $(-2, 2)$, use a closed dot for each point. See Figure 75.

   $f(-2) = 1 - 2(-2) = 5, (-2, 5); f(1) = 1 - 2(1) = -1, (1, -1);$ graph a segment from

   $(-2, 5)$ to $(1, -1)$, use an open dot at $(-2, 5)$ and a closed dot for $(1, -1)$. See Figure 75.

   $f(1) = \dfrac{2}{3}(1) + \dfrac{4}{3} = 2, (1, 2); f(4) = \dfrac{2}{3}(4) + \dfrac{4}{3} = 4, (4, 4);$ graph a segment from $(1, 2)$ to $(4, 4)$, use an

   open dot at $(1, 2)$ and a closed dot for $(4, 4)$. See Figure 75.

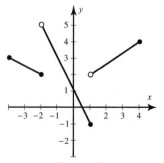

Figure 75

77. (a) $f(-3) = 3(-3) - 1 = -10, f(1) = 4, f(2) = 4$, and $f(5) = 6 - 5 = 1$

(b) The function f is constant with a value of 4 on the interval $[1, 3]$.

(c) See Figure 77. f is not continuous.

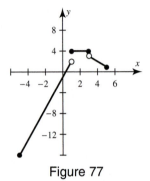

Figure 77

79. (a) Graph $Y_1 = \text{int}(2X - 1)$ as shown in Figure 79.

(b) $f(-3.1) = [2(-3.1) - 1] = [-7.2] = -8$ and $f(1.7) = [2(1.7) - 1] = [2.4] = 2$

81. (a) Graph $Y_1 = 2(\text{int}(X)) + 1$ as shown in Figure 81.

(b) $f(-3.1) = 2[-3.1] + 1 = 2(-4) + 1 = -7$ and $f(1.7) = 2[1.7] + 1 = 2(1) + 1 = 3$

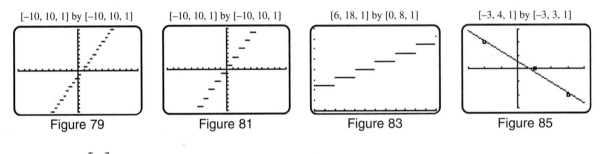

| [−10, 10, 1] by [−10, 10, 1] | [−10, 10, 1] by [−10, 10, 1] | [6, 18, 1] by [0, 8, 1] | [−3, 4, 1] by [−3, 3, 1] |

Figure 79   Figure 81   Figure 83   Figure 85

83. (a) $f(x) = 0.8\left[\dfrac{x}{2}\right]$ for $6 \le x \le 18$

(b) Graph $Y_1 = 0.8(\text{int}(X/2))$ as shown in Figure 83.

(c) $f(8.5) = 0.8\left[\dfrac{8.5}{2}\right] = 0.8[4.25] = 0.8(4) = \$3.20;\ f(15.2) = 0.8\left[\dfrac{15.2}{2}\right] = 0.8[7.6] = 0.8(7) = \$5.60$

85. Enter the x-values into the list L_1 and the y-values into the list L_2. Use the statistical feature of your graphing

calculator to find the correlation coefficient r and the regression equation. $r \approx -0.993;\ y \approx -0.789x + 0.526$

See Figure 85.

87. (a) Enter the x-values into the list L_1 and the y-values into the list L_2 in the statistical feature of your graphing calculator; the scatterplot of the data indicates that the correlation coefficient will be positive (and very close to 1).

(b) $y = ax + b$, where $a \approx 3.25$ and $b \approx -2.45$; $r \approx 0.9994$

(c) $y \approx 3.25(2.4) - 2.45 = 5.35$

89. (a) Enter the x-values into the list L_1 and the y-values into the list L_2 in the statistical feature of your graphing calculator; the scatterplot of the data indicates that the correlation coefficient will be negative (and very close to –1).

(b) $y = ax + b$, where $a \approx -3.8857$ and $b \approx 9.3254$; $r \approx -0.9996$

(c) $y \approx -3.8857(2.4) + 9.3254 = -0.00028$. *Due to rounding answers may very slightly.*

91. (a) The data points (50, 990), (650, 9300), (950, 15000) and (1700, 25000) are plotted in Figure 91. The data appears to have a linear relationship.

(b) Use the linear regression feature on your graphing calculator to find the values of a and b in the equation $y = ax + b$. In this instance, $a \approx 14.680$ and $b \approx 277.82$.

(c) We must find the x-value when $y = 37,000$. This can be done by solving the equation

$37,000 = 14.680x + 277.82 \Rightarrow 14.680x = 36,722.18 \Rightarrow x \approx 2500$ light years away. One could also solve the equation graphically to obtain the same approximation.

[–100, 1800, 100] by [–1000, 28000, 1000]   [–5, 40, 5] by [0, 6, 1]   [–5, 40, 5] by [0, 6, 1]

Figure 91   Figure 93a   Figure 93c

93. (a) Positive. See Figure 93a.

(b) Enter the x-values into the list L_1 and the y-values into the list L_2 in the statistical feature of your graphing calculator. $f(x) \approx 0.0854x + 2.078$

(c) See Figure 93c. The slope indicates the number of miles traveled by passengers per year.

(d) Year 2010 $\Rightarrow x = 40$; $f(40) \approx 0.0854(40) + 2.078 \approx 5.5$; 5.5 trillion miles

2.2: Equations of Lines

1. Find slope: $m = \dfrac{-2 - 2}{3 - 1} = \dfrac{-4}{2} = -2$. Using $(x_1, y_1) = (1, 2)$ and point-slope form $y = m(x - x_1) + y_1$,

   we get $y = -2(x - 1) + 2$. See Figure 1.

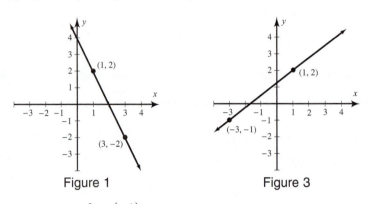

Figure 1   Figure 3

3. Find slope: $m = \dfrac{2 - (-1)}{1 - (-3)} = \dfrac{3}{4}$. Using $(x_1, y_1) = (-3, -1)$ and point-slope form $y = m(x - x_1) + y_1$,

   we get $y = \dfrac{3}{4}(x + 3) - 1$. See Figure 3.

5. The point-slope form is given by $y = m(x - x_1) + y_1$. Thus, $m = -2.4$ and $(x_1, y_1) = (4, 5) \Rightarrow$

   $y = -2.4(x - 4) + 5 \Rightarrow y = -2.4x + 9.6 + 5 \Rightarrow y = -2.4x + 14.6$.

7. First find the slope between the points $(1, -2)$ and $(-9, 3)$: $m = \dfrac{3 - (-2)}{-9 - 1} = -\dfrac{1}{2}$.

   $y = -\dfrac{1}{2}(x - 1) - 2 \Rightarrow y = -\dfrac{1}{2}x + \dfrac{1}{2} - 2 \Rightarrow y = -\dfrac{1}{2}x - \dfrac{3}{2}$.

9. $(4, 0), (0, -3)$; $m = \dfrac{-3 - 0}{0 - 4} = \dfrac{3}{4}$. Thus, $y = \dfrac{3}{4}(x - 4) + 0$ or $y = \dfrac{3}{4}x - 3$.

11. Using the points $(0, -1)$ and $(3, 1)$, we get $m = \dfrac{1 - (-1)}{3 - 0} = \dfrac{2}{3}$ and $b = -1$; $y = mx + b \Rightarrow y = \dfrac{2}{3}x - 1$.

13. Using the points $(-2, 1.8)$ and $(1, 0)$, we get $m = \dfrac{0 - 1.8}{1 - (-2)} = \dfrac{-1.8}{3} = -\dfrac{18}{30} = -\dfrac{3}{5}$; to find b, we use $(1, 0)$

    in $y = mx + b$ and solve for b: $0 = -\dfrac{3}{5}(1) + b \Rightarrow b = \dfrac{3}{5}$; $y = -\dfrac{3}{5}x + \dfrac{3}{5}$.

15. c

17. b

19. e

21. $m = \dfrac{2 - (-4)}{1 - (-1)} = 3$; $y = 3(x + 1) - 4 = 3x + 3 - 4 = 3x - 1$

23. $m = \dfrac{-3 - 5}{1 - 4} = \dfrac{8}{3}$; $y = \dfrac{8}{3}(x - 4) + 5 = \dfrac{8}{3}x - \dfrac{32}{3} + 5 = \dfrac{8}{3}x - \dfrac{17}{3}$

25. $b = 5$ and $m = -7.8 \Rightarrow y = -7.8x + 5$.

27. The line passes through the points $(0, 45)$ and $(90, 0)$.

    $m = \dfrac{0 - 45}{90 - 0} = -\dfrac{1}{2}$; $b = 45$ and $m = -\dfrac{1}{2} \Rightarrow y = -\dfrac{1}{2}x + 45$

29. $m = -3$ and $b = 5 \Rightarrow y = -3x + 5$

31. $m = \dfrac{0 - (-6)}{4 - 0} = \dfrac{6}{4} = \dfrac{3}{2}$ and $b = -6$; $y = mx + b \Rightarrow y = \dfrac{3}{2}x - 6$

33. $m = \dfrac{\frac{2}{3} - \frac{3}{4}}{\frac{1}{5} - \frac{1}{2}} = \dfrac{-\frac{1}{12}}{-\frac{3}{10}} = \dfrac{5}{18}$; using the point-slope form with $m = \dfrac{5}{18}$ and $\left(\dfrac{1}{2}, \dfrac{3}{4}\right)$, we get

$$y = \dfrac{5}{18}\left(x - \dfrac{1}{2}\right) + \dfrac{3}{4} \Rightarrow y = \dfrac{5}{18}x - \dfrac{5}{36} + \dfrac{3}{4} \Rightarrow y = \dfrac{5}{18}x + \dfrac{11}{18}.$$

35. The line has a slope of 4 and passes through the point $(-4, -7)$; $y = 4(x + 4) - 7 \Rightarrow y = 4x + 9$.

37. The slope of the perpendicular line is equal to $\dfrac{3}{2}$ and the line passes through the point $(1980, 10)$;

$$y = \dfrac{3}{2}(x - 1980) + 10 \Rightarrow y = \dfrac{3}{2}x - 2960$$

39. $y = \dfrac{2}{3}x + 3 \Rightarrow m = \dfrac{2}{3}$; the parallel line has slope $\dfrac{2}{3}$; since it passes through $(0, -2.1)$,

the y-intercept $= -2.1$; $y = mx + b \Rightarrow y = \dfrac{2}{3}x - 2.1$.

41. $y = -2x \Rightarrow m = -2$; the perpendicular line has slope $\dfrac{1}{2}$; since it passes through $(-2, 5)$, the equation is

$$y = \dfrac{1}{2}(x + 2) + 5 = \dfrac{1}{2}x + 1 + 5 = \dfrac{1}{2}x + 6.$$

43. $y = -x + 4 \Rightarrow m = -1$; the perpendicular line has slope 1; since it passes through $(15, -5)$, the equation is

$$y = 1(x - 15) - 5 = x - 15 - 5 = x - 20.$$

45. $m = \dfrac{1 - 3}{-3 - 1} = \dfrac{-2}{-4} = \dfrac{1}{2}$; a line parallel to this line also has slope $m = \dfrac{1}{2}$. Using

$(x_1, y_1) = (5, 7)$, $m = \dfrac{1}{2}$, and point-slope form $y = m(x - x_1) + y_1$, we get $y = \dfrac{1}{2}(x - 5) + 7 \Rightarrow$

$y = \dfrac{1}{2}x + \dfrac{9}{2}$.

47. $m = \dfrac{\frac{2}{3} - \frac{1}{2}}{-3 - (-5)} = \dfrac{\frac{1}{6}}{2} = \dfrac{1}{12}$; a line perpendicular to this line has slope $m = -\dfrac{12}{1} = -12$.

Using $(x_1, y_1) = (-2, 4)$, $m = -12$, and point-slope form $y = m(x - x_1) + y_1$, we get

$y = -12(x + 2) + 4 \Rightarrow y = -12x - 24 + 4 \Rightarrow y = -12x - 20$.

49. $x = -5$

51. $y = 6$

53. Since the line $y = 15$ is horizontal, the perpendicular line through $(4, -9)$ is vertical and has equation $x = 4$.

55. The line through $(19, 5.5)$ and parallel to $x = 4.5$ is also vertical and has equation $x = 19$.

57. Let $4x - 5y = 20$.

   x-intercept: Substitute $y = 0$ and solve for x. $4x - 5(0) = 20 \Rightarrow 4x = 20 \Rightarrow x = 5$; x-intercept: 5

   y-intercept: Substitute $x = 0$ and solve for y. $4(0) - 5y = 20 \Rightarrow -5y = 20 \Rightarrow y = -4$; y-intercept: -4

   See Figure 57.

59. Let $x - y = 7$.

   x-intercept: Substitute $y = 0$ and solve for x. $x - 0 = 7 \Rightarrow x = 7$; x-intercept: 7

   y-intercept: Substitute $x = 0$ and solve for y. $0 - y = 7 \Rightarrow -y = 7 \Rightarrow y = -7$; y-intercept: -7

   See Figure 59.

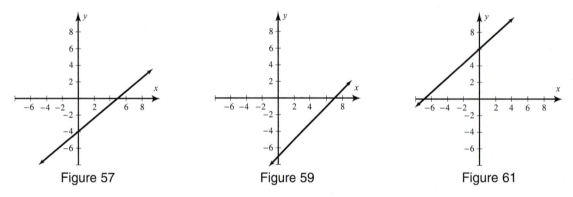

Figure 57   Figure 59   Figure 61

61. Let $6x - 7y = -42$.

   x-intercept: Substitute $y = 0$ and solve for x. $6x - 7(0) = -42 \Rightarrow 6x = -42 \Rightarrow x = -7$; x-intercept: -7

   y-intercept: Substitute $x = 0$ and solve for y. $6(0) - 7y = -42 \Rightarrow -7y = -42 \Rightarrow y = 6$; y-intercept: 6

   See Figure 61.

63. Let $y - 3x = 7$.

   x-intercept: Substitute $y = 0$ and solve for x. $0 - 3x = 7 \Rightarrow -3x = 7 \Rightarrow x = -\dfrac{7}{3}$; x-intercept: $-\dfrac{7}{3}$

   y-intercept: Substitute $x = 0$ and solve for y. $y - 3(0) = 7 \Rightarrow y - 0 = 7 \Rightarrow y = 7$; y-intercept: 7

   See Figure 63.

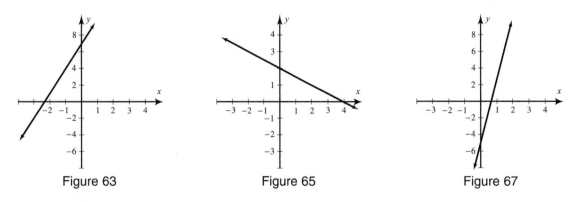

Figure 63   Figure 65   Figure 67

65. Let $0.2x + 0.4y = 0.8$.

   x-intercept: Substitute $y = 0$ and solve for x. $0.2x + 0.4(0) = 0.8 \Rightarrow 0.2x = 0.8 \Rightarrow x = 4$; x-intercept: 4

   y-intercept: Substitute $x = 0$ and solve for y. $0.2(0) + 0.4y = 0.8 \Rightarrow 0.4y = 0.8 \Rightarrow y = 2$; y-intercept: 2

   See Figure 65.

67. Let $y = 8x - 5$.

x-intercept: Substitute $y = 0$ and solve for x. $0 = 8x - 5 \Rightarrow 5 = 8x \Rightarrow x = \dfrac{5}{8}$; x-intercept: $\dfrac{5}{8}$

y-intercept: Substitute $x = 0$ and solve for y. $y = 8(0) - 5 \Rightarrow y = -5$; y-intercept: -5

See Figure 67.

69. Let $\dfrac{x}{5} + \dfrac{y}{7} = 1$.

x-intercept: Substitute $y = 0$ and solve for x. $\dfrac{x}{5} + \dfrac{0}{7} = 1 \Rightarrow \dfrac{x}{5} = 1 \Rightarrow x = 5$; x-intercept: 5

y-intercept: Substitute $x = 0$ and solve for y. $\dfrac{0}{5} + \dfrac{y}{7} = 1 \Rightarrow \dfrac{y}{7} = 1 \Rightarrow y = 7$; y-intercept: 7

a and b represent the x- and y-intercepts, respectively.

71. Let $\dfrac{2x}{3} + \dfrac{4y}{5} = 1$.

x-intercept: Substitute $y = 0$ and solve for x. $\dfrac{2x}{3} + \dfrac{4(0)}{5} = 1 \Rightarrow \dfrac{2x}{3} = 1 \Rightarrow x = \dfrac{3}{2}$; x-intercept: $\dfrac{3}{2}$

y-intercept: Substitute $x = 0$ and solve for y. $\dfrac{2(0)}{3} + \dfrac{4y}{5} = 1 \Rightarrow \dfrac{4y}{5} = 1 \Rightarrow y = \dfrac{5}{4}$; y-intercept: $\dfrac{5}{4}$

a and b represent the x- and y-intercepts, respectively.

73. $\dfrac{x}{a} + \dfrac{y}{b} = 1$; x-intercept: $5 \Rightarrow a = 5$, y-intercept: $9 \Rightarrow b = 9$; $\dfrac{x}{5} + \dfrac{y}{9} = 1$

75. (a) Since the point $(0, -3.2)$ is on the graph, the y-intercept is -3.2. The data is exactly linear, so one can use

any two points to determine the slope. Using the points $(0, -3.2)$ and $(1, -1.7)$, $m = \dfrac{-1.7 - (-3.2)}{1 - 0} = 1.5$.

The slope-intercept form of the line is $y = 1.5x - 3.2$.

(b) When $x = -2.7$, $y = 1.5(-2.7) - 3.2 = -7.25$. This calculation involves interpolation.

When $x = 6.3$, $y = 1.5(6.3) - 3.2 = 6.25$. This calculation involves extrapolation.

77. (a) Since the data is exactly linear, one can use any two points to determine the slope. Using the points

$(5, 94.7)$ and $(23, 56.9)$, $m = \dfrac{56.9 - 94.7}{23 - 5} = -2.1$. The point-slope form of the line is

$y = -2.1(x - 5) + 94.7$ and the slope-intercept form of the line is $y = -2.1x + 105.2$.

(b) When $x = -2.7$, $y = -2.1(-2.7) + 105.2 = 110.87$. This calculation involves extrapolation.

When $x = 6.3$, $y = -2.1(6.3) + 105.2 = 91.97$. This calculation involves interpolation.

79. (a) The slope between $(1998, 3305)$ and $(1999, 3185)$ is -120, and the slope between $(1999, 3185)$ and

$(2000, 3089)$ is -96. Using the average of -120 and -96, we will let $m = -108$.

$f(x) = -108(x - 1998) + 3305$, or $f(x) = -108x + 219{,}089$ approximately models the data.

Answers may vary.

(b) $f(2005) = -108(2005) + 219{,}089 = 2549$; this estimated value is too low (compared to the actual value

of 3450); this estimate involved extrapolation.

(c) Numbers were decreasing but increased after 911.

81. (a) Find the slope: $m = \dfrac{37{,}000 - 25{,}000}{2010 - 2003} = \dfrac{12{,}000}{7}$. Using the first point $(2003, 25000)$ for (x_1, y_1) and

$m = \dfrac{12{,}000}{7}$, we get $y = \dfrac{12{,}000}{7}(x - 2003) + 25{,}000$. The cost of attending a private college or

university is increasing by $\dfrac{12{,}000}{7} \approx \$1714$ per year on average.

(b) $y = \dfrac{12{,}000}{7}(2007 - 2003) + 25{,}000 \Rightarrow y = \dfrac{12{,}000}{7}(4) + 25{,}000 \Rightarrow y \approx 6857 + 25{,}000 \Rightarrow$

$y \approx \$31{,}857$; interpolation

(c) $y = \dfrac{12{,}000}{7}(x - 2003) + 25{,}000 \Rightarrow y \approx \dfrac{12{,}000}{7}x - 3{,}433{,}714 + 25{,}000 \Rightarrow$

$y \approx 1714x - 3{,}408{,}714$ (approximate)

83. (a) Find the slope: $m = \dfrac{3.6 - 1.6}{2005 - 2002} = \dfrac{2}{3}$. Using the first point $(2002, 1.6)$ for (x_1, y_1) and $m = \dfrac{2}{3}$, we get

$y = \dfrac{2}{3}(x - 2002) + 1.6$; online music sales increased by $\dfrac{2}{3}$ billion dollars $\approx \$0.67$ billion per year on average.

(b) $y = \dfrac{2}{3}(2008 - 2002) + 1.6 \Rightarrow y = \dfrac{2}{3}(6) + 1.6 \Rightarrow y = 4 + 1.6 \Rightarrow y = 5.6$ or $\$5.6$ billion;

extrapolation

(c) $y = \dfrac{2}{3}(x - 2002) + 1.6 \Rightarrow y = \dfrac{2}{3}x - \dfrac{4004}{3} + \dfrac{8}{5} \Rightarrow y = \dfrac{2}{3}x - \dfrac{19{,}996}{15}$

85. (a) See Figure 85.

(b) Use the first and last points to find slope $m = \dfrac{8.8 - 1.0}{2004 - 1999} = \dfrac{7.8}{5} = 1.56$. Now using the first point

$(1999, 1.0)$ for (x_1, y_1) and $m = 1.56$, we get $y = 1.56(x - 1999) + 1.0$. The daily worldwide spam

message numbers increased 1.56 billion per year on average. *Answers may vary.*

(c) $y = 1.56(2007 - 1999) + 1.0 \Rightarrow y = 1.56(8) + 1.0 \Rightarrow y = 12.48 + 1 \Rightarrow y \approx 13.5$ billion.

Answers may vary.

87. (a) See Figure 87.

(b) Use the first and last points to find slope $m = \dfrac{2 - 1.4}{2004 - 1998} = \dfrac{0.6}{6} = 0.1$. Now using the first point

and slope $m = 0.1$, we get $y = 0.1(x - 1998) + 1.4$. U.S. sales of Toyota vehicles has increased

by 0.1 million per year.

(c) $f(x)$ is an exact model for the listed data.

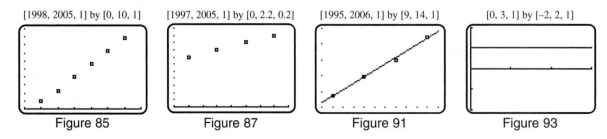

[1998, 2005, 1] by [0, 10, 1]   [1997, 2005, 1] by [0, 2.2, 0.2]   [1995, 2006, 1] by [9, 14, 1]   [0, 3, 1] by [−2, 2, 1]

Figure 85   Figure 87   Figure 91   Figure 93

89. (a) The annual fixed cost would be $350 \times 12 = \$4200$. The variable cost of driving x miles is $0.29x$. Thus,

$f(x) = 0.29x + 4200$.

(b) The y-intercept is 4200, which represents the annual fixed costs. This means that even if the car is not driven, it will still cost \$4200 each year to own it.

91. (a) Scatterplot the data in the table.

(b) Start by picking a data point for the line to pass through. If we choose $(1996, 9.7)$, f is represented by

$f(x) = m(x - 1996) + 9.7$. Using trial and error, the slope m is between 0 and 1. Let $m = 0.42$. The graph of f together with the scatterplot is shown in Figure 91. *Answers may vary.*

(c) A slope of $m \approx 0.42$ means that Asian-American population is predicted to increase by approximately 0.42 million (420,000) people each year.

(d) To predict the population in the year 2008, evaluate $f(2008) = 0.4167(2008 - 1996) + 9.7 \approx 14.7$ million people.

93. (a) Graph $Y_1 = X/1024 + 1$ in $[0, 3, 1]$ by $[-2, 2, 1]$ as in Figure 93. The line appears to be horizontal in this viewing rectangle, however, we know that the graph of the line is not horizontal because its slope

is $\dfrac{1}{1024} \neq 0$.

(b) The resolution of most graphing calculator screens is not good enough to show the slight increase in the y-values. Since the x-axis is 3 units long, this increase in y-values amounts to only $\dfrac{1}{1024} \times 3 \approx 0.003$ units, which does not show up on the screen.

95. (a) From Figure 95a, one can see that the lines do not appear to be perpendicular.

(b) The lines are graphed in the specified viewing rectangles and shown in Figures 95b-d, respectively. In the windows $[-15, 15, 1]$ by $[-10, 10, 1]$ and $[-3, 3, 1]$ by $[-2, 2, 1]$ the lines appear to be perpendicular.

(c) The lines appear perpendicular when the distance shown along the x-axis is approximately 1.5 times the distance along the y-axis. For example, in window $[-12, 12, 1]$ by $[-8, 8, 1]$, the lines will appear perpendicular. The distance along the x-axis is 24 while the distance along the y-axis is 16. Notice that $1.5 \times 16 = 24$. This is called a "square window" and can be set automatically on some graphing calculators.

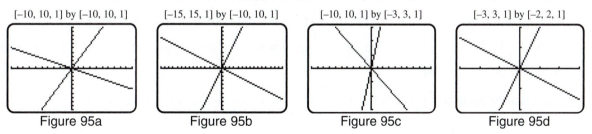

$[-10, 10, 1]$ by $[-10, 10, 1]$	$[-15, 15, 1]$ by $[-10, 10, 1]$	$[-10, 10, 1]$ by $[-3, 3, 1]$	$[-3, 3, 1]$ by $[-2, 2, 1]$
Figure 95a	Figure 95b	Figure 95c	Figure 95d

97. (i)   The slope of the line connecting $(0, 0)$ and $(2, 2)$ is 1. Let $y_1 = x$.

(ii)   A second line passing through $(0, 0)$ has a slope of -1. Let $y_2 = -x$.

(iii)   A third line passing through $(1, 3)$ has a slope of 1. Let $y_3 = (x - 1) + 3 = x + 2$.

(iv)   A fourth line passing through $(2, 2)$ has a slope of -1. Let $y_4 = -(x - 2) + 2 = -x + 4$.

99. (i)   The slope of the line connecting $(-4, 0)$ and $(0, 4)$ is 1. Let $y_1 = x + 4$.

    (ii)   A second line passing through $(4, 0)$ and $(0, -4)$ has a slope of 1. Let $y_2 = x - 4$.

    (iii) A third line passing through $(0, -4)$ and $(-4, 0)$ has a slope of -1. Let $y_3 = -x - 4$.

    (iv)   A fourth line passing through $(0, 4)$ and $(4, 0)$ is -1. Let $y_4 = -x + 4$.

101. Since y is directly proportional to x, the variation equation $y = kx$ must hold. To find the value of k, use the

    value $y = 7$ when $x = 14$. Solve the equation $7 = k(14) \Rightarrow k = \dfrac{1}{2}$. Then $y = \dfrac{1}{2}(5) = \dfrac{5}{2} = 2.5$.

103. Since y is directly proportional to x, the variation equation $y = kx$ must hold. To find the value of k, use the

    value $y = \dfrac{3}{2}$ when $x = \dfrac{2}{3}$. Solve the equation $\dfrac{3}{2} = k\left(\dfrac{2}{3}\right) \Rightarrow k = \dfrac{9}{4}$. Then $y = \dfrac{9}{4}\left(\dfrac{1}{2}\right) = \dfrac{9}{8}$.

105. Since y is directly proportional to x, the variation equation $y = kx$ must hold. To find the value of k use the

    value $y = 7.5$ when $x = 3$ from the table. Solve the equation $7.5 = k(3) \Rightarrow k = 2.5$. The variation equation

    is $y = 2.5x$ and hence $y = 2.5(8) = 20$ when $x = 8$. A graph of $Y_1 = 2.5X$ together with the data points is

    shown in Figure 105.

107. Since y is directly proportional to x, the variation equation $y = kx$ must hold. To find the value of k use the

    value $y = 1.50$ when $x = 25$ from the table. Solve the equation $1.50 = k(25) \Rightarrow k = 0.06$. The variation

    equation is $y = 0.06x$ and hence when $y = 5.10$, $x = \dfrac{5.1}{0.06} = 85$. A graph of $Y_1 = 0.06X$ together with the

    data points is shown in Figure 107.

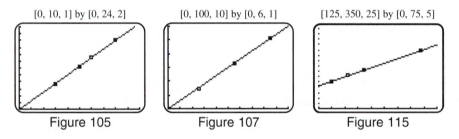

[0, 10, 1] by [0, 24, 2]   [0, 100, 10] by [0, 6, 1]   [125, 350, 25] by [0, 75, 5]

Figure 105   Figure 107   Figure 115

109. Let y represent the cost of tuition and x represent the number of credits taken. Since the cost of tuition is

    directly proportional to the number of credits taken, the variation equation $y = kx$ must hold. If cost

    $y = \$720.50$ when the number of credits $x = 11$, we find the constant of proportionality k by solving

    $720.50 = k(11) \Rightarrow k = 65.50$. The variation equation is $y = 65.50x$. Therefore, the cost of taking 16 credits

    is $y = 65.50(16) = \$1048$.

111. (a) Since the points $(0, 0)$ and $(300, 3)$ lie on the graph of $y = kx$, the slope of the graph is $\dfrac{3 - 0}{300 - 0} = 0.01$

       and $y = 0.01x$, so $k = 0.01$.

    (b) $y = 0.01(110) = 1.1$ millimeters.

113. (a) Using $F = kx \Rightarrow 15 = k(8) \Rightarrow k = \dfrac{15}{8}$

    (b) The variation equation is $y = \dfrac{15}{8}x$; $25 = \dfrac{15}{8}(x) \Rightarrow x = 13\dfrac{1}{3}$ inches.

115. (a) For $(150, 26)$, $\dfrac{F}{x} = \dfrac{26}{150} \approx 0.173$; for $(180, 31)$, $\dfrac{F}{x} = \dfrac{31}{180} \approx 0.172$; for $(210, 36)$, $\dfrac{F}{x} = \dfrac{36}{210} \approx 0.171$;

for $(320, 54)$, $\dfrac{F}{x} = \dfrac{54}{320} \approx 0.169$; the ratios give the force needed to push a 1 lb box.

(b) From the table it appears that approximately 0.17 lb of force is needed to push a 1 lb cargo box $\Rightarrow$

$k = 0.17$.

(c) See Figure 115.

(d) $F \approx 0.17(275) \Rightarrow F = 46.75$ lbs of force.

2.3: Linear Equations

1. $ax + b = 0 \Rightarrow ax = -b \Rightarrow x = \dfrac{-b}{a}$. This shows that the equation $ax + b = 0$ has only one solution.

3. $4 - (5 - 4x) = 4 - 5 + 4x = -1 + 4x = 4x - 1$

5. The zero of f and the x-intercept of the graph of f are equal. The zero of f and the x-intercept of the graph of f are both found by finding the value of x when $y = 0$.

7. $3x - 1.5 = 7 \Rightarrow 3x - 1.5 - 7 = 0 \Rightarrow 3x - 8.5 = 0$; the equation is linear.

9. $2\sqrt{x} + 2 = 1$; since the equation cannot be written in the form $ax + b = 0$, it is nonlinear.

11. $7x - 5 = 3(x - 8) \Rightarrow 7x - 5 = 3x - 24 \Rightarrow 4x + 19 = 0$; the equation is linear.

13. $2x - 8 = 0 \Rightarrow 2x = 8 \Rightarrow x = 4$ Check: $2(4) - 8 = 0 \Rightarrow 8 - 8 = 0 \Rightarrow 0 = 0$

15. $-5x + 3 = 23 \Rightarrow -5x = 20 \Rightarrow x = -4$ Check: $-5(-4) + 3 = 23 \Rightarrow 20 + 3 = 23 \Rightarrow 23 = 23$

17. $4(z - 8) = z \Rightarrow 4z - 32 = z \Rightarrow 3z = 32 \Rightarrow z = \dfrac{32}{3}$ Check: $4\left(\dfrac{32}{3} - 8\right) = \dfrac{32}{3} \Rightarrow 4\left(\dfrac{8}{3}\right) = \dfrac{32}{3} \Rightarrow$

$\dfrac{32}{32} = \dfrac{32}{32}$

19. $-5(3 - 4t) = 65 \Rightarrow -15 + 20t = 65 \Rightarrow 20t = 80 \Rightarrow t = 4$ Check: $-5[3 - 4(4)] = 65 \Rightarrow$

$-5(3 - 16) = 65 \Rightarrow -5(-13) = 65 \Rightarrow 65 = 65$

21. $k + 8 = 5k - 4 \Rightarrow -4k = -12 \Rightarrow k = 3$ Check: $3 + 8 = 5(3) - 4 \Rightarrow 11 = 15 - 4 \Rightarrow 11 = 11$

23. $2(1 - 3x) + 1 = 3x \Rightarrow 2 - 6x + 1 = 3x \Rightarrow -6x + 3 = 3x \Rightarrow -9x = -3 \Rightarrow x = \dfrac{1}{3}$

Check: $2\left[1 - 3\left(\dfrac{1}{3}\right)\right] + 1 = 3\left(\dfrac{1}{3}\right) \Rightarrow 2(1 - 1) + 1 = 1 \Rightarrow 0 + 1 = 1 \Rightarrow 1 = 1$

25. $-5(3 - 2x) - (1 - x) = 4(x - 3) \Rightarrow -15 + 10x - 1 + x = 4x - 12 \Rightarrow 11x - 16 = 4x - 12 \Rightarrow$

$7x = 4 \Rightarrow x = \dfrac{4}{7}$ Check: $-5\left[3 - 2\left(\dfrac{4}{7}\right)\right] - \left(1 - \dfrac{4}{7}\right) = 4\left(\dfrac{4}{7} - 3\right) \Rightarrow$

$-5\left(\dfrac{13}{7}\right) - \dfrac{3}{7} = 4\left(-\dfrac{17}{7}\right) \Rightarrow -\dfrac{65}{7} - \dfrac{3}{7} = -\dfrac{68}{7} \Rightarrow -\dfrac{68}{7} = -\dfrac{68}{7}$

27. $-4(5x - 1) = 8 - (x + 2) \Rightarrow -20x + 4 = 8 - x - 2 \Rightarrow -20x + 4 = 6 - x \Rightarrow -19x = 2 \Rightarrow$

    $x = -\dfrac{2}{19}$ Check: $-4\left[5\left(-\dfrac{2}{19}\right) - 1\right] = 8 - \left(-\dfrac{2}{19} + 2\right) \Rightarrow -4\left(-\dfrac{10}{19} - 1\right) = 8 + \dfrac{2}{19} - 2 \Rightarrow$

    $\dfrac{40}{19} + 4 = 6 + \dfrac{2}{19} \Rightarrow \dfrac{116}{19} = \dfrac{116}{19}$

29. $\dfrac{2}{7}n + \dfrac{1}{5} = \dfrac{4}{7} \Rightarrow \dfrac{2}{7}n = \dfrac{13}{7} \Rightarrow n = \dfrac{13}{35} \Rightarrow n = \dfrac{13}{10}$ Check: $\dfrac{2}{7}\left(\dfrac{13}{10}\right) + \dfrac{1}{5} = \dfrac{4}{7} \Rightarrow \dfrac{26}{70} + \dfrac{1}{5} = \dfrac{4}{7} \Rightarrow \dfrac{40}{70} = \dfrac{4}{7} \Rightarrow$

    $\dfrac{4}{7} = \dfrac{4}{7}$

31. $\dfrac{1}{2}(d - 3) - \dfrac{2}{3}(2d - 5) = \dfrac{5}{12} \Rightarrow \dfrac{1}{2}d - \dfrac{3}{2} - \dfrac{4}{3}d + \dfrac{10}{3} = \dfrac{5}{12} \Rightarrow -\dfrac{5}{6}d + \dfrac{11}{6} = \dfrac{5}{12} \Rightarrow$

    $-\dfrac{5}{6}d = -\dfrac{17}{12} \Rightarrow d = \dfrac{17}{10}$

    Check: $\dfrac{1}{2}\left(\dfrac{17}{10} - 3\right) - \dfrac{2}{3}\left[2\left(\dfrac{17}{10}\right) - 5\right] = \dfrac{5}{12} \Rightarrow \dfrac{1}{2}\left(-\dfrac{13}{10}\right) - \dfrac{2}{3}\left(\dfrac{34}{10} - 5\right) = \dfrac{5}{12} \Rightarrow$

    $\dfrac{1}{2}\left(-\dfrac{13}{10}\right) - \dfrac{2}{3}\left(-\dfrac{16}{10}\right) = \dfrac{5}{12} \Rightarrow -\dfrac{13}{20} + \dfrac{32}{30} = \dfrac{5}{12} \Rightarrow \dfrac{25}{60} = \dfrac{5}{12} \Rightarrow \dfrac{5}{12} = \dfrac{5}{12}$

33. $\dfrac{x - 5}{3} + \dfrac{3 - 2x}{2} = \dfrac{5}{4} \Rightarrow 12\left(\dfrac{x - 5}{3} + \dfrac{3 - 2x}{2}\right) = 12\left(\dfrac{5}{4}\right) \Rightarrow 4x - 20 + 18 - 12x = 15 \Rightarrow$

    $-8x - 2 = 15 \Rightarrow -8x = 17 \Rightarrow x = -\dfrac{17}{8}$ Check: $\dfrac{-\frac{17}{8} - 5}{3} + \dfrac{3 - 2(-\frac{17}{8})}{2} = \dfrac{5}{4} \Rightarrow$

    $\dfrac{-\frac{57}{8}}{3} + \dfrac{3 + \frac{34}{8}}{2} = \dfrac{5}{4} \Rightarrow -\dfrac{19}{8} + \dfrac{\frac{58}{8}}{2} = \dfrac{5}{4} \Rightarrow -\dfrac{19}{8} + \dfrac{29}{8} = \dfrac{5}{4} \Rightarrow \dfrac{10}{8} = \dfrac{5}{4} \Rightarrow \dfrac{5}{4} = \dfrac{5}{4}$

35. $0.1z - 0.05 = -0.07z \Rightarrow 0.17z = 0.05 \Rightarrow z = \dfrac{0.05}{0.17} \Rightarrow z = \dfrac{5}{17}$ Check:

    $0.1\left(\dfrac{5}{17}\right) - 0.05 = -0.07\left(\dfrac{5}{17}\right) \Rightarrow \dfrac{5}{170} - \dfrac{5}{100} = -\dfrac{35}{1700} \Rightarrow \dfrac{50}{1700} - \dfrac{85}{1700} = -\dfrac{35}{1700} \Rightarrow$

    $-\dfrac{35}{1700} = -\dfrac{35}{1700}$

37. $0.15t + 0.85(100 - t) = 0.45(100) \Rightarrow 0.15t + 85 - 0.85t = 45 \Rightarrow -0.7t = -40 \Rightarrow t = \dfrac{40}{0.7} \Rightarrow$

    $t = \dfrac{400}{7}$ Check: $0.15\left(\dfrac{400}{7}\right) + 0.85\left(100 - \dfrac{400}{7}\right) = 0.45(100) \Rightarrow \dfrac{6000}{700} + 85 - \dfrac{34{,}000}{700} = 45 \Rightarrow$

    $\dfrac{60}{7} + 85 - \dfrac{340}{7} = 45 \Rightarrow 85 - \dfrac{280}{7} = 45 \Rightarrow 85 - 40 = 45 \Rightarrow 45 = 45$

39. (a) $5x - 1 = 5x + 4 \Rightarrow -1 = 4 \Rightarrow$ there is no solution.

    (b) Since no x-value satisfies the equation, it is contradiction.

41. (a) $3(x - 1) = 5 \Rightarrow 3x - 3 = 5 \Rightarrow 3x = 8 \Rightarrow x = \dfrac{8}{3}$

    (b) Since one x-value is a solution and other x-values are not, the equation is conditional.

43. (a) $0.5(x - 2) + 5 = 0.5x + 4 \Rightarrow 0.5x - 1 + 5 = 0.5x + 4 \Rightarrow 0.5x + 4 = 0.5x + 4 \Rightarrow$ every x-value satisfies this equation.

    (b) Since every x-value satisfies the equation, it is an identity.

45. (a) $\dfrac{t+1}{2} = \dfrac{3t-2}{6} \Rightarrow 6\left(\dfrac{t+1}{2} = \dfrac{3t-2}{6}\right) \Rightarrow 3t+3 = 3t-2 \Rightarrow 3 = -2 \Rightarrow$ there is no solution.

(b) Since no x-value satisfies the equation, it is contradiction.

47. (a) $\dfrac{1-2x}{4} = \dfrac{3x-1.5}{-6} \Rightarrow -6(1-2x) = 4(3x-1.5) \Rightarrow -6+12x = 12x-6 \Rightarrow 0 = 0 \Rightarrow$ every

x-value satisfies this equation.

(b) Since every x-value satisfies the equation, it is an identity.

49. In the graph, the lines intersect at $(3, -1)$. The solution is the x-value, 3.

51. (a) From the graph, when $f(x)$ or $y = -1$, $x = 4$; the solution is the x-value, 4.

(b) From the graph, when $f(x)$ or $y = 0$, $x = 2$; the solution is the x-value, 2.

(c) From the graph, when $f(x)$ or $y = 2$, $x = -2$; the solution is the x-value, -2.

53. Graph $Y_1 = X + 4$ and $Y_2 = 1 - 2X$. See Figure 53. The lines intersect at $x = -1$.

$x + 4 = 1 - 2x \Rightarrow 3 = -3x \Rightarrow x = -1$

55. Graph $Y_1 = -X + 4$ and $Y_2 = 3X$. See Figure 55. The lines intersect at $x = 1$.

$-x + 4 = 3x \Rightarrow 4 = 4x \Rightarrow x = 1$

Figure 53	Figure 55	Figure 57

57. Graph $Y_1 = 2(X - 1) - 2$ and $Y_2 = X$. See Figure 57. The lines intersect at $x = 4$.

$2(x - 1) - 2 = x \Rightarrow 2x - 2 - 2 = x \Rightarrow 2x - 4 = x \Rightarrow -4 = -x \Rightarrow x = 4$

59. Graph $Y_1 = 5X - 1.5$ and $Y_2 = 5$. Their graphs intersect at $(1.3, 5)$. The solution is 1.3. See Figure 59.

61. Graph $Y_1 = 3X - 1.7$ and $Y_2 = 1 - X$. Their graphs intersect at $(0.675, 0.325)$. The solution is 0.675.
See Figure 61.

63. Graph $Y_1 = 3.1(X - 5)$ and $Y_2 = X/5 - 5$. Their graphs intersect near $(3.621, -4.276)$. The solution is
approximately 3.621. See Figure 63.

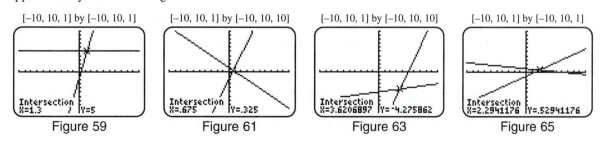

[−10, 10, 1] by [−10, 10, 1]	[−10, 10, 1] by [−10, 10, 10]	[−10, 10, 1] by [−10, 10, 10]	[−10, 10, 1] by [−10, 10, 1]
Figure 59	Figure 61	Figure 63	Figure 65

65. Graph $Y_1 = (6 - X)/7$ and $Y_2 = (2X - 3)/3$. Their graphs intersect near $(2.294, 0.529)$. The solution is
approximately 2.294. See Figure 65.

67. One way to solve this equation is to table $Y_1 = 2X - 7$ and determine the x-value where $Y_1 = -1$. See Figure 67. This occurs when $x = 3$, so the solution is 3.

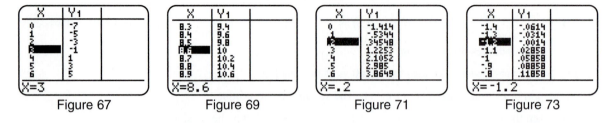

Figure 67   Figure 69   Figure 71   Figure 73

69. Table $Y_1 = 2X - 7.2$ and determine the x-value where $Y_1 = 10$. See Figure 69. This occurs when $x = 8.6$, so the solution is 8.6.

71. Table $Y_1 = \sqrt{(2)}(4X - 1) + \pi X$ and determine the x-value where $Y_1 = 0$. See Figure 71. This occurs when $x \approx 0.2$, so the solution is 0.2.

73. Table $Y_1 = 0.5 - 0.1(\sqrt{(2)} - 3X)$ and determine the x-value where $Y_1 = 0$. See Figure 73. This occurs when $x \approx -1.2$, so the solution is -1.2.

75. (a) $5 - (x + 1) = 3 \Rightarrow 5 - x - 1 = 3 \Rightarrow 4 - x = 3 \Rightarrow x = 1$

    (b) Using the intersection of graphs method, graph $Y_1 = 5 - (X + 1)$ and $Y_2 = 3$. Their point of intersection is shown in Figure 75b as $(1, 3)$. The solution is the x-value, 1.

    (c) Table $Y_1 = 5 - (X + 1)$ and $Y_2 = 3$. Figure 75c shows a table where $Y_1 = Y_2$ at $x = 1$.

[−10, 10, 1] by [−10, 10, 1]   [−10, 10, 1] by [−10, 10, 1]

Figure 75b   Figure 75c   Figure 77b   Figure 77c

77. (a) $\sqrt{3}(2 - \pi x) + x = 0 \Rightarrow 2\sqrt{3} - \sqrt{3}\pi x + x = 0 \Rightarrow -\sqrt{3}\pi x + x = -2\sqrt{3} \Rightarrow$

    $x(-\sqrt{3}\pi + 1) = -2\sqrt{3} \Rightarrow x = \dfrac{-2\sqrt{3}}{(-\sqrt{3}\pi + 1)} \approx 0.8.$

    (b) Using the intersection of graphs method, graph $Y_1 = \sqrt{(3)}(2 - \pi X) + X$ and $Y_2 = 0$. Their point of intersection is shown in Figure 77b as approximately $(0.8, 0)$. The solution is the x-value, 0.8.

    (c) Table $Y_1 = \sqrt{(3)}(2 - \pi X) + X$ and $Y_2 = 0$. Figure 77c shows a table where $Y_1 = Y_2$ at $x \approx 0.8$.

79. (a) $x - 3 = 2x + 1 \Rightarrow -x = 4 \Rightarrow x = -4$

    (b) Using the intersection-of-graphs method, graph $Y_1 = X - 3$ and $Y_2 = 2X + 1$. Their point of intersection is shown in Figure 79b as $(-4, -7)$. The solution is the x-value, -4.

    (c) Table $Y_1 = X - 3$ and $Y_2 = 2X + 1$, starting at $x = -7$, incrementing by 1. Figure 79c shows a table where $Y_1 = Y_2$ at $x = -4$.

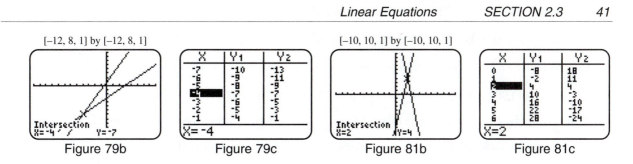

Figure 79b   Figure 79c   Figure 81b   Figure 81c

81. (a) $6x - 8 = -7x + 18 \Rightarrow 13x = 26 \Rightarrow x = 2$

(b) Using the intersection-of-graphs method, graph $Y_1 = 6X - 8$ and $Y_2 = -7X + 18$. Their point of intersection is shown in Figure 81b as $(2, 4)$. The solution is the x-value, 2.

(c) Table $Y_1 = 6X - 8$ and $Y_2 = -7X + 18$, starting at $x = 0$, incrementing by 1. Figure 81c shows a table where $Y_1 = Y_2$ at $x = 2$.

83. $A = LW \Rightarrow W = \dfrac{A}{L}$

85. $P = 2L + 2W \Rightarrow P - 2W = 2L \Rightarrow L = \dfrac{P - 2W}{2} \Rightarrow L = \dfrac{1}{2}P - W$

87. $3x + 2y = 8 \Rightarrow 2y = 8 - 3x \Rightarrow y = 4 - \dfrac{3}{2}x$

89. $y = 3(x - 2) + x \Rightarrow y = 3x - 6 + x \Rightarrow y = 4x - 6 \Rightarrow 4x = y + 6 \Rightarrow x = \dfrac{1}{4}y + \dfrac{3}{2}$

91. $f(x) = 19,000$ and $f(x) = 1000(x - 1980) + 10,000 \Rightarrow 1000(x - 1980) + 10,000 = 19,000 \Rightarrow$ $1000(x - 1980) = 9000 \Rightarrow x - 1980 = 9 \Rightarrow x = 1989$. The per capita income was \$19,000 in 1989.

93. Using the intersection of graphs method, graph $Y_1 = 51.6(X - 1985) + 9.1$ and $Y_2 = -31.9(X - 1985) + 167.7$. Their approximate point of intersection is shown in Figure 93 as $(1987, 107)$. In approximately 1987 the sales of LP records and compact discs were equal.

[1980, 1990, 2] by [0, 200, 20]

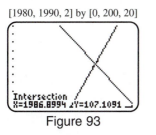

Figure 93

95. The graph of f must pass through the points $(1980, 64)$ and $(2000, 80)$. Its slope is $m = \dfrac{80 - 64}{2000 - 1980} = 0.8$. Thus, $f(x) = 0.8(x - 2000) + 80$. Find x when $f(x) = 87 \Rightarrow 87 = 0.8(x - 2000) + 80 \Rightarrow$ $7 = 0.8(x - 2000) \Rightarrow (x - 2000) = 8.75 \Rightarrow x = 2008.75$. The US population density reached 87 people per square mile in about 2009.

97. To calculate the sale price subtract 25% of the regular price from the regular price. $f(x) = x - 0.25x \Rightarrow f(x) = 0.75x$. An item which normally costs \$56.24 will be on sale for $f(56.24) = 0.75(56.24) = \$42.18$.

99. (a) The number of skin cancer cases can is given by 0.045x.

   (b) There were 65,000 cases of skin cancer diagnosed in 2007. So, $65,000 = 0.045x \Rightarrow x = \dfrac{65,000}{0.045} \Rightarrow$

   $x = 1,444,000$. There were about 1,444,000 cancer cases in 2007.

101. (a) It would take a little less time than the faster gardener, who can rake the lawn alone in 3 hours. It would

   take both gardeners about 2 hours working together. *Answers may vary.*

   (b) Let $x =$ time to rake the lawn working together. In 1 hour thge first gardener can rake $\dfrac{1}{3}$ of the lawn,

   whereas the second gardener can rake $\dfrac{1}{5}$ of the lawn; in x hours both gardeners working together can rake

   $\dfrac{x}{3} + \dfrac{x}{5}$ of the lawn; $\dfrac{x}{3} + \dfrac{x}{5} = 1 \Rightarrow 5x + 3x = 15 \Rightarrow 8x = 15 \Rightarrow x = \dfrac{15}{8} = 1.875$ hours.

103. Let $t =$ time spent traveling at 55 mph and $6 - t =$ time spent traveling at 70 mph. Using $d = rt$, we get

   $d = 55t + 70(6 - t) \Rightarrow 372 = 55t + 420 - 70t \Rightarrow -48 = -15t \Rightarrow t = 3.2$ and $6 - t = 2.8$; the car

   traveled 3.2 hours at 55 mph and 2.8 hours at 70 mph.

105. Let $t =$ time traveled by car at 55 mph and $t + \dfrac{1}{2} =$ time traveled by runner at 10 mph; since $d = rt$ and the

   distance is the same for both runner and driver, we get $55t = 10\left(t + \dfrac{1}{2}\right) \Rightarrow 55t = 10t + 5 \Rightarrow 45t = 5 \Rightarrow$

   $t = \dfrac{1}{9}$; it takes the driver $\dfrac{1}{9}$ hour or $6\dfrac{2}{3}$ minutes to catch the runner.

107. The follow sketch illustrates the situation, where $x =$ height of streetlight. See Figure 107.

   Using similar triangles, we get $\dfrac{x}{15 + 7} = \dfrac{5.5}{7} \Rightarrow x = \dfrac{(5.5)(22)}{7} \Rightarrow x \approx 17.29$. The streetlight is about

   17.29 feet high.

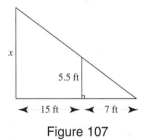

Figure 107

109. Use similar triangles to find the radius of the cone when the water is 7 feet deep: $\dfrac{r}{3.5} = \dfrac{7}{11} \Rightarrow r \approx \dfrac{49}{22}$ ft.

   Use $V = \dfrac{1}{3}\pi r^2 h$ to find the volume of the water in the cone at $h = 7$ ft.: $V = \dfrac{1}{3}\pi\left(\dfrac{49}{22}\right)^2 (7) \approx 36.4$ ft^3.

111. Let $x =$ amount of pure water to be added and $x + 5 =$ final amount of the 15% solution. Since pure water

   is 0% sulfuric acid, we get $0\%x + 40\%(5) = 15\%(x + 5) \Rightarrow 0.40(5) = 0.15(x + 5) \Rightarrow$

   $40(5) = 15(x + 5) \Rightarrow 200 = 15x + 75 \Rightarrow 15x = 125 \Rightarrow x = \dfrac{125}{15} \approx 8.333$; about 8.33 liters of pure

   water should be added.

113. $P = 2w + 2l \Rightarrow 180 = 2w + 2(w + 18) \Rightarrow 180 = 4w + 36 \Rightarrow 4w = 144 \Rightarrow w = 36$ and

   $w + 18 = 54$; the window is 36 inches by 54 inches.

115. (a) Using $(2002, 75)$ and $(2006, 45)$, find slope: $m = \dfrac{45 - 75}{2006 - 2002} \Rightarrow \dfrac{-30}{4} \Rightarrow -7.5$; therefore,

   $C(x) = -7.5(x - 2002) + 75 \Rightarrow C(x) = -7.5x + 15{,}090$. Now using $(2002, 29)$ and $(2006, 88)$,

   find slope: $m = \dfrac{88 - 29}{2006 - 2002} = \dfrac{59}{4} = 14.75$. Therefore, $L(x) = 14.75(x - 2002) + 29 \Rightarrow$

   $L(x) = 14.75x - 29{,}500.5$.

   (b) Sales of CRT monitors decreased by 7.5 million per year, on average. Sales of LCD monitors increased

   by 14.75 million monitors per year, on average.

   (c) Graph $C(x) = -7.5x + 15{,}090$ and $L(x) = 14.75x - 29{,}500$. See Figure 115c. The lines of the

   functions intersect at $x = 2004$ or year 2004.

   (d) Set $L(x) = C(x)$ and solve: $14.75x - 29{,}500 = -7.5x + 15{,}090 \Rightarrow 22.25x = 44{,}590 \Rightarrow x \approx 2004$

   (e) Table $Y_1 = -7.5X + 15{,}090$ and $Y_2 = 14.75X - 29{,}500$, starting at 2000, incrementing by 1.

   Figure 115e shows a table where $Y_1 = Y_2$ and $x = 2004$.

[2000, 2008, 1] by [0, 80, 10]

Figure 115c   Figure 115e

117. $C = \dfrac{5}{9}(F - 32)$ and $F = C \Rightarrow F = \dfrac{5}{9}(F - 32) \Rightarrow F = \dfrac{5}{9}F - \dfrac{160}{9} \Rightarrow \dfrac{4}{9}F = -\dfrac{160}{9} \Rightarrow F = -40$;

   $-40° F$ is equivalent to $-40° C$.

119. (a) It is reasonable to expect that f is linear because if the number of gallons of gas doubles so should the

   amount of oil. Five gallons of gasoline requires five times the oil that one gallon of gasoline would. The

   increase in oil is always equal to 0.16 pint for each additional gallon of gasoline. Oil is mixed at a

   constant rate, so a linear function describes this amount.

   (b) $f(3) = 0.16(3) = 0.48$; 0.48 pint of oil should be added to 3 gallons of gasoline to get the correct

   mixture.

   (c) $0.16x = 2 \Rightarrow x = 12.5$; 12.5 gallons of gasoline should be mixed with 2 pints of oil.

121. Linear regression gives the model:

   $y = 0.36x - 0.21.$ $y = 2.99 \Rightarrow 2.99 = 0.36x - 0.21 \Rightarrow 3.2 = 0.36x \Rightarrow x \approx 8.89$

123. (a) Linear regression gives the model:

   $S(x) \approx 3.974x - 14{,}479.$ *Answers may vary.*

   (b) $S(x) = 6 \Rightarrow 6 = 3.974x - 14.479 \Rightarrow 20.479 = 3.974x \Rightarrow x \approx 5.15$; The circumference of a

   finger with ring size 6 is approximately 5.15 cm.

125. (a) Linear regression gives the model:

$f(x) \approx 0.10677x - 211.69$. *Answers may vary.*

(b) $f(1987) = 0.10677(1987) - 211.69 \Rightarrow f(1987) \approx 0.462$.

The cost of a 30-scond Super Bowl ad in 1987 was approximated to be \$0.5 million. The estimate that was found involved extrapolation.

(c) $f(x) = 3.2 \Rightarrow 3.2 = 0.10677x - 211.69 \Rightarrow 214.89 = 0.10677x \Rightarrow x \approx 2012.644$.

Thus, the cost for a 30-second Super Bowl ad could reach \$3.2 million in 2013.

2.4: Linear Inequalities

1. $(-\infty, 2)$

3. $[-1, \infty)$

5. $[1, 8)$

7. $(-\infty, 1]$

9. $2x + 6 \geq 10 \Rightarrow 2x \geq 4 \Rightarrow x \geq 2$; $[2, \infty)$; set-builder notation the interval is $\{x \mid x \geq 2\}$.

11. $-2(x - 10) + 1 > 0 \Rightarrow -2x + 21 > 0 \Rightarrow -2x > -21 \Rightarrow x < 10.5$; $(-\infty, 10.5)$; set-builder notation the interval is $\{x \mid x < 10.5\}$.

13. $\dfrac{t + 2}{3} \geq 5 \Rightarrow t + 2 \geq 15 \Rightarrow t \geq 13$; $[13, \infty)$; set-builder notation the interval is $\{t \mid t \geq 13\}$.

15. $4x - 1 < \dfrac{3 - x}{-3} = 7 \Rightarrow -12x + 3 > 3 - x \Rightarrow -11x > 0 \Rightarrow x < 0$; $(-\infty, 0)$; set-builder notation the interval is $\{x \mid x < 0\}$.

17. $-3(z - 4) \geq 2(1 - 2z) \Rightarrow -3z + 12 \geq 2 - 4z \Rightarrow z \geq -10$; $[-10, \infty)$; set-builder notation the interval is $\{z \mid z \geq -10\}$.

19. $\dfrac{1 - x}{4} < \dfrac{2x - 2}{3} \Rightarrow 3(1 - x) < 4(2x - 2) \Rightarrow 3 - 3x < 8x - 8 \Rightarrow -11x < -11 \Rightarrow x > 1$; $(1, \infty)$; set-builder notation the interval is $\{x \mid x > 1\}$.

21. $2x - 3 > \dfrac{1}{2}(x + 1) \Rightarrow 2x - 3 > \dfrac{1}{2}x + \dfrac{1}{2} \Rightarrow \dfrac{3}{2}x > \dfrac{7}{2} \Rightarrow x > \dfrac{7}{3}$; $\left(\dfrac{7}{3}, \infty\right)$; set-builder notation the interval is $\left\{x \mid x > \dfrac{7}{3}\right\}$.

23. $5 < 4t - 1 \leq 11 \Rightarrow 6 < 4t \leq 12 \Rightarrow \dfrac{3}{2} < t \leq 3$; $\left(\dfrac{3}{2}, 3\right]$; set-builder notation the interval is $\left\{t \mid \dfrac{3}{2} < t \leq 3\right\}$.

25. $3 \leq 4 - x \leq 20 \Rightarrow -1 \leq -x \leq 16 \Rightarrow 1 \geq x \geq -16$; $[-16, 1]$; set-builder notation the interval is $\{x \mid -16 \leq x \leq 1\}$.

27. $-7 \leq \dfrac{1 - 4x}{7} < 12 \Rightarrow -49 \leq 1 - 4x < 84 \Rightarrow -50 \leq -4x < 83 \Rightarrow 12.5 \geq x > -20.75$; $(-20.75, 12.5]$; set-builder notation the interval is $\{x \mid -20.75 < x \leq 12.5\}$.

29. $5 > 2(x + 4) - 5 > -5 \Rightarrow 5 > 2x + 8 - 5 > -5 \Rightarrow 2 > 2x > -8 \Rightarrow 1 > x > -4;\ (-4, 1);$

    set-builder notation the interval is $\{x \mid -4 < x < 1\}$.

31. $3 \le \dfrac{1}{2}x + \dfrac{3}{4} \le 6 \Rightarrow 12 \le 2x + 3 \le 24 \Rightarrow 9 \le 2x \le 21 \Rightarrow \dfrac{9}{2} \le x \le \dfrac{21}{2};\ \left[\dfrac{9}{2}, \dfrac{21}{2}\right];$ set-builder

    notation the interval is $\left\{x \mid \dfrac{9}{2} \le x \le \dfrac{21}{2}\right\}$.

33. $5x - 2(x + 3) \ge 4 - 3x \Rightarrow 5x - 2x - 6 \ge 4 - 3x \Rightarrow 6x \ge 10 \Rightarrow x \ge \dfrac{5}{3};\ \left[\dfrac{5}{3}, \infty\right);$ set-builder

    notation the interval is $\left\{x \mid x \ge \dfrac{5}{3}\right\}$.

35. $\dfrac{1}{2} \le \dfrac{1 - 2t}{3} < \dfrac{2}{3} \Rightarrow \dfrac{3}{2} \le 1 - 2t < 2 \Rightarrow \dfrac{1}{2} \le -2t < 1 \Rightarrow -\dfrac{1}{4} \ge t > -\dfrac{1}{2} \Rightarrow -\dfrac{1}{2} < t \le -\dfrac{1}{4};$

    $\left(-\dfrac{1}{2}, -\dfrac{1}{4}\right];$ set-builder notation the interval is $\left\{t \mid -\dfrac{1}{2} < t \le -\dfrac{1}{4}\right\}$.

37. $\dfrac{1}{2}z + \dfrac{2}{3}(3 - z) - \dfrac{5}{4}z \ge \dfrac{3}{4}(z - 2) + z \Rightarrow \dfrac{1}{2}z + 2 - \dfrac{2}{3}z - \dfrac{5}{4}z \ge \dfrac{3}{4}z - \dfrac{3}{2} + z \Rightarrow$

    $-\dfrac{17}{12}z + 2 \ge \dfrac{7}{4}z - \dfrac{3}{2} \Rightarrow -\dfrac{38}{12}z \ge -\dfrac{7}{2} \Rightarrow z \le \dfrac{21}{19};\ \left(-\infty, \dfrac{21}{19}\right];$ set-builder notation the interval is

    $\left\{z \mid z \le \dfrac{21}{19}\right\}$.

39. Graph $y_1 = x + 2$ and $y_2 = 2x$. See Figure 39. $y_1 \ge y_2$ when the graph of y_1 is above the graph of y_2, which

    is left of the intersection point $(2, 4)$ and includes point $(2, 4) \Rightarrow \{x \mid x \le 2\}$.

41. Graph $y_1 = \dfrac{2}{3}x - 2$ and $y_2 = -\dfrac{4}{3}x + 4$. See Figure 41. $y_1 > y_2$ when the graph of y_1 is above the graph of y_2,

    which is right of the intersection point $(3, 0)$ and does not include point $(3, 0) \Rightarrow \{x \mid x > 3\}$.

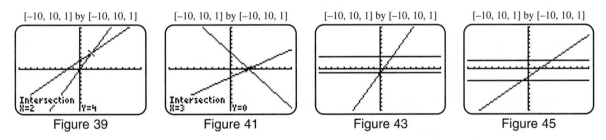

| [–10, 10, 1] by [–10, 10, 1] | [–10, 10, 1] by [–10, 10, 1] | [–10, 10, 1] by [–10, 10, 1] | [–10, 10, 1] by [–10, 10, 1] |

Figure 39   Figure 41   Figure 43   Figure 45

43. Graph $y_1 = -1$, $y_2 = 2x - 1$, and $y_3 = 3$. See Figure 43. $y_1 \le y_2 \le y_3$ when the graph of y_2 is in between the

    graphs of y_1 and y_3, which is in between the intersection points $(0, -1)$ and $(2, 3)$ and it does include each

    point $\Rightarrow \{x \mid 0 \le x \le 2\}$.

45. Graph $y_1 = -3$, $y_2 = x - 2$, and $y_3 = 2$. See Figure 45. $y_1 < y_2 \le y_3$ when the graph of y_2 is in between the

    graphs of y_1 and y_3, which is in between the intersection points $(-1, -3)$ and $(4, 2)$ and it does not include the

    point $(-1, -3)$ but does include $(4, 2) \Rightarrow \{x \mid -1 < x \le 4\}$.

47. (a) $y = \frac{3}{2}x - 3$, then $ax + b = 0$ gives us $\frac{3}{2}x - 3 = 0 \Rightarrow \frac{3}{2}x = 3 \Rightarrow x = 2$

    (b) $ax + b < 0$ gives us $\frac{3}{2}x - 3 < 0 \Rightarrow x < 2 \Rightarrow (-\infty, 2)$ or in set builder notation, $\{x \mid x < 2\}$.

    (c) $ax + b \geq 0$ gives us $\frac{3}{2}x - 3 \geq 0 \Rightarrow x \geq 2 \Rightarrow [2, \infty)$ or in set builder notation, $\{x \mid x \geq 2\}$.

49. (a) $y = -x - 2$, then $ax + b = 0$ gives us $-x - 2 = 0 \Rightarrow -x = 2 \Rightarrow x = -2$

    (b) $ax + b < 0$ gives us $-x - 2 < 0 \Rightarrow x > -2 \Rightarrow (-2, \infty)$ or in set builder notation, $\{x \mid x > -2\}$.

    (c) $ax + b \geq 0$ gives us $-x - 2 \geq 0 \Rightarrow x \leq -2 \Rightarrow (-\infty, -2]$ or in set builder notation, $\{x \mid x \leq -2\}$.

51. $x - 3 \leq \frac{1}{2}x - 2 \Rightarrow x - 3 - \frac{1}{2}x + 2 \leq 0 \Rightarrow \frac{1}{2}x - 1 \leq 0$. Figure 51 shows the graph of $y_1 = \frac{1}{2}x - 1$.

    The solution set for $y_1 \leq 0$ occurs when the graph is on or below the x-axis, or when $x \leq 2$. The solution set

    is $(-\infty, 2]$. Solving symbolically, $x - 3 \leq \frac{1}{2}x - 2 \Rightarrow \frac{1}{2}x \leq 1 \Rightarrow x \leq 2 \Rightarrow$

    $(-\infty, 2]$. In set-builder notation the interval is $\{x \mid x \leq 2\}$.

53. $2 - x < 3x - 2 \Rightarrow 2 - x - 2x + 2 < 0 \Rightarrow -4x + 4 < 0$. Figure 53 shows the graph of $y_1 = -4x + 4$.

    The solution set for $y_1 < 0$ occurs when the graph is below the x-axis, or when $x > 1$. The solution set is

    $(1, \infty)$. Solving symbolically, $2 - x < 3x - 2 \Rightarrow -4x < -4 \Rightarrow x > 1 \Rightarrow (1, \infty)$. In set-builder notation

    the interval is $\{x \mid x > 1\}$.

[−10, 10, 1] by [−10, 10, 1]	[−10, 10, 1] by [−10, 10, 1]	[−15, 15, 2] by [−15, 15, 2]	[1980, 2000, 1] by [0, 100, 10]
Figure 51	Figure 53	Figure 55	Figure 57

55. Graph $Y_1 = 5X - 4$ and $Y_2 = 10$. The graphs intersect at the point $(2.8, 10)$. The graph of Y_1 is above

    the graph of Y_2 for x-values to the right of this intersection point or where $x > 2.8$, $\{x \mid x > 2.8\}$.

    See Figure 55.

57. Graph $Y_1 = -2(X - 1990) + 55$ and $Y_2 = 60$. The graphs intersect at the point $(1987.5, 60)$. The graph

    of Y_1 is above the graph of Y_2 for x-values to the left of this intersection point, so $y_1 \geq y_2$ when $x \leq 1987.5$,

    $\{x \mid x \leq 1987.5\}$. See Figure 57.

59. Graph $Y_1 = \sqrt{(5)}(X - 1.2) - \sqrt{(3)}X$ and $Y_2 = 5(X + 1.1)$. The graphs intersect near the point

    $(-1.820, -3.601)$. The graph of Y_1 is below the graph of Y_2 for x-values to the right of this intersection

    point or when $x > k$, where $k \approx -1.82$, $\{x \mid x > -1.82\}$. See Figure 59.

61. Graph $Y_1 = 3$, $Y_2 = 5X - 17$ and $Y_3 = 15$, as shown in Figure 61. The graphs intersect at the points $(4, 3)$

    and $(6.4, 15)$. The solutions to $Y_1 \leq Y_2 < Y_3$ are the x-values between 4 and 6.4, including $4 \Rightarrow [4, 6.4)$. In

    set-builder notation the interval is $\{x \mid 4 \leq x < 6.4\}$.

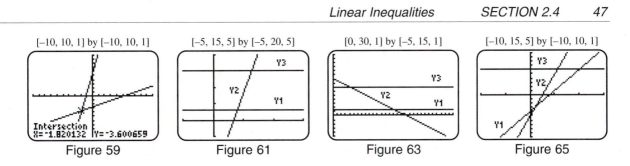

[–10, 10, 1] by [–10, 10, 1]   [–5, 15, 5] by [–5, 20, 5]   [0, 30, 1] by [–5, 15, 1]   [–10, 15, 5] by [–10, 10, 1]

Figure 59   Figure 61   Figure 63   Figure 65

63. Graph $Y_1 = 1.5$, $Y_2 = 9.1 - 0.5X$ and $Y_3 = 6.8$, as shown in Figure 63. The graphs intersect at the points $(4.6, 6.8)$ and $(15.2, 1.5)$. The solutions to $Y_1 \le Y_2 \le Y_3$ are the x-values between 4.6 and 15.2 (inclusive) or $4.6 \le x \le 15.2 \Rightarrow [4.6, 15.2]$. In set builder notation the interval is $\{x \,|\, 4.6 \le x \le 15.2\}$.

65. Graph $Y_1 = X - 4$, $Y_2 = 2X - 5$ and $Y_3 = 6$, as shown in Figure 65. The graph of y_2 intersects the graphs of y_1 and y_3 at $(1, -3)$ and $(5.5, 6)$. The solutions to $Y_1 < Y_2 < Y_3$ are the x-values between 1 and 5.5 or $1 < x < 5.5 \Rightarrow (1, 5.5)$. In set-builder notation the interval is $\{x \,|\, 1 < x < 5.5\}$.

67. (a) The graphs intersect at the point $(8, 7)$. Therefore, $g(x) = f(x)$ is satisfied when $x = 8$. The solution is 8.

(b) $g(x) > f(x)$ whenever the y-values on the graph of g are above the y-values on the graph of f. This occurs to the left of the point of intersection. Therefore the x-values that satisfy this inequality are $x < 8$. In set-builder notation the interval is $\{x \,|\, x < 8\}$.

69. From the table,

$Y_1 = 0$ when $x = 4$. $Y_1 > 0$ when $x < 4 \Rightarrow \{x \,|\, x < 4\}$; $Y_1 \le 0$ when $x \ge 4 \Rightarrow \{x \,|\, x \ge 4\}$.

71. Let $Y_1 = -4X - 6$. From the table shown in Figure 71, $Y_1 = 0$ when $x = -1.5$ or $-\dfrac{3}{2}$. $Y_1 > 0$ when $x < -\dfrac{3}{2}$.

In set-builder notation the interval is $\left\{ x \,\middle|\, x < -\dfrac{3}{2} \right\}$.

73. Let $Y_1 = 3X - 2$. From the table shown in Figure 73, Y_1 is between 10 and 4 (inclusive) for x-values between 1 and 4 (inclusive) $\Rightarrow [1, 4]$. In set-builder notation the interval is $\{x \,|\, 1 \le x \le 4\}$.

Figure 71   Figure 73   Figure 75   Figure 77

75. Let $Y_1 = (2 - 5X)/3$. From the table shown in Figure 75, Y_1 is between -0.75 and 0.75 for x-values between -0.05 and 0.85 and $Y_1 = 0.75$ when $x = -0.05 \Rightarrow \left[-\dfrac{1}{20}, \dfrac{17}{20} \right)$. In set-builder notation the interval is

$\left\{ x \,\middle|\, -\dfrac{1}{20} \le x < \dfrac{17}{20} \right\}$.

77. Let $Y_1 = (\sqrt{11} - \pi)X - 5.5$. From the table shown in Figure 77, $Y_1 \approx 0$ when $x = 31.4$. $Y_1 \le 0$ when $x \le 31.4$; $(-\infty, 31.4]$. In set-builder notation the interval is $\{x \,|\, x \le 31.4\}$.

79. Symbolically: $2x - 8 > 5 \Rightarrow 2x > 13 \Rightarrow x > \dfrac{13}{2}$. The solution set is $\left(\dfrac{13}{2}, \infty\right)$. In set-builder notation

the interval is $\left\{ x \,\middle|\, x > \dfrac{13}{2} \right\}$.

81. Graphically: Let $Y_1 = \pi X - 5.12$ and $Y_2 = \sqrt{(2)}X - 5.7(X - 1.1)$. Graph Y_1 and Y_2 as shown in Figure 81. The graphs intersect near $(1.534, -0.302)$. The graph of Y_1 is below Y_2 for $x < 1.534$, so $Y_1 \leq Y_2$ when $x \leq 1.534$. The solution set is $(-\infty, 1.534]$.

85. (a) Graph $Y_1 = 65 - 19X$ and $Y_2 = 50 - 5.8X$. These graphs intersect near the point $(1.14, 43.4)$ as shown in Figure 85. At an altitude of approximately 1.14 miles the temperature and the dew point are both equal to 43.4°F. The air temperature is greater than the dew point below 1.14 miles. The region where the clouds will not form is below 1.14 miles or when $0 \leq x < 1.14$.

(b) $65 - 19x > 50 - 5.8x \Rightarrow 15 > 13.2x \Rightarrow \dfrac{15}{13.2} > x$ or $0 \leq x < \dfrac{15}{13.2} \approx 1.14$

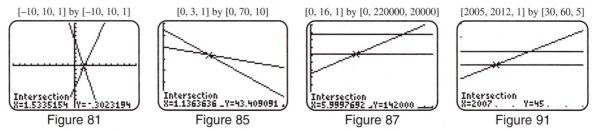

| [−10, 10, 1] by [−10, 10, 1] | [0, 3, 1] by [0, 70, 10] | [0, 16, 1] by [0, 220000, 20000] | [2005, 2012, 1] by [30, 60, 5] |
| Figure 81 | Figure 85 | Figure 87 | Figure 91 |

83. (a) Car A is traveling faster since it passes Car B. Its graph has the greater slope.

(b) The cars are the same distance from St. Louis when their graphs intersect. This point of intersection occurs at $(2.5, 225)$. The cars are both 225 miles from St. Louis after 2.5 hours.

(c) Car B is ahead of Car A when $0 \leq x < 2.5$.

87. (a) The slope of the graph of P is 8667. This means that the median price of a single-family home has increased by approximately \$8667 per year.

(b) Graph $Y_1 = 8667X + 90000$, $Y_2 = 142000$ and $Y_3 = 194000$. These graphs are shown in Figure 87. The points of intersection are located near $(5.99, 142{,}000)$ and $(11.99, 194{,}000)$. For approximately $5.99 \leq x \leq 11.99$ or, rounded to the nearest year, between 1996 and 2002, the median price was between \$142,000 and \$194,000. (Note that $x = 0$ corresponds to 1990.)

89. (a) Using $(2000, 6)$ and $(2004, 30)$, find slope: $m = \dfrac{30 - 6}{2004 - 2000} = \dfrac{24}{4} = 6 \Rightarrow$

$B(x) = 6(x - 2000) + 6$ or $B(x) = 6(x - 2004) + 30$.

(b) $6(x - 2000) + 6 \geq 24 \Rightarrow 6x - 11{,}994 \geq 24 \Rightarrow 6x \geq 12{,}018 \Rightarrow x \geq 2003$, from 2003 to 2006.

91. (a) The graph of linear function P will contain the points $(2005, 40)$ and $(2011, 55)$.

$m = \dfrac{55 - 40}{2011 - 2005} = \dfrac{15}{6} = 2.5 \Rightarrow P(x) = 2.5(x - 2005) + 40 \Rightarrow P(x) = 2.5x - 4972.5$

(b) Let $Y_1 = 2.5X - 4972.5$, $Y_2 = 45$ and $Y_3 = 50$. The points of intersection are located at $(2007, 45)$ and $(2009, 50)$. The percentage was between 45% and 50% from 2007 to 2009. See Figure 91.

93. The graph of linear function will intersect the points (90, 6.5) and (129, 5.5).

$$m = \frac{6.5 - 5.5}{90 - 129} = -\frac{1}{39} \Rightarrow f(x) = -\frac{1}{39}(x - 129) + 5.5 \Rightarrow 5.75 < -\frac{1}{39}(x - 129) + 5.5 < 6 \Rightarrow$$

$$0.25 < -\frac{1}{39}(x - 129) < 0.5 \Rightarrow -9.75 > x - 129 > 79.5 \Rightarrow 119.25 > x > 109.5$$

95. $r = \dfrac{C}{2\pi}$ and $1.99 \le r \le 2.01 \Rightarrow 1.99 \le \dfrac{C}{2\pi} \le 2.01 \Rightarrow 3.98\pi \le C \le 4.02\pi$

97. (a) $m = \dfrac{4.5 - (-1.5)}{2 - 0} = \dfrac{6}{2} = 3$ and y-intercept $= -1.5$; $f(x) = 3x - 1.5$ models the data.

(b) $f(x) > 2.25 \Rightarrow 3x - 1.5 > 2.25 \Rightarrow 3x > 3.75 \Rightarrow x > 1.25$

99. (a) Using the linear regression function on the calculator the function f is found to be

$$f(x) = 20.1x - 40,0096.7.$$

(b) Let $Y_1 = 20.1X - 40096.7$, $Y_2 = 43$ and $Y_3 = 83$. The points of intersection are at (1997, 43) and near

(1999, 83). Therefore, the number of cell phone subscribers was between 43 and 83 million between the

years 1997 and 1999.

(c) The answer was a result of extrapolation.

2.5: Absolute Value Equations and Inequalities

1. $|x| = 3 \Rightarrow x = 3$ or $x = -3$

3. $|x| > 3 \Rightarrow x > 3$ or $x < -3$; $(-\infty, -3) \cup (3, \infty)$

5. The graph of $y = |ax + b|$ is V-shaped with the vertex on the x-axis.

7. $\sqrt{36a^2} = |6a|$ since 36 and a^2 are always positive values.

9. (a) $x + 1 = 0 \Rightarrow x = -1 \Rightarrow$ the vertex is $(-1, 0)$. Find any other point, $x = 0 \Rightarrow (0, 1)$; graph the

absolute value graph with the vertex $(-1, 0)$, point $(0, 1)$ and its reflection through $x = -1$.

See Figure 9.

(b) $y = |x + 1|$ is increasing on $x \ge -1$ or $[-1, \infty)$ and decreasing on $x \le -1$ or $(-\infty, -1]$.

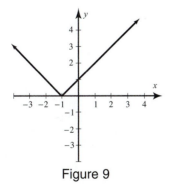

Figure 9

11. (a) $2x - 3 = 0 \Rightarrow 2x = 3 \Rightarrow x = \dfrac{3}{2} \Rightarrow$ the vertex is $\left(\dfrac{3}{2}, 0\right)$. Find another point such as $(0, 3)$; graph the

absolute value function. See Figure 11.

(b) $y = |2x - 3|$ is increasing on $x \geq \dfrac{3}{2}$ or $\left[\dfrac{3}{2}, \infty\right)$ and decreasing on $x \leq \dfrac{3}{2}$ or $\left(-\infty, \dfrac{3}{2}\right]$.

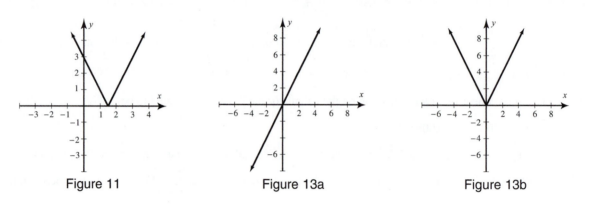

| Figure 11 | Figure 13a | Figure 13b |

13. (a) The graph of $y_1 = 2x$ is shown in Figure 13a.

(b) The graph of $y = |2x|$ is similar to the graph of $y = 2x$ except that it is reflected across the x-axis

whenever $2x < 0$. The graph of $y_1 = |2x|$ is shown in Figure 13b.

(c) The x-intercept occurs when $2x = 0$ or when $x = 0$. The x-intercept is 0.

15. (a) The graph of $y_1 = 3x - 3$ is shown in Figure 15a.

(b) The graph of $y = |3x - 3|$ is similar to the graph of $y = 3x - 3$ except that it is reflected across the x-axis

whenever $3x - 3 < 0$ or $x < 1$. The graph of $y_1 = |3x - 3|$ is shown in Figure 15b.

(c) The x-intercept occurs when $3x - 3 = 0$ or when $x = 1$. The x-intercept is located at 1.

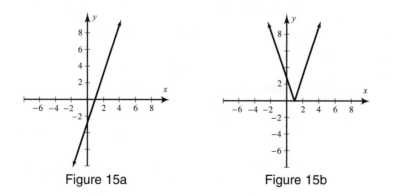

| Figure 15a | Figure 15b |

17. (a) The graph of $y_1 = 6 - 2x$ is shown in Figure 17a.

   (b) The graph of $y = |6 - 2x|$ is similar to the graph of $y = 6 - 2x$ except that it is reflected across the *x*-axis whenever $6 - 2x < 0$ or $x > 3$. The graph of $y_1 = |6 - 2x|$ is shown in Figure 17b.

   (c) The *x*-intercept occurs when $6 - 2x = 0$ or when $x = 3$. The *x*-intercept is located at 3.

Figure 17a   Figure 17b

19. $|-2x| = 4 \Rightarrow -2x = 4$ or $-2x = -4 \Rightarrow 2x = -4$ or $2x = 4 \Rightarrow x = -2$ or 2.

21. $|5x - 7| = 2 \Rightarrow 5x - 7 = -2$ or $5x - 7 = 2$. If $5x - 7 = -2$, then $5x = 5 \Rightarrow x = 1$;

   If $5x - 7 = 2$, then $5x = 9 \Rightarrow x = \dfrac{9}{5};\ 1, \dfrac{9}{5}$.

23. $|3 - 4x| = 5 \Rightarrow 3 - 4x = -5$ or $3 - 4x = 5$. If $3 - 4x = -5$, then $-4x = -8 \Rightarrow x = 2$;

   If $3 - 4x = 5$ then $-4x = 2 \Rightarrow x = -\dfrac{1}{2};\ -\dfrac{1}{2}, 2$.

25. $|-6x - 2| = 0 \Rightarrow -6x - 2 = 0 \Rightarrow -6x = 2 \Rightarrow x = -\dfrac{1}{3}$.

27. $|7 - 16x| = 0 \Rightarrow 7 - 16x = 0 \Rightarrow 7 = 16x \Rightarrow x = \dfrac{7}{16}$.

29. $|17x - 6| = -3$ has no solutions since the absolute value of any quantity is always greater than or equal to 0.

31. $|1.2x - 1.7| - 1 = 3 \Rightarrow |1.2x - 1.7| = 4$, then $1.2x - 1.7 = -4$ or $1.2x - 1.7 = 4$.

   If $1.2x - 1.7 = -4$ then, $1.2x = -2.3 \Rightarrow x = -\dfrac{2.3}{1.2} \Rightarrow x = -\dfrac{23}{12}$; if $1.2x - 1.7 = 4$ then

   $1.2x = 5.7 \Rightarrow x = \dfrac{5.7}{1.2} \Rightarrow x = \dfrac{19}{4};\ \Rightarrow -\dfrac{23}{12}, \dfrac{19}{4}$.

33. $|4x - 5| + 3 = 2 \Rightarrow |4x - 5| = -1$ has no solution since the absolute value of any quantity is always greater than or equal to 0.

35. $|2x - 9| = |8 - 3x| \Rightarrow 2x - 9 = 8 - 3x$ or $2x - 9 = -(8 - 3x) \Rightarrow$

   $2x + 3x = 8 + 9$ or $2x - 3x = -8 + 9 \Rightarrow 5x = 17$ or $-x = 1 \Rightarrow x = \dfrac{17}{5}, -1$

37. $\left|\dfrac{3}{4}x - \dfrac{1}{4}\right| = \left|\dfrac{3}{4} - \dfrac{1}{4}x\right| \Rightarrow \dfrac{3}{4}x - \dfrac{1}{4} = \dfrac{3}{4} - \dfrac{1}{4}x$ or $\dfrac{3}{4}x - \dfrac{1}{4} = -\left(\dfrac{3}{4} - \dfrac{1}{4}x\right) \Rightarrow$

   $\dfrac{3}{4}x + \dfrac{1}{4}x = \dfrac{3}{4} + \dfrac{1}{4}$ or $\dfrac{3}{4}x - \dfrac{1}{4}x = -\dfrac{3}{4} + \dfrac{1}{4} \Rightarrow x = 1$ or $\dfrac{1}{2}x = -\dfrac{1}{2} \Rightarrow x = -1, 1$

39. $\left| 15x - 5 \right| = \left| 35 - 5x \right| \Rightarrow 15x - 5 = 35 - 5x$ or $15x - 5 = -(35 - 5x) \Rightarrow$

    $15x - 5 = 35 - 5x \Rightarrow 20x = 40 \Rightarrow x = 2$ or $15x - 5 = -(35 - 5x) \Rightarrow 15x - 5 = -35 + 5x \Rightarrow$

    $10x = -30 \Rightarrow x = -3 \Rightarrow x = -3$ or 2

41. (a) $f(x) = g(x)$ when $x = -1$ or 7.

    (b) $f(x) < g(x)$ between these x-values or when $-1 < x < 7$; $(-1, 7)$

    (c) $f(x) > g(x)$ outside of these x-values or when $x < -1$ or $x > 7$; $(-\infty, -1) \cup (7, \infty)$

43. (a) $|2x - 3| = 1 \Rightarrow 2x - 3 = 1$ or $2x - 3 = -1$. If $2x - 3 = 1$, then $2x = 4 \Rightarrow x = 2$; If

    $2x - 3 = -1$ then $2x = 2 \Rightarrow x = 1$; $x = 1$ or $x = 2$

    (b) $|2x - 3| < 1 \Rightarrow -1 < 2x - 3 < 1 \Rightarrow 2 < 2x < 4 \Rightarrow 1 < x < 2$; $(1, 2)$

    (c) $|2x - 3| > 1 \Rightarrow 2x - 3 > 1$ or $2x - 3 < -1$. If $2x - 3 > 1$, then $2x > 4 \Rightarrow x > 2$.

    If $2x - 3 < -1$, then $2x < 2 \Rightarrow x < 1$. $x < 1$ or $x > 2$ or $(-\infty, 1) \cup (2, \infty)$

45. (a) Graph $Y_1 = \text{abs}(2X - 5)$ and $Y_2 = 10$. See Figures 45a and 45b. The solutions are -2.5 and 7.5.

    (b) Table $Y_1 = \text{abs}(2X - 5)$ starting at -5, incrementing by 2.5. See Figure 45c. The solutions are -2.5 and 7.5.

    (c) $\left| 2x - 5 \right| = 10 \Rightarrow 2x - 5 = -10$ or $2x - 5 = 10 \Rightarrow x = -\dfrac{5}{2}$ or $\dfrac{15}{2}$

    From each method, the solution to $\left| 2x - 5 \right| < 10$ lies between -2.5 and 7.5, exclusively: $-2.5 < x < 7.5$

    or $\left(-\dfrac{5}{2}, \dfrac{15}{2} \right)$.

[−10, 10, 1] by [−5, 15, 1]   [−10, 10, 1] by [−5, 15, 1]

Figure 45a   Figure 45b   Figure 45c

47. (a) Graph $Y_1 = \text{abs}(5 - 3X)$ and $Y_2 = 2$. See Figures 47a and 47b. The solutions are 1 and $\dfrac{7}{3}$.

    (b) Table $Y_1 = \text{abs}(5 - 3X)$ starting at $-\dfrac{1}{3}$, incrementing by $\dfrac{2}{3}$. See Figure 47c. The solutions are 1 and $\dfrac{7}{3}$.

    (c) $\left| 5 - 3x \right| = 2 \Rightarrow 5 - 3x = -2$ or $5 - 3x = 2 \Rightarrow x = \dfrac{7}{3}$ or 1

    From each method, the solution to $\left| 5 - 3x \right| > 2$ lies outside of 1 and $\dfrac{7}{3}$, exclusively: $x < 1$ or $x > \dfrac{7}{3}$ or

    $(-\infty, 1) \cup \left(\dfrac{7}{3}, \infty \right)$.

[−1, 4, 1] by [−1, 3, 1]   [−1, 4, 1] by [−1, 3, 1]

Figure 47a   Figure 47b   Figure 47c

49. $|2.1x - 0.7| = 2.4 \Rightarrow 2.1x - 0.7 = -2.4$ or $2.1x - 0.7 = 2.4 \Rightarrow x = -\dfrac{17}{21}$ or $\dfrac{31}{21}$

The solution to $|2.1x - 0.7| \geq 2.4$ lies outside of $-\dfrac{17}{21}$ and $\dfrac{31}{21}$, inclusively: $x \leq -\dfrac{17}{21}$ or $x \geq \dfrac{31}{21}$ or

$\left(-\infty, -\dfrac{17}{21}\right] \cup \left[\dfrac{31}{21}, \infty\right)$.

51. $|3x| + 5 = 6 \Rightarrow |3x| = 1 \Rightarrow 3x = -1$ or $3x = 1 \Rightarrow x = -\dfrac{1}{3}$ or $\dfrac{1}{3}$.

The solution to $|3x| + 5 > 6$ lies outside of $-\dfrac{1}{3}$ and $\dfrac{1}{3}$, exclusively: $x < -\dfrac{1}{3}, x > \dfrac{1}{3}$,

$\left(-\infty, -\dfrac{1}{3}\right) \cup \left(\dfrac{1}{3}, \infty\right)$.

53. $\left|\dfrac{2}{3}x - \dfrac{1}{2}\right| = -\dfrac{1}{4}$ has no solutions since the absolute value of any quantity is always greater than or equal to 0.

There are no solutions to $\left|\dfrac{2}{3}x - \dfrac{1}{2}\right| \leq -\dfrac{1}{4}$.

55. The solutions to $|3x - 1| < 8$ satisfy $s_1 < x < s_2$ where s_1 and s_2, are the solutions to $|3x - 1| = 8$.

$|3x - 1| = 8$ is equivalent to $3x - 1 = -8 \Rightarrow x = -\dfrac{7}{3}$ or $3x - 1 = 8 \Rightarrow x = 3$.

The interval is $\left(-\dfrac{7}{3}, 3\right)$.

57. The solutions to $|7 - 4x| \leq 11$ satisfy $s_1 \leq x \leq s_2$, where s_1 and s_2 are the solutions to $|7 - 4x| = 11$.

$|7 - 4x| = 11$ is equivalent to $7 - 4x = -11 \Rightarrow x = \dfrac{9}{2}$ or $7 - 4x = 11 \Rightarrow x = -1$.

The interval is $\left[-1, \dfrac{9}{2}\right]$.

59. The solutions to $|0.5x - 0.75| < 2$ satisfy $s_1 < x < s_2$, where s_1 and s_2 are the solutions to $|0.5x - 0.75| = 2$.

$|0.5x - 0.75| = 2$ is equivalent to $0.5x - 0.75 = -2 \Rightarrow x = -\dfrac{5}{2}$ or $0.5x - 0.75 = 2 \Rightarrow x = \dfrac{11}{2}$.

The interval is $\left(-\dfrac{5}{2}, \dfrac{11}{2}\right)$.

61. The solutions to $|2x - 3| > 1$ satisfy $x < s_1$ or $x > s_2$, where s_1 and s_2 are the solutions to $|2x - 3| = 1$.

$|2x - 3| = 1$ is equivalent to $2x - 3 = -1 \Rightarrow x = 1$ or $2x - 3 = 1 \Rightarrow x = 2$.

The solution set is $(-\infty, 1) \cup (2, \infty)$.

63. The solutions to $|-3x + 8| \geq 3$ satisfy $x \leq s_1$ or $x \geq s_2$, where s_1 and s_2 are the solutions to $|-3x + 8| = 3$.

$|-3x + 8| = 3$ is equivalent to $-3x + 8 = -3 \Rightarrow x = \dfrac{11}{3}$ or $-3x + 8 = 3 \Rightarrow x = \dfrac{5}{3}$.

The solution set is $\left(-\infty, \dfrac{5}{3}\right] \cup \left[\dfrac{11}{3}, \infty\right)$.

65. The solutions to $|0.25x - 1| > 3$ satisfy $x < s_1$ or $x > s_2$, where s_1 and s_2 are the solutions to

$|0.25x - 1| = 3$.

$|0.25x - 1| = 3$ is equivalent to $0.25x - 1 = -3 \Rightarrow x = -8$ or $0.25x - 1 = 3 \Rightarrow x = 16$.

The solution set is $(-\infty, -8) \cup (16, \infty)$.

67. $|-6| = 6$

69. Since the inputs of absolute values can be positive or negative, the domain of $|f(x)|$ is also $[-2, 4]$.

71. Since all solutions or the range of absolute values must be non-negative, all negative solutions will change to

positive solutions; therefore, if the range of $f(x)$ is $(-\infty, 0]$, the range of $|f(x)|$ is $[0, \infty)$.

73. $|S - 57.5| = 17.5 \Rightarrow S - 57.5 = 17.5$ or $S - 57.5 = -17.5$. If $S - 57.5 = 17.5$, then $S = 75$. If

$S - 57.5 = -17.5$, then $S = 40$. Therefore, the maximum speed limit is 75 mph and the minimum speed limit

is 40 mph.

75. (a) $0 \leq 80 - 19x \leq 32 \Rightarrow -80 \leq -19x \leq -48 \Rightarrow \dfrac{80}{19} \geq x \geq \dfrac{48}{19} \Rightarrow \dfrac{48}{19} \leq x \leq \dfrac{80}{19}$. The air temperature

is between 0°F and 32°F when the altitudes are between $\dfrac{48}{19}$ and $\dfrac{80}{19}$ miles inclusively.

(b) The air temperature is between 0°F and 32°F inclusively when the altitude is within $\dfrac{16}{19}$ mile of $\dfrac{64}{19}$ miles.

$\left| x - \dfrac{64}{19} \right| \leq \dfrac{16}{19}$.

77. (a) $|T - 43| = 24 \Rightarrow T - 43 = -24$ or $T - 43 = 24 \Rightarrow T = 19$ or 67. The average monthly temperature

range is $19°F \leq T \leq 67°F$.

(b) The monthly average temperatures in Marquette vary between a low of 19°F and a high of 67°F.

The monthly averages are always within 24° of 43°F.

79. (a) $|T - 50| = 22 \Rightarrow T - 50 = -22$ or $T - 50 = 22 \Rightarrow T = 28$ or 72. The average monthly temperature

range is $28°F \leq T \leq 72°F$.

(b) The monthly average temperatures in Boston vary between a low of 28°F and a high of 72°F.

The monthly averages are always within 22° of 50°F.

81. (a) $|T - 61.5| = 12.5 \Rightarrow T - 61.5 = -12.5$ or $T - 61.5 = 12.5 \Rightarrow T = 49$ or 74. The average monthly

temperature range is $49°F \leq T \leq 74°F$.

(b) The monthly average temperatures in Buenos Aires vary between a low of 49°F and a high of 74°F.

The monthly averages are always within 12.5° of 61.5°F.

83. The solutions to $|d - 3| \leq 0.004$ satisfy $s_1 \leq d \leq s_2$ where s_1 and s_2 are the solutions to $|d - 3| = 0.004$.

$|d - 3| = 0.004$ is equivalent to $d - 3 = -0.004 \Rightarrow d = 2.996$ and $d - 3 = 0.004 \Rightarrow d = 3.004$.

The solution set is $\{d \mid 2.996 \leq d \leq 3.004\}$. The acceptable diameters are from 2.994 and 3.004 inches.

85. $\left| \dfrac{Q - A}{A} \right| \leq 0.02 \Rightarrow \left| \dfrac{Q - 35}{35} \right| \leq 0.02$, so $-0.02 \leq \dfrac{Q - 35}{35} \leq 0.02 \Rightarrow -0.7 \leq Q - 35 \leq 0.7 \Rightarrow$

$34.3 \leq Q \leq 35.7$

Chapter 2 Review Exercises

1. (a) Using the points $(0, 6)$ and $(2, 2)$, $m = \dfrac{2 - 6}{2 - 0} = \dfrac{-4}{2} = -2$; y-intercept: 6; x-intercept: 3.

    (b) $f(x) = -2x + 6$

    (c) The zeros of f are the same as the x-intercepts. That is $x = 3$.

3. $m = \dfrac{0 - 2.5}{2 - 1} = \dfrac{-2.5}{1} = -2.5$; $(1, 2.5) \Rightarrow (1 - 1, 2.5 - (-2.5)) = (0, 5)$, so $b = 5$; $f(x) = -2.5x + 5$

5. See Figure 5.

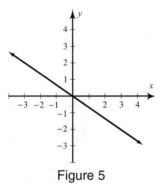

Figure 5

7. Using point-slope form $y = m(x - x_1) + y_1$, we get $y = -2(x + 2) + 3 \Rightarrow f(x) = -2x - 1$

9. $y = 7(x + 3) + 9 \Rightarrow y = 7x + 21 + 9 \Rightarrow y = 7x + 30$

11. Let $m = -3$. Then, $y = -3(x - 1) - 1 \Rightarrow y = -3x + 2$.

13. The line segment has slope $m = \dfrac{0 - 3.1}{5.7 - 0} = -\dfrac{31}{57}$; the parallel line has slope $m = -\dfrac{31}{57}$;

    $y = -\dfrac{31}{57}(x - 1) - 7 \Rightarrow y = -\dfrac{31}{57}x - \dfrac{368}{57}$

15. The line is vertical passing through $(6, -7)$, so the equation is $x = 6$.

17. The line is horizontal passing through $(1, 3)$, so the equation is $y = 3$.

19. The equation of the vertical line with x-intercept 2.7 is $x = 2.7$.

21. For x-intercept: $y = 0 \Rightarrow 5x - 4(0) = 20 \Rightarrow 5x = 20 \Rightarrow x = 4$; for y-intercept: $x = 0 \Rightarrow$

    $5(0) - 4y = 20 \Rightarrow -4y = 20 \Rightarrow y = -5$; use $(4, 0)$ and $(0, -5)$ to graph the equation. See Figure 21.

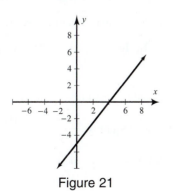

Figure 21

23. Graphical: Graph $Y_1 = 5X - 22$ and $Y_2 = 10$. Their graphs intersect at (6.4, 10) as shown in Figure 23. The solution is $x = 6.4$.

    Symbolic: $5x - 22 = 10 \Rightarrow 5x = 32 \Rightarrow x = \dfrac{32}{5} = 6.4$

25. Graphical: Graph $Y_1 = -2(3X - 7) + X$ and $Y_2 = 2X - 1$. Their graphs intersect near (2.143, 3.286) as shown in Figure 25. The solution is approximately 2.143.

    Symbolic: $-2(3x - 7) + x = 2x - 1 \Rightarrow -6x + 14 + x = 2x - 1 \Rightarrow -7x = -15 \Rightarrow x = \dfrac{15}{7} \approx 2.143$

27. Graphical: Graph $Y_1 = \pi X + 1$ and $Y_2 = 6$. Their graphs intersect near (1.592, 6) as shown in Figure 27. The solution is approximately 1.592.

    Symbolic: $\pi x + 1 = 6 \Rightarrow \pi x = 5 \Rightarrow x = \dfrac{5}{\pi} \approx 1.592$

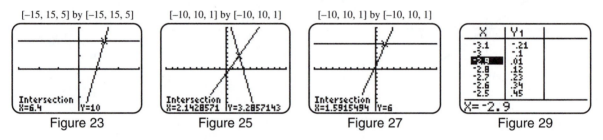

[−15, 15, 5] by [−15, 15, 5] [−10, 10, 1] by [−10, 10, 1] [−10, 10, 1] by [−10, 10, 1]

Figure 23   Figure 25   Figure 27   Figure 29

29. Let $Y_1 = 3.1X - 0.2 - 2(X - 1.7)$ and approximate where $Y_1 = 0$. From Figure 29 this occurs when $x \approx -2.9$.

31. (a) $4(6 - x) = -4x + 24 \Rightarrow 24 - 4x = -4x + 24 \Rightarrow 0 = 0 \Rightarrow$ all real numbers are solutions.

    (b) Because all real numbers are solutions, the equation is an identity.

33. (a) $5 - 2(4 - 3x) + x = 4(x - 3) \Rightarrow 5 - 8 + 6x + x = 4x - 12 \Rightarrow 7x - 3 = 4x - 12 \Rightarrow$

    $3x = -9 \Rightarrow x = -3$

    (b) Because there are finitely many solutions, the equation in condtional.

35. $(-3, \infty)$

37. $\left[-2, \dfrac{3}{4}\right)$

39. Graphical: Graph $Y_1 = 3X - 4$ and $Y_2 = 2 + X$. Their graphs intersect at (3, 5). The graph of Y_1 is below the graph of Y_2 to the left of the point of intersection. Thus, $3x - 4 \le 2 + x$ holds when $x \le 3$ or $(-\infty, 3]$. See Figure 39. Symbolic: $3x - 4 \le 2 + x \Rightarrow 2x \le 6 \Rightarrow x \le 3$ or $(-\infty, 3]$. In set-builder notation, the interval is $\{x \mid x \le 3\}$.

41. Graphical: Graph $Y_1 = (2X - 5)/2$ and $Y_2 = (5X + 1)/5$. Their graphs are parallel and never intersect. The graph of Y_1 is always below the graph of Y_2, so $Y_1 < Y_2$ for all values of x; the inequality $\dfrac{2x - 5}{2} < \dfrac{5x + 1}{5}$ holds when $-\infty < x < \infty$, or $(-\infty, \infty)$. See Figure 41. In set-builder notation the interval is $\{x \mid -\infty < x < \infty\}$.

[-10, 10, 1] by [-10, 10, 1]   [-10, 10, 1] by [-10, 10, 1]   [-10, 10, 1] by [-10, 10, 1]   [-10, 10, 1] by [-10, 10, 1]

Figure 39   Figure 41   Figure 43   Figure 45

43. Graphical: Graph $Y_1 = -2$, $Y_2 = 5 - 2X$ and $Y_3 = 7$. See Figure 43. Their graphs intersect at the points $(-1, 7)$ and $(3.5, -2)$. The graph of Y_2 is between the graphs of Y_1 and Y_3 when $-1 < x \le 3.5$. In interval notation the solution is $(-1, 3.5]$.

Symbolic: $-2 \le 5 - 2x < 7 \Rightarrow -7 \le -2x < 2 \Rightarrow \dfrac{7}{2} \ge x > -1 \Rightarrow -1 < x \le \dfrac{7}{2}$ or $\left(-1, \dfrac{7}{2}\right]$

In set-builder notation, the interval is $\left\{ x \mid -1 < x \le \dfrac{7}{2} \right\}$.

45. Graph $Y_1 = 2x$ and $Y_2 = x - 1$. See Figure 45. The lines intersect at $(-1, -2)$. $Y_1 > Y_2$ when the graph of Y_1 is above the graph of Y_2; this happens when $x > -1 \Rightarrow (-1, \infty)$. In set-builder notation the interval is $\{x \mid x > -1\}$.

47. (a) The graphs intersect at $(2, 1)$. The solution to $f(x) = g(x)$ is 2.

    (b) The graph of f is below the graph of g to the right of $(2, 1)$.

    Thus, $f(x) < g(x)$ when $x > 2$ or on $(2, \infty)$.

    (c) The graph of f is above the graph of g to the left of $(2, 1)$. Thus, $f(x) > g(x)$ when $x < 2$ or on $(-\infty, 2)$.

49. (a) $f(-2) = 8 + 2(-2) = 4$; $f(-1) = 8 + 2(-1) = 6$; $f(2) = 5 - 2 = 3$; $f(3) = 3 + 1 = 4$.

    (b) The graph of f is shown in Figure 49. It is essentially a piecewise line graph with the points $(-3, 2)$, $(-1, 6)$, $(2, 3)$, and $(5, 6)$. Since there are no breaks in the graph, f is continuous.

    (c) From the graph we can see that there are two x-values where $f(x) = 3$. They occur when $8 + 2x = 3 \Rightarrow x = -2.5$ and when $5 - x = 3 \Rightarrow x = 2$. The solutions are $x = -2.5$ or 2.

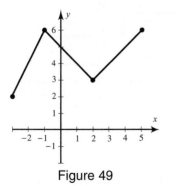

Figure 49

51. $|2x - 5| - 1 = 8 \Rightarrow |2x - 5| = 9 \Rightarrow 2x - 5 = -9$ or $2x - 5 = 9$; $2x - 5 = -9 \Rightarrow 2x = -4 \Rightarrow$ $x = -2$; $2x - 5 = 9 \Rightarrow 2x = 14 \Rightarrow x = 7 \Rightarrow -2, 7$

53. $|6 - 4x| = -2$ has no solutions since the absolute value of any quantity is always greater than or equal to 0.

55. $|x| = 3 \Rightarrow x = \pm 3$. The solutions to $|x| > 3$ lie to the left of -3 and to the right of 3. That is, $x < -3$ or $x > 3$. This can be supported by graphing $Y_1 = |x|$ and $Y_2 = 3$ and determining where the graph of Y_1 is above the graph of Y_2. To support this result numerically, table $Y_1 = \text{abs}(X)$ starting at -9 and incrementing by 3.

57. $|3x - 7| = 10 \Rightarrow 3x - 7 = 10$ or $3x - 7 = -10 \Rightarrow x = \dfrac{17}{3}$ or $x = -1$. The solutions to $|3x - 7| > 10$ lie to the left of -1 or to the right of $x = \dfrac{17}{3}$; that is, $x < -1$ or $x > \dfrac{17}{3}$. This can be supported by graphing $Y_1 = |3x - 7|$ and $Y_2 = 10$ and determining where the graph of Y_1 is above the graph of Y_2. To support this result numerically, table $Y_1 = \text{abs}(3X - 7)$ starting at -3 and incrementing by $\dfrac{1}{3}$.

59. The solutions to $|3 - 2x| < 9$ satisfy $s_1 < x < s_2$ where s_1 and s_2, are the solutions to $|3 - 2x| = 9$.

$|3 - 2x| = 9$ is equivalent to $3 - 2x = -9 \Rightarrow x = 6$ or $3 - 2x = 9 \Rightarrow x = -3$.

The solutions are $-3 < x < 6$ or $(-3, 6)$.

61. The solutions to $\left| \dfrac{1}{3}x - \dfrac{1}{6} \right| \geq 1$ satisfy $x \leq s_1$ or $x \geq s_2$ where s_1 and s_2, are the solutions to $\left| \dfrac{1}{3}x + \dfrac{1}{6} \right| = 1$.

$\left| \dfrac{1}{3}x - \dfrac{1}{6} \right| = 1$ is equivalent to $\dfrac{1}{3}x - \dfrac{1}{6} = -1 \Rightarrow x = -\dfrac{5}{2}$ or $\dfrac{1}{3}x - \dfrac{1}{6} = 1 \Rightarrow x = \dfrac{7}{2}$.

The solutions are $x \leq -\dfrac{5}{2}$ or $x \geq \dfrac{7}{2}$ or $\left(-\infty, -\dfrac{5}{2} \right] \cup \left[\dfrac{7}{2}, \infty \right)$.

63. (a) Graph $Y_1 = 1450(x - 1980) + 20{,}000$ and $Y_2 = 34{,}500$. Their graphs intersect at the point $(1990, 34{,}500)$. See Figure 63. This means that in 1990 the median income was $\$34{,}500$.

(b) $1450(x - 1980) + 20{,}000 = 34{,}500 \Rightarrow 1450(x - 1980) = 14{,}500 \Rightarrow x = \dfrac{14{,}500}{1450} + 1980 = 1990$

[1980, 2005, 2] by [20000, 40000, 1000] [1995, 2007, 1] by [200, 400, 50] [1995, 2007, 1] by [200, 400, 50] [0, 4, 1] by [0, 30, 5]

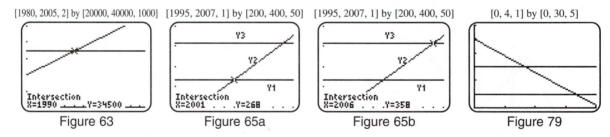

Figure 63   Figure 65a   Figure 65b   Figure 79

65. $268 \leq 18x - 35{,}750 \leq 358 \Rightarrow 36{,}018 \leq 18x \leq 36{,}108 \Rightarrow 2001 \leq x \leq 2006$. Medicare costs will be between 268 and 358 billion dollars from 2001 to 2006.

Graph $Y_1 = 268$, $Y_2 = 18X - 35{,}750$, and $Y_3 = 358$. Their graphs intersect at the points $(2001, 268)$ and $(2006, 358)$. See Figures 65a & 65b. Medicare costs will be between 268 and 358 billion dollars from 2001 to 2006.

67. Since the graph is piecewise linear, the slope each line segment represents a constant speed. Initially, the car is home. After 1 hour it is 30 miles from home and has traveled at a constant speed of 30 mph. After 2 hours it is 50 miles away. During the second hour the car travels 20 mph. During the third hour the car travels toward home at 30 mph until it is 20 miles away. During the fourth hour the car travels away from home at 40 mph until it is 60 miles away from home. The last hour the car travels 60 miles at 60 mph until it arrives back at home.

69. The midpoint is computed by $\left(\dfrac{2004 + 2008}{2}, \dfrac{143{,}247 + 167{,}933}{2} \right) = (2006, 155{,}590)$.

The population was about 155,590.

71. Let $x = $ time it takes for both working together; the first worker can shovel $\dfrac{1}{50}$ of the sidewalk in 1 minute,

and the second worker can shovel $\dfrac{1}{30}$ of the sidewalk in 1 minute; for the entire job, we get the equation

$\dfrac{x}{50} + \dfrac{x}{30} = 1 \Rightarrow 3x + 5x = 150 \Rightarrow 8x = 150 \Rightarrow x = 18.75$; it takes the two workers 18.75 minutes

to shovel the sidewalk together.

73. Let $t = $ time spent jogging at 7 mph; then $1.8 - t = $ time spent jogging at 8 mph; since $d = rt$ and the total distance jogged is 13.5 miles, we get the equation $7t + 8(1.8 - t) = 13.5 \Rightarrow 7t + 14.4 - 8t = 13.5 \Rightarrow$

$-t = -0.9 \Rightarrow t = 0.9$ and $1.8 - t = 0.9$; the runner jogged 0.9 hour at 7 mph and 0.9 hour at 8 mph.

75. (a) Begin by selecting any two points to determine the equation of the line. For example, if we use $(-1, 4.2)$

and $(2, 0.6)$, then $m = \dfrac{4.2 - 0.6}{-1 - 2} = \dfrac{3.6}{-3} = -1.2$. $y - y_1 = m(x - x_1) \Rightarrow y - 0.6 = -1.2(x - 2) \Rightarrow$

$y - 0.6 = -1.2x + 2.4 \Rightarrow y = -1.2x + 3$.

(b) When $x = -1.5$, then $y = -1.2(-1.5) + 3 = 4.8$. This involves interpolation.

When $x = 3.5$, then $y = -1.2(3.5) + 3 = -1.2$. This involves extrapolation.

(c) $1.3 = -1.2x + 3 \Rightarrow -1.7 = -1.2x \Rightarrow x = \dfrac{17}{12}$.

77. The tank is initially empty. When $0 \leq x \leq 3$, the slope is 5. The inlet pipe is open; the outlet pipe is closed. When $3 < x \leq 5$, the slope is 2. Both pipes are open. When $5 < x \leq 8$, the slope is 0. Both pipes are closed. When $8 < x \leq 10$, the slope is –3. The inlet pipe is closed; the outlet pipe is open.

79. Let x represent the distance above the ground and let y represent the temperature. Since the ground temperature is 25°C, the point $(0, 25)$ is on the graph of the function which models the situation. Since the rate of change is a constant -6°C per kilometer, the model is linear with a slope of $m = -6$. Therefore, the equation of the linear model is $y = -6x + 25$.

Graphically: Graph $Y_1 = 15$, $Y_2 = -6x + 25$, and $Y_3 = 5$ in $[0, 4, 1]$ by $[0, 30, 5]$. See Figure 79. The

intersection points are $\left(1\dfrac{2}{3}, 15 \right)$ and $\left(3\dfrac{1}{3}, 5 \right)$. The distance above the ground is between $1\dfrac{2}{3}$ km and $3\dfrac{1}{3}$ km.

Symbolically: Solve $5 \leq -6x + 25 \leq 15 \Rightarrow -20 \leq -6x \leq -10 \Rightarrow \dfrac{20}{6} \geq x \geq \dfrac{10}{6} \Rightarrow 1\dfrac{2}{3} \leq x \leq 3\dfrac{1}{3}$.

The solution interval is the same for either method, $\left[1\dfrac{2}{3}, 3\dfrac{1}{3} \right]$. The distance above the ground is between

$1\dfrac{2}{3}$ km and $3\dfrac{1}{3}$ km.

81. $\left| \dfrac{C - A}{A} \right| \le 0.003 \Rightarrow -0.003 \le \dfrac{C - 52.3}{52.3} \le 0.003 \Rightarrow -0.1569 \le C - 52.3 \le 0.1569 \Rightarrow$

   $52.1431 \le C \le 52.4569 \Rightarrow$ between 52.1431 and 52.4569 ft.

Chapters 1-2 Cumulative Review Exercises

1. Move the decimal point five places to the left; $123{,}000 = 1.23 \times 10^5$

   Move the decimal point three places to the right; $0.005 = 5.1 \times 10^{-3}$

3. $\dfrac{4 + \sqrt{2}}{4 - \sqrt{2}} \approx 2.09$

5. The standard equation of a circle must fit the form $(x - h)^2 + (y - k)^2 = r^2$, where (h, k) is the center and the

   radius r. The equation of the circle with center $(-2, 3)$ and radius 7 is $(x + 2)^2 + (y - 3)^2 = 49$.

7. $d = \sqrt{[2 - (-3)]^2 + ((-3) - 5)^2} = \sqrt{25 + 64} = \sqrt{89}$

9. (a) $D =$ all real numbers $\Rightarrow \{x \mid -\infty < x < \infty\}$; $R = \{y \mid y \ge -2\}$; $f(-1) = -1$

   (b) $D = \{x \mid -3 \le x \le 3\}$; $R = \{y \mid -3 \le y \le 2\}$; $f(-1) = -\dfrac{1}{2}$

11. (a) $f(2) = 5(2) - 3 = 7$; $f(a - 1) = 5(a - 1) - 3 = 5a - 5 - 3 = 5a - 8$

   (b) The domain of f includes all real numbers $\Rightarrow D = \{x \mid -\infty \le x \le \infty\}$

13. No, this is not a graph of a function because some vertical lines intersect the graph twice.

15. $f(1) = (1)^2 - 2(1) + 1 = 1 - 2 + 1 = 0 \Rightarrow (1, 0)$; $f(2) = (2)^2 - 2(2) + 1 = 4 - 4 + 1 = 1 \Rightarrow$

   $(2, 1)$. The slope $m = \dfrac{1 - 0}{2 - 1} = \dfrac{1}{1} = 1$, so the average rate of change is 1.

17. (a) $m = \dfrac{2}{3}$; y-intercept: -2, x-intercept: 3

   (b) $f(x) = mx + b \Rightarrow f(x) = \dfrac{2}{3}x - 2$

   (c) 3

19. Using point-slope form

   $y = m(x - x_1) + y_1 \Rightarrow y = -3\left(x - \dfrac{2}{3}\right) - \dfrac{2}{3} \Rightarrow y = -3x + \dfrac{4}{3} \Rightarrow f(x) = -3x + \dfrac{4}{3}$

21. $m = \dfrac{\frac{1}{2} - (-5)}{-3 - 1} = \dfrac{\frac{11}{2}}{-4} = -\dfrac{11}{8}$; using $(1, -5)$ and point-slope form: $y = -\dfrac{11}{8}(x - 1) - 5 \Rightarrow$

   $y = -\dfrac{11}{8}x + \dfrac{11}{8} - 5 \Rightarrow y = -\dfrac{11}{8}x - \dfrac{29}{8}$

23. All lines parallel to the y-axis have undefined slope $\Rightarrow y$ changes but x remains constant $\Rightarrow x = -1$.

25. For the points $(2.4, 5.6)$ and $(3.9, 8.6)$ we get $m = \dfrac{8.6 - 5.6}{3.9 - 2.4} = \dfrac{3}{1.5} = 2$. A line parallel to this has

   the same slope. Using point-slope form: $y = 2(x + 3) + 5 \Rightarrow y = 2x + 11$.

27. For $-2x + 3y = 6$: x-intercept, then $y = 0 \Rightarrow -2x + 3(0) = 6 \Rightarrow -2x = 6 \Rightarrow x = -3$;

y-intercept, then $x = 0 \Rightarrow -2(0) + 3y = 6 \Rightarrow 3y = 6 \Rightarrow y = 2$. See Figure 27.

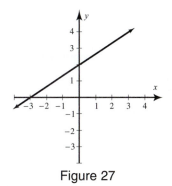

Figure 27

29. $4x - 5 = 1 - 2x \Rightarrow 6x = 6 \Rightarrow x = 1$; 1

31. $\dfrac{2}{3}(x - 2) - \dfrac{4}{5}x = \dfrac{4}{15} + x \Rightarrow \dfrac{2}{3}x - \dfrac{4}{3} - \dfrac{4}{5}x = \dfrac{4}{15} + x \Rightarrow 15\left(\dfrac{2}{3}x - \dfrac{4}{3} - \dfrac{4}{5}x = \dfrac{4}{15} + x\right) \Rightarrow$

$10x - 20 - 12x = 4 + 15x \Rightarrow -17x = 24 \Rightarrow x = -\dfrac{24}{17};\ -\dfrac{24}{17}$

33. Graph $Y_1 = X + 1$ and $Y_2 = 2X - 2$. See Figure 33a. The lines intersect at point $(3, 4) \Rightarrow x = 3$. Make a

table of $Y_1 = X + 1$ and $Y_2 = 2X - 2$ for x values from 0 to 5. See Figure 33b.

Both equations have y-value 4 at $x = 3$.

$[-10, 10, 1]$ by $[-10, 10, 1]$

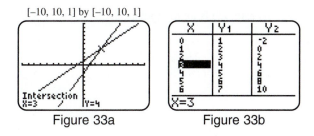

Figure 33a   Figure 33b

35. $(-\infty, 5)$

37. $(-\infty, -2) \cup (2, \infty)$

39. $-3(1 - 2x) + x \le 4 - (x + 2) \Rightarrow -3 + 6x + x \le 4 - x - 2 \Rightarrow 7x - 3 \le -x + 2 \Rightarrow 8x \le 5 \Rightarrow$

$x \le \dfrac{5}{8} \Rightarrow \left(-\infty, \dfrac{5}{8}\right]$. In set builder notation the interval is $\left\{x \mid x \le \dfrac{5}{8}\right\}$.

41. (a) 2

(b) The graph of $f(x)$ is above the graph of $g(x)$ to the left of $x = 2 \Rightarrow x < 2$

(c) The graph of $f(x)$ intersects or is below the graph of $g(x)$ to the right of $x = 2 \Rightarrow f(x) \le g(x)$ when $x \ge 2$.

43. $|d + 1| = 5 \Rightarrow d + 1 = 5$ or $d + 1 = -5$. If $d + 1 = -5, d = -6$; if $d + 1 = 5, d = 4 \Rightarrow -6, 4$.

45. $|2t| - 4 = 10 \Rightarrow |2t| = 14 \Rightarrow 2t = 14$ or $2t = -14$. If $2t = -14, t = -\dfrac{14}{2} \Rightarrow t = -7$; if

$2t = 14, t = 7 \Rightarrow -7, 7$.

47. The solutions to $|2t - 5| \le 5$ satisfy $s_1 \le t \le s_2$ where s_1 and s_2, are the solutions to $|2t - 5| = 5$.

    $|2t - 5| = 5$ is equivalent to $2t - 5 = -5 \Rightarrow t = 0$ or $2t - 5 = 5 \Rightarrow t = 5$.

    The interval is $[0, 5]$. In set-builder notation the interval is $\{t \,|\, 0 \le t \le 5\}$.

49. $V = \pi r^2 h \Rightarrow 24 = \pi (1.5)^2 h \Rightarrow 24 = \pi(2.25)h \Rightarrow h \approx 3.40$ inches

51. (a) $C(1500) = 500(1500) + 20{,}000 = 770{,}000$; it costs \$770,000 to manufacture 1500 computers.

    (b) 500; each additional computer costs \$500 to manufacture.

53. (a) $T(2) = 70 + \dfrac{3}{2}(2)^2 = 70 + 6 = 76$; $T(4) = 70 + \dfrac{3}{2}(4)^2 = 70 + 24 = 94$

    Using $(2, 76)$ and $(4, 94)$: $m = \dfrac{94 - 76}{4 - 2} = \dfrac{18}{2} = 9°F$ increase per hour.

    (b) On average the temperature increased by 9°F per hour over this 2-hour period.

55. Let $t =$ time for the two to mow the lawn together. Then the first person mows $\dfrac{1}{5}t$ of the lawn and the second

    person mows $\dfrac{1}{12}t$ of the lawn $\Rightarrow \dfrac{1}{5}t + \dfrac{1}{12}t = 1 \Rightarrow \dfrac{12}{60}t + \dfrac{5}{60}t = 1 \Rightarrow \dfrac{17}{60}t = 1 \Rightarrow t = \dfrac{60}{17} = 3.53$ hours.

57. (a) Using $(2001, 56)$ and $(2012, 61)$, $m = \dfrac{61 - 56}{2012 - 2001} = \dfrac{5}{11}$; $f(x) = \dfrac{5}{11}(x - 2001) + 56$

    (b) $f(2007) = \dfrac{5}{11}(2007 - 2001) + 56 = \dfrac{5}{11}(6) + 56 = \dfrac{30}{11} + 56 \approx 58.7$ lbs.

59. (a) Enter the data $(1970, 4095)$, $(1980, 10{,}182)$, $(1990, 19{,}572)$ and $(2000, 29{,}760)$;

    $f(x) = 863.84x - 1{,}698{,}819.9$

    (b) $f(1995) \approx 863.84(1995) - 1{,}698{,}819.9 \approx \$24{,}541$; this estimate is an interpolation.

Chapter 3: Quadratic Functions and Equations

3.1: Quadratic Functions and Models

1. $f(x) = 1 - 2x + 3x^2$ is quadratic; $a = 3$; $f(-2) = 1 - 2(-2) + 3(-2)^2 = 17$.

3. $f(x) = \dfrac{1}{x^2 - 1}$ is neither linear nor quadratic.

5. $f(x) = \dfrac{1}{2} - \dfrac{3}{10}x$ is linear.

7. (a) $a > 0$

   (b) vertex: $(1, 0)$

   (c) axis of symmetry: $x = 1$

   (d) f is increasing for $x \geq 1$ and decreasing for $x \leq 1$.

9. (a) $a < 0$

   (b) vertex: $(-3, -2)$

   (c) axis of symmetry: $x = -3$

   (d) f is increasing for $x \leq -3$ and decreasing for $x \geq -3$.

11. The graph of g is narrower than the graph of f.

13. The graph of g is wider than the graph of f and opens downward rather than upward.

15. $f(x) = -3(x - 1)^2 + 2 \Rightarrow$ vertex: $(1, 2)$; leading coefficient: -3; $f(x) = -3x^2 + 6x - 1$

17. $f(x) = 5 - 2(x - 4)^2 \Rightarrow$ vertex: $(4, 5)$; leading coefficient: -2; $f(x) = -2x^2 + 16x - 27$

19. $f(x) = \dfrac{3}{4}(x + 5)^2 - \dfrac{7}{4} \Rightarrow$ vertex: $\left(-5, -\dfrac{7}{4}\right)$; leading coefficient: $\dfrac{3}{4}$; $f(x) = \dfrac{3}{4}x^2 + \dfrac{15}{2}x + 17$

21. The vertex of the parabola is $(2, -2)$, so $f(x) = a(x - h)^2 + k \Rightarrow f(x) = a(x - 2)^2 - 2$; since $(0, 2)$ is a point on the parabola, $f(0) = a(0 - 2)^2 - 2 \Rightarrow 2 = a(-2)^2 - 2 \Rightarrow 4 = 4a \Rightarrow a = 1$; $f(x) = (x - 2)^2 - 2$.

23. The vertex of the parabola is $(2, -3)$, so $f(x) = a(x - h)^2 + k \Rightarrow f(x) = a(x - 2)^2 - 3$; since $(0, -1)$ is a point on the parabola, $f(0) = a(0 - 2)^2 - 3 \Rightarrow -1 = a(4) - 3 \Rightarrow -1 = 4a - 3 \Rightarrow 2 = 4a \Rightarrow a = \dfrac{1}{2}$; $f(x) = \dfrac{1}{2}(x - 2)^2 - 3$.

25. The vertex of the parabola is $(-1, 3)$, so $f(x) = a(x - h)^2 + k \Rightarrow f(x) = a(x + 1)^2 + 3$; since $(0, 1)$ is a point on the parabola, $f(0) = a(0 + 1)^2 + 3 \Rightarrow 1 = a(1) + 3 \Rightarrow 1 = a + 3 \Rightarrow a = -2$; $f(x) = -2(x + 1)^2 + 3$.

27. The vertex of the parabola is $(2, 6)$, so $f(x) = a(x - h)^2 + k \Rightarrow f(x) = a(x - 2)^2 + 6$; since $(0, -6)$ is a point on the parabola, $f(0) = a(0 - 2)^2 + 6 \Rightarrow -6 = 4a + 6 \Rightarrow -12 = 4a \Rightarrow a = -3$; $f(x) = -3(x - 2)^2 + 6$.

29. $f(x) = x^2 + 4x - 5 \Rightarrow x = -\dfrac{b}{2a} = -\dfrac{4}{2(1)} = -2$ and

$f(-2) = (-2)^2 + 4(-2) - 5 = -9 \Rightarrow$ vertex: $(-2, -9)$; since $a = 1$, $f(x) = 1(x - (-2))^2 - 9$, or $f(x) = (x + 2)^2 - 9$

31. $f(x) = x^2 - 3x \Rightarrow x = -\dfrac{b}{2a} = -\dfrac{(-3)}{2(1)} = \dfrac{3}{2}$ and

$f\!\left(\dfrac{3}{2}\right) = \left(\dfrac{3}{2}\right)^2 - 3\left(\dfrac{3}{2}\right) = -\dfrac{9}{4} \Rightarrow$ vertex: $\left(\dfrac{3}{2}, -\dfrac{9}{4}\right)$; since $a = 1$, $f(x) = 1\left(x - \dfrac{3}{2}\right)^2 - \dfrac{9}{4}$, or

$f(x) = \left(x - \dfrac{3}{2}\right)^2 - \dfrac{9}{4}$

33. $f(x) = 2x^2 - 5x + 3 \Rightarrow x = -\dfrac{b}{2a} = -\dfrac{(-5)}{2(2)} = \dfrac{5}{4}$ and

$f\!\left(\dfrac{5}{4}\right) = 2\left(\dfrac{5}{4}\right)^2 - 5\left(\dfrac{5}{4}\right) + 3 = -\dfrac{1}{8} \Rightarrow$ vertex: $\left(\dfrac{5}{4}, -\dfrac{1}{8}\right)$; since $a = 2$, $f(x) = 2\left(x - \dfrac{5}{4}\right)^2 - \dfrac{1}{8}$

35. $f(x) = \dfrac{1}{3}x^2 + x + 1 \Rightarrow x = -\dfrac{b}{2a} = -\dfrac{1}{2(\frac{1}{3})} = -\dfrac{1}{\frac{2}{3}} = -\dfrac{3}{2}$ and

$f\!\left(-\dfrac{3}{2}\right) = \dfrac{1}{3}\left(-\dfrac{3}{2}\right)^2 - \dfrac{3}{2} + 1 = \dfrac{1}{4} \Rightarrow$ vertex: $\left(-\dfrac{3}{2}, \dfrac{1}{4}\right)$; since $a = \dfrac{1}{3}$, $f(x) = \dfrac{1}{3}\left(x + \dfrac{3}{2}\right)^2 + \dfrac{1}{4}$

37. $f(x) = 2x^2 - 8x - 1 \Rightarrow x = -\dfrac{b}{2a} = -\dfrac{(-8)}{2(2)} = 2$ and

$f(2) = 2(2)^2 - 8(2) - 1 = -9 \Rightarrow$ vertex: $(2, -9)$; since $a = 2$, $f(x) = 2(x - 2)^2 - 9$

39. $f(x) = 2 - 9x - 3x^2 \Rightarrow x = -\dfrac{b}{2a} = -\dfrac{(-9)}{2(-3)} = -1.5$ and

$f(-1.5) = 2 - 9(-1.5) - 3(-1.5)^2 = 8.75 \Rightarrow$ vertex: $(-1.5, 8.75)$; since $a = -3$, $f(x) = -3(x - (-1.5))^2 + 8.75$, or $f(x) = -3(x + 1.5)^2 + 8.75$

41. (a) It may be helpful to write f in standard form as follows: $f(x) = -x^2 + 0x + 6$.

To find the vertex symbolically, use the vertex formula with $a = -1$ and $b = 0$.

$x = -\dfrac{b}{2a} = -\dfrac{0}{2(-1)} = \dfrac{0}{2} = 0$. The x-coordinate of the vertex is 0.

$y = f\!\left(-\dfrac{b}{2a}\right) = f(0) = 6 - (0)^2 = 6$. The y-coordinate of the vertex is 6. Thus, the vertex is $(0, 6)$.

The graph of $Y_1 = 6 - X\char`^2$ shows graphical support for this vertex. See Figure 41.

(b) f is increasing on $x \le 0$ or $(-\infty, 0]$ and decreasing on $x \ge 0$ or $[0, \infty)$.

43. (a) It may be helpful to write f in standard form as follows: $f(x) = x^2 - 6x + 0$.

To find the vertex symbolically, use the vertex formula with $a = 1$ and $b = -6$.

$x = -\dfrac{b}{2a} = -\dfrac{(-6)}{2(1)} = \dfrac{6}{2} = 3$. The x-coordinate of the vertex is 3.

$y = f\!\left(-\dfrac{b}{2a}\right) = f(3) = (3)^2 - 6(3) = -9$. The y-coordinate of the vertex is –9. Thus, the vertex is

$(3, -9)$. The graph of $Y_1 = X\char`^2 - 6X$ shows graphical support for this vertex. See Figure 43.

(b) f is increasing on $x \ge 3$ or $[3, \infty)$ and decreasing on $x \le 3$ or $(-\infty, 3]$.

[−10, 10, 1] by [−10, 10, 1]   [−10, 10, 1] by [−15, 5, 1]   [−10, 10, 1] by [−10, 10, 1]   [−40, 40, 5] by [−40, 40, 5]

Figure 41   Figure 43   Figure 45   Figure 47

45. (a) The function is already in standard form: $f(x) = 2x^2 - 4x + 1$.

   To find the vertex symbolically, use the vertex formula with $a = 2$ and $b = -4$.

   $$x = -\frac{b}{2a} = -\frac{(-4)}{2(2)} = \frac{4}{4} = 1.$$ The x-coordinate of the vertex is 1.

   $$y = f\left(-\frac{b}{2a}\right) = f(1) = 2(1)^2 - 4(1) + 1 = -1.$$ The y-coordinate of the vertex is -1. Thus, the vertex

   is $(1, -1)$. The graph of $Y_1 = 2X^2 - 4X + 1$ shows graphical support for this vertex. See Figure 45.

   (b) f is increasing on $x \geq 1$ or $[1, \infty)$ and decreasing on $x \leq 1$ or $(-\infty, 1]$.

47. (a) It may be helpful to write f in standard form as follows: $f(x) = \frac{1}{2}x^2 + 0x + 10$.

   To find the vertex symbolically, use the vertex formula with $a = \frac{1}{2}$ and $b = 0$.

   $$x = -\frac{b}{2a} = -\frac{0}{2(\frac{1}{2})} = 0.$$ The x-coordinate of the vertex is 0.

   $$y = f\left(-\frac{b}{2a}\right) = f(0) = \frac{1}{2}(0)^2 + 0(0) + 10 = 10.$$ The y-coordinate of the vertex is 10. Thus, the

   vertex is $(0, 10)$. The graph of $Y_1 = (1/2)X^2 + 10$ shows graphical support for this vertex. See Figure 47.

   (b) f is increasing on $x \geq 0$ or $[0, \infty)$ and decreasing on $x \leq 0$ or $(-\infty, 0]$.

49. (a) The function is already in standard form: $f(x) = -\frac{3}{4}x^2 + \frac{1}{2}x - 3$.

   To find the vertex symbolically, use the vertex formula with $a = -\frac{3}{4}$ and $b = \frac{1}{2}$.

   $$x = -\frac{b}{2a} = -\frac{\frac{1}{2}}{2(-\frac{3}{4})} = \frac{1}{3}.$$ The x-coordinate of the vertex is $\frac{1}{3}$.

   $$y = f\left(-\frac{b}{2a}\right) = f\left(\frac{1}{3}\right) = -\frac{3}{4}\left(\frac{1}{3}\right)^2 + \frac{1}{2}\left(\frac{1}{3}\right) - 3 = -\frac{35}{12}.$$ The y-coordinate of the vertex is $-\frac{35}{12}$.

   Thus, the vertex is $\left(\frac{1}{3}, -\frac{35}{12}\right)$. The graph of $Y_1 = (-3/4)X^2 + (1/2)X - 3$ shows graphical support for

   this vertex. See Figure 49.

   (b) f is increasing on $x \leq \frac{1}{3}$ or $\left(-\infty, \frac{1}{3}\right]$ and decreasing on $x \geq \frac{1}{3}$ or $\left[\frac{1}{3}, \infty\right)$.

[−10, 10, 1] by [−10, 10, 1]

Figure 49

51. (a) It may be helpful to write f in standard form as follows: $f(x) = -6x^2 - 3x + 1.5$.

   To find the vertex symbolically, use the vertex formula with $a = -6$ and $b = -3$.

   $$x = -\frac{b}{2a} = -\frac{(-3)}{2(-6)} = -\frac{3}{12} = -\frac{1}{4}.$$ The x-coordinate of the vertex is $-\frac{1}{4}$.

   $$y = f\left(-\frac{b}{2a}\right) = f\left(-\frac{1}{4}\right) = 1.5 - 3\left(-\frac{1}{4}\right) - 6\left(-\frac{1}{4}\right)^2 = 1\frac{7}{8}.$$ The y-coordinate of the vertex

   is $1\frac{7}{8}$. Thus, the vertex is $\left(-\frac{1}{4}, 1\frac{7}{8}\right)$. The graph of $Y_1 = 1.5 - 3X - 6X^2$ shows graphical support

   for this vertex. See Figure 51.

   (b) f is increasing on $x \le -\frac{1}{4}$ or $\left(-\infty, -\frac{1}{4}\right]$ and decreasing on $x \ge -\frac{1}{4}$ or $\left[-\frac{1}{4}, \infty\right)$.

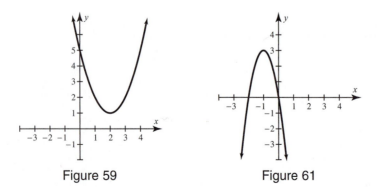

[-3, 3, 1] by [-5, 5, 1]

Maximum
X=-.2500007 Y=1.875

Figure 51

53. The graph of $f(x) = x^2$ is shown in Figure 53.

55. The graph of $f(x) = -\frac{1}{2}x^2$ is shown in Figure 55.

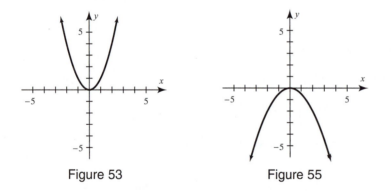

Figure 53

Figure 55

Figure 57

57. The graph of $f(x) = x^2 - 3$ is shown in Figure 57.

59. The graph of $f(x) = (x - 2)^2 + 1$ is shown in Figure 59.

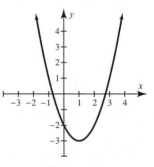

Figure 59

Figure 61

Figure 63

61. The graph of $f(x) = -3(x + 1)^2 + 3$ is shown in Figure 61.

63. The graph of $f(x) = x^2 - 2x - 2$ is shown in Figure 63.

65. The graph of $f(x) = -x^2 + 4x - 2$ is shown in Figure 65.

67. The graph of $f(x) = 2x^2 - 4x - 1$ is shown in Figure 67.

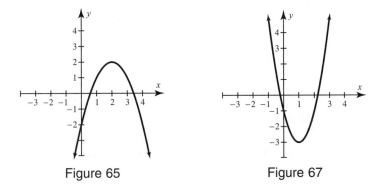

Figure 65　　　　　Figure 67　　　　　Figure 69

69. The graph of $f(x) = -3x^2 - 6x + 1$ is shown in Figure 69.

71. The graph of $f(x) = -\dfrac{1}{2}x^2 + x + 1$ is shown in Figure 71.

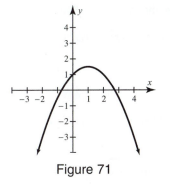

Figure 71

73. The average rate of change from 1 to 3 for $f(x) = -3x^2 + 5x$ is $\dfrac{f(3) - f(1)}{3 - 1} = \dfrac{-14}{2} = -7$.

75. $\dfrac{f(x + h) - f(x)}{h} = \dfrac{3(x + h)^2 - 2(x + h) - [3x^2 - 2x]}{h} =$

$\dfrac{3x^2 + 6xh + 3h^2 - 2x - 2h - 3x^2 + 2x}{h} = \dfrac{6xh + 3h^2 - 2h}{h} = 6x + 3h - 2$

77. Since the axis of symmetry is $x = 3$ and the graph passes through the point $(3, 1)$, the vertex is $(3, 1)$.

$f(x) = a(x - h)^2 + k \Rightarrow f(x) = a(x - 3)^2 + 1$. Using the point $(1, 9)$ to find a gives

$9 = a(1 - 3)^2 + 1 \Rightarrow a = 2$; $f(x) = 2(x - 3)^2 + 1$.

79. A stone thrown from ground level would first rise to some maximum height and then fall back to the ground. This is represented in figure d.

81. When the furnace first fails to work, the temperature begins to drop. Then, after the furnace is repaired, the temperature begins to rise. This is represented in figure a.

83. Perimeter of fence $= 2l + 2w = 1000 \Rightarrow l = \dfrac{1000 - 2w}{2} \Rightarrow l = 500 - w.$

   If $A = lw$ then $A = (500 - w)w \Rightarrow A = 500w - w^2 \Rightarrow A = -w^2 + 500w.$ This is a parabola opening

   downward and by the vertex formula, the maximum area occurs when $w = -\dfrac{b}{2a} = -\dfrac{500}{2(-1)} = 250.$

   The dimensions that maximize area are 250 ft. by 250 ft.

85. (a) $R(x) = x(40 - 2x) \Rightarrow R(2) = 2(40 - 2(2)) = 2(36) = 72;$ the company receives \$72,000 for

   producing 2000 CD players.

   (b) $R(x) = 40x - 2x^2$ is a quadratic function; to find the value at which the maximum value occurs, we need

   to find the vertex: $x = -\dfrac{b}{2a} = -\dfrac{40}{2(-2)} = 10$ and $R(10) = 40(10) - 2(10)^2 = 200 \Rightarrow (10, 200)$ is the

   vertex, thus, the company needs to produce 10,000 CD players to maximize its revenue.

   (c) Since $R(10) = 200,$ the maximum revenue for the company is \$200,000.

87. (a) $s(1) = -16(1)^2 + 44(1) + 4 = 32;$ the baseball is 32 feet high after 1 second.

   (b) For $s(t) = -16t^2 + 44t + 4,$ the vertex formula gives $t = -\dfrac{b}{2a} = -\dfrac{44}{2(-16)} = 1.375$ and

   $f(1.375) = -16(1.375)^2 + 44(1.375) + 4 = 34.25;$ the maximum height 34.25 feet. See Figure 87.

89. (a) The initial velocity $v_0 = -66$ and the initial height is $h_0 = 120,$ so $s(t) = -16t^2 + v_0t + h_0 \Rightarrow$

   $s(t) = -16t^2 - 66t + 120.$

   (b) When the stone hits the water, $s(t) = 0;$ $0 = -16t^2 - 66t + 120 \Rightarrow 0 = -2(8t^2 + 33t - 60) \Rightarrow$

   $0 = 8t^2 + 33t - 60;$ using the quadratic formula or graphing the parabola, we find that $s(t) = 0$ when

   $t \approx 1.37$ seconds, so the stone hits the water within the first 2 seconds.

91. Using the wall of the barn for one side gives us $W + L + W = 160$ for the three sides $\Rightarrow L = 160 - 2W.$

   If $A = LW$ then $A = (160 - 2W)W \Rightarrow A = 160W - 2W^2.$ Since this is a parabola opening downward,

   maximum area occurs when $W = -\dfrac{b}{2a} \Rightarrow W = \dfrac{-160}{2(-2)} \Rightarrow W = 40.$ Then $L = 160 - 2(40) = 80.$ Thus,

   the dimensions that yield maximum area are 40 feet by 80 feet.

93. (a) When $g = 32,$ $v_0 = 88,$ and $h_0 = 25,$ then $f(x) = -\dfrac{1}{2}(32)x^2 + 88x + 25 = -16x^2 + 88x + 25.$

   Graph $Y_1 = -16X^{\wedge}2 + 88X + 25.$ By using the calculator, the maximum height is found to be

   approximately 146 feet. The maximum height of 146 ft occurs when $x = 2.75$ seconds. See Figure 93.

   (b) To find the maximum height symbolically, use the vertex formula with $a = -16$ and $b = 88.$

   $x = -\dfrac{b}{2a} = -\dfrac{88}{2(-16)} = \dfrac{88}{32} = 2.75.$ The maximum height occurs at $x = 2.75$ seconds.

   $y = f\left(-\dfrac{b}{2a}\right) = f(2.75) = -16(2.75)^2 + 88(2.75) + 25 = 146.$ The maximum height is 146 feet.

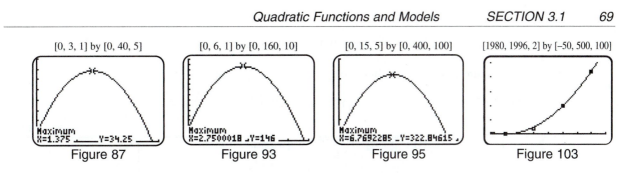

[0, 3, 1] by [0, 40, 5]   [0, 6, 1] by [0, 160, 10]   [0, 15, 5] by [0, 400, 100]   [1980, 1996, 2] by [−50, 500, 100]

Maximum
X=1.375   Y=34.25   Maximum
X=2.7500018 Y=146   Maximum
X=6.7692285 Y=322.84615

Figure 87   Figure 93   Figure 95   Figure 103

95. (a) When $g = 13$, $v_0 = 88$, and $h_0 = 25$, then $f(x) = -\dfrac{1}{2}(13)x^2 + 88x + 25 = -6.5x^2 + 88x + 25$.

Graph $Y_1 = -6.5X^2 + 88X + 25$. By using the calculator, the maximum height is found to be

approximately 323 feet. The maximum height of about 323 ft occurs when $x \approx 6.77$ seconds. See Figure 95.

(b) To find the maximum height symbolically, use the vertex formula with $a = -6.5$ and $b = 88$.

$x = -\dfrac{b}{2a} = -\dfrac{88}{2(-6.5)} = \dfrac{88}{13} \approx 6.77$. The maximum height occurs at $x \approx 6.77$ seconds.

$y = f\left(-\dfrac{b}{2a}\right) = f(6.77) = -6.5(6.77)^2 + 88(6.77) + 25 \approx 323$. The maximum height is about 323 feet.

97. The smallest y-value is −3 when $x = 1$; the symmetry in the y-values about $x = 1$ indicates that the axis of

symmetry is $x = 1$ and the vertex is $(1, -3)$, so $f(x) = a(x - 1)^2 - 3$. Since $(0, -1)$ is a data point,

$f(0) = -1$; $f(0) = a(0 - 1)^2 - 3 \Rightarrow -1 = a - 3 \Rightarrow a = 2$; the function $f(x) = 2(x - 1)^2 - 3$

models the data exactly.

99. The cables drawn as a parabola can be placed in the coordinate plane. Use $(0, 20)$ as the vertex.

$f(x) = a(x - h)^2 + k \Rightarrow f(x) = a(x - 0)^2 + 20 = f(x) = ax^2 + 20$; since the parabola passes through

$(150, 120)$, $f(150) = 120$; $f(150) = a(150)^2 + 20 \Rightarrow 120 = 22{,}500a + 20 \Rightarrow 100 = 22{,}500a \Rightarrow$

$a = \dfrac{1}{225}$; $f(x) = \dfrac{1}{225}x^2 + 20$, or $f(x) \approx 0.0044x^2 + 20$.

101. (a) Since the minimum occurs at $t = 4$, let $(4, 90)$ be the vertex and write $H(t) = a(t - 4)^2 + 90$. Use

the data point $(0, 122)$ to find a: $H(0) = a(0 - 4)^2 + 90 \Rightarrow 122 = a(0 - 4)^2 + 90 \Rightarrow$

$122 = 16a + 90 \Rightarrow 16a = 32 \Rightarrow a = 2$; $H(t) = 2(t - 4)^2 + 90$; Domain of $H = \{0 \le t \le 4\}$

(b) $H(1.5) = 2(1.5 - 4)^2 + 90 \Rightarrow H(1.5) = 2(-2.5)^2 + 90 \Rightarrow H(1.5) = 12.5 + 90 \Rightarrow$

$H(1.5) = 102.5$ beats per minute.

103. (a) Using the coordinates $(1982, 1.6)$, $f(x) = a(x - 1982)^2 + 1.6$. To find a, use $(1994, 442) \Rightarrow$

$442 = a(1994 - 1982)^2 + 1.6 \Rightarrow 440.4 = a(144) \Rightarrow a \approx 3.06$; $f(x) = 3.06(x - 1982)^2 + 1.6$.

Answers may vary.

(b) See Figure 103.

(c) $f(1991) = 3.06(1991 - 1982)^2 + 1.6 = 3.06(81) + 1.6 = 249.46$. By 1991, a total of about 250,000

AIDS cases had been reported.

105. Enter the data into your calculator. See Figure 105a. Select quadratic regression from the STAT menu. See

Figure 105b. In Figure 105c, the modeling function is given (approximately) by

$f(x) = 3.125x^2 + 2.05x - 0.9$. $f(3.5) = 3.125(3.5)^2 + 2.05(3.5) - 0.9 \approx 44.56$.

[1935, 2005, 5] by [1.9, 2.6, 0.1]

| Figure 105a | Figure 105b | Figure 105c | Figure 107 |

107. (a) $f(x) = 0.00019838x^2 - 0.79153x + 791.46$. See Figure 107.

(b) $f(1975) = 0.00019838(1975)^2 - 0.79153(1975) + 791.46 \approx 1.99$, which compares favorably with

the actual value of 2.01.

109. (a) Enter data into your calculator. See Figure 109a. Select quadratic regression from the STAT menu. See

Figure 109b. In Figure 109c, the modeling function is given by approximately

$f(x) = 0.59462x^2 - 2350.82x + 2,323,895$.

(b) $f(1985) = 0.59462(1985)^2 - 2350.82(1985) + 2,323,895 = 453.8895$. In 1985, the approximate

enrollment was 454 thousand.

| Figure 109a | Figure 109b | Figure 109c |

3.2: Quadratic Equations and Problem Solving

1. $x^2 + x - 11 = 1 \Rightarrow x^2 + x - 12 = 0$. Factoring this we get: $(x + 4)(x - 3) = 0$. Then $x + 4 = 0 \Rightarrow$

$x = -4$ or $x - 3 = 0 \Rightarrow x = 3$. Therefore $x = -4, 3$.

Check: $(-4)^2 + (-4) - 11 = 1 \Rightarrow 16 - 4 - 11 = 1 \Rightarrow 12 - 11 = 1 \Rightarrow 1 = 1$;

$(3)^2 + (3) - 11 = 1 \Rightarrow 9 + 3 - 11 = 1 \Rightarrow 12 - 11 = 1 \Rightarrow 1 = 1$

3. $t^2 = 2t \Rightarrow t^2 - 2t = 0$. Factoring this we get: $t(t - 2) = 0$. Then $t = 0$ or $t - 2 = 0 \Rightarrow t = 2$.

Therefore $t = 0, 2$. Check: $0^2 = 2(0) \Rightarrow 0 = 0$; $2^2 = 2(2) \Rightarrow 4 = 4$

5. $3x^2 - 7x = 0$. Factoring we get $x(3x - 7) = 0$. Then $x = 0$ or $3x - 7 = 0 \Rightarrow x = \dfrac{7}{3}$. Therefore $x = 0, \dfrac{7}{3}$.

 Check: $3(0)^2 - 7(0) = 0 \Rightarrow 0 = 0$; $3\left(\dfrac{7}{3}\right)^2 - 7\left(\dfrac{7}{3}\right) = 0 \Rightarrow \dfrac{49}{3} - \dfrac{49}{3} = 0 \Rightarrow 0 = 0$

7. $2z^2 = 13z + 15 \Rightarrow 2z^2 - 13z - 15 = 0$. Factoring we get $(2z - 15)(z + 1) = 0$. Then $2z - 15 = 0 \Rightarrow$

 $z = \dfrac{15}{2}$ or $z + 1 = 0 \Rightarrow z = -1$. Check: $2\left(\dfrac{15}{2}\right)^2 = 13\left(\dfrac{15}{2}\right) + 15 \Rightarrow \dfrac{225}{2} = \dfrac{195}{2} + \dfrac{30}{2} \Rightarrow \dfrac{225}{2} = \dfrac{225}{2}$;

 $2(-1)^2 = 13(-1) + 15 \Rightarrow 2 = -13 + 15 \Rightarrow 2 = 2$.

9. $x(3x + 14) = 5 \Rightarrow 3x^2 + 14x - 5 = 0$. Factoring this we get: $(3x - 1)(x + 5) = 0$.

 Then $3x - 1 = 0 \Rightarrow 3x = 1 \Rightarrow x = \dfrac{1}{3}$ or $x + 5 = 0 \Rightarrow x = -5$. Therefore $x = -5, \dfrac{1}{3}$.

 Check: $-5(3(-5) + 14) = 5 \Rightarrow -5(-1) = 5 \Rightarrow 5 = 5$; $\dfrac{1}{3}\left(3\left(\dfrac{1}{3}\right) + 14\right) = 5 \Rightarrow \dfrac{1}{3}\left(15\right) = 5 \Rightarrow 5 = 5$

11. $6x^2 + \dfrac{5}{2} = 8x \Rightarrow 6x^2 - 8x + \dfrac{5}{2} = 0 \Rightarrow 12x^2 - 16x + 5 = 0$. Factoring this we get:

 $(2x - 1)(6x - 5) = 0$.

 Then $2x - 1 = 0 \Rightarrow 2x = 1 \Rightarrow x = \dfrac{1}{2}$ or $6x - 5 = 0 \Rightarrow 6x = 5 \Rightarrow x = \dfrac{5}{6}$. Therefore $x = \dfrac{1}{2}, \dfrac{5}{6}$.

 Check: $6\left(\dfrac{1}{2}\right)^2 + \dfrac{5}{2} = 8\left(\dfrac{1}{2}\right) \Rightarrow \dfrac{3}{2} + \dfrac{5}{2} = 4 \Rightarrow 4 = 4$;

 $6\left(\dfrac{5}{6}\right)^2 + \dfrac{5}{2} = 8\left(\dfrac{5}{6}\right) \Rightarrow \dfrac{25}{6} + \dfrac{15}{6} = \dfrac{40}{6} \Rightarrow \dfrac{40}{6} = \dfrac{40}{6}$

13. $(t + 3)^2 = 5 \Rightarrow t^2 + 6t + 4 = 0$. Using the quadratic formula: $t = \dfrac{-b \pm \sqrt{b^2 - 4ac}}{2a} \Rightarrow$

 $t = \dfrac{-6 \pm \sqrt{6^2 - 4(1)(4)}}{2(1)} \Rightarrow t = \dfrac{-6 \pm \sqrt{20}}{2} \Rightarrow t = \dfrac{-6 \pm 2\sqrt{5}}{2} \Rightarrow t = -3 \pm \sqrt{5}$

15. $4x^2 - 13 = 0 \Rightarrow 4x^2 + 0x - 13 = 0$. Using the quadratic formula: $x = \dfrac{-b \pm \sqrt{b^2 - 4ac}}{2a} \Rightarrow$

 $x = \dfrac{0 \pm \sqrt{0^2 - 4(4)(-13)}}{2(4)} \Rightarrow x = \dfrac{\pm\sqrt{208}}{8} \Rightarrow x = \dfrac{\pm 4\sqrt{13}}{8} \Rightarrow x = \dfrac{\pm\sqrt{13}}{2}$

17. $2(x - 1)^2 + 4 = 0 \Rightarrow 2(x^2 - 2x + 1) + 4 = 0 \Rightarrow 2x^2 - 4x + 6 = 0$. Using the quadratic formula:

 $x = \dfrac{-b \pm \sqrt{b^2 - 4ac}}{2a} \Rightarrow x = \dfrac{-(-4) \pm \sqrt{(-4)^2 - 4(2)(6)}}{2(2)} \Rightarrow x = \dfrac{4 \pm \sqrt{-32}}{4} = \dfrac{4 \pm 4\sqrt{-2}}{4} =$

 $1 \pm \sqrt{-2}$. Since $\sqrt{-2}$ is not real, there is no real solution to the equation.

19. $\dfrac{1}{2}x^2 - 3x + \dfrac{1}{2} = 0 \Rightarrow x^2 - 6x + 1 = 0$. Using the quadratic formula: $x = \dfrac{-b \pm \sqrt{b^2 - 4ac}}{2a} \Rightarrow$

 $x = \dfrac{-(-6) \pm \sqrt{(-6)^2 - 4(1)(1)}}{2(1)} \Rightarrow x = \dfrac{6 \pm \sqrt{32}}{2} \Rightarrow x = \dfrac{6 \pm 4\sqrt{2}}{2} \Rightarrow x = 3 \pm 2\sqrt{2}$

21. $-3z^2 - 2z + 4 = 0$. Using the quadratic formula: $x = \dfrac{-b \pm \sqrt{b^2 - 4ac}}{2a} \Rightarrow$

$$x = \frac{-(-2) \pm \sqrt{(-2)^2 - 4(-3)(4)}}{2(-3)} \Rightarrow x = \frac{2 \pm \sqrt{52}}{-6} \Rightarrow x = \frac{-2 \pm 2\sqrt{13}}{6} \Rightarrow x = \frac{-1 \pm \sqrt{13}}{3}$$

23. $25k^2 + 1 = 10k \Rightarrow 25k^2 - 10k + 1 = 0$. Using the quadratic formula: $x = \dfrac{-b \pm \sqrt{b^2 - 4ac}}{2a} \Rightarrow$

$$x = \frac{-(-10) \pm \sqrt{(-10)^2 - 4(25)(1)}}{2(25)} \Rightarrow x = \frac{10 \pm \sqrt{0}}{50} \Rightarrow x = \frac{1}{5}$$

25. $-0.3x^2 + 0.1x = -0.02 \Rightarrow -30x^2 + 10x + 2 = 0$.

Using the quadratic formula: $x = \dfrac{-b \pm \sqrt{b^2 - 4ac}}{2a} \Rightarrow$

$$x = \frac{-10 \pm \sqrt{10^2 - 4(-30)(2)}}{2(-30)} \Rightarrow x = \frac{-10 \pm \sqrt{340}}{-60} \Rightarrow x = \frac{10 \pm 2\sqrt{85}}{60} \Rightarrow x = \frac{5 \pm \sqrt{85}}{30}$$

27. $2x(x + 2) = (x - 1)(x + 2) \Rightarrow 2x^2 + 4x = x^2 + x - 2 \Rightarrow x^2 + 3x + 2 = 0$. Factoring we get

$(x + 2)(x + 1) = 0$. Then $x + 2 = 0 \Rightarrow x = -2$ or $x + 1 = 0 \Rightarrow x = -1$.

29. For the x-intercepts, $y = 0 \Rightarrow 6x^2 + 13x - 5 = 0$. Factoring this we get $(2x + 5)(3x - 1) = 0$. Then

$2x + 5 = 0 \Rightarrow 2x = -5 \Rightarrow x = -\dfrac{5}{2}$ or $3x - 1 = 0 \Rightarrow 3x = 1 \Rightarrow x = \dfrac{1}{3}$. Therefore $x = -\dfrac{5}{2}, \dfrac{1}{3}$.

31. For the x-intercept, $y = 0 \Rightarrow -4x^2 + 12x - 9 = 0$. Using the quadratic formula:

$$x = \frac{-12 \pm \sqrt{12^2 - 4(-4)(-9)}}{2(-4)} \Rightarrow x = \frac{-12 \pm \sqrt{144 - 144}}{-8} \Rightarrow x = \frac{-12 \pm \sqrt{0}}{-8} \Rightarrow x = \frac{3}{2}$$

33. For the x-intercepts, $y = 0 \Rightarrow -3x^2 + 11x - 6 = 0$. Using the quadratic formula:

$$x = \frac{-11 \pm \sqrt{11^2 - 4(-3)(-6)}}{2(-3)} \Rightarrow x = \frac{-11 \pm \sqrt{121 - 72}}{-6} \Rightarrow x = \frac{-11 \pm \sqrt{49}}{-6} \Rightarrow x = \frac{11 \pm 7}{6} \Rightarrow$$

$x = \dfrac{2}{3}, 3$

35. (a) Graphical: Graph $Y_1 = X^2 + 2X$ and locate the x-intercepts. See Figure 35a and Figure 35b; the

solutions to the equation are -2 and 0.

(b) Numerical: Table $Y_1 = X^2 + 2X$ starting at $x = -4$, incrementing by 1. $Y_1 = 0$ when $x = -2$ or 0.

See Figure 35c.

(c) Symbolic: $x^2 + 2x = 0 \Rightarrow x(x + 2) = 0$. Then $x = 0$ or $x + 2 = 0 \Rightarrow x = -2$. Therefore $x = -2, 0$.

[−10, 10, 1] by [−10, 10, 1]   [−10, 10, 1] by [−10, 10, 1]

Figure 35a   Figure 35b   Figure 35c

37. (a) Graphical: Graph $Y_1 = X^2 - X - 6$ and locate the x-intercepts. See Figures 37a and 37b; the solutions to

 the equation are -2 and 3.

 (b) Numerical: Table $Y_1 = X^2 - X - 6$ starting at $x = -3$, incrementing by 1. $Y_1 = 0$ when $x = -2$ or 3.

 See Figure 37c.

 (c) Symbolic: $x^2 - x - 6 = 0 \Rightarrow (x - 3)(x + 2) = 0 \Rightarrow x = 3$ or -2.

$[-10, 10, 1]$ by $[-10, 10, 1]$ $[-10, 10, 1]$ by $[-10, 10, 1]$

Figure 37a   Figure 37b   Figure 37c

39. (a) Graphical: Graph $Y_1 = 2X^2 - 6$ and locate the x-intercepts. See Figures 39a and 39b; the solutions to the

 equation are $x \approx \pm 1.7$.

 (b) Numerical: Table $Y_1 = 2X^2 - 6$ starting at $x = -2\sqrt{3}$, incrementing by $\sqrt{3}$. $Y_1 \approx 0$ when $x \approx \pm 1.7$.

 See Figure 39c.

 (c) Symbolic: $2x^2 = 6 \Rightarrow x^2 = 3 \Rightarrow x = \pm\sqrt{3} \approx \pm 1.7$.

$[-10, 10, 1]$ by $[-10, 10, 1]$ $[-10, 10, 1]$ by $[-10, 10, 1]$

Figure 39a   Figure 39b   Figure 39c

41. (a) Graphical: Graph $Y_1 = 4X^2 - 12X + 9$ and locate the x-intercept. See Figure 41a; the solution to

 the equation is $x = 1.5$.

 (b) Numerical: Table $Y_1 = 4X^2 - 12X + 9$ starting at $x = 0$, incrementing by 0.5. $Y_1 = 0$ when $x = 1.5$.

 See Figure 41b.

 (c) Symbolic: $4x^2 - 12x + 9 = 0 \Rightarrow (2x - 3)(2x - 3) = 0 \Rightarrow x = \dfrac{3}{2}$.

$[-5, 5, 1]$ by $[-5, 5, 1]$

Figure 41a   Figure 41b

43. $20x^2 + 11x = 3 \Rightarrow 20x^2 + 11x - 3 = 0$. Graph $Y_1 = 20X{\wedge}2 + 11X - 3$ and locate the x-intercepts.

See Figures 43a and 43b; the solutions to the equation are $x = -0.75$ or 0.2.

Figure 43a   Figure 43b   Figure 45a   Figure 45b

45. $2.5x^2 = 4.75x - 2.1 \Rightarrow 2.5x^2 - 4.75x + 2.1 = 0$. Graph $Y_1 = 2.5X{\wedge}2 - 4.75X + 2.1$ and locate the

x-intercepts. See Figures 45a and 45b; the solutions to the equation are $x = 0.7$ or 1.2.

47. $x^2 + 4x - 6 = 0 \Rightarrow x^2 + 4x = 6 \Rightarrow x^2 + 4x + 4 = 6 + 4 \Rightarrow (x + 2)^2 = 10 \Rightarrow$

$x + 2 = \pm\sqrt{10} \Rightarrow x = -2 \pm \sqrt{10}$

49. $x^2 + 5x = 4 \Rightarrow x^2 + 5x + \dfrac{25}{4} = 4 + \dfrac{25}{4} \Rightarrow \left(x + \dfrac{5}{2}\right)^2 = \dfrac{41}{4} \Rightarrow$

$x + \dfrac{5}{2} = \pm\sqrt{\dfrac{41}{4}} \Rightarrow x = -\dfrac{5}{2} \pm \dfrac{\sqrt{41}}{2} \Rightarrow x = -\dfrac{5}{2} \pm \dfrac{1}{2}\sqrt{41}$

51. $3x^2 - 6x = 2 \Rightarrow x^2 - 2x = \dfrac{2}{3} \Rightarrow x^2 - 2x + 1 = \dfrac{2}{3} + 1 \Rightarrow (x - 1)^2 = \dfrac{5}{3} \Rightarrow$

$x - 1 = \pm\sqrt{\dfrac{5}{3}} \Rightarrow x = 1 \pm \sqrt{\dfrac{5}{3}} \Rightarrow 1 \pm \dfrac{\sqrt{15}}{3}$

53. $x^2 - 8x = 10 \Rightarrow x^2 - 8x + 16 = 10 + 16 \Rightarrow (x - 4)^2 = 26 \Rightarrow x - 4 = \pm\sqrt{26} \Rightarrow x = 4 \pm \sqrt{26}$

55. $\dfrac{1}{2}t^2 - \dfrac{3}{2}t = 1 \Rightarrow t^2 - 3t = 2 \Rightarrow t^2 - 3t + \dfrac{9}{4} = 2 + \dfrac{9}{4} \Rightarrow \left(t - \dfrac{3}{2}\right)^2 = \dfrac{17}{4} \Rightarrow t - \dfrac{3}{2} = \pm\sqrt{\dfrac{17}{4}} \Rightarrow$

$t = \dfrac{3}{2} \pm \dfrac{\sqrt{17}}{2} \Rightarrow t = \dfrac{3 \pm \sqrt{17}}{2}$

57. $-2z^2 + 3z + 1 = 0 \Rightarrow z^2 - \dfrac{3}{2}z = \dfrac{1}{2} \Rightarrow z^2 - \dfrac{3}{2}z + \dfrac{9}{16} = \dfrac{1}{2} + \dfrac{9}{16} \Rightarrow \left(z - \dfrac{3}{4}\right)^2 = \dfrac{17}{16} \Rightarrow$

$z - \dfrac{3}{4} = \pm\sqrt{\dfrac{17}{16}} \Rightarrow z = \dfrac{3}{4} \pm \dfrac{\sqrt{7}}{4} \Rightarrow z = \dfrac{3 \pm \sqrt{17}}{4}$

59. $-\dfrac{3}{2}z^2 - \dfrac{1}{4}z + 1 = 0 \Rightarrow z^2 + \dfrac{1}{6}z = \dfrac{2}{3} \Rightarrow z^2 + \dfrac{1}{6}z + \dfrac{1}{144} = \dfrac{2}{3} + \dfrac{1}{144} \Rightarrow \left(z + \dfrac{1}{12}\right)^2 = \dfrac{97}{144} \Rightarrow$

$z + \dfrac{1}{12} = \pm\sqrt{\dfrac{97}{144}} \Rightarrow z = -\dfrac{1}{12} \pm \dfrac{\sqrt{97}}{12} \Rightarrow z = \dfrac{-1 \pm \sqrt{97}}{12}$

61. $D = $ all real numbers except when the denominator $x^2 - 5 = 0 \Rightarrow x^2 = 5 \Rightarrow x = \pm\sqrt{5} \Rightarrow$

$D = \{x \mid x \neq \sqrt{5}, x \neq -\sqrt{5}\}$

63. $D = $ all real numbers except when the denominator

$t^2 - t - 2 = 0 \Rightarrow (t - 2)(t + 1) = 0 \Rightarrow t = -1, 2 \Rightarrow D = \{t \mid t \neq -1, t \neq 2\}$

65. $4x^2 + 3y = \dfrac{y+1}{3} \Rightarrow 3y - \dfrac{y+1}{3} = -4x^2 \Rightarrow 9y - y - 1 = -12x^2 \Rightarrow 8y - 1 = -12x^2 \Rightarrow$

$8y = -12x^2 + 1 \Rightarrow y = \dfrac{-12x^2 + 1}{8}$; yes, y is a function of x since each x-input produces only one y-output.

67. $3y = \dfrac{2x - y}{3} \Rightarrow 9y = 2x - y \Rightarrow 10y = 2x \Rightarrow y = \dfrac{x}{5}$; yes, y is a function of x since each x-input produces

only one y-output.

69. $x^2 + (y-3)^2 = 9 \Rightarrow (y-3)^2 = 9 - x^2 \Rightarrow y - 3 = \pm\sqrt{9 - x^2} \Rightarrow y = 3 \pm \sqrt{9 - x^2}$; no, y is not a

function of x since some x-inputs produce two y-outputs.

71. $3x^2 + 4y^2 = 12 \Rightarrow 4y^2 = 12 - 3x^2 \Rightarrow 2y = \pm\sqrt{12 - 3x^2} \Rightarrow y = \pm\dfrac{\sqrt{12 - 3x^2}}{2}$;

no, y is not a function of x because some x-inputs produce two y-outputs.

73. $V = \dfrac{1}{3}\pi r^2 h$ for $r \Rightarrow \dfrac{3V}{\pi h} = r^2 \Rightarrow r = \pm\sqrt{\dfrac{3V}{\pi h}}$

75. $K = \dfrac{1}{2}mv^2$ for $v \Rightarrow \dfrac{2K}{m} = v^2 \Rightarrow v = \pm\sqrt{\dfrac{2K}{m}}$

77. $a^2 + b^2 = c^2$ for $b \Rightarrow b^2 = c^2 - a^2 \Rightarrow b = \pm\sqrt{c^2 - a^2}$

79. $s = -16t^2 + 100t$ for $t \Rightarrow -s = 16t^2 - 100t \Rightarrow \dfrac{-s}{16} = t^2 - \dfrac{25}{4}t \Rightarrow \dfrac{-s}{16} + \dfrac{625}{64} = t^2 - \dfrac{25}{4}t + \dfrac{625}{64} \Rightarrow$

$\left(t - \dfrac{25}{8}\right)^2 = \dfrac{625 - 4s}{64} \Rightarrow t - \dfrac{25}{8} = \pm\sqrt{\dfrac{625 - 4s}{64}} \Rightarrow t = \dfrac{25}{8} \pm \sqrt{\dfrac{625 - 4s}{64}} \Rightarrow$

$t = \dfrac{25 \pm \sqrt{625 - 4s}}{8}$

81. (a) $3x^2 = 12 \Rightarrow 3x^2 - 12 = 0$

(b) $b^2 - 4ac = 0^2 - 4(3)(-12) = 144 > 0$. There are two real solutions.

(c) $3x^2 = 12 \Rightarrow x^2 = 4 \Rightarrow x = \pm 2$

83. (a) $x^2 - 2x = -1 \Rightarrow x^2 - 2x + 1 = 0$

(b) $b^2 - 4ac = (-2)^2 - 4(1)(1) = 0$. There is one real solution.

(c) $x^2 - 2x + 1 = 0 \Rightarrow (x-1)^2 = 0 \Rightarrow x = 1$

85. (a) $4x = x^2 \Rightarrow x^2 - 4x = 0$

(b) $b^2 - 4ac = (-4)^2 - 4(1)(0) = 16 > 0$. There are two real solutions.

(c) $x^2 - 4x = 0 \Rightarrow x(x-4) = 0 \Rightarrow x = 0$ or $x - 4 = 0 \Rightarrow x = 4$. Therefore, $x = 0, 4$.

87. (a) $x^2 + 1 = x \Rightarrow x^2 - x + 1 = 0$

(b) $b^2 - 4ac = (-1)^2 - 4(1)(1) = -3 < 0$. There are no real solutions.

(c) There are no real solutions.

89. (a) $2x^2 + 3x = 12 - 2x \Rightarrow 2x^2 + 5x - 12 = 0$.

(b) $b^2 - 4ac = (5)^2 - 4(2)(-12) = 121 > 0$. There are two real solutions.

(c) $2x^2 + 5x - 12 = 0 \Rightarrow (2x - 3)(x + 4) = 0 \Rightarrow 2x = 3 \Rightarrow x = 1.5$ or $x = -4$. So, $x = 1.5, -4$.

91. (a) $9x(x - 4) = -36 \Rightarrow 9x^2 - 36x + 36 = 0$

    (b) $b^2 - 4ac = (-36)^2 - 4(9)(36) = 0$. There is one real solution.

    (c) $9x^2 - 36x + 36 = 0 \Rightarrow x^2 - 4x + 4 = 0 \Rightarrow (x - 2)^2 = 0 \Rightarrow x = 2$

93. (a) $x\left(\dfrac{1}{2}x + 1\right) = -\dfrac{13}{2} \Rightarrow \dfrac{1}{2}x^2 + x + \dfrac{13}{2} = 0$

    (b) $b^2 - 4ac = (1)^2 - 4\left(\dfrac{1}{2}\right)\left(\dfrac{13}{2}\right) = -12 < 0$. There are no real solutions.

    (c) There are no real solutions.

95. (a) $3x^2 = 1 - x \Rightarrow 3x^2 + x - 1 = 0$

    (b) $b^2 - 4ac = 1^2 - 4(3)(-1) = 13 > 0$. There are two real solutions.

    (c) $x = \dfrac{-1 \pm \sqrt{1^2 - 4(3)(-1)}}{2(3)} = \dfrac{-1 \pm \sqrt{13}}{2(3)} = \dfrac{-1 \pm \sqrt{13}}{6}$

97. (a) Since the parabola opens upward, $a > 0$.

    (b) Since the zeros of f are -6 and 2, the solutions to $ax^2 + bx + c = 0$ are also -6 and 2.

    (c) Since there are two real solutions, the discriminant is positive.

99. (a) Since the parabola opens upward, $a > 0$.

    (b) Since the only zero of f is -4, the solution is also -4.

    (c) Since there is one real solutions, the discriminant is equal to zero.

101. The height when it hits the ground will be $0 \Rightarrow 75 - 16t^2 = 0 \Rightarrow -16t^2 = -75 \Rightarrow t^2 = \dfrac{75}{16} \Rightarrow$

    $t = \sqrt{\dfrac{75}{16}} \Rightarrow t = \dfrac{\sqrt{75}}{4} \Rightarrow t \approx 2.2$ seconds.

103. $90{,}000 = 2375x^2 + 5134x + 5020 \Rightarrow 2375x^2 + 5134x - 84{,}980 = 0$. Using the quadratic formula we get:

    $x = \dfrac{-5134 \pm \sqrt{(5134)^2 - 4(2375)(-84{,}980)}}{2(2375)} \Rightarrow x \approx -7.2$ or 5. Since -7.2 does not apply, the number

    of years is 5. Therefore $1984 + 5 = 1989$.

105. Graphical: First, we must determine a formula for the area. Let x represent the height of the computer screen.

    Then, $x + 2.5$ is the width. The area of the screen is height times width, computed $A(x) = x(x + 2.5)$. We

    must solve the quadratic equation $x(x + 2.5) = 93.5$ or $x^2 + 2.5x - 93.5 = 0$. Graph

    $Y_1 = X^2 + 2.5X - 93.5$ and determine any zeros. Figure 105a shows that the equation has two zeros, one

    negative and one positive. The positive zero is located at $x = 8.5$. The height is 8.5 inches and the width is

    $8.5 + 2.5 = 11$ inches.

    Symbolic: The quadratic equation $x^2 + 2.5x - 93.5 = 0$ can be solved by the quadratic formula.

    $x = \dfrac{-b \pm \sqrt{b^2 - 4ac}}{2a} = \dfrac{-2.5 \pm \sqrt{2.5^2 - 4(1)(-93.5)}}{2(1)} = \dfrac{-2.5 \pm 19.5}{2} = 8.5, -11$. The positive

    answer gives a height of 8.5 inches. It follows that the width is 11 inches.

    The numerical solution is shown in Figure 105b. Yes, the symbolic, graphical and numerical answers agree.

[-15, 15, 1] by [-100, 100, 10]

Figure 105a

Figure 105b

Figure 107

[0, 150, 10] by [-20, 50, 10]

Figure 113

107. Let x = width of the metal sheet in inches and $x + 10$ = length of the metal sheet in inches. Make a sketch to find expressions for the dimensions of the box. See Figure 107. The width of the box is $x - 8$ inches, the length of the box is $x + 2$ inches, and the height of the box is 4 inches; the volume of the box, which is given as 476 cubic inches, is determined by the length times the width times the height: $4(x - 8)(x + 2) = 476 \Rightarrow$ $(x - 8)(x + 2) = 119 \Rightarrow x^2 - 6x - 16 = 119 \Rightarrow x^2 - 6x - 135 = 0 \Rightarrow (x + 9)(x - 15) = 0 \Rightarrow$ $x = -9$ or $x = 15$. Since width cannot be negative, $x = 15$ inches; thus, the dimensions of the metal sheet is 15 inches by 25 inches.

109. $V = \pi r^2 h$, $V = 28$ cubic inches, and $h = 4$ inches $\Rightarrow 28 = \pi r^2(4) \Rightarrow r^2 = \dfrac{28}{4\pi} \Rightarrow r = \sqrt{\dfrac{28}{4\pi}} \approx 1.49$; the radius of the cylinder is approximately 1.49 inches.

111. Since the diameter of the semicircle is x, the radius is $\dfrac{x}{2}$; the area of the semicircle

$= \dfrac{1}{2}\pi\left(\dfrac{x}{2}\right)^2 = \dfrac{1}{2}\pi\left(\dfrac{x^2}{4}\right) = \dfrac{1}{8}\pi x^2$. The area of the square $= x^2$; thus, the total area of the window, which

is 463 square inches, is $x^2 + \dfrac{1}{8}\pi x^2$; $463 = x^2 + \dfrac{1}{8}\pi x^2 \Rightarrow 463 = \left(1 + \dfrac{\pi}{8}\right)x^2 \Rightarrow x^2 = \dfrac{463}{\left(1 + \dfrac{\pi}{8}\right)} \Rightarrow$

$x = \sqrt{\dfrac{463}{1 + \frac{\pi}{8}}} \approx 18.23$ inches.

113. Let x be the number of shirts ordered. Revenue equals the number of shirts sold times the price of each shirt. If x shirts are sold, then the price in dollars of each shirt is $20 - 0.10(x - 1)$. The revenue $R(x)$ is given by $R(x) = x(20 - 0.10(x - 1))$. We must solve the equation $989 = x(20 - 0.10(x - 1)) \Rightarrow$ $-0.10x^2 + 20.1x - 989 = 0$. Let $y_1 = -0.10x^2 + 20.1x - 989$ and use the graphing calculator to find the x-intercept. Because the discount applies to orders up to 100 shirts, we find that $x = 86$ shirts. See Figure 113.

115. (a) Since $s(0) = 32$, $s(t) = -16t^2 + v_0 t + h_0 \Rightarrow 32 = -16(0)^2 + v_0(0) + h_0 \Rightarrow h_0 = 32$

and so $s(t) = -16t^2 + v_0 t + 32$; since $s(1) = 176$, $176 = -16(1)^2 + v_0(1) + 32 \Rightarrow v_0 = 160$;

thus, $s(t) = -16t^2 + 160t + 32$ models the data.

(b) When the projectile strikes the ground, $s(t) = 0$; $0 = -16t^2 + 160t + 32 \Rightarrow 0 = t^2 - 10t - 2 \Rightarrow$

$t = \dfrac{10 \pm \sqrt{10^2 - 4(1)(-2)}}{2} \approx 10.2$ or -0.2; since t cannot be negative, $t \approx 10.2$; the projectile strikes

the ground after about 10.2 seconds.

117. (a) If the speed doubles, the radius of the curve quadruples, or increases by a factor of 4.

(b) Any ordered pair in the table may be used to determine the constant a. $R(x) = ax^2$ and

$$R(10) = 50 \Rightarrow a(10)^2 = 50 \Rightarrow a = \frac{50}{10^2} = 0.5.$$

(c) If $R = 500$, $500 = 0.5x^2 \Rightarrow 1000 = x^2 \Rightarrow x = \sqrt{1000} \approx 31.6$. 31.6 mph is the maximum safe speed on a curve with a 500 foot radius.

119. (a) $E(15) = 1.4$, $1987 + 15 = 2002$; in 2002 there were 1.4 million Wal-Mart employees.

(b) Use the quadratic regression function on your graphing calculator to find

$$f(x) = 0.00474x^2 + 0.00554x + 0.205.$$ *Answers may vary.*

(c) See Figure 119c.

(d) Let $Y_1 = (0.00474)X\char`\^2 + 0.00554X + 0.205$ and $Y_2 = 3$. Use the CALC function on your calculator to find the intersection points of the Y_1 and Y_2. See Figure 119d. When $x \approx 24$, $f(x) = 3$. Therefore, in the year 2011 the number of employees may reach 3 million. *Answers may vary.*

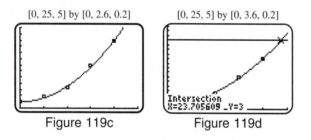

[0, 25, 5] by [0, 2.6, 0.2]   [0, 25, 5] by [0, 3.6, 0.2]

Figure 119c   Figure 119d

3.3: Complex Numbers

1. $\sqrt{-4} = i\sqrt{4} = 2i$

3. $\sqrt{-100} = i\sqrt{100} = 10i$

5. $\sqrt{-23} = i\sqrt{23}$

7. $\sqrt{-12} = i\sqrt{12} = i\sqrt{4}\sqrt{3} = 2i\sqrt{3}$

9. $\sqrt{-54} = \sqrt{(9)(6)(-1)} = \sqrt{9(-1)}\sqrt{6} = 3i\sqrt{6}$

11. $\dfrac{4 \pm \sqrt{-16}}{2} = \dfrac{4 \pm \sqrt{(16)(-1)}}{2} = \dfrac{4 \pm 4i}{2} = 2 \pm 2i$

13. $\dfrac{-6 \pm \sqrt{-72}}{3} = \dfrac{-6 \pm \sqrt{(36)(-1)}\sqrt{2}}{3} = \dfrac{-6 \pm 6i\sqrt{2}}{3} = -2 \pm 2i\sqrt{2}$

15. $\sqrt{-5} \cdot \sqrt{-5} = \sqrt{5}\sqrt{-1} \cdot \sqrt{5}\sqrt{-1} = \sqrt{5}i \cdot \sqrt{5}i = \sqrt{25}i^2 = 5(-1) = -5.$

17. $\sqrt{-18} \cdot \sqrt{-2} = (\sqrt{9}\sqrt{2}\sqrt{-1})(\sqrt{2}\sqrt{-1}) = (3\sqrt{2}\,i)(\sqrt{2}\,i) = 3\sqrt{4}i^2 = (3)(2)(-1) = -6$

19. $\sqrt{-3} \cdot \sqrt{-6} = (\sqrt{3}\sqrt{-1})(\sqrt{6}\sqrt{-1}) = (\sqrt{3}i)(\sqrt{6}i) = \sqrt{18}i^2 = \sqrt{9}\sqrt{2}(-1) = -3\sqrt{2}$

21. $3i + 5i = (3 + 5)i = 8i$

23. $(3 + i) + (-5 - 2i) = (3 + (-5)) + (1 - 2)i = -2 - i$

25. $2i - (-5 + 23i) = (-5) + (2 - 23)i = 5 - 21i$

27. $3 - (4 - 6i) = (3 - 4) + i = -1 + 6i$

29. $(2)(2 + 4i) = 4 + 8i$

31. $(1 + i)(2 - 3i) = (1)(2) + (1)(-3i) + (i)(2) + (i)(-3i) = 2 - i - 3i^2 = 2 - i - 3(-1) = 5 - i$

33. $(-3 + 2i)(-2 + i) = (-3)(-2) + (-3)(i) + (2i)(-2) + (2i)(i) = 6 - 7i + 2i^2 =$

$6 - 7i + 2(-1) = 4 - 7i$

35. $(-2 + 3i)^2 = (-2 + 3i)(-2 + 3i) = (-2)(-2) + (-2)(3i) + (-2)(3i) + (3i)(3i) =$

$4 + (-12i) + (9i^2) = 4 - 12i + (-9) = -5 - 12i$

37. $2i(1 - i)^2 = 2i(1 - i)(1 - i) = 2i[(1)(1) + (1)(-i) + (1)(-i) + (-i)(-i)] =$

$2i[1 - 2i + (-1)] = 2i(-2i) = -4i^2 = 4$

39. $\dfrac{1}{1 + i} = \dfrac{1}{1 + i} \cdot \dfrac{1 - i}{1 - i} = \dfrac{1 - i}{(1 + i)(1 - i)} = \dfrac{1 - i}{1 - i^2} = \dfrac{1 - i}{2} = \dfrac{1}{2} - \dfrac{1}{2}i$

41. $\dfrac{4 + i}{5 - i} = \dfrac{4 + i}{5 - i} \cdot \dfrac{5 + i}{5 + i} = \dfrac{(4 + i)(5 + i)}{(5 - i)(5 + i)} = \dfrac{20 + 9i + i^2}{25 - i^2} = \dfrac{19 + 9i}{26} = \dfrac{19}{26} + \dfrac{9}{26}i$

43. $\dfrac{2i}{10 - 5i} = \dfrac{2i}{10 - 5i} \cdot \dfrac{10 + 5i}{10 + 5i} = \dfrac{20i + 10i^2}{(10 - 5i)(10 + 5i)} = \dfrac{-10 + 20i}{100 - 25i^2} = \dfrac{-10 + 20i}{125} = -\dfrac{2}{25} + \dfrac{4}{25}i$

45. $\dfrac{3}{-i} \cdot \dfrac{i}{i} = \dfrac{3i}{-i^2} = \dfrac{3i}{-i^2} = \dfrac{3i}{-(-1)} = 3i$

47. $\dfrac{-2 + i}{(1 + i)^2} = \dfrac{-2 + i}{(1 + i)(1 + i)} = \dfrac{-2 + i}{(1)(1) + (1)(i) + (1)(i) + (i)(i)} = \dfrac{-2 + i}{1 + 2i + i^2} = \dfrac{-2 + i}{1 + 2i - 1} =$

$\dfrac{-2 + i}{2i} \cdot \dfrac{i}{i} = \dfrac{-2i + i^2}{2i^2} = \dfrac{-2i - 1}{2(-1)} = \dfrac{-2i - 1}{-2} = \dfrac{1}{2} + i$

49. $(23 - 5.6i) + (-41.5 + 93i) = -18.5 + 87.4i$

51. $(17.1 - 6i) - (8.4 + 0.7i) = 8.7 - 6.7i$

53. $(-12.6 - 5.7i)(5.1 - 9.3i) = -117.27 + 88.11i$

55. $\dfrac{17 - 135i}{18 + 142i} \approx -0.921 - 0.236i$

57. $x^2 + 5 = 0 \Rightarrow x^2 = -5 \Rightarrow x = \pm\sqrt{-5} = \pm i\sqrt{5}$

59. $5x^2 + 1 = 3x^2 \Rightarrow 2x^2 = -1 \Rightarrow x^2 = -\dfrac{1}{2} \Rightarrow x = \pm i\sqrt{\dfrac{1}{2}}$

61. $3x = 5x^2 + 1 \Rightarrow 5x^2 - 3x + 1 = 0$; use the quadratic formula with $a = 5, b = -3$, and $c = 1$.

$$x = \frac{-(-3) \pm \sqrt{(-3)^2 - 4(5)(1)}}{2(5)} = \frac{3 \pm \sqrt{9 - 20}}{10} = \frac{3 \pm \sqrt{-11}}{10} = \frac{3 \pm i\sqrt{11}}{10} = \frac{3}{10} \pm \frac{i\sqrt{11}}{10}$$

63. $x(x - 4) = -5 \Rightarrow x^2 - 4x = -5 \Rightarrow x^2 - 4x + 5 = 0$; use the quadratic formula to solve

$x^2 - 4x + 5 = 0$ with $a = 1, b = -4$, and $c = 5$.

$$x = \frac{-(-4) \pm \sqrt{(-4)^2 - 4(1)(5)}}{2(1)} = \frac{4 \pm \sqrt{16 - 20}}{2} = \frac{4 \pm \sqrt{-4}}{2} = \frac{4 \pm 2i}{2} = 2 \pm i$$

65. Use the quadratic formula to solve $x^2 - 3x + 5 = 0$ with $a = 1, b = -3$, and $c = 5$.

$$x = \frac{-(-3) \pm \sqrt{(-3)^2 - 4(1)(5)}}{2(1)} = \frac{3 \pm \sqrt{9 - 20}}{2} = \frac{3 \pm \sqrt{-11}}{2} = \frac{3 \pm i\sqrt{11}}{2} = \frac{3}{2} \pm \frac{i\sqrt{11}}{2}$$

67. Use the quadratic formula to solve $x^2 + 2x + 4 = 0$ with $a = 1, b = 2$, and $c = 4$.

$$x = \frac{-2 \pm \sqrt{2^2 - 4(1)(4)}}{2(1)} = \frac{-2 \pm \sqrt{4 - 16}}{2} = \frac{-2 \pm \sqrt{-12}}{2} = \frac{-2 \pm 2i\sqrt{3}}{2} = -1 \pm i\sqrt{3}$$

69. $3x^2 - 4x = x^2 - 3 \Rightarrow 2x^2 - 4x + 3 = 0$. Using the quadratic formula we get:

$$\frac{4 \pm \sqrt{16 - 4(2)(3)}}{2(2)} = \frac{4 \pm \sqrt{-8}}{4} = \frac{4 \pm 2\sqrt{-2}}{4} = 1 + \frac{\sqrt{-2}}{2} = 1 \pm \frac{i\sqrt{2}}{2}$$

71. $2x(x - 2) = x - 4 \Rightarrow 2x^2 - 4x = x - 4 \Rightarrow 2x^2 - 5x + 4 = 0$.

Using the quadratic formula we get: $\dfrac{5 \pm \sqrt{(5)^2 - 4(2)(4)}}{2(2)} = \dfrac{5 \pm \sqrt{-7}}{4} = \dfrac{5}{4} \pm \dfrac{i\sqrt{7}}{4}$

73. $3x(3 - x) - 8 = x(x - 2) \Rightarrow 9x - 3x^2 - 8 = x^2 - 2x \Rightarrow -4x^2 + 11x - 8 = 0$.

Using the quadratic formula we get: $\dfrac{-11 \pm \sqrt{(11)^2 - 4(-4)(-8)}}{2(-4)} = \dfrac{-11 \pm \sqrt{-7}}{-8} = \dfrac{11}{8} \pm \dfrac{i\sqrt{7}}{8}$

75. (a) The graph of $y = 2x^2 - x - 3$ intersects the x-axis twice, so there are two real zeros.

   (b) $x = \dfrac{-b \pm \sqrt{b^2 - 4ac}}{2a} = \dfrac{1 \pm \sqrt{(-1)^2 - 4(2)(-3)}}{2(2)} = \dfrac{1 \pm 5}{4} = \dfrac{3}{2}, -1$

77. (a) The graph of $f(x) = x^2 + x + 2$ does not intersect the x-axis. Both of its zeros are imaginary.

   (b) $x = \dfrac{-b \pm \sqrt{b^2 - 4ac}}{2a} = \dfrac{-1 \pm \sqrt{1^2 - 4(1)(2)}}{2(1)} = \dfrac{-1 \pm \sqrt{-7}}{2} = -\dfrac{1}{2} \pm \dfrac{i\sqrt{7}}{2}$

79. (a) The graph of $y = -x^2 - 2$ does not intersect the x-axis. Both of its zeros are imaginary.

   (b) $x = \dfrac{-b \pm \sqrt{b^2 - 4ac}}{2a} = \dfrac{0 \pm \sqrt{0^2 - 4(-1)(-2)}}{2(-1)} = \dfrac{\pm\sqrt{-8}}{-2} = \dfrac{\pm 2i\sqrt{2}}{-2} = \pm i\sqrt{2}$

3.4: Quadratic Inequalities

1. (a) The inequality $f(x) < 0$ is satisfied when the graph of f is below the x-axis. This occurs when

    $-3 < x < 2$.

    (b) The inequality $f(x) \geq 0$ is satisfied when the graph of f is above the x-axis or intersects it. This occurs

    when $x \leq -3$ or $x \geq 2$.

3. (a) The inequality $f(x) \leq 0$ is satisfied when the graph of f is below the x-axis or intersects it. This occurs

    only when $x = -2$.

    (b) The inequality $f(x) > 0$ is satisfied when the graph of f is above the x-axis. This occurs when $x \neq -2$.

5. (a) The inequality $f(x) > 0$ is satisfied when the graph of f is above the x-axis. Since the graph of f is always

    below the x-axis, this inequality has no solutions.

    (b) The inequality $f(x) < 0$ is satisfied when the graph of f is below the x-axis. This occurs for all real

    values of x.

7. (a) The x-intercepts are the solutions to $f(x) = 0$: $0 = 2x^2 + 6x + \dfrac{5}{2} \Rightarrow 4x^2 + 12x + 5 = 0$.

    $$x = \frac{-12 \pm \sqrt{(12)^2 - 4(4)(5)}}{2(4)} \Rightarrow x = \frac{-12 \pm \sqrt{64}}{8} \Rightarrow x = \frac{-12 \pm 8}{8} \Rightarrow x = \frac{-4}{8} \Rightarrow$$

    $$x = -\frac{1}{2} \text{ or } x = -\frac{20}{8} \Rightarrow x = -2\frac{1}{2}, x = -2\frac{1}{2}, -\frac{1}{2}$$

    (b) The inequality $f(x) < 0$ is satisfied when the graph of f is above the x-axis. This occurs in the interval

    $$\left(-\frac{5}{2}, -\frac{1}{2}\right) \text{ or } \left\{x \,\middle|\, -\frac{5}{2} < x < -\frac{1}{2}\right\}.$$

    (c) The inequality $f(x) > 0$ is satisfied when the graph of f is above the x-axis. This occurs in the interval

    $$\left(-\infty, -\frac{5}{2}\right) \cup \left(-\frac{1}{2}, \infty\right) \text{ or } \left\{x \,\middle|\, x < -\frac{5}{2} \text{ or } x > -\frac{1}{2}\right\}.$$

9. (a) The x-intercepts are the solutions to $f(x) = 0$: $0 = -5x^2 + 2x + 7$.

    $$x = \frac{-2 \pm \sqrt{(2)^2 - 4(-5)(7)}}{2(-5)} \Rightarrow x = \frac{-2 \pm \sqrt{144}}{-10} \Rightarrow x = \frac{-2 \pm 12}{-10} \Rightarrow x = -1, \frac{7}{5}$$

    (b) The inequality $f(x) < 0$ is satisfied when the graph of f is above the x-axis. This occurs in the interval

    $$(-\infty, -1) \cup \left(\frac{7}{5}, \infty\right) \text{ or } \left\{x \,\middle|\, x < -1 \text{ or } x > \frac{7}{5}\right\}$$

    (c) The inequality $f(x) > 0$ is satisfied when the graph of f is above the x-axis. This occurs in the interval

    $$\left(-1, \frac{7}{5}\right) \text{ or } \left\{x \,\middle|\, -1 < x < \frac{7}{5}\right\}$$

11. (a) $x^2 - x - 12 = 0 \Rightarrow (x - 4)(x + 3) = 0 \Rightarrow x = -3, 4.$

    (b) Using test intervals, $x^2 - x - 12 < 0$ on $(-3, 4)$ or $\{x \,|\, -3 < x < 4\}.$

    (c) Using test intervals, $x^2 - x - 12 > 0$ on $(-\infty, -3) \cup (4, \infty)$ or $\{x \,|\, x < -3 \text{ or } x > 4\}.$

13. (a) $k^2 - 5 = 0 \Rightarrow k^2 = 5 \Rightarrow k = \pm\sqrt{5}.$

    (b) Using test intervals, $k^2 - 5 \leq 0$ on $[-\sqrt{5}, \sqrt{5}]$ or $\{k \,|\, -\sqrt{5} \leq k \leq \sqrt{5}\}.$

    (c) Using test intervals, $k^2 - 5 \geq 0$ on $(-\infty, -\sqrt{5}] \cup [\sqrt{5}, \infty)$ or $\{k \,|\, k \leq -\sqrt{5} \text{ or } k \geq \sqrt{5}\}.$

15. (a) $3x^2 + 8x = 0 \Rightarrow x(3x + 8) = 0 \Rightarrow x = -\dfrac{8}{3}, 0.$ Using $x = -1$, which is inbetween $-\dfrac{8}{3}$ and 0

    produces $3(-1)^2 + 8(-1) = -5 \Rightarrow$ less than 0.

    (b) Using test intervals, $3x^2 + 8x \leq 0$ on $\left[-\dfrac{8}{3}, 0\right]$ or $\left\{x \,\middle|\, -\dfrac{8}{3} \leq x \leq 0\right\}.$

    (c) Using test intervals, $3x^2 + 8x \geq 0$ on $\left(-\infty, -\dfrac{8}{3}\right] \cup [0, \infty)$ or $\left\{x \,\middle|\, x \leq -\dfrac{8}{3} \text{ or } x \geq 0\right\}.$

17. (a) $-4x^2 + 12x - 9 = 0 \Rightarrow (2x - 3)(-2x + 3) = 0 \Rightarrow x = \dfrac{3}{2}.$

    (b) Using test intervals, $-4x^2 + 12x - 9 < 0$ on $x \neq \dfrac{3}{2}$ on $\left(-\infty, \dfrac{3}{2}\right) \cup \left(\dfrac{3}{2}, \infty\right)$ or $\left\{x \,\middle|\, x < \dfrac{3}{2} \text{ or } x > \dfrac{3}{2}\right\}$

    (c) Using test intervals, $-4x^2 + 12x - 9 > 0$ has no solution

19. (a) $12z^2 - 23z + 10 = 0 \Rightarrow (3z - 2)(4z - 5) = 0 \Rightarrow z = \dfrac{2}{3}, \dfrac{5}{4}.$

    (b) Using test intervals, $12z^2 - 23z + 10 \leq 0$ on $\left[\dfrac{2}{3}, \dfrac{5}{4}\right]$ or $\left\{z \,\middle|\, \dfrac{2}{3} \leq z \leq \dfrac{5}{4}\right\}$

    (c) Using test intervals, $12z^2 - 23z + 10 \geq 0$ on $\left(-\infty, \dfrac{2}{3}\right] \cup \left[\dfrac{5}{4}, \infty\right)$ or $\left\{z \,\middle|\, z \leq \dfrac{2}{3} \text{ or } z \geq \dfrac{5}{4}\right\}$

21. (a) $x^2 + 2x - 1 = 0$, using the quadratic formula we get: $x = \dfrac{-2 \pm \sqrt{(2)^2 - 4(1)(-1)}}{2(1)} \Rightarrow$

    $x = \dfrac{-2 \pm \sqrt{8}}{2} \Rightarrow x = \dfrac{-2 \pm 2\sqrt{2}}{2} \Rightarrow x = -1 \pm \sqrt{2}.$

    (b) Using test intervals, $x^2 + 2x - 1 < 0$ on $(-1 - \sqrt{2}, -1 + \sqrt{2})$ or $\{x \,|\, -1 - \sqrt{2} < x < -1 + \sqrt{2}\}$

    (c) Using test intervals, $x^2 + 2x - 1 > 0$ on $(-\infty, -1 - \sqrt{2}) \cup (-1 + \sqrt{2}, \infty)$ or

    $\{x \,|\, x < -1 - \sqrt{2} \text{ or } x > -1 + \sqrt{2}\}$

23. (a) $f(x) > 0$ when $x < -1$ or $x > 1$

    (b) $f(x) \leq 0$ when $-1 \leq x \leq 1$

25. (a) $f(x) > 0$ when $-6 < x < -2$

    (b) $f(x) \leq 0$ when $x \leq -6$ or $x \geq -2$

27. Start by solving $2x^2 + 5x + 2 = 0 \Rightarrow (2x + 1)(x + 2) = 0 \Rightarrow x = -2$ or -0.5; thus,

    $2x^2 + 5x + 2 \leq 0$ when $-2 \leq x \leq -0.5.$

29. Start by solving $x^2 + x - 6 = 0 \Rightarrow (x + 3)(x - 2) = 0 \Rightarrow x = -3$ or 2; thus,

    $x^2 + x - 6 > 0$ when $x < -3$ or $x > 2$.

31. Start by solving the equation $x^2 = 4 \Rightarrow x = \pm 2$. Next write the quadratic inequality with $a > 0$.

    $x^2 \leq 4 \Rightarrow x^2 - 4 \leq 0$. The graph of $y = x^2 - 4$ is a parabola opening up. It will be less than or equal to 0

    between (and including) the solutions to the equation. The solutions are $-2 \leq x \leq 2$.

33. Since $x(x - 4) \geq 4 \Rightarrow x^2 - 4x + 4 \geq 0 \Rightarrow (x - 2)^2 \geq 0$, the solutions are all real numbers.

35. Start by solving the equation $-x^2 + x + 6 = 0 \Rightarrow x^2 - x - 6 = 0 \Rightarrow (x + 2)(x - 3) = 0 \Rightarrow$

    $x = -2, 3$. The graph of $y = -x^2 + x + 6$ is a parabola opening downward. It will be below the x-axis or

    intersect it outside of the solutions of the equation. The solutions are $x \leq -2$ or $x \geq 3$.

37. Start by solving the equation $6x^2 - x = 1 \Rightarrow 6x^2 - x - 1 = 0 \Rightarrow (3x + 1)(2x - 1) = 0 \Rightarrow$

    $x = -\dfrac{1}{3}, \dfrac{1}{2}$. The graph of $y = 6x^2 - x - 1$ is a parabola opening up. It will be below the x-axis between the

    solutions of the equation. The solutions to $6x^2 - x < 1$ are $-\dfrac{1}{3} < x < \dfrac{1}{2}$.

39. Start by solving the equation $(x + 4)(x - 10) = 0 \Rightarrow x = -4, 10$. The graph of

    $y = (x + 4)(x - 10) = x^2 - 6x - 40$ is a parabola opening up. It will intersect or be below the x-axis

    between (and including) the solutions of the equation. The solutions to $(x + 4)(x - 10) \leq 0$ are $-4 \leq x \leq 10$.

41. Since $2x^2 + 4x + 3 > 0$ for all values of x, there are no solutions to $2x^2 + 4x + 3 < 0$.

43. Start by solving the equation $9x^2 - 12x + 4 = 0 \Rightarrow (3x - 2)(3x - 2) = 0 \Rightarrow x = \dfrac{2}{3}$. The graph of

    $y = 9x^2 - 12x + 4 = (3x - 2)^2$ is a parabola opening up and intersecting the x-axis at $x = \dfrac{2}{3}$. It will always

    be above the x-axis except at $x = \dfrac{2}{3}$. The solutions to $9x^2 + 4 > 12x$ are all real numbers except $\dfrac{2}{3}$.

45. Start by solving the equation $x^2 = x \Rightarrow x^2 - x = 0 \Rightarrow x(x - 1) = 0 \Rightarrow x = 0, 1$. The graph of

    $y = x^2 - x$ is a parabola opening up. It will be above or intersect the x-axis outside (and including) the

    solutions of the equation. The solutions to $x^2 \geq x$ are $x \leq 0$ or $x \geq 1$.

47. Start by solving the equation $x(x - 1) - 6 = 0 \Rightarrow x^2 - x - 6 = 0 \Rightarrow (x + 2)(x - 3) = 0 \Rightarrow$

    $x = -2, 3$. The graph of $y = x^2 - x - 6$ is a parabola opening up. It will be above or intersect the x-axis

    outside (and including) the solutions of the equation. The solutions to $x(x - 1) \geq 6$ are $x \leq -2$ or $x \geq 3$.

49. Start by solving the equation $x^2 - 5 = 0 \Rightarrow x^2 = 5 \Rightarrow x = \pm\sqrt{5}$. The graph of $y = x^2 - 5$ is a parabola

    opening up. It will intersect or be below the x-axis between (and including) the solutions of the equation.

    The solutions are $-\sqrt{5} \leq x \leq \sqrt{5}$.

51. Start by graphing $Y_1 = 7X^2 - 179.8X + 515.2$ in $[-4, 30, 2]$ by $[-750, 250, 50]$. The inequality given by

    $7x^2 - 179.8x + 515.2 \geq 0$ is satisfied when the graph Y_1 is above or intersects the x-axis. This occurs when

    $x \leq k$ or $x \geq 22.4$, where $k = \dfrac{23}{7}$.

53. For $x^2 - 9x + 14 \le 0$, first solve $x^2 - 9x + 14 = 0 \Rightarrow (x - 7)(x - 2) = 0 \Rightarrow x = 2, 7$.

These boundary numbers give us the disjoint intervals: $(-\infty, 2), (2, 7), (7, \infty)$. See Figure 53.

$x^2 - 9x + 14 \le 0$ when $2 \le x \le 7$.

Interval	Test Value x	$x^2 - 9x + 14$	Positive or Negative?
$(-\infty, 2)$	0	14	Positive
$(2, 7)$	4	-6	Negative
$(7, \infty)$	10	24	Positive

Figure 53

55. For $x^2 \ge 3x + 10 \Rightarrow x^2 - 3x - 10 \ge 0$, first solve $x^2 - 3x - 10 = 0 \Rightarrow (x - 5)(x + 2) = 0 \Rightarrow$

$x = -2, 5$. These boundary numbers give us the disjoint intervals: $(-\infty, -2), (-2, 5), (5, \infty)$. See Figure 55.

$x^2 - 3x - 10 \ge 0$ when $x \le -2$ or $x \ge 5$.

Interval	Test Value x	$x^2 - 3x - 10$	Positive or Negative?
$(-\infty, -2)$	-3	8	Positive
$(-2, 5)$	0	-10	Negative
$(5, \infty)$	6	8	Positive

Figure 55

57. For $\dfrac{1}{8}x^2 + x + 2 \ge 0 \Rightarrow x^2 + 8x + 16 \ge 0$, first solve $x^2 + 8x + 16 = 0 \Rightarrow (x + 4)(x + 4) = 0 \Rightarrow$

$x = -4$. This boundary gives us the disjoint intervals: $(-\infty, -4), (-4, \infty)$. See Figure 57.

$\dfrac{1}{8}x^2 + x + 2 \ge 0$ for all real numbers.

Interval	Test Value x	$1/8x^2 + x + 2$	Positive or Negative?
$(-\infty, -4)$	-8	2	Positive
$(-4, \infty)$	0	2	Positive

Figure 57

59. For $x^2 > 3 - 4x \Rightarrow x^2 + 4x - 3 > 0$, first solve $x^2 + 4x - 3 = 0$. Using the quadratic formula we get:

$$x = \frac{-4 \pm \sqrt{(4)^2 - 4(1)(-3)}}{2(1)} \Rightarrow x = \frac{-4 \pm \sqrt{28}}{2} \Rightarrow x = -2 + \sqrt{7} \text{ or } x = -2 - \sqrt{7}.$$ These

boundary numbers give us the disjoint intervals:

$(-\infty, -2 - \sqrt{7}), (-2 - \sqrt{7}, -2 + \sqrt{7}), (-2 + \sqrt{7}, \infty)$. See Figure 59. $x^2 + 4x - 3 > 0$ when

$x < -2 - \sqrt{7}$ or $x > -2 + \sqrt{7}$.

Interval	Test Value x	$x^2 + 4x - 3$	Positive or Negative?
$(-\infty, -4.6)$	-5	2	Positive
$(-4.6, 0.6)$	0	-3	Negative
$(0.6, \infty)$	1	2	Positive

Figure 59

61. Graph $Y_1 = (1/9)X^2 + (11/3)X$, $Y_2 = 300$ in $[30, 40, 1]$ by $[0, 400, 100]$ (Not Shown). The intersection is

   (38.018, 300). Thus, the safe stopping speed is to be no more than 38 mph. The speed limit might be 35 mph.

63. To find the possible values of the radius of the can, solve the following inequality: $24\pi \leq \pi(r^2)6 \leq 54\pi \Rightarrow$

   $24 \leq 6r^2 \leq 54 \Rightarrow 4 \leq r^2 \leq 9 \Rightarrow 2 \leq r \leq 3$. The possible values of r range between 2 and 3 inches.

65. (a) $f(0) = \dfrac{4}{5}(0 - 10)^2 + 80 = 160$ and $f(2) = \dfrac{4}{5}(2 - 10)^2 + 80 = 131.2$. Initially when the person stops

   exercising the heart rate is 160 beats per minute, and after 2 minutes the heart rate has dropped to about 131

   beats per minute.

   (b) Graph $Y_1 = 100$, $Y_2 = 0.8(X - 10)^2 + 80$ and $Y_3 = 120$ (Not shown). To find the points of intersection,

   use the calc function on the calculator. The person's heart rate is between 100 and 120 when the graph of

   Y_2 is between the graphs of Y_1 and Y_3. This occurs between approximately 2.9 minutes and 5 minutes after

   the person stops exercising.

67. Graph $Y_1 = 2375X^2 + 5134X - 5020$, $Y_2 = 90,000$ and $Y_3 = 200,000$ in $[-1, 11, 1]$ by

   $[5000, 210000, 10000]$ (Not shown). The intersection points are approximately (4.9977492, 90000) and

   (8.044127, 200,000). From the graphs, we see that $90,000 \leq Y_1 \leq 200,000$ when $4.998 \leq X \leq 8.044$. Since

   1984 corresponds to $x = 0$, the number of AIDS deaths was from 90,000 to 200,000 for the years from 1989 to 1992.

69. (a) The data cannot be modeled by a linear function because the rate at which the height of the water decreases

   is not constant. When the tank first starts to drain, the large quantity of water in the tank will cause the

   water to drain rapidly. As the tank empties, there will be less water in the tank and the draining rate will

   decrease. This is evident in the data.

   (b) From the table, it appears that the height of the water was between 5 and 10 centimeters from approximately

   43 to 95 seconds after draining began.

   (c) The graph of $Y_1 = 0.00036706X^2 - 0.1517X + 15.94$ with the data shown in Figure 69c.

   (d) To find the times when the water height was between 5 and 10 centimeters, graph the function

   $Y_1 = 5$, $Y_2 = 0.00036706X^2 - 0.1517X + 15.94$, and $Y_3 = 10$ in the same window. The intersection

   points are approximately (93.1, 5) and (43.8, 10). See Figure 69d. The water level was between 5 and 10

   centimeters from approximately 43.8 to 93.1 seconds after draining started.

$[-25, 200, 25]$ by $[-5, 20, 5]$ $[-25, 200, 25]$ by $[-5, 20, 5]$

Figure 69c   Figure 69d

3.5: Transformations of Graphs

1. The vertex is $(-2, 0)$. The parabola $y = x^2$ has been shifted 2 units to the left. The equation is $y = (x + 2)^2$.

3. The endpoint is $(-3, 0)$. The curve $y = \sqrt{x}$ has been shifted 3 units to the left. The equation is $y = \sqrt{x + 3}$.

5. The vertex is $(-2, -1)$. The graph of $y = |x|$ has been shifted 2 units to the left and 1 unit down.

    The equation is $y = |x + 2| - 1$.

7. The endpoint is $(-2, -3)$. The curve $y = \sqrt{x}$ has been shifted 2 units to the left and 3 units down.

    The equation is $y = \sqrt{x + 2} - 3$.

9. To shift the graph of $f(x) = x^2$ to the right 2 units, replace x with $(x - 2)$ in the formula for $f(x)$.

    This results in $y = f(x - 2) = (x - 2)^2$. To shift the graph of this new equation down 3 units, subtract 3

    from the formula to obtain $y = f(x - 2) - 3 = (x - 2)^2 - 3$. See Figure 9.

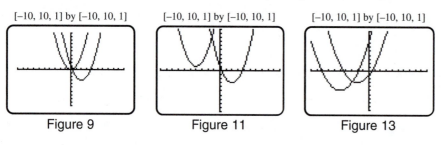

$[-10, 10, 1]$ by $[-10, 10, 1]$ $[-10, 10, 1]$ by $[-10, 10, 1]$ $[-10, 10, 1]$ by $[-10, 10, 1]$

Figure 9   Figure 11   Figure 13

11. To shift the graph of $f(x) = x^2 - 4x + 1$ to the left 6 units, replace x with $(x + 6)$ in the formula for $f(x)$.

    This results in $y = f(x + 6) = (x + 6)^2 - 4(x + 6) + 1$. To shift the graph of this new equation up 4

    units, add 4 to the formula to obtain

    $$y = f(x + 6) + 4 = (x + 6)^2 - 4(x + 6) + 1 + 4 = (x + 6)^2 - 4(x + 6) + 5. \text{ See Figure 11.}$$

13. To shift the graph of $f(x) = \frac{1}{2}x^2 + 2x - 1$ to the left 3 units, replace x with $(x + 3)$ in the formula for $f(x)$.

    The result is $y = f(x + 3) = \frac{1}{2}(x + 3)^2 + 2(x + 3) - 1$. To shift the graph of this new equation down 2

    units, subtract 2 to obtain $y = f(x + 3) - 2 = \frac{1}{2}(x + 3)^2 + 2(x + 3) - 1 - 2 \Rightarrow$

    $y = \frac{1}{2}(x + 3)^2 + 2(x + 3) - 3$. See Figure 13.

15. (a) $g(x) = 3(x + 3)^2 + 2(x + 3) - 5$

    (b) $g(x) = 3x^2 + 2x - 9$

17. (a) $g(x) = 2(x - 2)^2 - 4(x - 2) + 5$

    (b) $g(x) = 2(x + 8)^2 - 4(x + 8) - 4$

19. (a) $g(x) = 3(x - 2000)^2 - 3(x - 2000) + 72$

    (b) $g(x) = 3(x + 300)^2 - 3(x + 300) - 28$

21. (a) $g(x) = -\sqrt{x} - 4$

    (b) $g(x) = \sqrt{-x} + 2$

23. $(x - 3)^2 + (y + 4)^2 = 4$; Center: $(3, -4)$; Radius $= 2$

25. $(x + 5)^2 + (y - 3)^2 = 5$; Center: $(-5, 3)$; Radius $= \sqrt{5}$

27. (a) See Figure 27a.

    (b) See Figure 27b.

    (c) See Figure 27c.

Figure 27a

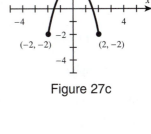

Figure 27b

Figure 27c

29. (a) See Figure 29a.

    (b) See Figure 29b.

    (c) See Figure 29c.

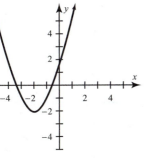

Figure 29a

Figure 29b

Figure 29c

31. (a) See Figure 31a.

    (b) See Figure 31b.

    (c) See Figure 31c.

Figure 31a

Figure 31b

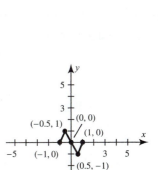

Figure 31c

33. (a) See Figure 33a.

    (b) See Figure 33b.

    (c) See Figure 33c.

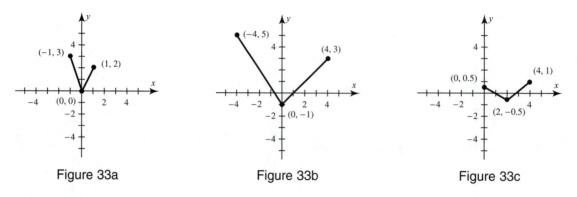

Figure 33a   Figure 33b   Figure 33c

35. Shift the graph of $y = x^2$ right 3 units and upward 1 unit.

37. Shift the graph of $y = x^2$ left 1 unit and vertically shrink it with factor $\frac{1}{4}$.

39. Reflect the graph of $y = \sqrt{x}$ across the x-axis and shift it left 5 units.

41. Reflect the graph of $y = \sqrt{x}$ across the y-axis and vertically stretch it with factor 2.

43. Reflect the graph of $y = |x|$ across the y-axis and shift it left 1 unit.

45. Shift the graph of $y = x^2$ down 3 units. The vertex is located at $(0, -3)$. See Figure 45.

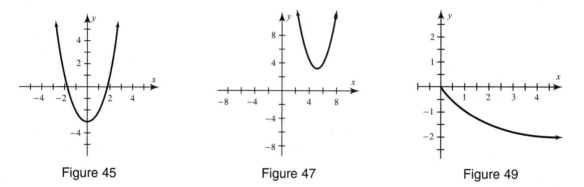

Figure 45   Figure 47   Figure 49

47. Shift the graph of $y = x^2$ to the right 5 units and up 3 units. The vertex is located at $(5, 3)$. See Figure 47.

49. Reflect the graph of $y = \sqrt{x}$ across the x-axis. See Figure 49.

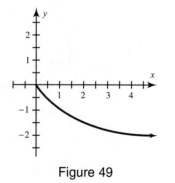

Figure 49

51. Reflect the graph of $y = x^2$ across the x-axis and shift the graph 4 units up. See Figure 51.

53. Shift the graph of $y = |x|$ down 4 units. See Figure 53.

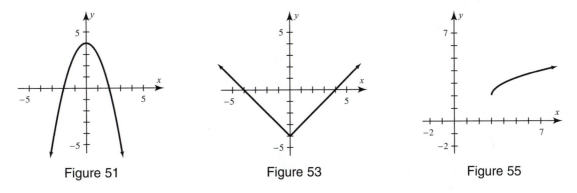

Figure 51   Figure 53   Figure 55

55. Shift the graph of $y = \sqrt{x}$ to the right 3 units and up 2 units. See Figure 55.

57. The graph is like $f(x) = |x|$ but narrower by a factor of 2, vertex $(0, 0)$. See Figure 57.

59. Reflect the graph of $f(x) = \sqrt{x}$ across the x-axis and then shift up 1 unit, the vertex is $(0, 1)$.

    See Figure 59.

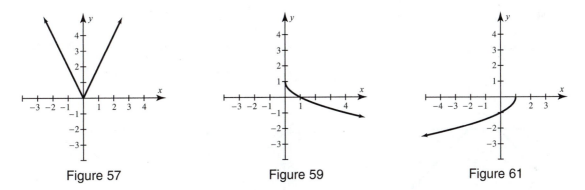

Figure 57   Figure 59   Figure 61

61. Shift the graph of $f(x) = \sqrt{x}$ left 1 unit, then reflect this graph across the y-axis and then reflect this graph

    across the x-axis. See Figure 61.

63. Reflect the graph of $f(x) = \sqrt{x}$ across the y-axis, then shift left 1 unit. See Figure 63.

65. Shift the graph of $f(x) = x^3$ right 1 unit. See Figure 65.

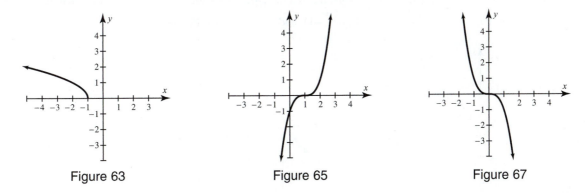

Figure 63   Figure 65   Figure 67

67. Reflect the graph of $f(x) = x^3$ across the x-axis. See Figure 67.

69. To reflect this function across the x-axis, graph $y = -f(x) = -(x^2 - 2x - 3) = -x^2 + 2x + 3$.

    See Figure 69a.

    To reflect this function across the y-axis, graph $y = f(-x) = (-x)^2 - 2(-x) - 3 = x^2 + 2x - 3$.

    See Figure 69b.

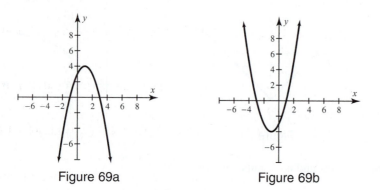

    Figure 69a   Figure 69b

71. To reflect this function across the x-axis, graph $y = -f(x) = -(|x + 1| - 1) = -|x + 1| + 1$.

    See Figure 71a.

    To reflect this function across the y-axis, graph $y = f(-x) = |-x + 1| - 1 = |-x + 1| - 1$.

    See Figure 71b.

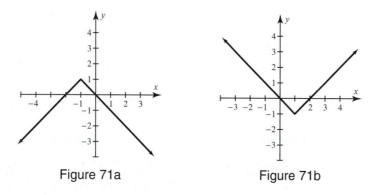

    Figure 71a   Figure 71b

73. (a) To reflect this line graph across the x-axis, make a table of values for $y = -f(x)$; that is, change each (x, y)

       to $(x, -y)$. Plot the points $(-3, -2)$, $(-1, -3)$, $(1, 1)$, and $(2, 2)$. See Figure 73a.

   (b) To reflect this line graph across the y-axis, make a table of values for $y = f(-x)$; that is, change each (x, y)

       to $(-x, y)$. Plot the points $(3, 2)$, $(1, 3)$, $(-1, -1)$, and $(-2, -2)$. See Figure 73b.

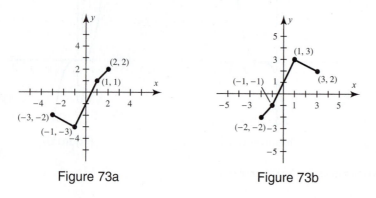

    Figure 73a   Figure 73b

75. Since $g(x) = f(x) + 7$, the output for $g(x)$ can be determined by adding 7 to the output of $f(x)$.

See Figure 75.

x	1	2	3	4	5	6
$g(x)$	12	8	13	9	14	16

Figure 75

77. Since $g(x) = f(x - 2)$, each input value of g is found by shifting each input value of f two units right.

For example, $g(4) = f(4 - 2) = f(2) = -5$ and $g(2) = f(2 - 2) = f(0) = -3$.

However, the value of $g(-4) = f(-4 - 2) = f(-6)$ is undefined because -6 is not in the domain of f.

See Figure 77.

x	-2	0	2	4	6
$g(x)$	5	2	-3	-5	-9

Figure 77

79. Since $g(x) = f(x + 1) - 2$, each input value of g is found by shifting each input value of f one unit left. Then each output of g is found by shifting the cooresponding output of f two units down.

For example, $g(1) = f(1 + 1) - 2 = f(2) - 2 = 4 - 2 = 2$ and

$g(1) = f(1 + 1) - 2 = f(2) - 2 = 4 - 2 = 2$ and $g(2) = f(2 + 1) - 2 = f(3) - 2 = 3 - 2 = 1$.

$g(5) = f(6) - 2 = 10 - 2 = 8$; $g(6) = f(7) - 2$ is undefined because 7 is not in the domain of f.

See Figure 79.

x	0	1	2	3	4	5
$g(x)$	0	2	1	5	6	8

Figure 79

81. Since $g(x) = f(-x) + 1$, each input value of g is the opposite of each input value of f.

Then each output of g is found by shifting the cooresponding output of f one unit up.

For example, $g(-2) = f(-(-2)) + 1 = f(2) + 1 = -1 + 1 = 0$ and

$g(-1) = f(-(-1)) + 1 = f(1) + 1 = 2 + 1 = 3$. See Figure 81.

x	-2	-1	0	1	2
$g(x)$	0	3	6	9	12

Figure 81

82. 83. The graph of $f(x)$ is shifted up 2 units $\Rightarrow (-12, 8), (0, 10)$, and $(8, -2)$.

85. The graph of $f(x)$ is shifted right 2 units and up 1 unit $\Rightarrow (-10, 7), (2, 9)$, and $(10, -3)$.

87. The graph is reflected across the x-axis $\Rightarrow$ the y-coordinate is $-\dfrac{1}{2}$ times the given y-coordinate $\Rightarrow$

$(-12, -3), (0, -4)$, and $(8, 2)$.

89. The graph is reflected across the y-axis $\Rightarrow$ the x-coordinate is the reciprocal of -2 or $-\dfrac{1}{2}$ times the given

x-coordinate $\Rightarrow (6, 6), (0, 8)$, and $(-4, -4)$.

91. Let $(1997, 3.8)$ be the vertex (h, k). Translate the graph of $y = x^2$ right 1997 units and upward 3.8 units. To determine a value for a, graph the data and $y = a(x - 1997)^2 + 3.8$ for different values of a. By trial and error a value of $a = \dfrac{1}{4}$ appears to fit the data. Thus $f(x) = \dfrac{1}{4}(x - 1997)^2 + 3.8$.

93. Let $(1970, 30)$ be the vertex (h, k). Translate the graph of $y = x^2$ right 1970 units and upward 30 units. To determine a value for a, graph the data and $y = a(x - 1970)^2 + 30$ for different values of a. By trial and error a value of $a = 0.22$ appears to fit the data. Thus $f(x) = 0.22(x - 1970)^2 + 30$.

95. We must determine a function g such that $g(1991) = P(1)$, $g(1992) = P(2)...$. Thus, the relationship $g(x) = P(x - 1990)$ holds for $x = 1990, 1991, 1992...$. It follows that the representation for $g(x)$ is $g(x) = 0.00075(x - 1990)^2 + 0.17(x - 1990) + 44$.

97. (a) When $x = 2$, the y-value is about 9. There are about 9 hours of daylight on February 21 at 60°N latitude.

   (b) Since February 21 is 2 months after the shortest day and October 21 is 2 months before the shortest day, the days should have about the same number of daylight hours. A reasonable conjecture would be 9 hours.

   (c) The amount of daylight would increase to the left or right of $x = 0$. There are approximately the same number of daylight hours either x months before or after December 21. the graph should be symmetric about the y-axis, as shown in Figure 97.

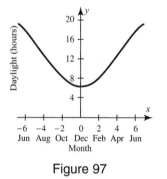

Figure 97

99. In 15 seconds, the plane has moved $15(0.2) = 3$ kilometers to the left. To show this movement, translate the graph of the mountain 3 kilometers (units) to the right; $y = -0.4(x - 3)^2 + 4$. See Figure 99.

Figure 99

101. (a) In 4 hours, the cold front has moved $4(40) = 160$ miles, which is $\dfrac{160}{100} = 1.6$ units on the graph. To show

this movement, shift the graph 1.6 units down; $y = \dfrac{1}{20}x^2 - 1.6$. See Figure 101a.

(b) The graph of the cold front has shifted $\dfrac{250}{100} = 2.5$ units down and $\dfrac{210}{100} = 2.1$ units to the right. To show

this movement, the new equation should be $y = \dfrac{1}{20}(x - 2.1)^2 - 2.5$. The location of Columbus, Ohio is

at $(5.5, -0.8)$ relative to the location of Des Moines at $(0, 0)$. Since the parabola representing the cold

front at midnight has past the point $(5.5, -0.8)$, the front reaches Columbus by midnight. See Figure 101b.

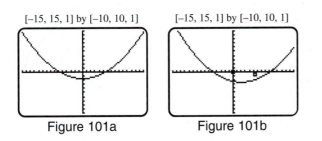

[−15, 15, 1] by [−10, 10, 1]   [−15, 15, 1] by [−10, 10, 1]

Figure 101a   Figure 101b

Chapter 3 Review Exercises

1. (a) The parabola opens downward, so $a < 0$.

(b) Vertex: $(2, 4)$

(c) Axis of symmetry: $x = 2$.

(d) f is increasing for $x \le 2$ and decreasing for $x \ge 2$.

3. $f(x) = -2(x - 5)^2 + 1 \Rightarrow f(x) = -2(x^2 - 10x + 25) + 1 \Rightarrow f(x) = -2x^2 + 20x - 49$

The leading coefficient is –2.

5. The graph of $y = x^2$ has been shifted 1 unit left, reflected across the x-axis, and shifed 2 units up $\Rightarrow$

$f(x) = -(x + 1)^2 + 2$.

7. $f(x) = x^2 + 6x - 1 \Rightarrow f(x) = x^2 + 6x + 9 - 9 - 1 \Rightarrow f(x) = (x^2 + 6x + 9) - 10 \Rightarrow$

$f(x) = (x + 3)^2 - 10$; the vertex is $(-3, -10)$.

9. $-\dfrac{b}{2a} = -\dfrac{2}{2(-3)} = \dfrac{1}{3}$ and $f\left(\dfrac{1}{3}\right) = -3\left(\dfrac{1}{3}\right)^2 + 2\left(\dfrac{1}{3}\right) - 4 = -\dfrac{11}{3}$; the vertex is $\left(\dfrac{1}{3}, -\dfrac{11}{3}\right)$.

11. Shift the graph of $f(x) = x^2$ up 3 units, reflect it across the x-axis, and make it narrower by the scale factor 3.

The vertex is located at $(0, 3)$. See Figure 11.

13. $h(t) = t^2 - 4t - 1 \Rightarrow h(t) = t^2 - 4t + 4 - 5 \Rightarrow h(t) = (t - 2)^2 - 5$. Shift the graph of $h(t) = t^2$

right 2 units, and down 5 units. The vertex is $(2, -5)$. See Figure 13.

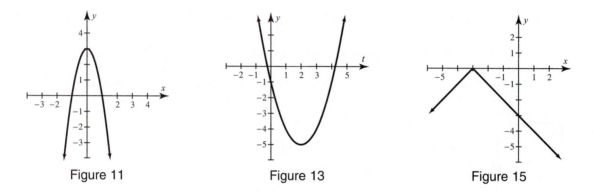

| Figure 11 | Figure 13 | Figure 15 |

15. Shift the graph of $f(x) = |x|$ left 3 units, and reflect it across the x-axis. The vertex is located at $(-3, 0)$.

See Figure 15.

17. Shift the graph of $f(x) = \sqrt{x}$ left 1 unit and reflect across the x-axis. The vertex is $(-1, 0)$. See Figure 17.

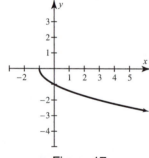

Figure 17

19. Average rate of change $= \dfrac{f(b) - f(a)}{b - a} \Rightarrow \dfrac{f(4) - f(2)}{4 - 2} = \dfrac{-63 - (-5)}{2} = \dfrac{-58}{2} = -29$

21. $x^2 - x - 20 = 0 \Rightarrow (x - 5)(x + 4) = 0 \Rightarrow x = -4, 5$

23. $x^2 = 4x \Rightarrow x^2 - 4x = 0 \Rightarrow x(x - 4) = 0 \Rightarrow x = 0, 4$

25. $4z^2 - 7 = 0 \Rightarrow 4z^2 = 7 \Rightarrow z^2 = \dfrac{7}{4} \Rightarrow z = \pm\sqrt{\dfrac{7}{4}} \Rightarrow z = \pm\dfrac{\sqrt{7}}{2}$

27. $-2t^2 - 3t + 14 = 0 \Rightarrow (2t + 7)(-t + 2) = 0 \Rightarrow 2t + 7 = 0 \Rightarrow 2t = -7 \Rightarrow$

$t = -\dfrac{7}{2}$ or $-t + 2 = 0 \Rightarrow -t = -2 \Rightarrow t = 2 \Rightarrow t = -\dfrac{7}{2}, 2$

29. $0.1x^2 - 0.3x = 1 \Rightarrow x^2 - 3x - 10 = 0 \Rightarrow (x - 5)(x + 2) = 0 \Rightarrow x = -2, 5$

31. $(k - 1)^2 = \dfrac{9}{4} \Rightarrow k^2 - 2k + 1 = \dfrac{9}{4} \Rightarrow k^2 - 2k - \dfrac{5}{4} = 0 \Rightarrow 4k^2 - 8k - 5 = 0 \Rightarrow$

$(2k - 5)(2k + 1) = 0 \Rightarrow 2k - 5 = 0 \Rightarrow 2k = 5 \Rightarrow k = \dfrac{5}{2}$ or $2k + 1 = 0 \Rightarrow 2k = -1 \Rightarrow$

$k = -\dfrac{1}{2} \Rightarrow k = -\dfrac{1}{2}, \dfrac{5}{2}$

33. (a) Since the parabola is opening upward, $a > 0$.

    (b) The real solutions to $ax^2 + bx + c = 0$ appear to be $-3, 2$, the x-intercepts.

    (c) The discriminant must be positive since $ax^2 + bx + c = 0$ has two real solutions.

35. $x^2 + 2x = 5 \Rightarrow x^2 + 2x + 1 = 5 + 1 \Rightarrow (x + 1)^2 = 6 \Rightarrow x + 1 = \pm\sqrt{6} \Rightarrow x = -1 \pm \sqrt{6}$

37. $2z^2 - 6z - 1 = 0 \Rightarrow z^2 - 3z = \dfrac{1}{2} \Rightarrow z^2 - 3z + \dfrac{9}{4} = \dfrac{1}{2} + \dfrac{9}{4} \Rightarrow \left(z - \dfrac{3}{2}\right)^2 = \dfrac{11}{4} \Rightarrow$

$z - \dfrac{3}{2} = \pm\sqrt{\dfrac{11}{4}} \Rightarrow z = \dfrac{3}{2} \pm \dfrac{\sqrt{11}}{2} \Rightarrow z = \dfrac{3 \pm \sqrt{11}}{2}$

39. $\dfrac{1}{2}x^2 - 4x + 1 = 0 \Rightarrow x^2 - 8x + 2 = 0 \Rightarrow x^2 - 8x + 16 = -2 + 16 \Rightarrow (x - 4)^2 = 14 \Rightarrow$

$x - 4 = \pm\sqrt{14} \Rightarrow x = 4 \pm \sqrt{14}$

41. $2x^2 - 3y^2 = 6 \Rightarrow -3y^2 = -2x^2 + 6 \Rightarrow y^2 = \dfrac{2x^2 - 6}{3} \Rightarrow y = \pm\sqrt{\dfrac{2x^2 - 6}{3}}$

    Since there are two output values for each input, y is not a function of x.

43. (a) $\sqrt{-16} = \sqrt{16}\sqrt{-1} = 4i$

    (b) $\sqrt{-48} = \sqrt{16}\sqrt{-1}\sqrt{3} = 4i\sqrt{3}$

    (c) $\sqrt{-5} \cdot \sqrt{-15} = i\sqrt{5} \cdot i\sqrt{5}\sqrt{3} = i^2(5)\sqrt{3} = 5(-1)\sqrt{3} = -5\sqrt{3}$

45. (a) The x-intercepts of f are $-\dfrac{5}{2}$ and $\dfrac{1}{2}$.

    (b) The complex zeros of f are $-\dfrac{5}{2}$ and $\dfrac{1}{2}$.

47. (a) There are no x-intercepts of f.

    (b) Use the quadratic formula to find the complex zeros. $x = \dfrac{-b \pm \sqrt{b^2 - 4ac}}{2} \Rightarrow$

$x = \dfrac{-1 \pm \sqrt{1^2 - 4(1)(2)}}{2(1)} \Rightarrow x = \dfrac{-1 \pm \sqrt{-7}}{2} \Rightarrow x = -\dfrac{1}{2} \pm \dfrac{i\sqrt{7}}{2}$

49. $4x^2 + 9 = 0 \Rightarrow 4x^2 = -9 \Rightarrow x^2 = -\dfrac{9}{4} \Rightarrow x = \pm\sqrt{-\dfrac{9}{4}} = \pm\dfrac{3}{2}i$

51. (a) The graph of $y = f(x)$ has x-intercepts at $x = -3, 2$. It is above the x-axis between these values, which

    are solutions to $f(x) = 0$, so $f(x) > 0$ when $-3 < x < 2$ or on $(-3, 2)$.

    (b) The graph of $y = f(x)$ is below or intersects the x-axis outside of and including the values where $f(x) = 0$,

    so $f(x) \leq 0$ when $x \leq -3$ or $x \geq 2$ or on $(-\infty, -3] \cup [2, \infty)$.

53. (a) $x^2 - 3x + 2 = 0 \Rightarrow (x - 2)(x - 1) = 0 \Rightarrow x = 1, 2$.

    (b) Using test intervals, $x^2 - 3x + 2 < 0$ on $(1, 2)$, (when $1 < x < 2$).

    (c) Using test intervals, $x^2 - 3x + 2 > 0$ on $(-\infty, 1) \cup (2, \infty)$, (when $x < 1$ or $x > 2$).

55. Start by solving $x^2 - 3x + 2 = 0 \Rightarrow (x - 1)(x - 2) = 0 \Rightarrow x = 1$ or 2; thus, $x^2 - 3x + 2 \leq 0$ on $[1, 2]$ or $\{x | 1 \leq x \leq 2\}$. Note that a graph of $y = x^2 - 3x + 2$ intersects or is below the x-axis for values of x between the x-intercepts.

57. Start by solving $2x^2 + 3x + 1 = 0 \Rightarrow x = -1$ or $-\dfrac{1}{2}$; thus, $2x^2 + 3x + 1 < 0$ on $\left(-1, -\dfrac{1}{2}\right)$ or $\left\{x | -1 < x < -\dfrac{1}{2}\right\}$. Note that a graph of $y = 2x^2 + 3x + 1$ is below the x-axis for values of x between the x-intercepts.

59. Start by solving $n(n - 2) = 15 \Rightarrow n^2 - 2n - 15 = 0 \Rightarrow (n - 5)(n + 3) = 0 \Rightarrow n = -3$ or 5; thus, $n(n - 2) \geq 15$ on $(-\infty, -3] \cup [5, \infty)$ or $\{x | x \leq -3 \text{ or } x \geq 5\}$. Note that a graph of $y = n^2 - 2n - 15$ intersects or is above the n-axis for values of n outside of the n-intercepts.

61. $y = -f(x) = -(2x^2 - 3x + 1) = -2x^2 + 3x - 1$. See Figure 61a.

    $y = f(-x) = 2(-x)^2 - 3(-x) + 1 = 2x^2 + 3x + 1$. See Figure 61b.

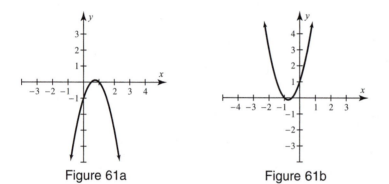

Figure 61a   Figure 61b

63. Shift the graph of $y = x^2$ four units down. See Figure 63.

65. Shift the graph of $y = \sqrt{x}$ two units left. See Figure 65.

67. Reflect the graph of $y = x^2$ across the x-axis, shift it two units right and three units up, then stretch it vertically. See Figure 67.

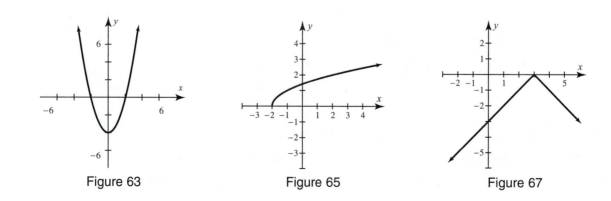

Figure 63   Figure 65   Figure 67

69. For example $g(2) = (-3)f(2 + 1) + 2 \Rightarrow -3f(3) + 2 = -3(4) + 2 = -10.$ See Figure 69.

x	0	1	2	3
$g(x)$	11	-7	-10	-19

Figure 69

71. Let $W + L + W = 44 \Rightarrow L = 44 - 2W.$ If $A = L \cdot W$ then $A = (44 - 2W)(W) \Rightarrow A = 44W - 2W^2$ or

$A = -2W^2 + 44W.$ Now use the vertex formula: $W = -\dfrac{44}{2(-2)} \Rightarrow W = 11$ and $L = 44 - 2(11) \Rightarrow$

$L = 22.$ So the dimensions are 11 feet by 22 feet.

73. (a) $h(t) = -16t^2 + 88t + 5 \Rightarrow h(0) = -16(0)^2 + 88(0) + 5 \Rightarrow h(0) = 5$

    The stone was 5 feet above the ground when it was released.

    (b) $h(2) = -16(2)^2 + 88(2) + 5 = 117;$ the stone was 117 feet high after 2 seconds.

    (c) $t = -\dfrac{b}{2a} = -\dfrac{88}{2(-16)} = 2.75$ and $h(2.75) = -16(2.75)^2 + 88(2.75) + 5 = 126;$ the maximum height of

    the stone was 126 feet.

    (d) $h(t) = 117 \Rightarrow 117 = -16t^2 + 88t + 5 \Rightarrow 16t^2 - 88t + 112 = 0 \Rightarrow$

    $t = \dfrac{88 \pm \sqrt{(-88)^2 - 4(16)(112)}}{2(16)} \Rightarrow t = 3.5 \text{ or } t = 2$

    The stone was 117 feet high after 2 seconds and after 3.5 seconds.

75. Use the sketch shown in Figure 75 to determine the dimensions of the box; let $x = $ width of the metal sheet.

    The dimensions of the box are 3 inches by $(x - 6)$ inches by $(x - 2)$ inches, and the volume is 135 cubic

    inches.

    $135 = 3(x - 6)(x - 2) \Rightarrow 135 = 3x^2 - 24x + 36 \Rightarrow 0 = 3x^2 - 24x - 99 \Rightarrow 0 = x^2 - 8x - 33 \Rightarrow$

    $0 = (x - 11)(x + 3) \Rightarrow x = 11$ or $x = -3.$ Since $x = -3$ is impossible, the width of the metal sheet is

    11 inches and the length is $11 + 4 = 15$ inches.

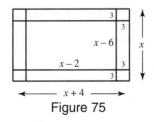

Figure 75

77. (a) Using least squares regression, we get the function $f(x) \approx 0.00051x^2 - 0.00604x + 1.6$. See Figure 77a.

(b) Graph $Y_1 = 0.00051X\text{^}2 - 0.00604X + 1.6$ and $Y_2 = 3.7$ in [0, 120, 20] by [1, 6.2, 1]. See Figure 77b.

The intersection point is approximately (70.363108, 3.7), so $x \approx 70.4$. The yield will be 3.7 tons per

hectare when 70.4% of the area is irrigated.

[0, 120, 10] by [1, 6.2, 1]

Figure 77a   Figure 77b

Chapter 4: Nonlinear Functions and Equations

4.1: Nonlinear Functions and Their Graphs

1. $f(x) = 2x^3 - x + 5$ is a polynomial; degree: 3; leading coefficient: 2.

3. $f(x) = \sqrt{x}$ is not a polynomial.

5. $f(x) = 1 - 4x - 5x^4$ is a polynomial; degree: 4; leading coefficient: –5.

7. $g(t) = \dfrac{1}{t^2 + 3t - 1}$ is not a polynomial.

9. $g(t) = 22$ is a polynomial; degree: 0; leading coefficient: 22.

11. (a) A local maximum of approximately 5.5 occurs when $x \approx -2$. A local minimum of approximately –5.5 occurs when $x \approx 2$. *Answers may vary slightly.*

    (b) There are no absolute extrema.

13. (a) Local maxima of approximately 17 and 27 occur when $x \approx -3$ and 2, respectively. Local minima of approximately –10 and 24 occur when $x \approx -1$ and 3, respectively. *Answers may vary slightly.*

    (b) There are no absolute extrema.

15. (a) Local maxima of approximately 0.5 and 2.8 occur when $x \approx 1$ and -2, respectively. A local minimum of approximately 0 occur when $x = 0$. *Answers may vary slightly.*

    (b) The absolute maximum is 2.8 when $x \approx -2$. There is no absolute minimum.

17. (a) A Local maximum of 0 occurs when $x = 0$. A local minimum of $-1{,}000$ occurs when $x = \pm 8$.

    (b) There is no absolute maximum. The absolute minimum of $-1{,}000$ occurs when $x = \pm 8$.

19. (a) Each local minimum is –1. Each local maximum is 1.

    (b) The absolute minimum is –1 and the absolute maximum is 1.

21. (a) A local minimum of approximately –3.2 occurs when $x \approx 0.5$. There are no local maxima. *Answers may vary slightly.*

    (b) The absolute minimum is approximately –3.2. The absolute maximum is 3 and occurs when $x = -2$.

23. (a) Local minima are approximately -0.5 and –2. Local maxima are approximately 0.5 and 2.

    (b) The absolute minimum is –2 and the absolute maximum is 2.

25. (a) There is no local maximum. A local minimum of -2 occurs when $x \approx 1$.

    (b) There is no absolute maximum. An absolute minimum of -2 occurs when $x \approx 1$.

27. (a) The graph of g is linear $\Rightarrow$ no local extrema.

    (b) The graph of g is linear $\Rightarrow$ no absolute extrema.

29. (a) The graph of g is a parabola opening up with a vertex $(0, 1)$ $\Rightarrow$ it has a local minimum: 1; and no local maximum.

    (b) The graph of g is a parabola opening up with a vertex $(0, 1)$ $\Rightarrow$ it has an absolute minimum: 1; and no absolute maximum.

31. (a) The graph of g is a parabola opening down with a vertex $(-3, 4)$ $\Rightarrow$ it has a local maximum: 4; and no local minimum.

  (b) The graph of g is a parabola opening down with a vertex $(-3, 4)$ $\Rightarrow$ it has an absolute maximum: 4; and no absolute minimum.

33. (a) The graph of g is a parabola opening up with a vertex $\left(\dfrac{3}{4}, -\dfrac{1}{8}\right)$ $\Rightarrow$ it has a local minimum: $-\dfrac{1}{8}$; and no local maximum.

  (b) The graph of g is a parabola opening up with a vertex $\left(\dfrac{3}{4}, -\dfrac{1}{8}\right)$ $\Rightarrow$ it has an absolute minimum: $-\dfrac{1}{8}$; and no absolute maximum.

35. (a) The absolute value graph opening up has a vertex $(-3, 0)$ $\Rightarrow$ it has a local minimum: 0; and no local maximum.

  (b) The absolute value graph opening up has a vertex $(-3, 0)$ $\Rightarrow$ it has an absolute minimum: 0; and no absolute maximum.

37. (a) The graph of g is the basic $\sqrt[3]{x}$ graph and has no local extrema.

  (b) The graph of g is the basic $\sqrt[3]{x}$ graph and has no absolute extrema.

39. (a) The cubic function graph of g has turning points $(-1, -2)$ and $(1, 2)$ $\Rightarrow$ it has a local minimum: -2; and a local maximum: 2.

  (b) The cubic function graph of g has turning points $(-1, -2)$ and $(1, 2)$ but then continues on to $-\infty$ and ∞ $\Rightarrow$ it has no absolute extrema.

41. (a) A graph of $f(x) = -3x^4 + 8x^3 + 6x^2 - 24x$ is shown in Figure 41. Local maxima of 19 and -8 occur at $x \approx -1$ and $x \approx 2$ respectively. A local minima of -13 occurs when $x \approx 1$.

  (b) The absolute maximum of 19 occurs at $x \approx -1$, and there is no absolute minimum.

[-5, 5, 1] by [-30, 30, 5]   [-10, 10, 1] by [-10, 10, 1]   [-10, 10, 1] by [-10, 10, 1]

Figure 41   Figure 43   Figure 45

43. (a) A graph of $f(x) = 0.5x^4 - 5x^2 + 4.5$ is shown in Figure 43. A local maximum of 4.5 occurs at $x = 0$. A local minimum of -8 occurs when $x \approx \pm 2.236$.

  (b) There is no absolute maximum. An absolute minimum of -8 occurs at $x \approx \pm 2.236$.

45. (a) A graph of $f(x) = \dfrac{8}{1 + x^2}$ is shown in Figure 45. A local maximum of 8 occurs at $x = 0$. There are no local minima.

  (b) An absolute maximum of 8 occurs at $x = 0$. There is no absolute minimum.

47. From the graph we can see that the function f is neither even or odd.

49. From the graph we can see that the function is even. Since the point $(1, -1)$ is on the graph of f, the point $(-1, -1)$ must also be located on the graph.

51. The function f is a monomial with only an odd power $\Rightarrow$ it is odd.

53. The function f is a polynomial with an odd power and a constant term, which is an even power $\Rightarrow$ it is neither.

55. The function f is a polynomial with an even power and a constant term, which is also even $\Rightarrow$ it is even.

57. The function f is a polynomial with all even powers and a constant term, which is also even $\Rightarrow$ it is even.

59. The function f is a polynomial with all odd powers $\Rightarrow$ it is odd.

61. The function f is a polynomial with both an odd powered term and an even powered term $\Rightarrow$ it is neither.

63. If $f(x) = \sqrt[3]{x^2}$, then $f(-x) = \sqrt[3]{(-x)^2} = \sqrt[3]{x^2}$, which is equal to $f(x) \Rightarrow$ it is even.

65. If $f(x) = \sqrt{1 - x^2}$, then $f(-x) = \sqrt{1 - (-x)^2} = \sqrt{1 - x^2}$, which is equal to $f(x) \Rightarrow$ it is even.

67. If $f(x) = \dfrac{1}{1 + x^2}$, then $f(-x) = \dfrac{1}{1 + (-x)^2} = \dfrac{1}{1 + x^2}$, which is equal to $f(x) \Rightarrow$ it is even.

69. If $f(x) = |x + 2|$, then $f(-x) = |-x + 2|$, and $-f(x) = -|x + 2|$. Since $f(x) \neq f(-x)$ and $f(-x) \neq -f(x)$. It is neither.

71. Notice that $f(-100) = 56$ and $f(100) = -56$. In general, since $f(-x) = -f(x)$ holds for each x in the table, f is odd.

73. Since f is even, the condition $f(-x) = f(x)$ must hold. For example, since $f(-3) = 21$, $f(3)$ must also equal 21. The value assigned to $f(0)$ makes no difference. See Figure 73.

x	-3	-2	-1	0	1	2	3
$f(x)$	21	-12	-25	1	-25	-12	21

Figure 73

75. Since f is an odd function and $(-5, -6)$ and $(-3, 4)$ are on the graph of f, then $(5, 6)$ and $(3, -4)$ are on the graph. Thus, $f(5) = 6$ and $f(3) = -4$.

77. Sketch any linear function that passes through the origin will work. For example, the graph of $y = x$ is shown in Figure 77.

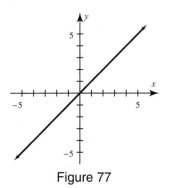

Figure 77

79. No. If (2, 5) is on the graph of an odd function f, then so is $(-2, -5)$. Since f would pass through $(-3, -4)$

    and then $(-2, -5)$, it could not always be increasing.

81. Plot points $(-2, -3)$ and $(2, -1)$ and make them minima. See Figure 81.

83. Plot point $(2, 3)$ and make it a maxima. See Figure 83. Yes, it could be quadratic, but it does not have to be

    quadratic.

Figure 81   Figure 83   Figure 85

85. Shift the graph of $y = f(x) = 4x - \dfrac{x^3}{3}$ to the left 1 unit to get the graph of

    $y = f(x + 1) = 4(x + 1) - \dfrac{(x + 1)^3}{3}$. See Figure 85.

87. Sketch the graph of $y = f(x) = 4x - \dfrac{x^3}{3}$ vertically by multiplying each y-value by 2, that is, each (x, y)

    point becomes $(x, 2y)$, to get the graph of $y = 2f(x) = 2\left(4x - \dfrac{x^3}{3}\right) = 8x - \dfrac{2}{3}x^3$. See Figure 87.

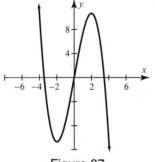

Figure 87

89. The first equation graph is transformed 1 unit left $\Rightarrow y = f(x + 1) - 2$ increases on $[0, 3]$. The second

    equation graph is transformed across the x-axis and transformed 2 units right $\Rightarrow y = -f(x - 2)$ decreases on

    $[3, 6]$.

91. (a) An absolute maximum of approximately 84°F occurs when $x \approx 5$. An absolute minumum of approximately 63°F

    occurs when $x \approx 1.6$. These are the high and low temperatures between 1 p.m. and 6 p.m. *Answers may vary slightly.*

    (b) Local maxima of approximately 78°F and 84°F occur when $x \approx 2.9$ and 5, respectively. Local mimima of

    approximately 63°F and 72°F occur when $x \approx 1.6$ and 3.8. *Answers may vary slightly.*

    (c) The temperature is increasing on the intervals $1.6 \leq x \leq 2.9$ and $3.8 \leq x \leq 5$. *Answers may vary*

    slightly.

93. (a) One might expect an absolute maximum to occur in January since it is the coldest month and requires the

    heating. An absolute minimum might occur in July since it is the warmest month.

    (b) The graph f is shown in both Figures 93a & 93b. One can see from the graph that there is one local

    maximum and one local minimum where $1 \leq x \leq 12$. The peak is near the point $(1.46, 140.06)$ and the

    valley is located near $(7.12, 14.78)$. This means that the peak heating expense of $140 occurs in January

    $(x \approx 1)$. The minimum heating cost is during July $(x \approx 7)$ when it is roughly $15.

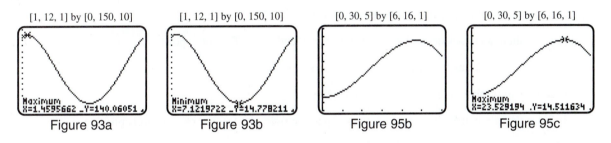

| [1, 12, 1] by [0, 150, 10] | [1, 12, 1] by [0, 150, 10] | [0, 30, 5] by [6, 16, 1] | [0, 30, 5] by [6, 16, 1] |

| Figure 93a | Figure 93b | Figure 95b | Figure 95c |

95. (a) $f(5) = -0.00113(5)^3 + 0.0408(5)^2 - 0.432(5) + 7.66 \approx 8.32$; in 1955 U.S. consumption of energy

    was about 8.32 quadrillion Btu.

    (b) The graph of f is shown in Figure 95b; the energy consumption increased, reached a maximum value,

    and the started to decrease.

    (c) The local maximum is approximately 14.5 when $x \approx 23.5$. In 1973 energy use peaked at 14.5 quadrillion

    Btu. See Figure 95c.

97. (a) Since the graph of f is symmetric with respect to the y-axis, it is a graphical representation of an even function.

    (b) If f represents the average temperature in the month x, $f(1) = f(-1)$. Thus, the average monthly

    temperature in August is also 83°F.

    (c) The average monthly temperatures in March and November are equal, since $f(-4) = f(4)$.

    (d) In Austin the average monthly temperature is symmetric about July. In July the highest average monthly

    temperature occurs. In January the lowest temperature occurs. The pairs June-August, May-September,

    April-October, March-November, and February-December have approximately the same average temperatures.

4.2: Polynomial Functions and Models

1. (a) Turning points occur near (1.6, 3.6), (3, 1.2), and (4.4, 3.6).

    (b) The turning points represent the times and distances where the runner turned around and jogged the other direction. In this example, after 1.6 minutes the runner is 360 feet from the starting line. The runner turns and jogs toward the starting line. After 3 minutes the runner is 120 feet from the starting line, turns, and jogs away from the starting line. After 4.4 minutes the runner is again 360 feet from the starting line. The runner turns and jogs back to the starting line.

3. (a) It has no turning points and one x-intercept of 0.5.

    (b) The leading coefficient is positive, since the slope of the line is positive.

    (c) Since the graph appears to be a line that is not horizontal, its minimum degree is 1.

5. (a) It has three turning points and three x-intercepts of $-6, -1, 6$.

    (b) Since the graph goes down for large values of x, the leading coefficient is negative.

    (c) Since the graph has three turning points, its minimum degree is 4.

7. (a) It has four turning points and five x-intercepts of $-3, -1, 0, 1$, and 2.

    (b) Since the graph goes up for large values of x, the leading coefficient is positive.

    (c) Since the graph has four turning points, its minimum degree is 5.

9. (a) It has two turning points and one x-intercept of -3.

    (b) Since the graph goes down to the left and up to the right, the leading coefficient is positive.

    (c) The graph has two turning points. The minimum degree of f is 3.

11. (a) It has one turning point and two x-intercepts of -1, and 2.

    (b) Since the graph is a parabola opening up, the leading coefficient is positive.

    (c) The graph has one turning point. The minimum degree of f is 2.

13. (a) Graph d.

    (b) There is one turning point at (1, 0).

    (c) It has one x-intercept: $x = 1$.

    (d) The only local minimum is 0.

    (e) The absolute minimum is 0. There is no absolute maximum.

15. (a) Graph b.

    (b) The turning points are at (–3, 27) and (1, –5).

    (c) The three x-intercepts are $x \approx -4.9$, $x = 0$, and $x \approx 1.9$.

    (d) The local minimum is –5, and the local maximum is 27.

    (e) There are no absolute extrema.

17. (a) Graph a.

    (b) The turning points are at $(-2, 16)$, $(0, 0)$, and $(2, 16)$.

    (c) The three x-intercepts are $x \approx -2.8$, $x = 0$, and $x \approx 2.8$.

    (d) The local minimum is 0, and the local maximum is 16.

    (e) The absolute maximum is 16. There is no absolute minimum.

19. (a) The graph of $f(x) = \dfrac{1}{9}x^3 - 3x$ is shown in Figure 19.

    (b) There are two turning points located at $(-3, 6)$ and $(3, -6)$.

    (c) At $x = -3$ there is a local maximum of 6 and at $x = 3$ there is a local minimum of -6.

21. (a) The graph of $f(x) = 0.025x^4 - 0.45x^2 - 3$ is shown in Figure 21.

    (b) There are three turning points located at $(-3, -7.025)$, $(0, -5)$, and $(3, -7.025)$.

    (c) At $x = \pm 3$ there is a local minimum of -7.025. At $x = 0$ there is a local maximum of -5.

| [–10, 10, 1] by [–10, 10, 1] | [–10, 10, 1] by [–10, 10, 1] | [–10, 10, 1] by [–10, 10, 1] | [–10, 10, 1] by [–10, 10, 1] |

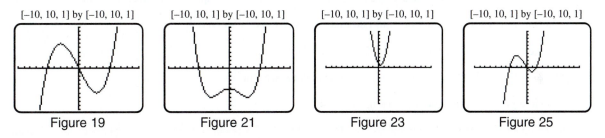

| Figure 19 | Figure 21 | Figure 23 | Figure 25 |

23. (a) The graph of $f(x) = 1 - 2x + 3x^2$ is shown in Figure 23.

    (b) There is one turning point located at $\left(\dfrac{1}{3}, \dfrac{2}{3}\right) \approx (0.333, 0.667)$.

    (c) At $x = \dfrac{1}{3}$ there is a local minimum of $\dfrac{2}{3} \approx 0.667$.

25. (a) The graph of $f(x) = \dfrac{1}{3}x^3 + \dfrac{1}{2}x^2 - 2x$ is shown in Figure 25.

    (b) The turning points are at $\left(-2, \dfrac{10}{3}\right) \approx (-2, 3.333)$ and $\left(1, -\dfrac{7}{6}\right) \approx (1, -1.167)$.

    (c) There is a local minimum of $-\dfrac{7}{6} \approx -1.167$, and there is a local maximum of $\dfrac{10}{3} \approx 3.333$.

27. (a) The degree is 1 and the leading coefficient is -2.

    (b) Since the degree of f is 1 and the lead coefficient is negative the graph is a line with a negative slope $\Rightarrow$ it goes up on the left end and down on the right.

29. (a) The degree is 2 and the leading coefficient is 1.

    (b) Since the degree of f is 2 and the lead coefficient is positive the graph is a parabola that opens up on each side.

31. (a) The degree is 3 and the leading coefficient is -2.

    (b) Since the degree of f is odd and the lead coefficient is negative its graph will go up on the left and down on the right.

33. (a) The degree is 3 and the leading coefficient is –1.

    (b) Since the degree of f is odd and the leading coefficient is negative, its graph will go up on the left and down on the right.

35. (a) The degree is 5 and the leading coefficient is 0.1.

    (b) Since the degree of f is odd and the leading coefficient is positive, its graph will go down on the left and up on the right.

37. (a) The degree is 2 and the leading coefficient is $-\dfrac{1}{2}$.

    (b) Since the degree of f is even and the leading coefficient is negative, its graph will go down on both sides.

39. (a) A line graph of the data is shown in Figure 39.

    (b) The data appears to decrease, increase, and decrease. The line graph crosses the x-axis three times. Neither a linear or quadratic function could do this. Thus, f must be degree 3.

    (c) Since f rises to the left and falls to the right $a < 0$.

    (d) Using the cubic regression function on the calculator we find $f(x) = -x^3 + 8x$.

41. (a) A line graph of the data is shown in Figure 41.

    (b) The data decreases and then increases. This could not be a linear function. The data appears to be parabolic and the line graph crosses the x-axis twice. Since all zeros are real and between -3 and 3, f must be degree 2 with two real zeros.

    (c) Since f is opening upward $a > 0$.

    (d) Using the quadratic regession function on the calclator we find $f(x) = x^2 - 2x - 1$.

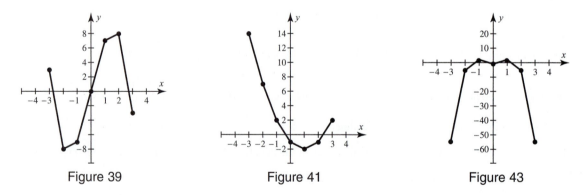

Figure 39   Figure 41   Figure 43

43. (a) See Figure 43.

    (b) The data increases then decreases then increases and the decreases. The graph crosses the x-axis 4 times. Therefore, f is degree 4.

    (c) Since both ends of f go down $a < 0$.

    (d) Using the quartic regression function on the calculator we find $f(x) = -x^4 + 3x^2 - 1$.

45. One possible graph is shown in Figure 45.

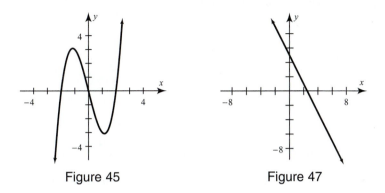

Figure 45   Figure 47

47. One possible graph is shown in Figure 47.

49. No such graph is possible.

51. No such graph is possible.

53. Start by plotting the points $(-1, 2)$ and $\left(1, \dfrac{2}{3}\right)$. Then sketch a cubic polynomial having these two turning

points. One possible graph is shown in Figure 53.

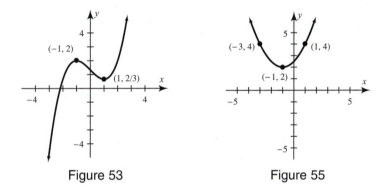

Figure 53   Figure 55

55. Since a quadratic polynomial has only one turning point, $(-1, 2)$ must be the vertex. Plot the points $(-1, 2)$, $(-3, 4)$, and $(1, 4)$. Sketch a parabola passing through these points with a vertex of $(-1, 2)$. The parabola must open up. See Figure 55.

57. The graphs of $f(x) = 2x^4$, $g(x) = 2x^4 - 5x^2 + 1$, and $h(x) = 2x^4 + 3x^2 - x - 2$ are shown in Figures 57a-c. As the viewing rectangle increases in size, the graphs begin to look alike. Each formula contains the term $2x^4$. This term determines the end behavior of the graph for large values of $|x|$. The other terms are relatively small in absolute value by comparison when $|x| \geq 100$. End behavior is determined by the leading term.

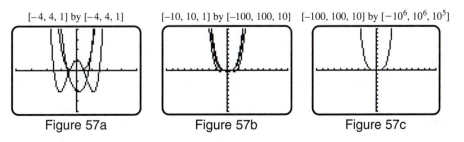

Figure 57a   Figure 57b   Figure 57c

59. For $f(x) = x$ we get $(0, 0)$ and $\left(\frac{1}{2}, \frac{1}{2}\right)$ $\Rightarrow$ the average rate of change is $\dfrac{\frac{1}{2} - 0}{\frac{1}{2} - 0} = \dfrac{\frac{1}{2}}{\frac{1}{2}} = 1$

   For $g(x) = x^2$ we get $(0, 0)$ and $\left(\frac{1}{2}, \frac{1}{4}\right)$ $\Rightarrow$ the average rate of change is $\dfrac{\frac{1}{4} - 0}{\frac{1}{2} - 0} = \dfrac{\frac{1}{4}}{\frac{1}{2}} = 0.5$

   For $h(x) = x^3$ we get $(0, 0)$ and $\left(\frac{1}{2}, \frac{1}{8}\right)$ $\Rightarrow$ the average rate of change is $\dfrac{\frac{1}{8} - 0}{\frac{1}{2} - 0} = \dfrac{\frac{1}{8}}{\frac{1}{2}} = 0.25$

   It decreases with higher degree.

61. (a) For $[1.9, 2.1]$ we get: $(1.9, 6.859)$ and $(2.1, 9.261)$ $\Rightarrow$ the average rate of change is $\dfrac{9.261 - 6.859}{2.1 - 1.9} = 12.01$

   (b) For $[1.99, 2.01]$ we get: $(1.99, 7.880599)$ and $(2.01, 8.120601)$ $\Rightarrow$ the average rate of change is

   $\dfrac{8.120601 - 7.880599}{2.01 - 1.99} = 12.0001$

   (c) For $[1.999, 2.001]$ we get: $(1.999, 7.988005999)$ and $(2.001, 8.012006001)$ $\Rightarrow$ the average rate of change is

   $\dfrac{8.012006001 - 7.988005999}{2.001 - 1.999} = 12.000001$

   As the interval decreases in length the average rate of change is approaching 12.

63. (a) For $[1.9, 2.1]$ we get: $\frac{1}{4}(1.9)^4 - \frac{1}{3}(1.9)^3 = 0.971691\overline{6}$ or $(1.9, 0.971691\overline{6})$ and $\frac{1}{4}(2.1)^4 - \frac{1}{3}(2.1)^3 =$

   1.775025 or $(2.1, 1.775025)$ $\Rightarrow$ the average rate of change is: $\dfrac{1.775025 - 0.971691\overline{6}}{2.1 - 1.9} = 4.01\overline{6}$

   (b) For $[1.99, 2.01]$ we get: $\frac{1}{4}(1.99)^4 - \frac{1}{3}(1.99)^3 = 1.293731\overline{6}$ or $(1.99, 1.293731\overline{6})$ and

   $\frac{1}{4}(2.01)^4 - \frac{1}{3}(2.01)^3 = 1.373735003$ or $(2.01, 1.373735003)$ $\Rightarrow$ the average rate of change is:

   $\dfrac{1.373735003 - 1.293731\overline{6}}{2.01 - 1.99} = 4.00016\overline{6}$

   (c) For $[1.999, 2.001]$ we get: $\frac{1}{4}(1.999)^4 - \frac{1}{3}(1.999)^3 = 1.329337332$ or $(1.999, 1.329337332)$ and

   $\frac{1}{4}(2.001)^4 - \frac{1}{3}(2.001)^3 = 1.337337335$ or $(2.001, 1.337337335)$ $\Rightarrow$ the average rate of change is:

   $\dfrac{1.337337335 - 1.329337332}{2.001 - 1.999} = 4.000001\overline{6}$

   As the interval decreases in length the average rate of change is approaching 4.

65. Use $\dfrac{f(x + h) - f(x)}{h}$, then $\dfrac{3(x + h)^3 - 3x^3}{h} = \dfrac{3x^3 + 9x^2h + 9xh^2 + 3h^3 - 3x^3}{h} = 9x^2 + 9xh + 3h^2$

67. Use $\dfrac{f(x + h) - f(x)}{h}$, then $\dfrac{1 + (x + h) - (x + h)^3 - (1 + x - x^3)}{h} =$

   $\dfrac{1 + x + h - x^3 - 3x^2h - 3xh^2 - h^3 - 1 - x + x^3}{h} = -3x^2 - 3xh - h^2 + 1$

69. $f(-2) \approx 5$ and $f(1) \approx 0$

71. $f(-1) \approx -1$, $f(1) \approx 1$, and $f(2) \approx -2$

73. $f(-3) = (-3)^3 - 4(-3)^2 = -27 - 36 = -63$, $f(1) = 3(1)^2 = 3$, and $f(4) = (4)^3 - 54 =$

   $64 - 54 = 10$

75. $f(-2) = (-2)^2 + 2(-2) + 6 = 6$, $f(1) = 1 + 6 = 7$, and $f(2) = 2^3 + 1 = 9$

77. (a) The graph of f is shown in Figure 77.

    (b) The function f is discontinuous at $x = 0$.

    (c) First solve: $4 - x^2 = 0 \Rightarrow -x^2 = -4 \Rightarrow x^2 = 4 \Rightarrow x = -2, 2$; only $x = -2$ is in the domain for the
    equation. Now solve: $x^2 - 4 = 0 \Rightarrow x^2 = 4 \Rightarrow x = -2, 2$; only $x = 2$ is in the domain for the
    equation $\Rightarrow x = -2, 2$.

79. (a) The graph of f is shown in Figure 79.

    (b) f is continuous.

    (c) First solve: $2x = 0 \Rightarrow x = 0$; this is not in the domain for the equation.
    $-2 \neq 0 \Rightarrow$ no solutions $-1 \le x \le 0$. Finally solve: $x^2 - 2 = 0 \Rightarrow x^2 = 2 \Rightarrow x = \pm\sqrt{2}$; only $\sqrt{2}$
    is in the domain for the equation $\Rightarrow x = \sqrt{2}$.

Figure 77   Figure 79   Figure 81

81. (a) The graph of f is shown in Figure 81.

    (b) f is continuous on its domain.

    (c) First solve: $x^3 + 3 = 0 \Rightarrow x^3 = -3 \Rightarrow x = \sqrt[3]{-3}$ or $-\sqrt[3]{3}$; this is in the domain for the equation.
    Now solve: $x + 3 = 0 \Rightarrow x = -3$; this is not in the domain for the equation.
    Finally solve: $4 + x - x^2 = 0 \Rightarrow$ using quadratic formula for $-x^2 + x + 4 = 0$ we get:
    $\dfrac{-1 \pm \sqrt{1 - 4(-1)(4)}}{2(-1)} = \dfrac{-1 \pm \sqrt{17}}{-2}$; only $\dfrac{1 + \sqrt{17}}{2}$ is in the domain of the equation $\Rightarrow$
    $x = -\sqrt[3]{3}, \dfrac{\sqrt{17} + 1}{2}$.

83. (a) $H(-2) = 0$, because $t < 0$; $H(0) = 1$, because $t \ge 0$; $H(3.5) = 1$, because $t \ge 0$

    (b) See Figure 83.

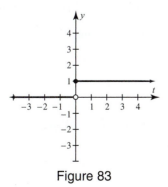

Figure 83

85. (a) There are two turning points that occur near (1, 13) and (7, 72).

(b) The point (1, 13) means that in January the average temperature is 13°F. This is the minimum average monthly temperature in Minneapolis. After this the average temperature begins to increase. The point (7, 72) means that in July the average temperature is 72°F. This is the maximum average monthly temperature in Minneapolis. After this the average temperature begins to decrease.

87. (a) Using the cubic regession function on the calculator we find

$f(x) \approx -0.15311111x^3 + 914.695238x^2 - 1,821,416.18x + 1,208,942,069.3$ *Answers may vary.*

(b) $f(1998) = -0.15311111(1998)^3 + 914.695238(1998)^2 - 1,821,416.18(1998) + 1,208,942,069.3 \approx 1704.$ The result that was found using interpolation.

(c) Using the graph of $f(x)$ we can see that in approximately 1996 the number of endangered and threatened species will be 1600.

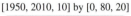

[1950, 2010, 10] by [0, 80, 20]

Figure 89

89. (a) Using the cubic regression function on the calculator we find

$f(x) = 0.000667x^3 - 3.9x^2 + 7604.7333x - 4,942725.$

(b) See Figure 89.

(c) $f(1994) = 0.000667(1994)^3 - 3.9(1994)^2 + 7604.7333(1994) - 4,942,725 \approx 50.05.$ This result is very close to the actual value of 50 thousand.

$f(2010) = 0.000667(1994)^3 - 3.9(2010)^2 + 7604.7333(2010) - 4,942,725 \approx 133.$ This result is 4 thousand higher than the actual result of 129 thousand.

(d) The estimate found for 1994 involved interpolation, and the estimate found for 2010 involved extrapolation.

91. (a) See Figure 91a.

(b) Water entering the container can be modeled with a linear formula and occurs in the first 2 minutes. The height increases 16 centimeters in 2 minutes. Thus, $m = 8$, $f(x) = 8x$ for $0 \le x \le 2$. The water leaks out over the time interval of 2 to 5 minutes and can be modeled with a nonlinear formula. Use quadratic regression to find $f(x) = 1.32x^2 - 14.4x + 39.39$ for $2 < x \le 5$.

Thus, $f(x) = \begin{cases} 8x, & \text{if } 0 \le x \le 2 \\ 1.32x^2 - 14.4 + 39.39, & \text{if } 2 < x \le 5 \end{cases}$

(c) When $x = 1.25$, water was flowing into the tank and $f(x) = 8x$. $f(1.25) = 8(1.25) = 10$ cm. When $x = 3.2$, water is flowing out of the tank and $f(x) = 1.32x^2 - 14.4x + 39.39.$

$f(3.2) = 1.32(3.2)^2 - 14.4(3.2) + 39.39 \approx 7$ cm.

(d) The water level in figure 91 equals 5 centimeters twice. This occurs when the water is entering the container and when it is leaking out. Solve $8x = 5$ and $1.32x^2 - 14.4x + 39.39 = 5$. Thus, after $\dfrac{5}{8}$ minute the water level was 5 centimeters. The equation $1.32x^2 - 14.4x + 39.39 = 5$ is solved graphically. See Figure 91d.

Intersection: $y = 5$ and $x = 3.5312577$. The solution satisfying $2 < x \le 5$ is $x \approx 3.53$ min. *Answers may vary.*

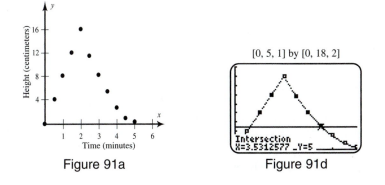

Figure 91a   Figure 91d

4.3: Division of Polynomials

1. $\dfrac{5x^4 - 15}{10x} = \dfrac{5x^4}{10x} - \dfrac{15}{10x} = \dfrac{x^3}{2} - \dfrac{3}{2x}$

3. $\dfrac{3x^4 - 2x^2 - 1}{3x^3} = \dfrac{3x^4}{3x^3} - \dfrac{2x^2}{3x^3} - \dfrac{1}{3x^3} = x - \dfrac{2}{3x} - \dfrac{1}{3x^3}$

5. $\dfrac{x^3 - 4}{4x^3} = \dfrac{x^3}{4x^3} - \dfrac{4}{4x^3} = \dfrac{1}{4} - \dfrac{1}{x^3}$

7. $\dfrac{5x(3x^2 - 6x + 1)}{3x^2} = \dfrac{15x^3 - 30x^2 + 5x}{3x^2} = \dfrac{15x^3}{3x^2} - \dfrac{30x^2}{3x^2} + \dfrac{5x}{3x^2} = 5x - 10 + \dfrac{5}{3x}$

9. $x^3 - 2x^2 - 5x + 6$ divided by $x - 3$ can be performed using synthetic division.

    $$\begin{array}{r|rrrr} 3 & 1 & -2 & -5 & 6 \\ & & 3 & 3 & -6 \\ \hline & 1 & 1 & -2 & 0 \end{array}$$

    The quotient is $x^2 + x - 2$ and the remainder is 0.

11. $2x^4 - 7x^3 - 5x^2 - 19x + 17$ divided by $x + 1$ can be performed using synthetic division.

    $$\begin{array}{r|rrrrr} -1 & 2 & -7 & -5 & -19 & 17 \\ & & -2 & 9 & -4 & 23 \\ \hline & 2 & -9 & 4 & -23 & 40 \end{array}$$

    The quotient is $2x^3 - 9x^2 + 4x - 23$ and the remainder is 40.

12. $x^4 - x^3 - 4x + 1$ divided by $x - 2$ can be performed using synthetic division.

13. $3x^3 - 7x + 10$ divided by $x - 1$ can be performed using synthetic division.

$$\begin{array}{r|rrrr} 1 & 3 & 0 & -7 & 10 \\ & & 3 & 3 & -4 \\ \hline & 3 & 3 & -4 & 6 \end{array}$$

The quotient is $3x^2 + 3x - 4$ and the remainder is 6.

This can also be found using long division as shown.

$$
\begin{array}{r}
3x^2 + 3x - 4 \\
x - 1 \overline{) 3x^3 - 0x^2 - 7x + 10} \\
\underline{3x^3 - 3x^2} \\
3x^2 - 7x \\
\underline{3x^2 - 3x} \\
-4x + 10 \\
\underline{-4x + 4} \\
6
\end{array}
$$

15. We can use synthetic division to divide $x^4 - 3x^3 - x + 3$ by $x - 3$.

$$\begin{array}{r|rrrrr} 3 & 1 & -3 & 0 & -1 & 3 \\ & & 3 & 0 & 0 & -3 \\ \hline & 1 & 0 & 0 & -1 & 0 \end{array}$$

The quotient is $x^3 - 1$ and the remainder is 0.

17. We can use long division to divide $4x^3 - x^2 - 5x + 6$ by $x - 1$.

$$
\begin{array}{r}
4x^2 + 3x - 2 \\
x - 1 \overline{) 4x^3 - x^2 - 5x + 6} \\
\underline{4x^3 - 4x^2} \\
3x^2 - 5x \\
\underline{3x^2 - 3x} \\
-2x + 6 \\
\underline{-2x + 2} \\
4
\end{array}
$$

The quotient is $4x^2 + 3x - 2 + \dfrac{4}{x - 1}$.

19. We can use synthetic division to divide $x^3 + 1$ by $x + 1$.

$$\begin{array}{r|rrrr} -1 & 1 & 0 & 0 & 1 \\ & & -1 & 1 & -1 \\ \hline & 1 & -1 & 1 & 0 \end{array}$$

The quotient is $x^2 - x + 1$ and the remainder is 0.

21. We can use long division to divide $6x^3 + 5x^2 - 8x + 4$ by $2x - 1$.

$$
\begin{array}{r}
3x^2 + 4x - 2 \\
2x - 1 \overline{) 6x^3 + 5x^2 - 8x + 4} \\
\underline{6x^3 - 3x^2} \\
8x^2 - 8x + 4 \\
\underline{8x^2 - 4x} \\
-4x + 4 \\
\underline{-4x + 2} \\
2
\end{array}
$$

The quotient is $3x^2 + 4x - 2 + \dfrac{2}{2x - 1}$.

23. We can use long division to divide $3x^4 - 7x^3 + 6x - 16$ by $3x - 7$.

$$
\begin{array}{r}
x^3 + 2 \\
3x - 7 \overline{)3x^4 - 7x^3 + 0x^2 + 6x - 16} \\
\underline{3x^4 - 7x^3 } \\
6x - 16 \\
\underline{6x - 14} \\
-2
\end{array}
$$

The quotient is $x^3 + 2 + \dfrac{-2}{3x - 7}$.

25. We can use long division to divide $5x^4 - 2x^2 + 6$ by $x^2 + 2$.

$$
\begin{array}{r}
5x^2 - 12 \\
x^2 + 2 \overline{)5x^4 - 2x^2 + 6} \\
\underline{5x^4 + 10x^2 } \\
-12x^2 + 6 \\
\underline{-12x^2 - 24} \\
30
\end{array}
$$

The quotient is $5x^2 - 12 + \dfrac{30}{x^2 + 2}$.

27. We can use long division to divide $8x^3 + 10x^2 - 12x - 15$ by $2x^2 - 3$.

$$
\begin{array}{r}
4x + 5 \\
2x^2 - 3 \overline{)8x^3 + 10x^2 - 12x - 15} \\
\underline{8x^3 - 12x } \\
10x^2 - 15 \\
\underline{10x^2 - 15} \\
0
\end{array}
$$

The quotient is $4x + 5$.

29. We can use long division to divide $2x^4 - x^3 + 4x^2 + 8x + 7$ by $2x^2 + 3x + 2$.

$$
\begin{array}{r}
x^2 - 2x + 4 \\
2x^2 + 3x + 2 \overline{)2x^4 - x^3 + 4x^2 + 8x + 7} \\
\underline{2x^4 + 3x^3 + 2x^2 } \\
-4x^3 + 2x^2 + 8x \\
\underline{-4x^3 - 6x^2 - 4x } \\
8x^2 + 12x + 7 \\
\underline{8x^2 + 12x + 8} \\
-1
\end{array}
$$

The quotient is $x^2 - 2x + 4 + \dfrac{-1}{2x^3 + 3x + 2}$.

31. The divisor times the quotient will be equal to the dividend, $(x - 2)(x^2 - 6x + 3) = x^3 - 8x^2 + 15x - 6$.

33.
$$
\begin{array}{r}
x - 1 \\
x - 2 \overline{)x^2 - 3x + 1} \\
\underline{x^2 - 2x } \\
-x + 1 \\
\underline{-x + 2} \\
-1
\end{array}
$$
$(x - 2)(x - 1) - 1$

35.
$$
\begin{array}{r}
x^2 - 1 \\
2x + 1 \overline{)2x^3 - x^2 - 2x} \\
\underline{2x^3 + x^2 } \\
-2x + 0 \\
\underline{-2x - 1} \\
1
\end{array}
$$
$(2x + 1)(x^2 - 1) + 1$

37.
$$
\begin{array}{r}
x - 1 \\
x^2 + 0x + 1 \overline{\smash{\big)}\ x^3 - x^2 + x + 1} \\
\underline{x^3 + 0x^2 + x} \\
-x^2 + 1 \\
\underline{-x^2 - 1} \\
2 \quad (x^2 + 1)(x - 1) + 2
\end{array}
$$

39.
$$
\begin{array}{r|rrrr}
-5 & 1 & 2 & -17 & -10 \\
   &   & -5 & 15 & 10 \\
\hline
   & 1 & -3 & -2 & 0
\end{array}
$$
The quotient is $x^2 - 3x - 2$.

41.
$$
\begin{array}{r|rrrr}
5 & 3 & -11 & -20 & 3 \\
  &   & 15 & 20 & 0 \\
\hline
  & 3 & 4 & 0 & 3
\end{array}
$$
The quotient is $3x^2 + 4x + \dfrac{3}{x - 5}$.

43.
$$
\begin{array}{r|rrrrr}
2 & 1 & -3 & -4 & 12 & 0 \\
  &   & 2 & -2 & -12 & 0 \\
\hline
  & 1 & -1 & -6 & 0
\end{array}
$$
The quotient is $x^3 + x^2 - 6x$.

45.
$$
\begin{array}{r|rrrrrr}
-\dfrac{1}{2} & 2 & -1 & -1 & 0 & 4 & 3 \\
 &   & -1 & 1 & 0 & 0 & -2 \\
\hline
 & 2 & -2 & 0 & 0 & 4 & 1
\end{array}
$$
The quotient is $2x^4 - 2x^3 + 4 + \dfrac{1}{x + 0.5}$.

47. Using the remainder theorem we find: $f(1)$, $\Rightarrow 5(1)^2 - 3(1) + 1 = 5 - 3 + 1 = 3$

49. Using the remainder theorem we find:

$f(-2)$, $\Rightarrow 4(-2)^3 - (-2)^2 + 4(-2) + 2 = -32 - 4 - 8 + 2 = -42$

51. If we divide the Area by the Width we will find the Length:

$$
\begin{array}{r}
4x + 3 \\
3x + 1 \overline{\smash{\big)}\ 12x^2 + 13x + 3} \\
\underline{12x^2 + 4x} \\
9x + 3 \\
\underline{9x + 3} \\
0
\end{array}
$$

The length is $4x + 3$. When $x = 10$, the Length is $4(10) + 3 = 43$ feet.

4.4: Real Zeros of Polynomial Functions

1. The x-intercepts of f are -1, 1, and 2. Since $f(-1) = 0$, the factor theorem states that $(x + 2)$ is a factor of $f(x)$. Similarly, $f(-1) = 0$ implies that $(x + 1)$ is a factor, and $f(1) = 0$ implies that $(x - 1)$ is a factor.

3. The x-intercepts of f are -2, -1, 1 and 2. Since $f(-2) = 0$, the factor theorem states that $(x + 2)$ is a factor of $f(x)$. Similarly, $f(-1) = 0$ implies that $(x + 1)$ is a factor, $f(1) = 0$ implies that $(x - 1)$ is a factor and $f(2) = 0$ implies that $(x - 2)$ is a factor.

5. $f(x) = 2x^2 - 25x + 77$ and zeros: $\dfrac{11}{2}$ and $7 \Rightarrow f(x) = 2\left(x - \dfrac{11}{2}\right)(x - 7)$

7. $f(x) = x^3 - 2x^2 - 5x + 6$ and zeros: -2, 1, and $3 \Rightarrow f(x) = (x + 2)(x - 1)(x - 3)$

9. $f(x) = -2x^3 + 3x^2 + 59x - 30$ and zeros: $-5, \frac{1}{2}$, and $6 \Rightarrow f(x) = -2(x + 5)\left(x - \frac{1}{2}\right)(x - 6)$

11. If $f(-3) = 0$ then the quadratic equation has a factor of $x - (-3)$ or $x + 3$, likewise if $f(2) = 0$ then the quadratic equation has a factor $x - 2$. If this quadratic equation has a leading coefficient 7, the complete factored form of $f(x)$ is $f(x) = 7(x + 3)(x - 2)$.

13. To factor $f(x)$ we need to determine the leading coefficient and zeros of f. The leading coefficient is -2 and the zeros are $-1, 0$, and 1. The complete factorization is $f(x) = -2x(x + 1)(x - 1)$.

15. From the graph the zeros are $-4, 2$, and 8. $f(x)$ is a cubic with a positive leading coefficient. Therefore, $f(x) = (x + 4)(x - 2)(x - 8)$.

17. From the graph the zeros are $-8, -4, -2$, and 4. $f(x)$ is a quartic polynomial with a negative leading coefficient. Therefore, $f(x) = -1(x + 8)(x + 4)(x + 2)(x - 4)$.

19. Since the polynomial has zeros of $-1, 2$, and 3, it has factors $(x + 1)(x - 2)(x - 3)$.

If f passes through $(0, 3)$ then $a(0 + 1)(0 - 2)(0 - 3) = 3 \Rightarrow a(1)(-2)(-3) = 3 \Rightarrow 6a = 3 \Rightarrow a = \frac{1}{2}$.

The complete factored form is: $f(x) = \frac{1}{2}(x + 1)(x - 2)(x - 3)$.

21. Since f has zeros $-1, 1$, and 2, it has factors $(x + 1)(x - 1)(x - 2)$. If $f(0) = 1$ then $a(1)(-1)(-2) = 1$ or $2a = 1 \Rightarrow a = \frac{1}{2}$. The complete factored form is: $f(x) = \frac{1}{2}(x + 1)(x - 1)(x - 2)$.

23. Since f has zeros $-2, -1, 1$, and 2, it has factors $(x + 2)(x + 1)(x - 1)(x - 2)$. If $f(0) = -8$ then $a(2)(1)(-1)(-2) = -8 \Rightarrow 4a = -8 \Rightarrow a = -2$.

The complete factored form is: $f(x) = -2(x + 2)(x + 1)(x - 1)(x - 2)$.

25. A graph of $Y_1 = 10X^2 + 17X - 6$ is shown in Figure 25. Its zeros are -2 and 0.3. Since the leading coefficient is 10, the complete factorization is $f(x) = 10(x + 2)\left(x - \frac{3}{10}\right)$.

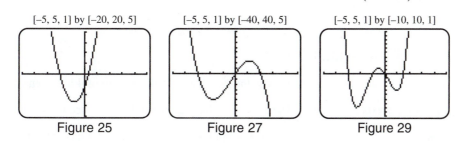

[-5, 5, 1] by [-20, 20, 5]   [-5, 5, 1] by [-40, 40, 5]   [-5, 5, 1] by [-10, 10, 1]

Figure 25   Figure 27   Figure 29

27. A graph of $Y_1 = -3X^3 - 3X^2 + 18X$ is shown in Figure 27. Its zeros are $-3, 0$, and 2. Since the leading coefficient is -3, the complete factorization is $f(x) = -3(x - 0)(x - 2)(x + 3) = -3x(x - 2)(x + 3)$.

29. A graph of $Y_1 = X^4 + (5/2)X^3 - 3X^2 - (9/2)X$ is shown in Figure 29. Its zeros are $-3, -1, 0$, and $\frac{3}{2}$. Since the leading coefficient is 1, the complete factorization is

$$f(x) = (x + 3)(x + 1)(x - 0)\left(x - \frac{3}{2}\right) = x(x + 3)(x + 1)\left(x - \frac{3}{2}\right).$$

31. By the factor theorem, since 1 is a zero, $(x - 1)$ is a factor. $x^3 - 9x^2 + 23x - 15$ divided by $x - 1$ can be performed using synthetic division.

$$\begin{array}{r|rrrr} 1 & 1 & -9 & 23 & -15 \\ & & 1 & -8 & 15 \\ \hline & 1 & -8 & 15 & 0 \end{array}$$

The quotient is $x^2 - 8x + 15$ and the remainder is 0.

Thus, $x^3 - 9x^2 + 23x - 15 = (x - 1)(x^2 - 8x + 15) = (x - 1)(x - 3)(x - 5)$.

The complete factored form of $f(x) = x^3 - 9x^2 + 23x - 15$ is $f(x) = (x - 1)(x - 3)(x - 5)$.

33. By the factor theorem, since -4 is a zero, $(x + 4)$ is a factor. $-4x^3 - x^2 + 51x - 36$ divided by $x + 4$ can be performed using synthetic division.

$$\begin{array}{r|rrrr} -4 & -4 & -1 & 51 & -36 \\ & & 16 & -60 & 36 \\ \hline & -4 & 15 & -9 & 0 \end{array}$$

The quotient is $-4x^2 + 15x - 9$ and the remainder is 0.

Thus, $-4x^3 - x^2 + 51x - 36 = (x + 4)(-4x^2 + 15x - 9) = (x + 4)(-4x + 3)(x - 3)$.

The complete factored form of $f(x) = -4x^3 - x^2 + 51x - 36$ is $f(x) = -4(x + 4)\left(x - \dfrac{3}{4}\right)(x - 3)$.

35. By the factor theorem, since -2 is a zero, $(x + 2)$ is a factor. $2x^4 - x^3 - 13x^2 - 6x$ divided by $x + 2$ can be performed using synthetic division.

$$\begin{array}{r|rrrrr} -2 & 2 & -1 & -13 & -6 & 0 \\ & & -4 & 10 & 6 & 0 \\ \hline & 2 & -5 & -3 & 0 & 0 \end{array}$$

The quotient is $2x^3 - 5x^2 - 3x$ and the remainder is 0. Thus,

$2x^4 - x^3 - 13x^2 - 6x = (x + 2)(2x^3 - 5x^2 - 3x) = (x + 2)x(2x^2 - 5x - 3) = x(x + 2)(2x + 1)(x - 3).$

The complete factored form of $f(x) = 2x^4 - x^3 - 13x^2 - 6x$ is $f(x) = 2x(x + 2)\left(x + \dfrac{1}{2}\right)(x - 3)$.

37. $f(2) = (2)^3 - 6(2)^2 + 11(2) - 6 = 0$; since $f(2) = 0$, by the factor theorem,

$x - 2$ is a factor of $f(x) = x^3 - 6x^2 + 11x - 6$.

39. $f(3) = (3)^4 - 2(3)^3 - 13(3)^2 - 10(3) = -120$; since $f(-3) \neq 0$, by the factor theorem,

$x - 3$ is not a factor of $x^4 - 2x^3 - 13x^2 - 10x$.

41. The zeros of $f(x)$ are 4 and -2. Since the graph does not cross the x-axis at $x = 4$, the zero of 4 has even multiplicity. The graph crosses the x-axis at $x = -2$. The zero of -2 has odd multiplicity. Since the graph levels off, crossing the x-axis at $x = -2$, this zero has at least multiplicity 3, and the zero of 4 has at least multiplicity 2. Thus, the minimum degree of $f(x)$ is $3 + 2$, or 5.

43. Degree: 3; zeros: -1 with multiplicity 2 and 6 with multiplicity 1. $f(x) = (x + 1)^2(x - 6)$

45. Degree: 4; zeros: 2 with multiplicity 3 and 6 with multiplicity 1. $f(x) = (x - 2)^3(x - 6)$

47. The graph shows a cubic polynomial with a positive leading coefficient and zeros of 4 and -2. The zero of 4 has odd multiplicity, whereas the zero of -2 has even multiplicity. Since the graph has a degree three, its complete factored form is $f(x) = (x - 4)(x + 2)^2$.

49. The graph shows a quartic polynomial with a negative leading coefficient and zeros of –3 and 3. Both zeros have even multiplicity. Since the graph has a degree four polynomial, its complete factored form is

    $f(x) = -1(x + 3)^2(x - 3)^2.$

51. The graph shows a fifth degree polynomial with a positive leading coefficient and zeros of –1 and 1. The –1 has an even multiplicity, whereas the zero of 1 has an odd multiplicity. Since the graph shows a fifth degree polynomial the factors are $(x + 1)^2(x - 1)^3$. To find the leading coefficient we use

    $f(0) = -2 \Rightarrow a(1)^2(-1)^3 = -2 \Rightarrow -a = -2 \Rightarrow a = 2.$ Its factored form is $f(x) = 2(x + 1)^2(x - 1)^3.$

53. (a) From the factors the x-intercepts are: $-2, -1$. To find the y-intercept set $x = 0 \Rightarrow$

    $2(0 + 2)(0 + 1)^2 = 2(2)(1)^2 = 4$

    (b) The zero -2 has multiplicity 1. The zero -1 has multiplicity 2.

    (c) See Figure 53.

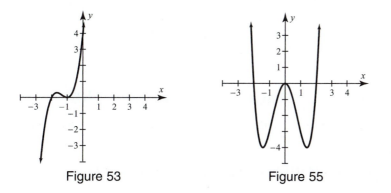

Figure 53   **Figure 55**

55. (a) From the factors the x-intercepts are: $-2, 0,$ and 2. To find the y-intercept set $x = 0 \Rightarrow$

    $0(0 + 2)(0 - 2) = 0(2)(-2) = 0$

    (b) The zero -2 has multiplicity 1. The zero 0 has multiplicity 2. The zero 2 has multiplicity 1.

    (c) See Figure 55.

57. $f(x) = 2x^3 + 3x^2 - 8x + 3$

    (a) If $\dfrac{p}{q}$ is a rational zero, then p is a factor of 3, which are ± 1 and ± 3 and q is a factor of 2, which are

    ± 1 or ± 2. Thus, any rational zero must be in the list $\pm\dfrac{1}{2}, \pm 1, \pm\dfrac{3}{2},$ or ± 3. From Figure 57 we see that

    there are three rational zeros of $\dfrac{1}{2}$, 1, and –3.

    (b) The complete factored form is $f(x) = 2\left(x - \dfrac{1}{2}\right)(x - 1)(x + 3).$

x	$\frac{1}{2}$	$-\frac{1}{2}$	1	-1	$\frac{3}{2}$	$-\frac{3}{2}$	3	-3
$f(x)$	0	$\frac{15}{2}$	0	12	$\frac{9}{2}$	15	60	0

Figure 57

59. $f(x) = 2x^4 + x^3 - 8x^2 - x + 6$

(a) If $\dfrac{p}{q}$ is a rational zero, then p is a factor of 6, which are ± 1, ± 2, ± 3, and ± 6 and q is a factor of 2, which

are ± 1 and ± 2. Thus, any rational zero must be in the list $\pm \dfrac{1}{2}$, ± 1, $\pm \dfrac{3}{2}$, ± 2, ± 3, or ± 6. By evaluating $f(x)$

at each of these values, we find that the zeros are $-2, -1, 1$, and $\dfrac{3}{2}$.

(b) The complete factored form is $f(x) = 2(x + 2)(x + 1)(x - 1)\left(x - \dfrac{3}{2}\right)$.

61. $f(x) = 3x^3 - 16x^2 + 17x - 4$

(a) If $\dfrac{p}{q}$ is a rational zero, then p is a factor of 4, which are ± 1, ± 2, and ± 4 and q is a factor of 3, which are

± 1 and ± 3. Thus, any rational zero must be in the list $\pm \dfrac{1}{3}$, ± 1, $\pm \dfrac{2}{3}$, ± 2, $\pm \dfrac{4}{3}$, or ± 4. By evaluating $f(x)$ at

each of these values, we find that the zeros are $\dfrac{1}{3}$, 1, and 4.

(b) The complete factored form is $f(x) = 3\left(x - \dfrac{1}{3}\right)(x - 1)(x - 4)$.

63. $f(x) = x^3 - x^2 - 7x + 7$

(a) If $\dfrac{p}{q}$ is a rational zero, then p is a factor of 7, which are ± 1, and ± 7 and q is a factor of 1, which are ± 1.

Thus, any rational zero must be in the list ± 1 or ± 7. By evaluating $f(x)$ at each of these values, we find

that the only rational zero is 1.

(b) In order to find the complete factored form of $f(x)$ we need to divide the factor $(x - 1)$ into

$x^3 - x^2 - 7x + 7$ using synthetic division.

$$
\begin{array}{r|rrrr}
1 & 1 & -1 & -7 & 7 \\
  &   & 1 & 0 & 7 \\
\hline
  & 1 & 0 & -7 & 0
\end{array}
\qquad \text{Thus } x^3 - x^2 - 7x + 7 = (x - 1)(x^2 - 7).
$$

The complete factored form is $f(x) = (x - 1)(x - \sqrt{7})(x + \sqrt{7})$.

65. $P(x) = 2x^3 - 4x^2 + 2x + 7$, $P(x)$ has two sign changes, therefore are 2 or $2 - 2 = 0$ possible positive

zeros. $P(-x) = 2(-x)^3 - 4(-x)^2 + 2(-x) + 7 = -2(x)^3 - 4x^2 - 2x + 7$, $P(-x)$ has one sign change,

therefore there is one possible negative zero. From the graph of $P(x)$ in Figure 65, we see that the actual number of positive and negative zeros are 0 and 1 repectively.

67. $P(x) = 5x^4 + 3x^2 + 2x - 9$, $P(x)$ has one sign change, therefore there is one possible positive zero.

$P(-x) = 5(-x)^4 + 3(-x)^2 + 2(-x) - 9 = 5x^4 + 3x - 2x - 9$, $P(-x)$ has one sign change, therefore

there is one possible negative zero. From the graph of $P(x)$ in Figure 67, we see that the actual number of positive and negative zeros are 1 and 1 respectively.

69. $P(x) = x^5 + 3x^4 - x^3 + 2x + 3$, $P(x)$ has two sign changes, therefore there are 2 or $2 - 2 = 0$ possible

positive zeros. $P(-x) = (-x)^5 + 3(-x)^4 - (-x)^3 + 2(-x) + 3 = -x^5 + 3x^4 + x^3 - 2x + 3$, $P(-x)$

has three sign changes, therefore there are 3 or $3 - 2 = 1$ possible negative zeros. From the graph of $P(x)$ in

Figure 69, we see that the actual number of positive and negative zeros are 0 and 1 respectively.

[−4, 4, 1] by [−25, 25, 5]   [−4, 4, 1] by [−25, 25, 5]   [−4, 4, 1] by [−50, 50, 10]

Figure 65   Figure 67   Figure 69

71. (a) $f(x) = x^3 + x^2 - 6x = 0 \Rightarrow x(x^2 + x - 6) = x(x + 3)(x - 2) = 0 \Rightarrow x = 0, -3,$ or 2.

(b) Graph $Y_1 = $ X^3 + X^2 − 6X in $[-5, 5, 1]$ by $[-10, 10, 1]$. The x-intercepts are −3, 0, and 2.

(c) Table $Y_1 = $ X^3 + X^2 − 6X starting at $x = -4$, incrementing by 1. The zeros or x-intercepts

are −3, 0, and 2.

73. (a) $f(x) = x^4 - 1 = 0 \Rightarrow x^4 = 1 \Rightarrow x = \pm 1$.

(b) Graph $Y_1 = $ X^4 − 1 in $[-2, 2, 1]$ by $[-10, 10, 1]$. The x-intercepts are −1 and 1.

(c) Table $Y_1 = $ X^4 − 1 starting at $x = -2$, incrementing by 1. The zeros or x-intercepts are ± 1.

75. (a) $f(x) = -x^3 + 4x = 0 \Rightarrow -x(x^2 - 4) = x(x + 2)(x - 2) = 0 \Rightarrow x = 0,$ or ± 2.

(b) Graph $Y_1 = $ −X^3 + 4X in $[-5, 5, 1]$ by $[-5, 5, 1]$. The x-intercepts are −2, 0, and 2.

(c) Table $Y_1 = $ −X^3 + 4X starting at $x = -3$, incrementing by 1. The zeros or x-intercepts are −2, 0, and 2.

77. $x^3 - 25x = 0 \Rightarrow x(x^2 - 25) = 0 \Rightarrow x(x - 5)(x + 5) = 0 \Rightarrow x = 0,$ or ± 5

79. $x^4 - x^2 = 2x^2 + 4 \Rightarrow x^4 - 3x^2 - 4 = 0 \Rightarrow (x^2 - 4)(x^2 + 1) = 0 \Rightarrow$

$(x - 2)(x + 2)(x^2 + 1) = 0 \Rightarrow x = \pm 2$

81. $x^3 - 3x^2 - 18x = 0 \Rightarrow x(x^2 - 3x - 18) = 0 \Rightarrow x(x - 6)(x + 3) = 0 \Rightarrow x = 0, 6,$ or -3

83. $2x^3 = 4x^2 - 2x \Rightarrow 2x^3 - 4x^2 + 2x = 0 \Rightarrow 2x(x^2 - 2x + 1) = 0 \Rightarrow$

$2x(x - 1)(x - 1) = 0 \Rightarrow x = 0$ or 1

85. $12x^3 = 17x^2 + 5x \Rightarrow 12x^3 - 17x^2 - 5x = 0 \Rightarrow x(12x^2 - 17x - 5) = 0 \Rightarrow$

$x(4x + 1)(3x - 5) = 0 \Rightarrow x = 0, -\dfrac{1}{4},$ or $\dfrac{5}{3}$

87. $9x^4 + 4 = 13x^2 \Rightarrow 9x^4 - 13x^2 + 4 = 0 \Rightarrow (9x^2 - 4)(x^2 - 1) = 0 \Rightarrow x = \pm\dfrac{2}{3}, x = \pm 1$

89. $4x^3 + 4x^2 - 3x - 3 = 0 \Rightarrow (4x^3 + 4x^2) - (3x + 3) = 0 \Rightarrow 4x^2(x + 1) - 3(x + 1) = 0 \Rightarrow$

$(4x^2 - 3)(x + 1) = 0.$ Set $4x^2 - 3 = 0 \Rightarrow 4x^2 = 3 \Rightarrow x^2 = \dfrac{3}{4} \Rightarrow x = \pm\dfrac{\sqrt{3}}{2}.$

The solutions are; $x = -1, \pm\dfrac{\sqrt{3}}{2}.$

91. $2x^3 + 4 = x(x + 8) \Rightarrow 2x^3 + 4 = x^2 + 8x \Rightarrow 2x^3 - x^2 - 8x + 4 = 0 \Rightarrow$

$(2x^3 - x^2) - (8x - 4) = 0 \Rightarrow x^2(2x - 1) - 4(2x - 1) = 0 \Rightarrow (x^2 - 4)(2x - 1) = 0 \Rightarrow$

$(x + 2)(x - 2)(2x - 1) = 0 \Rightarrow x = -2, 2, \dfrac{1}{2}$

93. $8x^4 - 30x^2 + 27 = 0 \Rightarrow (4x^2 - 9)(2x^2 - 3) = 0 \Rightarrow (2x + 3)(2x - 3)(2x^2 - 3) = 0.$

Set $2x^2 - 3 = 0 \Rightarrow 2x^2 = 3 \Rightarrow x^2 = \frac{3}{2} \Rightarrow x = \pm\sqrt{\frac{3}{2}} \Rightarrow x = \pm\frac{\sqrt{6}}{2}.$ The solutions are;

$x = \pm\frac{3}{2}, \pm\frac{\sqrt{6}}{2}$

95. $x^6 - 19x^3 - 216 = 0 \Rightarrow (x^3 + 8)(x^3 - 27) = 0.$ set $x^3 + 8 = 0 \Rightarrow x^3 = -8 \Rightarrow x = -2.$

Also set $x^3 - 27 = 0 \Rightarrow x^3 = 27 \Rightarrow x = 3.$ The solutions are; $x = -2, 3.$

97. The graph of $f(x) = x^3 - 1.1x^2 - 5.9x + 0.7$ is shown in Figure 97. Its zeros are approximately -2.0095, 0.11639, and 2.9931. The solutions are $x \approx -2.01, 0.12,$ or $2.99.$

99. The graph of $f(x) = -0.7x^3 - 2x^2 + 4x + 2.5$ is shown in Figure 99. Its zeros are approximately -4.0503, -0.51594, and 1.7091. The solutions are $x \approx -4.05, -0.52,$ or $1.71.$

[–10, 10, 1] by [–10, 10, 1]   [–10, 10, 1] by [–10, 10, 1]   [–10, 10, 1] by [–120, 120, 20]   [–20, 40, 5] by [–2500, 2500, 500]

Figure 97   Figure 99   Figure 101   Figure 111

101. The graph of $f(x) = 2x^4 - 1.5x^3 - 24x^2 - 10x + 13$ is shown in Figure 101. Its zeros are approximately $-2.6878, -1.0957, 0.55475,$ and 3.9787. The solutions are $x \approx -2.69, -1.10, 0.55$ or $3.98.$

103. $f(x) = x^2 - 5 \Rightarrow f(2) = 2^2 - 5 = -1$ and $f(3) = 3^2 - 5 = 4$; because $f(2) < 0$ and $f(3) > 0$, by the intermediate value property, there exists an x-value between 2 and 3 such that $f(x) = 0.$

105. $f(x) = 2x^3 - 1 \Rightarrow f(0) = 2(0)^3 - 1 = -1$ and $f(1) = 2(1)^3 - 1 = 1$; because $f(0) < 0$ and $f(1) > 0$, by the intermediate value property, there exists an x-value between 0 and 1 such that $f(x) = 0.$

107. $f(x) = x^5 - x^2 + 4 \Rightarrow f(1) = 1^5 - 1^2 + 4 = 4$ and $f(2) = 2^5 - 2^2 + 4 = 32.$ Because $f(1) < 20$ and $f(2) > 20$, by the intermediate value property, there exists a number K such that $f(K) = 20.$

109. $T(x) = x^3 - 6x^2 + 8x$ when $0 \le x \le 4.$ Graph $Y_1 = X^3 - 6X^2 + 8X$ in the window $[0, 4, 1]$ by $[-5, 5, 1].$ The zeros are $x = 0, x = 2,$ and $x = 4.$ Since $x = 0, 2,$ and 4 correspond to the hours after midnight, then the temperature was $0°F$ at 12 am, 2 am and 4 am.

111. The graph of $Y_1 = (\pi/3)X^3 - 10\pi X^2 + ((4000\pi)(0.6))/3$ and the smallest positive zero are shown in Figure 111. The ball with a 20-centimeter diameter will sink approximately 11.34 centimeters into the water.

[–20, 40, 5] by [–2500, 2500, 500]

113. $f(x) = x^3 - 66x^2 + 1052x + 652$ and $f(x) = 2500 \Rightarrow 2500 = x^3 - 66x^2 + 1052x + 652 \Rightarrow$ $x^3 - 66x^2 + 1052x - 1848 = 0$; graph $Y_1 = X^3 - 66X^2 + 1052X - 1848$ in the window $[0, 45, 5]$ by $[-5000, 5000, 1000].$ The zeros are at $x = 2, x = 22,$ and $x = 42.$ Since $x = 1$ corresponds to June 1, there were 2500 birds on approximately June 2, June 22, and July 12.

115. (a) Graph f in $[-10, 15, 1]$ by $[-70, 70, 10]$. It has three zeros of approximately -6.01, 2.15, and 11.7. The approximate complete factored form is $-0.184(x + 6.01)(x - 2.15)(x - 11.7)$.

(b) The zeros represent the months when the average temperature is 0°F. The zero of -6.01 has no significance since it does not correspond to a month. The zeros of 2.15 and 11.7 mean that in approximately February and November the average temperature in Trout Lake is 0°F.

117. (a) The greater the distance downstream from the plant the less the concentration of copper. This agrees with intuition.

(b) Using the cubic regression function on your calculator we find
$$f(x) \approx -0.000068x^3 + 0.0099x^2 - 0.653x + 23.$$

(c) See Figure 117c.

(d) We must approximate the distance where the concentration of copper first drops to 10. Graph $Y_1 = C(x)$ and $Y_2 = 10$. The point of intersection is near $(32.1, 10)$ as shown in Figure 117d. Mussels would not be expected to live between the plant and approximately 32.1 miles downstream, that is when $0 \le x < 32.1$ (approximately).

[0, 70, 10] by [0, 22, 5]   [0, 70, 10] by [0, 22, 5]

Figure 117c   Figure 117d

4.5: The Fundamental Theorem of Algebra

1. The graph of $f(x)$ does not intersect the x-axis. Therefore, f has no real zeros. Since f is degree 2, there must be two imaginary zeros.

3. The graph of $f(x)$ intersects the x-axis once. Therefore, f has one real zero. Since f is degree 3, there must be two imaginary zeros.

5. The graph of $f(x)$ intersects the x-axis twice. Therefore, f has two real zeros. Since f is degree 4, there must be two imaginary zeros.

7. The graph of $f(x)$ intersects the x-axis three times. Therefore, f has three real zeros. Since f is degree 5, there must be two imaginary zeros.

9. Degree: 2; leading coefficient: 1; zeros: $6i$ and $-6i$

(a) $f(x) = (x - 6i)(x + 6i)$

(b) $(x - 6i)(x + 6i) = x^2 - 36i^2 = x^2 + 36$. Thus, $f(x) = x^2 + 36$.

11. Degree: 3; leading coefficient: –1; zeros: –1, $2i$, and $-2i$

   (a) $f(x) = -1(x + 1)(x - 2i)(x + 2i)$

   (b) $-1(x + 1)(x - 2i)(x + 2i) = -1(x + 1)(x^2 - 4i^2) = -1(x + 1)(x^2 + 4) =$

   $-(x^3 + 4x + x^2 + 4) = -x^3 - x^2 - 4x - 4$. Thus, $f(x) = -x^3 - x^2 - 4x - 4$.

13. Degree: 4; leading coefficient: 10; zeros: 1, –1, $3i$, and $-3i$

   (a) $f(x) = 10(x - 1)(x + 1)(x - 3i)(x + 3i)$

   (b) $10(x - 1)(x + 1)(x - 3i)(x + 3i) = 10(x^2 - 1)(x^2 + 9) = 10(x^4 + 8x^2 - 9) = 10x^4 + 80x^2 - 90$.

   Thus, $f(x) = 10x^4 + 80x^2 - 90$.

15. Degree: 4; leading coefficient: $\dfrac{1}{2}$; zeros: $-i$ and $2i$

   (a) Since $f(x)$ has real coefficients, it must also have a third and fourth zero of i and $-2i$ the conjugate of

   $-i$ and $2i$. Therefore the complete factored form is: $f(x) = \dfrac{1}{2}(x + i)(x - i)(x + 2i)(x - 2i)$

   (b) $\dfrac{1}{2}(x + i)(x - i)(x + 2i)(x - 2i) = \dfrac{1}{2}(x^2 - i^2)(x^2 - 4i^2) = \dfrac{1}{2}(x^2 + 1)(x^2 + 4) =$

   $\dfrac{1}{2}(x^4 + 4x^2 + x^2 + 4) = \dfrac{1}{2}x^4 + \dfrac{5}{2}x^2 + 2$. Thus, $f(x) = \dfrac{1}{2}x^4 + \dfrac{5}{2}x^2 + 2$.

17. Degree: 3; leading coefficient: -2; zeros: $1 - i$ and 3

   (a) Since $f(x)$ has real coefficients, it must also have a third zero of $1 + i$ the conjugate of

   $1 - i$. Therefore the complete factored form is: $f(x) = -2(x - (1 + i))(x - (1 - i))(x - 3)$

   (b) $-2(x - (1 + i))(x - (1 - i))(x - 3) = -2(x^2 - x + xi - x - xi + 1 - i^2)(x - 3) =$

   $-2(x^2 - 2x + 2)(x - 3) = -2(x^3 - 2x^2 + 2x - 3x^2 + 6x - 6) = -2(x^3 - 5x^2 + 8x - 6) =$

   $-2x^3 + 10x^2 - 16x + 12$. Thus, $f(x) = -2x^3 + 10x^2 - 16x + 12$.

19. First divide:

$$
\begin{array}{r}
3x^2 + 75 \\
x - \tfrac{5}{3} \overline{)\, 3x^3 - 5x^2 + 75x - 125} \\
\underline{3x^3 - 5x^2 } \\
75x - 125 \\
\underline{75x - 125} \\
0
\end{array}
$$

Now set $3x^2 + 75 = 0$ and solve. $3x^2 + 75 = 0 \Rightarrow 3x^2 = -75 \Rightarrow x^2 = -25 \Rightarrow x = \pm\sqrt{-25} \Rightarrow$

$x = \pm 5i$. The solutions are $x = \dfrac{5}{3}, \pm 5i$.

21. If $-3i$ is a zero then $3i$ is also a zero $\Rightarrow (x + 3i)(x - 3i) = x^2 - 9i^2 = x^2 + 9$. Now divide this into the

    equation:

$$
\begin{array}{r}
2x^2 - x + 1 \\
x^2 + 9\overline{)2x^4 - x^3 + 19x^2 - 9x + 9} \\
\underline{2x^4 \quad\quad + 18x^2} \\
-x^3 + \quad x^2 - 9x + 9 \\
\underline{-x^3 \quad\quad\quad - 9x} \\
x^2 \quad\quad + 9 \\
\underline{x^2 \quad\quad + 9} \\
0
\end{array}
$$

Then use the quadratic formula to solve $2x^2 - x + 1$. $\dfrac{1 \pm \sqrt{1 - 4(2)(1)}}{2(2)} = \dfrac{1 \pm \sqrt{-7}}{4} =$

$\dfrac{1}{4} \pm \dfrac{i\sqrt{7}}{4}$. Therefore the solutions are: $x = \pm 3i, \dfrac{1}{4} \pm \dfrac{i\sqrt{7}}{4}$.

23. $x^2 + 25 = 0 \Rightarrow x^2 = -25 \Rightarrow x = \pm 5i$. Thus, $f(x) = (x - 5i)(x + 5i)$.

25. $3x^3 + 3x = 0 \Rightarrow 3x(x^2 + 1) = 0 \Rightarrow x = 0$ or $\pm i$. Thus, $f(x) = 3(x - 0)(x - i)(x + i)$.

27. $x^4 + 5x^2 + 4 = 0 \Rightarrow (x^2 + 1)(x^2 + 4) = 0 \Rightarrow x = \pm i$ or $\pm 2i$.

    Thus, $f(x) = (x - i)(x + i)(x - 2i)(x + 2i)$.

29. The graph of $y = x^3 + 2x^2 + 16x + 32$ is shown in Figure 29. The x-intercept appears to be -2. We can use

    synthetic division to help factor f.

$$
\begin{array}{r}
-2|\quad 1 \quad 2 \quad 16 \quad 32 \\
\underline{\quad\quad -2 \quad 0 \; -32} \\
1 \quad 0 \quad 16 \quad 0
\end{array}
$$

$x^3 + 2x^2 + 16x + 32 = (x + 2)(x^2 + 16) = (x + 2)(x + 4i)(x - 4i)$.

Thus, $f(x) = (x + 2)(x + 4i)(x - 4i)$.

[–10, 10, 1] by [–50, 50, 10]

Figure 29

31. $x^3 + x = 0 \Rightarrow x(x^2 + 1) = 0 \Rightarrow x(x + i)(x - i) = 0 \Rightarrow x = 0, \pm i$

33. Factor or find the zero of 2 graphically.

    $x^3 = 2x^2 - 7x + 14 \Rightarrow x^3 - 2x^2 + 7x - 14 = 0 \Rightarrow x^2(x - 2) + 7(x - 2) = 0 \Rightarrow$

    $(x - 2)(x^2 + 7) = 0 \Rightarrow (x - 2)(x + i\sqrt{7})(x - i\sqrt{7}) = 0 \Rightarrow x = 2, \pm i\sqrt{7}$

35. $x^4 + 5x^2 = 0 \Rightarrow x^2(x^2 + 5) = x^2(x + i\sqrt{5})(x - i\sqrt{5}) = 0 \Rightarrow x = 0, \pm i\sqrt{5}$

37. $x^4 = x^3 - 4x^2 \Rightarrow x^4 - x^3 + 4x^2 = 0 \Rightarrow x^2(x^2 - x + 4) = 0$

    Use the quadratic formula to find the zeros of $x^2 - x + 4$. The solutions are $x = 0, \dfrac{1}{2} \pm \dfrac{i\sqrt{15}}{2}$.

39. Find the zeros of 1 and 2 using the rational zero test or a graph. See Figures 39a & 39b.

$$x^4 + x^3 = 16 - 8x - 6x^2 \implies x^4 + x^3 + 6x^2 + 8x - 16 = 0$$

The graph of $x^4 + x^3 + 6x^2 + 8x - 16$ shows that -2 and 1 are zeros, so $(x + 2)$ and $(x - 1)$ are factors; using synthetic division twice gives the missing factor.

$$\begin{array}{r|rrrrr} -2 & 1 & 1 & 6 & 8 & -16 \\ & & -2 & 2 & -16 & 16 \\ \hline & 1 & -1 & 8 & -8 & 0 \end{array}$$

$$\begin{array}{r|rrrr} 1 & 1 & -1 & 8 & -8 \\ & & 1 & 0 & 8 \\ \hline & 1 & 0 & 8 & 0 \end{array}$$

$$x^4 + x^3 + 6x^2 + 8x - 16 = (x + 2)(x - 1)(x^2 + 8) = (x + 2)(x - 1)(x + i\sqrt{8})(x - i\sqrt{8})$$

The solutions are $x = -2, 1, \pm i\sqrt{8}$.

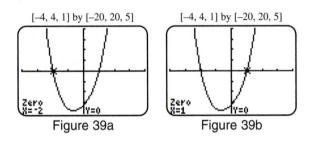

[−4, 4, 1] by [−20, 20, 5]   [−4, 4, 1] by [−20, 20, 5]

Figure 39a   Figure 39b

41. Find the zero of -2 using the rational zero test or a graph of $y = 3x^3 + 4x^2 - x + 6$. Use synthetic division to factor $3x^3 + 4x^2 - x + 6$.

$$\begin{array}{r|rrrr} -2 & 3 & 4 & -1 & 6 \\ & & -6 & 4 & -6 \\ \hline & 3 & -2 & 3 & 0 \end{array}$$

$$3x^3 + 4x^2 - x + 6 = (x + 2)(3x^2 - 2x + 3)$$

Use the quadratic formula to find the zeros of $3x^2 - 2x + 3$. The solutions are $x = -2, \dfrac{1}{3} \pm \dfrac{i\sqrt{8}}{3}$.

43. $Z = \dfrac{V}{I} = \dfrac{50 + 98i}{8 + 5i} = 10 + 6i$

45. $V = IZ = (1 + 2i)(3 - 4i) = 11 + 2i$

47. $I = \dfrac{V}{Z} = \dfrac{27 + 17i}{22 - 5i} = 1 + i$

4.6: Rational Functions and Models

1. Yes, since the numerator and denominator are both polynomials. Since $4x - 5 \neq 0$, $D = \left\{ x \mid x \neq \dfrac{5}{4} \right\}$.

3. Yes, since f can be written as $f(x) = \dfrac{x^2 - x - 2}{1}$, and $g(x) = 1$ is a polynomial. $D =$ all real numbers.

5. No, since the numerator is not a polynomial. $D = \{ x \mid x \neq -1 \}$

7. Yes, since the numerator and denominator are both polynomials. Since $x^2 + 1 \neq 0$, $D =$ all real numbers.

9. No, since the numerator is not a polynomial. Since $x^2 + x \neq 0 \Rightarrow x(x + 1) \neq 0, D = \{x \mid x \neq -1, x \neq 0\}$.

11. Yes, since f can be written as $f(x) = \dfrac{4(x + 1) - 3}{x + 1} \Rightarrow f(x) = \dfrac{4x + 1}{x + 1}$.

   Since $x + 1 \neq 0, D = \{x \mid x \neq -1\}$.

13. There is a horizontal asymptote of $y = 4$ and a vertical asymptote of $x = 2$; $D = \{x \mid x \neq 2\}$.

15. There is a horizontal asymptote of $y = -4$ and a vertical asymptote of $x = \pm 2$; $D = \{x \mid x \neq \pm 2\}$.

17. There is a horizontal asymptote of $y = 0$ and there is no vertical asymptote; $D =$ all real numbers.

19. Since the output of Y_1 gets closer to 3 as the input x gets larger, it is reasonable to conjecture that the equation for the horizontal asymptote is $y = 3$.

21. Horizontal Asymptotes: The degree of the numerator is equal to the degree of the denominator and the ratio of the leading coefficients is $\dfrac{4}{2} = 2$. Therefore, $y = 2$ is a horizontal asymptote.

   Vertical Asymptotes: Find the zeros of the denominator, $2x - 6 = 0 \Rightarrow 2x = 6 \Rightarrow x = 3$. Therefore $x = 3$ is a vertical asymptote. (Note that 3 is not a zero of the numerator.)

23. Horizontal Asymptotes: The degree of the numerator is less than the degree of the denominator, therefore, the x-axis, $y = 0$, is a horizontal asymptote.

   Vertical Asymptotes: Find the zeros of the denominator, $x^2 - 5 = 0 \Rightarrow x^2 = 5 \Rightarrow x = \pm \sqrt{5}$. Therefore $x = \pm \sqrt{5}$ are vertical asymptotes. (Note that $\pm \sqrt{5}$ are not zeros of the numerator.)

25. Horizontal Asymptotes: The degree of the numerator is greater than the degree of the denominator, therefore there is no horizontal asymptote.

   Vertical Asymptotes: Find the zeros of the denominator,

   $x^2 + 3x - 10 = 0 \Rightarrow (x + 5)(x - 2) = 0 \Rightarrow x = -5, 2$. Therefore $x = -5$ and $x = 2$ are the vertical asymptotes. (Note that -5 and 2 are not a zeros of the numerator.)

27. Horizontal Asymptotes: The degree of the numerator is equal to the degree of the denominator and the ratio of the leading coefficients is $\dfrac{1}{2}$. Therefore, $y = \dfrac{1}{2}$ is a horizontal asymptote.

   Vertical Asymptotes: Find the zeros of the denominator, $(2x - 5)(x + 1) = 0 \Rightarrow x = \dfrac{5}{2}$ and $x = -1$.

   Therefore $x = \dfrac{5}{2}$ is a vertical asymptote. (Note that $\dfrac{5}{2}$ is not a zero of the numerator and $x = -1$ is a zero of the numerator).

29. Horizontal Asymptotes: The degree of the numerator is equal to the degree of the denominator and the ratio of the leading coefficients is $\dfrac{3}{1} = 3$. Therefore, $y = 3$ is a horizontal asymptote.

   Vertical Asymptotes: Find the zeros of the denominator, $(x + 2)(x - 1) = 0 \Rightarrow x = -2, 1$. Here, only $x = 1$ is a vertical asymptote. (Note that -2 is a zero of the numerator.)

31. Horizontal Asymptotes: The degree of the numerator is greater than the degree of the denominator, therefore there is no horizontal asymptote.

Vertical Asymptotes: Find the zeros of the denominator, $x + 3 = 0 \Rightarrow x = -3$. But $x = -3$ is not a vertical asymptote since -3 is a zero of the numerator and $f(x) = x - 3$ for $x \neq -3$. There are no vertical asymptotes.

33. $f(x) = \dfrac{a}{x - 1}$ has a vertical asymptote of $x = 1$. It has a horizontal asymptote of $y = 0$, since the degree of the numerator is less than the degree of the denominator. The best choice is graph b.

35. $f(x) = \dfrac{x - a}{x + 2}$ has a vertical asymptote of $x = -2$ and a horizontal asymptote of $y = 1$, since the degree of the numerator equals the degree of the denominator and the ratio of the leading coefficients is $\dfrac{1}{1}$.

The best choice is graph d.

37. One example of a symbolic representation of a rational function with a vertical asymptote of $x = -3$ and a horizontal asymptote of $y = 1$ is $f(x) = \dfrac{x + 1}{x + 3}$. *Answers may vary.*

39. One example of a symbolic representation of a rational function with vertical asymptotes of $x = \pm 3$ and a horizontal asymptote of $y = 0$ is $f(x) = \dfrac{1}{x^2 - 9}$. *Answers may vary.*

41. See Figure 41. Since the degree of the numerator is less than the degree of the denominator, the horizontal asymptote is $y = 0$. To find the vertical asymptote we find the zero of the denominator, $x^2 = 0 \Rightarrow x = 0$.

43. See Figure 43. Since the degree of the numerator is less than the degree of the denominator, the horizontal asymptote is $y = 0$. To find the vertical asymptote we find the zero of the denominator, $2x = 0 \Rightarrow x = 0$.

Figure 41   Figure 43   Figure 45

45. $g(x) = \dfrac{1}{x - 3}$ is $y = \dfrac{1}{x}$ transformed 3 units right $\Rightarrow$ it has a vertical asymptote of $x = 3$. It still has a horizontal asymptote of $y = 0$. See Figure 45. Then $g(x) = f(x - 3)$.

47. $g(x) = \dfrac{1}{x} + 2$ is $y = \dfrac{1}{x}$ transformed 2 units up $\Rightarrow$ it has a horizontal asymptote of $y = 2$. It still has a vertical asymptote of $x = 0$. See Figure 47. Then $g(x) = f(x) + 2$.

49. $g(x) = \dfrac{1}{x + 1} - 2$ is $y = \dfrac{1}{x}$ transformed 1 unit left and 2 units down. It now has a vertical asymptote of $x = -1$, and a horizontal asymptote of $y = -2$. See Figure 49. Then $g(x) = f(x + 1) - 2$.

Figure 47   Figure 49   Figure 51

51. $g(x) = -\dfrac{2}{(x-1)^2}$ is $y = \dfrac{1}{x^2}$ transformed by reflecting it across the x-axis, then shifting 1 unit right, and

vertically stretching by a factor of 2. It now has a vertical asymptote of $x = 1$, and it still has a horizontal

asymptote of $y = 0$. See Figure 51. Then $g(x) = -2h(x-1)$.

53. $g(x) = \dfrac{1}{(x+1)^2} - 2$ is $y = \dfrac{1}{x^2}$ transformed 1 unit left and 2 units down. It now has a vertical asymptote of

$x = -1$, and a horizontal asymptote of $y = -2$. See Figure 53. Then $g(x) = h(x+1) - 2$.

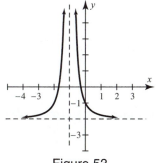

Figure 53

55. (a) $x - 2 = 0 \Rightarrow x = 2; D = \{x \mid x \neq 2\}$

(b) The graph of f using dot mode is shown in Figure 55b.

(c) Since the degree of the numerator equals the degree of the denominator and the ratio of the leading

coefficients is $\dfrac{1}{1} = 1$, the horizontal asymptote is $y = 1$. There is a vertical asymptote at $x = 2$.

(d) First, sketch the vertical and horizontal asymptotes found in part (c). Then use Figure 55b as a guide to a

more complete graph of f. The sketch is shown in Figure 55d.

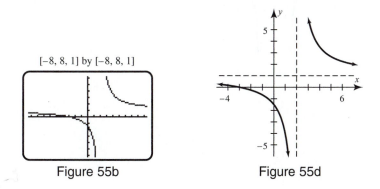

[−8, 8, 1] by [−8, 8, 1]

Figure 55b   Figure 55d

57. (a) $x^2 - 4 = 0 \Rightarrow x^2 = 4 \Rightarrow x = \pm 2; D = \{x \mid x \neq \pm 2\}$

    (b) The graph of f using dot mode is shown in Figure 57b.

    (c) Since the degree of the numerator is less than the degree of the denominator there is a horizontal asymptote at $y = 0$. There are vertical asymptotes at $x = \pm 2$.

    (d) First, sketch the vertical and horizontal asymptotes found in part (c). Then use Figure 57b as a guide to a more complete graph of f. The sketch is shown in Figure 57d.

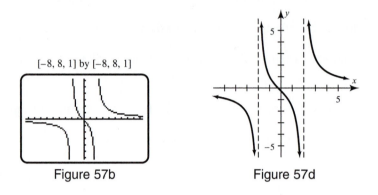

[−8, 8, 1] by [−8, 8, 1]

Figure 57b   Figure 57d

59. (a) $1 - 0.25x^2 = 0 \Rightarrow 0.25x^2 = 1 \Rightarrow x^2 = 4 \Rightarrow x = \pm 2; D = \{x \mid x \neq \pm 2\}$

    (b) The graph of f using dot mode is shown in Figure 59b.

    (c) Since the degree of the numerator is less than the degree of the denominator there is a horizontal asymptote at $y = 0$. There are vertical asymptotes at $x = \pm 2$.

    (d) First, sketch the vertical and horizontal asymptotes found in part (c). Then use Figure 59b as a guide to a more complete graph of f. The sketch is shown in Figure 59d.

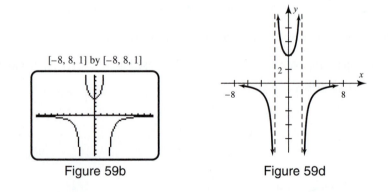

[−8, 8, 1] by [−8, 8, 1]

Figure 59b   Figure 59d

61. (a) $x - 2 = 0 \Rightarrow x = 2; D = \{x \mid x \neq 2\}$

    (b) The graph of f using dot mode is shown in Figure 61b.

    (c) Since the degree of the numerator is greater than the degree of the denominator there is no horizontal
    asymptote. There are no vertical asymptotes. (The numerator equals 2 when $x = 2$).

    (d) First, sketch the vertical and horizontal asymptotes found in part (c). Then use Figure 61b as a guide to a
    more complete graph of f. The sketch is shown in Figure 61d.

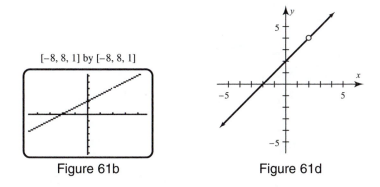

$[-8, 8, 1]$ by $[-8, 8, 1]$

Figure 61b   Figure 61d

63. First divide:

$$\begin{array}{r} x - 1 \\ x - 1 \overline{)x^2 - 2x + 1} \\ \underline{x^2 - x} \\ -x + 1 \\ \underline{-x + 1} \\ 0 \end{array}$$

Now graph $x - 1$; since the denominator $x - 1 \neq 0$ there is a hole at $x = 1$ in the graph. See Figure 63.

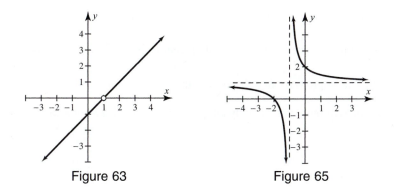

Figure 63   Figure 65

65. First we find the asymptotes. Since the degree of the numerator and denominator are the same, the ratio of the
leading coefficients $\frac{1}{1}$ or $y = 1$ is the horizontal asymptote. To find the vertical asymptote we find the zero of
the denominator. $x + 1 = 0 \Rightarrow x = -1$. Since $x = -1$ is a vertical asymptote it has no holes.

See Figure 65.

67. First divide: $g(x) = \dfrac{(2x + 1)(x - 2)}{(x - 2)(x - 2)} = \dfrac{2x + 1}{x - 2}$. Now graph $\dfrac{2x + 1}{x - 2}$ by finding it's asymptotes. The ratio

of the leading coefficients $\dfrac{2}{1}$ or $y = 2$ is the horizontal asymptote. To find the vertical asymptote we find the

zero of the denominator. $x - 2 = 0 \Rightarrow x = 2$. Since $x = 2$ is an asymptote there are no holes.

See Figure 67.

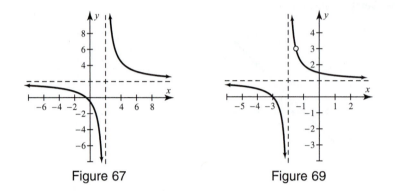

Figure 67   Figure 69

69. Factor the numerator and denominator, then divide: $f(x) = \dfrac{2x^2 + 9x + 9}{2x^2 + 7x + 6} = \dfrac{(2x + 3)(x + 3)}{(2x + 3)(x + 2)} = \dfrac{x + 3}{x + 2}$.

Now graph $\dfrac{x + 3}{x + 2}$ by finding it's asymptotes. The ratio of the leading coefficients $\dfrac{1}{1}$ or $y = 1$ is the horizontal

asymptote. To find the vertical asymptote we find the zero of the denominator. $x + 2 = 0 \Rightarrow x = -2$.

Since $x = -2$ is an asymptote there is only a hole is at $2x + 3 = 0$ or $x = \dfrac{-3}{2}$. See Figure 69.

71. Factor the numerator and denominator, then divide:

$$f(x) = \dfrac{-2x^2 + 11x - 14}{x^2 - 5x + 6} = \dfrac{(-2x + 7)(x - 2)}{(x - 3)(x - 2)} = \dfrac{-2x + 7}{x - 3}.$$

Now graph $\dfrac{-2x + 7}{x - 3}$ by finding it's asymptotes. The ratio of the leading coefficients $\dfrac{-2}{1}$ or $y = -2$ is the

horizontal asymptote. To find the vertical asymptote we find the zero of the denominator.

$x - 3 = 0 \Rightarrow x = 3$. Since $x = 3$ is an asymptote there is only a hole is at $x = 2$. See Figure 71.

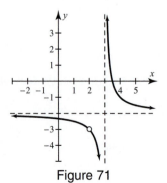

Figure 71

73. There is a vertical asymptote at $x = -1$ since -1 is a zero of the denominator but is not a zero of the numerator.

To find any slant asymptote we will divide the numerator by the denominator using synthetic division:

$$\begin{array}{r|rrr} -1 & 1 & 0 & 1 \\ & & -1 & 1 \\ \hline & 1 & -1 & 2 \end{array}$$

Therefore, $f(x) = x - 1 + \dfrac{2}{x + 1}$ and as $|x|$ becomes large, $f(x)$ approaches $y = x - 1$. There is a slant

asymptote at $y = x - 1$. The graph of f using dot mode is shown in Figure 73a. To sketch this graph, first

sketch the vertical and slant asymptotes found above. Then use Figure 73a as a guide to a more complete graph

of f. The sketch is shown in Figure 73b.

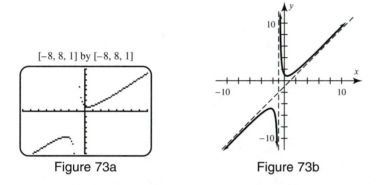

[−8, 8, 1] by [−8, 8, 1]

Figure 73a   Figure 73b

75. There is a vertical asymptote at $x = -2$ since -2 is a zero of the denominator but is not a zero of the numerator.

To find any slant asymptote we will divide the numerator by the denominator using synthetic division:

$$\begin{array}{r|rrr} -2 & 0.5 & -2 & 2 \\ & & -1 & 6 \\ \hline & 0.5 & -3 & 8 \end{array}$$

Therefore, $f(x) = 0.5x - 3 + \dfrac{8}{x + 2}$ and as $|x|$ becomes large, $f(x)$ approaches $y = 0.5x - 3$. There is a

slant asymptote at $y = 0.5x - 3$. The graph of f using dot mode is shown in Figure 75a. To sketch this graph,

first sketch the vertical and slant asymptotes found above. Then use Figure 75a as a guide to a more complete

graph of f. The sketch is shown in Figure 75b.

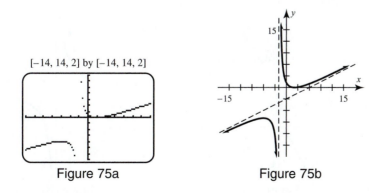

[−14, 14, 2] by [−14, 14, 2]

Figure 75a   Figure 75b

77. There is a vertical asymptote at $x = 1$ since 1 is a zero of the denominator but is not a zero of the numerator.

To find any slant asymptote we will divide the numerator by the denominator using synthetic division:

$$
\begin{array}{r|rrr}
1 & 1 & 2 & 1 \\
  &   & 1 & 3 \\
\hline
  & 1 & 3 & 4
\end{array}
$$

Therefore, $f(x) = x + 3 + \dfrac{4}{x-1}$ and as $|x|$ becomes large, $f(x)$ approaches $y = x + 3$. There is a slant asymptote at $y = x + 3$. The graph of f using dot mode is shown in Figure 77a. To sketch this graph, first sketch the vertical and slant asymptotes found above. Then use Figure 77a as a guide to a more complete graph of f. The sketch is shown in Figure 77b.

[−12, 12, 2] by [−20, 20, 4]

Figure 77a   Figure 77b

79. There is a vertical asymptote at $x = \dfrac{1}{2}$ since $\dfrac{1}{2}$ is a zero of the denominator but is not a zero of the numerator. To find any slant asymptote we will divide the numerator by the denominator.

$$
\begin{array}{r}
2x + 1 + \frac{1}{2x-1} \\
2x - 1 \overline{)\, 4x^2 + 0x + 0} \\
\underline{4x^2 - 2x} \\
2x + 0 \\
\underline{2x - 1} \\
1
\end{array}
$$

Therefore, $f(x) = 2x + 1 + \dfrac{1}{2x-1}$ and as $|x|$ becomes large, $f(x)$ approaches $y = 2x + 1$. There is a slant asymptote at $y = 2x + 1$. The graph of f using dot mode is shown in Figure 79a. To sketch this graph, first sketch the vertical and slant asymptotes found above. Then use Figure 79a as a guide to a more complete graph of f. The sketch is shown in Figure 79b.

[−3, 7, 1] by [−3, 7, 1]

Figure 79a   Figure 79b   Figure 87

81. $\dfrac{4}{x+2} = -4 \Rightarrow 4 = -4(x+2) \Rightarrow 4 = -4x - 8 \Rightarrow 12 = -4x \Rightarrow x = -3$

83. $\dfrac{x+1}{x} = 2 \Rightarrow x + 1 = 2x \Rightarrow x = 1$

85. $\dfrac{1-x}{3x-1} = -\dfrac{3}{5} \Rightarrow -3(3x-1) = 5(1-x) \Rightarrow -9x + 3 = 5 - 5x \Rightarrow -2 = 4x \Rightarrow x = -\dfrac{1}{2}$

87. $f(x) = \dfrac{2x-4}{x-1}$ is in lowest terms.

 1) To find the vertical asymptote we find the zero of the denominator, $x - 1 = 0 \Rightarrow x = 1$.

 2) Since the degree of the numerator is equal to the degree of the denominator the equation of the horizontal

 asymptote is $y = \dfrac{2}{1} = 2$.

 3) $f(0) = \dfrac{2(0)-4}{0-1} = 4;\ (0,4)$

 4) $\dfrac{2x-4}{x-1} = 0 \Rightarrow 2x - 4 = 0 \Rightarrow 2x = 4 \Rightarrow x = 2;\ (2,0)$

 5) $\dfrac{2x-4}{x-1} = 2 \Rightarrow 2x - 4 = 2(x-1) \Rightarrow 2x - 4 = 2x - 2 \Rightarrow 0 = 2 \Rightarrow f$ does not cross the horizontal

 asymptote.

 6) See Figure 87.

 7) See Figure 87.

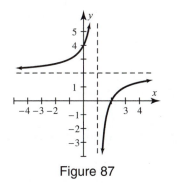

Figure 87

89. $f(x) = \dfrac{x^2 - 2x}{x^2 + 6x + 9}$ is in lowest terms.

1) To find the vertical asymptote we find the zero of the denominator, $x^2 + 6x + 9 = 0 \Rightarrow$

   $(x + 3)^2 = 0 \Rightarrow x + 3 = 0 \Rightarrow x = -3.$

2) Since the degree of the numerator is equal to the degree of the denominator the equation of the horizontal

   asymptote is $y = \dfrac{1}{1} = 1.$

3) $f(0) = \dfrac{0^2 - 2(0)}{0^2 + 6(0) + 9} = \dfrac{0}{9} = 0;\ (0, 0)$

4) $\dfrac{x^2 - 2x}{x^2 + 6x + 9} = 0 \Rightarrow x^2 - 2x = 0 \Rightarrow x(x - 2) = 0 \Rightarrow x = 0, x = 2;\ (0, 0), (2, 0)$

5) $\dfrac{x^2 - 2x}{x^2 + 6x + 9} = 1 \Rightarrow x^2 - 2x = x^2 + 6x + 9 \Rightarrow -8x = 9 \Rightarrow x = -\dfrac{9}{8} \Rightarrow f$ crosses the horizontal

   asymptote at $\left(-\dfrac{9}{8}, 1\right)$

6) See Figure 89.

7) See Figure 89.

Figure 89   Figure 91

91. $f(x) = \dfrac{x^2 + 2x + 1}{x^2 - x - 6}$ is in lowest terms.

1) To find the vertical asymptotes we find the zeros of the denominator, $x^2 - x - 6 = 0 \Rightarrow$

   $(x - 3)(x + 2) = 0 \Rightarrow x = 3, -2.$

2) Since the degree of the numerator is equal to the degree of the denominator the equation of the horizontal

   asymptote is $y = \dfrac{1}{1} = 1.$

3) $f(0) = \dfrac{0^2 + 2(0) + 1}{0^2 - 0 - 6} = -\dfrac{1}{6} = 0;\ \left(0, -\dfrac{1}{6}\right)$

4) $\dfrac{x^2 + 2x + 1}{x^2 - x - 6} = 0 \Rightarrow x^2 + 2x + 1 = 0 \Rightarrow (x + 1)^2 = 0 \Rightarrow x = -1;\ (-1, 0)$

5) $\dfrac{x^2 + 2x + 1}{x^2 - x - 6} = 1 \Rightarrow x^2 + 2x + 1 = x^2 - x - 6 \Rightarrow 3x = -7 \Rightarrow x = -\dfrac{7}{3} \Rightarrow f$ crosses the

   horizontal asymptote at $-\dfrac{7}{3}.$

6) See Figure 91.

7) See Figure 91.

93. (a) $T(4) = -\dfrac{1}{4-8} \Rightarrow T(4) = \dfrac{1}{4} \Rightarrow T(4) = 0.25$; when vehicles leave the ramp at an average rate

    of 4 vehicles per minute, the wait is 0.25 minutes or 15 seconds.

    $T(7.5) = \dfrac{1}{7.5-8} \Rightarrow T(7.5) = \dfrac{1}{0.5} \Rightarrow T(7.5) = 2.0$; when vehicles leave the ramp at an average rate of

    7.5 vehicles per minute, the wait is 2 minutes.

    (b) The wait increases dramatically.

95. (a) $N(20) = \dfrac{20^2}{1600-40(20)} = \dfrac{400}{800} = \dfrac{1}{2}$; $N(39) = \dfrac{39^2}{1600-40(39)} = \dfrac{1521}{40} = 38.025$

    (b) The wait increases dramatically.

    (c) We find the vertical asymptote by finding the zero of the denominator,

    $1600 - 40x = 0 \Rightarrow 1600 = 40x \Rightarrow x = 40$.

97. (a) Graph $Y_1 = (10X + 1)/(X + 1)$ and $Y_2 = 10$ in [0, 14, 1] by [0, 14, 1]. Since the degree of the

    numerator and denominator are equal, there is a horizontal asymptote at $y = \dfrac{10}{1} = 10$. See Figure 97.

    (b) The initial population would be $f(0) = 1$ million insects.

    (c) After many months the population starts to level off at 10 million.

    (d) The horizontal asymptote $y = 10$ represents the limiting population after a very long time.

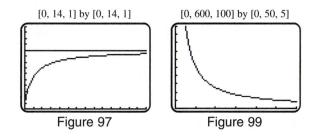

[0, 14, 1] by [0, 14, 1]   [0, 600, 100] by [0, 50, 5]

Figure 97   Figure 99

99. (a) $f(x) = \dfrac{2540}{x} \Rightarrow f(400) = \dfrac{2540}{400} = 6.35$ inches. A curve designed for 60 mph with a radius of 400 feet

    should have the outer rail elevated 6.35 inches.

    (b) Figure 99 shows the graph of $Y_1 = 2540/X$. This means that a sharper curve must be banked more. As

    the radius increases, the curve is not as sharp and the elevation of the outer rail decreases.

    (c) Since the degree of the numerator is less than the degree of the denominator, the graph of $f(x) = \dfrac{2540}{x}$

    has a horizontal asymptote of $y = 0$. As the radius of the curve increases without bound $(x \to \infty)$, the

    tracks become straight and no elevation or banking $(y \to 0)$ of the outer rail is necessary.

    (d) $f(x) = 12.7 \Rightarrow 12.7 = \dfrac{2540}{x} \Rightarrow x = \dfrac{2540}{12.7} = 200$; a radius of 200 feet requires an elevation of 12.7

    inches.

101. (a) At $x = 0$, the denominator of $\dfrac{2500}{30x}$ is equal to zero, whereas the numerator is nonzero. There is a vertical

asymptote at $x = 0$. As the coefficient of friction x becomes smaller and approaches 0, the stopping

distance becomes larger and larger without bound. This means that as the road surface becomes very

slippery, the distance required to stop becomes very large. If it were possible for the roadway to be

covered with glare ice having a coefficient of friction of $x = 0$, then the car would continue indefinitely.

A situation similar to this occurs in space where a satellite will travel indefinitely without slowing down.

(b) $340 = \dfrac{2500}{30x} \Rightarrow x = \dfrac{2500}{30(340)} \Rightarrow x = \dfrac{2500}{10{,}200} = \dfrac{25}{102}$

The coefficient of friction associated with a braking distance of 340 feet is $\dfrac{25}{102} \approx 0.245$.

4.7: More Equations and Inequalities

1. (a) $\dfrac{2x}{x + 2} = 6 \Rightarrow 2x = 6(x + 2) \Rightarrow 2x = 6x + 12 \Rightarrow 4x = -12 \Rightarrow x = -3$

(b) Graph $Y_1 = (2X)/(X + 2)$ and $Y_2 = 6$ in $[-10, 10, 1]$ by $[-10, 10, 1]$ using dot mode. See Figure 1b.

The intersection point is $(-3, 6)$. The solution is $x = -3$.

(c) Table $Y_1 = (2X)/(X + 2)$ starting at $x = -5$ and incrementing by 1. See Figure 1c. The solution is

$x = -3$.

$[-10, 10, 1]$ by $[-10, 10, 1]$

Figure 1b   Figure 1c

3. (a) $2 - \dfrac{5}{x} + \dfrac{2}{x^2} = 0 \Rightarrow \dfrac{2x^2}{x^2} - \dfrac{5x}{x^2} + \dfrac{2}{x^2} = 0 \Rightarrow \dfrac{2x^2 - 5x + 2}{x^2} = 0 \Rightarrow 2x^2 - 5x + 2 = 0 \Rightarrow$

$(2x - 1)(x - 2) = 0 \Rightarrow x = \dfrac{1}{2}, 2$

(b) Graph $Y_1 = 2 - 5/X + 2/X^2$ in $[-10, 10, 1]$ by $[-10, 10, 1]$ using dot mode. The x-intercepts are the

points is $(0.5, 0)$ and $(2, 0)$. The solutions are $x = \dfrac{1}{2}, 2$.

(c) Table $Y_1 = 2 - 5/X + 2/X^2$ starting at $x = -0.5$ and incrementing by 0.5. The solutions are $x = \dfrac{1}{2}, 2$.

5. (a) $\dfrac{1}{x+1} + \dfrac{1}{x-1} = \dfrac{1}{x^2-1} \Rightarrow \dfrac{x-1}{x^2-1} + \dfrac{x+1}{x^2-1} = \dfrac{1}{x^2-1} \Rightarrow \dfrac{2x}{x^2-1} = \dfrac{1}{x^2-1} \Rightarrow 2x = 1 \Rightarrow$

$x = \dfrac{1}{2}$

(b) Graph $Y_1 = 1/(X+1) + 1/(X-1)$ and $Y_2 = 1/(X^2-1)$ in $[-2, 2, 1]$ by $[-5, 5, 1]$ using dot mode.

The intersection point is $\left(\dfrac{1}{2}, -\dfrac{4}{3}\right)$. The solution is $x = \dfrac{1}{2}$.

(c) Table $Y_1 = 1/(X+1) + 1/(X-1) - 1/(X^2-1)$ starting at $x = -1$ and incrementing by 0.5. Find the

x-value where $Y_1 = 0$. The solution is $x = \dfrac{1}{2}$.

7. $\dfrac{x+1}{x-5} = 0 \Rightarrow x+1 = 0(x-5) \Rightarrow x+1 = 0 \Rightarrow x = -1$; Check: $\dfrac{-1+1}{-1-5} = \dfrac{0}{-6} = 0$

9. $\dfrac{6(1-2x)}{x-5} = 4 \Rightarrow 6(1-2x) = 4(x-5) \Rightarrow 6-12x = 4x - 20 \Rightarrow -16x = -26 \Rightarrow x = \dfrac{13}{8}$

Check: $\dfrac{6(1-2(\frac{13}{8}))}{\frac{13}{8}-5} = \dfrac{6(-\frac{9}{4})}{-\frac{27}{8}} = \dfrac{-\frac{27}{2}}{-\frac{27}{8}} = -\dfrac{27}{2} \cdot \left(-\dfrac{8}{27}\right) = 4$

11. $\dfrac{1}{x+2} + \dfrac{1}{x} = 1 \Rightarrow x + (x+2) = x(x+2) \Rightarrow 2x + 2 = x^2 + 2x \Rightarrow x^2 = 2 \Rightarrow x = \pm\sqrt{2}$

Check: $\dfrac{1}{\sqrt{2}+2} + \dfrac{1}{\sqrt{2}} = \dfrac{\sqrt{2}}{\sqrt{2}(\sqrt{2}+2)} + \dfrac{\sqrt{2}+2}{\sqrt{2}(\sqrt{2}+2)} = \dfrac{2\sqrt{2}+2}{2+2\sqrt{2}} = 1$

Check: $\dfrac{1}{-\sqrt{2}+2} + \dfrac{1}{\sqrt{2}} = \dfrac{-\sqrt{2}}{-\sqrt{2}(-\sqrt{2}+2)} + \dfrac{-\sqrt{2}+2}{-\sqrt{2}(-\sqrt{2}+2)} = \dfrac{2-2\sqrt{2}}{2-2\sqrt{2}} = 1$

13. $\dfrac{1}{x} - \dfrac{2}{x^2} = 5 \Rightarrow x - 2 = 5x^2 \Rightarrow 5x^2 - x + 2 = 0$; This quadratic equation has no solutions since the

discriminant is negative: $(-1)^2 - 4(5)(2) = 1 - 40 = -39 < 0$. No real solution.

15. $\dfrac{x^3 - 4x}{x^2+1} = 0 \Rightarrow x^3 - 4x = 0(x^2+1) \Rightarrow x^3 - 4x = 0 \Rightarrow x(x+2)(x-2) = 0 \Rightarrow x = 0, -2, 2$

Check: $\dfrac{(0)^3 - 4(0)}{(0)^2 + 1} = \dfrac{0}{1} = 0$; Check: $\dfrac{(-2)^3 - 4(-2)}{(-2)^2 + 1} = \dfrac{0}{5} = 0$; Check: $\dfrac{(2)^3 - 4(2)}{(2)^2 + 1} = \dfrac{0}{5} = 0$

17. $\dfrac{35}{x^2} = \dfrac{4}{x} + 15 \Rightarrow 35 = 4x + 15x^2 \Rightarrow 15x^2 + 4x - 35 = 0 \Rightarrow (5x-7)(3x+5) = 0 \Rightarrow x = \dfrac{-5}{3}, \dfrac{7}{5}$

Check: $\dfrac{35}{(\frac{-5}{3})^2} = \dfrac{4}{\frac{-5}{3}} + 15 \Rightarrow \dfrac{63}{5} = \dfrac{-12}{5} + \dfrac{75}{5} \Rightarrow \dfrac{63}{5} = \dfrac{63}{5}$;

Check: $\dfrac{35}{(\frac{7}{5})^2} = \dfrac{4}{\frac{7}{5}} + 15 \Rightarrow \dfrac{125}{7} = \dfrac{20}{7} + \dfrac{105}{7} \Rightarrow \dfrac{125}{7} = \dfrac{125}{7}$

19. $\dfrac{x+5}{x+2} = \dfrac{x-4}{x-10} \Rightarrow (x+5)(x-10) = (x-4)(x+2) \Rightarrow x^2 - 5x - 50 = x^2 - 2x - 8 \Rightarrow$

$3x + 42 = 0 \Rightarrow x = -14$; Check: $\dfrac{(-14)+5}{(-14)+2} = \dfrac{(-14)-4}{(-14)-10} \Rightarrow \dfrac{-9}{-12} = \dfrac{-18}{-24} \Rightarrow \dfrac{3}{4} = \dfrac{3}{4}$

21. $\dfrac{1}{x-2} - \dfrac{2}{x-3} = \dfrac{-1}{x^2 - 5x + 6} \Rightarrow \dfrac{1}{x-2} - \dfrac{2}{x-3} = \dfrac{-1}{(x-2)(x-3)} \Rightarrow$

$x - 3 - 2(x-2) = -1 \Rightarrow -x + 1 = -1 \Rightarrow x = 2$; No solution. Check: There is no solution since

$x = 2$ is not definrd in the original equation.

23. $\dfrac{2}{x-1} + 1 = \dfrac{4}{x^2-1} \Rightarrow \dfrac{2}{x-1} + 1 = \dfrac{4}{(x+1)(x-1)} \Rightarrow 2(x+1) + (x+1)(x-1) = 4 \Rightarrow$

$2x + 2 + x^2 - 1 = 4 \Rightarrow x^2 + 2x - 3 = 0 \Rightarrow (x+3)(x-1) = 0 \Rightarrow x = -3, 1.$ Since $x = 1$ is not

defined in the original equation $x = -3.$ Check: $\dfrac{2}{-3-1} + 1 = \dfrac{4}{(-3)^2-1} \Rightarrow -\dfrac{1}{2} + 1 = \dfrac{4}{8} \Rightarrow \dfrac{1}{2} = \dfrac{1}{2}$

25. $\dfrac{1}{x+2} = \dfrac{4}{4-x^2} - 1 \Rightarrow \dfrac{1}{x+2} = \dfrac{4}{(2+x)(2-x)} - 1 \Rightarrow 2 - x = 4 - (2+x)(2-x) \Rightarrow$

$2 - x = 4 - (4 - x^2) \Rightarrow 2 - x = 4 - 4 + x^2 \Rightarrow x^2 + x - 2 = 0 \Rightarrow (x+2)(x-1) = 0 \Rightarrow$

$x = -2, 1.$ Since $x = -2$ is not defined in the original equation $x = 1.$

Check: $\dfrac{1}{1+2} = \dfrac{4}{4-(1)^2} - 1 \Rightarrow \dfrac{1}{3} = \dfrac{4}{3} - 1 \Rightarrow \dfrac{1}{3} = \dfrac{1}{3}$

27. $\dfrac{1}{x-1} + \dfrac{1}{x+1} = \dfrac{2}{x^2-1} \Rightarrow \dfrac{1}{x-1} + \dfrac{1}{x+1} = \dfrac{2}{(x+1)(x-1)} \Rightarrow x + 1 + x - 1 = 2 \Rightarrow$

$2x = 2 \Rightarrow x = 1.$ Since $x = 1$ is not defined in the original equation, there is no solution to the equation.

29. (a) The boundary numbers, the x-values for which $f(x) = 0$, are -4, -2, and 2.

    (b) $f(x) > 0$ on the interval for which the graph of f is above the x-axis. That is $(-4, -2) \cup (2, \infty)$.

    In set builder notation the intervals are $\{x \mid -4 < x < -2 \text{ or } x > 2\}$.

    (c) $f(x) < 0$ on the interval for which the graph of f is below the x-axis. That is $(-\infty, -4) \cup (-2, 2)$.

    In set builder notation the intervals are $\{x \mid x < -4 \text{ or } -2 < x < 2\}$.

31. (a) The boundary numbers, the x-values for which $f(x) = 0$, are -4, -2, 0, and 2.

    (b) $f(x) > 0$ on the interval for which the graph of f is above the x-axis. That is $(-4, -2) \cup (0, 2)$.

    In set builder notation the intervals are $\{x \mid -4 < x < -2 \text{ or } 0 < x < 2\}$.

    (c) $f(x) < 0$ on the interval for which the graph of f is below the x-axis. That is $(-\infty, -4) \cup (-2, 0) \cup (2, \infty)$.

    In set builder notation the intervals are $\{x \mid x < -4 \text{ or } -2 < x < 0 \text{ or } x > 2\}$.

33. (a) The boundary numbers, the x-values for which $f(x) = 0$, are -2, 1, and 2.

    (b) $f(x) > 0$ on the interval for which the graph of f is above the x-axis. That is $(-\infty, -2) \cup (-2, 1)$.

    In set builder notation the intervals are $\{x \mid x < -2 \text{ or } -2 < x < 1\}$.

    (c) $f(x) < 0$ on the interval for which the graph of f is below the x-axis. That is $(1, 2) \cup (2, \infty)$.

    In set builder notation the intervals are $\{x \mid 1 < x < 2 \text{ or } x > 2\}$.

35. (a) $f(x)$ is undefined at $x = 0$.

    (b) $f(x) > 0$ on the interval for which the graph of f is above the x-axis. That is $(-\infty, 0) \cup (0, \infty)$.

    In set builder notation the intervals are $\{x \mid x < 0 \text{ or } x > 0\}$.

    (c) $f(x) < 0$ on the interval for which the graph of f is below the x-axis. There are no solutions.

37. (a) $f(x)$ is undefined at $x = 1$, $f(x) = 0$ at $x = 0$.

    (b) $f(x) > 0$ on the interval for which the graph of f is above the x-axis. That is $(-\infty, 0) \cup (1, \infty)$.

    In set builder notation the intervals are $\{x \mid x < 0 \text{ or } x > 1\}$.

    (c) $f(x) < 0$ on the interval for which the graph of f is below the x-axis. That is $(0, 1)$.

    In set builder notation the interval is $\{x \mid 0 < x < 1\}$.

39. (a) $f(x)$ is undefined at $x = \pm 2$, $f(x) = 0$ at $x = 0$.

(b) $f(x) > 0$ on the interval for which the graph of f is above the x-axis. That is $(-\infty, -2) \cup (2, \infty)$.

In set builder notation the intervals are $\{x \mid x < -2 \text{ or } x > 2\}$.

(c) $f(x) < 0$ on the interval for which the graph of f is below the x-axis. That is $(-2, 0) \cup (0, 2)$.

In set builder notation the intervals are $\{x \mid -2 < x < 0 \text{ or } 0 < x < 2\}$.

41. (a) First find the boundary numbers by solving $x^3 - x = 0$.

$x^3 - x = 0 \Rightarrow x(x^2 - 1) = 0 \Rightarrow x(x + 1)(x - 1) = 0 \Rightarrow x = 0, -1, \text{ or } 1$

The boundary values divide the number line into four intervals. Choose a test value from each of these intervals and evaluate $x^3 - x$ for these values. See Figure 41a. The solution is $(-1, 0) \cup (1, \infty)$.

In set builder notation the intervals are $\{x \mid -1 < x < 0 \text{ or } x > 1\}$.

(b) Graph $Y_1 = X^3 - X$ as shown in Figure 41b. The x-intercepts are -1, 0, and 1. The graph is above the x-axis on the interval $(-1, 0) \cup (1, \infty)$.

Interval	Test Value x	$x^3 - x$	Positive or Negative?
$(-\infty, -1)$	-2	-6	Negative
$(-1, 0)$	-0.5	0.375	Positive
$(0, 1)$	0.5	-0.375	Negative
$(1, \infty)$	2	6	Positive

Figure 41a

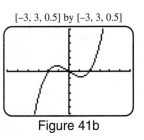

$[-3, 3, 0.5]$ by $[-3, 3, 0.5]$

Figure 41b

43. (a) Write the inequality as $x^3 + x^2 - 2x \geq 0$. Then find the boundary numbers by solving $x^3 + x^2 - 2x = 0$.

$x^3 + x^2 - 2x = 0 \Rightarrow x(x^2 + x - 2) = 0 \Rightarrow x(x + 2)(x - 1) = 0 \Rightarrow x = 0, -2, \text{ or } 1$

The boundary values divide the number line into four intervals. Choose a test value from each of these intervals and evaluate $x^3 + x^2 - 2x$ for these values. See Figure 43a. The solution is $[-2, 0] \cup [1, \infty)$.

In set builder notation the intervals are $\{x \mid -2 \leq x \leq 0 \text{ or } x \geq 1\}$.

(b) Graph $Y_1 = X^3 + X^2 - 2X$ as shown in Figure 43b. The x-intercepts are -1, 0, and 1. The graph intersects or is above the x-axis on the interval $[-2, 0] \cup [1, \infty)$.

Interval	Test Value x	$x^3 + x^2 - 2x$	Positive or Negative?
$(-\infty, -2)$	-3	-12	Negative
$(-2, 0)$	-1	2	Positive
$(0, 1)$	0.5	-0.625	Negative
$(1, \infty)$	2	8	Positive

Figure 43a

$[-3, 3, 1]$ by $[-5, 5, 1]$

Figure 43b

45. (a) First find the boundary numbers by solving $x^4 - 13x^2 + 36 = 0$.

$x^4 - 13x^2 + 36 = 0 \Rightarrow (x^2 - 4)(x^2 - 9) = 0 \Rightarrow (x + 2)(x - 2)(x + 3)(x - 3) = 0 \Rightarrow$

$x = -3, -2, 2$ or 3 The boundary values divide the number line into five intervals. Choose a test value from each of these intervals and evaluate $x^4 - 13x^2 + 36$ for these values. See Figure 45a. The solution is $(-3, -2) \cup (2, 3)$. In set builder notation the intervals are $\{x \mid -3 < x < -2 \text{ or } 2 < x < 3\}$.

(b) Graph $Y_1 = X^4 - 13X^2 + 36$ as shown in Figure 45b. The x-intercepts are $-3, -2, 2$ and 3. The graph is below the x-axis on the interval $(-3, -2) \cup (2, 3)$.

Interval	Test Value x	$x^4 - 13x^2 + 36$	Positive or Negative?
$(-\infty, -3)$	-4	84	Positive
$(-3, -2)$	-2.5	-6.1875	Negative
$(-2, 2)$	0	36	Positive
$(2, 3)$	2.5	-6.1875	Negative
$(3, \infty)$	4	84	Positive

Figure 45a

$[-5, 5, 1]$ by $[-10, 100, 10]$

Figure 45b

47. Write the inequality as $7x^4 - 14x^2 \geq 0$. Then find the boundary numbers by solving $7x^4 - 14x^2 = 0$.

$7x^4 - 14x^2 = 0 \Rightarrow 7x^2(x^2 - 2) = 0 \Rightarrow x = 0$ or $\pm\sqrt{2}$. The boundary values divide the number line into four intervals. Choose a test value from each of these intervals and evaluate $7x^4 - 14x^2$ for these values. See Figure 47. The solution is $(-\infty, -\sqrt{2}) \cup (\sqrt{2}, \infty)$. In set builder notation the intervals are $\{x \mid x < -\sqrt{2} \text{ or } x > \sqrt{2}\}$.

Interval	Test Value x	$7x^4 - 14x^2$	Positive or Negative?
$(-\infty, -\sqrt{2})$	-2	56	Positive
$(-\sqrt{2}, 0)$	-1	-7	Negative
$(0, \sqrt{2})$	1	-7	Negative
$(\sqrt{2}, \infty)$	2	56	Positive

Figure 47

49. First find the boundary numbers by solving $(x - 1)(x - 2)(x + 2) = 0$.

$(x - 1)(x - 2)(x + 2) = 0 \Rightarrow x = -2, 1$ or 2

The boundary values divide the number line into four intervals. Choose a test value from each of these intervals and evaluate $(x - 1)(x - 2)(x + 2)$ for these values. See Figure 49. The solution is $[-2, 1] \cup [2, \infty)$. In set builder notation the intervals are $\{x \mid 2 \leq x \leq 1 \text{ or } x \geq 2\}$.

Interval	Test Value x	$(x - 1)(x - 2)(x + 2)$	Positive or Negative?
$(-\infty, -2)$	-3	-20	Negative
$(-2, 1)$	0	4	Positive
$(1, 2)$	1.5	-0.875	Negative
$(2, \infty)$	3	10	Positive

Figure 49

51. Write the inequality as $2x^4 + 2x^3 - 12x^2 \leq 0$. Find the boundary numbers by solving $2x^4 + 2x^3 - 12x^2 = 0$.

    $2x^4 + 2x^3 - 12x^2 = 0 \Rightarrow 2x^2(x^2 + x - 6) = 0 \Rightarrow 2x^2(x + 3)(x - 2) = 0 \Rightarrow x = -3, 0 \text{ or } 2$

    The boundary values divide the number line into four intervals. Choose a test value from each of these intervals

    and evaluate $2x^4 + 2x^3 - 12x^2$ for these values. See Figure 51. The solution is $[-3, 2]$.

    In set builder notation the interval is $\{x \mid -3 \leq x \leq 2\}$.

Interval	Test Value x	$2x^4 + 2x^3 - 12x^2$	Positive or Negative?
$(-\infty, -3)$	-4	192	Positive
$(-3, 0)$	-1	-12	Negative
$(0, 2)$	1	-8	Negative
$(2, \infty)$	3	108	Positive

Figure 51

53. Write the inequality as $x^3 - 7x^2 + 14x - 8 \leq 0$.

    Then Graph $Y_1 = X^3 - 7X^2 + 14X - 8$ as shown in Figure 53. The x-intercepts are 1, 2, and 4.

    The graph intersects or is below the x-axis on the interval $(-\infty, 1] \cup [2, 4]$. In set builder notation the intervals

    are $\{x \mid x \leq 1 \text{ or } 2 \leq x \leq 4\}$.

[-2, 5, 1] by [-10, 10, 1]

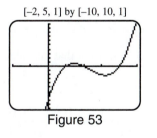

Figure 53

55. Graph $Y_1 = 3X^4 - 7X^3 - 2X^2 + 8X$ as shown in Figure 55. The x-intercepts are $-1, 0, \dfrac{4}{3}$ and 2.

    The graph is above the x-axis on the interval $(-\infty, -1) \cup \left(0, \dfrac{4}{3}\right) \cup (2, \infty)$. In set builder notation the

    intervals are $\left\{ x \mid x < -1 \text{ or } 0 < x < \dfrac{4}{3} \text{ or } x > 2 \right\}$.

[-4, 4, 1] by [-5, 5, 1]

Figure 55

57. (a) Find the zeros of the numerator and the denominator.

Numerator: No zero; Denominator: $x = 0$

This boundary number divides the number line into two intervals. Choose a test value from each of these intervals and evaluate $\dfrac{1}{x}$ for these values. See Figure 57a. The solution is $(-\infty, 0)$. In set builder notation the interval is $\{x \mid x < 0\}$.

(b) Graph $Y_1 = 1/X$ as shown in Figure 57b. The graph has a vertical asymptote at $x = 0$ and no x-intercepts. The graph is below the x-axis on the interval $(-\infty, 0)$.

$[-4, 4, 1]$ by $[-2, 2, 0.5]$

Interval	Test Value x	$1/x$	Positive or Negative?
$(-\infty, 0)$	-1	-1	Negative
$(0, \infty)$	1	1	Positive

Figure 57a

Figure 57b

59. (a) Find the zeros of the numerator and the denominator.

Numerator: No zero; Denominator: $x + 3 = 0 \Rightarrow x = -3$

This boundary number divides the number line into two intervals. Choose a test value from each of these intervals and evaluate $\dfrac{4}{x + 3}$ for these values. See Figure 59a. The solution is $(-3, \infty)$. In set builder notation the interval is $\{x \mid x > -3\}$.

(b) Graph $Y_1 = 4/(X + 3)$ using dot mode as shown in Figure 59b. The graph has a vertical asymptote at $x = -3$ and no x-intercepts. The graph is above the x-axis on the interval $(-3, \infty)$. Note that the value $x = -3$ can not be included in the solution set since $\dfrac{4}{x + 3}$ is undefined for $x = -3$.

$[-6, 1, 1]$ by $[-4, 4, 1]$

Interval	Test Value x	$4/(x + 3)$	Positive or Negative?
$(-\infty, -3)$	-4	-4	Negative
$(-3, \infty)$	1	1	Positive

Figure 59a

Figure 59b

61. (a) Find the zeros of the numerator and the denominator.

Numerator: No zero; Denominator: $x^2 - 4 = 0 \Rightarrow x = \pm 2$

These boundary numbers divide the number line into three intervals. Choose a test value from each of these intervals and evaluate $\dfrac{5}{x^2 - 4}$ for these values. See Figure 61a. The solution is $(-2, 2)$. In set builder notation the interval is $\{x \mid -2 < x < 2\}$.

(b) Graph $Y_1 = 5/(X^2 - 4)$ using dot mode as shown in Figure 61b. The graph has vertical asymptotes at $x = \pm 2$ and no x-intercepts. The graph is below the x-axis on the interval $(-2, 2)$.

Interval	Test Value x	$5/(x^2 - 4)$	Positive or Negative?
$(-\infty, -2)$	-3	1	Positive
$(-2, 2)$	0	-1.25	Negative
$(2, \infty)$	3	1	Positive

Figure 61a

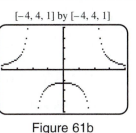
[−4, 4, 1] by [−4, 4, 1]

Figure 61b

63. The graph of $Y_1 = (X + 1)^{\wedge}2/(X - 2)$ has a vertical asymptote at $x = 2$ and an x-intercept at $x = -1$.

The graph of Y_1 intersects or is below the x-axis on the interval $(-\infty, 2)$. Note that 2 can not be included.

In set builder notation the interval is $\{x \mid x > 2\}$.

65. The graph of $Y_1 = (3 - 2X)/(1 + X)$ has a vertical asymptote at $x = -1$ and an x-intercept at $x = \dfrac{3}{2}$.

The graph of Y_1 is below the x-axis on the interval $(-\infty, -1) \cup \left(\dfrac{3}{2}, \infty\right)$.

In set builder notation the intervals are $\left\{x \mid x < -1 \text{ or } x > \dfrac{3}{2}\right\}$.

67. The graph of $Y_1 = (X + 1)(X - 2)/(X + 3)$ has a vertical asymptote of at $x = -3$ and x-intercepts at

$x = -1$ and $x = 2$. The graph of Y_1 is below the x-axis on the interval $(-\infty, -3) \cup (-1, 2)$.

In set builder notation the intervals are $\{x \mid x < -3 \text{ or } -1 < x < 2\}$.

69. The graph of $Y_1 = 2X - 5/((X + 1)(X - 1))$ has vertical asymptotes at $x = -1$ and $x = 1$ and an x-intercept

at $x = \dfrac{5}{2}$. The graph of Y_1 is on or above the x-axis on the interval $(-1, 1) \cup \left[\dfrac{5}{2}, \infty\right)$.

In set builder notation the intervals are $\left\{x \mid -1 < x < 1 \text{ or } x \geq \dfrac{5}{2}\right\}$.

71. First, rewrite the inequality:

$$\dfrac{1}{x - 3} \leq \dfrac{5}{x - 3} \Rightarrow \dfrac{1}{x - 3} - \dfrac{5}{x - 3} \leq 0 \Rightarrow \dfrac{-4}{x - 3} \leq 0$$

The graph of $Y_1 = -4/(X - 3)$ has a vertical asymptote at $x = 3$ and no x-intercept.

The graph of Y_1 intersects or is below the x-axis on the interval $(3, \infty)$. Note that 3 can not be included.

In set builder notation the interval is $\{x \mid x > 3\}$.

73. First, rewrite the inequality:

$$2 - \dfrac{5}{x} + \dfrac{2}{x^2} \geq 0 \Rightarrow \dfrac{2x^2 - 5x + 2}{x^2} \geq 0 \Rightarrow \dfrac{(2x - 1)(x - 2)}{x^2} \geq 0$$

The graph of $Y_1 = (2X - 1)(X - 2)/X^{\wedge}2$ has a vertical asymptote at $x = 0$ and x-intercepts

$x = \dfrac{1}{2}$ and $x = 2$. The graph of Y_1 intersects or is above the x-axis on the interval $(-\infty, 0) \cup \left(0, \dfrac{1}{2}\right] \cup [2, \infty)$.

In set builder notation the intervals are $\left\{x \mid x < 2 \text{ or } 0 < x \leq \dfrac{1}{2} \text{ or } x \geq 2\right\}$.

75. First, rewrite the inequality:

$\dfrac{1}{x} \le \dfrac{2}{x+2} \Rightarrow \dfrac{1}{x} - \dfrac{2}{x+2} \le 0 \Rightarrow \dfrac{x+2-2x}{x(x+2)} \le 0 \Rightarrow \dfrac{2-x}{x(x+2)} \le 0$. The graph of

$Y_1 = (2-X)/((X(+2))$ has vertical asymptotes at $x = 0$ and $x = -2$, and x-intercept of $x = 2$. The graph

of Y_1 is below the x-axis on the interval $(-2, 0) \cup [2, \infty)$. In set builder notation the intervals are

$\left\{ x \mid -2 < x < 0 \text{ or } x \ge 2 \right\}$.

77. (a) Since 15 seconds equals 0.25 minute, graph $Y_1 = (X-5)/(X^2-10X)$ and $Y_2 = 0.25$ as shown in

Figure 77. The intersection point is near $(12.4, 0.25)$. The gate should admit 12.4 cars per minute on

average to keep the wait less than 15 seconds. Note: The reason the answer is greater than 10 cars per

minute is because cars are arriving randomly. For some minutes, more than 10 vehicles might arrive at

the gate.

(b) $\dfrac{12.4}{5} = 2.48$ or 3 parking attendants must be on duty to keep the average wait less than 15 seconds.

[10, 15, 1] by [0, 1, 0.1]

Intersection
X=12.385165   Y=.25

Figure 77

79. Let y = height, x = width, and $2x$ = length of the box. To relate the variables, we use the volume formula for

a box and the fact that $V = 196$ cubic inches. $V = (2x)xy = 2x^2 y \Rightarrow y = \dfrac{V}{2x^2} = \dfrac{196}{2x^2} = \dfrac{98}{x^2}$.

The surface area of the box is the sum of the areas of the 6 rectangular sides. We are given that $A = 280 \text{ in}^2$.

$A = 2(x \cdot y) + 2(2x \cdot y) + 2(x \cdot 2x) \Rightarrow A = 6xy + 4x^2 = 280$; and since $y = \dfrac{98}{x^2}$ we get the equation

$6x\left(\dfrac{98}{x^2}\right) + 4x^2 = 280 \Rightarrow \dfrac{588}{x} + 4x^2 = 280$.

Figures 79a and 79b show the intersection points when graphing $Y_1 = 588/X + 4X^2$ and $Y_2 = 280$.

There are two possible solutions (in inches):

width $= 7$, length $= 14$, height $= \dfrac{98}{7^2} = 2$ or width ≈ 2.266, length ≈ 4.532, height $\approx \dfrac{98}{2.266^2} \approx 19.086$.

[0, 10, 2] by [100, 400, 20]   [0, 10, 2] by [100, 400, 20]

Intersection
X=2.2662813   Y=280   Intersection
X=7   Y=280

Figure 79a   Figure 79b

81. (a) $V = L \cdot W \cdot H \Rightarrow V = x \cdot x \cdot h \Rightarrow 108 = x^2 h \Rightarrow \dfrac{108}{x^2} = h.$ The surface area $A(x) = x^2 + 4xh \Rightarrow$

$A(x) = x^2 + 4x\left(\dfrac{108}{x^2}\right) \Rightarrow A(x) = \left(x^2 + \dfrac{432}{x}\right).$ This finds surface area in square inches. To find

square feet we must divide by 144 (the number of square inches in a square foot) $\Rightarrow$

$A(x) = \left(x^2 + \dfrac{432}{x}\right) \div 144 \Rightarrow A(x) = \dfrac{x^2}{144} + \dfrac{3}{x}.$

(b) $C = 0.10\left(\dfrac{x^2}{144} + \dfrac{3}{x}\right)$

(c) Graph the equation: $C = 0.10\left(\dfrac{x^2}{144} + \dfrac{3}{x}\right).$ The minimum cost is when $x \approx 6 \Rightarrow h = \dfrac{108}{(6)^2} \Rightarrow h = 3.$

The dimensions are $6 \times 6 \times 3$ inches.

83. (a) $D(0.05) = \dfrac{2500}{30(0.3 + 0.05)} \approx 238$

The braking distance far a car traveling at 50 mph on a 5% uphill grade is about 238 feet.

(b) Table $Y_1 = 2500/(30(0.3 + X))$ starting at $x = 0$ and incrementing by 0.05 as shown in Figure 83.

The table indicates that as the grade x increases, the braking distance decreases. This agrees with driving experience.

(c) $220 = \dfrac{2500}{30(0.3 + x)} \Rightarrow 0.3 + x = \dfrac{2500}{30(220)} \Rightarrow x = \dfrac{2500}{30(220)} - 0.3 \approx 0.079,$ or 7.9%

X	Y1
0	277.78
.05	238.1
.1	208.33
.15	185.19
.2	166.67
.25	151.52
.3	138.89

X=0

Figure 83

85. (a) Graph $Y_1 = X^2/(1600 - 40X)$ and $Y_2 = 8$ (not shown). The graphs intersect approximately at $(36, 8).$

The graph of Y_1 is equal to or is below the graph of Y_2 when $x \leq 36$ (approximately).

(b) The average line length is less than or equal to 8 cars when the average arrival rate is 36 cars per hour or less.

87. (a) The graph of $D(x) = \dfrac{120}{x}$ increases as x decreases, which means the braking distance increases as the

coefficient of friction becomes smaller.

(b) $\dfrac{120}{x} \geq 400 \Rightarrow \dfrac{120}{400} \geq x \Rightarrow 0.3 \geq x;$ the braking distance is 400 feet or more when $0 < x \leq 0.3.$

89. The volume of a cube is $V = x^3,$ where x is the length of a side.

$212.8 \leq V \leq 213.2 \Rightarrow 212.8 \leq x^3 \leq 213.2 \Rightarrow \sqrt[3]{212.8} \leq x \leq \sqrt[3]{213.2}$ inches.

This is approximately $5.97022 \leq x \leq 5.97396$ inches.

91. If $y = \dfrac{k}{x}$ and $y = 2$ when $x = 3,$ then $2 = \dfrac{k}{3} \Rightarrow k = 6.$

93. If $y = kx^3$ and $y = 64$ when $x = 2,$ then $64 = k(8) \Rightarrow 8k = 64 \Rightarrow k = 8.$

95. If $T = kx^{3/2}$ and $T = 20$ when $x = 4$, then $20 = k(4)^{3/2} \Rightarrow 20 = 8k \Rightarrow k = 2.5$.

   The variation equation becomes $T = 2.5x^{3/2}$. When $x = 16$, $T = 2.5(16)^{3/2} = 2.5(64) = 160$.

97. If $y = \dfrac{k}{x}$ and $y = 5$ when $x = 6$, then $5 = \dfrac{k}{6} \Rightarrow k = 30$.

   The variation equation becomes $y = \dfrac{30}{x}$. When $x = 15$, $y = \dfrac{30}{15} = 2$.

99. If y is inversely proportional to x, then $y = \dfrac{k}{x}$ must hold. So if the value of x is doubled, the right side of this

   equation becomes $\dfrac{k}{2x} = \dfrac{1}{2} \cdot \dfrac{k}{x} = \dfrac{1}{2}y$. Therefore y becomes half its original value.

101. If y is directly proportional to x^3, then $y = kx^3$ must hold. So if the value of x is tripled, the right side of this

   equation becomes $k(3x)^3 = 27 \cdot kx^3 = 27y$. Therefore y becomes 27 times its original value.

103. Since $y = kx^n$, we know that $k = \dfrac{y}{x^n}$ where k is a constant. Using trial and error, let $n = 1$ while calculating

   various values for k using x- and y-values from the table. For example, when $x = 2$ and $y = 2$, the value of k is

   1. But for $x = 4$ and $y = 8$, the value of k is 2. Since the value of k did not remain constant we know that the

   value of n is not 1. Repeat this process for $n = 2$. For each x and y pair in the table, the value of k is 0.5.

   Therefore $k = 0.5$ and $n = 2$.

105. Since $y = \dfrac{k}{x^n}$, we know that $k = yx^n$ where k is a constant. Using trial and error, let $n = 1$ while calculating

   various values for k using x- and y-values from the table. For example, when $x = 2$ and $y = 1.5$, the value of

   k is 3. And for $x = 3$ and $y = 1$, the value of k is also 3. Since the value of k remained constant we know

   that the value of n is 1. For each x and y pair in the table, the value of k is 3. Therefore $k = 3$ and $n = 1$.

107. If y is directly proportional to $x^{1.25}$, then $y = kx^{1.25}$ must hold. Given that $y = 1.9$ when $x = 1.1$, we may

   calculate the value of $k = \dfrac{y}{x^{1.25}} = \dfrac{1.9}{1.1^{1.25}} \approx 1.69$. The variation equation can be written as $y \approx 1.69x^{1.25}$.

   When a fiddler crab has claws weighing 0.75 grams, its body weight will be $y \approx 1.69(0.75)^{1.25} \approx 1.18$ grams.

109. Let $I = $ the intensity of the light, and let $d = $ the distance from a star to the earth. If I is inversely

   proportional to d^2, then $I = \dfrac{k}{d^2}$ must hold. Suppose a ground-based telescope can see a star with intensity I_g.

   Then the Hubble Telescope can see stars of intensity $\dfrac{1}{50} \cdot I_g = \dfrac{1}{50} \cdot \dfrac{k}{d^2} = \dfrac{k}{(d\sqrt{50})^2}$. That is, the Hubble

   Telescope can see $\sqrt{50} \approx 7$ times as far as ground-based telescopes.

111. Since the resistance varies inversely as the square of the diameter of the wire, increasing the diameter of the

   wire by a factor of 1.5 will decrease the resistance by a factor of $\dfrac{1}{1.5^2} = \dfrac{4}{9}$. Hence, a 25 foot wire with a

   diameter of 3 millimeters (1.5 times 2 millimeters) will have a resistance of $\dfrac{4}{9}$ (0.5 ohm) $= \dfrac{2}{9}$ ohm.

113. In this exercise $F = \dfrac{K\sqrt{T}}{L}$. If both T and L are doubled, $\dfrac{K\sqrt{2T}}{2L} = \dfrac{\sqrt{2}}{2} \cdot \dfrac{K\sqrt{T}}{L} = \dfrac{\sqrt{2}}{2}F$. Therefore F

   decreases by a factor of $\dfrac{\sqrt{2}}{2}$.

4.8: Power Functions and Radical Equations

1. $8^{2/3} = (8^{1/3})^2 = 2^2 = 4$

3. $16^{-3/4} = (16^{1/4})^{-3} = (2\ )^{-3} = \dfrac{1}{2^3} = \dfrac{1}{8}$

5. $-81^{0.5} = -81^{1/2} = -\sqrt{81} = -9$

7. $(9^{3/4})^2 = 9^{3/2} = (\sqrt{9}\ )^3 = 3^3 = 27$

9. $\dfrac{8^{5/6}}{8^{1/2}} = 8^{5/6-1/2} = 8^{1/3} = 2$

11. $27^{5/6} \cdot 27^{-1/6} = 27^{5/6+(-1/6)} = 27^{2/3} = \sqrt[3]{27^2} = 3^2 = 9$

13. $(-27)^{-5/3} = \dfrac{1}{(-27)^{5/3}} = \dfrac{1}{\sqrt[3]{(-27)^5}} = \dfrac{1}{(-3)^5} = -\dfrac{1}{243}$

15. $(0.5^{-2})^2 = \left(\dfrac{1}{2}\right)^{-4} = 2^4 = 16$

17. $\left(\dfrac{2}{3}\right)^{-2} = \left(\dfrac{3}{2}\right)^2 = \dfrac{9}{4}$

19. $\sqrt{2x} = (2x)^{1/2}$

21. $\sqrt[3]{z^5} = z^{5/3}$

23. $(\sqrt[4]{y}\ )^{-3} = (y^{1/4})^{-3} = y^{-3/4} = \dfrac{1}{y^{3/4}}$

25. $\sqrt{x} \cdot \sqrt[3]{x} = x^{1/2} \cdot x^{1/3} = x^{1/2+1/3} = x^{5/6}$

27. $\sqrt{y \cdot \sqrt{y}} = (y \cdot y^{1/2})^{1/2} = (y^{3/2})^{1/2} = y^{3/4}$

29. $a^{-3/4}b^{1/2} = \dfrac{b^{1/2}}{a^{3/4}} = \dfrac{\sqrt{b}}{\sqrt[4]{a^3}}$

31. $(a^{1/2} + b^{1/2})^{1/2} = \sqrt{\sqrt{a} + \sqrt{b}}$

33. $\sqrt{x+2} = x - 4 \Rightarrow x + 2 = x^2 - 8x + 16 \Rightarrow x^2 - 9x + 14 = 0 \Rightarrow (x-2)(x-7) = 0 \Rightarrow$
    $x = 2$ or 7.
    Check: $\sqrt{2+2} = 2 \neq 2 - 4$ (not a solution); $\sqrt{7+2} = 3 = 7 - 4$. The only solution is $x = 7$.

35. $\sqrt{3x+7} = 3x + 5 \Rightarrow 3x + 7 = 9x^2 + 30x + 25 \Rightarrow 9x^2 + 27x + 18 = 0 \Rightarrow 9(x^2 + 3x + 2) = 0 \Rightarrow$
    $9(x+2)(x+1) = 0 \Rightarrow x = -2$ or -1
    Check: $\sqrt{3(-2)+7} \neq 3(-2) + 5$ (not a solution); $\sqrt{3(-1)+7} = 2 = 3(-1) + 5$. The only solution is
    $x = -1$.

37. $\sqrt{5x-6} = x \Rightarrow 5x - 6 = x^2 \Rightarrow x^2 - 5x + 6 = 0 \Rightarrow (x-2)(x-3) = 0 \Rightarrow x = 2$ or 3
    Check: $\sqrt{5(2)-6} = 2$; $\sqrt{5(3)-6} = 3$

39. $\sqrt{x+5} + 1 = x \Rightarrow \sqrt{x+5} = x - 1 \Rightarrow x + 5 = (x-1)^2 \Rightarrow x + 5 = x^2 - 2x + 1 \Rightarrow$
    $x^2 - 3x - 4 = 0 \Rightarrow (x-4)(x+1) = 0 \Rightarrow x = -1$ or $x = 4$
    Check: $\sqrt{-1+5} + 1 = -1 \Rightarrow \sqrt{4} + 1 = -1 \Rightarrow 3 \neq -1$ (not a solution).
    Check: $\sqrt{4+5} + 1 = 4 \Rightarrow \sqrt{9} + 1 = 4 \Rightarrow 4 = 4$. The only solution is $x = 4$.

41. $\sqrt{x+1} + 3 = \sqrt{3x+4} \Rightarrow (\sqrt{x+1}+3)^2 = 3x+4 \Rightarrow x+1+6\sqrt{x+1}+9 = 3x+4 \Rightarrow$

$6\sqrt{x+1} = 2x - 6 \Rightarrow 3\sqrt{x+1} = x - 3 \Rightarrow (3\sqrt{x+1})^2 = (x-3)^2 \Rightarrow$

$9(x+1) = x^2 - 6x + 9 \Rightarrow x^2 - 15x = 0 \Rightarrow x(x-15) = 0 \Rightarrow x = 0$ or 15

Check: $\sqrt{(0)+1} + 3 \neq \sqrt{3(0)+4}$ (not a solution); $\sqrt{(15)+1} + 3 = 7 = \sqrt{3(15)+4}$.

The only solution is $x = 15$.

43. $\sqrt{2x} - \sqrt{x+1} = 1 \Rightarrow \sqrt{2x} = 1 + \sqrt{x+1} \Rightarrow 2x = (1 + \sqrt{x+1})^2 \Rightarrow$

$2x = 1 + 2\sqrt{x+1} + x + 1 \Rightarrow x - 2 = 2\sqrt{x+1} \Rightarrow x^2 - 4x + 4 = 4(x+1) \Rightarrow$

$x^2 - 4x + 4 = 4x + 4 \Rightarrow x^2 - 8x = 0 \Rightarrow x(x-8) = 0 \Rightarrow x = 0$ or 8.

Check: $\sqrt{2(0)} - \sqrt{0+1} = 1 \Rightarrow \sqrt{0} - \sqrt{1} = 1 \Rightarrow -1 = 1$ (not a solution);

$\sqrt{2(8)} - \sqrt{8+1} = 1 \Rightarrow \sqrt{16} - \sqrt{9} = 1 \Rightarrow 4 - 3 = 1 \Rightarrow 1 = 1$ The only solution is 8.

45. $\sqrt[3]{z+1} = -3 \Rightarrow z+1 = (-3)^3 \Rightarrow z+1 = -27 \Rightarrow z = -28$

Check: $\sqrt[3]{-28-1} = \sqrt[3]{-27} = -3$

47. $\sqrt[3]{x+1} = \sqrt[3]{2x-1} \Rightarrow x+1 = 2x-1 \Rightarrow x = 2$

Check: $\sqrt[3]{2+1} = \sqrt[3]{3} = \sqrt[3]{2(2)-1}$

49. $\sqrt[4]{x-2} + 4 = 20 \Rightarrow \sqrt[4]{x-2} = 16 \Rightarrow x-2 = (16)^4 \Rightarrow x-2 = 65{,}536 \Rightarrow x = 65{,}538$

Check: $\sqrt[4]{65{,}538 - 2} + 4 = 20 \Rightarrow \sqrt[4]{65{,}536} + 4 = 20 \Rightarrow 20 = 20$

51. $f(x) = x^{3/2} - x^{1/2} \Rightarrow f(50) = 50^{3/2} - 50^{1/2} \approx 346.48$

53. The graph of $f(x) = x^a$ is increasing, since $a > 0$. However, since $a < b$ its graph increases more slowly than the graph of $f(x) = x^b$. The best choice is graph b.

55. Shift the graph of $f(x) = \sqrt{x}$ up 1 unit to get the graph of $f(x) = \sqrt{x} + 1$. See Figure 55.

57. Shift the graph of $f(x) = x^{2/3}$ down 1 unit to get the graph of $f(x) = x^{2/3} - 1$. See Figure 57.

Figure 55   Figure 57   Figure 59

59. Shift the graph of $f(x) = \sqrt{x}$ left 2 units and down 1 unit to get the graph of $f(x) = \sqrt{x+2} - 1$.

See Figure 59.

61. $x^3 = 8 \Rightarrow x = \sqrt[3]{8} \Rightarrow x = 2$;   Check: $2^3 = 8$

63. $x^{1/4} = 3 \Rightarrow x = 3^4 \Rightarrow x = 81$;   Check: $81^{1/4} = \sqrt[4]{81} = 3$

65. $x^{2/5} = 4 \Rightarrow (x^{2/5})^{5/2} = 4^{5/2} \Rightarrow x = (\sqrt{4})^5 \Rightarrow x = (\pm 2)^5 \Rightarrow x = \pm 32$

Check: $(\pm 32)^{2/5} = (\sqrt[5]{\pm 32})^2 = (\pm 2)^2 = 4$

67. $2(x^{1/5} - 2) = 0 \Rightarrow x^{1/5} - 2 = 0 \Rightarrow x^{1/5} = 2 \Rightarrow (x^{1/5})^5 = 2^5 \Rightarrow x = 32$

    Check: $2(32^{1/5} - 2) = 2(2 - 2) = 2(0) = 0$

69. $4x^{3/2} + 5 = 21 \Rightarrow 4x^{3/2} = 16 \Rightarrow x^{3/2} = 4 \Rightarrow (x^{3/2})^{2/3} = 4^{2/3} \Rightarrow x = \sqrt[3]{4^2} \Rightarrow x = \sqrt[3]{16}$

    Check: $4(\sqrt[3]{16})^{3/2} + 5 = 4(16^{1/3})^{3/2} + 5 = 4(16^{1/2}) + 5 = 4(\sqrt{16}) + 5 = 4(4) + 5 = 16 + 5 = 21$

71. $n^{-2} + 3n^{-1} + 2 = 0 \Rightarrow (n^{-1})^2 + 3(n^{-1}) + 2 = 0$, let $u = n^{-1}$, then $u^2 + 3u + 2 = 0 \Rightarrow$

    $(u + 2)(u + 1) = 0 \Rightarrow u = -2$ or $u = -1$. Because $u = n^{-1}$, it follows that $n = u^{-1}$.

    Thus $n = (-2)^{-1} \Rightarrow n = \dfrac{1}{(-2)} \Rightarrow n = -\dfrac{1}{2}$ or $n = (-1)^{-1} \Rightarrow n = \dfrac{1}{(-1)} = -1$. Therefore, $n = -1, -\dfrac{1}{2}$.

73. $5n^{-2} + 13n^{-1} = 28 \Rightarrow 5(n^{-1})^2 + 13(n^{-1}) - 28 = 0$, let $u = n^{-1}$, then $5u^2 + 13u - 28 = 0 \Rightarrow$

    $(5u - 7)(u + 4) = 0 \Rightarrow u = \dfrac{7}{5}$ or $u = -4$. Because $u = n^{-1}$, it follows that $n = u^{-1}$.

    Thus $n = \left(\dfrac{7}{5}\right)^{-1} \Rightarrow n = \dfrac{5}{7}$ or $n = (-4)^{-1} \Rightarrow n = -\dfrac{1}{4}$. Therefore, $n = -\dfrac{1}{4}, \dfrac{5}{7}$.

75. $x^{2/3} - x^{1/3} - 6 = 0 \Rightarrow (x^{1/3})^2 - x^{1/3} - 6 = 0$, let $u = x^{1/3}$, then $u^2 - u - 6 = 0 \Rightarrow$

    $(u - 3)(u + 2) = 0 \Rightarrow u = -2$ or $u = 3$. Because $u = x^{1/3}$, it follows that $x = u^3$.

    Thus $x = (-2)^3 \Rightarrow x = -8$ or $x = 3^3 \Rightarrow x = 27$. Therefore, $x = -8, 27$.

77. $6x^{2/3} - 11x^{1/3} + 4 = 0 \Rightarrow 6(x^{1/3})^2 - 11(x^{1/3}) + 4 = 0$, let $u = x^{1/3}$, then $6u^2 - 11u + 4 = 0$.

    Using the quadratic formula to solve we get: $u = \dfrac{11 \pm \sqrt{121 - 4(6)(4)}}{2(6)} = \dfrac{11 \pm \sqrt{25}}{12} = \dfrac{11 \pm 5}{12} \Rightarrow$

    $u = \dfrac{16}{12} \Rightarrow u = \dfrac{4}{3}$ or $u = \dfrac{6}{12} \Rightarrow u = \dfrac{1}{2}$. Because $u = x^{1/3}$, it follows that $x = u^3$.

    Thus $x = \left(\dfrac{4}{3}\right)^3 \Rightarrow x = \dfrac{64}{27}$ or $x = \left(\dfrac{1}{2}\right)^3 \Rightarrow x = \dfrac{1}{8}$. Therefore, $x = \dfrac{1}{8}, \dfrac{64}{27}$.

79. $x^{3/4} - x^{1/2} - x^{1/4} + 1 = 0 \Rightarrow (x^{1/4})^3 - (x^{1/4})^2 - (x^{1/4}) + 1 = 0$, let $u = x^{1/4}$,

    then $u^3 - u^2 - u + 1 = 0 \Rightarrow (u^3 - u^2) - (u - 1) = 0 \Rightarrow u^2(u - 1) - 1(u - 1) = 0 \Rightarrow$

    $(u^2 - 1)(u - 1) = 0 \Rightarrow (u + 1)(u - 1)(u - 1) = 0 \Rightarrow u = -1$ or $u = 1$.

    Because $u = x^{1/4}$ it follows that $x = u^4$.

    Thus $x = (-1)^4 \Rightarrow x = 1$ or $x = (1)^4 \Rightarrow x = 1$. Therefore, $x = 1$.

81. $x^{-2/3} - 2x^{-1/3} - 3 = 0 \Rightarrow (x^{-1/3})^2 - 2x^{-1/3} - 3 = 0$, let $u = x^{-1/3}$, then $u^2 - 2u - 3 = 0 \Rightarrow$

    $(u - 3)(u + 1) = 0 \Rightarrow u = 3$ or -1. Because $u = x^{-1/3}$ it follows that $x = u^{-3}$.

    Thus $x = 3^{-3} = \dfrac{1}{27}$, $x = -1^{-3} = -1$.

83. Average rate of change $= \dfrac{s\left(\frac{9}{2}\right) - s\left(\frac{1}{2}\right)}{\frac{9}{2} - \frac{1}{2}} = \dfrac{\sqrt{96\left(\frac{9}{2}\right)} - \sqrt{96\left(\frac{1}{2}\right)}}{4} = \dfrac{\sqrt{432} - \sqrt{48}}{4} \approx 3.5$. The average speed

    over the time interval is about 3.5 mph.

85. $S(w) = 3 \Rightarrow 1.27w^{2/3} = 3 \Rightarrow w^{2/3} = \dfrac{3}{1.27} \Rightarrow w = \left(\dfrac{3}{1.27}\right)^{3/2} \Rightarrow w \approx 3.63$

    The bird weighs about 3.63 pounds

87. $f(15) = 15^{1.5} \approx 58.1$; The planet would take about 58.1 years to orbit the sun.

89. (a) Following the hint, $f(1) = a(1)^b = 1960 \Rightarrow a = 1960$

(b) Plot the four data points (0.5, 4500), (1, 1960), (2, 850), and (3, 525) together with $f(x) = 1960x^b$ for different values of b. Through trial and error, a value of $b \approx -1.2$ can be found. Figure 89 shows a graph of $Y_1 = 1960X^\wedge(-1.2)$ along with the data.

(c) $f(x) = 1960x^{-1.2} \Rightarrow f(4) = 1960(4)^{-1.2} \approx 371$. This means that if the zinc ion concentration in the water reaches 371 milligrams per liter, the rainbow trout will live only 4 minutes on average.

91. (a) $f(2) = 0.445(2)^{1.25} \approx 1.06$ grams

(b) We must solve the equation $0.445x^{1.25} = 0.5$ for x. Graph $Y_1 = 0.445X^\wedge 1.25$ and $Y_2 = 0.5$. Their graphs intersect near (1.1, 0.5). See Figure 91. When the claw weighs 0.5 grams, the crab weighs about 1.1 grams.

(c) $0.445x^{1.25} = 0.5 \Rightarrow 0.445x^{5/4} = 0.5 \Rightarrow x^{5/4} = \dfrac{0.5}{0.445} \Rightarrow x = \left(\dfrac{0.5}{0.445}\right)^{4/5} \Rightarrow x \approx 1.1$ grams

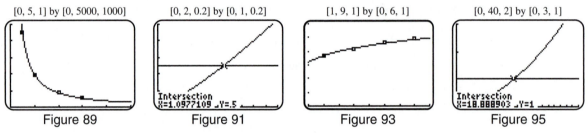

[0, 5, 1] by [0, 5000, 1000]   [0, 2, 0.2] by [0, 1, 0.2]   [1, 9, 1] by [0, 6, 1]   [0, 40, 2] by [0, 3, 1]

Figure 89   Figure 91   Figure 93   Figure 95

93. Use the power regression feature of your graphing calculator to find the values of a and b in the equation $f(x) = ax^b$. For this exercise, $a \approx 3.20$ and $b \approx 0.20$.

The graph of $Y_1 = 3.20X^\wedge 0.20$ is shown together with the data in Figure 93.

95. (a) Using the power regression function on the graphing calculator we see that $f(x) \approx 0.005192x^{1.7902}$.

Answers may vary.

(b) The year 2012 is 32 years after 1980. $f(32) \approx 0.005192(32)^{1.7902} \approx 2.6$ million. The estimate involved extrapolation.

(c) See Figure 95. We can see from the graph of $f(x)$ that the number of employees reached 1.million in about 1999.

97. Use the power regression feature of your graphing calculator to find the values of a and b in the equation $f(x) = ax^b$. For this exercise, $a \approx 874.54$ and $b \approx -0.49789$.

Chapter 4 Review Exercises

1. First put in standard order:

$f(x) = -7x^3 - 2x^2 + x + 4$, now the degree is: 3, and the leading coefficient is: -7.

3. (a) There is a local minimum of -2. There is a local maximum of 4.

(b) There is neither an absolute minimum nor an absolute maximum.

5. The graph of $f(x) = x^4 + 2x^3 - 9x^2 - 2x + 20$ is shown in Figure 5.

   There are two local minima, one local maximum and two x-intercepts.

   [−10, 10, 1] by [−100, 100, 10]

   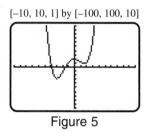

   Figure 5

7. $f(x) = 2x^6 - 5x^4 - x^2 \Rightarrow f$ is even since each power of x is even.

9. $f(x) = 7x^5 + 3x^3 - x \Rightarrow f$ is odd since each power of x is odd.

11. f is odd since $f(-x) = -f(x)$ for all x-values in the table.

13. See Figure 13. *Answers may vary.*

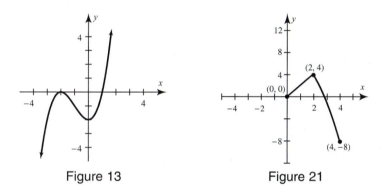

Figure 13   Figure 21

15. (a) 2 turning points; 3 x-intercepts at: $x \approx -2, 0, 1$.

   (b) This is a an odd degreed function going down to the right $\Rightarrow$ negative.

   (c) Since it has 2 turning points, it is a minimum of 3rd degree.

17. A negative cubic function $\Rightarrow$ up on the left end and down on the right end;

   $f(x) \to \infty$ as $x \to -\infty$; $f(x) \to -\infty$ as $x \to \infty$.

19. $f(-2) = (-2)^3 + 1 \Rightarrow f(-2) = -7 \Rightarrow (-2, -7)$ and $f(-1) = (-1)^3 + 1 \Rightarrow f(-1) = 0 \Rightarrow (-1, 0)$.

   To find the average rate of change we use: $\dfrac{y_2 - y_1}{x_2 - x_1} \Rightarrow \dfrac{0 - (-7)}{(-1) - (-2)} \Rightarrow \dfrac{7}{1} \Rightarrow 7$

21. (a) The graph of f has of two different pieces: $y = 2x$ when $0 \le x < 2$ and $y = 8 - x^2$ when $2 \le x \le 4$.

   The graph is shown in Figure 21. The graph of f is continuous.

   (b) To evaluate $f(1)$ we must use the first piece of the function: $f(1) = 2(1) = 2$

   To evaluate $f(3)$ we must use the second piece of the function: $f(3) = 8 - (3)^2 = 8 - 9 = -1$

   (c) We must solve the equation $f(x) = 2$ for each piece of the piecewise-defined function f.

   $2x = 2 \Rightarrow x = 1$ and $8 - x^2 = 2 \Rightarrow x^2 = 6 \Rightarrow x = \pm\sqrt{6}$

   Since $x = -\sqrt{6}$ is not in the domain of f, the solutions are $x = 1$ or $\sqrt{6}$.

23. $\dfrac{14x^3 - 21x^2 - 7x}{7x} = 2x^2 - 3x - 1$

25.

$$
\begin{array}{r}
2x^2 - 3x + 1 \\
2x + 3{\overline{\smash{\big)}\,4^3 \qquad - 7x + 4}} \\
\underline{4x^3 + 6x^2} \\
-6x^2 - 7x + 4 \\
\underline{-6x^2 - 9x} \\
2x + 4 \\
\underline{2x + 3} \\
1
\end{array}
$$

Therefore the quotient is: $2x^2 - 3x + 1 + \dfrac{1}{2x + 3}$

27. Since the leading coefficient of $f(x)$ is $\dfrac{1}{2}$ and the degree of the polynomial is 3, all of the zeros are given, thus:

$$f(x) = \frac{1}{2}(x - 1)(x - 2)(x - 3).$$

29. The graph of $Y_1 = 2X^3 + 3X^2 - 18X + 8$ is shown in Figure 29. From the graph, the x-intercepts of f are

located at $-4, \dfrac{1}{2}$, and 2. Three x-intercepts is the maximum number possible for a cubic polynomial. The

leading coefficient is $a_n = 2$. Thus $f(x) = 2(x + 4)\left(x - \dfrac{1}{2}\right)(x - 2)$.

[−5, 5, 1] by [−20, 50, 5]   [−10, 10, 1] by [−10, 10, 1]

Figure 29   Figure 39

31. The graph has x-intercepts or zeros of $x = -2, 1, 3 \Rightarrow$ it has factors $(x + 2)(x - 1)(x - 3)$.

Since $f(0) = 3$ we can solve for the leading coefficient a by setting $a(0 + 2)(0 - 1)(0 - 3) = 3$

and solving $\Rightarrow a(2)(-1)(-3) = 3 \Rightarrow 6a = 3 \Rightarrow a = \dfrac{1}{2}$.

The complete factored form is: $f(x) = \dfrac{1}{2}(x + 2)(x - 1)(x - 3)$.

33. By the rational zeros test, any rational zero of $2x^3 + x^2 - 13x + 6$ must be one of the following:

$\pm 6, \ \pm 3, \ \pm 2, \ \pm 1, \ \pm \dfrac{3}{2}, \ \pm \dfrac{1}{2}$. By evaluating f at each of these values we find that the three rational zeros are

$-3, \dfrac{1}{2}$, and 2.

35. $9x = 3x^3 \Rightarrow 3x^3 - 9x = 0 \Rightarrow 3x(x^2 - 3) = 0 \Rightarrow x = 0, \pm\sqrt{3}$

37. $x^4 - 3x^2 + 2 = 0 \Rightarrow (x^2 - 1)(x^2 - 2) = 0 \Rightarrow (x + 1)(x - 1)(x^2 - 2) = 0 \Rightarrow x = \pm 1, \pm\sqrt{2}$

39. The x-intercepts on the graph of $Y_1 = X^3 - 3X + 1$ are approximately $-1.88, 0.35$ and 1.53. See Figure 39.

41. $(2 - 2i) + (-3 + 2i) = (2 + (-3)) + (-2 + 2)i = -1$

43. $(3 + 2i)^2 = (3 + 2i)(3 + 2i) = 9 + 12i + 4i^2 = 9 + 12i - 4 = 5 + 12i$

45. $x^3 + x = 0 \Rightarrow x(x^2 + 1) = 0 \Rightarrow x = 0$ or $x = \pm i$

47. Graph $Y_1 = X^3 - 3X^2 + 3X - 9$ in $[-5, 5, 1]$ by $[-25, 25, 5]$ (not shown). Its graph crosses the x-axis once. Therefore, f has one real zero. Since f is degree 3, there must be two imaginary zeros.

49. The easiest way to factor a polynomial is to determine its zeros. Then, we can find its complete factorization. If one graphs $f(x) = 2x^2 + 4$, it does not intersect the x-axis. Therefore, there are no real zeros. However, the Fundamental Theorem of Algebra tells us that it must have complex zeros. These can be found by solving for x. $2x^2 + 4 = 0 \Rightarrow 2x^2 = -4 \Rightarrow x^2 = -2 \Rightarrow x = \pm\sqrt{-2} \Rightarrow x = \pm i\sqrt{2}$

    Since $a_n = 2$, the complete factorization is

    $$f(x) = 2(x - i\sqrt{2})(x - (-i\sqrt{2})) = 2(x - i\sqrt{2})(x + i\sqrt{2}).$$

51. If i is a zero then $-i$ is also a zero $\Rightarrow (x + i)(x - i)$ or $x^2 + 1$ is a factor. By polynomial long division:

    $$
    \begin{array}{r}
    x^2 + x + 1 \\
    x^2 + 1 \overline{)\, x^4 + x^3 + 2x^2 + x + 1} \\
    \underline{x^4 + x^2} \\
    x^3 + x^2 + x + 1 \\
    \underline{x^3 + x} \\
    x^2 + 1 \\
    \underline{x^2 + 1} \\
    0
    \end{array}
    $$

    Solve for $x^2 + x + 1 = 0$ using the quadratic formula: $x = \dfrac{-1 \pm \sqrt{1 - 4(1)(1)}}{2(1)} = \dfrac{-1 \pm \sqrt{-3}}{2} = -\dfrac{1}{2} \pm \dfrac{i\sqrt{3}}{2}$. Therefore the zeros are $\pm i$, $-\dfrac{1}{2} \pm \dfrac{i\sqrt{3}}{2}$ and the complete factored form is:

    $$f(x) = (x + i)(x - i)\left(x - \left(-\frac{1}{2} + \frac{i\sqrt{3}}{2}\right)\right)\left(x - \left(-\frac{1}{2} - \frac{i\sqrt{3}}{2}\right)\right)$$

53. Horizontal asymptote: since the degree of the numerator and denominator is the same the horizontal asymptote is the leading coefficient ratio $y = \dfrac{2}{3}$. Vertical asymptotes: they can be found by finding the zero's of the denominator that are not zeros of the numerator

    $\Rightarrow$ solve: $3x^2 + 8x - 3 \Rightarrow (3x - 1)(x + 3) = 0 \Rightarrow x = -3, \dfrac{1}{3}$. However, $x = -3$ is not a vertical asymptote because -3 is a zero of the numerator.

    The horizontal asymptote is $y = \dfrac{2}{3}$ and the vertical asymptote is $x = \dfrac{1}{3}$.

55. $g(x)$ is the function $f(x) = \dfrac{1}{x}$ shifted 1 left and 2 down. See Figure 55.

57. $g(x) = \dfrac{x^2 - 1}{x^2 + 2x + 1} \Rightarrow g(x) = \dfrac{(x + 1)(x - 1)}{(x + 1)(x + 1)} \Rightarrow g(x) = \dfrac{(x - 1)}{(x + 1)} \Rightarrow g(x) = \dfrac{x - 1}{x + 1}$. This is similar

to $f(x) = \dfrac{1}{x}$ with a horizontal asymptote of $y = 1$ and a vertical asymptote of $x = -1$. See Figure 57.

Figure 55   Figure 57   Figure 59

59. One example of a function that has a vertical asymptote at $x = -2$ and a horizontal asymptote at $y = 2$ is shown in Figure 59.

61. $\dfrac{5x + 1}{x + 3} = 3 \Rightarrow 5x + 1 = 3(x + 3) \Rightarrow 5x + 1 = 3x + 9 \Rightarrow 2x = 8 \Rightarrow x = 4$

This answer is supported graphically by graphing $Y_1 = (5X + 1)/(X + 3)$ and $Y_2 = 3$. The intersection point is $(4, 3)$. Making a table of Y_1 starting at 0 and incrementing by 1 will support this result numerically.

63. $\dfrac{1}{x + 2} + \dfrac{1}{x - 2} = \dfrac{4}{x^2 - 4} \Rightarrow x - 2 + x + 2 = 4 \Rightarrow 2x - 4 \Rightarrow x = 2$. Since 2 is defined in the original equation, there is no solution.

65. $\dfrac{x + 5}{x - 2} = \dfrac{x - 1}{x + 1} \Rightarrow (x + 5)(x + 1) = (x - 1)(x - 2) \Rightarrow x^2 + 6x + 5 = x^2 - 3x + 2 \Rightarrow$

$9x = -3 \Rightarrow x = -\dfrac{1}{3}$; Check: $\dfrac{-\frac{1}{3} + 5}{-\frac{1}{3} - 2} = \dfrac{-\frac{1}{3} - 1}{-\frac{1}{3} + 1} \Rightarrow \dfrac{\frac{14}{3}}{-\frac{7}{3}} = \dfrac{-\frac{4}{3}}{\frac{2}{3}} \Rightarrow -2 = -2$

67. (a) $(-\infty, -4)\cup(-2, 3)$; In set builder notation the interval is $\{x \mid x < -4 \text{ or } -2 < x < 3\}$.

(b) $(-4, -2)\cup(3, \infty)$; In set builder notation the interval is $\{x \mid -4 < x < -2 \text{ or } x > 3\}$.

69. Find the boundary numbers by solving $x^3 + x^2 - 6x = 0$.

$x^3 + x^2 - 6x = 0 \Rightarrow x(x^2 + x - 6) = 0 \Rightarrow x(x + 3)(x - 2) = 0 \Rightarrow x = -3, 0 \text{ or } 2$.

The boundary values divide the number line into four intervals. Choose a test value from each of these intervals and evaluate $x^3 + x^2 - 6x$ for these values. See Figure 69. The solution is $(-3, 0)\cup(2, \infty)$; In set builder notation the interval is $\{x \mid -3 < x < 0 \text{ or } x > 2\}$.

Interval	Test Value x	$x^3 + x^2 - 6x$	Positive or Negative?
$(-\infty, -3)$	-4	-24	Negative
$(-3, 0)$	-1	6	Positive
$(0, 2)$	1	-4	Negative
$(2, \infty)$	3	18	Positive

Figure 69

71. Graph $Y_1 = (2X - 1)/(X + 2)$ using dot mode as shown in Figure 71. The graph has a vertical asymptote

at $x = -2$ and an x-intercept at $x = \dfrac{1}{2}$. The graph is above the x-axis on the interval $(-\infty, -2) \cup \left(\dfrac{1}{2}, \infty\right)$;

In set builder notation the interval is $\left\{ x \,\middle|\, x < -2 \text{ or } x > \dfrac{1}{2} \right\}$

[-8, 4, 2] by [-5, 5, 1]

Figure 71

73. $(36^{3/4})^2 = 36^{3/2} = (\sqrt{36}\,)^3 = 6^3 = 216$

75. $(2^{-3/2} \cdot 2^{1/2})^{-3} = (2^{-3/2 + 1/2})^{-3} = (2^{-1})^{-3} = 2^3 = 8$

77. $\sqrt[3]{x^4} = x^{4/3}$

79. $\sqrt[3]{y \cdot \sqrt{y}} = (y \cdot y^{1/2})^{1/3} = (y^{3/2})^{1/3} = y^{1/2}$

81. $D = \{x \mid x \geq 0\}; f(3) = 3^{5/2} \approx 15.59$

83. $x^5 = 1024 \Rightarrow x = \sqrt[5]{1024} = 4$　　Check: $4^5 = 1024$

85. $\sqrt{x - 2} = x - 4 \Rightarrow x - 2 = (x - 4)^2 \Rightarrow x - 2 = x^2 - 8x + 16 \Rightarrow x^2 - 9x + 18 = 0 \Rightarrow$

$(x - 6)(x - 3) = 0 \Rightarrow x = 3 \text{ or } 6$

Check: $\sqrt{3 - 2} \neq 3 - 4$ (not a solution); $\sqrt{6 - 2} = 2 = 6 - 4$. The only solution is $x = 6$.

87. $2x^{1/4} + 3 = 6 \Rightarrow 2x^{1/4} = 3 \Rightarrow x^{1/4} = \dfrac{3}{2} \Rightarrow x = \left(\dfrac{3}{2}\right)^4 \Rightarrow x = \dfrac{81}{16}$

Check: $2\left(\dfrac{81}{16}\right)^{1/4} + 3 = 2\left(\dfrac{3}{2}\right) + 3 = 3 + 3 = 6$

89. $\sqrt[3]{2x - 3} + 1 = 4 \Rightarrow \sqrt[3]{2x - 3} = 3 \Rightarrow 2x - 3 = 3^3 \Rightarrow 2x - 3 = 27 \Rightarrow 2x = 30 \Rightarrow x = 15$

Check: $\sqrt[3]{2(15) - 3} + 1 = \sqrt[3]{27} + 1 = 3 + 1 = 4$

91. $2n^{-2} - 5n^{-1} = 3 \Rightarrow 2(n^{-1})^2 - 5(n^{-1}) = 3$, let $u = n^{-1}$, then $2u^2 - 5u - 3 = 0 \Rightarrow$

$(2u + 1)(u - 3) = 0 \Rightarrow u = -\dfrac{1}{2}, 3$. If $u = n^{-1}$, then it follows that $n = u^{-1}$, thus $n = \left(-\dfrac{1}{2}\right)^{-1} \Rightarrow$

$n = -2 \text{ or } n = (3)^{-1} \Rightarrow n = \dfrac{1}{3}$

Check: $2(-2)^{-2} - 5(-2)^{-1} = 3 \Rightarrow 2\left(\dfrac{1}{4}\right) - 5\left(-\dfrac{1}{2}\right) = 3 \Rightarrow \dfrac{1}{2} + \dfrac{5}{2} = 3 \Rightarrow 3 = 3$

Check: $2\left(\dfrac{1}{3}\right)^{-2} - 5\left(\dfrac{1}{3}\right)^{-1} = 3 \Rightarrow 2(9) - 5(3) = 3 \Rightarrow 18 - 15 = 3 \Rightarrow 3 = 3$

The solution are: $n = -2, \dfrac{1}{3}$.

93. $k^{2/3} - 4k^{1/3} - 5 = 0 \Rightarrow (k^{1/3})^2 - 4(k^{1/3}) - 5 = 0$, let $u = k^{1/3}$, then $u^2 - 4u - 5 = 0 \Rightarrow$

$(u - 5)(u + 1) = 0 \Rightarrow u = -1, 5$. If $u = k^{1/3}$, then it follows that $k = u^3$, thus $k = (-1)^3 \Rightarrow$

$k = -1$ or $k = (5)^3 \Rightarrow k = 125$.

Check: $(-1)^{2/3} - 4(-1)^{1/3} - 5 = 0 \Rightarrow 1 - (-4) - 5 = 0 \Rightarrow 0 = 0$

Check: $(125)^{2/3} - 4(125)^{1/3} - 5 = 0 \Rightarrow 25 - 4(5) - 5 = 0 \Rightarrow 0 = 0$. The solutions are: $k = -1, 125$.

95. $\sqrt{x + 1} + 1 = \sqrt{2x} \Rightarrow (\sqrt{x + 1} + 1)^2 = 2x \Rightarrow x + 1 + 2\sqrt{x + 1} + 1 = 2x \Rightarrow$

$2\sqrt{x + 1} = x - 2 \Rightarrow 4(x + 1) = x^2 - 4x + 4 \Rightarrow 4x + 4 = x^2 - 4x + 4 \Rightarrow x^2 - 8x = 0 \Rightarrow$

$x(x - 8) = 0 \Rightarrow x = 0, 8$.

Check: $\sqrt{0 + 1} + 1 = \sqrt{2(0)} \Rightarrow 1 + 1 = 0 \Rightarrow 2 \neq 0 \Rightarrow 0$ is not a solution.

Check: $\sqrt{8 + 1} + 1 = \sqrt{2(8)} \Rightarrow 4 = 4$. The solution is: $x = 8$.

97. (a) Dog: $f(24) = 1607(24)^{-0.75} \approx 148$ bpm; Person: $f(66) = 1607(66)^{-0.75} \approx 69$ bpm

(b) We must approximate the length of an animal with a heart rate of 400 bpm. To do this, solve the

equation $f x) = 400$ or $1607x^{-0.75} = 400$ bpm.

$1607x^{-0.75} = 400 \Rightarrow x^{-0.75} = \dfrac{400}{1607} \Rightarrow x^{0.75} = \dfrac{1607}{400} \Rightarrow x^{3/4} = \dfrac{1607}{400} \Rightarrow x = \left(\dfrac{1607}{400}\right)^{4/3} \approx 6.4$

A six-inch animal such as a large bird or smaller rodent might have a heart rate of 400 bpm.

[0, 13, 1] by [0, 100, 10]

Figure 99

99. (a) May is $T(5) \Rightarrow T(5) = -0.064(5)^3 + 0.56(5)^2 + 2.9(5) + 61 \Rightarrow$

$T(5) = -8 + 14 + 14.5 + 61 \Rightarrow T(5) = 81.5$. The temperature in May is 81.5°F.

(b) Graph the function. See Figure 99. From the graph the ocean reaches a maximum temperature of about

87.3°F in July.

101. Since the time is directly proportional to the square root of the height, the equation $t = k\sqrt{h}$ must hold. If it

takes 1 second to drop 16 feet then $1 = k\sqrt{16} \Rightarrow k = 0.25$. Therefore, from a height of 256 feet the object

will take $t = 0.25\sqrt{256} = 4$ seconds to strike the ground.

Chapters 1-4 Cumulative Review Exercises

1. Percent change is: $\dfrac{\text{change}}{\text{original quantity}} = \dfrac{54 - 45}{45} = \dfrac{9}{45} = 0.20$. Therefore a 20% change.

3. (a) $D = \{-3, -1, 0, 1\}$; $R = \{4, -2, 5\}$

   (b) No, all input must have unique outputs and -1 has an output of -2 and 5.

5. $D = \{x \mid -2 \le x \le 2\}$; $R = \{x \mid 0 \le x \le 3\}$

7. (a) $D = \{x \mid x \le -2 \text{ or } x \ge 2\}$, all numbers between -2 and 2 give the unreal solutions, square roots of negative numbers.

   (b) $f(2) = \sqrt{(2)^2 - 4} \Rightarrow f(2) = \sqrt{0} \Rightarrow f(2) = 0$

9. $c(x) = 0.25x + 200$; $c(2000) = 0.25(2000) + 200 \Rightarrow c(2000) = 700$. The monthly cost of driving and maintaining a car driven 2000 miles is \$700.

11. $f(-3) = (-3)^3 - (-3) \Rightarrow f(-3) = -24$ or $(-3, -24)$; $f(-2) = (-2)^3 - (-2) \Rightarrow$

    $f(-2) = -6$ or $(-2, -6)$. The average rate of change $= \dfrac{-6 - (-24)}{-2 - (-3)} = \dfrac{18}{1} = 18$.

13. Slope $= \dfrac{(-4) - 5}{3 - (-2)} = \dfrac{-9}{5}$. Using point slope form $y = m(x - x_1) + y$ and point $(-2, 5)$ we get:

    $y = \dfrac{-9}{5}(x + 2) + 5 \Rightarrow y = \dfrac{-9}{5}x - \dfrac{18}{5} + 5 \Rightarrow y = \dfrac{-9}{5}x + \dfrac{7}{5}$

15. All lines parallel to the x-axis have 0 slope and will have graphs of $y = $ (the y-coordinate) $\Rightarrow y = -5$.

17. Each radio costs \$15 to manufacture. The fixed cost for this company to produce radios is \$2000.

19. $-3(2 - 3x) - (-x - 1) = 1 \Rightarrow -6 + 9x + x + 1 = 1 \Rightarrow 10x - 5 = 1 \Rightarrow 10x = 6 \Rightarrow x = \dfrac{3}{5}$

21. $|3x - 4| + 1 = 5 \Rightarrow |3x - 4| = 4 \Rightarrow 3x - 4 = 4 \Rightarrow 3x = 8 \Rightarrow x = \dfrac{8}{3}$ or

    $3x - 4 = -4 \Rightarrow 3x = 0 \Rightarrow x = 0$. Therefore $x = 0, \dfrac{8}{3}$.

23. $7x^2 + 9x = 10 \Rightarrow 7x^2 + 9x - 10 = 0 \Rightarrow (7x - 5)(x + 2) = 0 \Rightarrow x = -2, \dfrac{5}{7}$

25. $2x^3 + 4x^2 = 6x \Rightarrow 2x^3 + 4x^2 - 6x = 0 \Rightarrow 2x(x^2 + 2x - 3) = 0 \Rightarrow 2x(x + 3)(x - 1) = 0 \Rightarrow$

    $x = -3, 0, 1$

27. $3x^{2/3} + 5x^{1/3} - 2 = 0 \Rightarrow 3(x^{1/3})^2 + 5(x^{1/3}) - 2 = 0$. Let $u = x^{1/3}$, then $3u^2 + 5u - 2 = 0 \Rightarrow$

    $(3u - 1)(u + 2) = 0 \Rightarrow u = -2, \dfrac{1}{3}$, If $u = x^{1/3}$ then it follows $x = u^3$, thus $x = (-2)^3 \Rightarrow x = -8$ or

    $x = \left(\dfrac{1}{3}\right)^3 \Rightarrow x = \dfrac{1}{27}$. Therefore $x = -8, \dfrac{1}{27}$

29. $\dfrac{2x - 3}{5 - x} = \dfrac{4x - 3}{1 - 2x} \Rightarrow (2x - 3)(1 - 2x) = (4x - 3)(5 - x) \Rightarrow$

    $2x - 4x^2 - 3 + 6x = 20x - 4x^2 - 15 + 3x \Rightarrow 15x = 12 \Rightarrow x = \dfrac{4}{5}$

31. $\frac{1}{2}x - (4 - x) + 1 = \frac{3}{2}x - 5 \Rightarrow \frac{1}{2}x - 4 + x + 1 = \frac{3}{2}x - 5 \Rightarrow \frac{3}{2}x - 3 = \frac{3}{2}x - 5 \Rightarrow$

$-3 = -5$, since this is false the equation has no solutions and is a contradiction.

33. $-\frac{1}{3}x - (1 + x) > \frac{2}{3}x \Rightarrow -\frac{1}{3}x - 1 - x > \frac{2}{3}x \Rightarrow -\frac{4}{3}x - 1 > \frac{2}{3}x \Rightarrow -2x > 1 \Rightarrow x < -\frac{1}{2} \Rightarrow$

$\left(-\infty, -\frac{1}{2}\right)$; In set builder notation the interval is $\left\{x \mid x < -\frac{1}{2}\right\}$.

35. The solutions to $|5x - 7| \geq 3$ satisfy $x \leq s_1$ or $x \geq s_2$ where s_1 and s_2 are the solutions to $|5x - 7| = 3$.

$|5x - 7| = 3$ is equivalent to $5x - 7 = -3 \Rightarrow x = \frac{4}{5}$ and $5x - 7 = 3 \Rightarrow x = 2$.

The interval is $\left(-\infty, \frac{4}{5}\right] \cup [2, \infty)$; In set builder notation the interval is $\left\{x \mid x \leq \frac{4}{5} \text{ or } x \geq 2\right\}$.

37. For $x^3 - 9x \leq 0$, first set $x^3 - 9x = 0 \Rightarrow x(x^2 - 9) = 0 \Rightarrow x(x + 3)(x - 3) = 0 \Rightarrow x = -3, 0, 3$.

These boundary numbers separate the number line into these intervals: $(-\infty, -3), (-3, 0), (0, 3)$ and $(3, \infty)$.

Checking an x value in each interval we get:

for $x = -5, (-5)^3 - 9(-5) \leq 0 \Rightarrow -125 + 45 \leq 0 \Rightarrow -80 \leq 0$, which is true $\Rightarrow$ a solution;

for $x = -1, (-1)^3 - 9(-1) \leq 0 \Rightarrow -1 + 9 \leq 0 \Rightarrow 8 \leq 0$, which is false $\Rightarrow$

not a solution; for $x = 1, (1)^3 - 9(1) \leq 0 \Rightarrow 1 - 9 \leq 0 \Rightarrow -8 \leq 0$, which is true $\Rightarrow$ a solution.

for $x = 5, (5)^3 - 9(5) \leq 0 \Rightarrow 125 - 45 \leq 0 \Rightarrow 80 \leq 0$, which is false $\Rightarrow$ not a solution.

Therefore $(-\infty, -3] \cup [0, 3]$. In set builder notation the intervals are $\{x \mid x \leq -3 \text{ or } 0 \leq x \leq 3\}$.

39. (a) $x = -3, -1, 1, 2$

(b) $(-\infty, -3) \cup (-1, 1) \cup (2, \infty)$; In set builder notation the intervals are $\{x \mid x < -3 \text{ or } -1 < x < 1 \text{ or } x > 2\}$.

(c) $[-3, -1] \cup [1, 2]$; In set builder notation the intervals are $\{x \mid -3 \leq x \leq -1 \text{ or } 1 \leq x \leq 2\}$.

41. $f(x) = -\frac{1}{2}x^2 + 3x - 2 \Rightarrow y = -\frac{1}{2}x^2 + 3x - 2 \Rightarrow y + 2 = -\frac{1}{2}x^2 + 3x \Rightarrow$

$y + 2 = -\frac{1}{2}(x^2 - 6x) \Rightarrow y + 2 - \frac{9}{2} = -\frac{1}{2}(x^2 - 6x + 9) \Rightarrow y - \frac{5}{2} = -\frac{1}{2}(x - 3)^2 \Rightarrow$

$y = -\frac{1}{2}(x - 3)^2 + \frac{5}{2} \Rightarrow f(x) = -\frac{1}{2}(x - 3)^2 + \frac{5}{2} \Rightarrow$ the vertex is $\left(3, \frac{5}{2}\right)$.

43. (a) Shift $f(x)$ 2 units left and 1 unit down. See Figure 43a.

(b) Multiply the y-coordinate of each point by -2 and plot. The new graph will open up. See Figure 43b.

(c) Shift $f(x)$ up one unit and reflect across the y-axis. See Figure 43c.

(d) Multiply the x-coordinate of each point by 2 and plot. See Figure 43d.

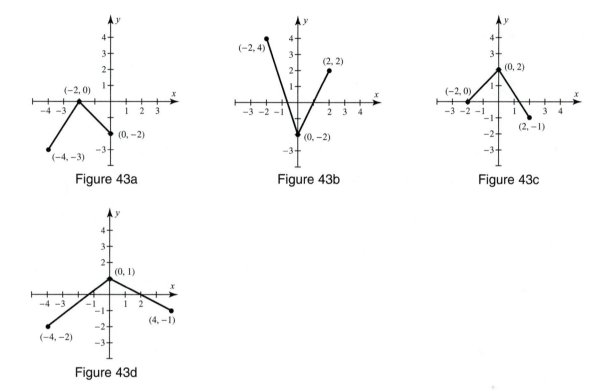

Figure 43a   Figure 43b   Figure 43c

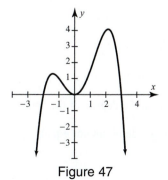

Figure 43d

45. (a) It is increasing: $(-\infty, -2]$ and $[1, \infty)$; it is decreasing: $[-2, 1]$. In set builder notation the intervals are $\{x \mid x \leq -2 \text{ or } x \geq 1\}$ and $\{x \mid -2 \leq x \leq 1\}$.

(b) The zeros are approximately $x = -3.3, 0$ and 1.8.

(c) The turning points are: $(-2, 3)$ and $(1, -1)$.

(d) It has local extrema, maximum: 3, and minimum: -1.

47. Answers may vary. See Figure 47.

Figure 47

49. (a)

$$\require{enclose}\begin{array}{r} a - 2 + \frac{12}{4a^2} \\ 4a^2 \enclose{longdiv}{4a^3 - 8a^2 + 12} \\ \underline{4a^3 } \\ -8a^2 + 12 \\ \underline{-8a^2 } \\ 12 \end{array}$$

Therefore the quotient is: $a - 2 + \dfrac{3}{a^2}$

(b)

$$\begin{array}{r} 2x^2 + 2x - 2 + \frac{-1}{x-1} \\ x - 1 \enclose{longdiv}{2x^3 - 4x + 1} \\ \underline{2x^3 - 2x^2 } \\ 2x^2 - 4x + 1 \\ \underline{2x^2 - 2x } \\ -2x + 1 \\ \underline{-2x + 2} \\ -1 \end{array}$$

Therefore the quotient is: $2x^2 + 2x - 2 + \dfrac{-1}{x - 1}$

(c)

$$\begin{array}{r} x^2 + 2x - 3 + \frac{x+4}{x^2+2} \\ x^2 + 2 \enclose{longdiv}{x^4 + 2x^3 - x^2 + 5x - 2} \\ \underline{x^4 + 2x^2 } \\ 2x^3 - 3x^2 + 5x - 2 \\ \underline{2x^3 + 4x } \\ -3x^2 + x - 2 \\ \underline{-3x^2 - 6} \\ x + 4 \end{array}$$

Therefore the quotient is: $x^2 + 2x - 3 + \dfrac{x + 4}{x^2 + 2}$

51. $f(x) = 4(x + 3)(x - 1)^2(x - 4)^3$

53. $2x^3 - x^2 - 6x + 3 = (2x^3 - x^2) - (6x - 3) = x^2(2x - 1) - 3(2x - 1) = (x^2 - 3)(2x - 1) =$

$(x + \sqrt{3})(x - \sqrt{3})(2x - 1) \Rightarrow f(x) = (x + \sqrt{3})(x - \sqrt{3})(2x - 1) \Rightarrow$

the complete factored form is: $f(x) = 2(x + \sqrt{3})(x - \sqrt{3})\left(x - \dfrac{1}{2}\right)$

55. $\dfrac{3 + 4i}{1 - i} \cdot \dfrac{1 + i}{1 + i} = \dfrac{3 + 3i + 4i + 4i^2}{1 - (i)^2} = \dfrac{3 + 7i + 4(-1)}{1 - (-1)} = \dfrac{-1 + 7i}{2}$ or $\dfrac{-1}{2} + \dfrac{7i}{2}$

57. For the domain the denominator cannot equal zero $\Rightarrow$ set $x^2 - 3x - 4 = 0 \Rightarrow (x + 1)(x - 4) = 0$.

Then $D = \{x \mid x \neq -1, x \neq 4\}$. Since the degree of the numerator is less then the degree of the denominator

the horizontal asymptote is $y = 0$. The vertical asymptotes are values when the demoninator equals $0 \Rightarrow$

the vertical asymptotes are: $x = -1, x = 4$.

59. For m_1 use $(0, 0)$ and $(2, 40{,}000)$ $\Rightarrow$ $\dfrac{40{,}000 - 0}{2 - 0}$ $\Rightarrow$ $\dfrac{40{,}000}{2}$ $\Rightarrow$ $m_1 = 20{,}000$; the pool is being filled at a

rate of 20,000 gallons per hour.

For m_2 use $(2, 40{,}000)$ and $(3, 50{,}000)$ $\Rightarrow$ $\dfrac{50{,}000 - 40{,}000}{3 - 2}$ $\Rightarrow$ $\dfrac{10{,}000}{1}$ $\Rightarrow$ $m_2 = 10{,}000$; now the pool is

being filled at a rate of 10,000 gallons per hour.

For m_3 use $(3, 50{,}000)$ and $(4, 50{,}000)$ $\Rightarrow$ $\dfrac{50{,}000 - 50{,}000}{4 - 3}$ $\Rightarrow$ $\dfrac{0}{1}$ $\Rightarrow$ $m_3 = 0$; the amount of water in the

pool is remaining constant.

For m_4 use $(4, 50{,}000)$ and $(6, 20{,}000)$ $\Rightarrow$ $\dfrac{20{,}000 - 50{,}000}{6 - 4}$ $\Rightarrow$ $\dfrac{-30{,}000}{2}$ $\Rightarrow$ $m_4 = -15{,}000$; the pool is

being drained at a rate of 15,000 gallons per hour.

61. $D(0) = 4(0)^2 = 0 \Rightarrow (0, 0)$; $D(2) = 4(2)^2 = 16 \Rightarrow (2, 16)$; $D(4) = 4(4)^2 = 64 \Rightarrow (4, 64)$

The average rate of change from 0 to 2 is: $\dfrac{16 - 0}{2 - 0} = \dfrac{16}{2} = 8$.

The average rate of change from 2 to 4 is: $\dfrac{64 - 16}{4 - 2} = \dfrac{48}{2} = 24$.

Between 0 and 2 seconds, the car travels at an average rate of 8 feet per second. Between 2 and 4 seconds,

the car travels at an average rate of 24 feet per second.

63. $0.35(2) + 0.12x = 0.20(x + 2) \Rightarrow 0.70 + 0.12x = 0.20x + 0.40 \Rightarrow 0.30 = 0.08x \Rightarrow x = 3.75$ liters

65. Graph $R(x) = x(800 - x)$ with x by the 100's from 0 to 800. See Figure 65. The maximum revenue is made

when 400 toy figures are sold.

[0, 1000, 100] by [0, 200000, 20,000]   [0, 300, 100] by [0, 50000, 10,000]   [1997, 2003, 1] by [0, 20, 2]

Figure 65   Figure 67   Figure 69

67. (a) $C(t) = t(805 - 5t)$

(b) $C(t) = 17{,}000 \Rightarrow 17{,}000 = t(805 - 5t) \Rightarrow 17{,}000 = 805t - 5t^2 \Rightarrow 5t^2 - 805t + 17{,}000 = 0 \Rightarrow$

$5(t^2 - 161t + 3400) = 0 \Rightarrow 5(t - 25)(t - 136) = 0 \Rightarrow t = 25, 136$. The cost is $17,000 when either

25 or 136 tickets are purchased.

(c) Graph $C(t) = t(805 - 5t)$. See Figure 67. The maximum cost of $32,400 is reached at 80 to 81 tickets

are sold.

69. (a) Using the lowest values $x = 1998$ and $y = 6.1$ as the vertex of the parabola we get:

$f(x) = a(x - 1998)^2 + 6.1$. Using the graph, see Figure 69, and trial and error, $a \approx 0.5$ produces a

reasonable fit for the data $\Rightarrow$ $f(x) = 0.5(x - 1998)^2 + 6.1$. *Answers may vary.*

(b) $f(2004) = 0.5(2004 - 1998)^2 + 6.1 \Rightarrow f(2004) = 0.5(6)^2 + 6.1 \Rightarrow f(2004) = 0.5(36) + 6.1 \Rightarrow$

$f(2004) = 24.1\%$. *Answers may vary.*

71. Since volume of a cylinder is: $V = \pi r^2 h$. Then $10\pi = \pi r^2 h \Rightarrow h = \dfrac{10}{r^2}$.

The surface area of the can is: $S(r) = 2\pi r^2 + 2\pi rh \Rightarrow S(r) = 2\pi r^2 + 2\pi r \cdot \dfrac{10}{r^2} \Rightarrow$

$S(r) = 2\pi r^2 + \dfrac{20\pi}{r}$. Graph this equation. See Figure 71. The minimum surface area occurs when

$r \approx 1.7$ inches $\Rightarrow$ then $h = \dfrac{10}{(1.7)^2}$ or $h \approx 3.5$ inches.

[0, 5, 1] by [0, 100, 10]

Figure 71

Chapter 5: Exponential and Logarithmic Functions

5.1: Combining Functions

1. $(f + g)(3) = f(3) + g(3) = 2 + 5 = 7$

3. $(fg)(x) = f(x)g(x) = (x^2)(4x) = 4x^3$

5. $(g \circ f)(x) = g(f(x))$. Therefore, $(g \circ f)(x)$ calculates the cost of x square yards of carpet.

7. (a) $f(3) = 2(3) - 3 = 3$ and $g(3) = 1 - 3^2 = -8$. Thus, $(f + g)(3) = f(3) + g(3) = 3 + (-8) = -5$.

   (b) $f(-1) = 2(-1) - 3 = -5$ and $g(-1) = 1 - (-1)^2 = 0$.

   Thus, $(f - g)(-1) = f(-1) - g(-1) = -5 - 0 = -5$.

   (c) $f(0) = 2(0) - 3 = -3$ and $g(0) = 1 - 0^2 = 1$. Thus, $(fg)(0) = f(0) \cdot g(0) = (-3)(1) = -3$.

   (d) $f(2) = 2(2) - 3 = 1$ and $g(2) = 1 - 2^2 = -3$. Thus, $(f/g)(2) = \dfrac{f(2)}{g(2)} = \dfrac{1}{-3} = -\dfrac{1}{3}$.

9. (a) $f(2) = 2(2) + 1 = 5$ and $g(2) = \dfrac{1}{2}$. Thus, $(f + g)(2) = f(2) + g(2) = 5 + \dfrac{1}{2} = \dfrac{11}{2}$.

   (b) $f\left(\dfrac{1}{2}\right) = 2\left(\dfrac{1}{2}\right) + 1 = 2$ and $g\left(\dfrac{1}{2}\right) = \dfrac{1}{\frac{1}{2}} = 2$. Thus, $(f - g)\left(\dfrac{1}{2}\right) = f\left(\dfrac{1}{2}\right) - g\left(\dfrac{1}{2}\right) = 2 - 2 = 0$.

   (c) $f(4) = 2(4) + 1 = 9$ and $g(4) = \dfrac{1}{4}$. Thus, $(fg)(4) = f(4) \cdot g(4) = (9)\left(\dfrac{1}{4}\right) = \dfrac{9}{4}$.

   (d) $f(0) = 2(0) + 1 = 1$ and $g(0) = \dfrac{1}{0} \Rightarrow$ undefined. Thus, $(f/g)(0) \Rightarrow$ undefined.

11. (a) $(f + g)(x) = f(x) + g(x) = 2x + x^2$; Domain: all real numbers.

    (b) $(f - g)(x) = f(x) - g(x) = 2x - x^2$; Domain: all real numbers.

    (c) $(fg)(x) = f(x) \cdot g(x) = (2x)(x^2) = 2x^3$; Domain: all real numbers.

    (d) $(f/g)(x) = \dfrac{f(x)}{g(x)} = \dfrac{2x}{x^2} = \dfrac{2}{x}$; Domain: $\{x \mid x \neq 0\}$.

13. (a) $(f + g)(x) = f(x) + g(x) = (x^2 - 1) + (x^2 + 1) = 2x^2$; Domain: all real numbers.

    (b) $(f - g)(x) = f(x) - g(x) = (x^2 - 1) - (x^2 + 1) = -2$; Domain: all real numbers.

    (c) $(fg)(x) = f(x) \cdot g(x) = (x^2 - 1)(x^2 + 1) = x^4 - 1$; Domain: all real numbers.

    (d) $(f/g)(x) = \dfrac{f(x)}{g(x)} = \dfrac{x^2 - 1}{x^2 + 1}$; Domain: all real numbers.

15. (a) $(f + g)(x) = f(x) + g(x) = (x - \sqrt{x - 1}) + (x + \sqrt{x - 1}) = 2x$; Domain: $\{x \mid x \geq 1\}$.

    (b) $(f - g)(x) = f(x) - g(x) = (x - \sqrt{x - 1}) - (x + \sqrt{x - 1}) = -2\sqrt{x - 1}$; Domain: $\{x \mid x \geq 1\}$.

    (c) $(fg)(x) = f(x) \cdot g(x) = (x - \sqrt{x - 1})(x + \sqrt{x - 1}) = x^2 - (x - 1) = x^2 - x + 1$;

    Domain: $\{x \mid x \geq 1\}$.

    (d) $(f/g)(x) = \dfrac{f(x)}{g(x)} = \dfrac{x - \sqrt{x - 1}}{x + \sqrt{x - 1}}$; Domain: $\{x \mid x \geq 1\}$.

17. (a) $(f + g)(x) = f(x) + g(x) = (\sqrt{x} - 1) + (\sqrt{x} + 1) = 2\sqrt{x}$; Domain: $\{x \mid x \geq 0\}$.

(b) $(f - g)(x) = f(x) - g(x) = (\sqrt{x} - 1) - (\sqrt{x} + 1) = -2$; Domain: $\{x \mid x \geq 0\}$.

(c) $(fg)(x) = f(x) \cdot g(x) = (\sqrt{x} - 1)(\sqrt{x} + 1) = x - 1$; Domain: $\{x \mid x \geq 0\}$.

(d) $(f/g)(x) = \dfrac{f(x)}{g(x)} = \dfrac{\sqrt{x} - 1}{\sqrt{x} + 1}$; Domain: $\{x \mid x \geq 0\}$.

19. (a) $(f + g)(x) = f(x) + g(x) = \dfrac{1}{x + 1} + \dfrac{3}{x + 1} = \dfrac{4}{x + 1}$; Domain: $\{x \mid x \neq -1\}$.

(b) $(f - g)(x) = f(x) - g(x) = \dfrac{1}{x + 1} - \dfrac{3}{x + 1} = -\dfrac{2}{x + 1}$; Domain: $\{x \mid x \neq -1\}$.

(c) $(fg)(x) = f(x) \cdot g(x) = \left(\dfrac{1}{x + 1}\right)\left(\dfrac{3}{x + 1}\right) = \dfrac{3}{(x + 1)^2}$; Domain: $\{x \mid x \neq -1\}$.

(d) $(f/g)(x) = \dfrac{f(x)}{g(x)} = \dfrac{\frac{1}{x + 1}}{\frac{3}{x + 1}} = \dfrac{1}{x + 1} \cdot \dfrac{x + 1}{3} = \dfrac{1}{3}$; Domain: $\{x \mid x \neq -1\}$.

21. (a) $(f + g)(x) = f(x) + g(x) = \dfrac{1}{2x - 4} + \dfrac{x}{2x - 4} = \dfrac{1 + x}{2x - 4}$; Domain: $\{x \mid x \neq 2\}$.

(b) $(f - g)(x) = f(x) - g(x) = \dfrac{1}{2x - 4} - \dfrac{x}{2x - 4} = \dfrac{1 - x}{2x - 4}$; Domain: $\{x \mid x \neq 2\}$.

(c) $(fg)(x) = f(x) \cdot g(x) = \left(\dfrac{1}{2x - 4}\right)\left(\dfrac{x}{2x - 4}\right) = \dfrac{x}{(2x - 4)^2}$; Domain: $\{x \mid x \neq 2\}$.

(d) $(f/g)(x) = \dfrac{f(x)}{g(x)} = \dfrac{\frac{1}{2x - 4}}{\frac{x}{2x - 4}} = \dfrac{1}{2x - 4} \cdot \dfrac{2x - 4}{x} = \dfrac{1}{x}$; Domain: $\{x \mid x \neq 0 \text{ and } x \neq 2\}$.

23. (a) $(f + g)(x) = f(x) + g(x) = (x^2 - 1) + (|x + 1|) = x^2 - 1 + |x + 1|$; Domain: all real numbers.

(b) $(f - g)(x) = f(x) - g(x) = (x^2 - 1) - (|x + 1|) = x^2 - 1 - |x + 1|$; Domain: all real numbers.

(c) $(fg)(x) = f(x) \cdot g(x) = (x^2 - 1)(|x + 1|) = (x^2 - 1)(|x + 1|)$; Domain: all real numbers.

(d) $(f/g)(x) = \dfrac{f(x)}{g(x)} = \dfrac{x^2 - 1}{|x + 1|}$; Domain: $\{x \mid x \neq -1\}$.

25. (a) $(f + g)(x) = f(x) + g(x) = \dfrac{(x - 1)(x - 2)}{x + 1} + \dfrac{(x + 1)(x - 1)}{x - 2} =$

$\dfrac{(x - 1)(x - 2)(x - 2)}{(x + 1)(x - 2)} + \dfrac{(x - 1)(x + 1)(x + 1)}{(x + 1)(x - 2)} =$

$\dfrac{(x - 1)(x^2 - 4x + 4)}{(x + 1)(x - 2)} + \dfrac{(x - 1)(x^2 + 2x + 1)}{(x + 1)(x - 2)} = \dfrac{(x - 1)(2x^2 - 2x + 5)}{(x + 1)(x - 2)}$;

Domain: $\{x \mid x \neq -1, x \neq 2\}$

(b) $(f - g)(x) = f(x) - g(x) = \dfrac{(x - 1)(x - 2)}{x + 1} - \dfrac{(x + 1)(x - 1)}{x - 2} =$

$\dfrac{(x - 1)(x - 2)(x - 2)}{(x + 1)(x - 2)} - \dfrac{(x - 1)(x + 1)(x + 1)}{(x + 1)(x - 2)} =$

$\dfrac{(x - 1)(x^2 - 4x + 4)}{(x + 1)(x - 2)} - \dfrac{(x - 1)(x^2 + 2x + 1)}{(x + 1)(x - 2)} = \dfrac{(x - 1)(-6x + 3)}{(x + 1)(x - 2)} = \dfrac{-3(x - 1)(2x - 1)}{(x + 1)(x - 2)}$;

Domain: $\{x \mid x \neq -1, x \neq 2\}$

(c) $(fg)(x) = f(x) \cdot g(x) = \dfrac{(x-1)(x-2)}{x+1} \cdot \dfrac{(x+1)(x-1)}{x-2} =$

$\dfrac{(x-2)(x-1)(x+1)(x-1)}{(x+1)(x-2)} = (x-1)(x-1) = (x-1)^2$; Domain: $\{x \mid x \neq -1, x \neq 2\}$

(d) $(f/g)(x) = \dfrac{f(x)}{g(x)} = \dfrac{(x-2)(x-1)}{x+1} \div \dfrac{(x+1)(x-1)}{x-2} = \dfrac{(x-2)(x-1)}{x+1} \cdot \dfrac{x-2}{(x+1)(x-1)} =$

$\dfrac{(x-2)^2}{(x+1)^2}$; Domain: $\{x \mid x \neq 1, x \neq -1 \text{ and } x \neq 2\}$

27. (a) $(f+g)(x) = f(x) + g(x) = \dfrac{2}{(x+1)(x-1)} + \dfrac{(x+1)}{(x-1)(x-1)} =$

$\dfrac{2(x-1)}{(x+1)(x-1)^2} + \dfrac{(x+1)^2}{(x+1)(x-1)^2} = \dfrac{2x-2}{(x+1)(x-1)^2} + \dfrac{x^2+2x+1}{(x+1)(x-1)^2} = \dfrac{x^2+4x-1}{(x+1)(x-1)^2}$;

Domain: $\{x \mid x \neq -1, x \neq 1\}$

(b) $(f-g)(x) = f(x) - g(x) = \dfrac{2}{(x+1)(x-1)} - \dfrac{(x+1)}{(x-1)(x-1)} =$

$\dfrac{2(x-1)}{(x+1)(x-1)^2} - \dfrac{(x+1)^2}{(x+1)(x-1)} = \dfrac{2x-2}{(x+1)(x-1)^2} - \dfrac{x^2+2x+1}{(x+1)(x-1)^2} = \dfrac{-x^2-3}{(x+1)(x-1)^2}$;

Domain: $\{x \mid x \neq -1, x \neq 1\}$

(c) $(fg)(x) = f(x) \cdot g(x) = \dfrac{2}{(x+1)(x-1)} \cdot \dfrac{(x+1)}{(x-1)(x-1)} = \dfrac{2}{(x-1)^3}$; Domain: $\{x \mid x \neq -1, x \neq 1\}$

(d) $(f/g)(x) = f(x) \div g(x) = \dfrac{2}{(x+1)(x-1)} \div \dfrac{(x+1)}{(x-1)(x-1)} =$

$\dfrac{2}{(x+1)(x-1)} \cdot \dfrac{(x-1)(x-1)}{(x+1)} = \dfrac{2(x-1)}{(x+1)^2}$; Domain: $\{x \mid x \neq -1, x \neq 1\}$

29. (a) $(f+g)(x) = f(x) + g(x) = (x^{5/2} - x^{3/2}) + (x^{1/2}) = x^{5/2} - x^{3/2} + x^{1/2} = x^{1/2}(x^2 - x + 1)$;

Domain: $\{x \mid x \geq 0\}$

(b) $(f-g)(x) = f(x) - g(x) = (x^{5/2} - x^{3/2}) - (x^{1/2}) = x^{5/2} - x^{3/2} - x^{1/2} = x^{1/2}(x^2 - x - 1)$;

Domain: $\{x \mid x \geq 0\}$

(c) $(fg)(x) = f(x) \cdot g(x) = (x^{5/2} - x^{3/2})(x^{1/2}) = x^3 - x^2 = x^2(x - 1)$; Domain: $\{x \mid x \geq 0\}$

(d) $(f/g)(x) = f(x) \div g(x) = \left(\dfrac{x^{5/2} - x^{3/2}}{x^{1/2}}\right) = x^2 - x = x(x-1)$; Domain: $\{x \mid x > 0\}$

31. (a) From the graph $f(2) = 4$ and $g(2) = -2$. Thus, $(f+g)(2) = f(2) + g(2) = 4 + (-2) = 2$.

(b) From the graph $f(1) = 1$ and $g(1) = -3$. Thus, $(f-g)(1) = f(1) - g(1) = 1 - (-3) = 4$.

(c) From the graph $f(0) = 0$ and $g(0) = -4$. Thus, $(fg)(0) = f(0) \cdot g(0) = (0)(-4) = 0$.

(d) From the graph $f(1) = 1$ and $g(1) = -3$. Thus, $(f/g)(1) = \dfrac{f(1)}{g(1)} = \dfrac{1}{-3} = -\dfrac{1}{3}$.

33. (a) From the graph $f(0) = 0$ and $g(0) = 2$. Thus, $(f + g)(0) = f(0) + g(0) = 0 + 2 = 2$.

    (b) From the graph $f(-1) = -2$ and $g(-1) = 1$. Thus, $(f - g)(-1) = f(-1) - g(-1) = -2 - 1 = -3$.

    (c) From the graph $f(1) = 2$ and $g(1) = 1$. Thus, $(fg)(1) = f(1) \cdot g(1) = (2)(1) = 2$.

    (d) From the graph $f(2) = 4$ and $g(2) = -2$. Thus, $(f/g)(2) = \dfrac{f(2)}{g(2)} = \dfrac{4}{-2} = -2$.

35. (a) $(f + g)(-1) = f(-1) + g(-1) = -3 + -2 = -5$.

    (b) $(g - f)(0) = g(0) - f(0) = 3 - 5 = -2$.

    (c) $(gf)(2) = g(2) \cdot f(2) = 0 \cdot 1 = 0$.

    (d) $(f/g)(2) = \dfrac{f(2)}{g(2)} = \dfrac{1}{0} \Rightarrow$ undefined.

37. (a) $(f + g)(2) = f(2) + g(2) = 7 + (-2) = 5$

    (b) $(f - g)(4) = f(4) - g(4) = 10 - 5 = 5$

    (c) $(fg)(-2) = f(-2) \cdot g(-2) = (0)(6) = 0$

    (d) $(f/g)(0) = \dfrac{f(0)}{g(0)} = \dfrac{5}{0} \Rightarrow$ undefined

39. For example: $(f - g)(2) = f(2) - g(2) = 7 - (-2) = 9$. See Figure 39. A dash (—) indicates that the value of the function is undefined.

x	-2	0	2	4
$(f + g)(x)$	6	5	5	15
$(f - g)(x)$	-6	5	9	5
$(fg)(x)$	0	0	-14	50
$(f/g)(x)$	0	—	-3.5	2

Figure 39

41. (a) $g(x) = 2x + 1 \Rightarrow g(-3) = 2(-3) + 1 \Rightarrow g(-3) = -5$

    (b) $g(x) = 2x + 1 \Rightarrow g(b) = 2(b) + 1 \Rightarrow g(b) = 2b + 1$

    (c) $g(x) = 2x + 1 \Rightarrow g(x^3) = 2(x^3) + 1 \Rightarrow g(x^3) = 2x^3 + 1$

    (d) $g(x) = 2x + 1 \Rightarrow g(2x - 3) = 2(2x - 3) + 1 \Rightarrow g(2x - 3) = 4x - 6 + 1 \Rightarrow$

    $g(2x - 3) = 4x - 5$

43. (a) $g(x) = 2(x + 3)^2 - 4 \Rightarrow g(-3) = 2((-3) + 3)^2 - 4 \Rightarrow g(-3) = 2(0) - 4 \Rightarrow g(-3) = -4$

    (b) $g(x) = 2(x + 3)^2 - 4 \Rightarrow g(b) = 2((b) + 3)^2 - 4 \Rightarrow g(b) = 2(b + 3)^2 - 4$

    (c) $g(x) = 2(x + 3)^2 - 4 \Rightarrow g(x^3) = 2((x^3) + 3)^2 - 4 \Rightarrow g(x^3) = 2(x^3 + 3)^2 - 4$

    (d) $g(x) = 2(x + 3)^2 - 4 \Rightarrow g(2x - 3) = 2((2x - 3) + 3)^2 - 4 \Rightarrow g(2x - 3) = 2(2x)^2 - 4 \Rightarrow$

    $g(2x - 3) = 2(4x^2) - 4 \Rightarrow g(2x - 3) = 8x^2 - 4$

45. (a) $g(x) = \frac{1}{2}x^2 + 3x - 1 \Rightarrow g(-3) = \frac{1}{2}(-3)^2 + 3(-3) - 1 \Rightarrow g(-3) = \frac{9}{2} - 9 - 1 \Rightarrow$

$g(-3) = \frac{9}{2} - \frac{20}{2} \Rightarrow g(-3) = -\frac{11}{2}$

(b) $g(x) = \frac{1}{2}x^2 + 3x - 1 \Rightarrow g(b) = \frac{1}{2}(b)^2 + 3(b) - 1 \Rightarrow g(b) = \frac{1}{2}b^2 + 3b - 1$

(c) $g(x) = \frac{1}{2}x^2 + 3x - 1 \Rightarrow g(x^3) = \frac{1}{2}(x^3)^2 + 3(x^3) - 1 \Rightarrow g(x^3) = \frac{1}{2}x^6 + 3x^3 - 1$

(d) $g(x) = \frac{1}{2}x^2 + 3x - 1 \Rightarrow g(2x - 3) = \frac{1}{2}(2x - 3)^2 + 3(2x - 3) - 1 \Rightarrow$

$g(2x - 3) = \frac{4x^2 - 12x + 9}{2} + 6x - 9 - 1 \Rightarrow g(2x - 3) = 2x^2 - 6x + \frac{9}{2} + 6x - \frac{20}{2} \Rightarrow$

$g(2x - 3) = 2x^2 - \frac{11}{2}$

47. (a) $g(x) = \sqrt{x + 4} \Rightarrow g(-3) = \sqrt{(-3) + 4} \Rightarrow g(-3) = \sqrt{1} \Rightarrow g(-3) = 1$

(b) $g(x) = \sqrt{x + 4} \Rightarrow g(b) = \sqrt{(b) + 4} \Rightarrow g(b) = \sqrt{b + 4}$

(c) $g(x) = \sqrt{x + 4} \Rightarrow g(x^3) = \sqrt{(x^3) + 4} \Rightarrow g(x^3) = \sqrt{x^3 + 4}$

(d) $g(x) = \sqrt{x + 4} \Rightarrow g(2x - 3) = \sqrt{(2x - 3) + 4} \Rightarrow g(2x - 3) = \sqrt{2x + 1}$

49. (a) $g(x) = |3x - 1| + 4 \Rightarrow g(-3) = |3(-3) - 1| + 4 \Rightarrow g(-3) = |-10| + 4 \Rightarrow g(-3) = 14$

(b) $g(x) = |3x - 1| + 4 \Rightarrow g(b) = |3(b) - 1| + 4 \Rightarrow g(b) = |3b - 1| + 4$

(c) $g(x) = |3x - 1| + 4 \Rightarrow g(x^3) = |3(x^3) - 1| + 4 \Rightarrow g(x^3) = |3x^3 - 1| + 4$

(d) $g(x) = |3x - 1| + 4 \Rightarrow g(2x - 3) = |3(2x - 3) - 1| + 4 \Rightarrow g(2x - 3) = |6x - 10| + 4$

51. (a) $g(x) = \frac{4x}{x + 3} \Rightarrow g(-3) = \frac{4(-3)}{(-3) + 3} \Rightarrow g(-3) = \frac{-12}{0} \Rightarrow g(-3) = $ undefined

(b) $g(x) = \frac{4x}{x + 3} \Rightarrow g(b) = \frac{4(b)}{b + 3} \Rightarrow g(b) = \frac{4b}{b + 3}$

(c) $g(x) = \frac{4x}{x + 3} \Rightarrow g(x^3) = \frac{4(x^3)}{x^3 + 3} \Rightarrow g(x^3) = \frac{4x^3}{x^3 + 3}$

(d) $g(x) = \frac{4x}{x + 3} \Rightarrow g(2x - 3) = \frac{4(2x - 3)}{(2x - 3) + 3} \Rightarrow g(2x - 3) = \frac{8x - 12}{2x} \Rightarrow$

$g(2x - 3) = \frac{4x - 6}{x} \Rightarrow g(2x - 3) = \frac{2(2x - 3)}{x}$

53. (a) $(f \circ g)(2) = f(g(2)) = f(2^2) = f(4) = \sqrt{4 + 5} = \sqrt{9} = 3$

(b) $(g \circ f)(-1) = g(f(-1)) = g(\sqrt{-1 + 5}) = g(2) = 2^2 = 4$

55. (a) $(f \circ g)(-4) = f(g(-4)) = f(|-4|) = f(4) = 5(4) - 2 = 18$

(b) $(g \circ f)(5) = g(f(5)) = g(5(5) - 2) = g(23) = |23| = 23$

57. (a) $(f \circ g)(x) = f(g(x)) = f(x^2 + 3x - 1) = (x^2 + 3x - 1)^3$; *D*: all real numbers

(b) $(g \circ f)(x) = g(f(x)) = g(x^3) = (x^3)^2 + 3(x^3) - 1 = x^6 + 3x^3 - 1$; *D*: all real numbers

(c) $(f \circ f)(x) = f(f(x)) = f(x^3) = (x^3)^3 = x^9$; *D*: all real numbers

59. (a) $(f \circ g)(x) = f(g(x)) = f(x^4 + x^2 - 3x - 4) = x^4 + x^2 - 3x - 2$; D: all real numbers

(b) $(g \circ f)(x) = g(f(x)) = g(x + 2) = (x + 2)^4 + (x + 2)^2 - 3(x + 2) - 4$; D: all real numbers

(c) $(f \circ f)(x) = f(f(x)) = f(x + 2) = (x + 2) + 2 = x + 4$; D: all real numbers

61. (a) $(f \circ g)(x) = f(g(x)) = f(x^3) = 2 - 3x^3$; D: all real numbers

(b) $(g \circ f)(x) = g(f(x)) = g(2 - 3x) = (2 - 3x)^3$; D: all real numbers

(c) $(f \circ f)(x) = f(f(x)) = f(2 - 3x) = 2 - 3(2 - 3x) = 9x - 4$; D: all real numbers

63. (a) $(f \circ g)(x) = f(g(x)) = f(5x) = \dfrac{1}{5x + 1}$; $D = \left\{ x \mid x \neq -\dfrac{1}{5} \right\}$

(b) $(g \circ f)(x) = g(f(x)) = g\left(\dfrac{1}{x + 1}\right) = 5\left(\dfrac{1}{x + 1}\right) = \dfrac{5}{x + 1}$; $D = \{x \mid x \neq -1\}$

(c) $(f \circ f)(x) = f(f(x)) = f\left(\dfrac{1}{x + 1}\right) = \dfrac{1}{\dfrac{1}{x + 1} + 1} = \dfrac{1}{\dfrac{1 + x + 1}{x + 1}} = \dfrac{x + 1}{x + 2}$;

$D = \{x \mid x \neq -1 \text{ and } x \neq -2\}$

65. (a) $(f \circ g)(x) = f(g(x)) = f(\sqrt{4 - x^2}) = \sqrt{4 - x^2} + 4$; $D = \{x \mid -2 \leq x \leq 2\}$

(b) $(g \circ f)(x) = g(f(x)) = g(x + 4) = \sqrt{4 - (x + 4)^2}$; $D = \{x \mid -6 \leq x \leq -2\}$

(c) $(f \circ f)(x) = f(f(x)) = f(x + 4) = (x + 4) + 4 = x + 8$; D: all real numbers

67. (a) $(f \circ g)(x) = f(g(x)) = f(3x) = \sqrt{3x - 1}$; $D = \left\{ x \mid x \geq \dfrac{1}{3} \right\}$

(b) $(g \circ f)(x) = g(f(x)) = g(\sqrt{x - 1}) = 3\sqrt{x - 1}$; $D = \{x \mid x \geq 1\}$

(c) $(f \circ f)(x) = f(f(x)) = f(\sqrt{x - 1}) = \sqrt{\sqrt{x - 1} - 1}$; $D = \{x \mid x \geq 2\}$

69. (a) $(f \circ g)(x) = f(g(x)) = f\left(\dfrac{1 - x}{5}\right) = 1 - 5\left(\dfrac{1 - x}{5}\right) = 1 - (1 - x) = x$; D: all real numbers

(b) $(g \circ f)(x) = g(f(x)) = g(1 - 5x) = \dfrac{1 - (1 - 5x)}{5} = \dfrac{5x}{5} = x$; D: all real numbers

(c) $(f \circ f)(x) = f(f(x)) = f(1 - 5x) = 1 - 5(1 - 5x) = 1 - 5 + 25x = 25x - 4$;

D: all real numbers

71. (a) $(f \circ g)(x) = f(g(x)) = f\left(\dfrac{1}{kx}\right) = \dfrac{1}{k\left(\frac{1}{kx}\right)} = \dfrac{1}{\frac{1}{x}} = x$; $D: \{x \mid x \neq 0\}$

(b) $(g \circ f)(x) = g(f(x)) = g\left(\dfrac{1}{kx}\right) = \dfrac{1}{k\left(\frac{1}{kx}\right)} = \dfrac{1}{\frac{1}{x}} = x$; $D: \{x \mid x \neq 0\}$

(c) $(f \circ f)(x) = f(f(x)) = f\left(\dfrac{1}{kx}\right) = \dfrac{1}{k\left(\frac{1}{kx}\right)} = \dfrac{1}{\frac{1}{x}} = x$; $D: \{x \mid x \neq 0\}$

73. (a) $(f \circ g)(4) = f(g(4)) = f(0) = -4$

(b) $(g \circ f)(3) = g(f(3)) = g(2) = 2$

(c) $(f \circ f)(2) = f(f(2)) = f(0) = -4$

75. (a) $(f \circ g)(1) = f(g(1)) = f(2) = -3$

  (b) $(g \circ f)(-2) = g(f(-2)) = g(-3) = -2$

  (c) $(g \circ g)(-2) = g(g(-2)) = g(-1) = 0$

77. (a) $(g \circ f)(1) = g(f(1)) = g(4) = 5$

  (b) $(f \circ g)(4) = f(g(4)) = f(5) \Rightarrow$ undefined

  (c) $(f \circ f)(3) = f(f(3)) = f(1) = 4$

79. $g(3) = 4;\ f(4) = 2$

81. $h(x) = \sqrt{x - 2} \Rightarrow g(x) = \sqrt{x}$ and $f(x) = x - 2$. *Answers may vary.*

83. $h(x) = \dfrac{1}{x + 2} \Rightarrow g(x) = \dfrac{1}{x}$ and $f(x) = x + 2$. *Answers may vary.*

85. $h(x) = 4(2x + 1)^3 \Rightarrow g(x) = 4x^3$ and $f(x) = 2x + 1$. *Answers may vary.*

87. $h(x) = (x^3 - 1)^2 \Rightarrow g(x) = x^2$ and $f(x) = x^3 - 1$. *Answers may vary.*

89. $h(x) = -4|x + 2| - 3 \Rightarrow g(x) = -4|x| - 3$ and $f(x) = x + 2$. *Answers may vary.*

91. $h(x) = \dfrac{1}{(x - 1)^2} \Rightarrow g(x) = \dfrac{1}{x^2}$ and $f(x) = x - 1$. *Answers may vary.*

93. $h(x) = x^{3/4} - x^{1/4} \Rightarrow g(x) = x^3 - x$ and $f(x) = x^{1/4}$. *Answers may vary.*

95. The cost function would not change but the revenue function would be $R(x) = 15x$. Therefore,

  $P(x) = R(x) - C(x) = 15x - (2x + 2000) = 13x - 2000;\ P(3000) = 13(3000) - 2000 = \$37{,}000$

97. (a) $I(x) = 36x$

  (b) $C(x) = 2.54x$

  (c) $F(x) = (C \circ I)(x)$

  (d) $F(x) = 36(2.54x) = 91.44x$

99. (a) In Example 1, the reaction distance for a reaction time of 2.5 seconds was $\dfrac{11}{3}x.$

  The reaction distance for a driver with a reaction time of 1.25 seconds would be half. Thus, $f(x) = \dfrac{11}{6}x.$

  (b) $d(x) = f(x) + g(x) = \dfrac{11}{6}x + \dfrac{1}{9}x^2$

  (c) $d(60) = \dfrac{11}{6}(60) + \dfrac{1}{9}(60)^2 = 510;$ this car requires 510 feet to stop at 60 mph.

101. (a) $(g \circ f)(1) = g(f(1)) = g(1.5) = 5.25;$

  A 1% decrease in the ozone layer could result in a 5.25% increase in skin cancer.

  (b) $(f \circ g)(21) = f(g(21))$, but $g(21)$ is not given by the table. So $(f \circ g)(21)$ is not possible using these tables.

103. (a) $(g \circ f)(1975) = g(f(1975)) = g(3) = 4.5\%$

  (b) $(g \circ f)(x)$ computes the percent increase in peak demand during year x.

105. (a) $(g \circ f)(1960) = g(f(1960)) = g(0.11(1960 - 1948)) = g(1.32) = 1.5(1.32) = 1.98$

In 1960 the temperature had risen $1.32°C$ which resulted in a 1.98% increase in peak-demand for electricity.

(b) $(g \circ f)(x) = g(f(x)) = g(0.11(x - 1948)) = 1.5(0.11(x - 1948)) = 0.165(x - 1948)$

(c) f, g, and $f \circ g$ are all linear functions.

107. (a) $(g \circ f)(2) = g(f(2)) = g(77) \approx 25°C$. After 2 hours the temperature is about $25°C$.

Answers may vary.

(b) $(g \circ f)(x)$ computes the Celsius temperature after x hours.

109. $A = \pi r^2 \Rightarrow A(t) = \pi(6t)^2 = 36\pi t^2$

111. (a) For $A(s) = \frac{\sqrt{3}}{4}s^2$, $A(4s) = \frac{\sqrt{3}}{4}(4s)^2 = 16\frac{\sqrt{3}}{4}s^2 = 16A(s)$; if the length of a side is quadrupled, the

area increases by a factor of 16.

(b) For $A(s) = \frac{\sqrt{3}}{4}s^2$, $A(s + 2) = \frac{\sqrt{3}}{4}(s + 2)^2 = \frac{\sqrt{3}}{4}(s^2 + 4s + 4) = \frac{\sqrt{3}}{4}s^2 + \sqrt{3}s + \sqrt{3} = $

$A(s) + \sqrt{3}(s + 1)$; if the length of a side increases by 2, the area increases by $\sqrt{3}(s + 1)$.

113. (a) $(c + o)(1970) = c(1970) + o(1970) = 32.4 + 17.6 = 50$

(b) $(c + o)(x)$ computes the total coal and oil SO_2 emissions in the year x.

(c) A tabular representation of $(c + o)(x)$ is given in Figure 113.

x	1860	1900	1940	1970	2000
$(C + O)(x)$	2.4	12.8	26.5	50.0	78.0

Figure 113

x	1990	2000	2010	2020	2030
$h(x)$	32	35.5	39	42.5	46

Figure 115

115. (a) To find the total emissions we must add the developed and developing countries' emissions for each year.

The resulting table is shown in Figure 115.

(b) $h(x) = f(x) + g(x)$

117. $h(x) = f(x) + g(x) = (0.1x - 172) + (0.25x - 492.5) = 0.35x - 664.5$

119. (a) $P(h_0) = mph_0 = \frac{1}{2}m \cdot 2gh_0 = \frac{1}{2}m(\sqrt{2gh_0})^2 = \frac{1}{2}m(V_f) = K(V_f)$

(b) The potential energy of the ball before it is dropped is equal to the kinetic energy of the ball when it lands.

121. Let $f(x) = ax + b$ and $g(x) = cx + d$. Then $f(x) + g(x) = (ax + b) + (cx + d) = $

$(a + c)x + (b + d)$, which is linear.

123. (a) $f(x) = k$ and $g(x) = ax + b$. Thus, $(f \circ g)(x) = f(g(x)) = f(ax + b) = k$; $f \circ g$ is a constant

function.

(b) $f(x) = k$ and $g(x) = ax + b$. Thus, $(g \circ f)(x) = g(f(x)) = g(k) = ak + b$; $g \circ f$ is a constant

function.

5.2: Inverse Functions and Their Representations

1. Closing a window.

3. Closing a book, standing up, and walking out of the classroom.

5. Subtract 2 from x; $x + 2$ and $x - 2$

7. Divide x by 3 and add 2; $3(x - 2)$ and $\dfrac{x}{3} + 2$

9. Subtract 1 from x and cube the result; $\sqrt[3]{x} + 1$ and $(x - 1)^3$

11. Take the reciprocal of x; $\dfrac{1}{x}$ and $\dfrac{1}{x}$

13. Since a horizontal line will intersect the graph at most once, f is one-to-one.

15. Since a horizontal line will intersect the graph two times, f is not one-to-one.

17. Since a horizontal line will intersect the graph two times, f is not one-to-one.

19. Since $f(2) = 3$ and $f(3) = 3$, different inputs result in the same output. Therefore f is not one-to-one. Does not have an inverse.

21. Since different inputs always result in different outputs, f is one-to-one. Does have an inverse.

23. Since the graph of f is a line sloping upward from left to right, a horizontal line can intersect it at most once. Therefore, f is one-to-one.

25. Since the graph of f is a parabola, a horizontal line can intersect it more than once. Therefore, f is not one-to-one.

27. Since the graph of f is a parabola, a horizontal line can intersect it more than once. Therefore, f is not one-to-one.

29. Since the graph of f is a V-shape, a horizontal line can intersect it more than once. Therefore, f is not one-to-one.

31. Since the graph of f is both increasing and decreasing, a horizontal line can intersect it more than once. Therefore, f is not one-to-one.

33. Since the graph of f is both increasing and decreasing, a horizontal line can intersect it more than once. Therefore, f is not one-to-one.

35. Since the graph of f is a line sloping upward from left to right, a horizontal line can intersect it at most once. Therefore f is one-to-one.

37. Since the person goes up and down, there would be several times when the person attained a particular height above the ground. A one-to-one function would not model this situation.

39. Since the population of the United States increased each of these years, a horizontal line would intersect a graph of this population at most once. A one-to-one function would model this situation.

41. The inverse operation of taking the cube root of x is cubing x. Therefore, $f^{-1}(x) = x^3$.

43. The inverse operations of multiplying by -2 and adding 10 are subtracting 10 and dividing by -2.
    Therefore, $f^{-1}(x) = \dfrac{x - 10}{-2} = -\dfrac{1}{2}x + 5$.

45. The inverse operations of multiplying by 3 and subtracting 1 are adding 1 and dividing by 3.
    Therefore, $f^{-1}(x) = \dfrac{x + 1}{3}$.

47. The inverse operations of cubing a number, multiplying by 2 and subtracting 5 are adding 5, dividing by 2 and taking the cube root of the result. Therefore, $f^{-1}(x) = \sqrt[3]{\dfrac{x+5}{2}}$.

49. The inverse operations of squaring a number and subtracting 1 are adding 1 and taking the square root. Therefore, $f^{-1}(x) = \sqrt{x+1}$.

51. The inverse operations of multiplying by 2 and taking the reciprocal are taking the reciprocal and dividing by 2. Therefore, $f^{-1}(x) = \dfrac{1}{x} \div 2 = \dfrac{1}{2x}$.

53. The inverse operations of multiplying by -5, adding 4, multiplying by $\dfrac{1}{2}$, and adding 1 are subtracting 1, dividing by $\dfrac{1}{2}$ (or multipying by 2), subtracting 4, and dividing by -5 $\Rightarrow$

$$f^{-1}(x) = \frac{2(x-1)-4}{-5} = \frac{2x-6}{-5} = -\frac{2(x-3)}{5}.$$

55. $y = \dfrac{x}{x+2} \Rightarrow y(x+2) = x \Rightarrow xy + 2y = x \Rightarrow 2y = x - xy \Rightarrow 2y = x(1-y) \Rightarrow x = \dfrac{2y}{1-y} \Rightarrow$

$f^{-1}(x) = \dfrac{2x}{1-x} \Rightarrow f^{-1}(x) = -\dfrac{2x}{x-1}$

57. $y = \dfrac{2x+1}{x-1} \Rightarrow y(x-1) = 2x+1 \Rightarrow xy - y = 2x + 1 \Rightarrow xy - 2x = y + 1 \Rightarrow$

$x(y-2) = y+1 \Rightarrow x = \dfrac{y+1}{y-2} \Rightarrow f^{-1}(x) = \dfrac{x+1}{x-2}$

59. $y = \dfrac{1}{x} - 3 \Rightarrow y + 3 = \dfrac{1}{x} \Rightarrow x = \dfrac{1}{y+3} \Rightarrow f^{-1}(x) = \dfrac{1}{x+3}$

61. $y = \dfrac{1}{x^3 - 1} \Rightarrow \dfrac{1}{y} = x^3 - 1 \Rightarrow \dfrac{1}{y} + 1 = x^3 \Rightarrow \dfrac{1}{y} + \dfrac{y}{y} = x^3 \Rightarrow \dfrac{y+1}{y} = x^3 \Rightarrow f^{-1}(x) = \sqrt[3]{\dfrac{x+1}{x}}$

63. $y = 4 - x^2, x \geq 0 \Rightarrow x^2 = 4 - y \Rightarrow x = \sqrt{4-y}$; therefore $f^{-1}(x) = \sqrt{4-x}$

65. $y = (x-2)^2 + 4, x \geq 2 \Rightarrow y - 4 = (x-2)^2 \Rightarrow \sqrt{y-4} = x - 2 \Rightarrow x = \sqrt{y-4} + 2$; therefore $f^{-1}(x) = \sqrt{x-4} + 2$.

67. $y = x^{2/3} + 1, x \geq 0 \Rightarrow x^{2/3} = y - 1 \Rightarrow x = (y-1)^{3/2}; \Rightarrow f^{-1}(x) = (x-1)^{3/2}$

69. $y = \sqrt{9 - 2x^2}, -\dfrac{3}{12} \leq x \leq \dfrac{3}{\sqrt{2}} \Rightarrow y^2 = 9 - 2x^2 \Rightarrow y^2 - 9 = -2x^2 \Rightarrow \dfrac{9-y^2}{2} = x^2 \Rightarrow$

$\sqrt{\dfrac{9-y^2}{2}} = x; \Rightarrow f^{-1}(x) = \sqrt{\dfrac{9-x^2}{2}}$

71. $y = 5x - 15 \Rightarrow y + 15 = 5x \Rightarrow \dfrac{y+15}{5} = x \Rightarrow f^{-1}(x) = \dfrac{x+15}{5}$; D and R are all real numbers.

73. $y = \sqrt[3]{x-5} \Rightarrow y^3 = x - 5 \Rightarrow y^3 + 5 = x \Rightarrow f^{-1}(x) = x^3 + 5$; D and R are all real numbers

75. $y = \dfrac{x-5}{4} \Rightarrow 4y = x - 5 \Rightarrow 4y + 5 = x \Rightarrow f^{-1}(x) = 4x + 5$; D and R are all real numbers.

77. $y = \sqrt{x-5}, x \geq 5 \Rightarrow y^2 = x - 5 \Rightarrow y^2 + 5 = x \Rightarrow f^{-1}(x) = x^2 + 5$;

$D = \{x \mid x \geq 0\}$ and $R = \{y \mid y \geq 5\}$

79. $y = \dfrac{1}{x+3} \Rightarrow \dfrac{1}{y} = x + 3 \Rightarrow \dfrac{1}{y} - 3 = x \Rightarrow f^{-1}(x) = \dfrac{1}{x} - 3; D = \{x \mid x \neq 0\}$ and $D = \{y \mid y \neq -3\}$

81. $y = 2x^3 \Rightarrow \dfrac{y}{2} = x^3 \Rightarrow \sqrt[3]{\dfrac{y}{2}} = x \Rightarrow f^{-1}(x) = \sqrt[3]{\dfrac{x}{2}}; D$ and R are all real numbers.

83. $y = x^2, x \geq 0 \Rightarrow \sqrt{y} = x \Rightarrow f^{-1}(x) = \sqrt{x}; D = \{x \mid x \geq 0\}$ and $D = \{y \mid y \geq 0\}$

85. The domain and range of f are $D = \{1, 2, 3\}$ and $R = \{5, 7, 9\}$. Interchange these to get the domain and range

    of $f^{-1}; D = \{5, 7, 9\}$ and $R = \{1, 2, 3\}$. See Figure 85.

x	5	7	9
$f^{-1}(x)$	1	2	3

Figure 85

x	0	4	16
$f^{-1}(x)$	0	2	4

Figure 87

87. The domain and range of f are $D = \{0, 2, 4\}$ and $R = \{0, 4, 16\}$. Interchange these to get the domain and

    range of $f^{-1}; D = \{0, 4, 16\}$ and $R = \{0, 2, 4\}$. See Figure 87.

89. Since f multiplies x by 4, f^{-1} divides x by 4. See Figure 89.

x	0	2	4	6
$f^{-1}(x)$	0	$\frac{1}{2}$	1	$\frac{3}{2}$

Figure 89

91. $f(1) = 3 \Rightarrow f^{-1}(3) = 1$

93. $g(3) = 4 \Rightarrow g^{-1}(4) = 3$

95. $(f \circ g^{-1})(1) = f(g^{-1}(1)) = f(2) = 5$. Note that $g(2) = 1 \Rightarrow g^{-1}(1) = 2$.

97. $(g \circ f^{-1})(5) = g(f^{-1}(5)) = g(2) = 1$. Note that $f(2) = 5 \Rightarrow f^{-1}(5) = 2$.

99. (a) $f(1) \approx \$110$

    (b) $f^{-1}(110) \approx 1$ year

    (c) $f^{-1}(160) \approx 5$ years

    The function f^{-1} computes the number of years it takes for this savings account to accumulate x dollars.

101. (a) $f(-1) = 2$

    (b) $f^{-1}(-2) = 3$

    (c) $f^{-1}(0) = 1$

    (d) $(f^{-1} \circ f)(3) = f^{-1}(f(3)) = f^{-1}(-2) = 3$

103. (a) $f(4) = 4$

    (b) $f^{-1}(0) = 0$

    (c) $f^{-1}(6) = 9$

    (d) $(f^{-1} \circ f)(4) = f^{-1}(f(4)) = f^{-1}(4) = 4$

105. The graph of f passes through the points $(-2, -4)$ and $(1, 2)$. Thus, the graph of f^{-1} passes through the points $(-4, -2)$ and $(2, 1)$. Plot these points and sketch the reflection of f in the line $y = x$ to obtain the graph of f^{-1}. Notice that since the graph of f is a line, its reflection will also be a line. See Figure 105.

107. The graph of f passes through the points $\left(-2, \frac{1}{4}\right)$, $(0, 1)$, $(1, 2)$ and $(2, 4)$. Thus, the graph of f^{-1} passes through the points $\left(\frac{1}{4}, -2\right)$, $(1, 0)$, $(2, 1)$ and $(4, 2)$. Plot these points and sketch the reflection of f in the line $y = x$ to obtain the graph of f^{-1}. See Figure 107.

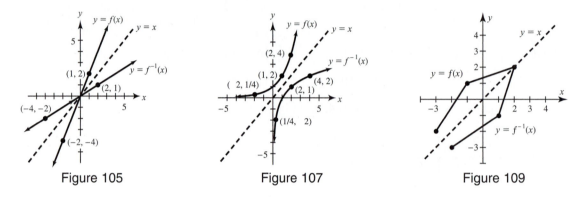

Figure 105   Figure 107   Figure 109

109. The graph of f passes through the points $(-3, -2)$, $(-1, 1)$ and $(2, 2)$. Thus, the graph of f^{-1} passes through the points $(-2, -3)$, $(1, -1)$ and $(2, 2)$. Plot these points and sketch the reflection of f in the line $y = x$ to obtain the graph of f^{-1}. See Figure 109.

111. The graphs of $y = 2x - 1$, $y = \dfrac{x + 1}{2}$, and $y = x$ are shown in Figure 111.

113. The graphs of $y = x^3 - 1$, $y = \sqrt[3]{x + 1}$, and $y = x$ are shown in Figure 113.

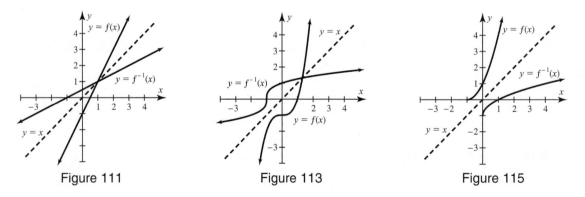

Figure 111   Figure 113   Figure 115

115. The graphs of $y = (x + 1)^2$, $y = \sqrt{x} - 1$, and $y = x$ are shown in Figure 115.

117. The graphs of $Y_1 = 3X - 1$, $Y_2 = (X + 1)/3$ and $Y_3 = X$ are shown in Figure 117.

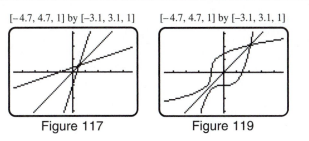

[−4.7, 4.7, 1] by [−3.1, 3.1, 1]   [−4.7, 4.7, 1] by [−3.1, 3.1, 1]

Figure 117   Figure 119

119. The graphs of $Y_1 = X^3/3 - 1$, $Y_2 = \sqrt[3]{(3X + 3)}$ and $Y_3 = X$ are shown in Figure 119.

121. (a) Since each volume value is the result of exactly one radius value, V represents a one-to-one function.

(b) The inverse of V computes the radius r of a sphere with volume V.

(c) $V = \dfrac{4}{3}\pi r^3 \Rightarrow r^3 = \dfrac{3V}{4\pi} \Rightarrow r = \sqrt[3]{\dfrac{3V}{4\pi}}$

(d) No; if V and r were interchanged, then r would represent the volume and V would represent the radius.

123. (a) $W = \dfrac{25}{7}(70) - \dfrac{800}{7} = \dfrac{950}{7} \approx 135.7$ pounds

(b) Yes, since no weight value corresponds to more than one height value.

(c) $W = \dfrac{25}{7}h - \dfrac{800}{7} \Rightarrow \dfrac{25}{7}h = W + \dfrac{800}{7} \Rightarrow h = \dfrac{7}{25}\left(W + \dfrac{800}{7}\right) = \dfrac{7}{25}W + 32 \Rightarrow W^{-1} = \dfrac{7}{25}W + 32$

(d) $W^{-1}(150) = \dfrac{7}{25}(150) + 32 \Rightarrow W^{-1}(150) = 74$; the maximum recommended height for a person weighing 150 pounds is 74 inches.

(e) The inverse computes the maximum recommended height of a person with a given weight.

125. (a) $(F \circ Y)(2) = F(Y(2)) = F(3520) = 10,560.$ $(F \circ Y)(2)$ computes the number of feet in 2 miles.

(b) $F^{-1}(26,400) = 8800.$ F^{-1} converts feet to yards. There are 8800 yards in 26,400 feet.

(c) $(Y^{-1} \circ F^{-1})(21,120) = Y^{-1}(F^{-1}(21,120)) = Y^{-1}(7040) = 4.$ $(Y^{-1} \circ F^{-1})(21,120)$ computes the number of miles in 21,120 feet.

127. (a) $(Q \circ C)(96) = Q(C(96)) = Q(6) = 1.5.$ $(Q \circ C)(96)$ computes the number of quarts in 96 tablespoons.

(b) $Q^{-1}(2) = 8.$ Q^{-1} converts quarts into cups. There are 8 cups in 2 quarts.

(c) $(C^{-1} \circ Q^{-1})(1.5) = C^{-1}(Q^{-1}(1.5)) = C^{-1}(6) = 96.$ $(C^{-1} \circ Q^{-1})(1.5)$ computes the number of tablespoons in 1.5 quarts.

129. (a) $f(x) = 0.06(x - 1930) + 62.5 \Rightarrow f(1930) = 0.06(1930 - 1930) + 62.5 = 62.5$ and $f(1980) = 0.06(1980 - 1930) + 62.5 = 65.5.$ In 1930 there was cloud cover 62.5% of the time, whereas in 1980 it increased to 65.5%. It increased by 3%.

(b) f^{-1} computes the year when the cloud cover was x percent.

(c) $f(1930) = 62.5 \Rightarrow f^{-1}(62.5) = 1930.$ Similarly, $f(1980) = 65.5 \Rightarrow f^{-1}(65.5) = 1980.$

(d) Using the technique shown in Example 6, we can find $f^{-1}(x)$.

$$y = 0.06(x - 1930) + 62.5 \Rightarrow y - 62.5 = 0.06(x - 1930) \dfrac{y - 62.5}{0.06} = x - 1930 \Rightarrow$$

$$1930 + \dfrac{y - 62.5}{0.06} = x \Rightarrow f^{-1}(x) = 1930 + \dfrac{x - 62.5}{0.06} \text{ or } f^{-1}(x) = \dfrac{50}{3}(x - 62.5) + 1930$$

5.3: Exponential Functions and Models

1. $2^{-3} = \dfrac{1}{2^3} = \dfrac{1}{8}$

3. $3(4)^{1/2} = 3\sqrt{4} = 3(2) = 6$

5. $-2(27)^{2/3} = -2(\sqrt[3]{27})^2 = -2(3)^2 = -2(9) = -18$

7. $4^{1/6}4^{1/3} = 4^{1/6+1/3} = 4^{1/2} = \sqrt{4} = 2$

9. $e^x e^x = e^{x+x} = e^{2x}$

11. $3^0 = 1$

13. $(5^{101})^{1/101} = 5^1 = 5$

15. For each unit increase in x, the y-values decrease by 1.2, so the data is linear. Since $y = 2$ when $x = 0$, the function $f(x) = -1.2x + 2$ can model the data.

17. For each unit increase in x, the y-values are multiplied by $\dfrac{1}{2}$, so the data is exponential. Since the initial value is $C = 8$ and $a = \dfrac{1}{2}$, the function $f(x) = 8\left(\dfrac{1}{2}\right)^x$ can model the data.

19. For each 2-unit increase in x, the y-values are multiplied by 4. That is, for each unit increase in x, the y-values are multiplied by 2, so the data is exponential. Since the initial value is $C = 5$ and $a = 2$, the function $f(x) = 5(2^x)$ can model the data.

21. For each 1-year increase, the salary is multiplied by 1.08 to get the new salary. This is an example of exponential growth; $C = \$40{,}000$ and $a = 1.08 \Rightarrow f(n) = 40{,}000(1.08)^{n-1}$, where $n = 1, 2, 3, \ldots$.

23. For $x > 4$, $f(x) > g(x)$; for example $f(10) = 1024$ whereas $g(10) = 100$. That is, $f(x) = 2^x$ becomes larger.

25. $f(x) = 2x + 1$, $f(0) = 1$, $f(10) = 21$. $g(x) = 2^{-x}$, $g(0) = 2$, $g(10) = 9.76 \times 10^{-4}$.
Therefore, $f(x)$ becomes larger for $0 \le x \le 10$.

27. $f(0) = 5 \Rightarrow C = 5$ and $a = 1.5$

29. $f(0) = 10$ and $f(1) = 20 \Rightarrow C = 10$ and $a = \dfrac{20}{10} = 2$

31. $f(1) = 9$ and $f(2) = 27 \Rightarrow a = \dfrac{27}{9} = 3$; $C = f(0) = 9 \div 3 = 3$

33. $f(-2) = \dfrac{9}{2}$ and $f(2) = \dfrac{1}{18} \Rightarrow Ca^{-2} = \dfrac{9}{2} \Rightarrow \dfrac{C}{a^2} = \dfrac{9}{2} \Rightarrow C = \dfrac{9a^2}{2}$; $Ca^2 = \dfrac{1}{18} \Rightarrow C = \dfrac{1}{18a^2}$ thus
$\dfrac{1}{18a^2} = \dfrac{9a^2}{2} \Rightarrow 162a^4 = 2 \Rightarrow a^4 = \dfrac{1}{81} \Rightarrow a = \dfrac{1}{3}$; then $C = \dfrac{1}{18(\frac{1}{3})^2} \Rightarrow C = \dfrac{1}{18(\frac{1}{9})} \Rightarrow C = \dfrac{1}{2}$.

35. $C = 5000$, $a = 2$; x represents time in hours

37. $C = 200{,}000$, $a = 0.95$; x represents the number of years after 2000

39. $f(9.5) = 30(0.9)^{9.5} \approx 11$; the tire's pressure is about 11 pounds per square inch after 9.5 minutes.

41. $f(x) = 4e^{-1.2x} \Rightarrow f(-2.4) = 4e^{-1.2(-2.4)} = 4e^{2.88} \approx 71.2571$

43. $f(x) = \dfrac{e^x - e^{-x}}{2} \Rightarrow f(-0.7) = \dfrac{e^{-0.7} - e^{0.7}}{2} \approx -0.7586$

45. See Figure 45.

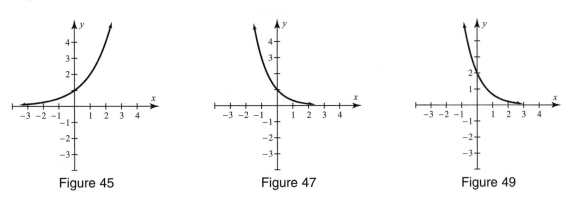

Figure 45   Figure 47   Figure 49

47. See Figure 47.

49. See Figure 49.

51. See Figure 51.

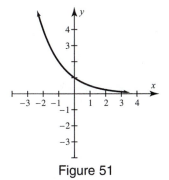

Figure 51

53. Since $y = 1$ when $x = 0$, $C = 1$ and so $y = a^x$.

Since $y = 4$ when $x = -2$, $4 = a^{-2} \Rightarrow \dfrac{1}{a^2} = 4 \Rightarrow a^2 = \dfrac{1}{4} \Rightarrow a = \dfrac{1}{2}$. That is $C = 1$ and $a = \dfrac{1}{2}$.

55. Since $y = \dfrac{1}{2}$ when $x = 0$, $C = \dfrac{1}{2}$ and so $y = \dfrac{1}{2}a^x$.

Since $y = 8$ when $x = 2$, $8 = \dfrac{1}{2}a^2 \Rightarrow a^2 = 16 \Rightarrow a = 4$. That is $C = \dfrac{1}{2}$ and $a = 4$.

57. (a) D: $(-\infty, \infty)$; R: $(0, \infty)$

(b) Decreasing, as x increases $\left(\dfrac{1}{8}\right)^x$ decreases.

(c) $y = 0$ as x increases, $7\left(\dfrac{1}{8}\right)^x$ gets closer and closer to 0.

(d) y-intercept: 7; no x-intercept.

(e) All horizontal lines intersect once $\Rightarrow$ yes; yes.

59. (i) The graph of $y = e^x$ increases faster than the graph of $y = 1.5^x$. The best choice is graph b.

(ii) The graph of $y = 3^{-x}$ decreases faster than the graph of $y = 0.99^x$. The best choice is graph d.

(iii) The graph of $y = 1.5^x$ increases slower than the graph of $y = e^x$. The best choice is graph a.

(iv) The graph of $y = 0.99^x$ is almost a horizontal line since $y = 1^x$ is horizontal. The best choice is graph c.

61. (a) To graph $y = 2^x - 2$, translate the graph of $y = 2^x$ down 2 units. See Figure 61a.

    (b) To graph $y = 2^{x-1}$, translate the graph of $y = 2^x$ right 1 unit. See Figure 61b.

    (c) To graph $y = 2^{-x}$, reflect the graph of $y = 2^x$ about the y-axis. See Figure 61c.

    (d) To graph $y = -2^x$, reflect the graph of $y = 2^x$ about the x-axis. See Figure 61d.

Figure 61a

Figure 61b

Figure 61c

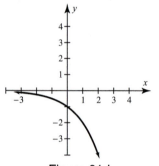

Figure 61d

63. $A_n = A_0(1 + r)^n \Rightarrow A_5 = 600(1 + 0.07)^5 \approx \$841.53$

65. $A_n = A_0\left(1 + \dfrac{r}{m}\right)^{mn} \Rightarrow A_{20} = 950\left(1 + \dfrac{0.03}{365}\right)^{365(20)} \approx \$1730.97$

67. $A_n = A_0 e^{rn} \Rightarrow A_8 = 2000 e^{0.10(8)} \approx \$4451.08$

69. $A_n = A_0\left(1 + \dfrac{r}{m}\right)^{mn} \Rightarrow A_{2.5} = 1600\left(1 + \dfrac{0.104}{12}\right)^{12(2.5)} \approx \$2072.76$

71. $A_{20} = 2000\left(1 + \dfrac{0.10}{12}\right)^{12(20)} \approx \$14{,}656.15; \ A_{20} = 2000\left(1 + \dfrac{0.13}{12}\right)^{12(20)} \approx \$26{,}553.58$

    A 13% interest rate results in considerably more interest than a 10% interest rate.

73. $A_{10} = 8000(1 + 0.06)^{10} \approx \$14{,}326.78$

75. (a) $P = 1500, r = 0.06$ and $t = 30 \Rightarrow A = Pe^{rt} \Rightarrow A = 1500 e^{0.06(30)} \approx 9074.47$

    (b) During the last 10 years. See Figure 75.

    [0, 30, 2] by [0, 10,000, 1000]

Figure 75

77. $A_n = A_o(1 + r)^n \Rightarrow A_{30} = 340(1 + 0.0454)^{30} \approx 1288.11$ billion or 1.288 trillion.

79. $A_{10} = 50\left[\dfrac{(1 + \frac{0.08}{26})^{260} - 1}{\frac{0.08}{26}}\right] \approx \$19,870.65$

81. $A(x) = Pe^{rt} \Rightarrow 0.85 = e^{5r} \Rightarrow \ln(0.85) = 5r \Rightarrow r = \dfrac{\ln(0.85)}{5} \approx -0.0325 \Rightarrow A(x) = Pe^{-0.0325x}$

    P is the initial value.

83. (a) The number of E. coli was modeled by $N(x) = N_0\, e^{0.014x}$, where x is in minutes and $N_0 = 500,000$.

    Since 3 hours is 180 minutes, $N(180) = 500,000\, e^{0.014(180)} \approx 6,214,000$ bacteria per milliliter.

    (b) Solve the equation $500,000\, e^{0.014x} = 10,000,000$. Graph $Y_1 = 500000e^{\wedge}(0.014X)$ and $Y_2 = 10\text{E}6$

    as shown in Figure 83. The point of intersection is near $(214, 10,000,000)$. Thus, there will be 10

    million E. coli after about 214 minutes, or about 3.6 hours.

[0, 300, 100] by [0, 20,000,000, 5,000,000]   [0, 5, 1] by [0, 1, 0.1]   [0, 6, 1] by [–2, 2, 1]

Figure 83   Figure 87   Figure 93

85. (a) $f(0) = 0.72 \Rightarrow C \approx 0.72$, so $f(x) = 0.72\,a^x$. One possible way to determine the value of a is to let the

    graph of f pass through the point $(20, 1.60)$. $f(20) = 1.60 \Rightarrow 1.60 = 0.72\,a^{20} \Rightarrow a^{20} = \dfrac{1.60}{0.72} \Rightarrow$

    $a = \left(\dfrac{1.60}{0.72}\right)^{1/20} \Rightarrow a \approx 1.041$; Thus $f(x) = 0.72(1.041)^x$. *Answers may vary slightly.*

    (b) Let $x = 13$ correspond to 2013, so ; $f(13) = 0.72(1.041)^{13} \approx 1.21$ the CFC-12 concentration in 2013 is

    about 1.21 ppb. *Answers may vary slightly.*

87. (a) $p(x) = 1 - e^{-5x/6} \Rightarrow p(3) = 1 - e^{-5(3)/6} \Rightarrow p(3) = 1 - e^{-15/6} \approx 0.92$, or 92%

    (b) Find x when $p(x) = 0.5$. That is, solve $0.5 = 1 - e^{-5x/6}$ for x.

    Graph $Y_1 = 0.5$ and $Y_2 = 1 - e^{\wedge}(-5X/6)$ as shown in Figure 87. There is a 50-50 chance of at least one

    car entering the intersection during an interval of about 0.83 minutes.

89. To find the age of the fossil solve $0.10 = 1\left(\dfrac{1}{2}\right)^{x/5700}$. Graph $Y_1 = 0.10$ and $Y_2 = 0.5^{\wedge}(X/5700)$. The graphs

    intersect near $(18934.99, 0.1)$, so the fossil is about 18,935 years old.

91. $P = 1\left(\dfrac{1}{2}\right)^{x/1600} \Rightarrow P = 1\left(\dfrac{1}{2}\right)^{3000/1600} \Rightarrow P \approx 0.273$, or 27.3%

93. (a) Since the initial amount is 2.5 ppm, $C = 2.5$. Since 30% of the chlorine dissipates each day, 70% remains.

    So, $a = 0.7$.

    (b) $f(2) = 2.5(0.7)^2 = 1.225$ parts per million.

    (c) Chlorine will need to be added after the level drops to 1.5 parts per million. Therefore, we must solve the

    equation $2.5(0.7)^x = 1.5 \Rightarrow 2.5(0.7)^x - 1.5 = 0$. Graph $Y_1 = 2.5(0.7)^{\wedge}X - 1.5$ and determine any

    x-intercepts or zeros. There is one zero located at $x \approx 1.4$ as shown in Figure 93. If the chlorine level

    should not drop below 1.5 parts per million, chlorine should be added before the middle of the second day.

95. (a) Using the exponential regression function on your calculator we find $W = 18.29(1.279)^x$.

(b) $W = 18.29(1.279)^9 \approx 168$ thousand pounds.

(c) Solve by graphing $Y_1 = 18.29(1.279)\wedge X$ and $Y_2 = 242$. The minimum thickness is about 10.5 inches. See Figure 95.

[0, 12, 1] by [0, 300, 50]

Figure 95

97. (a) $H(30) = 0.157(1.033)^{30} \approx 0.42$. This means that approximately 0.42 horsepower are required for each ton that the locomotive is pulling at 30 mph.

(b) To pull a 5000-ton train at 30 mph, approximately a $0.42(5000) = 2100$ horsepower engine is needed.

(c) $\dfrac{2100}{1350} \approx 1.56$. This value must be rounded up. Two locomotives having 1350 horsepower would move a 5000-ton train at 30 mph.

5.4: Logarithmic Functions and Models

1. See Figure 1.

x	10^0	10^4	10^{-8}	$10^{1.26}$
$\log x$	0	4	−8	1.26

Figure 1

3. (a) $\log(-3)$ is undefined

(b) $\log \dfrac{1}{100} = \log \dfrac{1}{10^2} = \log 1^{-2} = -2$

(c) $\log \sqrt{0.1} = \log 10^{-1/2} = -\dfrac{1}{2}$

(d) $\log 5^0 = 0$

5. (a) $\log 10 = \log 10^1 = 1$

(b) $\log 10{,}000 = \log 10^4 = 4$

(c) $20 \log 0.1 = 20 \log 10^{-1} = 20(-1) = -20$

(d) $\log 10 + \log 0.001 = \log 10^1 + \log 10^{-3} = 1 + (-3) = -2$

7. (a) $2 \log 0.1 + 4 = 2 \log 10^{-1} + 4 = 2(-1) + 4 = -2 + 4 = 2$

(b) $\log 10^{1/2} = \dfrac{1}{2}$

(c) $3 \log 100 - \log 1000 = 3 \log 10^2 - \log 10^3 = 3(2) - 3 = 6 - 3 = 3$

(d) $\log(-10)$ is undefined

9. (a) Since $10^1 \le 79 \le 10^2$, $\log 10^1 \le \log 79 \le \log 10^2 \Rightarrow 1 \le \log 79 \le 2 \Rightarrow n = 1$; $\log 79 \approx 1.898$

   (b) Since $10^2 \le 500 \le 10^3$, $\log 10^2 \le \log 500 \le \log 10^3 \Rightarrow 2 \le \log 500 \le 3 \Rightarrow n = 2$; $\log 500 \approx 2.699$

   (c) Since $10^0 \le 5 \le 10^1$, $\log 10^0 \le \log 5 \le \log 10^1 \Rightarrow 0 \le \log 5 \le 1 \Rightarrow n = 0$; $\log 5 \approx 0.6990$

   (d) Since $10^{-1} \le 0.5 \le 10^0$, $\log 10^{-1} \le \log 0.5 \le \log 10^0 \Rightarrow -1 \le \log 0.5 \le 0 \Rightarrow n = -1$;

   $\log 0.5 \approx -0.3010$

11. (a) $\dfrac{3}{2}$ since $\sqrt{1000} = 1000^{1/2} = (10^3)^{1/2} = 10^{3/2}$

    (b) $\dfrac{1}{3}$ since $\log \sqrt[3]{10} = \log 10^{1/3}$

    (c) $\log \sqrt[5]{0.1} = \log (10^{-1})^{1/5} = \log (10)^{-1/5} = -\dfrac{1}{5}$

    (d) $\log \sqrt{0.01} = \log (10^{-2})^{1/2} = \log 10^{-1} = -1$

13. The input to a logarithmic function must be positive. Thus, any element of the domain of f must satisfy
    $x + 3 > 0$, or equivalently, $x > -3$. Thus D: $(-3, \infty)$, or $\{x \mid x > -3\}$.

15. The input to a logarithmic function must be positive. Thus, any element of the domain of f must satisfy
    $x^2 - 1 > 0 \Rightarrow x^2 > 1 \Rightarrow x < -1$ or $x > 1$ Thus D: $(-\infty, -1) \cup (1, \infty)$, or $\{x \mid x < -1, \text{ or } x > 1\}$.

17. The input to a logarithmic function must be positive. Thus, any element of the domain of f must satisfy
    $4^x > 0 \Rightarrow x$ can be any real number. Thus D: $(-\infty, \infty)$, or $\{x \mid -\infty < x < \infty\}$.

19. The input to a logarithmic function must be positive. Thus, any element of the domain of f must satisfy
    $\sqrt{3 - x} - 1 > 0 \Rightarrow \sqrt{3 - x} > 1 \Rightarrow x < 2$. Thus D: $(-\infty, 2)$, or $\{x \mid x < 2\}$.

21. $\log_8 8^{-5.7} = -5.7$

23. $7^{\log_7 2x} = 2x$ for $x > 0$

25. $\log_{1/3} \left(\dfrac{1}{3}\right)^{64} = 64$

27. $\ln e^{-4} = -4$

29. $\log_5 5^\pi = \pi$

31. $3^{\log_3 (x-1)} = x - 1$, for $x > 1$

33. $\log_2 64 = \log_2 2^6 = 6$

35. $\log_4 2 = \log_4 \sqrt{4} = \log_4 4^{1/2} = \dfrac{1}{2}$

37. $\ln e^{-3} = -3$

39. $\log_8 64 = \log_8 8^2 = 2$

41. $\log_{1/2} \left(\dfrac{1}{4}\right) = \log_{1/2} \left(\dfrac{1}{2}\right)^2 = 2$

43. $\log_{1/6} 36 = \log_{1/6} \left(\dfrac{1}{6}\right)^{-2} = -2$

45. $\log_a \dfrac{1}{a} = \log_a a^{-1} = -1$

47. $\log_5 5^0 = 0$

49. $\log_2 \dfrac{1}{16} = \log_2 2^{-4} = -4$

51. See Figure 51.

x	6	7	21
$f(x)$	0	2	8

Figure 51

53. (a) $10^x = 0.01 \implies \log 10^{-2} = x \implies x = -2$

    (b) $10^x = 7 \implies \log 7 = x \implies x \approx 0.85$

    (c) $10^x = -4 \implies \log(-4) = x \implies$ no solution

55. (a) $4^x = \dfrac{1}{16} \implies 4^x = 4^{-2} \implies x = -2$

    (b) $e^x = 2 \implies \ln(2) = x \implies x \approx 0.69$

    (c) $5^x = 125 \implies 5^x = 5^3 \implies x = 3$

57. (a) $9^x = 1 \implies 9^x = 9^0 \implies x = 0$

    (b) $10^x = \sqrt{10} \implies 10^x = 10^{1/2} \implies x = \dfrac{1}{2}$

    (c) $4^x = \sqrt[3]{4} \implies 4^x = 4^{1/3} \implies x = \dfrac{1}{3}$

59. $e^{-x} = 3 \implies \ln e^{-x} = \ln 3 \implies -x = \ln 3 \implies x = -\ln 3 \approx -1.10$

61. $10^x - 5 = 95 \implies 10^x = 100 \implies 10^x = 10^2 \implies x = 2$

63. $10^{3x} = 100 \implies 10^{3x} = 10^2 \implies 3x = 2 \implies x = \dfrac{2}{3} \approx 0.67$

65. $5(10^{4x}) = 65 \implies 10^{4x} = 13 \implies \log(13) = 4x \implies x = \dfrac{\log(13)}{4} \approx 0.28$

67. $4(3^x) - 3 = 13 \implies 4(3^x) = 16 \implies 3^x = 4 \implies \log_3(4) = \dfrac{\log(4)}{\log(3)} \approx 1.26$

69. $e^x + 1 = 24 \implies e^x = 23 \implies \ln e^x = \ln 23 \implies x = \ln 23 \approx 3.14$

71. $2^x + 1 = 15 \implies 2^x = 14 \implies \log_2 2^x = \log_2 14 \implies x = \log_2 14$ or $x \approx 3.81$

73. $5e^x + 2 = 20 \implies 5e^x = 18 \implies e^x = \dfrac{18}{5} \implies \ln e^x = \ln \dfrac{18}{5} \implies x = \ln\left(\dfrac{18}{5}\right) \approx 1.28$

75. $8 - 3(2)^{0.5x} = -40 \implies -3(2)^{0.5x} = -48 \implies 2^{0.5x} = 16 \implies 2^{0.5x} = 2^4 \implies 0.5x = 4 \implies x = \dfrac{4}{0.5} \implies$

    $x = 8$

77. (a) $\log x = 2 \implies 10^2 = x \implies x = 100$

    (b) $\log x = -3 \implies 10^{-3} = x \implies x = \dfrac{1}{1000}$

    (c) $\log x = 1.2 \implies 10^{1.2} = x \implies x \approx 15.8489$

79. (a) $\log_2 x = 6 \implies x = 2^6 = x \implies 64$

    (b) $\log_3 x = -2 \implies x = 3^{-2} = x \implies \dfrac{1}{9}$

    (c) $\ln x = 2 \implies x = e^2 \implies x \approx 7.3891$

81. $\log_2 x = 1.2 \implies 2^{\log_2 x} = 2^{1.2} \implies x = 2^{1.2} \approx 2.2974$

83. $5\log_7 (2x) = 10 \implies \log_7 (2x) = 2 \implies 2x = 7^2 \implies 2x = 49 \implies x = \dfrac{49}{2}$

85. $2 \log x = 6 \implies \log x = 3 \implies 10^{\log x} = 10^3 \implies x = 10^3 = 1000$

87. $2 \log 5x = 4 \implies \log 5x = 2 \implies 10^{\log 5x} = 10^2 \implies 5x = 100 \implies x = 20$

89. $4 \ln x = 3 \implies \ln x = \dfrac{3}{4} \implies e^{\ln x} = e^{3/4} \implies x = e^{3/4} \approx 2.1170$

91. $5 \ln x - 1 = 6 \implies 5 \ln x = 7 \implies \ln x = \dfrac{7}{5} \implies e^{\ln x} = e^{7/5} \implies x = e^{7/5} \approx 4.0552$

93. $4 \log_2 x = 16 \implies \log_2 x = 4 \implies 2^{\log_2 x} = 2^4 \implies x = 2^4 = 16$

95. $5 \ln (2x) + 6 = 12 \implies 5 \ln (2x) = 6 \implies \ln (2x) = \dfrac{6}{5} \implies e^{\ln 2x} = e^{6/5} \implies 2x = e^{6/5} \implies$

   $x = \dfrac{e^{6/5}}{2} \approx 1.6601$

97. $9 - 3 \log_4 2x = 3 \implies -3 \log_4 2x = -6 \implies \log_4 2x = 2 \implies 4^{\log_4 2x} = 4^2 \implies 2x = 16 \implies x = 8$

99. $f(x) = a + b \log x$ and $f(1) = 5 \implies 5 = a + b \log 1 \implies 5 = a + b(0) \implies a = 5$

   $f(x) = 5 + b \log x$ and $f(10) = 7 \implies 7 = 5 + b \log 10 \implies 7 = 5 + b(1) \implies b = 2$

   The function is $f(x) = 5 + 2 \log x$.

101. Since $f(x) = e^x$, $f^{-1}(x) = \ln x$. See Figure 101.

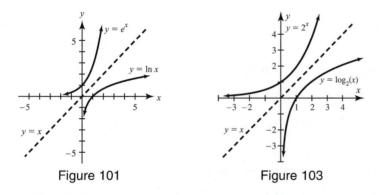

Figure 101   Figure 103

103. See Figure 103.

105. The graph is shown in Figure 105. $D = \{x \,|\, x > -1\}$

107. The graph is shown in Figure 107. $D = \{x \,|\, x < 0\}$

$[-6, 6, 1]$ by $[-4, 4, 1]$ $[-6, 6, 1]$ by $[-4, 4, 1]$

Figure 105   Figure 107

109. Decreasing. See Figure 109.

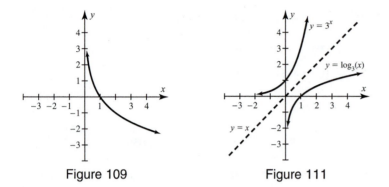

Figure 109   Figure 111

111. (a) See Figure 111.

(b) f is increasing on $(0, \infty)$; f^{-1} is increasing on $(-\infty, \infty)$.

113. (a) $7^{4x} = 4 \Rightarrow \log_7 4 = 4x$

(b) $e^x = 7 \Rightarrow \ln(7) = x$

(c) $c^x = b \Rightarrow \log_c(b) = x$

115. (a) $\log_8(x) = 3 \Rightarrow x = 8^3$

(b) $\log_9(2 + x) = 5 \Rightarrow 2 + x = 9^5$

(c) $\log_k b = c \Rightarrow b = k^c$

117. $D(x) = 10 \log(10^{16}x),\ x = 10^{-11/2} \Rightarrow D(10^{-11/2}) = 10 \log(10^{16}(10^{-11/2})) \Rightarrow 10 \log(10^{21/2}) \Rightarrow$

$10\left(\dfrac{21}{2}\right) \Rightarrow 105$ decibels.

119. $f(x) = a + b \log x$ and $f(1) = 7 \Rightarrow 7 = a + b \log 1 \Rightarrow 7 = a + b(0) \Rightarrow a = 7$

$f(x) = 7 + b \log x$ and $f(10) = 11 \Rightarrow 11 = 7 + b \log 10 \Rightarrow 11 = 7 + b(1) \Rightarrow b = 4$

The function that models the given data is $f(x) = 7 + 4 \log x$.

$f(x) = 16 \Rightarrow 16 = 7 + 4 \log x \Rightarrow 9 = 4 \log x \Rightarrow \log x = \dfrac{9}{4} \Rightarrow 10^{\log x} = 10^{9/4} \Rightarrow$

$x = 10^{9/4} \approx 178$. The island would be about 178 square kilometers.

121. (a) $f(x) = Ca^x$ and $f(0) = 3 \Rightarrow 3 = Ca^0 \Rightarrow 3 = C(1) \Rightarrow C = 3$

$f(x) = 3a^x$ and $f(1) = 6 \Rightarrow 6 = 3a^1 \Rightarrow 2 = a^1 \Rightarrow a = 2$; thus $f(x) = 3(2^x)$ models the data.

(b) $f(x) = 16 \Rightarrow 16 = 3(2^x) \Rightarrow 2^x = \dfrac{16}{3}$. Graph $Y_1 = 2^\wedge X$ and $Y_2 = 16/3$ in $[0, 5, 1]$ by $[4, 20, 5]$.

The graphs (not shown) intersect near $(2.41, 5.33)$, so there were 16 million bacteria after about 2.4 days.

123. (a) The graph of L is shown in Figure 123. It is an increasing function. The implication is that heavier planes generally require longer runways.

(b) Since the weight is measured in 1000-pound units, start by evaluating $L(10)$ and $L(100)$.

$L(10) = 3 \log 10 = 3$ and $L(100) = 3 \log 100 = 6$. Thus, a 10,000-pound plane requires approximately 3000 feet of runway, whereas as 100,000-pound plane requires approximately 6000 feet. The distance does not increase by a factor of 10.

(c) If the weight increases tenfold, the runway length increases by 3000 feet.

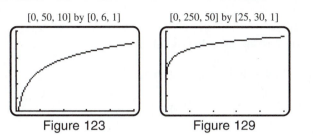

[0, 50, 10] by [0, 6, 1]   [0, 250, 50] by [25, 30, 1]

Figure 123   Figure 129

125. (a) $pH = -\log(x) \Rightarrow pH = -\log(10^{-4.7}) \Rightarrow pH = -(-4.7) = 4.7$

(b) $-\log(x) = 8.2 \Rightarrow \log(x) = -8.2 \Rightarrow 10^{\log(x)} = 10^{-8.2} \Rightarrow x = 10^{-8.2}$. Since $\dfrac{10^{-4.7}}{10^{-8.2}} = 10^{3.5}$ the

hydrogen concentration in the rainwater is $10^{3.5} \approx 3162$ times greater than it is in seawater.

127. (a) Since $I_0 = 1$, $R(x) = 6.0 \Rightarrow \log x = 6.0 \Rightarrow 10^{\log x} = 10^{6.0} \Rightarrow x = 10^6 = 1{,}000{,}000$

Similarly $R(x) = 8.0 \Rightarrow \log x = 8.0 \Rightarrow 10^{\log x} = 10^{8.0} \Rightarrow x = 10^8 = 100{,}000{,}000$

(b) $\dfrac{10^8}{10^6} = 10^{8-6} = 10^2 = 100$. The Indonesian earthquake was 100 times more intense (powerful) than the

Yugoslovian earthquake.

129. (a) $f(x) = 0.48 \ln(x+1) + 27 \Rightarrow f(0) = 0.48 \ln(0+1) + 27 = 0.48 \ln 1 + 27 =$

$0.48(0) + 27 = 27$ and $f(100) = 0.48 \ln(100+1) + 27 \approx 29.2$ inches. At the center, or eye, of the

hurricane the pressure is 27 inches of mercury, while 100 miles from the eye the air pressure has risen to

29.2 inches of mercury.

(b) Graph $Y_1 = 0.48 \ln(X+1) + 27$ as shown in Figure 129. At first, the air pressure rises rapidly as one

moves away from the eye. Then, the air pressure starts to level off and does not increase significantly for

distances greater than 200 miles.

(c) $f(x) = 28 \Rightarrow 28 = 0.48 \ln(x+1) + 27 \Rightarrow \ln(x+1) = \dfrac{1}{0.48} \Rightarrow e^{\ln(x+1)} = e^{1/0.48} \Rightarrow$

$x + 1 = e^{1/0.48} \Rightarrow x = e^{1/0.48} - 1 \approx 7.03$; the air pressure is 28 inches of mercury about 7 miles from

the eye of the hurricane.

131. (a) $T(x) = 20 + 80e^{-x} \Rightarrow T(1) = 20 + 80e^{-1} \approx 49.4$; the temperature is about 49.4°C after 1 hour.

(b) $T(x) = 60 \Rightarrow 60 = 20 + 80e^{-x} \Rightarrow 40 = 80e^{-x} \Rightarrow e^{-x} = \dfrac{1}{2} \Rightarrow \ln e^{-x} = \ln\left(\dfrac{1}{2}\right) \Rightarrow$

$-x = \ln\left(\dfrac{1}{2}\right) \Rightarrow x = -\ln\left(\dfrac{1}{2}\right) \approx 0.693$; the water took about 0.69 hours, or 41.4 minutes to cool to 60°C.

133. (a) $f(x) = e^{-x/3} \Rightarrow f(5) = e^{-5/3} \approx 0.189$

The probability that no car enters the intersection during a 5-minute period is about 0.189 or 18.9%.

(b) $f(x) = 0.30 \Rightarrow 0.30 = e^{-x/3} \Rightarrow \ln e^{-x/3} = \ln 0.3 \Rightarrow -\dfrac{x}{3} = \ln 0.3 \Rightarrow x = -3 \ln 0.3 \approx 3.6$

The probability that no car enters the intersection will be 30% during a period of about 3.6 minutes.

5.5: Properties of Logarithms

1. $\log 4 + \log 7 \approx 1.447$; $\log 28 \approx 1.447$; $\log 4 + \log 7 = \log 28 = \log (4 \cdot 7)$; Property 2

3. $\ln 72 - \ln 8 \approx 2.197$; $\ln 9 \approx 2.197$; $\ln 72 - \ln 8 = \ln 9 = \ln \left(\dfrac{72}{8} \right)$; Property 3

5. $10 \log 2 \approx 3.010$; $\log 1024 \approx 3.010$; $10 \log 2 = \log 1024 = \log 2^{10}$; Property 4

7. $\log_2 ab = \log_2 a + \log_2 b$

9. $\ln 7a^4 = \ln 7 + 4 \ln a$

11. $\log \dfrac{6}{z} = \log 6 - \log z$

13. $\log \dfrac{x^2}{3} = \log x^2 - \log 3 = 2 \log x - \log 3$

15. $\ln \dfrac{2x^7}{3k} = \ln 2x^7 - \ln 3k = \ln 2 + \ln x^7 - (\ln 3 + \ln k) = \ln 2 + 7 \ln x - \ln 3 - \ln k$

17. $\log_2 4k^2 x^3 = \log_2 4k^2 + \log_2 x^3 = \log_2 4 + \log_2 k^2 + \log_2 x^3 = 2 + 2 \log_2 k + 3 \log_2 x$

19. $\log_5 \dfrac{25x^3}{y^4} = \log_5 25x^3 - \log_5 y^4 = 2 + 3 \log_5 x - 4 \log_5 y$

21. $\ln \dfrac{x^4}{y^2 \sqrt{z^3}} = \ln (x^4) - (\ln y^2 + \ln (z^{3/2})) = 4 \ln (x) - 2 \ln (y) - \dfrac{3}{2} \ln (z)$

23. $\log_4 0.25(x + 2)^3 = \log_4 0.25 + \log_4 (x + 2)^3 = -1 + 3 \log_4 (x + 2)$

25. $\log_5 \dfrac{x^3}{(x - 4)^4} = \log_5 x^3 - \log_5 (x - 4)^4 = 3 \log_5 x - 4 \log_5 (x - 4)$

27. $\log_2 \dfrac{\sqrt{x}}{z^2} = \log_2 \sqrt{x} - \log_2 z^2 = \log_2 x^{1/2} - \log_2 z^2 = \dfrac{1}{2} \log_2 x - 2 \log_2 z$

29. $\ln \sqrt[3]{\dfrac{2x + 6}{(x + 1)^5}} = \ln \left(\dfrac{2x + 6}{(x + 1)^5} \right)^{1/3} = \dfrac{1}{3} \left[\ln (2x + 6) - 5 \ln (x + 1) \right] = \dfrac{1}{3} \ln (2x + 6) - \dfrac{5}{3} \ln (x + 1)$

31. $\log_2 \dfrac{\sqrt[3]{x^2 - 1}}{\sqrt{1 + x^2}} = \log_2 \sqrt[3]{x^2 - 1} - \log_2 \sqrt{1 + x^2} = \log_2 (x^2 - 1)^{1/3} - \log_2 (1 + x^2)^{1/2} =$

    $\dfrac{1}{3} \log_2 (x^2 - 1) - \dfrac{1}{2} \log_2 (1 + x^2)$

33. $\log 2 + \log 3 = \log (2 \cdot 3) = \log 6$

35. $\ln \sqrt{5} - \ln 25 = \ln \left(\dfrac{5^{1/2}}{5^2} \right) = \ln 5^{-3/2} = -\dfrac{3}{2} \ln 5$

37. $\log 20 + \log \dfrac{1}{10} = \log (20) \left(\dfrac{1}{10} \right) = \log \dfrac{20}{10} = \log 2$

39. $\log 4 + \log 3 - \log 2 = \log \dfrac{4(3)}{2} = \log \dfrac{12}{2} = \log 6$

41. $\log_7 5 + \log_7 k^2 = \log_7 (5)(k^2) = \log_7 5k^2$

43. $\ln x^6 - \ln x^3 = \ln \dfrac{x^6}{x^3} = \ln x^{6-3} = \ln x^3$

45. $\log \sqrt{x} + \log x^2 - \log x = \log \left(\dfrac{x^{1/2} \cdot x^2}{x} \right) = \log x^{3/2} = \dfrac{3}{2} \log x$

47. $3 \ln (x) - \dfrac{3}{2} \ln (y) + 4 \ln (z) = \ln (x^3) - \ln (\sqrt{y^3}) + \ln (z^2) = \ln \left(\dfrac{x^3 z^4}{\sqrt{y^3}} \right)$

49. $\ln \dfrac{1}{e^2} + \ln 2e = \ln \left(\dfrac{1}{e^2} \cdot 2e \right) = \ln \dfrac{2}{e}$

51. $2 \ln x - 4 \ln y + \dfrac{1}{2} \ln z = \ln x^2 - \ln y^4 + \ln z^{1/2} = \ln \dfrac{x^2}{y^4} + \ln \sqrt{z} = \ln \dfrac{x^2 \sqrt{z}}{y^4}$

53. $\log 4 - \log x + 7 \log \sqrt{x} = \log \dfrac{4}{x} + \log (\sqrt{x})^7 = \log \left(\dfrac{4 x^{7/2}}{x} \right) = \log 4 x^{5/2} = \log 4 \sqrt{x^5}$

55. $2 \log (x^2 - 1) + 4 \log (x - 2) - \dfrac{1}{2} \log y = \log (x^2 - 1)^2 + \log (x - 2)^4 - \log y^{1/2} =$

    $\log \dfrac{(x^2 - 1)^2 (x - 2)^4}{\sqrt{y}}$

57. (a) Table $Y_1 = \log (3X) + \log (2X)$ and $Y_2 = \log (6X^2)$ starting at $x = 1$, incrementing by 1.

    See Figure 57. From the table we see that $Y_1 = Y_2$, so $f(x) = g(x)$.

    (b) By property 2: $\log 3x + \log 2x = \log (3x \cdot 2x) = \log 6x^2$

59. (a) Table $Y_1 = \ln (2X^2) - \ln (X)$ and $Y_2 = \ln (2X)$ starting at $x = 1$, incrementing by 1. See Figure 59.

    From the table we see that $Y_1 = Y_2$, so $f(x) = g(x)$.

    (b) By property 3: $\ln 2x^2 - \ln x = \ln \left(\dfrac{2x^2}{x} \right) = \ln 2x$

Figure 57   Figure 59   Figure 61

61. (a) Table $Y_1 = \ln (X^4) - \ln (X^2)$ and $Y_2 = 2 \ln (X)$ starting at $x = 1$, incrementing by 1. See Figure 61.

    From the table we see that $Y_1 = Y_2$, so $f(x) = g(x)$.

    (b) By property 4: $\ln x^4 - \ln x^2 = 4 \ln x - 2 \ln x = 2 \ln x$

63. See Figure 63.

65. See Figure 65.

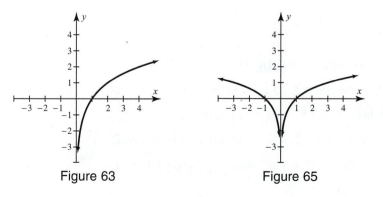

Figure 63   Figure 65

67. $\log_2 25 = \dfrac{\log 25}{\log 2} \approx 4.644$

69. $\log_5 130 = \dfrac{\log 130}{\log 5} \approx 3.024$

71. $\log_2 5 + \log_2 7 = \dfrac{\log 5}{\log 2} + \dfrac{\log 7}{\log 2} \approx 5.129$

73. $\sqrt{\log_4 46} = \sqrt{\dfrac{\log 46}{\log 4}} \approx 1.662$

75. $\dfrac{\log_2 12}{\log_2 3} = \dfrac{\frac{\log 12}{\log 2}}{\frac{\log 3}{\log 2}} = \dfrac{\log 12}{\log 2} \cdot \dfrac{\log 2}{\log 3} = \dfrac{\log 12}{\log 3} \approx 2.262$

77. Graph $Y_1 = \log (X{\wedge}3 + X{\wedge}2 + 1)/\log 2$ and $Y_2 = 7$ in $[0, 10, 2]$ by $[0, 10, 2]$. The graphs intersect near the point $(4.714, 7)$, so the solution to $\log_2 (x^3 + x^2 + 1) = 7$ is $x \approx 4.714$.

79. Graph $Y_1 = \log (X{\wedge}2 + 1)/\log 2$ and $Y_2 = 5 - \log (X{\wedge}4 + 1)/\log 3$ in $[-5, 5, 2]$ by $[-5, 10, 2]$. The graphs intersect near the points $(-2.035, 2.362)$ and $(2.035, 2.362)$, so the solutions to the equation $\log_2 (x^2 + 1) = 5 - \log_3 (x^4 + 1)$ are $x \approx \pm 2.035$.

81. Change base 10 to base e: $L(x) = 3 \log x = \dfrac{3 \cdot \ln x}{\ln 10}$;

$L(50) = 3 \log 50 \approx 5.097$ and $L(50) = \dfrac{3 \ln 50}{\ln 10} \approx 5.097$. Yes, the answers agree.

83. $f(x) = 160 + 10 \log x \Rightarrow f(10x) = 160 + 10 \log (10x) = 160 + 10 (\log 10 + \log x) = 160 + 10(1 + \log x) = 160 + 10 + 10 \log x = 160 + 10 \log x + 10$, thus $f(10x) = f(x) + 10$. In other words, the decibel level increases by 10 decibels.

85. $\ln I - \ln I_0 = -kx \Rightarrow \ln I = \ln I_0 - kx \Rightarrow e^{\ln I} = e^{\ln I_0 - kx} \Rightarrow e^{\ln I} = e^{\ln I_0} \cdot e^{-kx} \Rightarrow I = I_0 e^{-kx}$

87. (a) $P = 34e^{0.013x} \Rightarrow \dfrac{P}{34} = e^{0.013x} \Rightarrow \ln \left(\dfrac{P}{34}\right) = 0.013x \Rightarrow x = \dfrac{1}{0.013} \ln \left(\dfrac{P}{34}\right)$

(b) $x = \dfrac{\ln \left(\frac{38}{34}\right)}{0.013} \approx 8.6$. The population is expected to reach 38 million during 2009.

89. $A = Pe^{rt} \Rightarrow \dfrac{A}{P} = e^{rt} \Rightarrow \ln \dfrac{A}{P} = \ln e^{rt} \Rightarrow \ln \dfrac{A}{P} = rt \Rightarrow \dfrac{\ln \frac{A}{P}}{r} = t \Rightarrow t = \dfrac{\ln \frac{A}{P}}{r}$

91. $\log 1 + 2 \log 2 + 3 \log 3 + 4 \log 4 + 5 \log 5 = \log 1 + \log 2^2 + \log 3^3 + \log 4^4 + \log 5^5 = \log (1 \cdot 2^2 \cdot 3^3 \cdot 4^4 \cdot 5^5) = \log 86{,}400{,}000$

5.6: Exponential and Logarithmic Equations

1. (a) The graphs appear to intersect near the point $(2, 7.5)$. Thus, the solution is $x \approx 2$.

(b) $f(x) = g(x) \Rightarrow e^x = 7.5 \Rightarrow \ln e^x = \ln 7.5 \Rightarrow x = \ln 7.5$ about 2.015

3. (a) The graphs appear to intersect near the point $(2, 2.5)$. Thus, the solution is $x \approx 2$.

(b) $f(x) = g(x) \Rightarrow 10^{0.2x} = 2.5 \Rightarrow \log 10^{0.2x} = \log 2.5 \Rightarrow 0.2x = \log 2.5 \Rightarrow x = \dfrac{\log 2.5}{0.2}$ about 1.990

5. $4e^x = 5 \Rightarrow e^x = \dfrac{5}{4} \Rightarrow \ln e^x = \ln \dfrac{5}{4} \Rightarrow x = \ln \dfrac{5}{4} \approx 0.2231$

7. $2(10^x) + 5 = 45 \Rightarrow 2(10^x) = 40 \Rightarrow 10^x = 20 \Rightarrow \log 10^x = \log 20 \Rightarrow x = \log 20 \approx 1.301$

9. $2.5e^{-1.2x} = 1 \Rightarrow e^{-1.2x} = \dfrac{1}{2.5} \Rightarrow \ln e^{-1.2x} = \ln \dfrac{1}{2.5} \Rightarrow -1.2x = \ln \dfrac{1}{2.5} \Rightarrow x = \dfrac{\ln \frac{1}{2.5}}{-1.2} \approx 0.7636$

11. $1.2(0.9^x) = 0.6 \Rightarrow 0.9^x = 0.5 \Rightarrow \ln 0.9^x = \ln 0.5 \Rightarrow x \ln 0.9 = \ln 0.5 \Rightarrow x = \dfrac{\ln 0.5}{\ln 0.9} \approx 6.579$

13. $4(1.1^{x-1}) = 16 \Rightarrow 1.1^{x-1} = 4 \Rightarrow \ln 1.1^{x-1} = \ln 4 \Rightarrow (x-1)\ln 1.1 = \ln 4 \Rightarrow x - 1 = \dfrac{\ln 4}{\ln 1.1} \Rightarrow$

$x = \dfrac{\ln 4}{\ln 1.1} + 1 \approx 15.55$

15. $5(1.2)^{3x-2} + 94 = 100 \Rightarrow 5(1.2)^{3x-2} = 6 \Rightarrow (1.2)^{3x-2} = 1.2 \Rightarrow \ln 1.2^{3x-2} = \ln 1.2 \Rightarrow$

$(3x - 2)\ln 1.2 = \ln 1.2 \Rightarrow 3x - 2 = \dfrac{\ln 1.2}{\ln 1.2} \Rightarrow 3x - 2 = 1 \Rightarrow 3x = 3 \Rightarrow x = 1$

17. $5^{3x} = 5^{1-2x} \Rightarrow \log_5 5^{3x} = \log_5 5^{1-2x} \Rightarrow 3x = 1 - 2x \Rightarrow 5x = 1 \Rightarrow x = \dfrac{1}{5}$

19. $10^{x^2} = 10^{3x-2} \Rightarrow \log 10^{x^2} = \log 10^{3x-2} \Rightarrow x^2 = 3x - 2 \Rightarrow x^2 - 3x + 2 = 0 \Rightarrow$

$(x - 1)(x - 2) = 0 \Rightarrow x = 1 \text{ or } x = 2$

21. No solution since no power of $\dfrac{1}{5}$ will result in a negative value.

23. $\left(\dfrac{2}{5}\right)^{x-2} = \dfrac{1}{3} \Rightarrow \log \left(\dfrac{2}{5}\right)^{x-2} = \log \left(\dfrac{1}{3}\right) \Rightarrow (x - 2)\log \left(\dfrac{2}{5}\right) = \log \left(\dfrac{1}{3}\right) \Rightarrow x - 2 = \dfrac{\log \left(\frac{1}{3}\right)}{\log \left(\frac{2}{5}\right)} \Rightarrow$

$x = 2 + \dfrac{\log \left(\frac{1}{3}\right)}{\log \left(\frac{2}{5}\right)} \approx 3.199$

25. $4^{x-1} = 3^{2x} \Rightarrow \log 4^{x-1} = \log 3^{2x} \Rightarrow (x - 1)\log 4 = 2x \log 3 \Rightarrow x \log 4 - \log 4 = 2x \log 3 \Rightarrow$

$x \log 4 - 2x \log 3 = \log 4 \Rightarrow x(\log 4 - 2 \log 3) = \log 4 \Rightarrow x = \dfrac{\log 4}{\log 4 - 2 \log 3} \approx -1.710$

27. $e^{x-3} = 2^{3x} \Rightarrow \ln e^{x-3} = \ln 2^{3x} \Rightarrow x - 3 = 3x \ln 2 \Rightarrow -3 = -x + 3x \ln 2 \Rightarrow$

$3 = x - 3x \ln 2 \Rightarrow 3 = x(1 - 3 \ln 2) \Rightarrow x = \dfrac{3}{1 - 3 \ln 2} \Rightarrow x = \dfrac{3}{1 - \ln 8} \Rightarrow x \approx -2.779$

29. $3(1.4)^x - 4 = 60 \Rightarrow 3(1.4)^x = 64 \Rightarrow 1.4^x = \dfrac{64}{3} \Rightarrow \log 1.4^x = \log \dfrac{64}{3} \Rightarrow x \log 1.4 = \log \dfrac{64}{3} \Rightarrow$

$x = \dfrac{\log \frac{64}{3}}{\log 1.4} \approx 9.095$

31. $5(1.015)^{x-1980} = 8 \Rightarrow 1.015^{x-1980} = \dfrac{8}{5} \Rightarrow \log 1.015^{x-1980} = \log \dfrac{8}{5} \Rightarrow (x - 1980)\log 1.015 = \log \dfrac{8}{5} \Rightarrow$

$x \log 1.015 - 1980 \log 1.015 = \log \dfrac{8}{5} \Rightarrow x \log 1.015 = \log \dfrac{8}{5} + 1980 \log 1.015 \Rightarrow$

$x = \dfrac{\log \frac{8}{5} + 1980 \log 1.015}{\log 1.015} = \dfrac{\log \frac{8}{5}}{\log 1.015} + 1980 \approx 2012$

33. $4\left(\dfrac{3}{4}\right)^{x+1} = \dfrac{1}{81}\left(\dfrac{3}{2}\right)^{5+x} \Rightarrow 324\left(\dfrac{3}{4}\right)^{x+1} = \left(\dfrac{3}{2}\right)^{5+x} \Rightarrow \ln\left[324\left(\dfrac{3}{4}\right)^{x+1}\right] = \ln\left(\dfrac{3}{2}\right)^{5+x} \Rightarrow$

$\ln 324 + \ln\left(\dfrac{3}{4}\right)^{x+1} = \ln\left(\dfrac{3}{2}\right)^{5+x} \Rightarrow \ln 324 + (x+1)\ln\dfrac{3}{4} = (5+x)\ln\dfrac{3}{2} \Rightarrow$

$\ln 324 + x\ln\dfrac{3}{4} + \ln\dfrac{3}{4} = 5\ln\dfrac{3}{2} + x\ln\dfrac{3}{2} \Rightarrow x\ln\dfrac{3}{4} - x\ln\dfrac{3}{2} = 5\ln\dfrac{3}{2} - \ln 324 - \ln\dfrac{3}{4} \Rightarrow$

$x\left(\ln\dfrac{3}{4} - \ln\dfrac{3}{2}\right) = 5\ln\dfrac{3}{2} - \ln 324 - \ln\dfrac{3}{4} \Rightarrow x = \dfrac{5\ln\frac{3}{2} - \ln 324 - \ln\frac{3}{4}}{\ln\frac{3}{4} - \ln\frac{3}{2}} \Rightarrow x = 5$

35. $3\log x = 2 \Rightarrow \log x = \dfrac{2}{3} \Rightarrow 10^{\log x} = 10^{2/3} \Rightarrow x = 10^{2/3} \approx 4.642$

37. $\ln 2x = 5 \Rightarrow e^{\ln 2x} = e^5 \Rightarrow 2x = e^5 \Rightarrow x = \dfrac{e^5}{2} \approx 74.207$

39. $\log 2x^2 = 2 \Rightarrow 10^{\log 2x^2} = 10^2 \Rightarrow 2x^2 = 100 \Rightarrow x^2 = 50 \Rightarrow x = \pm\sqrt{50} \approx \pm 7.071$

41. $\log_2 (3x - 2) = 4 \Rightarrow 2^{\log_2 (3x-2)} = 2^4 \Rightarrow 3x - 2 = 16 \Rightarrow 3x = 18 \Rightarrow x = 6$

43. $\log_5 (8 - 3x) = 3 \Rightarrow 5^{\log_5 (8-3x)} = 5^3 \Rightarrow 8 - 3x = 125 \Rightarrow -3x = 117 \Rightarrow x = -39$

45. $160 + 10\log x = 50 \Rightarrow 10\log x = -110 \Rightarrow \log x = -11 \Rightarrow 10^{\log x} = 10^{-11} \Rightarrow x = 10^{-11}$

47. $\ln x + \ln x^2 = 3 \Rightarrow \ln x + 2\ln x = 3 \Rightarrow 3\ln x = 3 \Rightarrow \ln x = 1 \Rightarrow e^{\ln x} = e^1 \Rightarrow x = e \approx 2.718$

49. $2\log_2 x = 4.2 \Rightarrow \log_2 x = 2.1 \Rightarrow 2^{\log_2 x} = 2^{2.1} \Rightarrow x = 2^{2.1} \approx 4.287$

51. $\log x + \log 2x = 2 \Rightarrow \log (x \cdot 2x) = 2 \Rightarrow \log 2x^2 = 2 \Rightarrow 10^{\log 2x^2} = 10^2 \Rightarrow 2x^2 = 100 \Rightarrow$

$x^2 = 50 \Rightarrow x = \pm\sqrt{50}$. When $x = -\sqrt{50} < 0$, $\log x$ is undefined. Therefore, the only solution is

$x = \sqrt{50} \approx 7.071.$

53. $\log (2 - 3x) = 3 \Rightarrow 10^3 = 2 - 3x \Rightarrow 1000 = 2 - 3x \Rightarrow 998 = -3x \Rightarrow x = -\dfrac{998}{3}$

55. $\ln (x) + \ln (3x - 1) = \ln (10) \Rightarrow \ln (x(3x - 1)) = \ln (10) \Rightarrow \ln (3x^2 - x) = \ln (10) \Rightarrow$

$3x^2 - x = 10 \Rightarrow 3x^2 - x - 10 = 0 \Rightarrow (3x + 5)(x - 2) = 0 \Rightarrow 3x + 5 = 0 \text{ or } x - 2 = 0 \Rightarrow$

$x = -\dfrac{5}{3} \text{ or } x = 2$. When $x = -\dfrac{5}{3}$, both $\ln (3x - 1)$ and $\ln (x)$ are undefined, therefore the only solution is $x = 2.$

57. $2\ln x = \ln (2x + 1) \Rightarrow \ln x^2 = \ln (2x + 1) \Rightarrow e^{\ln x^2} = e^{\ln (2x+1)} \Rightarrow x^2 = 2x + 1 \Rightarrow$

$x^2 - 2x - 1 = 0 \Rightarrow x = \dfrac{2 \pm \sqrt{2^2 - 4(1)(-1)}}{2(1)} = \dfrac{2 \pm \sqrt{8}}{2} = \dfrac{2 \pm 2\sqrt{2}}{2} = 1 \pm \sqrt{2} \Rightarrow$

$x = 1 + \sqrt{2} \text{ or } x = 1 - \sqrt{2}.$

When $x = 1 - \sqrt{2} \approx -0.414$, $\ln x$ is undefined. The only solution is $x = 1 + \sqrt{2}.$

59. $\log (x + 1) + \log (x - 1) = \log 3 \Rightarrow \log [(x + 1)(x - 1)] = \log 3 \Rightarrow \log (x^2 - 1) = \log 3 \Rightarrow$

$10^{\log (x^2-1)} = 10^{\log 3} \Rightarrow x^2 - 1 = 3 \Rightarrow x^2 = 4 \Rightarrow x = -2 \text{ or } x = 2$

When $x = -2$ both $\log (x - 1)$ and $\log (x + 1)$ are undefined. The only solution is $x = 2.$

61. $\log_2 2x = 4 - \log_2 (x + 2) \Rightarrow \log_2 2x + \log_2 (x + 2) = 4 \Rightarrow \log_2 [2x(x+2)] = 4 \Rightarrow$

$\log_2 (2x^2 + 4x) = 4 \Rightarrow 2^{\log_2 (2x^2+4x)} = 4 \Rightarrow 2x^2 + 4x - 16 = 0 \Rightarrow 2(x^2 + 2x - 8) = 0 \Rightarrow$

$2(x + 4)(x - 2) = 0 \Rightarrow x = -4, x = 2$, since we cannot take the log of a negative number, $x = 2$.

63. $\log_5 (x + 1) + \log_5 (x - 1) = \log_5 15 \Rightarrow \log_5 (x + 1)(x - 1) = \log_5 15 \Rightarrow$

$\log_5 (x^2 - 1) = \log_5 15 \Rightarrow 5^{\log_5 (x^2-1)} = 5^{\log_5 15} \Rightarrow x^2 - 1 = 15 \Rightarrow x^2 - 16 = 0 \Rightarrow$

$(x + 4)(x - 4) = 0$, since we cannot take the log of a negative number, $x = 4$.

65. Graph as shown in Figure 65.

The graphs intersect near the point $(0.31, 2)$, so the solution to $2x + e^x = 2$ is $x \approx 0.31$.

67. Graph $Y_1 = X^2 + X \ln (X)$ and $Y_2 = 2$ as shown in Figure 67.

The graphs intersect near the point $(1.71, 2)$, so the solution to $x^2 - x \ln x = 2$ is $x \approx 1.71$.

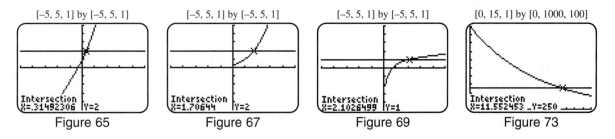

[-5, 5, 1] by [-5, 5, 1]   [-5, 5, 1] by [-5, 5, 1]   [-5, 5, 1] by [-5, 5, 1]   [0, 15, 1] by [0, 1000, 100]

Figure 65   Figure 67   Figure 69   Figure 73

69. Graph $Y_1 = Xe^{\wedge}(-X) + \ln (X)$ and $Y_2 = 1$ as shown in Figure 69.

The graphs intersect near the point $(2.10, 2)$, so the solution to $xe^{-x} + \ln x = 1$ is $x \approx 2.10$.

71. $4 = 3(1.017)^{x-1960} \Rightarrow \dfrac{4}{3} = (1.017)^{x-1960} \Rightarrow \ln \left(\dfrac{4}{3}\right) = \ln (1.017)^{x-1960} \Rightarrow$

$\ln \left(\dfrac{4}{3}\right) = (x - 1960) \ln (1.017) \Rightarrow \dfrac{\ln \left(\frac{4}{3}\right)}{\ln (1.017)} = x - 1960 \Rightarrow x = 1960 + \dfrac{\ln \left(\frac{4}{3}\right)}{\ln (1.017)} \Rightarrow x = 1977$

73. $250 = 1000e^{-0.12x} \Rightarrow 0.25 = e^{-0.12x} \Rightarrow \ln (0.25) = -0.12x \Rightarrow x = \dfrac{\ln (0.25)}{-0.12} \approx 11.55$ ft. Graph

$Y_1 = 250$ and $Y_2 = 1000 e^{-0.12x}$, See Figure 73. The graph intersects at approximately $(11.55, 250)$.

75. $0.5e^{0.014t} = 25 \Rightarrow e^{0.014t} = 50 \Rightarrow 0.014t = \ln 50 \Rightarrow t = \dfrac{\ln 50}{0.014} \approx 279.4$ min

77. We must solve the equation $f(x) = 95$.

$230(0.881)^x = 95 \Rightarrow 0.881^x = \dfrac{95}{230} \Rightarrow \log 0.881^x = \log \dfrac{95}{230} \Rightarrow x \log 0.881 = \log \dfrac{95}{230}$

$x = \dfrac{\log \frac{95}{230}}{\log 0.881} \approx 7$

Since $x = 0$ corresponds to 1974, $x \approx 7$ represents $1974 + 7 = 1981$, rounded to the nearest year.

79. (a) $P(x) = Ca^{x-2000} \Rightarrow P(2000) = Ca^{2000-2000} \Rightarrow 1 - Ca^0 \Rightarrow C = 1$, so $P(x) = a^{x-2000}$

$P(2025) = a^{2025-2000} \Rightarrow 1.4 = a^{25} \Rightarrow a = 1.4^{1/25} \approx 1.01355$

(b) $P(x) = (1.01355)^{x-2000} \Rightarrow P(2010) = (1.01355)^{10} \approx 1.144$;

in 2010 the population if India will be about 1.14 billion.

(c) $P(x) = 1.5 \Rightarrow 1.01355^{x-2000} = 1.5 \Rightarrow \ln 1.01355^{x-2000} = \ln 1.5 \Rightarrow (x - 2000)\ln 1.01355 = \ln 1.5 \Rightarrow$

$x \ln 1.01355 - 2000 \ln 1.01355 = \ln 1.5 \Rightarrow x \ln 1.01355 = \ln 1.5 + 2000 \ln 1.01355 \Rightarrow$

$x = \dfrac{\ln 1.5 + 2000 \ln 1.01355}{\ln 1.01355} = \dfrac{\ln 1.5}{\ln 1.01355} + 2000 \approx 2030.13$;

India's population might reach 1.5 billion in 2030.

81. Using the formula from the example, let $T(x) = 140$.

$60(1.066)^{x-1999} = 140 \Rightarrow 1.066^{x-1999} = \dfrac{7}{3} \Rightarrow \ln 1.066^{x-1999} = \ln \dfrac{7}{3} \Rightarrow (x - 1999)\ln 1.066 = \ln \dfrac{7}{3}$

$x - 1999 = \dfrac{\ln (7/3)}{\ln 1.066} \Rightarrow x = \dfrac{\ln (7/3)}{\ln 1.066} + 1999 \approx 2012.3$

The number of individuals waiting for organ transplants may reach 140,000 in about 2012.

83. (a) $T_0 = 32, D = 212 - 32 \Rightarrow D = 180$. Solving for a use $70 = 32 + 180a^{1/2} \Rightarrow 38 = 180a^{1/2} \Rightarrow$

$\dfrac{38}{180} = a^{1/2} \Rightarrow a = 0.045$

(b) $T(t) = 32 + 180(0.045)^t \Rightarrow T\left(\dfrac{1}{6}\right) = 32 + 180(0.045)^{1/6} \Rightarrow T\left(\dfrac{1}{6}\right) \approx 139°\text{F}$

(c) $40 = 32 + 180(0.045)^t \Rightarrow 8 = 180(0.045)^t \Rightarrow \dfrac{8}{180} = 0.045^t \Rightarrow \log \dfrac{8}{180} = \log 0.045^t \Rightarrow$

$\log \dfrac{8}{180} = t \log 0.045 \Rightarrow t = \dfrac{\log \dfrac{8}{180}}{\log 0.045} \Rightarrow t \approx 1$

Graph $Y_1 = 32 + 180(0.045)^t$ and $Y_2 = 40$. The lines intersect at $t \approx 1$. See Figure 83.

[0, 1.5, 0.5] by [0, 225, 25]

Figure 83

85. (a) $f(1.5) = 20 - 15(0.365)^{1.5} \approx 16.7°\text{C}$

(b) $15 = 20 - 15(0.365)^t \Rightarrow 15(0.365)^t = 5 \Rightarrow 0.365^t = \dfrac{1}{3} \Rightarrow \ln 0.365^t = \ln \dfrac{1}{3} \Rightarrow t \ln 0.365 = \ln \dfrac{1}{3} \Rightarrow$

$t = \dfrac{\ln \frac{1}{3}}{\ln 0.365} \approx 1.09$; the soda can warmed to 15°C after about 1.09 minutes or 1 hour 5.4 minutes.

87. $280 \ln (x + 1) + 1925 = 2300 \Rightarrow 280 \ln (x + 1) = 375 \Rightarrow \ln (x + 1) = \dfrac{375}{280} \Rightarrow e^{\ln (x+1)} = e^{375/280} \Rightarrow$

$x + 1 = e^{375/280} \Rightarrow x = e^{375/280} - 1 \approx 2.8$; a person who consumes 2300 calories daily owns about 2.8 acres.

89. $\dfrac{2 - \log(100 - x)}{0.42} = 2 \Rightarrow 2 - \log(100 - x) = 0.84 \Rightarrow \log(100 - x) = 1.16 \Rightarrow$

$10^{\log(100-x)} = 10^{1.16} \Rightarrow 100 - x = 10^{1.16} \Rightarrow x = 100 - 10^{1.16} \approx 85.546$; after 2 years approximately 85.5%

of the robins have died.

91. (a) A concentration increase of 15% implies $B(t) = 1.15$ when $B_0 = 1$.

$1.15 = e^{6k} \Rightarrow \ln(1.15) = 6k \Rightarrow k = \dfrac{\ln(1.15)}{6} \approx 0.0233.$

(b) $B(t) = 1.2e^{0.0233(8.2)} \approx 1.45$ billion per liter.

(c) $B(t) = 1e^{0.0233(1)} \approx 1.0236 \Rightarrow$ the concentration increases by about 2.36% per hour.

93. $A = P\left(1 + \dfrac{r}{n}\right)^{nt} \Rightarrow 2000 = 1000\left(1 + \dfrac{0.085}{4}\right)^{0.4t} \Rightarrow 2 = (1.02125)^{4t} \Rightarrow \ln(2) = \ln(1.02125)^{4t} \Rightarrow$

$\dfrac{\ln(2)}{\ln(1.02125)} = 4t \Rightarrow t = \dfrac{\frac{\ln(2)}{\ln(1.02125)}}{4} \approx 8.25$ yrs.

95. (a) $P(t) = 750 \Rightarrow 500e^{0.09x} = 750 \Rightarrow e^{0.09x} = 1.5 \Rightarrow \ln e^{0.09x} = \ln 1.5 \Rightarrow 0.09x = \ln 1.5 \Rightarrow$

$t = \dfrac{\ln 1.5}{0.09} \approx 4.505$

(b) If \$500 is invested at 9% compounded continuously, it will grow to approximately \$750 after 4.5 years.

97. $P = 100\left(\dfrac{1}{2}\right)^{t/5700} \Rightarrow 35 = 100\left(\dfrac{1}{2}\right)^{t/5700} \Rightarrow \left(\dfrac{1}{2}\right)^{t/5700} = 0.35 \Rightarrow \ln\left(\dfrac{1}{2}\right)^{t/5700} = \ln 0.35 \Rightarrow$

$\dfrac{t}{5700}\ln\dfrac{1}{2} = \ln 0.35 \Rightarrow \dfrac{t}{5700} = \dfrac{\ln 0.35}{\ln\frac{1}{2}} \Rightarrow t = 5700\left(\dfrac{\ln 0.35}{\ln\frac{1}{2}}\right) \approx 8633.$ The fossil is about 8633 years old.

99. $0.5 = 1 - e^{-0.5x} \Rightarrow -0.5 = -e^{-0.5x} \Rightarrow 0.5 = e^{-0.5x} \Rightarrow \ln(0.5) = -0.5x \Rightarrow x = \dfrac{\ln(0.5)}{-0.5} \Rightarrow$

$x = -2\ln(0.5) \approx 1.39$ min.

101. $N(t) = 100{,}000e^{rt} \Rightarrow 200{,}000 = 100{,}000e^{r(2)} \Rightarrow 2 = e^{2r} \Rightarrow \ln 2 = \ln e^{2r} \Rightarrow \ln 2 = 2r \Rightarrow$

$\dfrac{\ln 2}{2} = r \Rightarrow r = 0.3466.$ Now $350{,}000 = 100{,}000e^{0.3466t} \Rightarrow 3.5 = e^{0.3466t} \Rightarrow \ln 3.5 = \ln e^{0.3466t} \Rightarrow$

$\ln 3.5 = 0.3466t \Rightarrow t = \dfrac{\ln 3.5}{0.3466} \Rightarrow t \approx 3.6$ hrs.

103. $A_n = A_0 e^{rn} \Rightarrow 2300 = 2000e^{r(4)} \Rightarrow 1.15 = e^{4r} \Rightarrow \ln 1.15 = \ln e^{4r} \Rightarrow \ln 1.15 = 4r \Rightarrow$

$r = \dfrac{\ln 1.15}{4} \Rightarrow r = 0.03494.$ Now $3200 = 2000e^{0.03494t} \Rightarrow 1.6 = e^{0.03494t} \Rightarrow \ln 1.6 = \ln e^{0.03494t} \Rightarrow$

$\ln 1.6 = 0.03494t \Rightarrow t = \dfrac{\ln 1.6}{0.03494} \Rightarrow t \approx 13.5$ years.

105. (a) The initial concentration of the drug is 11 milligrams per liter since $C(0) = 11(0.72)^0 = 11$ milligrams

per liter.

(b) 50% of $11 = 5.5$ milligrams per liter.

$11(0.72)^t = 5.5 \Rightarrow 0.72^t = 0.5 \Rightarrow \ln 0.72^t = \ln 0.5 \Rightarrow t\ln 0.72 = \ln 0.5 \Rightarrow t = \dfrac{\ln 0.5}{\ln 0.72} \approx 2.11$

The drug concentration decreases to 50% of its initial level after about 2 hours.

107. $P\left(1 + \dfrac{r}{n}\right)^{nt} = A \Rightarrow \left(1 + \dfrac{r}{n}\right)^{nt} = \dfrac{A}{P} \Rightarrow \log\left(1 + \dfrac{r}{n}\right)^{nt} = \log\left(\dfrac{A}{P}\right) \Rightarrow$

$nt\ \log\left(1 + \dfrac{r}{n}\right) = \log\left(\dfrac{A}{P}\right) \Rightarrow t = \dfrac{\log\,(A/P)}{n \log\,(1 + r/n)}$

5.7: Constructing Nonlinear Models

1. Growth slows down as x increases; logarithmic.

3. Growth rate increases as x increases; exponential.

5. Exponential; least-squares regression gives $f(x) = 1.2(1.7)^x$.

7. Logarithmic; least-squares regression gives $f(x) = 1.088 + 2.937 \ln x$.

9. Logistic; least-squares regression gives $f(x) = \dfrac{9.96}{1 + 30.6e^{-1.51x}}$.

11. (a) See Figure 11.

    (b) $f(x) = 1.568(1.109)^x$ deaths per 100,000.

    (c) $f(80) = 1.568(1.109)^{80} \approx 6164$; about 6164 deaths per 100,000.

[25, 75, 5] by [−100, 2100, 200]

Figure 11

13. $a \approx 9.02, b \approx 1.03$, or $f(x) = 9.02 + 1.03 \ln x$.

15. (a) $a \approx 1.4734, b \approx 0.99986$, or $f(x) = 1.4734(0.99986)^x$.

    (b) $f(7000) = 1.4734(0.99986)^{7000} \approx 0.55$; approximately 0.55 kg/m^3.

17. (a) $f(x) = \dfrac{4.9955}{1 + 49.7081e^{-0.6998x}}$

    (b) For large x, $f(x) \approx \dfrac{4.9955}{1} \approx 5$; the density after a long time is about 5 thousand per acre.

19. (a) Linear: least-squares regression gives $N(x) \approx 213.78x - 414{,}673$.

    (b) See Figure 19.

    (c) $9500 = 213.78x - 414{,}673 \Rightarrow 424{,}173 = 213.78x \Rightarrow x \approx 1984.2$. The number of radio stations reached 9500 during 1984.

21. (a) Power: least-squares regession gives $A(w) \approx 101x^{0.662}$.

    (b) See Figure 21.

    (c) $500 = 101w^{0.662} \Rightarrow \dfrac{500}{101} = w^{0.662} \Rightarrow \left(\dfrac{500}{101}\right)^{1/0.662} = (w^{0.662})^{1/0.662} \Rightarrow w \approx 11.2$. lbs. A weight of about 11.2 lbs. corresponds to a bird with a wing area of 500 in^2.

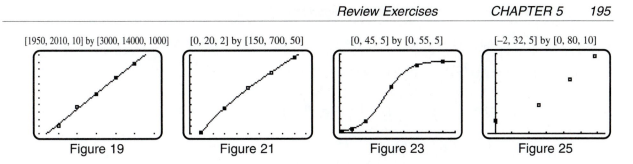

[1950, 2010, 10] by [3000, 14000, 1000]   [0, 20, 2] by [150, 700, 50]   [0, 45, 5] by [0, 55, 5]   [–2, 32, 5] by [0, 80, 10]

Figure 19   Figure 21   Figure 23   Figure 25

23. (a) From the table we see that $H(5) = 3$. After 5 years the tree is 3 feet tall.

(b) Logistic: least-squares regressions gives $H(x) \approx \dfrac{50.1}{1 + 47.4e^{-0.221x}}$.

(c) See Figure 23.

(d) $25 = \dfrac{50.1}{1 + 47.4e^{-0.221x}} \Rightarrow \dfrac{50.1}{25} = 1 + 47.4e^{-0.221x} \Rightarrow \dfrac{50.1}{25} - 1 = 47.4e^{-0.221x} \Rightarrow$

$\dfrac{\frac{50.1}{25} - 1}{47.4} = e^{-0.221x} \Rightarrow \ln\left(\dfrac{\frac{50.1}{25} - 1}{47.4}\right) = -0.221x \Rightarrow x = \dfrac{\ln\left(\dfrac{\frac{50.1}{25} - 1}{47.4}\right)}{-0.221} \approx 17.4$ years. After about 17.4

years the tree is 25 feet tall.

(e) The answer involved interpolation.

25. (a) The data is not linear. See Figure 25.

(b) $a \approx 12.42, b \approx 1.066$, or $f(x) = 12.42(1.066)^x$.

(c) $f(39) = 12.42(1.066)^{39} \approx 150$; 150 kilograms per hectare. Chemical fertilizer use increased, but at a slower rate than predicted by f.

Chapter 5 Review Exercises

1. (a) $(f + g)(1) = f(1) + g(1) = 7 + 1 = 8$

(b) $(f - g)(3) = f(3) - g(3) = 9 - 9 = 0$

(c) $(fg)(-1) = f(-1)g(-1) = 3(-2) = -6$

(d) $(f/g)(0) = \dfrac{f(0)}{g(0)} = \dfrac{5}{0}$; undefined.

3. (a) $f(x) = x^2 \Rightarrow f(3) = 9$ and $g(x) = 1 - x \Rightarrow g(3) = 1 - 3 = -2$.

Thus, $(f + g)(3) = f(3) + g(3) = 9 + (-2) = 7$.

(b) $f(-2) = 4$ and $g(-2) = 1 - (-2) = 3$. Thus, $(f - g)(-2) = f(-2) - g(-2) = 4 - 3 = 1$.

(c) $f(1) = 1$ and $g(1) = 1 - 1 = 0$. Thus, $(fg)(1) = f(1)g(1) = (1)(0) = 0$.

(d) $f(3) = 9$ and $g(3) = 1 - 3 = -2$. Thus, $(f/g)(3) = \dfrac{f(3)}{g(3)} = \dfrac{9}{-2} = -\dfrac{9}{2}$.

5. (a) $(g \circ f)(-2) = g(f(-2)) = g(1) = 2$

(b) $(f \circ g)(3) = f(g(3)) = f(-2) = 1$

(c) $f^{-1}(3) = 2$ since $f(2) = 3$

7. (a) $f(x) = \sqrt{x}$ and $g(x) = x^2 + x \Rightarrow (f \circ g)(2) = f(g(2)) = f(6) = \sqrt{6}$

    (b) $(g \circ f)(9) = g(f(9)) = g(\sqrt{9}) = g(3) = 3^2 + 3 = 12$

9. $(f \circ g) = f(g(x)) = f\left(\dfrac{1}{x}\right) = \left(\dfrac{1}{x}\right)^3 - \left(\dfrac{1}{x}\right)^2 + 3\left(\dfrac{1}{x}\right) - 2; \ D = \{x \mid x \ne 0\}$

11. $(f \circ g) = f(g(x)) = f\left(\dfrac{1}{2}x^3 + \dfrac{1}{2}\right) = \sqrt[3]{2\left(\dfrac{1}{2}x^3 + \dfrac{1}{2}\right) - 1} = \sqrt[3]{x^3 + 1 - 1} = \sqrt[3]{x^3} = x;$

    $D = $ all real numbers

13. $h(x) = (g \circ f)(x) \Rightarrow h(x) = \sqrt{x^2 + 3} \Rightarrow f(x) = x^2 + 3, g(x) = \sqrt{x}.$ *Answers may vary.*

15. Subtract 6 from x and then multiply the results by 10. $\dfrac{x}{10} + 6$ and $10(x - 6)$.

17. Since the graph of $f(x) = 3x - 1$ is a line sloping upward from left to right, a horizontal line can intersect it at

    most once. Therefore, f is one-to-one.

19. f is not one-to-one. It does not pass the Horizontal line test.

21. See Figure 21. The domain of f is $D = \{-1, 0, 4, 6\}$ and its range is $R = \{1, 3, 4, 6\}$. The domain

    and range of f^{-1} are $D = \{1, 3, 4, 6\}$ and $R = \{-1, 0, 4, 6\}$, repectively.

x	6	4	3	1
$f^{-1}(x)$	-1	0	4	6

Figure 21

23. The inverse operations are add 5 to x and divide by 3. Therefore, $f^{-1}(x) = \dfrac{x + 5}{3}.$

    $x = \dfrac{7y}{3 - y}; \Rightarrow f^{-1}(x) = \dfrac{7x}{3 - x}$

25. $(f \circ f^{-1})(x) = f(f^{-1}(x)) = f\left(\dfrac{x + 1}{2}\right) = 2\left(\dfrac{x + 1}{2}\right) - 1 = (x + 1) - 1 = x$

    $(f^{-1} \circ f)(x) = f^{-1}(f(x)) = f^{-1}(2x - 1) = \dfrac{(2x - 1) + 1}{2} = \dfrac{2x}{2} = x$

27. $(f \circ g^{-1})(4) = f(g^{-1}(4)) = f(3) = 1$

29. $y = \sqrt{x + 1} \Rightarrow y^2 = x + 1 \Rightarrow y^2 - 1 = x; \Rightarrow f^{-1}(x) = x^2 - 1, x \ge 0$

    For $f \colon D = \{x \mid x \ge -1\}$ and $R = \{y \mid y \ge 0\}$; for $f^{-1} \colon D = \{x \mid x \ge 0\}$ and $R = \{y \mid y \ge -1\}$

31. If $f(0) = 3 \Rightarrow Ca^x = 3 \Rightarrow Ca^0 = 3 \Rightarrow a^0 = 1; \ C(1) = 3 \Rightarrow C = 3;$ then $f(3) = 24 \Rightarrow$

    $Ca^x = 24 \Rightarrow 3a^3 = 24 \Rightarrow a^3 = 8 \Rightarrow a = 2.$ Therefore $C = 3$ and $a = 2.$

    $C = 8\left(\dfrac{1}{2}\right) \Rightarrow C = 4.$

33. See Figure 33. $D = $ all real numbers.

35. See Figure 35. $D = \{x \mid x > 0\}$.

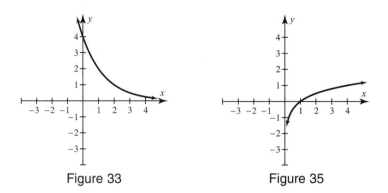

Figure 33   Figure 35

37. $y = Ca^x \Rightarrow y = 2(2)^x \Rightarrow C = 2;\ a = 2$

39. $A = 1200\left(1 + \dfrac{0.09}{2}\right)^{2(3)} \approx \$1562.71$

41. $e^x = 19 \Rightarrow \ln e^x = \ln 19 \Rightarrow x = \ln 19 \approx 2.9444$; the result can be supported graphically using the intersection method with the graphs $Y_1 = e^\wedge X$ and $Y_2 = 19$. The intersection point is near (2.9444, 19). The result may also be supported numerically with a table of $Y_1 = e^\wedge X$ starting at 2.9 and incrementing by 0.01. Here $Y_1 = 19$ when $x \approx 2.94$.

43. $\log 1000 = \log 10^3 = 3$

45. $10 \log 0.01 + \log \dfrac{1}{10} = 10 \log 10^{-2} + \log 10^{-1} = 10(-2) + (-1) = -21$

47. $\log_3 9 = \log_3 3^2 = 2$

49. $\ln e = \ln e^1 = 1$

51. $\log_3 18 = \dfrac{\log 18}{\log 3} \approx 2.631$

53. $10^x = 125 \Rightarrow \log 10^x = \log 125 \Rightarrow x = \log 125 \approx 2.097$

55. $e^{0.1x} = 5.2 \Rightarrow \ln e^{0.1x} = \ln 5.2 \Rightarrow 0.1x = \ln 5.2 \Rightarrow x = 10 \ln 5.2 \approx 16.49$

57. $5^{-x} = 10 \Rightarrow \log 5^{-x} = \log 10 \Rightarrow -x \log 5 = 1 \Rightarrow -x = \dfrac{1}{\log 5} \Rightarrow x = -\dfrac{1}{\log 5} \approx -1.431$

59. $50 - 3(0.78)^{x-10} = 21 \Rightarrow -3(0.78)^{x-10} = -29 \Rightarrow (0.78)^{x-10} = \dfrac{29}{3} \Rightarrow$

    $(x - 10) \log (0.78) = \log \left(\dfrac{29}{3}\right) \Rightarrow x - 10 = \dfrac{\log \left(\frac{29}{3}\right)}{\log (0.78)} \Rightarrow x = 10 + \dfrac{\log \left(\frac{29}{3}\right)}{\log (0.78)} \approx 0.869$

61. For each unit increase in x, the y-values are multiplied by 2, so the data are exponential. Since $y = 1.5$ when $x = 0$, the initial value is 1.5, so $f(x) = Ca^x \Rightarrow f(x) = 1.5a^x$. Since $y = 3$ when $x = 1$, $f(1) = 1.5a^1 \Rightarrow$ $3 = 1.5a \Rightarrow a = 2$. Thus the function $f(x) = 1.5(2^x)$ can model the data.

63. $\log x = 1.5 \Rightarrow 10^{\log x} = 10^{1.5} \Rightarrow x = 10^{1.5} \approx 31.62$

65. $\ln x = 3.4 \Rightarrow e^{\ln x} = e^{3.4} \Rightarrow x = e^{3.4} \approx 29.96$

67. $\log 6 + \log 5x = \log (6 \cdot 5x) = \log 30x$

69. $\ln \dfrac{y}{x^2} = \ln (y) - \ln (x^2) = \ln (y) - 2 \ln (x)$

71. $8 \log x = 2 \Rightarrow \log x = \dfrac{1}{4} \Rightarrow 10^{\log x} = 10^{1/4} \Rightarrow x = 10^{1/4} \Rightarrow x = \sqrt[4]{10} \approx 1.778$

73. $2 \log 3x + 5 = 15 \Rightarrow 2 \log 3x = 10 \Rightarrow \log 3x = 5 \Rightarrow 10^{\log 3x} = 10^5 \Rightarrow 3x = 100{,}000 \Rightarrow$

$x = \dfrac{100{,}000}{3} \approx 33{,}333$

75. $2 \log_5 (x + 2) = \log_5 (x + 8) \Rightarrow \log_5 (x + 2)^2 = \log_5 (x + 8) \Rightarrow 10^{\log_5 (x+2)^2} = 10^{\log_5 (x+8)} \Rightarrow$

$(x + 2)^2 = (x + 8) \Rightarrow x^2 + 4x + 4 = x + 8 \Rightarrow x^2 + 3x - 4 = 0 \Rightarrow (x + 4)(x - 1) = 0 \Rightarrow$

$x = -4 \text{ or } x = 1;$ since $x = -4$ is undefined in $\log (x + 2)$, the only solution is $x = 1$.

77. If b is the y-intercept, then the point $(0, b)$ is on the graph of f. Therefore, the point $(b, 0)$ is on the graph of f^{-1}. The point $(b, 0)$ lies on the x-axis and is the x-intercept of the graph of f^{-1}.

79. (a) $N(t) = N_0 e^{rt} \Rightarrow 6000 = 4000 e^{r(1)} \Rightarrow 1.5 = e^r \Rightarrow \ln 1.5 = \ln e^r \Rightarrow \ln 1.5 = r \Rightarrow r = 0.4055$

Now $N(2.5) = 4000 e^{0.4055(2.5)} \Rightarrow N(2.5) \approx 11{,}022$

(b) $8500 = 4000 e^{0.4055t} \Rightarrow 2.125 = e^{0.4055t} \Rightarrow \ln 2.125 = \ln e^{0.4055t} \Rightarrow \ln 2.125 = 0.4055t \Rightarrow$

$\dfrac{\ln 2.125}{0.4055} = t \Rightarrow t \approx 1.86$ hours.

81. $h(x) = f(x) + g(x) = 10x + 5x = 15x$

83. Use the intersection method by graphing $Y_1 = 175.6(1 - 0.66e\text{^}(-0.24X))\text{^}3$ and $Y_2 = 50$ in the viewing rectangle $[0, 14, 1]$ by $[0, 175, 25]$. The intersection is near the point $(2.74, 50)$. See Figure 83. This means that at approximately 3 weeks the fish weighed 50 milligrams.

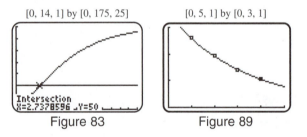

Figure 83   Figure 89

85. $15 = 32e^{-0.2t} \Rightarrow \dfrac{15}{32} = e^{-0.2t} \Rightarrow -0.2t = \ln\left(\dfrac{15}{32}\right) \Rightarrow t = \left(-\dfrac{1}{0.2}\right)\ln\left(\dfrac{15}{32}\right) \approx (-5)(-0.7577) \approx 3.8;$

after about 3.8 minutes.

87. Logistic: least-squares regression gives $f(x) \approx \dfrac{171.4}{1 + 18.4e^{-0.0744x}}$

89. See Figure 89. $a \approx 3.50, b \approx 0.74,$ or $f(x) = 3.50(0.74)^x$

Chapter 6: Trigonometric Functions

6.1: Angles and Their Measure

1. (a) See Figure 1a.

    (b) See Figure 1b.

Figure 1a

Figure 1b

Figure 1c

Figure 1d

    (c) See Figure 1c.

    (d) See Figure 1d.

3. Any angle with degree measure between $0°$ and $90°$. See Figure 3. *Answers may vary.*

5. See Figure 5.

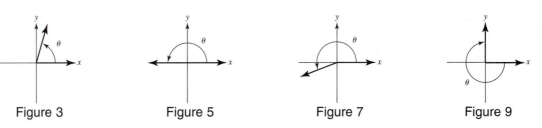

Figure 3   Figure 5   Figure 7   Figure 9

7. Any angle with degree measure between $180°$ and $270°$. See Figure 7. *Answers may vary.*

9. Any angle with degree measure of the form $90° - n(360°)$ where $n = 1, 2, 3, \ldots$. See Figure 9. *Answers may vary.*

11. (a) $\dfrac{90°}{360°} = \dfrac{1}{4}$

    (b) $\dfrac{30°}{360°} = \dfrac{1}{12}$

    (c) $\dfrac{\frac{\pi}{3}}{2\pi} = \dfrac{1}{6}$

    (d) $\dfrac{\frac{\pi}{4}}{2\pi} = \dfrac{1}{8}$

13. $150° + 360° = 510°$, $150° - 360° = -210°$ *Answers may vary.*

15. $-72° + 360° = 288°$, $-72° - 360° = -432°$ *Answers may vary.*

17. $\dfrac{\pi}{2} + 2\pi = \dfrac{\pi}{2} + \dfrac{4\pi}{2} = \dfrac{5\pi}{2}$, $\dfrac{\pi}{2} - 2\pi = \dfrac{\pi}{2} - \dfrac{4\pi}{2} = -\dfrac{3\pi}{2}$ *Answers may vary.*

19. $-\dfrac{\pi}{5} + 2\pi = -\dfrac{\pi}{5} + \dfrac{10\pi}{5} = \dfrac{9\pi}{5}$, $-\dfrac{\pi}{5} - 2\pi = -\dfrac{\pi}{5} - \dfrac{10\pi}{5} = -\dfrac{11\pi}{5}$ *Answers may vary.*

21. $125°15' = 125° + \left(\dfrac{15}{60}\right)^\circ = 125.25°$

23. $108°45'36'' = 108° + \left(\dfrac{45}{60}\right)^\circ + \left(\dfrac{36}{3600}\right)^\circ = 108.76°$

25. $125.3° = 125° + \left(\dfrac{3}{10}\right)^\circ = 125° + \left(\dfrac{18}{60}\right)^\circ = 125°18'$

27. $51.36° = 51° + \left(\dfrac{36}{100}\right)^\circ = 51° + \left(\dfrac{35}{100}\right)^\circ + \left(\dfrac{1}{100}\right)^\circ = 51° + \left(\dfrac{21}{60}\right)^\circ + \left(\dfrac{36}{3600}\right)^\circ = 51°21'36''$

29. $\alpha = 90° - 55.9° = 34.1°, \beta = 180° - 55.9° = 124.1°$

31. $\alpha = 90° - 85°23'45'' = 89°59'60'' - 85°23'45'' = 4°36'15''$

 $\beta = 180° - 85°23'45'' = 179°59'60'' - 85°23'45'' = 94°36'15''$

33. $\alpha = 90° - 23°40'35'' = 89°59'60'' - 23°40'35'' = 66°19'25''$

 $\beta = 180° - 23°40'35'' = 179°59'60'' - 23°40'35'' = 156°19'25''$

35. $\theta = \dfrac{4}{2} = 2$ radians. The degree measure of θ is $2\left(\dfrac{180°}{\pi}\right) = \dfrac{360°}{\pi} \approx 114.6°$.

37. $\theta = \dfrac{13}{10} = 1.3$ radians. The degree measure of θ is $\dfrac{13}{10}\left(\dfrac{180°}{\pi}\right) = \dfrac{234°}{\pi} \approx 74.5°$.

39. (a) $45°\left(\dfrac{\pi}{180°}\right) = \dfrac{45\pi}{180} = \dfrac{\pi}{4}$

 (b) $135°\left(\dfrac{\pi}{180°}\right) = \dfrac{135\pi}{180} = \dfrac{3\pi}{4}$

 (c) $-120°\left(\dfrac{\pi}{180°}\right) = -\dfrac{120\pi}{180} = -\dfrac{2\pi}{3}$

 (d) $-210°\left(\dfrac{\pi}{180°}\right) = -\dfrac{210\pi}{180} = -\dfrac{7\pi}{6}$

41. (a) $37°\left(\dfrac{\pi}{180°}\right) = \dfrac{37\pi}{180}$

 (b) $123.4°\left(\dfrac{\pi}{180°}\right) = \dfrac{123.4\pi}{180} \approx 2.15$

 (c) $-92°25' = -\left[92° + \left(\dfrac{25}{60}\right)^\circ\right] \approx -92.42°\left(\dfrac{\pi}{180°}\right) = -\dfrac{92.42\pi}{180} \approx -1.61$

 (d) $230°17' = 230 + \left(\dfrac{17}{60}\right)^\circ \approx 230.28\left(\dfrac{\pi}{180°}\right) = \dfrac{230.28\pi}{180} \approx 4.02$

43. (a) $\dfrac{\pi}{6}\left(\dfrac{180°}{\pi}\right) = \dfrac{180°}{6} = 30°$

 (b) $\dfrac{\pi}{15}\left(\dfrac{180°}{\pi}\right) = \dfrac{180°}{15} = 12°$

 (c) $-\dfrac{5\pi}{3}\left(\dfrac{180°}{\pi}\right) = -\dfrac{900°}{3} = -300°$

 (d) $-\dfrac{7\pi}{6}\left(\dfrac{180°}{\pi}\right) = -\dfrac{1260°}{6} = -210°$

45. (a) $\dfrac{\pi}{4}\left(\dfrac{180°}{\pi}\right) = \dfrac{180°}{4} = 45°$

   (b) $\dfrac{\pi}{7}\left(\dfrac{180°}{\pi}\right) = \dfrac{180°}{7} \approx 25.71°$

   (c) $3.1\left(\dfrac{180°}{\pi}\right) = \dfrac{558°}{\pi} \approx 177.62°$

   (d) $-\dfrac{5}{2}\left(\dfrac{180°}{\pi}\right) = -\dfrac{900°}{2\pi} \approx -143.24°$

47. Convert 60° to radians. $60°\left(\dfrac{\pi}{180°}\right) = \dfrac{60\pi}{180} = \dfrac{\pi}{3} \Rightarrow s = 2\left(\dfrac{\pi}{3}\right) = \dfrac{2\pi}{3}$ in.

49. $12 = 5\theta \Rightarrow \theta = \dfrac{12}{5}$ radians

51. $5 = r\pi \Rightarrow r = \dfrac{5}{\pi}$ ft.

53. $s = 3\left(\dfrac{\pi}{12}\right) = \dfrac{3\pi}{12} = \dfrac{\pi}{4}$ m.

55. Convert 15° to radians. $15°\left(\dfrac{\pi}{180°}\right) = \dfrac{15\pi}{180} = \dfrac{\pi}{12} \Rightarrow s = 12\left(\dfrac{\pi}{12}\right) = \dfrac{12\pi}{12} = \pi$ ft.

57. Note that $1°45' = 1° + \left(\dfrac{45}{60}\right)° = 1.75°$.

   Convert 1.75° to radians. $1.75°\left(\dfrac{\pi}{180°}\right) = \dfrac{1.75\pi}{180} = \dfrac{7\pi}{720} \Rightarrow s = 2\left(\dfrac{7\pi}{720}\right) = \dfrac{14\pi}{720} = \dfrac{7\pi}{360}$ mi.

59. A total of 15 minutes passes between 10:15 A.M. and 10:30 A.M. The minute hand has moved through the

   angle of $\dfrac{15}{60}\cdot 2\pi = \dfrac{30\pi}{60} = \dfrac{\pi}{2}$ radians. Therefore, the tip of the minute hand travels

   $s = 4\left(\dfrac{\pi}{2}\right) = \dfrac{4\pi}{2} = 2\pi$ inches. The linear speed of the tip of the minute hand is $\dfrac{2\pi}{15}$ in./min.

61. A total of 75 minutes passes between 3:00 P.M. and 4:15 P.M. The minute hand has moved through the

   angle of $\dfrac{75}{60}\cdot 2\pi = \dfrac{150\pi}{60} = \dfrac{5\pi}{2}$ radians. Therefore, the tip of the minute hand travels

   $s = 4\left(\dfrac{5\pi}{2}\right) = 10\pi$ inches. The linear speed of the tip of the minute hand is $\dfrac{10\pi}{75} = \dfrac{2\pi}{15}$ in./min.

63. $A = \dfrac{1}{2}\cdot 3^2 \cdot \left(\dfrac{\pi}{3}\right) = \dfrac{3\pi}{2} = 1.5\pi$ in^2

65. Convert 45° to radians. $45°\left(\dfrac{\pi}{180°}\right) = \dfrac{\pi}{4}$. Then, $A = \dfrac{1}{2}\cdot 6^2 \cdot \left(\dfrac{\pi}{4}\right) = \dfrac{9\pi}{2} = 4.5\pi$ in^2

67. $A = \dfrac{1}{2}\cdot 13.1^2 \cdot \left(\dfrac{\pi}{15}\right) = \dfrac{17{,}161\pi}{3000} \approx 5.72\pi$ cm^2

69. Convert 30° to radians. $30°\left(\dfrac{\pi}{180°}\right) = \dfrac{\pi}{6}$. Then, $A = \dfrac{1}{2}\cdot 1.5^2 \cdot \left(\dfrac{\pi}{6}\right) = \dfrac{3\pi}{16} \approx 0.59$ ft^2

71. When rotating from $-45°$ to $90°$ the arm swings through an angle of $135°$. Convert $135°$ to radians:

$135°\left(\dfrac{\pi}{180°}\right) = \dfrac{3\pi}{4}$. To find the area of the work space, we subtract the area of the inner circular sector from

the area of the outer circular sector. $A = \dfrac{1}{2}\left(\dfrac{3\pi}{4}\right)(26^2 - 6^2) = \dfrac{3\pi}{8}(640) = 240\pi$ in^2

73. When rotating from $15°$ to $195°$ the arm swings through an angle of $180°$. Convert $180°$ to radians:

$180°\left(\dfrac{\pi}{180°}\right) = \pi$. To find the area of the work space, we subtract the area of the inner circular sector from

the area of the outer circular sector. $A = \dfrac{1}{2}(\pi)(95^2 - 21^2) = \dfrac{\pi}{2}(8584) = 4292\pi$ cm^2

75. The angular velocity is $\omega = 15$ radians/sec. The radius is $r = \dfrac{26}{2} = 13$ inches.

The linear speed is given by $v = r\omega = 13(15) = 195$ in/sec. That is $\dfrac{195 \text{ in}}{\text{sec}} \cdot \dfrac{1 \text{ ft}}{12 \text{ in}} = 16.25$ ft/sec.

And $\dfrac{16.25 \text{ ft}}{\text{sec}} \cdot \dfrac{1 \text{ mi}}{5280 \text{ ft}} \cdot \dfrac{3600 \text{ sec}}{1 \text{ hr}} \approx 11.1$ mi/hr.

77. Since the celestial north pole traces out a circular path every 26,000 years, its angular speed is

$\omega = \dfrac{360°}{26,000} \approx 0.013846° \approx 0°0'49.''/$yr. The celestial north pole moves approximately $0°0'49.8''/$yr.

79. (a) The angular velocity is $\omega = 500(2\pi) = 1000\pi \approx 3141.6$ radians/min.

(b) The linear speed is $v = r\omega$; $r = \dfrac{30}{2} = 15$ in. and $\omega = 1000\pi$

therefore $v = 15(1000\pi) \approx 47,123.89$ in./min or about 65.5 ft/sec.

81. Convert $40°55' - 29°11' = 11°44'$ to radians: $\left[11° + \left(\dfrac{44}{60}\right)°\right]\left(\dfrac{\pi}{180°}\right) \approx 0.2048$.

Then, $s = 3955(0.2048) \approx 809.98 \approx 810$ mi.

83. Angular speed ω measures the speed of rotation and is measured by $\omega = \dfrac{\theta}{t}$, $85° = \dfrac{17}{36}\pi$ and $t = 0.05 \Rightarrow$

$\omega = \dfrac{\frac{17\pi}{36}}{0.05} \approx 29.7$ rad/sec. The linear speed v at which the tip of the rocket travels is given by $v = r\omega$.

$r = 28$ inches or $2\dfrac{1}{3}$ ft and $\omega \approx 29.7$ rad/sec $\Rightarrow v = \left(2\dfrac{1}{3}\right)(29.7) = 69.3$ or about 69 ft/sec.

85. Convert $75.3°$ to radians: $75.3°\left(\dfrac{\pi}{180°}\right) \approx 1.314$. Then, $s = 11(1.314) \approx 14.5$ in.

87. (a) In one revolution of the pedals, the chain will travel $s = 3.75(2\pi) = 7.5\pi$ inches. The circumference of

the small gear is $C = 1.5(2\pi) = 3\pi$ inches. When the chain moves 7.5π inches, the smaller gear will

rotate $\dfrac{7.5\pi}{3\pi} = \dfrac{5}{2} = 2.5$ times. Thus, the bicycle tire will rotate 2.5 times for each revolution of the pedals.

(b) When the pedals turn through 2 revolutions per second, the angular velocity of the tire is

$\omega = 2.5(2)(2\pi) = 10\pi$ radians/sec. Thus, the tire of radius 13 inches (diameter = 26 inches) will travel

$v = 13(10\pi) = 130\pi$ in./sec. That is $\dfrac{130\pi \text{ in.}}{\text{sec}} \cdot \dfrac{1 \text{ ft}}{12 \text{ in.}} \approx 34$ ft/sec.

89. (a) $\omega = \dfrac{2\pi}{0.615}$ radians/yr $\Rightarrow \dfrac{\frac{2\pi}{0.615} \text{ rad}}{1 \text{ yr}} \cdot \dfrac{1 \text{ yr}}{365 \text{ days}} \cdot \dfrac{1 \text{ day}}{24 \text{ hr}} \approx 1.1662741 \times 10^{-3}$ radians/hr.

Thus, the orbital velocity of Venus is $v = 67.2 \times 10^6(1.1662741 \times 10^{-3}) \approx 78{,}370$ mi/hr.

(b) $\omega = \dfrac{2\pi}{1} = 2\pi$ radians/yr $\Rightarrow \dfrac{2\pi \text{ rad}}{1 \text{ yr}} \cdot \dfrac{1 \text{ yr}}{365 \text{ days}} \cdot \dfrac{1 \text{ day}}{24 \text{ hr}} \approx 7.172585967 \times 10^{-4}$ radians/hr.

Thus, the orbital velocity of Earth is $v = 92.9 \times 10^6(7.172585967 \times 10^{-4}) \approx 66{,}630$ mi/hr.

(c) $\omega = \dfrac{2\pi}{11.86}$ radians/yr $\Rightarrow \dfrac{\frac{2\pi}{11.86} \text{ rad}}{1 \text{ yr}} \cdot \dfrac{1 \text{ yr}}{365 \text{ days}} \cdot \dfrac{1 \text{ day}}{24 \text{ hr}} \approx 6.047711608 \times 10^{-5}$ radians/hr.

Thus, the orbital velocity of Jupiter is $v = 483.6 \times 10^6(6.047711608 \times 10^{-5}) \approx 29{,}250$ mi/hr.

(d) $\omega = \dfrac{2\pi}{164.8}$ radians/yr $\Rightarrow \dfrac{\frac{2\pi}{164.8} \text{ rad}}{1 \text{ yr}} \cdot \dfrac{1 \text{ yr}}{365 \text{ days}} \cdot \dfrac{1 \text{ day}}{24 \text{ hr}} \approx 4.35229731 \times 10^{-6}$ radians/hr.

Thus, the orbital velocity of Neptune is $v = 2794 \times 10^6(4.35229731 \times 10^{-6}) \approx 12{,}160$ mi/hr.

91. Since the bar is 2 meters in length, we may use 2 meters as an estimate of the arc length from P to Q.

Convert $0.835°$ to radians: $0.835°\left(\dfrac{\pi}{180°}\right) = \dfrac{167\pi}{36{,}000}$. Then, $2 = r\left(\dfrac{167\pi}{36{,}000}\right) \Rightarrow r = 2\left(\dfrac{36{,}000}{167\pi}\right) \approx 137.2$ m.

A reasonable approximation for d is 137.2 meters.

93. (a) $475{,}000 = r^2\pi \Rightarrow r = \sqrt{\dfrac{475{,}000}{\pi}} \approx 388.8$ m

(b) Convert $70°$ to radians: $70°\left(\dfrac{\pi}{180°}\right) = \dfrac{7\pi}{18}$. So

$$475{,}000 = \dfrac{1}{2}\left(\dfrac{7\pi}{18}\right)r^2 \Rightarrow r = \sqrt{2(475{,}000)\left(\dfrac{18}{7\pi}\right)} \approx 881.8 \text{ m}$$

6.2: Right Triangle Trigonometry

1. See Figure 1.

Figure 1   Figure 3

3. See Figure 3.

5. $\sin 60° = \dfrac{\text{side opposite}}{\text{hypotenuse}} = \dfrac{\sqrt{3}}{2}$

7. $\cos 30° = \dfrac{\text{side adjacent}}{\text{hypotenuse}} = \dfrac{\sqrt{3}}{2}$

9. $\sec 60° = \dfrac{\text{hypotenuse}}{\text{side adjacent}} = \dfrac{2}{1} = 2$

11. $\tan 45° = \dfrac{\text{side opposite}}{\text{side adjacent}} = \dfrac{1}{1} = 1$

13. $\cot 45° = \dfrac{\text{side adjacent}}{\text{side opposite}} = \dfrac{1}{1} = 1$

15. $\sin 45° = \dfrac{\text{side opposite}}{\text{hypotenuse}} = \dfrac{1}{\sqrt{2}}$

17. Use the Pythagorean theorem to calculate the length of the hypotenuse:

$$h^2 = 3^2 + 4^2 \Rightarrow h^2 = 9 + 16 \Rightarrow h^2 = 25 \Rightarrow h = 5$$

$\sin \theta = \dfrac{4}{5}$ $\qquad \cos \theta = \dfrac{3}{5}$ $\qquad \tan \theta = \dfrac{4}{3}$ $\qquad \csc \theta = \dfrac{5}{4}$ $\qquad \sec \theta = \dfrac{5}{3}$ $\qquad \cot \theta = \dfrac{3}{4}$

19. Use the Pythagorean theorem to calculate the length of the missing leg:

$$13^2 = 12^2 + l^2 \Rightarrow l^2 = 169 - 144 \Rightarrow l^2 = 25 \Rightarrow l = 5$$

$\sin \theta = \dfrac{12}{13}$ $\qquad \cos \theta = \dfrac{5}{13}$ $\qquad \tan \theta = \dfrac{12}{5}$ $\qquad \csc \theta = \dfrac{13}{12}$ $\qquad \sec \theta = \dfrac{13}{5}$ $\qquad \cot \theta = \dfrac{5}{12}$

21. $\sin 60° \approx 0.866$ $\qquad\qquad \cos 60° \approx 0.5$ $\qquad\qquad \tan 60° \approx 1.732$

   $\csc 60° \approx 1.155$ $\qquad\qquad \sec 60° \approx 2$ $\qquad\qquad \cot 60° \approx 0.577$

23. $\sin 25° \approx 0.423$ $\qquad\qquad \cos 25° \approx 0.906$ $\qquad\qquad \tan 25° \approx 0.466$

   $\csc 25° \approx 2.366$ $\qquad\qquad \sec 25° \approx 1.103$ $\qquad\qquad \cot 25° \approx 2.145$

25. $\sin 5°35' \approx 0.097$ $\qquad\qquad \cos 5°35' \approx 0.995$ $\qquad\qquad \tan 5°35' \approx 0.098$

   $\csc 5°35' \approx 10.278$ $\qquad\qquad \sec 5°35' \approx 1.005$ $\qquad\qquad \cot 5°35' \approx 10.229$

27. $\sin 13°45'30'' \approx 0.238$ $\qquad \cos 13°45'30'' \approx 0.971$ $\qquad \tan 13°45'30'' \approx 0.245$

   $\csc 13°45'30'' \approx 4.205$ $\qquad \sec 13°45'30'' \approx 1.030$ $\qquad \cot 13°45'30'' \approx 4.084$

29. $\sin 1.05° \approx 0.018$ $\qquad\qquad \cos 1.05° \approx 1.000$ $\qquad\qquad \tan 1.05° \approx 0.018$

   $\csc 1.05° \approx 54.570$ $\qquad\qquad \sec 1.05° \approx 1.000$ $\qquad\qquad \cot 1.05° \approx 54.561$

31. $\sec \theta = \dfrac{1}{\cos \theta} = \dfrac{1}{\frac{1}{3}} = 3$

33. $\csc \theta = \dfrac{1}{\sin \theta} = \dfrac{1}{\frac{12}{13}} = \dfrac{13}{12}$

35. $\tan \theta = \dfrac{1}{\cot \theta} = \dfrac{1}{\frac{7}{24}} = \dfrac{24}{7}$

37. Triangle ABC is a $30° \sim 60°$ right triangle, thus $b = 8$. Find a using the Pythagorean theorem:

$$8^2 + a^2 = 16^2 \Rightarrow a^2 = 256 - 64 \Rightarrow a^2 = 192 \Rightarrow a = 8\sqrt{3} \approx 13.86.$$

39. To find b note that $\tan 50° = \dfrac{6}{b} \Rightarrow b = \dfrac{6}{\tan 50°} \Rightarrow b \approx 5.03.$

   To find c note that $\sin 50° = \dfrac{6}{c} \Rightarrow c = \dfrac{6}{\sin 50°} \Rightarrow c \approx 7.83.$

41. To find a note that $\sin 41° = \dfrac{a}{8} \Rightarrow a = 8\sin 41° \Rightarrow a \approx 5.25.$

   To find b note that $\cos 41° = \dfrac{b}{8} \Rightarrow b = 8\cos 41° \Rightarrow b \approx 6.04.$

43. To find a note that $\tan 46°15' = \dfrac{a}{16.1} \Rightarrow a = 16.1 \tan 46°15' \Rightarrow a \approx 16.82$.

To find c note that $\cos 46°15' = \dfrac{16.1}{c} \Rightarrow c = \dfrac{16.1}{\cos 46°15'} \Rightarrow c \approx 23.28$.

45. To find a note that $\cos 60° = \dfrac{a}{24} \Rightarrow a = 24 \cos 60° \Rightarrow a = 24\left(\dfrac{1}{2}\right) = 12$.

To find b note that $\sin 60° = \dfrac{b}{24} \Rightarrow b = 24 \sin 60° \Rightarrow b = 24\left(\dfrac{\sqrt{3}}{2}\right) = 12\sqrt{3}$.

To find c note that $\cos 45° = \dfrac{24 \sin 60°}{c} \Rightarrow c = \dfrac{24 \sin 60°}{\cos 45°} \Rightarrow c = 24\sqrt{2} \sin 60° \Rightarrow c = 24\sqrt{2}\left(\dfrac{\sqrt{3}}{2}\right) \Rightarrow$

$c = 12\sqrt{6}$.

To find d note that $\sin 45° = \dfrac{d}{24\sqrt{2}\sin 60°} \Rightarrow d = \sin 45°(24\sqrt{2}\sin 60°) \Rightarrow d = \dfrac{1}{\sqrt{2}}(24\sqrt{2}\sin 60°) \Rightarrow$

$d = 24 \sin 60° \Rightarrow d = 24\left(\dfrac{\sqrt{3}}{2}\right) \Rightarrow d = 12\sqrt{3}$.

47. To find a note that $\sin 60° = \dfrac{7}{a} \Rightarrow a = \dfrac{7}{\sin 60°} \Rightarrow a = \dfrac{7}{\frac{\sqrt{3}}{2}} \Rightarrow a = \dfrac{14\sqrt{3}}{3}$.

To find b note that $\tan 60° = \dfrac{7}{b} \Rightarrow b = \dfrac{7}{\tan 60°} \Rightarrow b = \dfrac{7}{\sqrt{3}} \Rightarrow b = \dfrac{7\sqrt{3}}{3}$.

To find c note that $\tan 45° = \left(\dfrac{7}{\sin 60°} \div c\right) \Rightarrow \tan 45° = \dfrac{7}{c \sin 60°} \Rightarrow (c \sin 60°)(\tan 45°) = 7$,

$\tan 45° = 1 \Rightarrow (c \sin 60°)(1) = 7 \Rightarrow c = \dfrac{7}{\sin 60°} \Rightarrow c = \dfrac{7}{\frac{\sqrt{3}}{2}} \Rightarrow c = \dfrac{14\sqrt{3}}{3}$.

To find d note that $\sin 45° = \left(\dfrac{7}{\sin 60°} \div d\right) \Rightarrow \sin 45° = \dfrac{7}{d \sin 60°} \Rightarrow (d \sin 60°)(\sin 45°) = 7 \Rightarrow$

$d = \dfrac{7}{(\sin 60°)(\sin 45°)} \Rightarrow d = \dfrac{7}{\left(\frac{\sqrt{3}}{2}\right)\left(\frac{\sqrt{2}}{2}\right)} \Rightarrow d = \dfrac{14\sqrt{6}}{3}$.

49. To find a note that $\tan 60° = \dfrac{a}{12} \Rightarrow a = 12 \tan 60° \Rightarrow a \approx 20.78$.

51. To find c note that $\cos 53°43' = \dfrac{100}{c} \Rightarrow c = \dfrac{100}{\cos 53°43'} \Rightarrow c \approx 168.98$.

53. (a) $\sin 70° = \cos (90° - 70°) = \cos 20° \approx 0.9397$

    (b) $\cos 40° = \sin (90° - 40°) = \sin 50° \approx 0.7660$

55. (a) $\csc 49° = \sec (90° - 49°) = \sec 41° \approx 1.3250$

    (b) $\sec 63° = \csc (90° - 63°) = \csc 27° \approx 2.2027$

57. $\tan 37°30' = \dfrac{h}{1500} \Rightarrow h = 1500 \tan 37°30' \Rightarrow h \approx 1151$ ft.

59. $\tan 34° = \dfrac{5 \text{ ft } 3 \text{ in.}}{a}$ or $\tan 34° = \dfrac{63 \text{ in.}}{a} \Rightarrow a = \dfrac{63 \text{ in.}}{\tan 34°} \Rightarrow a \approx 93.4$ in. or about 7.8 ft.

61. $\tan 35.3° = \dfrac{h}{48} \Rightarrow h = 48 \tan 35.3° \Rightarrow h \approx 34$ ft.

63. Using the 13° angle of depression, we may find the distance between the airplane and the stadium as follows:

$$\tan 13° = \frac{12,000}{a} \Rightarrow a = \frac{12,000}{\tan 13°} \Rightarrow a \approx 51,977.7 \text{ or } 52,000 \text{ ft.}$$

65. Using the 20° angle of depression, we may find the distance between the buildings as follows:

$$\tan 20° = \frac{30}{d} \Rightarrow d = \frac{30}{\tan 20°} \Rightarrow d \approx 82.4 \text{ ft. Using this distance, we may now calculate the height } h \text{ of the}$$

upper part of the building on the right: $\tan 50° = \dfrac{h}{82.4} \Rightarrow h = 82.4 \tan 50° \Rightarrow h \approx 98.2$ ft. The total height

of the building is about $98.2 + 30 = 128.2$ ft.

67. From the diagram in the text we see that: $\tan 23.76° = \dfrac{PR}{PQ} \Rightarrow PQ = \dfrac{85.62}{\tan 23.76°} \Rightarrow PQ = 194.5$ ft.

69. See Figure 69. From right triangle TCA, $\tan 13.7° = \dfrac{h}{d} \Rightarrow d = \dfrac{h}{\tan 13.7°}$. From right triangle TCB,

$$\tan 10.4° = \frac{h}{d+5} \Rightarrow d = \frac{h}{\tan 10.4°} - 5. \text{ Thus, we calculate } h \text{ as follows:}$$

$$\frac{h}{\tan 10.4°} - 5 = \frac{h}{\tan 13.7°} \Rightarrow h \tan 13.7° - 5 \tan 13.7° \tan 10.4° = h \tan 10.4° \Rightarrow$$

$$h(\tan 13.7° - \tan 10.4°) = 5 \tan 13.7° \tan 10.4° \Rightarrow h = \frac{5 \tan 13.7° \tan 10.4°}{\tan 13.7° - \tan 10.4°} \Rightarrow h = 3.71 \text{ mi} \approx 19,600 \text{ ft.}$$

Figure 69

71. let y be the base of the smaller triangle, then $\tan(21°10') = \dfrac{x}{135+y} \Rightarrow x = (135+y)\tan 21°10'$

and $\tan 35°30' = \dfrac{x}{y} \Rightarrow x = y(\tan 35°30')$.

Now $(135 + y)\tan 21°10' = y(\tan 35°30') \Rightarrow 135(\tan 21°10') = y(\tan 35°30') - y(\tan 21°10') \Rightarrow$

$135(\tan 21°10') = y(\tan 35°30' - \tan 21°10') \Rightarrow y = \dfrac{135(\tan(21°10'))}{\tan(35°30') - \tan(21°10')} \Rightarrow y \approx 160.303 \Rightarrow$

Back substituting we find $x \approx 114.343$ or about 114 ft.

73. From the example we know that the distance d to a star with parallax θ is given by $d = \dfrac{93,000,000}{\sin \theta}$.

For Barnard's star: $d = \dfrac{93,000,000}{\sin 0.000152} \approx 3.5 \times 10^{13}$ mi or $\dfrac{3.5 \times 10^{13}}{5.9 \times 10^{12}} \approx 5.9$ light-years.

For Sirus: $d = \dfrac{93,000,000}{\sin 0.000105} \approx 5.1 \times 10^{13}$ mi or $\dfrac{5.1 \times 10^{13}}{5.9 \times 10^{12}} \approx 8.6$ light-years.

For 61 Cygni: $d = \dfrac{93,000,000}{\sin 0.0000811} \approx 6.6 \times 10^{13}$ mi or $\dfrac{6.6 \times 10^{13}}{5.9 \times 10^{12}} \approx 11.1$ light-years.

For Procyon: $d = \dfrac{93,000,000}{\sin 0.0000797} \approx 6.7 \times 10^{13}$ mi or $\dfrac{6.7 \times 10^{13}}{5.9 \times 10^{12}} \approx 11.3$ light-years.

75. Using the Earth - Sun distance of 93,000,000 miles, the minimum distance d is given by:

$$\sin 18° = \frac{d}{93,000,000} \Rightarrow d = 93,000,000 \sin 18° \Rightarrow d \approx 2.9 \times 10^7 \text{ mi.}$$

The maximum distance is given by: $\sin 28° = \dfrac{d}{93,000,000} \Rightarrow d = 93,000,000 \sin 28° \Rightarrow d \approx 4.4 \times 10^7 \text{ mi.}$

77. $d = r\left(\dfrac{1}{\cos \theta} - 1\right) = 3963\left(\dfrac{1}{\cos 76.1°} - 1\right) \approx 12{,}533.8 \approx 12{,}534 \text{ mi.}$

79. (a) $r = \dfrac{66^2}{4.5 + 32.2 \tan 3°} \approx 704 \text{ ft.}$

(b) $r = \dfrac{66^2}{4.5 + 32.2 \tan 5°} \approx 595 \text{ ft.}$

(c) Larger values of θ correspond to a smaller safe radius. A table of $Y_1 = 66^2/(4.5 + 32.2 \tan (X))$ starting

at $x = 0$ and incrementing by 1 is shown in Figure 79.

Figure 79

81. From the example $d = r\left(\dfrac{1}{\cos \theta} - 1\right)$. Thus, $r = 625$ and $\theta = 54° \Rightarrow d = 625\left(\dfrac{1}{\cos 54°} - 1\right) \approx 438 \text{ ft.}$

83. The area of a triangle is $\dfrac{b \cdot h}{2}$ and the height is unknown. Height is found by using the following:

$$\cos 30° = \frac{h}{s} \Rightarrow s \cos 30° = h, \cos 30° = \frac{\sqrt{3}}{2} \Rightarrow s \cdot \frac{\sqrt{3}}{2} = \text{ height or } h = \frac{s\sqrt{3}}{2}.$$

Since area equals $\dfrac{b \cdot h}{2}$ and $b = s$, then area $= \left(s \cdot \dfrac{s\sqrt{3}}{2}\right) \div 2$ therefore area $= \dfrac{\sqrt{3}}{4}s^2$.

6.3: The Sine and Cosine Functions and Their Graphs

1. (a) $r = \sqrt{12^2 + 5^2} = \sqrt{144 + 25} = \sqrt{169} = 13$

(b) $\sin \theta = \dfrac{y}{r} = \dfrac{5}{13} \qquad \cos \theta = \dfrac{x}{r} = \dfrac{12}{13}$

3. (a) $r = \sqrt{(-15)^2 + 8^2} = \sqrt{225 + 64} = \sqrt{289} = 17$

(b) $\sin \theta = \dfrac{y}{r} = \dfrac{8}{17} \qquad \cos \theta = \dfrac{x}{r} = -\dfrac{15}{17}$

5. $r = \sqrt{4^2 + 3^2} = \sqrt{16 + 9} = \sqrt{25} = 5, \qquad \sin \theta = \dfrac{y}{r} = \dfrac{3}{5} \qquad \cos \theta = \dfrac{x}{r} = \dfrac{4}{5}$

7. $r = \sqrt{1^2 + (-2)^2} = \sqrt{1 + 4} = \sqrt{5}, \qquad \sin \theta = \dfrac{y}{r} = -\dfrac{2}{\sqrt{5}} \qquad \cos \theta = \dfrac{x}{r} = \dfrac{1}{\sqrt{5}}$

9. The point $(1, 2)$ lies on the line $y = 2x$ in the first quadrant. This is shown in Figure 9.

So, $r = \sqrt{1^2 + 2^2} = \sqrt{1 + 4} = \sqrt{5}$. It follows that: $\sin \theta = \dfrac{2}{\sqrt{5}}$ and $\cos \theta = \dfrac{1}{\sqrt{5}}$.

11. The point $(1, -3)$ lies on the line $y = -3x$ in the fourth quadrant. This is shown in Figure 11.

So, $r = \sqrt{1^2 + (-3)^2} = \sqrt{1 + 9} = \sqrt{10}$. It follows that: $\sin \theta = -\dfrac{3}{\sqrt{10}}$ and $\cos \theta = \dfrac{1}{\sqrt{10}}$.

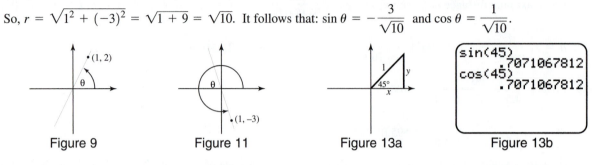

| Figure 9 | Figure 11 | Figure 13a | Figure 13b |

13. The length of the legs in a $45° \sim 45°$ right triangle are equal. For convenience we let $r = 1$ as shown in Figure 13a.

Next, find x and y: $x^2 + y^2 = 1^2 \Rightarrow x^2 + x^2 = 1^2 \Rightarrow 2x^2 = 1 \Rightarrow x^2 = \dfrac{1}{2} \Rightarrow x = \dfrac{1}{\sqrt{2}}$. Thus,

$x = y = \dfrac{1}{\sqrt{2}}$. $\sin 45° = \dfrac{1}{\sqrt{2}} \approx 0.7071$ $\cos 45° = \dfrac{1}{\sqrt{2}} \approx 0.7071$ The results are supported in Figure 13b.

15. The length of the shorter leg in a $30° \sim 60°$ right triangle is equal to half the length of the hypotenuse. For convenience we let $r = 2$ as shown in Figure 15a. Note that y is negative so $y = -1$.

Next, find x: $x^2 = r^2 - 1^2 \Rightarrow x^2 = 2^2 - 1^2 \Rightarrow x^2 = 3 \Rightarrow x = \sqrt{3}$.

$\sin (-30°) = -\dfrac{1}{2}$ $\cos (-30°) = \dfrac{\sqrt{3}}{2} \approx 0.8660$ The results are supported in Figure 15b.

| Figure 15a | Figure 15b | Figure 17a | Figure 17b |

17. The lengths of the legs in a $45° \sim 45°$ right triangle are equal. For convenience we let $r = 1$ as shown in

Figure 17a. Next, find x and y: $x^2 + y^2 = 1^2 \Rightarrow x^2 + x^2 = 1^2 \Rightarrow 2x^2 = 1 \Rightarrow x^2 = \dfrac{1}{2} \Rightarrow x = \pm\dfrac{1}{\sqrt{2}}$.

Thus, $x = y = -\dfrac{1}{\sqrt{2}}$; $\sin (225°) = -\dfrac{1}{\sqrt{2}}$, $\cos (225°) = -\dfrac{1}{\sqrt{2}}$. The results are supported in Figure 17b.

19. The length of the shorter leg in a $30° \sim 60°$ right triangle is equal to half the length of the hypotenuse. For

convenience we let $r = 2$ as shown in Figure 19a. Note that x is positive, so $x = 1$.

$y^2 = r^2 - x^2 \Rightarrow y^2 = 2^2 - 1^2 \Rightarrow y^2 = 3 \Rightarrow y = -\sqrt{3}$. Thus, $\sin (-420°) = -\dfrac{\sqrt{3}}{2}$, $\cos (-420°) = \dfrac{1}{2}$.

The results are supported in Figure 19b.

21. Note that $\dfrac{\pi}{3} = 60°$. The length of the shorter leg in a $30° \sim 60°$ right triangle is equal to half the length of the

hypotenuse. For convenience we let $r = 2$ as shown in Figure 21a. Note that x is positive so $x = 1$.

Next, find y: $y^2 = r^2 - 1^2 \Rightarrow y^2 = 2^2 - 1^2 \Rightarrow y^2 = 3 \Rightarrow y = \sqrt{3}$.

$\sin \dfrac{\pi}{3} = \dfrac{\sqrt{3}}{2} \approx 0.8660$ $\cos \dfrac{\pi}{3} = \dfrac{1}{2}$ The results are supported in Figure 21b.

Figure 19a   Figure 19b   Figure 21a   Figure 21b

23. A $-\dfrac{\pi}{2}$ angle is quadrantal. Its terminal side lies on the *y*-axis. For convenience we let $r = 1$ as shown in

Figure 23a. Note that if $r = 1$, then $x = 0$ and $y = -1$.

$\sin\left(-\dfrac{\pi}{2}\right) = -1$ $\cos\left(-\dfrac{\pi}{2}\right) = 0$ The results are supported in Figure 23b.

Figure 23a   Figure 23b   Figure 25a   Figure 25b

25. Note that $\dfrac{7\pi}{6} = 210°$ lies in the third quadrant and it forms a $30° \sim 60°$ right triangle with the *x*-axis.

For convenience we let $r = 2$ as shown in Figure 25a. Note that *y* is negative so $y = -1$.

Next find $x^2 = r^2 - y^2 \Rightarrow x^2 = 2^2 - (-1)^2 \Rightarrow x^2 = 3 \Rightarrow x = -\sqrt{3}$.

$\sin\dfrac{7\pi}{6} = -\dfrac{1}{2}$ $\cos\dfrac{7\pi}{6} = -\dfrac{\sqrt{3}}{2} \approx -0.8660$ These results are supported in Figure 25b.

27. Note that $-\dfrac{9\pi}{4} = -405°$. The length of the legs in a $45° \sim 45°$ right triangle are equal. For convenience we

let $r = 1$ as shown in Figure 27a. Next, find *x* and *y*:

$x^2 + y^2 = 1^2 \Rightarrow x^2 + x^2 = 1^2 \Rightarrow 2x^2 = 1 \Rightarrow x^2 = \dfrac{1}{2} \Rightarrow x = \pm\dfrac{1}{\sqrt{2}}$.

$\sin\left(-\dfrac{9\pi}{4}\right) = -\dfrac{1}{\sqrt{2}}$ $\cos\left(-\dfrac{9\pi}{4}\right) = \dfrac{1}{\sqrt{2}}$ The results are supported in Figure 27b.

Figure 27a   Figure 27b

29. $\sin 93.2° \approx 0.9984$ $\cos 93.2° \approx -0.0558$

31. $\sin 123°50' \approx 0.8307$ $\cos 123°50' \approx -0.5568$

33. $\sin(-4) \approx 0.7568$ $\cos(-4) \approx -0.6536$

35. $\sin\dfrac{11\pi}{7} \approx -0.9749$ $\cos\dfrac{11\pi}{7} \approx 0.2225$

37. $\sin \theta = y = \dfrac{3}{5}$ $\qquad\qquad$ $\cos \theta = x = \dfrac{4}{5}$

39. $\sin \theta = y = -\dfrac{5}{13}$ $\qquad\quad$ $\cos \theta = x = \dfrac{12}{13}$

41. An angle θ of $\dfrac{\pi}{2}$ radians in standard position has a terminal side that intersects the unit circle at the point $(0, 1)$.

Therefore $\sin \dfrac{\pi}{2} = 1$ and $\cos \dfrac{\pi}{2} = 0$.

43. An angle θ of $\dfrac{7\pi}{6}$ radians in standard position has a terminal side that intersects the unit circle in the third quadrant.

To find the point of intersection, the hypotenuse of a $30° \sim 60°$ right triangle has length 1 and the shorter leg has

length $y = \dfrac{1}{2}$. Since $x^2 + y^2 = 1$, we can solve for x. $x^2 + \left(\dfrac{1}{2}\right)^2 = 1 \Rightarrow x^2 = \dfrac{3}{4} \Rightarrow x = \dfrac{\sqrt{3}}{2}$.

Since the point (x, y) is located in the third quadrant, the point becomes $\left(-\dfrac{\sqrt{3}}{2}, -\dfrac{1}{2}\right)$.

Therefore $\sin\left(\dfrac{7\pi}{6}\right) = -\dfrac{1}{2}$ and $\cos\left(\dfrac{7\pi}{6}\right) = -\dfrac{\sqrt{3}}{2}$.

45. An angle θ of $-\dfrac{3\pi}{4}$ radians in standard position has a terminal side that intersects the unit circle in the third

quadrant at the point $\left(-\dfrac{1}{\sqrt{2}}, -\dfrac{1}{\sqrt{2}}\right)$. Therefore $\sin\left(-\dfrac{3\pi}{4}\right) = -\dfrac{1}{\sqrt{2}}$ and $\cos\left(-\dfrac{3\pi}{4}\right) = -\dfrac{1}{\sqrt{2}}$.

47. An angle θ of $\dfrac{5\pi}{2}$ radians in standard position has a terminal side that intersects the unit circle at the point

$(0, 1)$. Therefore $\sin \dfrac{5\pi}{2} = 1$ and $\cos \dfrac{5\pi}{2} = 0$.

49. An angle θ of $-\dfrac{\pi}{3}$ radians in standard position has a terminal side that intersects the unit circle in the fourth

quadrant. To find the point of intersection, the hypotenuse of a $30°/60°$ right triangle has length 1 and the shorter

leg has length $x = \dfrac{1}{2}$. Since $x^2 + y^2 = 1$, we can solve for y. $\left(\dfrac{1}{2}\right)^2 + y^2 = 1 \Rightarrow y^2 = \dfrac{3}{4} \Rightarrow y = \dfrac{\sqrt{3}}{2}$.

Since the point (x, y) is located in the fourth quadrant, the point becomes $\left(\dfrac{1}{2}, -\dfrac{\sqrt{3}}{2}\right)$.

Therefore $\sin\left(-\dfrac{\pi}{3}\right) = -\dfrac{\sqrt{3}}{2}$ and $\cos\left(-\dfrac{\pi}{3}\right) = \dfrac{1}{2}$.

51. If a string of $s = 3\pi$ is wrapped counter-clockwise around the unit circle, it will make $1\dfrac{1}{2}$

revolutions and the terminal point will coincide with the point $(-1, 0)$. Thus, $W(3\pi) = (-1, 0)$.

53. If a string of $s = -2\pi$ is wrapped clockwise around the unit circle, it will make 1 complete

revolution and the terminal point will coincide with the point $(1, 0)$. Thus, $W(-2\pi) = (1, 0)$.

55. If a string of $s = \dfrac{3\pi}{2}$ is wrapped counter-clockwise around the unit circle, it will make $\dfrac{3}{4}$ of a revolution and the

terminal point will coincide with the point $(0, -1)$. Thus, $W\left(\dfrac{3\pi}{2}\right) = (0, -1)$.

57. If a string of $s = -\dfrac{5\pi}{2}$ is wrapped clockwise around the unit circle, it will make $1\dfrac{1}{4}$ revolutions and the

terminal point will coincide with the point $(0, -1)$. Thus, $W\left(-\dfrac{5\pi}{2}\right) = (0, -1)$.

59. A distance of $\dfrac{\pi}{4}$ represents an eighth of a revolution in the unit circle. The $45° \sim 45°$ triangle formed has a

hypotenuse with length 1 and legs both length a. $a^2 + a^2 = 1 \Rightarrow 2a^2 = 1 \Rightarrow a^2 = \dfrac{1}{2} \Rightarrow a = \pm\dfrac{1}{\sqrt{2}}$.

Since the terminal point of $\dfrac{5\pi}{4}$ lies in quadrant III, the x-coordinate is $-\dfrac{1}{\sqrt{2}}$, and the y-coordinate is $-\dfrac{1}{\sqrt{2}}$.

Thus, $W\left(\dfrac{5\pi}{4}\right) = \left(-\dfrac{1}{\sqrt{2}}, -\dfrac{1}{\sqrt{2}}\right)$.

61. A distance of $\dfrac{\pi}{4}$ represents an eighth of a revolution in the unit circle. The $45° \sim 45°$ triangle formed has a

hypotenuse with length 1 and legs both length a. $a^2 + a^2 = 1 \Rightarrow 2a^2 = 1 \Rightarrow a^2 = \dfrac{1}{2} \Rightarrow a = \pm\dfrac{1}{\sqrt{2}}$.

Since the terminal point of $\dfrac{-5\pi}{4}$ lies in quadrant II, the x-coordinate is $-\dfrac{1}{\sqrt{2}}$, and the y-coordinate is $\dfrac{1}{\sqrt{2}}$.

Thus, $W\left(-\dfrac{5\pi}{4}\right) = \left(-\dfrac{1}{\sqrt{2}}, \dfrac{1}{\sqrt{2}}\right)$.

63. $W\left(\dfrac{11\pi}{6}\right)$ is illustrated in Figure 63. $W\left(\dfrac{11\pi}{6}\right) = \left(\dfrac{\sqrt{3}}{2}, -\dfrac{1}{2}\right)$

Figure 63

65. Use a similar procedure as is shown in solution 63. $W\left(-\dfrac{7\pi}{3}\right) = \left(\dfrac{1}{2}, -\dfrac{\sqrt{3}}{2}\right)$

67. Note that the values for s are the same as those used in exercises 51-66. Thus, if $W = (x, y)$, then we apply the
fact that $\sin s = y$ and $\cos s = x$. $\sin 3\pi = 0$ $\cos 3\pi = -1$

69. Note that the values for s are the same as those used in exercises 51-66. Thus, if $W = (x, y)$, then we apply the
fact that $\sin s = y$ and $\cos s = x$. $\sin(-2\pi) = 0$ $\cos(-2\pi) = 1$

71. Note that the values for s are the same as those used in exercises 51-66. Thus, if $W = (x, y)$, then we apply the
fact that $\sin s = y$ and $\cos s = x$. $\sin\left(\dfrac{3\pi}{2}\right) = -1$ $\cos\left(\dfrac{3\pi}{2}\right) = 0$

73. Note that the values for s are the same as those used in exercises 51-66. Thus, if $W = (x, y)$, then we apply the
fact that $\sin s = y$ and $\cos s = x$. $\sin\left(-\dfrac{5\pi}{2}\right) = -1$ $\cos\left(-\dfrac{5\pi}{2}\right) = 0$

75. Note that the values for s are the same as those used in exercises 51-66. Thus, if $W = (x, y)$, then we apply the
fact that $\sin s = y$ and $\cos s = x$. $\sin\left(\dfrac{5\pi}{4}\right) = \cos\left(\dfrac{5\pi}{4}\right) = -\dfrac{1}{\sqrt{2}}$

77. Note that the values for *s* are the same as those used in exercises 51-66. Thus, if $W = (x, y)$, then we apply the fact that $\sin s = y$ and $\cos s = x$. $\sin\left(-\dfrac{5\pi}{4}\right) = \dfrac{1}{\sqrt{2}}$ $\cos\left(-\dfrac{5\pi}{4}\right) = -\dfrac{1}{\sqrt{2}}$

79. Note that the values for *s* are the same as those used in exercises 51-66. Thus, if $W = (x, y)$, then we apply the fact that $\sin s = y$ and $\cos s = x$. $\sin\left(\dfrac{11\pi}{6}\right) = -\dfrac{1}{2}$ $\cos\left(\dfrac{11\pi}{6}\right) = \dfrac{\sqrt{3}}{2}$

81. Note that the values for *s* are the same as those used in exercises 51-66. Thus, if $W = (x, y)$, then we apply the fact that $\sin s = y$ and $\cos s = x$. $\sin\left(-\dfrac{7\pi}{3}\right) = -\dfrac{\sqrt{3}}{2}$ $\cos\left(-\dfrac{7\pi}{3}\right) = \dfrac{1}{2}$

83. (a) $R = \{y \mid -3 \le y \le 3\}$. See Figure 83a. $f\left(\dfrac{3\pi}{2}\right) = 3\sin\left(\dfrac{3\pi}{2}\right) = -3$

   (b) $R = \{y \mid -1 \le y \le 1\}$. See Figure 83b. $f\left(\dfrac{3\pi}{2}\right) = \sin\left(\dfrac{9\pi}{2}\right) = 1$

$[-2\pi, 2\pi, \pi/2]$ by $[-4, 4, 1]$ $[-2\pi, 2\pi, \pi/2]$ by $[-4, 4, 1]$ $[-2\pi, 2\pi, \pi/2]$ by $[-4, 4, 1]$ $[-2\pi, 2\pi, \pi/2]$ by $[-4, 4, 1]$

Figure 83a   Figure 83b   Figure 85a   Figure 85b

85. (a) $R = \{y \mid -1 \le y \le 3\}$. See Figure 85a. $f\left(\dfrac{3\pi}{2}\right) = 2\cos\left(\dfrac{3\pi}{2}\right) + 1 = 1$

   (b) $R = \{y \mid -2 \le y \le 0\}$. See Figure 85b. $f\left(\dfrac{3\pi}{2}\right) = \cos(3\pi) - 1 = -2$

87. (a) The slope of the line is 0.03, so when *x* increases by 100 feet, *y* increases by 3 feet. The hill has a 3% grade.

   (b) First we must find $\sin\theta$: $r = \sqrt{100^2 + 3^2} = \sqrt{10{,}009}$ and so $\sin\theta = \dfrac{y}{r} = \dfrac{3}{\sqrt{10{,}009}}$.

   The grade resistance is $R = W\sin\theta = 25{,}000\left(\dfrac{3}{\sqrt{10{,}009}}\right) \approx 750$ lb.

89. The first quarter phase corresponds to half the face on the moon being visible. This occurs when

   $\theta = \dfrac{\pi}{2} + 2\pi n$ for $n = 0, 1, 2, 3, \ldots$.

91. (a) The graph is sinusoidal and varies between -310 volts and 310 volts. The voltage is changing direction.

   (b) $V\left(\dfrac{1}{120}\right) = 310\sin\left(120\pi \cdot \dfrac{1}{120}\right) = 310\sin(\pi) = 310(0) = 0$. After $\dfrac{1}{120}$ second the voltage is 0.

   (c) The maximum voltage is 310 volts. The root mean square voltage is $\dfrac{310}{\sqrt{2}} \approx 219$ volts. Note that the common electrical rating for electric ranges and ovens is 220 volts.

93. (a) $D = \dfrac{1.05(88^2 - 44^2)}{27 + 64.4\sin 3°} \approx 201$ ft

   (b) $D = \dfrac{1.05(88^2 - 44^2)}{27 + 64.4\sin(-3°)} \approx 258$ ft

   (c) When the slope is negative (down hill) the car will require a greater stopping distance.

95. (a) When $S = 390$ and $r = 600$, $\beta = \dfrac{S}{r} = \dfrac{390}{600} = \dfrac{13}{20}$ radians. Then, $d = 600\left(1 - \cos\dfrac{\frac{13}{20}}{2}\right) \approx 31.4$ ft.

(b) When $S = 620$ and $r = 600$, $\beta = \dfrac{S}{r} = \dfrac{620}{600} = \dfrac{31}{30}$ radians. Then, $d = 600\left(1 - \cos\dfrac{\frac{31}{30}}{2}\right) \approx 78.3$ ft.

(c) Higher speed limits require more land to be cleared on the inside of the curve.

97. (a) The scatterplot is shown in Figure 97a.

(b) The graph of $Y_1 = 1.9 \sin(0.42X - 1.2) + 5.7$ in $[0, 13, 1]$ by $[3, 8, 1]$ is shown with the data in Figure 97b.

From the close data fit shown, we may conclude that the flying squirrels become active near sunset.

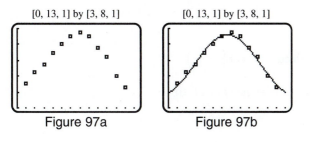

[0, 13, 1] by [3, 8, 1]   [0, 13, 1] by [3, 8, 1]

Figure 97a   Figure 97b

6.4: Other Trigonometric Functions and Their Graphs

1. $r = \sqrt{5^2 + 12^2} = \sqrt{25 + 144} = \sqrt{169} = 13$

$\sin\theta = \dfrac{12}{13}$ $\cos\theta = \dfrac{5}{13}$ $\tan\theta = \dfrac{12}{5}$

$\csc\theta = \dfrac{13}{12}$ $\sec\theta = \dfrac{13}{5}$ $\cot\theta = \dfrac{5}{12}$

3. $r = \sqrt{7^2 + (-24)^2} = \sqrt{49 + 576} = \sqrt{625} = 25$

$\sin\theta = -\dfrac{24}{25}$ $\cos\theta = \dfrac{7}{25}$ $\tan\theta = -\dfrac{24}{7}$

$\csc\theta = -\dfrac{25}{24}$ $\sec\theta = \dfrac{25}{7}$ $\cot\theta = -\dfrac{7}{24}$

5. The terminal side of a $90°$ angle intersects the unit circle at the point $(0, 1)$.

$\sin 90° = 1$ $\cos 90° = 0$ $\tan 90° \rightarrow$ undefined

$\csc 90° = 1$ $\sec 90° \rightarrow$ undefined $\cot 90° = 0$

7. The terminal side of a $-45°$ angle intersects the unit circle at the point $\left(\dfrac{1}{\sqrt{2}}, -\dfrac{1}{\sqrt{2}}\right)$.

$\sin(-45°) = -\dfrac{1}{\sqrt{2}}$ $\cos(-45°) = \dfrac{1}{\sqrt{2}}$ $\tan(-45°) = -1$

$\csc(-45°) = -\sqrt{2}$ $\sec(-45°) = \sqrt{2}$ $\cot(-45°) = -1$

9. The terminal side of a π radian angle intersects the unit circle at the point $(-1, 0)$.

$\sin\pi = 0$ $\cos\pi = -1$ $\tan\pi = 0$

$\csc\pi \rightarrow$ undefined $\sec\pi = -1$ $\cot\pi \rightarrow$ undefined

11. The terminal side of a $-\dfrac{\pi}{3}$ radian angle intersects the unit circle at the point $\left(\dfrac{1}{2}, -\dfrac{\sqrt{3}}{2}\right)$.

$$\sin\left(-\frac{\pi}{3}\right) = -\frac{\sqrt{3}}{2} \qquad \cos\left(-\frac{\pi}{3}\right) = \frac{1}{2} \qquad \tan\left(-\frac{\pi}{3}\right) = -\sqrt{3}$$

$$\csc\left(-\frac{\pi}{3}\right) = -\frac{2}{\sqrt{3}} \qquad \sec\left(-\frac{\pi}{3}\right) = 2 \qquad \cot\left(-\frac{\pi}{3}\right) = -\frac{1}{\sqrt{3}}$$

13. The terminal side of a $-\dfrac{\pi}{2}$ radian angle intersects the unit circle at the point $(0, -1)$.

$$\sin\left(-\frac{\pi}{2}\right) = -1 \qquad \cos\left(-\frac{\pi}{2}\right) = 0 \qquad \tan\left(-\frac{\pi}{2}\right) \rightarrow \text{undefined}$$

$$\csc\left(-\frac{\pi}{2}\right) = -1 \qquad \sec\left(-\frac{\pi}{2}\right) \rightarrow \text{undefined} \qquad \cot\left(-\frac{\pi}{2}\right) = 0$$

15. The terminal side of a $360°$ angle intersects the unit circle at the point $(1, 0)$.

$$\sin 360° = 0 \qquad \cos 360° = 1 \qquad \tan 360° = 0$$

$$\csc 360° \rightarrow \text{undefined} \qquad \sec 360° = 1 \qquad \cot 360° \rightarrow \text{undefined}$$

17. The terminal side of a $\dfrac{\pi}{6}$ radian angle intersects the unit circle at the point $\left(\dfrac{\sqrt{3}}{2}, \dfrac{1}{2}\right)$.

$$\sin\frac{\pi}{6} = \frac{1}{2} \qquad \cos\frac{\pi}{6} = \frac{\sqrt{3}}{2} \qquad \tan\frac{\pi}{6} = \frac{1}{\sqrt{3}}$$

$$\csc\frac{\pi}{6} = 2 \qquad \sec\frac{\pi}{6} = \frac{2}{\sqrt{3}} \qquad \cot\frac{\pi}{6} = \sqrt{3}$$

19. The terminal side of a $\dfrac{4\pi}{3}$ radian angle intersects the unit circle at the point $\left(-\dfrac{1}{2}, -\dfrac{\sqrt{3}}{2}\right)$.

$$\sin\left(\frac{4\pi}{3}\right) = -\frac{\sqrt{3}}{2} \qquad \cos\left(\frac{4\pi}{3}\right) = -\frac{1}{2} \qquad \tan\left(\frac{4\pi}{3}\right) = \sqrt{3}$$

$$\csc\left(\frac{4\pi}{3}\right) = -\frac{2}{\sqrt{3}} \qquad \sec\left(\frac{4\pi}{3}\right) = -2 \qquad \cot\left(\frac{4\pi}{3}\right) = \frac{1}{\sqrt{3}}$$

21. The terminal side of a $-225°$ radian angle intersects the unit circle at the point $\left(-\dfrac{1}{\sqrt{2}}, \dfrac{1}{\sqrt{2}}\right)$.

$$\sin(-225°) = \frac{1}{\sqrt{2}} \qquad \cos(-225°) = -\frac{1}{\sqrt{2}} \qquad \tan(-225°) = -1$$

$$\csc(-225°) = \sqrt{2} \qquad \sec(-225°) = -\sqrt{2} \qquad \cot(-225°) = -1$$

23. The terminal side of a $\dfrac{-13\pi}{6}$ radian angle intersects the unit circle at the point $\left(\dfrac{\sqrt{3}}{2}, -\dfrac{1}{2}\right)$.

$$\sin\left(\frac{-13\pi}{6}\right) = -\frac{1}{2} \qquad \cos\left(\frac{-13\pi}{6}\right) = \frac{\sqrt{3}}{2} \qquad \tan\left(\frac{-13\pi}{6}\right) = -\frac{1}{\sqrt{3}}$$

$$\csc\left(\frac{-13\pi}{6}\right) = -2 \qquad \sec\left(\frac{-13\pi}{6}\right) = \frac{2}{\sqrt{3}} \qquad \cot\left(\frac{-13\pi}{6}\right) = -\sqrt{3}$$

25. The point $(-1, 4)$ lies on the line $y = -4x$ in the second quadrant. This is shown in Figure 25.

So $r = \sqrt{(-1)^2 + 4^2} = \sqrt{1 + 16} = \sqrt{17}$.

$$\sin \theta = \frac{4}{\sqrt{17}} \qquad\qquad \cos \theta = -\frac{1}{\sqrt{17}} \qquad\qquad \tan \theta = -4$$

$$\csc \theta = \frac{\sqrt{17}}{4} \qquad\qquad \sec \theta = -\sqrt{17} \qquad\qquad \cot \theta = -\frac{1}{4}$$

The slope of the line is equal to $\tan \theta$.

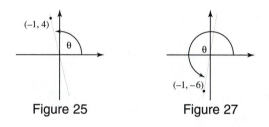

Figure 25   Figure 27

27. The point $(-1, -6)$ lies on the line $y = 6x$ in the third quadrant. This is shown in Figure 27.

So $r = \sqrt{(-1)^2 + (-6)^2} = \sqrt{1 + 36} = \sqrt{37}$.

$$\sin \theta = -\frac{6}{\sqrt{37}} \qquad\qquad \cos \theta = -\frac{1}{\sqrt{37}} \qquad\qquad \tan \theta = 6$$

$$\csc \theta = -\frac{\sqrt{37}}{6} \qquad\qquad \sec \theta = -\sqrt{37} \qquad\qquad \cot \theta = \frac{1}{6}$$

The slope of the line is equal to $\tan \theta$.

29. $\sin \theta = \dfrac{3}{5}$ $\qquad\qquad\qquad$ $\cos \theta = \dfrac{4}{5}$ $\qquad\qquad\qquad$ $\tan \theta = \dfrac{\sin \theta}{\cos \theta} = \dfrac{\frac{3}{5}}{\frac{4}{5}} = \dfrac{3}{4}$

$\csc \theta = \dfrac{1}{\sin \theta} = \dfrac{1}{\frac{3}{5}} = \dfrac{5}{3}$ $\qquad$ $\sec \theta = \dfrac{1}{\cos \theta} = \dfrac{1}{\frac{4}{5}} = \dfrac{5}{4}$ $\qquad$ $\cot \theta = \dfrac{\cos \theta}{\sin \theta} = \dfrac{\frac{4}{5}}{\frac{3}{5}} = \dfrac{4}{3}$

31. $\sin \theta = \dfrac{1}{\csc \theta} = \dfrac{1}{-\frac{17}{15}} = -\dfrac{15}{17}$ $\qquad$ $\cos \theta = \dfrac{1}{\sec \theta} = \dfrac{1}{-\frac{17}{8}} = -\dfrac{8}{17}$ $\qquad$ $\tan \theta = \dfrac{\sin \theta}{\cos \theta} = \dfrac{-\frac{15}{17}}{-\frac{8}{17}} = \dfrac{15}{8}$

$\csc \theta = -\dfrac{17}{15}$ $\qquad\qquad\qquad$ $\sec \theta = -\dfrac{17}{8}$ $\qquad\qquad\qquad$ $\cot \theta = \dfrac{\cos \theta}{\sin \theta} = \dfrac{-\frac{8}{17}}{-\frac{15}{17}} = \dfrac{8}{15}$

33. $\sin \theta = \tan \theta \cos \theta = \left(\dfrac{5}{12}\right)\left(\dfrac{12}{13}\right) = \dfrac{5}{13}$ $\qquad$ $\cos \theta = \dfrac{12}{13}$ $\qquad\qquad\qquad$ $\tan \theta = \dfrac{5}{12}$

$\csc \theta = \dfrac{1}{\sin \theta} = \dfrac{1}{\frac{5}{13}} = \dfrac{13}{5}$ $\qquad\qquad$ $\sec \theta = \dfrac{1}{\cos \theta} = \dfrac{1}{\frac{12}{13}} = \dfrac{13}{12}$ $\qquad$ $\cot \theta = \dfrac{\cos \theta}{\sin \theta} = \dfrac{\frac{12}{13}}{\frac{5}{13}} = \dfrac{12}{5}$

35. $\sin \theta = -\dfrac{3}{5}$ $\qquad$ $\cos \theta = \sqrt{1 - \left(-\frac{3}{5}\right)^2} = \dfrac{4}{5}$ $\qquad$ $\tan \theta = \dfrac{\sin \theta}{\cos \theta} = \dfrac{-\frac{3}{5}}{\frac{4}{5}} = -\dfrac{3}{4}$

$\csc \theta = \dfrac{1}{\sin \theta} = \dfrac{1}{-\frac{3}{5}} = -\dfrac{5}{3}$ $\qquad$ $\sec \theta = \dfrac{1}{\cos \theta} = \dfrac{1}{\frac{4}{5}} = \dfrac{5}{4}$ $\qquad$ $\cot \theta = \dfrac{\cos \theta}{\sin \theta} = \dfrac{\frac{4}{5}}{-\frac{3}{5}} = -\dfrac{4}{3}$

37. $\sin \theta = -\sqrt{1 - (-\frac{4}{5})^2} = -\frac{3}{5}$ $\cos \theta = -\frac{4}{5}$ $\tan \theta = \frac{\sin \theta}{\cos \theta} = \frac{-\frac{3}{5}}{-\frac{4}{5}} = \frac{3}{4}$

$\csc \theta = \frac{1}{\sin \theta} = \frac{1}{-\frac{3}{5}} = -\frac{5}{3}$ $\sec \theta = \frac{1}{\cos \theta} = \frac{1}{-\frac{4}{5}} = -\frac{5}{4}$ $\cot \theta = \frac{\cos \theta}{\sin \theta} = \frac{-\frac{4}{5}}{-\frac{3}{5}} = \frac{4}{3}$

39. The terminal side of -7π corresponds to the point $(-1, 0)$. Therefore,

$\sin(-7\pi) = 0$ $\cos(-7\pi) = -1$ $\tan(-7\pi) = 0$

$\csc(-7\pi) \rightarrow$ undefined $\sec(-7\pi) = -1$ $\cot(-7\pi) \rightarrow$ undefined

41. The terminal side of $\dfrac{7\pi}{2}$ corresponds to the point $(0, -1)$.

$\sin\left(\dfrac{7\pi}{2}\right) = -1$ $\cos\left(\dfrac{7\pi}{2}\right) = 0$ $\tan\left(\dfrac{7\pi}{2}\right) \rightarrow$ undefined

$\csc\left(\dfrac{7\pi}{2}\right) = -1$ $\sec\left(\dfrac{7\pi}{2}\right) \rightarrow$ undefined $\cot\left(\dfrac{7\pi}{2}\right) = 0$

43. The terminal side of $-\dfrac{3\pi}{4}$ corresponds to the point $\left(-\dfrac{1}{\sqrt{2}}, -\dfrac{1}{\sqrt{2}}\right)$.

$\sin\left(-\dfrac{3\pi}{4}\right) = -\dfrac{1}{\sqrt{2}}$ $\cos\left(-\dfrac{3\pi}{4}\right) = -\dfrac{1}{\sqrt{2}}$ $\tan\left(-\dfrac{3\pi}{4}\right) = 1$

$\csc\left(-\dfrac{3\pi}{4}\right) = -\sqrt{2}$ $\sec\left(-\dfrac{3\pi}{4}\right) = -\sqrt{2}$ $\cot\left(-\dfrac{3\pi}{4}\right) = 1$

45. The terminal side of $\dfrac{7\pi}{6}$ corresponds to the point $\left(-\dfrac{\sqrt{3}}{2}, -\dfrac{1}{2}\right)$.

$\sin\left(\dfrac{7\pi}{6}\right) = -\dfrac{1}{2}$ $\cos\left(\dfrac{7\pi}{2}\right) = -\dfrac{\sqrt{3}}{2}$ $\tan\left(\dfrac{7\pi}{6}\right) = \dfrac{1}{\sqrt{3}}$

$\csc\left(\dfrac{7\pi}{6}\right) = -2$ $\sec\left(\dfrac{7\pi}{6}\right) = -\dfrac{2}{\sqrt{3}}$ $\cot\left(\dfrac{7\pi}{6}\right) = \sqrt{3}$

47. $\sin \theta = \dfrac{1}{\sqrt{2}}$ $\cos \theta = \dfrac{1}{\sqrt{2}}$ $\tan \theta = 1$ $\csc \theta = \sqrt{2}$ $\sec \theta = \sqrt{2}$ $\cot \theta = 1$

49. $\sin \theta = -\dfrac{12}{13}$ $\cos \theta = \dfrac{5}{13}$ $\tan \theta = -\dfrac{12}{5}$ $\csc \theta = -\dfrac{13}{12}$ $\sec \theta = \dfrac{13}{5}$

51. See Figure 51. $\tan\left(\dfrac{-5\pi}{4}\right) = \dfrac{\frac{1}{\sqrt{2}}}{-\frac{1}{\sqrt{2}}} = -1$

Figure 51

Figure 53

Figure 55

53. See Figure 53. $\csc\left(\dfrac{\pi}{6}\right) = \dfrac{1}{\frac{1}{2}} = 2$

55. See Figure 55. $\sec\left(\dfrac{\pi}{6}\right) = \dfrac{1}{\frac{\sqrt{3}}{2}} = \dfrac{2}{\sqrt{3}}$

57. See Figure 57.

$\sin\left(\dfrac{9\pi}{2}\right) = 1$ $\cos\left(\dfrac{9\pi}{2}\right) = 0$ $\tan\left(\dfrac{9\pi}{2}\right) \rightarrow$ undefined

$\csc\left(\dfrac{9\pi}{2}\right) = 1$ $\sec\left(\dfrac{9\pi}{2}\right) \rightarrow$ undefined $\cot\left(\dfrac{9\pi}{2}\right) = 0$

59. See Figure 59.

$\sin(-\pi) = 0$ $\cos(-\pi) = -1$ $\tan(-\pi) = 0$

$\csc(-\pi) \rightarrow$ undefined $\sec(-\pi) = -1$ $\cot(-\pi) \rightarrow$ undefined

Figure 57

Figure 59

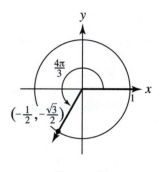

Figure 61

61. See Figure 61.

$\sin\left(\dfrac{4\pi}{3}\right) = -\dfrac{\sqrt{3}}{2}$ $\cos\left(\dfrac{4\pi}{3}\right) = -\dfrac{1}{2}$ $\tan\left(\dfrac{4\pi}{3}\right) = \sqrt{3}$

$\csc\left(\dfrac{4\pi}{3}\right) = -\dfrac{2}{\sqrt{3}}$ $\sec\left(\dfrac{4\pi}{3}\right) = -2$ $\cot\left(\dfrac{4\pi}{3}\right) = \dfrac{1}{\sqrt{3}}$

63. (a) $\sin 93.2° \approx 0.9984$

    (b) $\csc 93.2° = \dfrac{1}{\sin 93.2°} \approx 1.0016$

65. (a) $\tan 234°33' \approx 1.4045$

    (b) $\cot 234°33' = \dfrac{1}{\tan 234°33'} \approx 0.7120$

67. (a) $\cot(-4) = \dfrac{1}{\tan(-4)} \approx -0.8637$

    (b) $\tan(-4) \approx -1.1578$

69. (a) $\cos\dfrac{11\pi}{7} \approx 0.2225$

    (b) $\sec\dfrac{11\pi}{7} = \dfrac{1}{\cos\frac{11\pi}{7}} \approx 4.4940$

71. See the sine graph in "Putting it All Together" at the end of section 6.4. $f(t) = \sin t$ has origin symmetry.

73. See the tangent graph in "Putting it All Together" at the end of section 6.4. $f(t) = \tan t$ has origin symmetry.

75. See the secant graph in "Putting it All Together" at the end of section 6.4. $f(t) = \sec t$ has y-axis symmetry.

77. The domain is all real numbers. The range is $\{y \mid -1 \leq y \leq 1\}$. The period is 2π.

79. The domain is $\left\{ t \mid t \neq \pm\dfrac{\pi}{2}, \pm\dfrac{3\pi}{2}, \pm\dfrac{5\pi}{2}, \ldots \right\}$. The range is all real numbers. The period is π.

81. The domain is $\left\{ t \mid t \neq \pm\dfrac{\pi}{2}, \pm\dfrac{3\pi}{2}, \pm\dfrac{5\pi}{2}, \ldots \right\}$. The range is $\{y \mid |y| \geq 1\}$. The period is 2π.

83. $\dfrac{\sin\left(\frac{\pi}{4}\right) - \sin(0)}{\frac{\pi}{4} - 0} \approx 0.900$

85. $\dfrac{\tan\left((\frac{1}{2})(\frac{\pi}{4})\right) - \tan(0)}{\frac{\pi}{4} - 0} \approx 0.527$

87. (a) See Figure 87.

(b) At sunrise $(t = 0)$ the stick would cast a very long shadow (longer than four meters). As time moves toward noon, the length of the shadow decreases until it has length zero at exactly noon $\left(t = \dfrac{\pi}{2} \right)$.

In the afternoon, the length of the shadow begins to increase again but now the shadow is on the other side of the stick (negative values). This continues until sunset $(t = \pi)$ when the shadow is again very long.

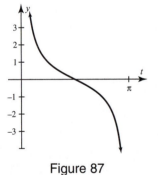

Figure 87

89. $\theta = 57.35 \tan 17°23'43'' \approx 17.95''$; Because $\tan\alpha = \dfrac{\sin\alpha}{\cos\alpha}$, $\theta = 57.3\dfrac{\sin\alpha}{\cos\alpha}$.

91. $L = \dfrac{2700}{h + 3\tan\alpha} = \dfrac{2700}{2 + 3\tan 1.5°} \approx 1300$ ft.

93. At noon $\theta = 90°$. The sun moves $15°$ per hour. Thus, three hours later $\theta = 135°$. Since $\csc 90° = 1$ and $\csc 135° = \dfrac{1}{\sin 135°} \approx 1.414$, we may conclude that sunlight travels through about 41% more atmosphere at 3:00 P.M. than at noon.

95. Table $Y_1 = 1/\sin(X)$ and $Y_2 = 1/(\sin(X) + 0.5(6 + X)^{\wedge}-1.64)$ starting at $x = 2$, incrementing by 1. See Figure 95. As θ increases, the values of Y_1 and Y_2 become closer together. When $\theta = 20°$ there is a difference of only about 0.02 or 2%.

[0, 20000, 5000] by [0, 4000, 1000]   [0, 20000, 5000] by [0, 4000, 1000]

Figure 95   Figure 97a   Figure 97b

97. Graph $Y_1 = -16X^2/(750^2\ (\cos\ (30))^2) + X \tan\ (30)$ in $[0, 20000, 5000]$ by $[0, 4000, 1000]$.

(a) The maximum height is shown to be about 2197 feet at (7612, 2197). See Figure 97a.

(b) The total distance traveled is shown to be about 15,223 feet. See Figure 97b.

6.5: Graphing Trigonometric Functions

1. Period is $\dfrac{2\pi}{\frac{1}{2}} = 4\pi$ and Amplitude is $|\,3\,| = 3$. See figure 1.

3. Period is $\dfrac{2\pi}{3}$ and Amplitude is $|-2| = 2$. See figure 3.

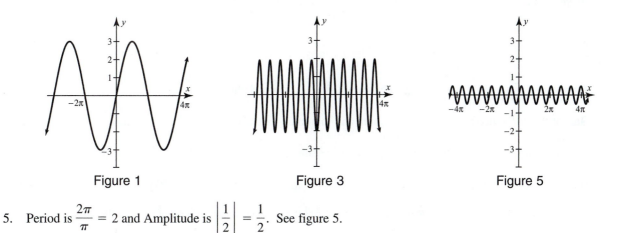

Figure 1   Figure 3   Figure 5

5. Period is $\dfrac{2\pi}{\pi} = 2$ and Amplitude is $\left|\dfrac{1}{2}\right| = \dfrac{1}{2}$. See figure 5.

7. Shorten the period of the sine graph to π, increase the amplitude to 3, and shift the graph downward 1 unit.

See Figure 7.

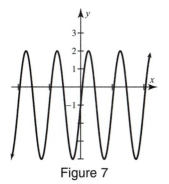

Figure 7

9. Shift the cosine graph $\frac{\pi}{2}$ units to the left, increase the amplitude to 2, and reflect the graph across the *x*-axis.

See Figure 9.

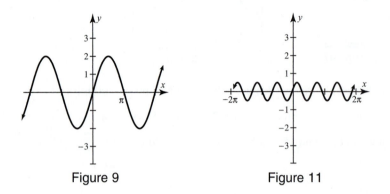

Figure 9   Figure 11

11. Shift the cosine graph $\frac{1}{\pi}$ unit to the right, shorten the period to 2, and decrease the amplitude to $\frac{1}{2}$.

See Figure 11.

13. Amplitude is 3; period is $\frac{2\pi}{4} = \frac{\pi}{2}$; phase shift is $\frac{\pi}{4}$; vertical shift is -4.

15. Amplitude is 4; period is $\frac{2\pi}{\frac{\pi}{2}} = 4$; phase shift is 1; vertical shift is 6.

17. Rewrite the equation as $y = \frac{2}{3}\sin\left(6\left(x + \frac{\pi}{2}\right)\right) - \frac{5}{2}$.

Amplitude is $\frac{2}{3}$; period is $\frac{2\pi}{6} = \frac{\pi}{3}$; phase shift is $-\frac{\pi}{2}$; vertical shift is $-\frac{5}{2}$.

19. Half of the difference between the maximum and minimum values is $0.5(3 - (-3)) = 3$.

Thus the amplitude is 3. The graph repeats every π units thus $\frac{2\pi}{b} = \pi \Rightarrow b = 2$. So $a = 3$ and $b = 2$.

21. Half of the difference between the maximum and minimum values is $0.5(3 - (-3)) = 3$. Thus the amplitude

is 3. The graph repeats every 4π units, thus the period is 4π. There is no phase shift for this graph.

23. Graph *c*.   amplitude $= 2$, period $= \frac{2\pi}{0.5} = 4\pi$, no phase shift

25. Graph *d*.   amplitude $= 3$, period $= \frac{2\pi}{\pi} = 2$, no phase shift

27. Graph *a*.   amplitude $= 1$, period $= \frac{2\pi}{1} = 2\pi$, phase shift $= -\frac{\pi}{2}$

29. $a = 3, \frac{2\pi}{b} = 4\pi \Rightarrow b = \frac{1}{2}$, and $c = 0$. Thus $y = 3\sin\left(\frac{1}{2}(x - 0)\right) \Rightarrow y = 3\sin\left(\frac{1}{2}x\right)$.

31. Amplitude $= 2$, period $= 2\pi$, phase shift $= 0$. See Figure 31.

33. Amplitude is 1; period is $\frac{2\pi}{\frac{1}{2}} = 4\pi$; phase shift is 0. See figure 33.

35. Amplitude is 1; period is 2π; phase shift is 0. See figure 35.

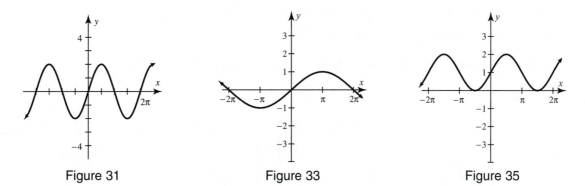

Figure 31   Figure 33   Figure 35

37. Amplitude $= 1$, period $= \dfrac{2\pi}{\pi} = 2$, phase shift $= 0$, vertical shift up 2 units. See Figure 37.

39. Amplitude $= 1$, period $= \dfrac{2\pi}{2} = \pi$, phase shift $= -\pi$. See Figure 39.

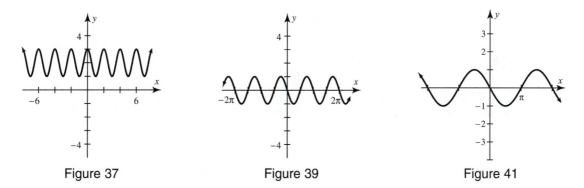

Figure 37   Figure 39   Figure 41

41. Amplitude is 1; period is 2π; phase shift is $\dfrac{\pi}{2}$. See figure 41.

43. Amplitude is $\left| -\dfrac{1}{2} \right| = \dfrac{1}{2}$; period is $\dfrac{2\pi}{2} = \pi$; phase shift is 0. See figure 43.44.

45. Rewrite the equation as $2\cos\left(2\left(x + \dfrac{\pi}{4} \right) \right) - 1$. Amplitude is 2; period is $\dfrac{2\pi}{2} = \pi$; phase shift is $-\dfrac{\pi}{4}$.

See Figure 45.

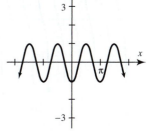

Figure 43   Figure 45   Figure 47

47. Amplitude $= 1$, period $= \dfrac{2\pi}{2} = \pi$, phase shift $= \dfrac{\pi}{2}$. See Figure 47.

49. Graph $Y_1 = 2 \sin (2X)$ in $[-2\pi, 2\pi, \pi/2]$ by $[-4, 4, 1]$. See Figure 49.

    amplitude $= 2$, period $= \pi$, phase shift $= 0$

51. Graph $Y_1 = 0.5 \cos (3(X + \pi/3))$ in $[-2\pi, 2\pi, \pi/2]$ by $[-4, 4, 1]$. See Figure 51.

    amplitude $= \dfrac{1}{2}$, period $= \dfrac{2\pi}{3}$, phase shift $= -\dfrac{\pi}{3}$

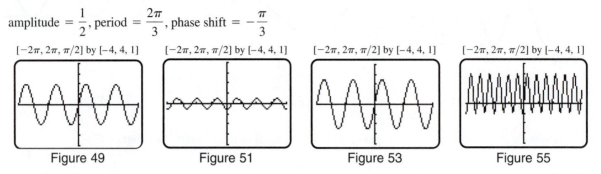

$[-2\pi, 2\pi, \pi/2]$ by $[-4, 4, 1]$ $[-2\pi, 2\pi, \pi/2]$ by $[-4, 4, 1]$ $[-2\pi, 2\pi, \pi/2]$ by $[-4, 4, 1]$ $[-2\pi, 2\pi, \pi/2]$ by $[-4, 4, 1]$

Figure 49   Figure 51   Figure 53   Figure 55

53. Graph $Y_1 = -2.5 \cos (2X + \pi/2))$ in $[-2\pi, 2\pi, \pi/2]$ by $[-4, 4, 1]$. See Figure 53.

    amplitude $= 2.5$, period $= \dfrac{2\pi}{2} = \pi$, phase shift $= -\dfrac{\pi}{4}$

55. Rewrite the equation as $f(x) = -2 \cos \left(2\pi \left(x + \dfrac{1}{8} \right) \right) + 1$.

    Graph $Y_1 = -2 \cos (2\pi (X + 1/8)) + 1$ in $[-2\pi, 2\pi, \pi/2]$ by $[-4, 4, 1]$. See Figure 55.

    Amplitude is $|-2| = 2$; period is $\dfrac{2\pi}{2\pi} = 1$; phase shift is $-\dfrac{1}{8}$.

57. If $f(t) = \tan 2t$, then $b = 2$ and $c = 0$. Also, period $= \dfrac{\pi}{b} = \dfrac{\pi}{2}$ and phase shift $= 0$. See Figure 57.

    The vertical asymptotes occur at $t = \pm \dfrac{\pi}{4}, \pm \dfrac{3\pi}{4}, \pm \dfrac{5\pi}{4}, \pm \dfrac{7\pi}{4}$.

59. If $f(t) = \tan \left(t - \dfrac{\pi}{2} \right)$, then $b = 1$ and $c = \dfrac{\pi}{2}$. Also, period $= \dfrac{\pi}{b} = \dfrac{\pi}{1} = \pi$ and phase shift $= \dfrac{\pi}{2}$. See

    Figure 59. The vertical asymptotes occur at $t = 0, \pm\pi, \pm 2\pi$.

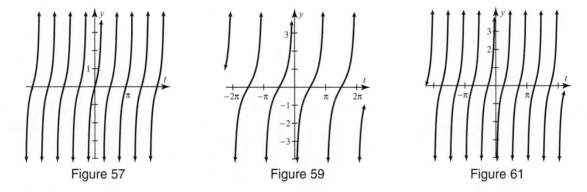

Figure 57   Figure 59   Figure 61

61. If $f(t) = -\cot 2t$, then $b = 2$ and $c = 0$. Also, period $= \dfrac{\pi}{b} = \dfrac{\pi}{2}$ and phase shift $= 0$. See Figure 61.

    The vertical asymptotes occur at $t = 0, \pm \dfrac{\pi}{2}, \pm\pi, \pm \dfrac{3\pi}{2}, \pm 2\pi$.

63. If $f(t) = \cot\left(2\left(x - \dfrac{\pi}{4}\right)\right) - 1$, then $b = 2$ and $c = \dfrac{\pi}{4}$. Also, period $= \dfrac{\pi}{2}$ and phase shift $= \dfrac{\pi}{4}$.

The graph is shown in Figure 63. The vertical asymptotes occur at $x = \pm\dfrac{\pi}{4}, \pm\dfrac{3\pi}{4}, \pm\dfrac{5\pi}{4}, \pm\dfrac{7\pi}{4}$.

65. If $f(t) = \sec\left(\dfrac{1}{2}t\right)$, then $b = \dfrac{1}{2}$ and $c = 0$. Also, period $= \dfrac{2\pi}{\frac{1}{2}} = 4\pi$ and phase shift $= 0$. See Figure 65.

The vertical asymptotes occur at $t = \pm\pi$.

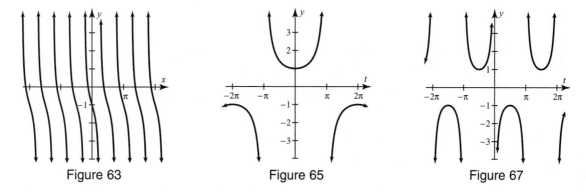

Figure 63   Figure 65   Figure 67

67. If $f(t) = \csc(t - \pi)$, then $b = 1$ and $c = \pi$. Also, period $= \dfrac{2\pi}{1} = 2\pi$ and phase shift $= \pi$. See Figure 67.

The vertical asymptotes occur at $t = 0, \pm\pi, \pm2\pi$.

69. If $f(x) = \sec\left(\dfrac{1}{3}\left(x - \dfrac{\pi}{6}\right)\right)$, then $b = \dfrac{1}{3}$ and $c = \dfrac{\pi}{6}$. Also, period $= \dfrac{2\pi}{\frac{1}{3}} = 6\pi$ and phase shift $= \dfrac{\pi}{6}$.

See Figure 69. The vertical asymptotes occur at $x = -\dfrac{4\pi}{3}, \dfrac{5\pi}{3}$.

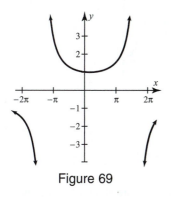

Figure 69

71. (a) The maximum monthly average temperature is found at the peak of the graph: $y = 40°$ F.

     The minimum monthly average temperature is found at the "valley" of the graph: $y = -40°$ F.

   (b) The amplitude is $0.5(40 - (-40)) = 40$. This represents a total temperature swing from $-40°$ F to $40°$ F.

     Since the average temperatures are cyclic, repeating every 12 months, the period is 12.

   (c) The x-intercepts represent the months in which the average temperature is $0°$ F (April and October).

73. (a) Graph $Y_1 = 34 \sin((\pi/6)(X - 4.3))$ in $[0, 25, 2]$ by $[-50, 50, 10]$. See Figure 73a.

    amplitude $= 34$, period $= \dfrac{2\pi}{\frac{\pi}{6}} = 12$, phase shift $= 4.3$

    (b) May corresponds to $x = 5$, $y \approx 12.2°$ F. See Figure 73b.

    December corresponds to $x = 12$, $y \approx -26.4°$ F. See Figure 73c.

    (c) Since half of the months have average temperatures above zero and half have average temperatures below

    zero, we would conjecture that the average yearly temperature is $0°$ F.

$[0, 25, 2]$ by $[-50, 50, 10]$	$[0, 25, 2]$ by $[-50, 50, 10]$	$[0, 25, 2]$ by $[-50, 50, 10]$
Figure 73a	Figure 73b	Figure 73c

75. (a) Plot the data is $[0, 25, 2]$ by $[0, 80, 10]$. See Figure 75a.

    (b) The maximum monthly average temperature is $64°$ F and the minimum is $36°$ F. The midpoint of these

    values is $0.5(64 + 36) = 50$, thus $d = 50$. Half the difference between these temperatures is

    $0.5(64 - 36) = 14$, thus $a = 14$. Since the temperatures cycle every 12 months, $b = \dfrac{2\pi}{12} = \dfrac{\pi}{6}$. The

    maximum of the $y = \sin x$ graph occurs when $x = \dfrac{\pi}{2}$ while the maximum in the table occurs when $x = 7$.

    Thus $\dfrac{\pi}{6}(7 - c) = \dfrac{\pi}{2} \Rightarrow 7 - c = 3 \Rightarrow c = 4$. The function is $f(x) = 14 \sin\left(\dfrac{\pi}{6}(x - 4)\right) + 50$.

    (c) See Figure 75c.

$[0, 25, 2]$ by $[0, 80, 10]$	$[0, 25, 2]$ by $[0, 80, 10]$	$[0, 25, 2]$ by $[40, 100, 10]$
Figure 75a	Figure 75c	Figure 77

77. Refer to the solution for exercise 75.

    (a) $a = 0.5(92 - 58) = 17$, $b = \dfrac{\pi}{6}$, $c = 7$, $d = 0.5(92 + 58) = 75$.

    Thus $f(x) = 17 \cos\left(\dfrac{\pi}{6}(x - 7)\right) + 75$. The graph of f and the actual data are shown together in

    $[0, 25, 2]$ by $[40, 100, 10]$. See Figure 77.

    (b) Yes, different values of c are possible of the form $c = 7 + 12n$ where n is an integer.

79. (a) The maximum number of daylight hours is about 18.5 hr. This occurs June 21st - summer solstice.

    (b) The minimum number of daylight hours is about 6 hr. This occurs December 21st - winter solstice.

    (c) The amplitude represents half the difference in daylight between the longest and shortest days. The period

    represents one year. *Answers may vary.*

81. (a) The maximum monthly average precipitation is about 8 inches.

    The minimum monthly average precipitation is about 0.5 inches.

    (b) The amplitude is $0.5(8 - 0.5) = 3.75$. The amplitude represents half of the difference between the maximum and minimum average monthly precipitation.

    (c) Since $a = 0.5(8 - 0.5) = 3.75$, $d = 0.5(8 + 0.5) = 4.25$. The maximum occurs at $x = 0$. Let $c = 0$.

    Since daylight hours cycle every 12 months, let $b = \dfrac{2\pi}{12} = \dfrac{\pi}{6}$. Thus, $f(x) = 3.75 \cos\left(\dfrac{\pi}{6}x\right) + 4.25$.

83. Refer to the solution for exercise 75.

    (a) $a = 0.5(6 - 2) = 2$, $b = \dfrac{\pi}{6}$, $c = 1$, $d = 0.5(6 + 2) = 4$. Thus $f(t) = 2 \cos\left(\dfrac{\pi}{6}(t - 1)\right) + 4$.

    (b) $f(1) = 2 \cos\left(\dfrac{\pi}{6}(1 - 1)\right) + 4 = 2 \cos(0) + 4 = 2 + 4 = 6$.

    $f(7) = 2 \cos\left(\dfrac{\pi}{6}(7 - 1)\right) + 4 = 2 \cos(\pi) + 4 = -2 + 4 = 2$.

85. Refer to the solution in exercise 75.

    (a) Plot the data is $[0, 25, 2]$ by $[60, 90, 5]$. See Figure 85.

    (b) $a = 0.5(85 - 72) = 6.5$, $b = \dfrac{\pi}{6}$, $c = 4$, $d = 0.5(85 + 72) = 78.5$.

    Thus $y = 6.5 \sin\left(\dfrac{\pi}{6}(x - 4)\right) + 78.5$.

87. Graph $Y_1 = 20 + 15 \sin(\pi X/12)$ in $[0, 12, 3]$ by $[0, 50, 10]$. See Figure 87.

    At 9:00 A.M. the outdoor temperature is $20°$ F. The temperature increases to a maximum of $35°$ F at 3:00 P.M. Then the temperature begins to fall until it reaches $20°$ F again at 9:00 P.M.

$[0, 25, 2]$ by $[60, 90, 5]$ $[0, 12, 3]$ by $[0, 50, 10]$ $[0, 1/110, 1/880]$ by $[-1.5, 1.5, 0.5]$

Figure 85   Figure 87   Figure 89

89. (a) Graph $Y_1 = \sin(880\pi X)$ in $[0, 1/110, 1/880]$ by $[-1.5, 1.5, 0.5]$. See Figure 89.

    (b) The period can be found by finding the change in the x-coordinate from one peak to the next.

    $x_1 \approx 0.00511371$, $x_2 \approx 0.00738592 \Rightarrow$

    $x_2 - x_1 = 0.00738592 - 0.00511371 \approx 0.00227$ sec. $\left(\text{actual: } \dfrac{1}{440} \approx 0.00227 \text{ sec.}\right)$

    (c) $F = \dfrac{1}{0.00227} \approx 440$ cycles per second $\left(\text{actual: } F = 440\right)$

91. (a) Since $s(0) = 2$ inches, $a = 2$. Since the period is 0.5 seconds, the frequency is $F = \dfrac{1}{0.5} = 2$.

Thus $s(t) = a \cos(2\pi F t) = 2 \cos(4\pi t)$.

(b) $s(1) = 2 \cos(4\pi) = 2$. The spring is compressed 2 inches one second after the weight is released. The weight is moving neither upward nor downward.

93. (a) Since $s(0) = -3$ inches, $a = -3$. Since the period is 0.8 seconds, the frequency is $F = \dfrac{1}{0.8} = \dfrac{5}{4}$.

Thus $s(t) = a \cos(2\pi F t) = -3 \cos(2.5\pi t)$.

(b) $s(1) = -3 \cos(2.5\pi) = -3(0) = 0$. The spring is at its natural length one second after the weight is released. The weight is moving upward.

95. First note that $b = 2\pi F = 2\pi(27.5) = 55\pi$. Thus $s(t) = a \cos(55\pi t)$.

Hence $s(0) = a \cos(0) = 0.21 \Rightarrow a = 0.21$. The equation is $s(t) = 0.21 \cos(55\pi t)$. See Figure 95.

97. First note that $b = 2\pi F = 2\pi(55) = 110\pi$. Thus $s(t) = a \cos(110\pi t)$.

Hence $s(0) = a \cos(0) = 0.14 \Rightarrow a = 0.14$. The equation is $s(t) = 0.14 \cos(110\pi t)$. See Figure 97.

$[0, 0.05, 0.01]$ by $[-0.3, 0.3, 0.1]$ $[0, 0.05, 0.01]$ by $[-0.3, 0.3, 0.1]$ $[0, 25, 2]$ by $[30, 80, 10]$

Figure 95   Figure 97   Figure 99a   Figure 99b

99. $y \approx 13.2 \sin(0.524x - 2.18) + 49.7$. See Figure 99a.

Graph $Y_1 = 13.2 \sin(0.524X - 2.18) + 49.7$ together with the data in $[0, 25, 2]$ by $[30, 80, 10]$.

See Figure 99b.

101. $y \approx 16.9 \sin(0.522x - 2.09) + 75.4$. See Figure 101a.

Graph $Y_1 = 16.9 \sin(0.522X - 2.09) + 75.4$ together with the data in $[0, 25, 2]$ by $[40, 100, 10]$.

See Figure 101b.

$[0, 25, 2]$ by $[40, 100, 10]$

Figure 101a   Figure 101b

6.6: Inverse Trigonometric Functions

1. one-to-one

3. π

5. 0

7. $\dfrac{\pi}{2}$

9. $-\dfrac{\pi}{4}$

11. $\dfrac{2\pi}{3}$

13. (a) The angle in $\left[-\dfrac{\pi}{2}, \dfrac{\pi}{2}\right]$ whose sine is 1 is $\dfrac{\pi}{2}$. Thus $\sin^{-1} 1 = \dfrac{\pi}{2}$ or $90°$.

   (b) No angle in $\left[-\dfrac{\pi}{2}, \dfrac{\pi}{2}\right]$ has a sine of 4. Thus, arcsin 4 is undefined.

   (c) The angle in $\left[-\dfrac{\pi}{2}, \dfrac{\pi}{2}\right]$ whose sine is $-\dfrac{\sqrt{3}}{2}$ is $-\dfrac{\pi}{3}$. Thus $\arcsin\left(-\dfrac{\sqrt{3}}{2}\right) = -\dfrac{\pi}{3}$ or $-60°$.

15. (a) The angle in $[0, \pi]$ whose cosine is 0 is $\dfrac{\pi}{2}$. Thus $\cos^{-1} 0 = \dfrac{\pi}{2}$ or $90°$.

   (b) The angle in $[0, \pi]$ whose cosine is -1 is π. Thus $\arccos(-1) = \pi$ or $180°$.

   (c) The angle in $[0, \pi]$ whose cosine is $\dfrac{1}{2}$ is $\dfrac{\pi}{3}$. Thus $\cos^{-1}\left(\dfrac{1}{2}\right) = \dfrac{\pi}{3}$ or $60°$.

17. (a) The angle in $\left(-\dfrac{\pi}{2}, \dfrac{\pi}{2}\right)$ whose tangent is 1 is $\dfrac{\pi}{4}$. Thus $\tan^{-1} 1 = \dfrac{\pi}{4}$ or $45°$.

   (b) The angle in $\left(-\dfrac{\pi}{2}, \dfrac{\pi}{2}\right)$ whose tangent is -1 is $-\dfrac{\pi}{4}$. Thus $\arctan(-1) = -\dfrac{\pi}{4}$ or $-45°$.

   (c) The angle in $\left(-\dfrac{\pi}{2}, \dfrac{\pi}{2}\right)$ whose tangent is $\sqrt{3}$ is $\dfrac{\pi}{3}$. Thus $\tan^{-1}\sqrt{3} = \dfrac{\pi}{3}$ or $60°$.

19. (a) $\sin^{-1} 1.5$ is undefined (Error). See Figure 19a.

   (b) $\tan^{-1} 10 \approx 1.47$ radians or $84.3°$. See Figure 19b.

   (c) $\arccos(-0.25) \approx 1.82$ radians or $104.5°$. See Figure 19c.

21. (a) Since $\tan\beta = \dfrac{4}{3}$, $\tan^{-1}\dfrac{4}{3} = \beta$.

   (b) Since $\sin\alpha = \dfrac{3}{5}$, $\sin^{-1}\dfrac{3}{5} = \alpha$.

   (c) Since $\cos\beta = \dfrac{3}{5}$, $\arccos\dfrac{3}{5} = \beta$.

23. $\sin(\sin^{-1} 1) = \sin\left(\dfrac{\pi}{2}\right) = 1$

25. $\cos^{-1}\left(\cos\dfrac{5\pi}{4}\right) = \cos^{-1}\left(-\dfrac{1}{\sqrt{2}}\right) = \dfrac{3\pi}{4}$

27. $\tan(\tan^{-1}(-3)) = -3$ since $-3 \in (-\infty, \infty)$

29. (a) $\sin^{-1}\dfrac{3}{5} + \cos^{-1}\dfrac{3}{5} = 90°$

   (b) $\sin^{-1}\dfrac{1}{3} + \cos^{-1}\dfrac{1}{3} = 90°$

   (c) $\sin^{-1}\dfrac{2}{7} + \cos^{-1}\dfrac{2}{7} = 90°$

   Conjecture: $\sin^{-1} x + \cos^{-1} x = 90°$ whenever $-1 \le x \le 1$.

31. Verbal: Determine the angle (or real number) θ such that $\sin\theta = 2x$ and $-\dfrac{\pi}{2} \le \theta \le \dfrac{\pi}{2}$.

Numerical: See Figure 31a.

Graphical: See Figure 31b.

	[-1, 1, 1] by [-2, 2, 1]		[-3, 3, 1] by [-1, 4, 1]
Figure 31a	Figure 31b	Figure 33a	Figure 33b

33. Verbal: Determine the angle (or real number) θ such that $\cos\theta = \dfrac{1}{2}x$ and $0 \le \theta \le \pi$.

Numerical: See Figure 33a.

Graphical: See Figure 33b.

35. Since $\tan\alpha = \dfrac{7}{24}$, $\alpha = \tan^{-1}\dfrac{7}{24} \approx 16.3°$. Since $\tan\beta = \dfrac{24}{7}$, $\beta = \tan^{-1}\dfrac{24}{7} \approx 73.7°$.

By the Pythagorean theorem: $c^2 = 7^2 + 24^2 \Rightarrow c^2 = 625 \Rightarrow c = 25$.

37. Since $\sin\alpha = \dfrac{6}{10}$, $\alpha = \sin^{-1}\dfrac{6}{10} \approx 36.9°$. Since $\cos\beta = \dfrac{6}{10}$, $\beta = \cos^{-1}\dfrac{6}{10} \approx 53.1°$.

By the Pythagorean theorem: $b^2 = 10^2 - 6^2 \Rightarrow b^2 = 64 \Rightarrow b = 8$.

39. β and $55°$ are complimentary, so $\beta = 90° - 55° = 35°$

Since $\tan 55° = \dfrac{5}{b}$, $b = \dfrac{5}{\tan 55°} \approx 3.5$. Since $\sin 55° = \dfrac{5}{c}$, $c = \dfrac{5}{\sin 55°} \approx 6.1$.

41. See Figure 41. By the Pythagorean theorem: $b^2 = 1^2 - x^2 \Rightarrow b = \sqrt{1 - x^2}$. Thus $\tan\theta = \dfrac{x}{\sqrt{1 - x^2}}$.

43. See Figure 43. By the Pythagorean theorem:

$$b^2 = (\sqrt{1 + x^2})^2 - x^2 \Rightarrow b^2 = 1 + x^2 - x^2 \Rightarrow b^2 = 1 \Rightarrow b = 1. \text{ Thus } \cos\theta = \dfrac{1}{\sqrt{1 + x^2}}.$$

Figure 41	Figure 43	Figure 45	Figure 47

45. Let θ be the angle whose cosine is u. That is $\theta = \cos^{-1}u$. See Figure 45. We find the unknown side x by using

the Pythagorean Theorem: $u^2 + x^2 = 1^2 \Rightarrow x^2 = 1 - u^2 \Rightarrow x = \sqrt{1 - u^2}$.

Therefore $\sin(\cos^{-1}u) = \sin\theta = \dfrac{\sqrt{1 - u^2}}{1} = \sqrt{1 - u^2}$.

47. Let θ be the angle whose cosine is u. That is $\theta = \cos^{-1}u$. See Figure 47. We find the unknown side x by using

the Pythagorean Theorem: $u^2 + x^2 = 1^2 \Rightarrow x^2 = 1 - u^2 \Rightarrow x = \sqrt{1 - u^2}$.

Therefore $\tan(\cos^{-1}u) = \tan\theta = \dfrac{\sqrt{1 - u^2}}{u}$.

49. Let θ be the angle whose tangent is $\dfrac{1}{u}$. That is $\theta = \tan^{-1}\dfrac{1}{u}$. See Figure 49. We find the unknown side x by

using the Pythagorean Theorem: $u^2 + 1^2 = x^2 \Rightarrow \sqrt{u^2 + 1^2} = x \Rightarrow x = \sqrt{u^2 + 1}$. Note that the

unknown side is not needed for the requested expression. Therefore $\cot\left(\tan^{-1}\dfrac{1}{u}\right) = \cot\theta = \dfrac{1}{\tan\theta} = \dfrac{1}{\frac{1}{u}} = u.$

51. $\sin\theta = 1 \Rightarrow \theta = \sin^{-1}1 = 90°$

53. $\tan\theta = 1 \Rightarrow \theta = \tan^{-1}1 = 45°$

55. $\cos\theta = 0 \Rightarrow \theta = \cos^{-1}0 = 90°$

57. $2\cos\theta = \dfrac{1}{4} \Rightarrow \cos\theta = \dfrac{1}{8} \Rightarrow \theta = \cos^{-1}\dfrac{1}{8} \approx 82.8°$

59. $\tan\theta - 1 = 5 \Rightarrow \tan\theta = 6 \Rightarrow \theta = \tan^{-1}6 \approx 80.5°$

61. $\sin^2\theta = 0.87 \Rightarrow \sin\theta \approx 0.9327 \Rightarrow \theta \approx \sin^{-1}0.9327 \approx 68.9°$

63. $\tan t = -\dfrac{1}{5} \Rightarrow t = \tan^{-1}\left(-\dfrac{1}{5}\right) \approx -0.197.$ See Figure 63.

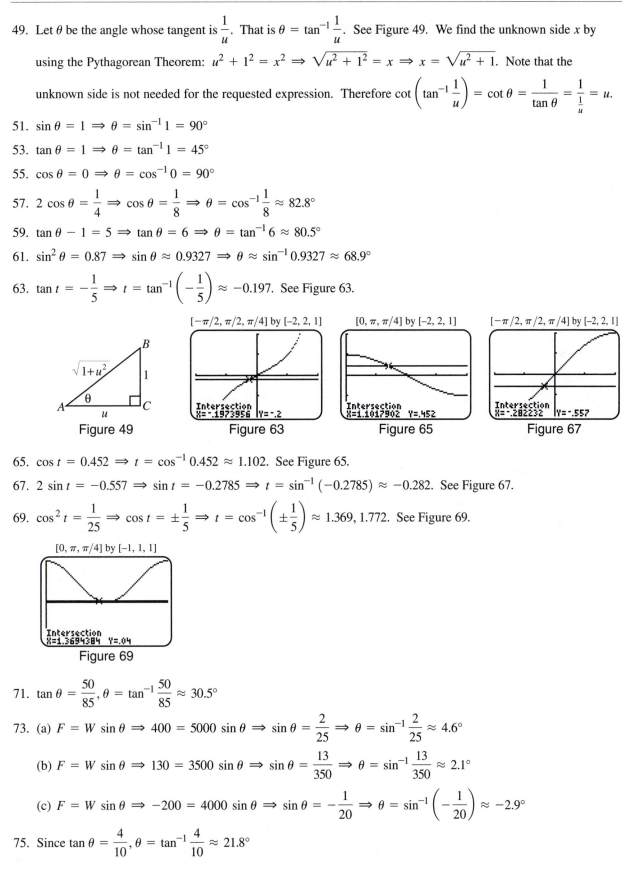

Figure 49   Figure 63   Figure 65   Figure 67

65. $\cos t = 0.452 \Rightarrow t = \cos^{-1}0.452 \approx 1.102.$ See Figure 65.

67. $2\sin t = -0.557 \Rightarrow \sin t = -0.2785 \Rightarrow t = \sin^{-1}(-0.2785) \approx -0.282.$ See Figure 67.

69. $\cos^2 t = \dfrac{1}{25} \Rightarrow \cos t = \pm\dfrac{1}{5} \Rightarrow t = \cos^{-1}\left(\pm\dfrac{1}{5}\right) \approx 1.369, 1.772.$ See Figure 69.

$[0, \pi, \pi/4]$ by $[-1, 1, 1]$

Figure 69

71. $\tan\theta = \dfrac{50}{85}, \theta = \tan^{-1}\dfrac{50}{85} \approx 30.5°$

73. (a) $F = W\sin\theta \Rightarrow 400 = 5000\sin\theta \Rightarrow \sin\theta = \dfrac{2}{25} \Rightarrow \theta = \sin^{-1}\dfrac{2}{25} \approx 4.6°$

(b) $F = W\sin\theta \Rightarrow 130 = 3500\sin\theta \Rightarrow \sin\theta = \dfrac{13}{350} \Rightarrow \theta = \sin^{-1}\dfrac{13}{350} \approx 2.1°$

(c) $F = W\sin\theta \Rightarrow -200 = 4000\sin\theta \Rightarrow \sin\theta = -\dfrac{1}{20} \Rightarrow \theta = \sin^{-1}\left(-\dfrac{1}{20}\right) \approx -2.9°$

75. Since $\tan\theta = \dfrac{4}{10}, \theta = \tan^{-1}\dfrac{4}{10} \approx 21.8°$

77. (a) $F = \dfrac{1}{2}(1 - \cos\theta) \Rightarrow \dfrac{1}{4} = \dfrac{1}{2}(1 - \cos\theta) \Rightarrow \dfrac{1}{2} = 1 - \cos\theta \Rightarrow \cos\theta = \dfrac{1}{2} \Rightarrow \theta = \cos^{-1}\dfrac{1}{2} = 60°$

(b) $F = \dfrac{1}{2}(1 - \cos\theta) \Rightarrow \dfrac{3}{4} = \dfrac{1}{2}(1 - \cos\theta) \Rightarrow \dfrac{3}{2} = 1 - \cos\theta \Rightarrow \cos\theta = -\dfrac{1}{2} \Rightarrow$

$\theta = \cos^{-1}\left(-\dfrac{1}{2}\right) = 120°$

79. $\theta = \sin^{-1}\left(\sqrt{\dfrac{43^2}{2(43^2) + 64.4(7)}}\right) \approx 41.9°$

81. If $\cos(0.1309H) = -0.4336 \tan L$, then $0.1309H = \cos^{-1}(-0.4336 \tan L) \Rightarrow H = \dfrac{\cos^{-1}(-0.4336 \tan L)}{0.1309}$.

(a) Convert $40°55'$ to radians: $\left(40 + \dfrac{55}{60}\right) \cdot \dfrac{\pi}{180} \approx 0.714$. Then $H \approx \dfrac{\cos^{-1}(-0.4336 \tan 0.714)}{0.1309} \approx 14.9$ hr.

(b) Convert $27°46'$ to radians: $\left(27 + \dfrac{46}{60}\right) \cdot \dfrac{\pi}{180} \approx 0.485$. Then $H \approx \dfrac{\cos^{-1}(-0.4336 \tan 0.485)}{0.1309} \approx 13.8$ hr.

(c) Convert $37°30'$ to radians: $\left(37 + \dfrac{30}{60}\right) \cdot \dfrac{\pi}{180} \approx 0.654$. Then $H \approx \dfrac{\cos^{-1}(-0.4336 \tan 0.654)}{0.1309} \approx 14.6$ hr.

83. After 5 seconds, $\theta_1 \approx 35.8°$ and so $\theta_2 \approx 90° - 35.8° \approx 54.2°$. See Figure 83.

[0, 15, 5] by [0, 100, 10]

Figure 83

85. Let α and β represent the angles of elevation from the shrub to the shorter and taller buildings respectively.

The distance from the shrub to the shorter building is $100 - x$, thus $\alpha = \arctan\left(\dfrac{75}{100 - x}\right)$. Similarly

$\beta = \arctan\left(\dfrac{150}{x}\right)$. Because the angles α, θ, and β form a straight angle, $\theta = \pi - \alpha - \beta$. That is

$\theta = \pi - \arctan\left(\dfrac{75}{100 - x}\right) - \arctan\left(\dfrac{150}{x}\right)$.

Chapter 6 Review Exercises

1. (a) See Figure 1a.

(b) See Figure 1b.

Figure 1a

Figure 1b

3. (a) $\dfrac{\pi}{3} \cdot \dfrac{180°}{\pi} = \dfrac{180°}{3} = 60°$

   (b) $\dfrac{\pi}{36} \cdot \dfrac{180°}{\pi} = \dfrac{180°}{36} = 5°$

   (c) $-\dfrac{5\pi}{6} \cdot \dfrac{180°}{\pi} = \dfrac{-5(180°)}{6} = -150°$

   (d) $-\dfrac{7\pi}{4} \cdot \dfrac{180°}{\pi} = \dfrac{-7(180°)}{4} = -315°$

5. Note that $60° = \dfrac{\pi}{3}$ radians, thus $s = \theta r = \dfrac{\pi}{3}(6) = 2\pi$ ft.

7. $\sin 30° = \dfrac{\text{opposite}}{\text{hypotenuse}} = \dfrac{1}{2}$

9. $\cot 60° = \dfrac{\text{adjacent}}{\text{opposite}} = \dfrac{1}{\sqrt{3}}$

11. $\sec \dfrac{\pi}{4} = \dfrac{\text{hypotenuse}}{\text{adjacent}} = \dfrac{\sqrt{2}}{1} = \sqrt{2}$

13. $c = \sqrt{8^2 + 9^2} = \sqrt{145}$

   $\sin \theta = \dfrac{8}{\sqrt{145}}$ $\cos \theta = \dfrac{9}{\sqrt{45}}$ $\tan \theta = \dfrac{8}{9}$ $\csc \theta = \dfrac{\sqrt{145}}{8}$ $\sec \theta = \dfrac{\sqrt{145}}{9}$ $\cot \theta = \dfrac{9}{8}$

15. $\csc \theta = \dfrac{1}{\sin \theta} = \dfrac{1}{\frac{1}{3}} = 3$

17. $\sin 25° \approx 0.423$ $\cos 25° \approx 0.906$ $\tan 25° \approx 0.466$

   $\csc 25° \approx 2.366$ $\sec 25° \approx 1.103$ $\cot 25° \approx 2.145$

19. $r = \sqrt{1^2 + (-2)^2} = \sqrt{5}$

   $\sin \theta = -\dfrac{2}{\sqrt{5}}$ $\cos \theta = \dfrac{1}{\sqrt{5}}$ $\tan \theta = -\dfrac{2}{1} = -2$ $\csc \theta = -\dfrac{\sqrt{5}}{2}$ $\sec \theta = \dfrac{\sqrt{5}}{1} = \sqrt{5}$ $\cot \theta = -\dfrac{1}{2}$

21. $\sin \theta = \dfrac{\sqrt{3}}{2}$ $\cos \theta = -\dfrac{1}{2}$ $\tan \theta = -\sqrt{3}$ $\csc \theta = \dfrac{2}{\sqrt{3}}$ $\sec \theta = -2$ $\cot \theta = -\dfrac{1}{\sqrt{3}}$

23. An angle $\theta = -\dfrac{\pi}{2}$ radians in standard position has terminal side that intersects the unit

   circle at the point $(0, -1)$. Therefore $\sin\left(-\dfrac{\pi}{2}\right) = -1$.

25. An angle $\theta = -3\pi$ radians in standard position has terminal side that intersects the unit

   circle at the point $(-1, 0)$. Therefore $\tan(-3\pi) = 0$.

27. $\sin \theta = -\dfrac{4}{5}$ $\cos \theta = \dfrac{3}{5}$ $\tan \theta = \dfrac{-\frac{4}{5}}{\frac{3}{5}} = -\dfrac{4}{3}$ $\csc \theta = -\dfrac{5}{4}$ $\sec \theta = \dfrac{5}{3}$ $\cot \theta = -\dfrac{3}{4}$

29. See Figure 29. Amplitude = 3; Period = $\dfrac{2\pi}{2} = \pi$; Phase shift = 0

31. See Figure 31. Amplitude = $|-3| = 3$; Period = $\dfrac{2\pi}{\frac{1}{2}} = 4\pi$; Phase shift = π

Figure 29

Figure 31

33. Amplitude = 2 $\Rightarrow$ a = 2   Period = $\dfrac{2\pi}{3} \Rightarrow \dfrac{2\pi}{3} = \dfrac{2\pi}{b} \Rightarrow b = 3$

35. See Figure 35. Period = $\dfrac{\pi}{b} = \dfrac{\pi}{2}$; Phase shift = 0

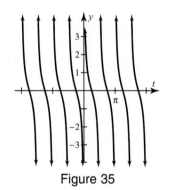

Figure 35

37. (a) $\sin^{-1}(-1) = -\dfrac{\pi}{2}$ or $-90°$

   (b) $\arccos \dfrac{1}{2} = \dfrac{\pi}{3}$ or $60°$

   (c) $\tan^{-1} 1 = \dfrac{\pi}{4}$ or $45°$

39. (a) $\sin^{-1}(-0.6) \approx -0.64$ radians or $-36.9°$

   (b) $\tan^{-1} 5 \approx 1.37$ radians or $78.7°$

   (c) $\arccos(0.12) \approx 1.45$ radians or $83.1°$

41. Since $\tan \alpha = \dfrac{5}{3}$, $\alpha = \tan^{-1} \dfrac{5}{3} \approx 59.0°$. And since $\tan \beta = \dfrac{3}{5}$, $\beta = \tan^{-1} \dfrac{3}{5} \approx 31.0°$.

   By the Pythagorean theorem: $c^2 = 5^2 + 3^2 \Rightarrow c^2 = 34 \Rightarrow c = \sqrt{34}$.

43. $\tan \theta = \dfrac{1}{\sqrt{3}} \Rightarrow \theta = \tan^{-1} \dfrac{1}{\sqrt{3}} \approx 30°$

45. $\cos \theta = \dfrac{1}{5} \Rightarrow \theta = \cos^{-1} \dfrac{1}{5} \approx 78.5°$

47. $\tan t = -\dfrac{3}{4} \Rightarrow t = \tan^{-1}\left(-\dfrac{3}{4}\right) \approx -0.6435$

Graph $Y_1 = \tan(X)$ together with $Y_2 = -0.75$ in $[-\pi/2, \pi/2, \pi/4]$ by $[-4, 4, 1]$. See Figure 47.

$[-\pi/2, \pi/2, \pi/4]$ by $[-4, 4, 1]$

Figure 47

49. (a) Angular velocity $= \dfrac{\text{angle of rotation}}{\text{time}} = \dfrac{2\pi}{50 \text{ sec}} = \dfrac{\pi}{25}$ radians/sec.

(b) Linear speed $= (\text{radius})(\text{angular velocity}) = (25)\left(\dfrac{\pi}{25}\right) = \pi$ ft/sec.

51. Let h represent the height of the tree. Then $\tan 48° = \dfrac{h}{80} \Rightarrow h = 80 \tan 48° \approx 89$ ft.

53. (a) $350 = 6000 \sin\theta \Rightarrow \sin\theta = \dfrac{7}{120} \Rightarrow \theta = \sin^{-1}\dfrac{7}{120} \approx 3.3°$

(b) $160 = 4500 \sin\theta \Rightarrow \sin\theta = \dfrac{8}{225} \Rightarrow \theta = \sin^{-1}\dfrac{8}{225} \approx 2.0°$

55. The two cities are $41°09' - 38°49' = 2°20'$ apart. Convert $2°20'$ to radians: $\left(2 + \dfrac{20}{60}\right)° \cdot \dfrac{\pi}{180°} \approx 0.0407$.

Using the arc length formula: $s \approx 0.0407(3955) \approx 161$ mi.

57. (a) Plot the data in $[0, 25, 2]$ by $[0, 70, 10]$. See Figure 57a.

(b) The maximum monthly average temperature is 58°F and the minimum is 6°F. The midpoint of these values is $0.5(58 + 6) = 32$, thus $d = 32$. Half the difference between these temperatures is $0.5(58 - 6) = 26$, thus $a = 26$. Since the temperature cycles every 12 months, $b = \dfrac{2\pi}{12} = \dfrac{\pi}{6}$. The maximum of the $y = \cos x$ graph occurs when $x = 0$, while the maximum in the table occurs when $x = 7$. Thus $c = 7$.

The function is $f(x) = 26 \cos\left(\dfrac{\pi}{6}(x - 7)\right) + 32$.

(c) Graph $Y_1 = 26 \cos\left((\pi/6)(X - 7)\right) + 32$ together with the data in $[0, 25, 2]$ by $[0, 70, 10]$.

See Figure 57c.

$[0, 25, 2]$ by $[0, 70, 10]$ $[0, 25, 2]$ by $[0, 70, 10]$

Figure 57a   Figure 57c

Chapters 1-6 Cumulative Review Exercises

1. Move the decimal point 5 place to the left, $125{,}000 = 1.25 \times 10^5$.

   Move the decimal point 3 places to the left, $4.67 \times 10^{-3} = 0.00467$

3. (a) See Figure 3a.

   (b) See Figure 3b.

   (c) See Figure 3c.

   (d) See Figure 3d.

   (e) See Figure 3e.

   (f) See Figure 3f.

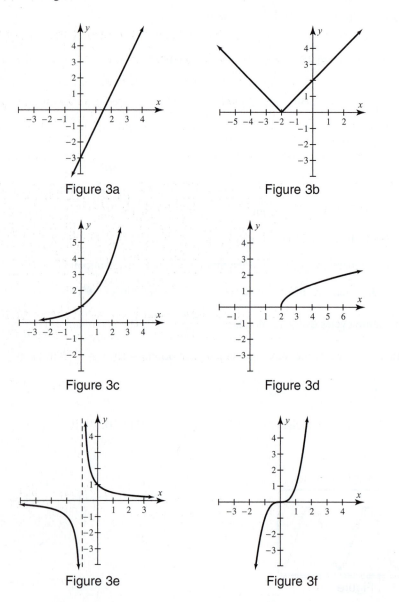

Figure 3a   Figure 3b

Figure 3c   Figure 3d

Figure 3e   Figure 3f

5. (a) Since $4x^2 - 16 \neq 0 \Rightarrow 4x^2 \neq 16 \Rightarrow x^2 \neq 4 \Rightarrow x \neq -2, 2 \Rightarrow D = \{x \mid x \neq -2, x \neq 2\}$.

   (b) $f(-1) = \dfrac{-1 - 2}{4(-1)^2 - 16} = \dfrac{-3}{-12} = \dfrac{1}{4}$

   For $a \neq 1$, $f(2a) = \dfrac{2a - 2}{4(2a)^2 - 16} = \dfrac{2a - 2}{16a^2 - 16} = \dfrac{2(a - 1)}{16(a^2 - 1)} = \dfrac{2(a - 1)}{16(a - 1)(a + 1)} = \dfrac{1}{8(a + 1)}$

7. $f(1) = 10^1 = 10 \Rightarrow (1, 10)$, $f(2) = 10^2 = 100 \Rightarrow (2, 100)$, the average rate of change is $\dfrac{100 - 10}{2 - 1} =$

   $\dfrac{90}{1} = 90$.

9. The line $2x + 3y = 6 \Rightarrow 3y = -2x + 6 \Rightarrow y = \dfrac{-2}{3}x + 2$ has a slope $\dfrac{-2}{3}$. A line parallel to this line also

   has $m = \dfrac{-2}{3}$. Using point-slope form we get: $y = \dfrac{-2}{3}(x - 2) + (-3) \Rightarrow y = \dfrac{-2}{3}x - \dfrac{5}{3}$.

11. (a) $|4 - 5x| = 8 \Rightarrow 4 - 5x = -8$ or $4 - 5x = 8$; $4 - 5x = -8 \Rightarrow -5x = -12 \Rightarrow x = \dfrac{12}{5}$ or

   $4 - 5x = 8 \Rightarrow -5x = 4 \Rightarrow x = -\dfrac{4}{5}$. Therefore $x = -\dfrac{4}{5}, \dfrac{12}{5}$.

   (b) $2e^x - 1 = 27 \Rightarrow 2e^x = 28 \Rightarrow e^x = 14 \Rightarrow \ln e^x = \ln 14 \Rightarrow x = \ln 14 \Rightarrow x \approx 2.64$

   (c) $x^3 - 3x^2 + 2x = 0 \Rightarrow x(x^2 - 3x + 2) = 0 \Rightarrow x(x - 2)(x - 1) = 0 \Rightarrow x = 0, 1, 2$

   (d) $x^4 + 8 = 6x^2 \Rightarrow x^4 - 6x^2 + 8 = 0 \Rightarrow (x^2 - 4)(x^2 - 2) = 0 \Rightarrow$

   $(x + 2)(x - 2)(x^2 - 2) = 0 \Rightarrow x = -2, 2, \pm\sqrt{2}$

   (e) $x^2 - x - 2 = 0 \Rightarrow (x - 2)(x + 1) = 0 \Rightarrow x = -1, 2$

   (f) $\sqrt{2x - 1} = x - 2 \Rightarrow 2x - 1 = x^2 - 4x + 4 \Rightarrow x^2 - 6x + 5 = 0 \Rightarrow (x - 5)(x - 1) = 0 \Rightarrow$

   $x = 1, 5$; Checking: $x = 1$ we get $\sqrt{2(1) - 1} = 1 - 2 \Rightarrow \sqrt{1} = -1$ which is false therefore $x = 5$.

   (g) $\dfrac{x}{x - 2} = \dfrac{2x - 1}{x + 1} \Rightarrow x(x + 1) = (2x - 1)(x - 2) \Rightarrow x^2 + x = 2x^2 - 5x + 2 \Rightarrow x^2 - 6x + 2 = 0$

   Using the quadratic formula: $\dfrac{6 \pm \sqrt{36 - 4(1)(2)}}{2(1)} = \dfrac{6 \pm \sqrt{28}}{2} = 3 \pm \sqrt{7}$

   (h) $\log_2(x + 1) = 16 \Rightarrow x + 1 = 2^{16} \Rightarrow x + 1 = 65{,}536 \Rightarrow x = 65{,}535$

13. (a) $-3(2-x) < 4 - (2x+1) \Rightarrow -6 + 3x < 3 - 2x \Rightarrow 5x < 9 \Rightarrow x < \dfrac{9}{5} \Rightarrow \left(-\infty, \dfrac{9}{5}\right)$

(b) $-3 \le 4 - 3x < 6 \Rightarrow -7 \le -3x < 2 \Rightarrow \dfrac{7}{3} \ge x > -\dfrac{2}{3} \Rightarrow \left(-\dfrac{2}{3}, \dfrac{7}{3}\right]$

(c) The solutions to $|4x - 3| \ge 9$ satisfy $x \le s_1$ or $x \ge s_2$ where s_1 and s_2 are the solutions to $|4x - 3| = 9$.

$|4x - 3| = 9$ is equivalent to $4x - 3 = -9 \Rightarrow x = -\dfrac{3}{2}$ and $4x - 3 = 9 \Rightarrow x = 3$.

The solution is $\left(-\infty, -\dfrac{3}{2}\right] \cup [3, \infty)$.

(d) For $x^2 - 5x + 4 \le 0$ first we solve $x^2 - 5x + 4 = 0 \Rightarrow (x - 4)(x - 1) = 0 \Rightarrow x = 1, 4$. These are

the boundary numbers and divide the number line into three sections: $(-\infty, 1], [1, 4],$ and $[4, \infty)$.

Testing a value in each section we get: $x = 0 \Rightarrow 0^2 - 5(0) + 4 \le 0 \Rightarrow 4 \le 0$, which is false;

$x = 2 \Rightarrow 2^2 - 5(2) + 4 \le 0 \Rightarrow 4 - 10 + 4 \le 0 \Rightarrow -2 \le 0$, which is true;

$x = 5 \Rightarrow 5^2 - 5(5) + 4 \le 0 \Rightarrow 25 - 25 + 4 \le 0 \Rightarrow 4 \le 0$, which is false.

Therefore the solution is; $[1, 4]$.

(e) For $t^3 - t > 0$ first solve $t^3 - t = 0 \Rightarrow t(t^2 - 1) = 0 \Rightarrow t(t + 1)(t - 1) = 0 \Rightarrow t = -1, 0, 1$. These

are the boundary numbers and divide the number line into four sections:

$(-\infty, -1), (-1, 0), (0, 1),$ and $(1, \infty)$.

Testing a value in each section we get: $t = -2 \Rightarrow -2^3 - (-2) > 0 \Rightarrow -6 > 0$, which is false;

$t = -\dfrac{1}{2} \Rightarrow \left(-\dfrac{1}{2}\right)^3 - \left(-\dfrac{1}{2}\right) > 0 \Rightarrow -\dfrac{1}{8} + \dfrac{1}{2} > 0 \Rightarrow \dfrac{3}{8} > 0$, which is true;

$t = \dfrac{1}{2} \Rightarrow \left(\dfrac{1}{2}\right)^3 - \dfrac{1}{2} > 0 \Rightarrow \dfrac{1}{8} - \dfrac{1}{2} > 0 \Rightarrow -\dfrac{3}{8} > 0$, which is false;

$t = 2 \Rightarrow 2^3 - 2 > 0 \Rightarrow 8 - 2 > 0 \Rightarrow 6 > 0$, which is true.

Therefore the solution is: $(-1, 0) \cup (1, \infty)$.

(f) One boundary number is: $t + 2 = 0 \Rightarrow t = -2$. For the other we solve: $\dfrac{1}{t + 2} - 3 = 0 \Rightarrow$

$1 - 3(t + 2) = 0 \Rightarrow 1 - 3t - 6 = 0 \Rightarrow -3t = 5 \Rightarrow t \le -\dfrac{5}{3}$. The two boundary numbers are:

-2 and $-\dfrac{5}{3} \Rightarrow$ the possible intervals are: $(-\infty, -2], \left[-2, -\dfrac{5}{3}\right],$ and $[-2, \infty)$.

Testing a value in each gives us: $t = -3 \Rightarrow \dfrac{1}{-3 + 2} - 3 = 0 \Rightarrow -4 = 0$, which is false;

$t = -\dfrac{11}{6} \Rightarrow \dfrac{1}{-\frac{11}{6} + 2} - 3 = 0 \Rightarrow 3 = 0$, which is true;

$t = 0 \Rightarrow \dfrac{1}{0 + 2} - 3 = 0 \Rightarrow -\dfrac{5}{2} = 0$, which is false. Therefore the solution is $\left(-2, -\dfrac{5}{3}\right]$.

15. Using the vertex formula: $x = -\dfrac{b}{2a}$, gives $x = -\dfrac{(-4)}{2(3)} = \dfrac{2}{3}$. Now substitute $x = \dfrac{2}{3}$ into $3x^2 - 4x + 1$;

$3\left(\dfrac{2}{3}\right)^2 - 4\left(\dfrac{2}{3}\right) + 1 = \dfrac{4}{3} - \dfrac{8}{3} + \dfrac{3}{3} = -\dfrac{1}{3}$. Therefore the vertex is: $\left(\dfrac{2}{3}, -\dfrac{1}{3}\right)$.

17. $2x^2 + 4x = 1 \Rightarrow 2(x^2 + 2x) = 1 \Rightarrow 2(x^2 + 2x + 1) = 1 + 2 \Rightarrow 2(x + 1)^2 = 3 \Rightarrow (x + 1)^2 = \dfrac{3}{2} \Rightarrow$

$x + 1 = \pm\sqrt{\dfrac{3}{2}} \Rightarrow x = -1 \pm \sqrt{\dfrac{3}{2}} \Rightarrow x = -1 \pm \dfrac{\sqrt{6}}{2} \Rightarrow \dfrac{-2 \pm \sqrt{6}}{2}$

19. (a) Increasing: $[-2, 0]$ and $[2, \infty)$; decreasing: $(-\infty, -2]$ and $[0, 2]$.

(b) The graph crosses or touches the x-axis at $x \approx -2.8, 0, 2.8$.

(c) The turning point coordinates: $(-2, -4)$, $(0, 0)$, and $(2, -4)$.

(d) Local minimum: -4; local maximum: 0.

21. (a) $\dfrac{5a^4 - 2a^2 + 4}{2a^2} = \dfrac{5a^4}{2a^2} - \dfrac{2a^2}{2a^2} + \dfrac{4}{2a^2} = \dfrac{5a^2}{2} - 1 + \dfrac{2}{a^2}$

(b)
$$
\begin{array}{r}
x - 3 + \frac{4}{x^2 + 1} \\[2pt]
x^2 + 1 \overline{)\,x^3 - 3x^2 + x + 1} \\
\underline{x^3 \qquad\quad + x} \\
-3x^2 \qquad + 1 \\
\underline{-3x^2 \qquad\quad - 3} \\
4
\end{array}
$$

The quotient is: $x - 3 + \dfrac{4}{x^2 + 1}$.

23. If $2i$ is a solution then $-2i$ is a solution $\Rightarrow (x - 2i)(x + 2i) = x^2 - 4i^2 = x^2 + 4$ divides into

$x^3 - x^2 + 4x - 4$.

$$
\begin{array}{r}
x - 1 \\[2pt]
x^2 + 4 \overline{)\,x^3 - x^2 + 4x - 4} \\
\underline{x^3 \qquad\; + 4x} \\
-x^2 \qquad - 4 \\
\underline{-x^2 \qquad - 4} \\
0
\end{array}
$$

$(x - 1)$ is a factor and therefore the zeros are: $\pm 2i, 1$.

25. $3x - 7 \neq 0 \Rightarrow 3x \neq 7 \Rightarrow x \neq \dfrac{7}{3}$. Then $D = \left\{ x \mid x \neq \dfrac{7}{3} \right\}$; the vertical asymptote is the denominator $=$

$0 \Rightarrow x = \dfrac{7}{3}$; and the horizontal asymptote is the ratio of the lead coefficients $\Rightarrow y = \dfrac{2}{3}$.

27. (a) $(f + g)(2) = f(2) + g(2) = 2 + 3 = 5$

(b) $(g/f)(4) = g(4)/f(4) = 1/0 \Rightarrow$ undefined

(c) $(f \circ g)(3) = f(g(3)) = f(2) = 2$

(d) $(f^{-1} \circ g)(1) = f^{-1}(g(1)) = f^{-1}(4) = 0$

29. (a) $f(2) = 2^2 + 3(2) - 2 = 4 + 6 - 2 = 8$; $g(2) = 2 - 2 = 0$; $(f + g)(2) = 8 + 0 = 8$

(b) $(g \circ f)(1) = g(f(1)) \Rightarrow f(1) = 1^2 + 3(1) - 2 = 1 + 3 - 2 = 2$. Then $g(2) = 2 - 2 = 0 \Rightarrow$

$(g \circ f)(1) = 0$

(c) $f(x) = x^2 + 3x - 2$ and $g(x) = x - 2 \Rightarrow (f - g)(x) = x^2 + 3x - 2 - (x - 2) \Rightarrow$

$(f - g)(x) = x^2 + 2x$

(d) $(f \circ g)(x) = f(g(x)) \Rightarrow (f \circ g)(x) = (x - 2)^2 + 3(x - 2) - 2 \Rightarrow$

$(f \circ g)(x) = x^2 - 4x + 4 + 3x - 6 - 2 \Rightarrow (f \circ g)(x) = x^2 - x - 4$

31. Graph: $Y_1 = 3X - 2$ and $Y_2 = X$ and reflect $Y_1 = 3X - 2$ across $Y_2 = X$ to graph $y = f^{-1}(x)$. See Figure 31.

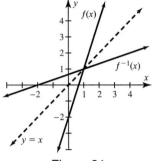

Figure 31

33. From the graph when $x = 0$, $y = \dfrac{1}{2}$, $\Rightarrow C = \dfrac{1}{2}$ and when $x = 1$, $y = 1 \Rightarrow$ using $y = Ca^x \Rightarrow 1 = \dfrac{1}{2}a^1 \Rightarrow$

    $a = 2$. So $C = \dfrac{1}{2}$ and $a = 2$.

35. (a) $\log 100 \Rightarrow 10^x = 100 \Rightarrow x = 2 \Rightarrow \log 100 = 2$

    (b) $\log_2 16 \Rightarrow 2^x = 16 \Rightarrow x = 4 \Rightarrow \log_2 16 = 4$

    (c) $\ln \dfrac{1}{e^2} \Rightarrow e^x = \dfrac{1}{e^2} \Rightarrow x = -2 \Rightarrow \ln \dfrac{1}{e^2} = -2$

    (d) $\log_4 \sqrt[3]{4} \Rightarrow 4^x = \sqrt[3]{4}$ or $4^x = 4^{1/3} \Rightarrow x = \dfrac{1}{3} \Rightarrow \log_4 \sqrt[3]{4} = \dfrac{1}{3}$

    (e) $\log 4 + \log 25 = \log 4(25) = \log 100 \Rightarrow 10^x = 100 \Rightarrow x = 2 \Rightarrow \log 4 + \log 25 = 2$

    (f) $\log_6 24 - \log_6 4 = \log_6 \dfrac{24}{4} = \log_6 6 \Rightarrow 6^x = 6 \Rightarrow x = 1 \Rightarrow \log_6 24 - \log_6 4 = 1$

37. $\log_3 125 = 4.395$

39. Complimentary angle $\alpha = 90° - 10° \, 34' = 89° \, 60' - 10° \, 34' = 79° \, 26'$

    Supplementary angle $\beta = 180° - 10° \, 34' = 179° \, 60' - 10° \, 34' = 169° \, 26'$

41. $150° \left(\dfrac{\pi}{180°} \right) = \dfrac{5\pi}{6}$

43. Convert $15°$ to radians $\Rightarrow 15° \left(\dfrac{\pi}{180°} \right) = \dfrac{15\pi}{180} = \dfrac{\pi}{12};\ s = 3 \left(\dfrac{\pi}{12} \right) = \dfrac{3\pi}{12} = \dfrac{\pi}{4} \approx 0.79$ ft.

45. $\tan 30° = \dfrac{9}{a} \Rightarrow a \tan 30° = 9 \Rightarrow a = \dfrac{9}{\tan 30°} \Rightarrow a \approx 15.6$

47. $r = \sqrt{(-3)^2 + (-3)^2} = \sqrt{18} = 3\sqrt{2}$

$$\sin \theta = \dfrac{-3}{3\sqrt{2}} = -\dfrac{1}{\sqrt{2}} \qquad \cos \theta = \dfrac{-3}{3\sqrt{2}} = -\dfrac{1}{\sqrt{2}} \qquad \tan \theta = \dfrac{-3}{-3} = 1$$

$$\csc \theta = \dfrac{3\sqrt{2}}{-3} = -\sqrt{2} \qquad \sec \theta = \dfrac{3\sqrt{2}}{-3} = -\sqrt{2} \qquad \cot \theta = \dfrac{-3}{-3} = 1$$

49. Refer to Figure 49 and use the Pythagorean theorem to find the opposite side.

$5^2 + x^2 = 13^2$, $25 + x^2 = 169$, $x^2 = 144$, $x = 12$.

$$\sin \theta = -\frac{12}{13} \qquad\qquad \cos \theta = \frac{5}{13} \qquad\qquad \tan \theta = -\frac{12}{5}$$

$$\csc \theta = -\frac{13}{12} \qquad\qquad \sec \theta = \frac{13}{5} \qquad\qquad \cot \theta = -\frac{5}{12}$$

Figure 49

51. (a) See Figure 51a.

(b) See Figure 51b.

(c) See Figure 51c.

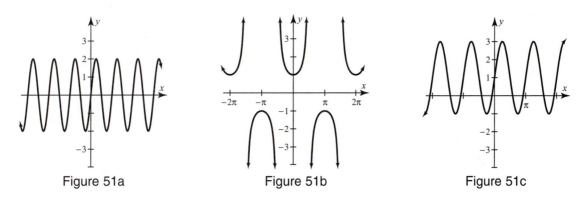

Figure 51a   Figure 51b   Figure 51c

53. $\sin \alpha = \frac{11}{20} \Rightarrow \alpha = \sin^{-1}\left(\frac{11}{20}\right) \Rightarrow \alpha \approx 33.4°$, $\cos \beta = \frac{11}{20} \Rightarrow \beta = \cos^{-1}\left(\frac{11}{20}\right) \Rightarrow \beta = 56.6$.

Using the Pythagorean theorem: $11^2 + b^2 = 20^2 \Rightarrow 121 + b^2 = 400 \Rightarrow b^2 = 279 \Rightarrow b \approx 16.7$.

55. (a) $t = 0 \Rightarrow D(t) = 16(0)^2 \Rightarrow (0, 0)$; $t = 1 \Rightarrow D(t) = 16(1)^2 \Rightarrow (1, 16)$

The average rate of change from 0 to 1 is: $\frac{16 - 0}{1 - 0} = 16$ ft./sec.

$t = 1 \Rightarrow D(t) = 16(1)^2 \Rightarrow (1, 16)$; $t = 2 \Rightarrow D(t) = 16(2)^2 \Rightarrow (2, 64)$

The average rate of change from 1 to 2 is: $\frac{64 - 16}{2 - 1} = 48$ ft./sec.

(b) During the first second, the average speed is 16 ft./sec. During the next second, the average speed is 48 ft./sec. The object is speeding up.

(c) The difference quotient is: $\frac{16(t + h)^2 - 16t^2}{h} = \frac{16(t^2 + 2th + h^2) - 16t^2}{h} =$

$\frac{16t^2 + 32th + 16h^2 - 16t^2}{h} = 32t + 16h$

57. $F = \dfrac{k}{x^2} \Rightarrow 150 = \dfrac{k}{(4000)^2} \Rightarrow k = 2,400,000,000.$ Now, $F = \dfrac{2,400,000,000}{(10,000)^2} \Rightarrow F = \dfrac{2,400,000,000}{100,000,000} \Rightarrow$

   $F = 24$ lbs.

59. (a) Use $(1980, 800)$ and $(2000, 3500)$, now find $m = \dfrac{3500 - 800}{2000 - 1980} \Rightarrow m = \dfrac{2700}{20} \Rightarrow m = 135.$

   Using $C(t) = m(t - t_1) + y_1$ we get: $C(t) = 135(t - 1980) + 800.$

   (b) $3000 = 135(t - 1980) + 800 \Rightarrow 2200 = 135(t - 1980) \Rightarrow 16.3 = t - 1980 \Rightarrow t = 1996.3$ or the year

   1996 which does agree with the actual, the formula gives the correct year.

61. (a) If $f(x) = \dfrac{5}{9}(x - 32)$, this is subtract 32 and multiply by $\dfrac{5}{9} \Rightarrow$ we divide by $\dfrac{5}{9}$, which is multiply by $\dfrac{9}{5}$,

   and add 32 $\Rightarrow f(x)^{-1} = \dfrac{9}{5}x + 32.$

   (b) Since $f(x)$ converted Fahrenheit to Celsius, $f(x)^{-1}$ converts Celsius to Fahrenheit.

63. Angular speed is 20 radians per second. The linear speed is the radius (12 inches or 1 foot) multiplied by the

   angular speed (20 radians per second). Therefore the speed is (1 ft)(20 radians/sec) = 20 ft/sec.

65. The maximum monthly average temperature is $90°$ F and the minimum is $61°$ F. The midpoint of these values

   is $0.5(90 + 61) = 75.5$, thus $d = 75.5$. Half the difference between these temperatures is

   $0.5(90 - 61) = 14.5$, thus $a = 14.5$. Since the temperatures cycle every 12 months,

   $b = \dfrac{2\pi}{12} = \dfrac{\pi}{6}$. The maximum of the $y = \sin x$ graph occurs when $x = \dfrac{\pi}{2}$, while the maximum in the table

   occurs when $x = 7$. Thus $\dfrac{\pi}{6}(7 - c) = \dfrac{\pi}{2} \Rightarrow 7 - c = 3 \Rightarrow c = 4.$

   The function $f(x) = 14.5 \sin\left(\dfrac{\pi}{6}(x - 4)\right) + 75.5.$

Chapter 7: Trigonometric Identities and Equations

7.1: Fundamental Identities

1. $\cot \theta = \dfrac{1}{\tan \theta} = \dfrac{1}{\frac{1}{2}} = 2$

3. $\sec \theta = \dfrac{1}{\cos \theta} = \dfrac{1}{\frac{2}{7}} = \dfrac{7}{2}$

5. $\cos \theta = \dfrac{1}{\sec \theta} = \dfrac{1}{-4} = -\dfrac{1}{4}$

7. $\dfrac{\tan \theta}{\cot \theta} = \dfrac{\tan \theta}{\frac{1}{\tan \theta}} = \tan^2 \theta = \dfrac{\sin^2 \theta}{\cos^2 \theta}$

9. $\dfrac{\cot^2 \theta}{\csc^2 \theta} = \dfrac{\frac{1}{\tan^2 \theta}}{\frac{1}{\sin^2 \theta}} = \dfrac{\frac{\cos^2 \theta}{\sin^2 \theta}}{\frac{1}{\sin^2 \theta}} = \cos^2 \theta$

11. $\dfrac{1}{\csc \theta} + \dfrac{1}{\sec \theta} = \sin \theta + \cos \theta$

13. $\tan \theta = \dfrac{\sin \theta}{\cos \theta} = \dfrac{\frac{3}{5}}{-\frac{4}{5}} = -\dfrac{3}{4}$

$\csc \theta = \dfrac{1}{\sin \theta} = \dfrac{1}{\frac{3}{5}} = \dfrac{5}{3}$

$\sec \theta = \dfrac{1}{\cos \theta} = \dfrac{1}{-\frac{4}{5}} = -\dfrac{5}{4}$

$\cot \theta = \dfrac{1}{\tan \theta} = \dfrac{1}{-\frac{3}{4}} = -\dfrac{4}{3}$

15. $\cos \theta = \sin \theta \cot \theta = -\dfrac{24}{25}\left(\dfrac{7}{24}\right) = -\dfrac{7}{25}$

$\tan \theta = \dfrac{1}{\cot \theta} = \dfrac{1}{\frac{7}{24}} = \dfrac{24}{7}$

$\csc \theta = \dfrac{1}{\sin \theta} = \dfrac{1}{-\frac{24}{25}} = -\dfrac{25}{24}$

$\sec \theta = \dfrac{1}{\cos \theta} = \dfrac{1}{-\frac{7}{25}} = -\dfrac{25}{7}$

17. $\tan \theta = \dfrac{\sin \theta}{\cos \theta} = \dfrac{-\frac{60}{61}}{-\frac{11}{61}} = \dfrac{60}{11}$

$\cot \theta = \dfrac{\cos \theta}{\sin \theta} = \dfrac{-\frac{11}{61}}{-\frac{60}{61}} = \dfrac{11}{60}$

$\csc \theta = \dfrac{1}{\sin \theta} = \dfrac{1}{-\frac{60}{61}} = -\dfrac{61}{60}$

$\sec \theta = \dfrac{1}{\cos \theta} = \dfrac{1}{-\frac{11}{61}} = -\dfrac{61}{11}$

19. $\sin \theta = \dfrac{1}{\csc \theta} = \dfrac{1}{\sqrt{2}}$

$\cos \theta = \dfrac{1}{\sec \theta} = -\dfrac{1}{\sqrt{2}}$

$\tan \theta = \dfrac{\sin \theta}{\cos \theta} = \dfrac{\frac{1}{\sqrt{2}}}{-\frac{1}{\sqrt{2}}} = -\dfrac{1}{1} = -1$

$\cot \theta = \dfrac{\cos \theta}{\sin \theta} = \dfrac{-\frac{1}{\sqrt{2}}}{\frac{1}{\sqrt{2}}} = -\dfrac{1}{1} = -1$

21. $\sec \theta \cos \theta = \dfrac{1}{\cos \theta} \cdot \cos \theta = 1$

23. $\sin \theta \csc \theta = \sin \theta \cdot \dfrac{1}{\sin \theta} = 1$

25. $\left(\sin^2 \theta + \cos^2 \theta\right)^3 = 1^3 = 1$

27. $1 - \sin^2 \theta = 1 - \left(1 - \cos^2 \theta\right) = \cos^2 \theta$

29. $\sec^2\theta - 1 = (1 + \tan^2\theta) - 1 = \tan^2\theta$

31. $\dfrac{\sin(-\theta)}{\cos(-\theta)} = \dfrac{-\sin\theta}{\cos\theta} = -\tan\theta$

33. $\dfrac{\sin^2\theta + \cos^2\theta}{\cos\theta} = \dfrac{1}{\cos\theta} = \sec\theta$

35. $\dfrac{\sec^2(-\theta)}{\csc^2\theta} = \dfrac{\sec^2\theta}{\csc^2\theta} = \left(\dfrac{\sec\theta}{\csc\theta}\right)^2 = \left(\dfrac{\frac{1}{\cos\theta}}{\frac{1}{\sin\theta}}\right)^2 = \left(\dfrac{1}{\cos\theta}\cdot\dfrac{\sin\theta}{1}\right)^2 = \left(\dfrac{\sin\theta}{\cos\theta}\right)^2 = \tan^2\theta$

37. $\dfrac{\cot x}{\csc x} = \dfrac{\frac{\cos x}{\sin x}}{\frac{1}{\sin x}} = \dfrac{\cos x}{\sin x}\cdot\dfrac{\sin x}{1} = \cos x$

39. $(\sin^2 x)(1 + \cot^2 x) = (\sin^2 x)(\csc^2 x) = (\sin^2 x)\left(\dfrac{1}{\sin^2 x}\right) = 1$

41. $\sec(-x) + \csc(-x) = \sec x + (-\csc x) = \sec x - \csc x$

43. Yes, since $\sec\theta\cot\theta = \left(\dfrac{1}{\cos\theta}\right)\left(\dfrac{\cos\theta}{\sin\theta}\right) = \dfrac{1}{\sin\theta} = \csc\theta$. For numerical support, table the following in
    degree mode: $Y_1 = (1/\cos(X))(1/\tan(X))$ and $Y_2 = (1/\sin(X))$ starting at $x = 0$, incrementing by 50. See
    Figure 43.

45. No, since $1 + \cot^2\theta = \csc^2\theta \Rightarrow \cot^2\theta - \csc^2\theta = -1 \neq 1$. For numerical support, table the following in
    degree mode: $Y_1 = (1/\tan(X))^2 - (1/\sin(X))^2$ and $Y_2 = 1$ starting at $x = 0$, incrementing by 50.
    See Figure 45.

| Figure 43 | Figure 45 | Figure 47 | Figure 49 |

47. If $\sin\theta < 0$ and $\cos\theta > 0$ then any point (x, y) on the terminal side of θ must satisfy $y < 0$ and $x > 0$. Thus,
    θ is contained in Quadrant IV. To support this result numerically, table the following in degree mode:
    $Y_1 = \sin(X)$ and $Y_2 = \cos(X)$ starting at $x = 270$, incrementing by 15. See Figure 47.

49. If $\sec\theta < 0$ and $\sin\theta < 0$ then any point (x, y) on the terminal side of θ must satisfy the following conditions:
    $y < 0$ and $\dfrac{r}{x} < 0 \Rightarrow x < 0$. Thus, θ is contained in quadrant III. For numerical support, table the following
    in degree mode: $Y_1 = 1/\cos(X)$ and $Y_2 = \sin(X)$ starting at $x = 180$, incrementing by 15. See Figure 49.

51. If $\cot\theta < 0$ and $\sin\theta > 0$ then any point (x, y) on the terminal side of θ must satisfy the following conditions:
    $y > 0$ and $\dfrac{x}{y} < 0 \Rightarrow x < 0$. Thus, θ is contained in quadrant II. For numerical support, table the following
    in degree mode: $Y_1 = 1/\tan(X)$ and $Y_2 = \sin(X)$ starting at $x = 90$, incrementing by 15. See Figure 51.

X	Y1	Y2
90	ERROR	1
105	-.2679	.96593
120	-.5774	.86603
135	-1	.70711
150	-1.732	.5
165	-3.732	.25882
180	ERROR	0

Y1⊟1/tan(X)

Figure 51

53. $\sin^2 \theta + \cos^2 \theta = 1 \Rightarrow \sin \theta = \pm\sqrt{1 - \cos^2 \theta}$ and since $\sin \theta < 0$ we know $\sin \theta = -\sqrt{1 - \cos^2 \theta}$.

$$\sin \theta = -\sqrt{1 - \cos^2 \theta} = -\sqrt{1 - \left(\frac{1}{2}\right)^2} = -\sqrt{1 - \frac{1}{4}} = -\sqrt{\frac{3}{4}} = -\frac{\sqrt{3}}{2} \qquad \tan \theta = \frac{\sin \theta}{\cos \theta} = \frac{-\frac{\sqrt{3}}{2}}{\frac{1}{2}} = -\sqrt{3}$$

$$\csc \theta = \frac{1}{\sin \theta} = \frac{1}{-\frac{\sqrt{3}}{2}} = -\frac{2}{\sqrt{3}} \qquad \sec \theta = \frac{1}{\cos \theta} = \frac{1}{\frac{1}{2}} = 2 \qquad \cot \theta = \frac{1}{\tan \theta} = \frac{1}{-\sqrt{3}} = -\frac{1}{\sqrt{3}}$$

55. $1 + \cot^2 \theta = \csc^2 \theta \Rightarrow \csc \theta = \pm\sqrt{1 + \cot^2 \theta}$ and since $\csc \theta < 0$, $\csc \theta = -\sqrt{1 + \cot^2 \theta}$

$$\cot \theta = \frac{1}{\tan \theta} = \frac{1}{-\frac{11}{60}} = -\frac{60}{11}$$

$$\csc \theta = -\sqrt{1 + \cot^2 \theta} = -\sqrt{1 + \left(-\frac{60}{11}\right)^2} = -\sqrt{1 + \frac{3600}{121}} = -\sqrt{\frac{3721}{121}} = -\frac{61}{11}$$

$$\sin \theta = \frac{1}{\csc \theta} = \frac{1}{-\frac{61}{11}} = -\frac{11}{61} \qquad \cos \theta = \frac{\sin \theta}{\tan \theta} = \frac{-\frac{11}{61}}{-\frac{11}{60}} = \frac{60}{61} \qquad \sec \theta = \frac{1}{\cos \theta} = \frac{1}{\frac{60}{61}} = \frac{61}{60}$$

57. $\sin^2 \theta + \cos^2 \theta = 1 \Rightarrow \cos \theta = \pm\sqrt{1 - \sin^2 \theta}$ and since $\cos \theta > 0$, $\cos \theta = \sqrt{1 - \sin^2 \theta}$

$$\cos \theta = \sqrt{1 - \sin^2 \theta} = \sqrt{1 - \left(\frac{7}{25}\right)^2} = \sqrt{1 - \frac{49}{625}} = \sqrt{\frac{576}{625}} = \frac{24}{25} \qquad \tan \theta = \frac{\sin \theta}{\cos \theta} = \frac{\frac{7}{25}}{\frac{24}{25}} = \frac{7}{24}$$

$$\csc \theta = \frac{1}{\sin \theta} = \frac{1}{\frac{7}{25}} = \frac{25}{7} \qquad \sec \theta = \frac{1}{\cos \theta} = \frac{1}{\frac{24}{25}} = \frac{25}{24} \qquad \cot \theta = \frac{1}{\tan \theta} = \frac{1}{\frac{7}{24}} = \frac{24}{7}$$

59. If $\sin \theta < 0$ and $\sec \theta < 0$, θ lies in quadrant III, $\sin^2 \theta + \cos^2 \theta = 1 \Rightarrow \left(-\frac{1}{3}\right)^2 + \cos^2 \theta = 1 \Rightarrow$

$$\cos^2 \theta = 1 - \frac{1}{9} = \frac{8}{9} \Rightarrow \cos \theta = \pm\frac{\sqrt{8}}{3} \Rightarrow \cos \theta = \pm\frac{2\sqrt{2}}{3} \Rightarrow \cos \theta = -\frac{2\sqrt{2}}{3} \text{ in quadrant III.}$$

$$\tan \theta = \frac{\sin \theta}{\cos \theta} = \frac{-\frac{1}{3}}{-\frac{2\sqrt{2}}{3}} = \frac{1}{2\sqrt{2}} \qquad \cot \theta = \frac{\cos \theta}{\sin \theta} = \frac{-\frac{2\sqrt{2}}{3}}{-\frac{1}{3}} = 2\sqrt{2} = \sqrt{8} \qquad \csc \theta = \frac{1}{\sin \theta} = \frac{1}{-\frac{1}{3}} = -3$$

$$\csc \theta = \frac{1}{\cos \theta} = \frac{1}{-\frac{2\sqrt{2}}{3}} = -\frac{3}{2\sqrt{2}} = -\frac{3}{\sqrt{8}}$$

61. $\sin^2 \theta + \cos^2 \theta = 1 \Rightarrow \sin \theta = \pm\sqrt{1 - \cos^2 \theta}$ and since θ is in quadrant IV, $\sin \theta = -\sqrt{1 - \cos^2 \theta}$.

$$\cos \theta = \frac{1}{\sec \theta} = \frac{12}{37} \qquad \sin \theta = -\sqrt{1 - \cos^2 \theta} = -\sqrt{1 - \left(\frac{12}{37}\right)^2} = -\frac{35}{37}$$

$$\tan \theta = \frac{\sin \theta}{\cos \theta} = \frac{-\frac{35}{37}}{\frac{12}{37}} = -\frac{35}{12} \qquad \csc \theta = \frac{1}{\sin \theta} = \frac{1}{-\frac{35}{37}} = -\frac{37}{35} \qquad \cot \theta = \frac{\cos \theta}{\sin \theta} = \frac{\frac{12}{37}}{-\frac{35}{37}} = -\frac{12}{35}$$

63. $\sin^2 \theta + \cos^2 \theta = 1 \Rightarrow \cos \theta = \pm\sqrt{1 - \sin^2 \theta}$ and since θ is in quadrant II, $\cos \theta = -\sqrt{1 - \sin^2 \theta}$.

$$\sin \theta = \frac{1}{\csc \theta} = \frac{1}{\frac{7}{3}} = \frac{3}{7} \qquad \cos \theta = -\sqrt{1 - \sin^2 \theta} = -\sqrt{1 - \left(\frac{3}{7}\right)^2} = -\sqrt{1 - \frac{9}{49}} = -\sqrt{\frac{40}{49}} = -\frac{\sqrt{40}}{7}$$

$$\tan \theta = \frac{\sin \theta}{\cos \theta} = \frac{\frac{3}{7}}{-\frac{\sqrt{40}}{7}} = -\frac{3}{\sqrt{40}} \qquad \sec \theta = \frac{1}{\cos \theta} = \frac{1}{-\frac{\sqrt{40}}{7}} = -\frac{7}{\sqrt{40}} \qquad \cot \theta = \frac{1}{\tan \theta} = \frac{1}{-\frac{3}{\sqrt{40}}} = -\frac{\sqrt{40}}{3}$$

65. If $\csc x > 0$, $\sin x > 0$. Then $\sin^2 x + \cos^2 x = 1 \Rightarrow \sin^2 x = 1 - \cos^2 x \Rightarrow \sin x = \sqrt{1 - \cos^2 x}$, thus

$$\tan x = \frac{\sin x}{\cos x} \Rightarrow \tan x = \frac{\sqrt{1 - \cos^2 \theta}}{\cos x}.$$

67. If $\cos x < 0$, $\sec x < 0$. Then $\sec^2 x = 1 + \tan^2 x \Rightarrow \sec x = -\sqrt{1 + \tan^2 x}$ and

$$\sin x = \frac{\tan x}{\sec x} \Rightarrow \sin x = -\frac{\tan x}{\sqrt{1 + \tan^2 x}}.$$

69. If $\cot x < 0$, $\cot^2 x = \csc^2 x - 1 \Rightarrow \cot x = -\sqrt{\csc^2 x - 1}$ and

$$\cos x = \frac{\cot x}{\csc x} \Rightarrow \cos x = -\frac{\sqrt{\csc^2 x - 1}}{\csc x}.$$

71. $\cos^2 \theta + \sin^2 \theta = 1 \Rightarrow \cos \theta = \pm\sqrt{1 - \sin^2 \theta}$ and since θ is acute, $\cos \theta = \sqrt{1 - \sin^2 \theta}$,

thus, $\cos \theta = \sqrt{1 - x^2} = \sqrt{1 - 0.5126^2} \approx 0.8586$.

73. $1 + \cot^2 \theta = \csc^2 \theta \Rightarrow \dfrac{1}{\sin^2 \theta} = 1 + \cot^2 \theta \Rightarrow \sin^2 \theta = \dfrac{1}{1 + \cot^2 \theta} \Rightarrow \sin \theta = \pm\sqrt{\dfrac{1}{1 + \cot^2 \theta}}$

Since θ is in quadrant III, $\sin \theta = -\sqrt{\dfrac{1}{1 + \cot^2 \theta}}$, $\sin \theta = -\sqrt{\dfrac{1}{1 + x^2}} = -\sqrt{\dfrac{1}{1 + 0.5126^2}} \approx -0.8899$.

75. If θ is acute, θ lies in quadrant I, $\sin^2 \theta + \cos^2 \theta = 1 \Rightarrow \cos^2 \theta = 1 - \sin^2 \theta \Rightarrow \cos \theta = \sqrt{1 - \sin^2 \theta}$,

$$\tan \theta = \frac{\sin \theta}{\sqrt{1 - \sin^2 \theta}}. \text{ Since } \sin \theta = x \Rightarrow \tan \theta = \frac{x}{\sqrt{1 - x^2}}.$$

77. Since sine is an odd function, $\sin(-13°) = -\sin 13°$.

79. Since tangent is an odd function, $\tan\left(-\dfrac{\pi}{11}\right) = -\tan\dfrac{\pi}{11}$.

81. Since secant is an even function, $\sec\left(-\dfrac{2\pi}{5}\right) = \sec\dfrac{2\pi}{5}$.

83. $d = 93{,}000{,}000 \cdot \dfrac{1}{\sin \theta} = 93{,}000{,}000 \csc \theta$

85. (a) Graph $Y_1 = 3(\cos(6000000X))^2$, $Y_2 = 3(\sin(6000000X))^2$ and $Y_3 = Y_1 + Y_2$

  in $[0, 10^{-6}, 10^{-7}]$ by $[-1, 4, 1]$. See Figure 85a. The total energy E (the sum of L and C) is always 3.

  (b) Table Y_1, Y_2 and Y_3 starting at $x = 0$, incrementing by 10^{-7}. See Figure 85b.

  (c) $E(t) = 3\cos^2(6000000t) + 3\sin^2(6000000t) = 3(\cos^2(6000000t) + \sin^2(6000000t)) =$

  $3(1) = 3$

$[0, 10^{-6}, 10^{-7}]$ by $[-1, 4, 1]$

Figure 85a   Figure 85b

87. (a) In radian mode, the graph has y-axis symmetry. The monthly high temperatures x months before and x months after July are equal. See Figure 87a.

(b) A table of Y_1 is shown in Figure 87b. f is an even function since the sign of the input does not affect the output.

(c) Symbolically this symmetry can be expressed as $f(-x) = f(x)$.

[−6, 6, 1] by [0, 100, 10]   [0, 2, 0.5] by [−1, 40, 8]

Figure 87a   Figure 87b   Figure 89

89. (a) $P = ky^2 = k(4 \cos (2\pi t))^2 = 16k \cos^2 (2\pi t)$

(b) Graph $Y_1 = 32 \, (\cos (2\pi X))$^2 in $[0, 2, 0.5]$ by $[-1, 40, 8]$. See Figure 89. P has a maximum value of 32 when $t = 0, 0.5, 1, 1.5, 2$ and P has a minimum value of 0 when $t = 0.25, 0.75, 1.25, 1.75$. The spring is either stretched or compressed the most when P is maximum.

(c) $P = 16k \cos^2 (2\pi t) = 16k (1 - \sin^2 (2\pi t))$

7.2: Verifying Identities

1. (a) $(1 + x)(1 - x) = 1 - x^2$

(b) $(1 + \sin \theta)(1 - \sin \theta) = 1 - \sin^2 \theta = \cos^2 \theta$

3. (a) $x(x - 1) = x^2 - x$

(b) $\sec \theta \, (\sec \theta - 1) = \sec^2 \theta - \sec \theta$

5. (a) $x^2 + 2x + 1 = (x + 1)(x + 1)$

(b) $\cos^2 \theta + 2 \cos \theta + 1 = (\cos \theta + 1)(\cos \theta + 1)$

7. (a) $x^2 - 2x = x(x - 2)$

(b) $\sec^2 t - 2 \sec t = \sec t \, (\sec t - 2)$

9. (a) $\dfrac{1}{1 - x} + \dfrac{1}{1 + x} = \dfrac{1 + x}{(1 - x)(1 + x)} + \dfrac{1 - x}{(1 - x)(1 + x)} = \dfrac{1 + x + 1 - x}{(1 - x)(1 + x)} = \dfrac{2}{1 - x^2}$

(b) $\dfrac{1}{1 - \cos \theta} + \dfrac{1}{1 + \cos \theta} = \dfrac{1 + \cos \theta}{(1 - \cos \theta)(1 + \cos \theta)} + \dfrac{1 - \cos \theta}{(1 - \cos \theta)(1 + \cos \theta)} =$

$\dfrac{1 + \cos \theta + 1 - \cos \theta}{(1 - \cos \theta)(1 + \cos \theta)} = \dfrac{2}{1 - \cos^2 \theta} = \dfrac{2}{\sin^2 \theta} = 2 \csc^2 \theta$

11. (a) $\dfrac{x}{y} + \dfrac{y}{x} = \dfrac{x^2}{xy} + \dfrac{y^2}{xy} = \dfrac{x^2 + y^2}{xy}$

(b) $\dfrac{\cos t}{\sin t} + \dfrac{\sin t}{\cos t} = \dfrac{\cos^2 t}{\cos t \, \sin t} + \dfrac{\sin^2 t}{\cos t \, \sin t} = \dfrac{\cos^2 t + \sin^2 t}{\cos t \, \sin t} = \dfrac{1}{\cos t \, \sin t} = \sec t \csc t$

13. (a) $\dfrac{1}{\frac{1}{y^2}} + \dfrac{1}{\frac{1}{x^2}} = y^2 + x^2$

   (b) $\dfrac{1}{\csc^2 t} + \dfrac{1}{\sec^2 t} = \sin^2 t + \cos^2 t = 1$

15. (a) $\dfrac{\frac{x}{y}}{\frac{1}{y}} = \dfrac{x}{y} \cdot \dfrac{y}{1} = x$

   (b) $\dfrac{\cot \theta}{\csc \theta} = \dfrac{\frac{\cos \theta}{\sin \theta}}{\frac{1}{\sin \theta}} = \dfrac{\cos \theta}{\sin \theta} \cdot \dfrac{\sin \theta}{1} = \cos \theta$

17. $\cos \theta \tan \theta = \cos \theta \cdot \dfrac{\sin \theta}{\cos \theta} = \sin \theta$

19. $\tan \theta \, (\cos \theta - \csc \theta) = \dfrac{\sin \theta}{\cos \theta}\left(\cos \theta - \dfrac{1}{\sin \theta}\right) = \sin \theta - \dfrac{1}{\cos \theta} = \sin \theta - \sec \theta$

21. $(1 + \tan t)^2 = 1 + 2 \tan t + \tan^2 t = 2 \tan t + (1 + \tan^2 t) = 2 \tan t + \sec^2 t$

23. $\dfrac{\csc^2 \theta - 1}{\csc^2 \theta} = \dfrac{\csc^2 \theta}{\csc^2 \theta} - \dfrac{1}{\csc^2 \theta} = 1 - \sin^2 \theta = \cos^2 \theta$

25. $1 - \tan^2 \theta = (1 - \tan \theta)(1 + \tan \theta)$

27. $\sec^2 t - \sec t - 6 = (\sec t - 3)(\sec t + 2)$

29. $\tan^4 \theta + 3 \tan^2 \theta + 2 = (\tan^2 \theta + 1)(\tan^2 \theta + 2) = \sec^2 \theta \, (\tan^2 \theta + 2)$

31. $\csc^2 \theta - \cot^2 \theta = (1 + \cot^2 \theta) - \cot^2 \theta = 1$

33. $(1 - \sin t)^2 = (1 - \sin t)(1 - \sin t) = 1 - \sin t - \sin t + \sin^2 t = 1 - 2 \sin t + \sin^2 t$

35. $\dfrac{\sin t + \cos t}{\sin t} = \dfrac{\sin t}{\sin t} + \dfrac{\cos t}{\sin t} = 1 + \cot t$

37. $\sec^2 \theta - 1 = (1 + \tan^2 \theta) - 1 = \tan^2 \theta$

39. $\dfrac{\tan^2 t}{\sec t} = \dfrac{\sec^2 t - 1}{\sec t} = \dfrac{\sec^2 t}{\sec t} - \dfrac{1}{\sec t} = \sec t - \cos t$

41. $\cot x + 1 = \dfrac{\cos x}{\sin x} + \dfrac{\sin x}{\sin x} = \dfrac{1}{\sin x}(\cos x + \sin x) = \csc x \, (\cos x + \sin x)$

43. $\dfrac{\sec t}{1 + \sec t} = \dfrac{\sec t}{1 + \sec t} \cdot \dfrac{\cos t}{\cos t} = \dfrac{1}{\cos t + 1}$

45. $(\sec t - 1)(\sec t + 1) = \sec^2 t - 1 = \tan^2 t$

47. $\dfrac{1 - \sin^2 \theta}{\cos \theta} = \dfrac{\cos^2 \theta}{\cos \theta} = \cos \theta$

49. $\dfrac{\sec t}{\tan t} - \dfrac{\tan t}{\sec t} = \dfrac{\sec^2 t - \tan^2 t}{\sec t \tan t} = \dfrac{(1 + \tan^2 t) - \tan^2 t}{\sec t \tan t} = \dfrac{1}{\sec t \tan t} = \cos t \cot t$

51. $\dfrac{\cot^2 t}{\csc t + 1} = \dfrac{\csc^2 t - 1}{\csc t + 1} = \dfrac{(\csc t - 1)(\csc t + 1)}{\csc t + 1} = \csc t - 1$

53. $\dfrac{\cot t}{\cot t + 1} = \dfrac{\cot t}{\cot t + 1} \cdot \dfrac{\tan t}{\tan t} = \dfrac{\cot t \tan t}{\cot t \tan t + \tan t} = \dfrac{1}{1 + \tan t}$

55. $\dfrac{1}{1 - \sin t} + \dfrac{1}{1 + \sin t} = \dfrac{(1 + \sin t) + (1 - \sin t)}{(1 - \sin t)(1 + \sin t)} = \dfrac{2}{1 - \sin^2 t} = \dfrac{2}{\cos^2 t} = 2 \sec^2 t$

57. $\dfrac{\csc t + \cot t}{\csc t - \cot t} = \dfrac{\csc t + \cot t}{\csc t - \cot t} \cdot \dfrac{\csc t + \cot t}{\csc + \cot t} = \dfrac{(\csc t + \cot t)^2}{\csc^2 t - \cot^2 t} = \dfrac{(\csc t + \cot t)^2}{\csc^2 t - (\csc^2 t - 1)} = (\csc t + \cot t)^2$

59. $\dfrac{\cos^2 t}{1 - \sin t} = \dfrac{1 - \sin^2 t}{1 - \sin t} = \dfrac{(1 + \sin t)(1 - \sin t)}{1 - \sin t} = 1 + \sin t$

61. $\dfrac{1}{1 + \sin \theta} = \dfrac{1}{1 + \sin \theta} \cdot \dfrac{1 - \sin \theta}{1 - \sin \theta} = \dfrac{1 - \sin \theta}{1 - \sin^2 \theta} = \dfrac{1 - \sin \theta}{\cos^2 \theta}$

63. $\sqrt{1 - \sin^2 \theta} = \sqrt{\cos^2 \theta} = \pm\cos \theta = \cos \theta$, since θ is acute.

65. $\dfrac{1 + 2 \sin x + \sin^2 x}{\cos^2 x} = \dfrac{(1 + \sin x)^2}{1 - \sin^2 x} = \dfrac{(1 + \sin x)(1 + \sin x)}{(1 - \sin x)(1 + \sin x)} = \dfrac{1 + \sin x}{1 - \sin x}$

67. $(1 - \cos^2 x)(1 + \cos^2 x) = \sin^2 x \,(1 + (1 - \sin^2 x)) = \sin^2 x \,(2 - \sin^2 x) = 2 \sin^2 x - \sin^4 x$

69. $\cot \theta \sin \theta = \dfrac{\cos \theta}{\sin \theta} \cdot \sin \theta = \cos \theta$

Graph $Y_1 = (\cos (X)/\sin (X)) \sin (X)$ and $Y_2 = \cos (X)$ in $[-2\pi, 2\pi, \pi/2]$ by $[-4, 4, 1]$.

Table Y_1 and Y_2 together in degree mode starting at $x = 0$, incrementing by 50.

Graph Y_1 is shown in Figure 69a. Graph Y_2 is shown in Figure 69b. The table is shown in Figure 69c.

Figure 69a   Figure 69b   Figure 69c

71. $(1 - \cos^2 \theta)(1 + \tan^2 \theta) = \sin^2 \theta \sec^2 \theta = \sin^2 \theta \cdot \dfrac{1}{\cos^2 \theta} = \dfrac{\sin^2 \theta}{\cos^2 \theta} = \tan^2 \theta$

Graph $Y_1 = (1 - (\cos (X))^2)(1 + (\tan (X))^2)$ and $Y_2 = (\tan (X))^2$ in $[-2\pi, 2\pi, \pi/2]$ by $[-4, 4, 1]$.

Table Y_1 and Y_2 together in degree mode starting at $x = 0$, incrementing by 50.

Graph Y_1 is shown in Figure 71a. Graph Y_2 is shown in Figure 71b. The table is shown in Figure 71c.

Figure 71a   Figure 71b   Figure 71c

73. $\cos t \, (\tan t - \sec t) = \cos t \left(\dfrac{\sin t}{\cos t} - \dfrac{1}{\cos t} \right) = \sin t - 1$

Graph $Y_1 = \cos (X)(\tan (X) - 1/\cos (X))$ and $Y_2 = \sin (X) - 1$ in $[-2\pi, 2\pi, \pi/2]$ by $[-4, 4, 1]$.

Table Y_1 and Y_2 together in degree mode starting at $x = 0$, incrementing by 50.

Graph Y_1 is shown in Figure 73a. Graph Y_2 is shown in Figure 73b. The table is shown in Figure 73c.

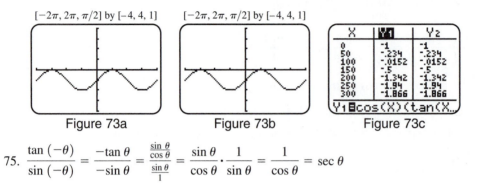

$[-2\pi, 2\pi, \pi/2]$ by $[-4, 4, 1]$ $[-2\pi, 2\pi, \pi/2]$ by $[-4, 4, 1]$

Figure 73a   Figure 73b   Figure 73c

75. $\dfrac{\tan (-\theta)}{\sin (-\theta)} = \dfrac{-\tan \theta}{-\sin \theta} = \dfrac{\frac{\sin \theta}{\cos \theta}}{\frac{\sin \theta}{1}} = \dfrac{\sin \theta}{\cos \theta} \cdot \dfrac{1}{\sin \theta} = \dfrac{1}{\cos \theta} = \sec \theta$

Graph $Y_1 = \tan (-X)/\sin (-X)$ and $Y_2 = 1/\cos (X)$ in $[-2\pi, 2\pi, \pi/2]$ by $[-4, 4, 1]$.

Table Y_1 and Y_2 together in degree mode starting at $x = 0$, incrementing by 50.

Graph Y_1 is shown in Figure 75a. Graph Y_2 is shown in Figure 75b. The table is shown in Figure 75c.

$[-2\pi, 2\pi, \pi/2]$ by $[-4, 4, 1]$ $[-2\pi, 2\pi, \pi/2]$ by $[-4, 4, 1]$ $[0, 1/15, 1/60]$ by $[-30, 30, 10]$

Figure 75a   Figure 75b   Figure 75c   Figure 77

77. (a) $W(t) = 5 \cos^2 (120\pi t) = 5(1 - \sin^2 (120\pi t))$

$V = 25 \sin (120\pi t)$ is maximum when $\sin (120\pi t) = 1$ and is minimum when $\sin (120\pi t) = -1$.

Thus, $W = 5(1 - (\pm 1)^2) = 0$.

(b) Graph $Y_1 = 5 \, (\cos (120\pi X))^2$ and $Y_2 = 25 \sin (120\pi X)$ in $[0, 1/15, 1/60]$ by $[-30, 30, 10]$.

See Figure 77. The wattage $Y_1 = 0$ whenever the voltage $Y_2 = \pm 25$.

7.3: Trigonometric Equations

1. $\theta = 120°$ is in quadrant II, $\theta_R = 180° - 120° = 60°$.

3. $\theta = 85°$ is in quadrant I, $\theta_R = 85°$.

5. $\theta = -65°$ is coterminal with $295°$ in quadrant IV, $\theta_R = 360° - 295° = 65°$.

7. $\theta = \dfrac{5\pi}{6}$ is in quadrant II, $\theta_R = \pi - \dfrac{5\pi}{6} = \dfrac{\pi}{6}$.

9. $\theta = -\dfrac{2\pi}{3}$ is coterminal with $\dfrac{4\pi}{3}$ in quadrant III, $\theta_R = \dfrac{4\pi}{3} - \pi = \dfrac{\pi}{3}$.

11. (a) Start by solving $\sin \theta_R = 1$: $\theta_R = \sin^{-1} 1 = 90°$. This angle is quadrantal, thus $\theta = 90°$ or $\theta = \dfrac{\pi}{2}$.

   (b) Start by solving $\sin \theta_R = 1$: $\theta_R = \sin^{-1} 1 = 90°$. This angle is quadrantal, thus $\theta = 270°$ or $\theta = \dfrac{3\pi}{2}$.

13. (a) Start by solving $\tan \theta_R = \sqrt{3}$: $\theta_R = \tan^{-1} \sqrt{3} = 60°$. The tangent function is positive in quadrants I and III.

   Angles in these quadrants with a $60°$ reference angle are $60°$ and $240°$ or $\dfrac{\pi}{3}$ and $\dfrac{4\pi}{3}$.

   (b) Start by solving $\tan \theta_R = \sqrt{3}$: $\theta_R = \tan^{-1} \sqrt{3} = 60°$. The tangent function is negative in quadrants II and IV.

   Angles in these quadrants with a $60°$ reference angle are $120°$ and $300°$ or $\dfrac{2\pi}{3}$ and $\dfrac{5\pi}{3}$.

15. First note that $\sec \theta = 2$ when $\cos \theta = \dfrac{1}{2}$ and $\sec \theta = -2$ when $\cos \theta = -\dfrac{1}{2}$.

   (a) Start by solving $\cos \theta_R = \dfrac{1}{2}$: $\theta_R = \cos^{-1} \dfrac{1}{2} = 60°$. The secant function is positive in quadrants I and IV.

   Angles in these quadrants with a $60°$ reference angle are $60°$ and $300°$ or $\dfrac{\pi}{3}$ and $\dfrac{5\pi}{3}$.

   (b) Start by solving $\cos \theta_R = \dfrac{1}{2}$: $\theta_R = \cos^{-1} \dfrac{1}{2} = 60°$. The secant function is negative in quadrants II and III.

   Angles in these quadrants with a $60°$ reference angle are $120°$ and $240°$ or $\dfrac{2\pi}{3}$ and $\dfrac{4\pi}{3}$.

17. (a) No solution since $\sin \theta \neq 3$ for any θ.

   (b) No solution since $\sin \theta \neq -3$ for any θ.

19. (a) Since $t_R = \sin^{-1}\left(\dfrac{1}{2}\right) = \dfrac{\pi}{6}$ and sine is positive in quadrants I and II, $t = \dfrac{\pi}{6} + 2\pi n$ and $\dfrac{5\pi}{6} + 2\pi n$.

   (b) Since $t_R = \sin^{-1}\left(\dfrac{1}{2}\right) = \dfrac{\pi}{6}$ and sine is negative in quadrants III and IV, $t = \dfrac{7\pi}{6} + 2\pi n$ and $\dfrac{11\pi}{6} + 2\pi n$.

21. (a) Since $t_R = \tan^{-1}(1) = \dfrac{\pi}{4}$ and tangent is positive in quadrants I and III, $t = \dfrac{\pi}{4} + \pi n$.

   (b) Since $t_R = \tan^{-1}(1) = \dfrac{\pi}{4}$ and tangent is negative in quadrants II and IV, $t = \dfrac{3\pi}{4} + \pi n$.

23. First note that $\cos t = \dfrac{1}{2}$ when $\sec t = 2$ and $\cos t = -\dfrac{1}{2}$ when $\sec t = -2$.

   (a) Since $t_R = \cos^{-1}\left(\dfrac{1}{2}\right) = \dfrac{\pi}{3}$ and the secant function is positive in quadrants I and IV,

   $t = \dfrac{\pi}{3} + 2\pi n$ and $\dfrac{5\pi}{3} + 2\pi n$.

   (b) Since $t_R = \cos^{-1}\left(\dfrac{1}{2}\right) = \dfrac{\pi}{3}$ and the secant function is negative in quadrants II and III,

   $t = \dfrac{2\pi}{3} + 2\pi n$ and $\dfrac{4\pi}{3} + 2\pi n$.

25. At the intersection points $t = \dfrac{\pi}{4}$ and $t = \dfrac{5\pi}{4}$ therefore, $\sin t = \cos t$ for $t = \dfrac{\pi}{4}, \dfrac{5\pi}{4}$.

   Symbolically: $\sin t = \cos t \Rightarrow \dfrac{\sin t}{\cos t} = 1 \Rightarrow \tan t = 1 \Rightarrow t = \dfrac{\pi}{4}, \dfrac{5\pi}{4}$.

27. At the intersection points $t = \dfrac{\pi}{3}$ and $t = \dfrac{5\pi}{3}$ therefore, $3 \cot t = 2 \sin t$ for $t = \dfrac{\pi}{3}, \dfrac{5\pi}{3}$.

    Symbolically: $3 \cot t = 2 \sin t \Rightarrow 3\dfrac{\cos t}{\sin t} = 2 \sin t \Rightarrow 3 \cos t = 2 \sin^2 t \Rightarrow 3 \cos t = 2(1 - \cos^2 t) \Rightarrow$

    $2 \cos^2 t + 3 \cos t - 2 = 0 \Rightarrow (2 \cos t - 1)(\cos t + 2) = 0 \Rightarrow \cos t = \dfrac{1}{2}$ or $\cos t = -2$.

    If $\cos t = \dfrac{1}{2}$, then $t = \dfrac{\pi}{3}, \dfrac{5\pi}{3}$. If $\cos t = -2$, then t is undefined.

29. (a) $2x - 1 = 0 \Rightarrow 2x = 1 \Rightarrow x = \dfrac{1}{2}$

    (b) $2 \sin \theta - 1 = 0 \Rightarrow 2 \sin \theta = 1 \Rightarrow \sin \theta = \dfrac{1}{2}$

    Since $\theta_R = \sin^{-1} \dfrac{1}{2} = 30°$ and sine is positive in quadrants I and II, $\theta = 30°, 150°$.

31. (a) $x^2 = x \Rightarrow x^2 - x = 0 \Rightarrow x(x - 1) = 0 \Rightarrow x = 0, 1$

    (b) $\sin^2 \theta = \sin \theta \Rightarrow \sin^2 \theta - \sin \theta = 0 \Rightarrow \sin \theta (\sin \theta - 1) = 0 \Rightarrow \sin \theta = 0, 1$

    Since $\theta_R = \sin^{-1} 0 = 0°$ or $\theta_R = \sin^{-1} 1 = 90°$ and sine is positive in quadrants I and II, $\theta = 0°, 90°, 180°$.

33. (a) $x^2 + 1 = 2 \Rightarrow x^2 = 1 \Rightarrow x = \pm 1$

    (b) $\tan^2 \theta + 1 = 2 \Rightarrow \tan^2 \theta = 1 \Rightarrow \tan \theta = \pm 1$

    Since $\theta_R = \tan^{-1} 1 = 45°$ and tangent is positive or negative in all quadrants, $\theta = 45°, 135°, 225°, 315°$.

35. (a) $x^2 + x = 2 \Rightarrow x^2 + x - 2 = 0 \Rightarrow (x + 2)(x - 1) = 0 \Rightarrow x = -2, 1$

    (b) $\cos^2 \theta + \cos \theta = 2 \Rightarrow \cos^2 \theta + \cos \theta - 2 = 0 \Rightarrow (\cos \theta + 2)(\cos \theta - 1) = 0 \Rightarrow \cos \theta = -2, 1$

    Since $\theta_R = \cos^{-1} 1 = 0°$ ($\cos \theta = -2$ is not possible) and cosine is positive in quadrants I and IV, $\theta = 0°$.

37. $\tan^2 t - 3 = 0 \Rightarrow \tan^2 t = 3 \Rightarrow \tan t = \pm\sqrt{3}$

    Since $t_R = \tan^{-1} \sqrt{3} = \dfrac{\pi}{3}$ and tangent is positive or negative in all quadrants, $t = \dfrac{\pi}{3}, \dfrac{2\pi}{3}, \dfrac{4\pi}{3}, \dfrac{5\pi}{3}$.

39. $3 \cos t + 4 = 0 \Rightarrow 3 \cos t = -4 \Rightarrow \cos t = -\dfrac{4}{3}$

    $\cos t = -\dfrac{4}{3}$ is not possible. No solution.

41. $\sin t \cos t = \cos t \Rightarrow \sin t \cos t - \cos t = 0 \Rightarrow \cos t (\sin t - 1) = 0 \Rightarrow \cos t = 0$ or $\sin t = 1$

    Since $t_R = \cos^{-1} 0 = \dfrac{\pi}{2}$ or $t_R = \sin^{-1} 1 = \dfrac{\pi}{2}$ and the angle is quadrantal, $t = \dfrac{\pi}{2}, \dfrac{3\pi}{2}$.

43. $\csc^2 t = 2 \cot t \Rightarrow 1 + \cot^2 t = 2 \cot t \Rightarrow \cot^2 t - 2 \cot t + 1 = 0 \Rightarrow (\cot t - 1)^2 = 0 \Rightarrow \cot t = 1$

    Since $t_R = \cot^{-1} 1 = \tan^{-1} 1 = \dfrac{\pi}{4}$ and tangent is positive in quadrants I and III,

    $t = \dfrac{\pi}{4}, \dfrac{5\pi}{4}$. $t = \dfrac{\pi}{4} + \pi n$ for $n = 0, \pm 1, \pm 2, \pm 3, \ldots$.

45. $\sin^2 t = \dfrac{1}{4} \Rightarrow \sin t = \pm \dfrac{1}{2}$

    Since $t_R = \sin^{-1} \dfrac{1}{2} = \dfrac{\pi}{6}$ and sine is positive or negative in all quadrants, $t = \dfrac{\pi}{6}, \dfrac{5\pi}{6}, \dfrac{7\pi}{6}, \dfrac{11\pi}{6}$.

47. $\sin t \cos t = 0 \Rightarrow \sin t = 0$ or $\cos t = 0$. Since $t_R = \sin^{-1} 0 = 0$ or $t_R = \cos^{-1} 0 = \dfrac{\pi}{2}, t = 0, \dfrac{\pi}{2}, \pi, \dfrac{3\pi}{2}$.

49. $2 \sec t = \tan^2 t + 1 \Rightarrow 2 \sec t = \sec^2 t \Rightarrow \sec^2 t - 2 \sec t = 0 \Rightarrow \sec t(\sec t - 2) = 0 \Rightarrow$

$\sec t = 0, \sec t - 2 = 0$. For $\sec t = 0$, t is undefined. For $\sec t - 2 = 0 \Rightarrow \sec t = 2 \Rightarrow t = \dfrac{\pi}{3}, \dfrac{5\pi}{3}$.

51. $\tan t + \sec t = 1 \Rightarrow \tan t = 1 - \sec t \Rightarrow \tan^2 t = (1 - \sec t)^2 \Rightarrow \sec^2 t - 1 = 1 - 2\sec t + \sec^2 t \Rightarrow$

$-2 = -2 \sec t \Rightarrow \sec t = 1 \Rightarrow t = 0.$

53. $2 \tan t - 1 = 1 \Rightarrow 2 \tan t = 2 \Rightarrow \tan t = 1 \Rightarrow t = \dfrac{\pi}{4}$ and tangent is positive in quadrants I and III,

$t = \dfrac{\pi}{4}, \dfrac{5\pi}{4}, \dfrac{9\pi}{4}, \dots .\ t = \dfrac{\pi}{4} + \pi n$ for $n = 0, \pm 1, \pm 2, \pm 3, \dots .$

55. $2 \sin t + 2 = 3 \Rightarrow 2 \sin t = 1 \Rightarrow \sin t = \dfrac{1}{2} \Rightarrow t = \dfrac{\pi}{6}, \dfrac{5\pi}{6} \Rightarrow t = \dfrac{\pi}{6} + 2\pi n,$

$t = \dfrac{5\pi}{6} + 2\pi n$ for $n = 0, \pm 1, \pm 2, \pm 3, \dots .$

57. $2 \sin^2 t - 3 \sin t = -1 \Rightarrow 2 \sin^2 t - 3 \sin t + 1 = 0 \Rightarrow (2 \sin t - 1)(\sin t - 1) = 0 \Rightarrow$

$(2 \sin t - 1) = 0, (\sin t - 1) = 0.\ 2 \sin t - 1 = 0 \Rightarrow \sin t = \dfrac{1}{2} \Rightarrow t = \dfrac{\pi}{6}, \dfrac{5\pi}{6}.$

$\sin t - 1 = 0 \Rightarrow \sin t = 1 \Rightarrow t = \dfrac{\pi}{2}.\ t = \dfrac{\pi}{6} + 2\pi n, \dfrac{5\pi}{6} + 2\pi n, \dfrac{\pi}{2} + 2\pi n$ for $n = 0, \pm 1, \pm 2, \pm 3, \dots .$

59. $\sec^2 t + 3 \sec t + 2 = 0 \Rightarrow (\sec t + 2)(\sec t + 1) = 0 \Rightarrow \sec t + 2 = 0, (\sec t + 1) = 0.$

$\sec t + 2 = 0 \Rightarrow \sec t = -2 \Rightarrow t = \dfrac{2\pi}{3}, \dfrac{4\pi}{3}.\ \sec t + 1 = 0 \Rightarrow \sec t = -1 \Rightarrow t = \pi.$

$t = \dfrac{2\pi}{3} + 2\pi n, \dfrac{4\pi}{3} + 2\pi n, \pi + 2\pi n$ for $n = 0, \pm 1, \pm 2, \pm 3, \dots .$

61. $\tan^2 t - 1 = 0 \Rightarrow \tan^2 t = 1 \Rightarrow \tan t = \pm 1$

Since $t_R = \tan^{-1} 1 = \dfrac{\pi}{4}$ and tangent is positive or negative in all quadrants, $t = \pm \dfrac{\pi}{4}, \pm \dfrac{3\pi}{4}, \pm \dfrac{5\pi}{4}, \dots .$

$t = \dfrac{\pi}{4} + \dfrac{\pi}{2} n$ for $n = 0, \pm 1, \pm 2, \pm 3, \dots .$

63. $\sin^2 t + \sin t - 20 = 0 \Rightarrow (\sin t + 5)(\sin t - 4) = 0 \Rightarrow \sin t = -5, 4$

$\sin t = -5$ and $\sin t = 4$ are both not possible. No solution.

65. $\cos t \sin t = \sin t \Rightarrow \cos t \sin t - \sin t = 0 \Rightarrow \sin t (\cos t - 1) = 0 \Rightarrow \sin t = 0$ or $\cos t = 1$

Since $t_R = \sin^{-1} 0 = 0$ or $t_R = \cos^{-1} 1 = 0$ and the angle is quadrantal, $t = 0, \pm \pi, \pm 2\pi, \dots .$

$t = \pi n$ for $n = 0, \pm 1, \pm 2, \pm 3, \dots .$

67. $\sec^2 t = 2 \tan t \Rightarrow 1 + \tan^2 t = 2 \tan t \Rightarrow \tan^2 t - 2 \tan t + 1 = 0 \Rightarrow (\tan t - 1)^2 = 0 \Rightarrow \tan t = 1$

Since $t_R = \tan^{-1} 1$ and tangent is positive in quadrants I and III, $t = \dfrac{\pi}{4}, \dfrac{5\pi}{4}.$

$t = \dfrac{\pi}{4} + \pi n$ for $n = 0, \pm 1, \pm 2, \pm 3, \dots .$

69. $\sin^2 t \cos^2 t = 0 \Rightarrow \sin^2 t = 0$ or $\cos^2 t = 0 \Rightarrow \sin t = 0$ or $\cos t = 0$

Since $t_R = \sin^{-1} 0 = 0$ or $t_R = \cos^{-1} 0 = \dfrac{\pi}{2}$ and the angle is quadrantal, $t = 0, \pm \dfrac{\pi}{2}, \pm \pi, \pm \dfrac{3\pi}{2}, \dots .$

$t = \dfrac{\pi}{2} n$ for $n = 0, \pm 1, \pm 2, \pm 3, \dots .$

71. $\sin t + \cos t = 1 \Rightarrow \sin^2 t + 2 \sin t \cos t + \cos^2 t = 1 \Rightarrow 2 \sin t \cos t = 0 \Rightarrow \sin t = 0$ or $\cos t = 0$

Since $t_R = \sin^{-1} 0 = 0$ or $t_R = \cos^{-1} 0 = \dfrac{\pi}{2}$ and the angle is quadrantal, $t = \ldots, -\dfrac{3\pi}{2}, 0, \dfrac{\pi}{2}, 2\pi, \ldots$.

$t = 2\pi n, \dfrac{\pi}{2} + 2\pi n$ for $n = 0, \pm 1, \pm 2, \pm 3, \ldots$.

73. $\sin 3t = \dfrac{1}{2}$, let $\theta = 3t$ so then the equation becomes $\sin \theta = \dfrac{1}{2}$ and in radian measure these solutions are

$\theta = \dfrac{\pi}{6}$ and $\dfrac{5\pi}{6}$. For all real number solutions $\theta = \dfrac{\pi}{6} + 2\pi n$ or $\theta = \dfrac{5\pi}{6} + 2\pi n$. Because $\theta = 3t$, we can

determine t by substituting $3t$ for $\theta \Rightarrow 3t = \dfrac{\pi}{6} + 2\pi n$ or $3t = \dfrac{5\pi}{6} + 2\pi n$. Therefore $t = \dfrac{\pi}{18} + \dfrac{2\pi}{3} n$ or

$t = \dfrac{5\pi}{18} + \dfrac{2\pi}{3} n$.

75. $\cos 4t = -\dfrac{\sqrt{3}}{2}$, let $\theta = 4t$ so then the equation becomes $\cos \theta = -\dfrac{\sqrt{3}}{2}$ and in radian measure these solutions

are $\theta = \dfrac{5\pi}{6}$ and $\dfrac{7\pi}{6}$. For all real number solutions $\theta = \dfrac{5\pi}{6} + 2\pi n$ or $\theta = \dfrac{7\pi}{6} + 2\pi n$. Because $\theta = 4t$, we can

determine t by substituting $4t$ for $\theta \Rightarrow 4t = \dfrac{5\pi}{6} + 2\pi n$ or $4t = \dfrac{7\pi}{6} + 2\pi n$. Therefore $t = \dfrac{5\pi}{24} + \dfrac{\pi}{2} n$ or

$t = \dfrac{7\pi}{24} + \dfrac{\pi}{2} n$.

77. $\tan 5t = 1$, let $\theta = 5t$ so then the equation becomes $\tan \theta = 1$ and in radian measure these solutions are

$\theta = \dfrac{\pi}{4}$ and $\dfrac{5\pi}{4}$. For all real number solutions $\theta = \dfrac{\pi}{4} + \pi n$ or $\theta = \dfrac{5\pi}{4} + \pi n$. Since these representations are

equivalent, we use $\theta = \dfrac{\pi}{4} + \pi n$. Because $\theta = 5t$, we can determine t by substituting

$5t$ for $\theta \Rightarrow 5t = \dfrac{\pi}{4} + \pi n$. Therefore $t = \dfrac{\pi}{20} + \dfrac{\pi}{5} n$.

79. $2 \sin 4t = -1 \Rightarrow \sin 4t = -\dfrac{1}{2}$, let $\theta = 4t$ so then the equation becomes $\sin \theta = -\dfrac{1}{2}$ and in radian measure

these solutions are $\theta = \dfrac{7\pi}{6}$ and $\dfrac{11\pi}{6}$. For all real number solutions $\theta = \dfrac{7\pi}{6} + 2\pi n$ or $\theta = \dfrac{11\pi}{6} + 2\pi n$.

Because $\theta = 4t$, we can determine t by substituting $4t$ for $\theta \Rightarrow 4t = \dfrac{7\pi}{6} + 2\pi n$ or $4t = \dfrac{11\pi}{6} + 2\pi n$.

Therefore $t = \dfrac{7\pi}{24} + \dfrac{\pi}{2} n$ or $t = \dfrac{11\pi}{24} + \dfrac{\pi}{2} n$.

81. $-\sec 4t = \sqrt{2} \Rightarrow \sec 4t = -\sqrt{2}$, let $\theta = 4t$ so then the equation becomes $\sec \theta = -\sqrt{2}$ or $\cos \theta = -\dfrac{1}{\sqrt{2}}$

and in radian measure these solutions are $\theta = \dfrac{3\pi}{4}$ and $\dfrac{5\pi}{4}$. For all real number solutions $\theta = \dfrac{3\pi}{4} + 2\pi n$ or

$\theta = \dfrac{5\pi}{4} + 2\pi n$. Because $\theta = 4t$, we can determine t by substituting $4t$ for $\theta \Rightarrow 4t = \dfrac{3\pi}{4} + 2\pi n$ or

$4t = \dfrac{5\pi}{4} + 2\pi n$. Therefore $t = \dfrac{3\pi}{16} + \dfrac{\pi}{2} n$ or $t = \dfrac{5\pi}{16} + \dfrac{\pi}{2} n$.

83. $2 \sin 8t - 3 = -1 \Rightarrow 2 \sin 8t = 2 \Rightarrow \sin 8t = 1$, let $\theta = 8t$ so then the equation becomes $\sin \theta = 1$ and in radian measure this solution is $\theta = \dfrac{\pi}{2}$. For all real number solutions $\theta = \dfrac{\pi}{2} + 2\pi n$. Because $\theta = 8t$, we can determine t by substituting $8t$ for $\theta \Rightarrow 8t = \dfrac{\pi}{2} + 2\pi n$. Therefore $t = \dfrac{\pi}{16} + \dfrac{\pi}{4}n$.

85. $\cot 4t + 5 = 6 \Rightarrow \cot 4t = 1$, let $\theta = 4t$ so then the equation becomes $\cot \theta = 1$ or $\tan \theta = 1$ and in radian measure this solution is $\theta = \dfrac{\pi}{4}$. For all real number solutions $\theta = \dfrac{\pi}{4} + \pi n$. Because $\theta = 4t$, we can determine t by substituting $4t$ for $\theta \Rightarrow 4t = \dfrac{\pi}{4} + \pi n$. Therefore $t = \dfrac{\pi}{16} + \dfrac{\pi}{4}n$.

87. $5 \cos 3t = 1 \Rightarrow \cos 3t = \dfrac{1}{5}$, let $\theta = 3t$ so then the equation becomes $\cos \theta = \dfrac{1}{5}$ and in radian measure these solutions are $\theta \approx 1.369$ and 4.914. For all real number solutions $\theta \approx 1.369 + 2\pi n$ or $\theta \approx 4.914 + 2\pi n$. Because $\theta = 3t$ we can determine t by substituting $3t$ for $\theta \Rightarrow 3t \approx 1.369 + 2\pi n$ or $3t \approx 4.914 + 2\pi n$. Therefore $t \approx 0.456 + \dfrac{2}{3}\pi n$ or $t \approx 1.638 + \dfrac{2}{3}\pi n$.

89. $\sin 2t = \dfrac{1}{3}$, let $\theta = 2t$ so then the equation becomes $\sin \theta = \dfrac{1}{3}$ and in radian measure these solutions are $\theta \approx 0.340$ and 2.802. For all real number solutions $\theta \approx 0.340 + 2\pi n$ or $\theta \approx 2.802 + 2\pi n$. Because $\theta = 2t$ we can determine t by substituting $2t$ for $\theta \Rightarrow 2t \approx 0.340 + 2\pi n$ or $2t \approx 2.802 + 2\pi n$. Therefore $t \approx 0.170 + \pi n$ or $t \approx 1.401 + \pi n$.

91. $2.1 \sec t - 4.5 = 0 \Rightarrow 2.1 \sec t = 4.5 \Rightarrow \sec t = \dfrac{4.5}{2.1} \Rightarrow \cos t = \dfrac{2.1}{4.5} \Rightarrow \cos^{-1}\left(\dfrac{2.1}{4.5}\right) = t \Rightarrow t \approx 1.085$.

The cosine function is positive in quadrants I and IV. For t in $[0, 2\pi)$, the two solutions are $t \approx 1.085$ and $2\pi - \cos^{-1}\left(\dfrac{2.1}{4.5}\right) \approx 5.198$.

93. $5.8 \sin t - 3.7 = 0.2 \Rightarrow 5.8 \sin t = 3.9 \Rightarrow \sin t = \dfrac{3.9}{5.8} \Rightarrow \sin^{-1}\left(\dfrac{3.9}{5.8}\right) = t \Rightarrow t \approx 0.737$.

The sine function is positive in quadrants I and II. For t in $[0, 2\pi)$, the two solutions are $t \approx 0.737$ and $\pi - \sin^{-1}\left(\dfrac{3.9}{5.8}\right) \approx 2.404$.

95. $5 \tan^2 t - 3 = 0 \Rightarrow 5 \tan^2 t = 3 \Rightarrow \tan^2 t = \dfrac{3}{5} \Rightarrow \tan t = \pm\sqrt{\dfrac{3}{5}} \Rightarrow \tan^{-1}\left(\pm\sqrt{\dfrac{3}{5}}\right) = t$. The reference number $t_R = \tan^{-1}\left(\sqrt{\dfrac{3}{5}}\right) \approx 0.659$. The tangent function is both positive and negative in all four quadrants.

For t in $[0, 2\pi)$, the solutions are $t \approx 0.659$, $\pi - \tan^{-1}\left(\sqrt{\dfrac{3}{5}}\right) \approx 2.483$, $\pi + \tan^{-1}\left(\sqrt{\dfrac{3}{5}}\right) \approx 3.801$ and $2\pi - \tan^{-1}\left(\sqrt{\dfrac{3}{5}}\right) \approx 5.624$.

97. Graph $Y_1 = \tan (X)$ and $Y_2 = X$ in dot mode in $[-\pi/6, 2\pi, \pi/2]$ by $[-2, 6, 1]$.

The intersections of the two graphs are shown in Figure 97a and Figure 97b.

Table $Y_1 = \tan (X)$ and $Y_2 = X$ starting at $x = 0$, incrementing by 0.9. See Figure 97c.

The solutions are $x = 0$, $x \approx 4.49$.

Figure 97a   Figure 97b   Figure 97c

99. Graph $Y_1 = \sin (X)$ and $Y_2 = (X - 1)^{\wedge}2$ in $[-\pi/6, 2\pi, \pi/2]$ by $[-3, 7, 1]$.

The intersections of the two graphs are shown in Figure 99a and Figure 99b.

Table $Y_1 = \sin (X)$ and $Y_2 = (X - 1)^{\wedge}2$ starting at $x = 0$, incrementing by 0.4. See Figure 99c.

The solutions are $x \approx 0.39$, $x \approx 1.96$.

Figure 99a   Figure 99b   Figure 99c

101. Graph $Y_1 = 2X \cos (X + 1)$ and $Y_2 = \sin (\cos (X))$ in $[-\pi/6, 2\pi, \pi/2]$ by $[-5, 5, 1]$.

The intersection of the two graphs is shown in Figure 101a.

Table $Y_1 = 2X \cos (X + 1)$ and $Y_2 = \sin (\cos (X))$ starting at $x = 3.0$, incrementing by 0.1.

See Figure 101b. The solution is $x \approx 3.60$.

Figure 101a   Figure 101b

103. $Y_1 = 0$ for $x = 30°, 210°$

We can write all solutions to this equation as $30° + 180°n$, where $n = 0, \pm 1, \pm 2, \ldots$.

105. $\sin^{-1} x = \dfrac{\pi}{2} \Rightarrow \sin (\sin^{-1} x) = \sin \dfrac{\pi}{2} \Rightarrow x = 1$

107. $2\cos^{-1} x = \dfrac{5\pi}{3} \Rightarrow \cos^{-1} x = \dfrac{5\pi}{6} \Rightarrow \cos (\cos^{-1} x) = \cos \dfrac{5\pi}{6} \Rightarrow x = -\dfrac{\sqrt{3}}{2} \approx -0.866$

109. $\pi + \tan^{-1} x = \frac{3\pi}{4} \Rightarrow \tan^{-1} x = -\frac{\pi}{4} \Rightarrow \tan(\tan^{-1} x) = \tan\left(-\frac{\pi}{4}\right) \Rightarrow x = -1$

111. $\tan^{-1}(3x + 1) = \frac{\pi}{4} \Rightarrow \tan(\tan^{-1}(3x + 1)) = \tan\left(\frac{\pi}{4}\right) \Rightarrow 3x + 1 = 1 \Rightarrow 3x = 0 \Rightarrow x = 0$

113. $\cos^{-1} x + 3 \cos^{-1} x = \pi \Rightarrow 4\cos^{-1} x = \pi \Rightarrow \cos^{-1} x = \frac{\pi}{4} \Rightarrow \cos(\cos^{-1} x) = \cos\frac{\pi}{4} \Rightarrow$

$x = \frac{1}{\sqrt{2}} \approx 0.707$

115. $0.25 = \frac{1}{2}(1 - \cos\theta) \Rightarrow 0.5 = 1 - \cos\theta \Rightarrow \cos\theta = 0.5.\ \theta_R = \cos^{-1}(0.5) = 60°.$ The cosine function is

positive in quadrants I and IV. Angles in these quadrants with $\theta_R = 60°$ are $60°$ and $300°$. Since the cosine

function has period $360°$, all solutions can be written as $60° + 360° \cdot n$ or $300° + 360° \cdot n$ where n is an integer.

117. $200 = \frac{44^2}{4.5 + 32.2\tan\theta} \Rightarrow 4.5 + 32.2\tan\theta = \frac{44^2}{200} \Rightarrow 32.2\tan\theta = \frac{44^2}{200} - 4.5 \Rightarrow \tan\theta = \frac{\frac{44^2}{200} - 4.5}{32.2} \Rightarrow$

$\tan\theta \approx 0.1609.\ \theta = \tan^{-1}(0.1609) \approx 9.1°.$ The superelevation should be about $9.1°$.

119. Graph $Y_1 = 6.5\sin(\pi/6(X - 3.65)) + 12.4$ and $Y_2 = 9$ in $[0, 13, 1]$ by $[0, 24, 2]$.

The intersections of the two graphs are shown in Figure 119a and Figure 119b.

Table $Y_1 = 6.5\sin(\pi/6(X - 3.65)) + 12.4$ starting at $x = 2.6$, incrementing by 1.35. See Figure 119c.

The solutions are $x \approx 2.6$ and $x \approx 10.7$; near February 17 and October 22.

$[0, 13, 1]$ by $[0, 24, 2]$	$[0, 13, 1]$ by $[0, 24, 2]$	
Figure 119a	Figure 119b	Figure 119c

121. $6.5\sin\left(\frac{\pi}{6}(x - 3.65)\right) + 12.4 = 9 \Rightarrow 6.5\sin\left(\frac{\pi}{6}(x - 3.65)\right) = -3.4 \Rightarrow$

$\sin\left(\frac{\pi}{6}(x - 3.65)\right) = -0.5231$

$\Rightarrow \frac{\pi}{6}(x - 3.65) = \sin^{-1}(-0.5231)$

$\Rightarrow \frac{\pi}{6}(x - 3.65) \approx -0.5505$ or $\frac{\pi}{6}(x - 3.65) \approx 3.6920$

$\Rightarrow x - 3.65 \approx -1.0513$ or $x - 3.65 \approx 7.0513$

$\Rightarrow x \approx 2.6$ or $x \approx 10.7$

123. (a) Using the sine regression function on the calculator we get $f(x) \approx 122.3 \sin (0.524x - 1.7) + 367$.

    Answers may vary.

    (b) Graph $Y_1 = 122.3 \sin (0.524X - 1.7) + 367$ and $Y_2 = 350$ together in $[0, 13, 1]$ by $[200, 600, 50]$.

    The intersections of the two graphs are shown in Figure 123a and Figure 123b.

    We see that $f(x) \geq 350$ for $k_1 \leq x \leq k_2$ where $k_1 \approx 2.98$ and $k_2 \approx 9.51$. At 50° N latitude the

    maximum monthly hours of sunshine is greater than or equal to 350 hours from roughly March through

    September.

$[0, 13, 1]$ by $[200, 600, 50]$ $[0, 13, 1]$ by $[200, 600, 50]$ $[0, 0.5, 0.1]$ by $[-0.2, 0.2, 0.1]$

    Figure 123a   Figure 123b   Figure 125a   Figure 125b

125. Use the intercept method to solve the equation.

    Graph $Y_1 = X - 0.26 - 0.017 \sin (X)$ in $[0, 0.5, 0.1]$ by $[-0.2, 0.2, 0.1]$. See Figure 125a.

    Table $Y_1 = X - 0.26 - 0.017 \sin (X)$ starting at $x = 0.20$, incrementing by 0.01. See Figure 125b.

    The solution is approximately 0.26.

127. (a) Let $Y_1 = 0.003 \sin (880\pi X - 0.7)$, $Y_2 = 0.002 \sin (880\pi X + 0.6)$.

    Graph Y_1, Y_2 and $Y_3 = Y_1 + Y_2$ seperately in $[0, 0.01, 0.005]$ by $[-0.005, 0.005, 0.001]$.

    Graph Y_1 is shown in Figure 127a. Graph Y_2 is shown in Figure 127b. Graph Y_3 is shown in Figure 127c.

    (b) The maximum pressure is $P \approx 0.004$. This can be found at any peak on the graph in Figure 127c.

    (c) No. The maximum of P_1 is 0.003 and the maximum of P_2 is 0.002.

$[0, 0.01, 0.005]$ by $[-0.005, 0.005, 0.001]$ $[0, 0.01, 0.005]$ by $[-0.005, 0.005, 0.001]$ $[0, 0.01, 0.005]$ by $[-0.005, 0.005, 0.001]$

    Figure 127a   Figure 127b   Figure 127c

7.4: Sum and Difference Identities

1. $\sin 15° = \sin (45° - 30°) = \sin 45° \cos 30° - \cos 45° \sin 30° = \left(\dfrac{\sqrt{2}}{2}\right)\left(\dfrac{\sqrt{3}}{2}\right) - \left(\dfrac{\sqrt{2}}{2}\right)\left(\dfrac{1}{2}\right) =$

    $\dfrac{\sqrt{6} - \sqrt{2}}{4}$

3. $\tan 15° = \tan (60° - 45°) = \dfrac{\tan 60° - \tan 45°}{1 + \tan 60° \tan 45°} = \dfrac{\sqrt{3} - 1}{1 + \sqrt{3}} = \dfrac{\sqrt{3} - 1}{\sqrt{3} + 1} \cdot \dfrac{\sqrt{3} - 1}{\sqrt{3} - 1} =$

$\dfrac{3 - 2\sqrt{3} + 1}{3 - 1} = 2 - \sqrt{3}$

5. $\cos 75° = \cos (135° - 60°) = \cos 135° \cos 60° + \sin 135° \sin 60° = \left(-\dfrac{\sqrt{2}}{2}\right)\left(\dfrac{1}{2}\right) + \left(\dfrac{\sqrt{2}}{2}\right)\left(\dfrac{\sqrt{3}}{2}\right) =$

$\dfrac{-\sqrt{2} + \sqrt{6}}{4} = \dfrac{\sqrt{6} - \sqrt{2}}{4}$

7. $\sin \dfrac{\pi}{12} = \sin \left(\dfrac{\pi}{3} - \dfrac{\pi}{4}\right) = \sin \dfrac{\pi}{3} \cos \dfrac{\pi}{4} - \cos \dfrac{\pi}{3} \sin \dfrac{\pi}{4} = \left(\dfrac{\sqrt{3}}{2}\right)\left(\dfrac{\sqrt{2}}{2}\right) - \left(\dfrac{1}{2}\right)\left(\dfrac{\sqrt{2}}{2}\right) = \dfrac{\sqrt{6} - \sqrt{2}}{4}$

9. $\sin \dfrac{5\pi}{12} = \sin \left(\dfrac{2\pi}{3} - \dfrac{\pi}{4}\right) = \sin \dfrac{2\pi}{3} \cos \dfrac{\pi}{4} - \cos \dfrac{2\pi}{3} \sin \dfrac{\pi}{4} = \left(\dfrac{\sqrt{3}}{2}\right)\left(\dfrac{\sqrt{2}}{2}\right) - \left(-\dfrac{1}{2}\right)\left(\dfrac{\sqrt{2}}{2}\right) =$

$\dfrac{\sqrt{6} + \sqrt{2}}{4}$

11. (a) Graphical: Graph $Y_1 = \sin (X + \pi/2)$ and $Y_2 = \cos (X)$ seperately in $[-2\pi, 2\pi, \pi/2]$ by $[-4, 4, 1]$.

   Graph Y_1 is shown in Figure 11a. Graph Y_2 is shown in Figure 11b. The graphs are the same.

   Verbal: If the sine graph is translated $\dfrac{\pi}{2}$ units left it coincides with the cosine graph.

   (b) $\sin \left(t + \dfrac{\pi}{2}\right) = \sin t \cos \dfrac{\pi}{2} + \cos t \sin \dfrac{\pi}{2} = \sin t \, (0) + \cos t \, (1) = \cos t$

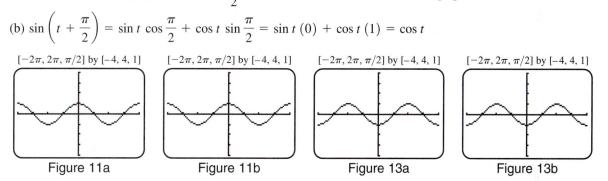

$[-2\pi, 2\pi, \pi/2]$ by $[-4, 4, 1]$ $[-2\pi, 2\pi, \pi/2]$ by $[-4, 4, 1]$ $[-2\pi, 2\pi, \pi/2]$ by $[-4, 4, 1]$ $[-2\pi, 2\pi, \pi/2]$ by $[-4, 4, 1]$

Figure 11a   Figure 11b   Figure 13a   Figure 13b

13. (a) Graphical: Graph $Y_1 = \cos (X + \pi)$ and $Y_2 = - \cos (X)$ seperately in $[-2\pi, 2\pi, \pi/2]$ by $[-4, 4, 1]$.

   Graph Y_1 is shown in Figure 13a. Graph Y_2 is shown in Figure 13b. The graphs are the same.

   Verbal: If the cosine graph is translated π units left it coincides with the cosine graph reflected across the *x*-axis.

   (b) $\cos (t + \pi) = \cos t \cos \pi - \sin t \sin \pi = \cos t \, (-1) - \sin t \, (0) = -\cos t$

15. (a) Graphical: Graph $Y_1 = 1/\cos (X - \pi/2)$ and $Y_2 = 1/\sin (X)$ seperately in $[-2\pi, 2\pi, \pi/2]$ by $[-4, 4, 1]$.

   Graph Y_1 is shown in Figure 15a. Graph Y_2 is shown in Figure 15b. The graphs are the same.

   Verbal: If the secant graph is translated $\dfrac{\pi}{2}$ units left it coincides with the cosecant graph.

   (b) $\sec \left(t - \dfrac{\pi}{2}\right) = \dfrac{1}{\cos (t - \frac{\pi}{2})} = \dfrac{1}{\cos t \cos \frac{\pi}{2} + \sin t \sin \frac{\pi}{2}} = \dfrac{1}{\cos t \, (0) + \sin t \, (1)} = \dfrac{1}{\sin t} = \csc t$

$[-2\pi, 2\pi, \pi/2]$ by $[-4, 4, 1]$ $[-2\pi, 2\pi, \pi/2]$ by $[-4, 4, 1]$

Figure 15a   Figure 15b

17. $\sin\left(\dfrac{\pi}{2} - t\right) = \sin\dfrac{\pi}{2}\cos t - \cos\dfrac{\pi}{2}\sin t = (1)\cos t - (0)\sin t = \cos t$

19. $\sec\left(\dfrac{\pi}{2} - t\right) = \dfrac{1}{\cos\left(\frac{\pi}{2} - t\right)} = \dfrac{1}{\cos\frac{\pi}{2}\cos t + \sin\frac{\pi}{2}\sin t} = \dfrac{1}{(0)\cos t + (1)\sin t} = \dfrac{1}{\sin t} = \csc t$

21. First sketch possible angles for α and β. See Figures 21a & 21b. Note: $\cos\alpha = \dfrac{4}{5}$ and $\cos\beta = \dfrac{12}{13}$.

(a) $\sin(\alpha + \beta) = \sin\alpha\cos\beta + \cos\alpha\sin\beta = \left(\dfrac{3}{5}\right)\left(\dfrac{12}{13}\right) + \left(\dfrac{4}{5}\right)\left(\dfrac{5}{13}\right) = \dfrac{36}{65} + \dfrac{20}{65} = \dfrac{56}{65}$

(b) $\cos(\alpha + \beta) = \cos\alpha\cos\beta - \sin\alpha\sin\beta = \left(\dfrac{4}{5}\right)\left(\dfrac{12}{13}\right) - \left(\dfrac{3}{5}\right)\left(\dfrac{5}{13}\right) = \dfrac{48}{65} - \dfrac{15}{65} = \dfrac{33}{65}$

(c) $\tan(\alpha + \beta) = \dfrac{\sin(\alpha + \beta)}{\cos(\alpha + \beta)} = \dfrac{\frac{56}{65}}{\frac{33}{65}} = \dfrac{56}{65}\cdot\dfrac{65}{33} = \dfrac{56}{33}$

(d) Since both $\sin(\alpha + \beta)$ and $\cos(\alpha + \beta)$ are positive, $\alpha + \beta$ is in quadrant I.

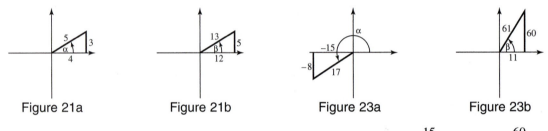

Figure 21a   Figure 21b   Figure 23a   Figure 23b

23. First sketch possible angles for α and β. See Figures 23a & 23b. Note: $\cos\alpha = -\dfrac{15}{17}$ and $\sin\beta = \dfrac{60}{61}$.

(a) $\sin(\alpha + \beta) = \sin\alpha\cos\beta + \cos\alpha\sin\beta = \left(-\dfrac{8}{17}\right)\left(\dfrac{11}{61}\right) + \left(-\dfrac{15}{17}\right)\left(\dfrac{60}{61}\right) = -\dfrac{88}{1037} - \dfrac{900}{1037} =$

$-\dfrac{988}{1037}$

(b) $\cos(\alpha + \beta) = \cos\alpha\cos\beta - \sin\alpha\sin\beta = \left(-\dfrac{15}{17}\right)\left(\dfrac{11}{61}\right) - \left(-\dfrac{8}{17}\right)\left(\dfrac{60}{61}\right) = -\dfrac{165}{1037} + \dfrac{480}{1037} = \dfrac{315}{1037}$

(c) $\tan(\alpha + \beta) = \dfrac{\sin(\alpha + \beta)}{\cos(\alpha + \beta)} = \dfrac{-\frac{988}{1037}}{\frac{315}{1037}} = -\dfrac{988}{1037}\cdot\dfrac{1037}{315} = -\dfrac{988}{315}$

(d) Since $\sin(\alpha + \beta)$ is negative and $\cos(\alpha + \beta)$ is positive, $\alpha + \beta$ is in quadrant IV.

25. First sketch possible angles for α and β. See Figures 25a & 25b. Note: $\sin\alpha = -\dfrac{4}{5}$ and $\sin\beta = -\dfrac{5}{13}$.

(a) $\sin(\alpha + \beta) = \sin\alpha\cos\beta + \cos\alpha\sin\beta = \left(-\dfrac{4}{5}\right)\left(\dfrac{12}{13}\right) + \left(-\dfrac{3}{5}\right)\left(-\dfrac{5}{13}\right) = -\dfrac{48}{65} + \dfrac{15}{65} = -\dfrac{33}{65}$

(b) $\cos(\alpha + \beta) = \cos\alpha\cos\beta - \sin\alpha\sin\beta = \left(-\dfrac{3}{5}\right)\left(\dfrac{12}{13}\right) - \left(-\dfrac{4}{5}\right)\left(-\dfrac{5}{13}\right) = -\dfrac{36}{65} - \dfrac{20}{65} = -\dfrac{56}{65}$

(c) $\tan(\alpha + \beta) = \dfrac{\sin(\alpha + \beta)}{\cos(\alpha + \beta)} = \dfrac{-\frac{33}{65}}{-\frac{56}{65}} = -\dfrac{33}{65}\cdot\left(-\dfrac{65}{56}\right) = \dfrac{33}{56}$

(d) Since both $\sin(\alpha + \beta)$ and $\cos(\alpha + \beta)$ are negative, $\alpha + \beta$ is in quadrant III.

Figure 25a

Figure 25b

Figure 27a

Figure 27b

27. First sketch possible angles for α and β. See Figures 27a & 27b.

Note: $\sin \alpha = \dfrac{5}{13}$, $\cos \alpha = -\dfrac{12}{13}$, $\sin \beta = \dfrac{60}{61}$, $\cos \beta = -\dfrac{11}{60}$

(a) $\sin (\alpha + \beta) = \sin \alpha \cos \beta + \cos \alpha \sin \beta = \left(\dfrac{5}{13}\right)\left(-\dfrac{11}{61}\right) + \left(-\dfrac{12}{13}\right)\left(\dfrac{60}{61}\right) =$

$\left(-\dfrac{55}{793}\right) + \left(-\dfrac{720}{793}\right) = -\dfrac{775}{793}$

(b) $\cos (\alpha + \beta) = \cos \alpha \cos \beta - \sin \alpha \sin \beta = \left(-\dfrac{12}{13}\right)\left(-\dfrac{11}{61}\right) - \left(\dfrac{5}{13}\right)\left(\dfrac{60}{61}\right) = \dfrac{132}{793} - \dfrac{300}{793} = -\dfrac{168}{793}$

(c) $\tan (\alpha + \beta) = \dfrac{\sin (\alpha + \beta)}{\cos (\alpha + \beta)} = \dfrac{-\frac{775}{793}}{-\frac{168}{793}} = \dfrac{775}{793} \cdot \left(-\dfrac{793}{168}\right) = \dfrac{775}{168}$

(d) Since both $\sin (\alpha + \beta)$ and $\cos (\alpha + \beta)$ are negative, $\alpha + \beta$ is in quadrant III.

29. $\cos\left(t - \dfrac{\pi}{4}\right) = \cos t \cos \dfrac{\pi}{4} + \sin t \sin \dfrac{\pi}{4} = \cos t \left(\dfrac{\sqrt{2}}{2}\right) + \sin t \left(\dfrac{\sqrt{2}}{2}\right) = \dfrac{\sqrt{2}}{2}(\cos t + \sin t)$

31. $\tan\left(t + \dfrac{\pi}{4}\right) = \dfrac{\tan t + \tan \frac{\pi}{4}}{1 - \tan t \tan \frac{\pi}{4}} = \dfrac{\tan t + 1}{1 - \tan t \,(1)} = \dfrac{1 + \tan t}{1 - \tan t}$

33. $\dfrac{\cos (x - y)}{\cos (x + y)} = \dfrac{\cos x \cos y + \sin x \sin y}{\cos x \cos y - \sin x \sin y} = \dfrac{\cos x \cos y + \sin x \sin y}{\cos x \cos y - \sin x \sin y} \cdot \dfrac{\sec x \sec y}{\sec x \sec y} = \dfrac{1 + \tan x \tan y}{1 - \tan x \tan y}$

35. $\dfrac{\cos (\alpha - \beta)}{\cos \alpha \sin \beta} = \dfrac{\cos \alpha \cos \beta + \sin \alpha \sin \beta}{\cos \alpha \sin \beta} = \dfrac{\cos \alpha \cos \beta}{\cos \alpha \sin \beta} + \dfrac{\sin \alpha \sin \beta}{\cos \alpha \sin \beta} = \cot \beta + \tan \alpha = \tan \alpha + \cot \beta$

37. $\sin 2t = \sin (t + t) = \sin t \cos t + \cos t \sin t = 2 \sin t \cos t$

39. $\sin (\alpha + \beta) + \sin (\alpha - \beta) = (\sin \alpha \cos \beta + \cos \alpha \sin \beta) + (\sin \alpha \cos \beta - \cos \alpha \sin \beta) = 2 \sin \alpha \cos \beta$

41. $\tan (\pi - \theta) = \dfrac{\tan \pi - \tan \theta}{1 + \tan \pi \tan \theta} = \dfrac{0 - \tan \theta}{1 + (0) \tan \theta} = - \tan \theta$

43. $\tan (x - y) - \tan (y - x) = \dfrac{\tan x - \tan y}{1 + \tan x \tan y} - \dfrac{\tan y - \tan x}{1 + \tan y \tan x} = \dfrac{\tan x - \tan y - \tan y + \tan x}{1 + \tan x \tan y} =$

$\dfrac{2 \tan x - 2 \tan y}{1 + \tan x \tan y} = \dfrac{2 (\tan x - \tan y)}{1 + \tan x \tan y}$

45. $\dfrac{\sin (x + y)}{\cos x \cos y} = \dfrac{\sin x \cos y + \cos x \sin y}{\cos x \cos y} = \dfrac{\sin x \cos y}{\cos x \cos y} + \dfrac{\cos x \sin y}{\cos x \cos y} = \dfrac{\sin x}{\cos x} + \dfrac{\sin y}{\cos y} = \tan x + \tan y$

47. $\tan \gamma = \tan (\alpha - \beta) = \dfrac{\tan \alpha - \tan \beta}{1 + \tan \alpha \tan \beta} = \dfrac{\left(\frac{6}{7}\right) - \left(\frac{5}{7}\right)}{1 + \left(\frac{6}{7}\right)\left(\frac{5}{7}\right)} = \dfrac{\frac{1}{7}}{1 + \frac{30}{49}} = \dfrac{\frac{1}{7}}{\frac{79}{49}} = \dfrac{1}{7} \cdot \dfrac{49}{79} = \dfrac{7}{79}$

49. $\tan \theta = \tan (\beta - \alpha) = \dfrac{\tan \beta - \tan \alpha}{1 + \tan \beta \tan \alpha} = \dfrac{m_2 - m_1}{1 + m_2 m_1} = \dfrac{m_2 - m_1}{1 + m_1 m_2}$

51. For $y = 2x - 3$, $m_2 = 2$ and for $y = \dfrac{3}{5}x + 1$, $m_1 = \dfrac{3}{5}$.

$$\tan \theta = \frac{2 - \frac{3}{5}}{1 + 2\left(\frac{3}{5}\right)} = \frac{\frac{7}{5}}{\frac{11}{5}} = \frac{7}{5} \cdot \frac{5}{11} = \frac{7}{11} \Rightarrow \theta = \tan^{-1}\frac{7}{11} \approx 32.5°$$

53. (a) $F = 2.89W \cos \theta \Rightarrow F = 2.89(200) \cos \dfrac{\pi}{4} \approx 409 \text{ lb}s$

(b) $2.89(200) \cos \theta = 400 \Rightarrow \cos \theta = \dfrac{400}{2.89(200)} \Rightarrow \theta = \cos^{-1}\left(\dfrac{2}{2.89}\right) \approx 0.81$ or about $46.2°$.

55. (a) Graph $Y_1 = 4 \cos (220\pi X) + 3 \sin (220\pi X)$ in $[0, 0.02, 0.001]$ by $[-6, 6, 1]$. See Figure 55.

(b) A maximum occurs at approximately $(0.00093114, 5)$. Thus $a = 5$. Since $\sin \theta$ is maximum when $\theta = \dfrac{\pi}{2}$,

we let $220\pi(0.00093114) + k = \dfrac{\pi}{2}$. Thus $k = \dfrac{\pi}{2} - 220\pi(0.00093114) \approx 0.9272$.

(c) $a \sin (220\pi t + k) \approx 5 \sin (220\pi t + 0.9272) =$

$5\left[\sin (220\pi t) \cos 0.9272 + \cos (220\pi t) \sin 0.9272\right] \approx$

$5\left[\sin (220\pi t)(0.6) + \cos (220\pi t)(0.8)\right] = 3 \sin (220\pi t) + 4 \cos (220\pi t)$

$[0, 0.02, 0.001]$ by $[-6, 6, 1]$

Figure 55

7.5: Multiple-Angle Identities

1. $\sin^2 10t \Rightarrow \theta = 10t \Rightarrow \sin^2 10t = \dfrac{1 - \cos (2(10t))}{2} = \dfrac{1 - \cos (20t)}{2}$

3. $\tan^2 5t \Rightarrow \theta = 5t \Rightarrow \tan^2 5t = \dfrac{1 - \cos (2(5t))}{1 + \cos (2(5t))} = \dfrac{1 - \cos (10t)}{1 + \cos (10t)}$

5. $\sin 20x \Rightarrow \theta = 10x \Rightarrow \sin 20x = 2 \sin 10x \cos 10x$

7. $\tan 5x \Rightarrow \theta = 10x \Rightarrow \tan \dfrac{\theta}{2} = \dfrac{1 - \cos (10x)}{\sin (10x)}$

9. (a) $\sin 30° + \sin 30° = \dfrac{1}{2} + \dfrac{1}{2} = 1$

(b) $\sin 60° = \dfrac{\sqrt{3}}{2}$

The results are not the same.

11. (a) $\cos 60° + \cos 60° = \dfrac{1}{2} + \dfrac{1}{2} = 1$

(b) $\cos 120° = -\dfrac{1}{2}$

The results are not the same.

13. (a) $\tan 45° + \tan 45° = 1 + 1 = 2$

(b) $\tan 90°$ is undefined

The results are not the same.

15. Graphical: Graph $Y_1 = \tan (2X)$ and $Y_2 = 2 \tan (X)$ in dot mode in $[-2\pi, 2\pi, \pi/2]$ by $[-4, 4, 1]$.

Graph Y_1 is shown in Figure 15a. Graph Y_2 is shown in Figure 15b. Note that the graphs are different.

Symbolic: $\tan 2\theta = \dfrac{2 \tan \theta}{1 - \tan^2 \theta} \neq 2 \tan \theta$, unless $\tan \theta = 0$

$[-2\pi, 2\pi, \pi/2]$ by $[-4, 4, 1]$ $[-2\pi, 2\pi, \pi/2]$ by $[-4, 4, 1]$

Figure 15a	Figure 15b	Figure 17a	Figure 17b

17. (a) $\sin 2\theta = 2 \sin \theta \cos \theta = 2\left(\dfrac{3}{5}\right)\left(\dfrac{4}{5}\right) = \dfrac{24}{25}$

$\cos 2\theta = \cos^2 \theta - \sin^2 \theta = \left(\dfrac{4}{5}\right)^2 - \left(\dfrac{3}{5}\right)^2 = \dfrac{16 - 9}{25} = \dfrac{7}{25}$

$\tan 2\theta = \dfrac{\sin 2\theta}{\cos 2\theta} = \dfrac{\frac{24}{25}}{\frac{7}{25}} = \dfrac{24}{25} \cdot \dfrac{25}{7} = \dfrac{24}{7}$

(b) Since θ is in quadrant I, $\theta = \cos^{-1} \dfrac{4}{5}$. Numerical support is shown in Figures 17a & 17b.

19. Since $\sin \theta$ is negative and $\cos \theta$ is positive, θ is in quadrant IV. One possibility is shown in Figure 19a.

We see that $\cos \theta = \dfrac{7}{25}$.

(a) $\sin 2\theta = 2 \sin \theta \cos \theta = 2\left(-\dfrac{24}{25}\right)\left(\dfrac{7}{25}\right) = -\dfrac{336}{625}$

$\cos 2\theta = \cos^2 \theta - \sin^2 \theta = \left(\dfrac{7}{25}\right)^2 - \left(-\dfrac{24}{25}\right)^2 = \dfrac{49 - 576}{625} = -\dfrac{527}{625}$

$\tan 2\theta = \dfrac{\sin 2\theta}{\cos 2\theta} = \dfrac{-\frac{336}{625}}{-\frac{527}{625}} = -\dfrac{336}{625} \cdot \left(-\dfrac{625}{527}\right) = \dfrac{336}{527}$

(b) Since θ is in quadrant IV, $\theta = \sin^{-1}\left(-\dfrac{24}{25}\right)$. Numerical support is shown in Figures 19b & 19c.

Figure 19a   Figure 19b   Figure 19c

21. Since $\sin \theta$ is negative and $\sec \theta$ is positive, θ is in quadrant IV. One possibility is shown in Figure 21a.

We see that $\cos \theta = \dfrac{60}{61}$.

(a) $\sin 2\theta = 2 \sin \theta \cos \theta = 2\left(-\dfrac{11}{61}\right)\left(\dfrac{60}{61}\right) = -\dfrac{1320}{3721}$

$\cos 2\theta = \cos^2 \theta - \sin^2 \theta = \left(\dfrac{60}{61}\right)^2 - \left(-\dfrac{11}{61}\right)^2 = \dfrac{3600 - 121}{3721} = \dfrac{3479}{3721}$

$\tan 2\theta = \dfrac{\sin 2\theta}{\cos 2\theta} = \dfrac{-\frac{1320}{3721}}{\frac{3479}{3721}} = -\dfrac{1320}{3721} \cdot \dfrac{3721}{3479} = -\dfrac{1320}{3479}$

(b) Since θ is in quadrant IV, $\theta = \sin^{-1}\left(-\dfrac{11}{61}\right)$. Numerical support is shown in Figures 21b & 21c.

Figure 21a   Figure 21b   Figure 21c

23. Since $\tan \theta$ is positive and $\cos \theta$ is negative, θ is in quadrant III. One possibility is shown in Figure 23a.

We see that $\sin \theta = -\dfrac{7}{25}$ and $\cos \theta = -\dfrac{24}{25}$.

(a) $\sin 2\theta = 2 \sin \theta \cos \theta = 2\left(-\dfrac{7}{25}\right)\left(-\dfrac{24}{25}\right) = \dfrac{336}{625}$

$\cos 2\theta = \cos^2 \theta - \sin^2 \theta = \left(-\dfrac{24}{25}\right)^2 - \left(-\dfrac{7}{25}\right)^2 = \dfrac{576 - 49}{625} = \dfrac{527}{625}$

$\tan 2\theta = \dfrac{\sin 2\theta}{\cos 2\theta} = \dfrac{\frac{336}{625}}{\frac{527}{625}} = \dfrac{336}{625} \cdot \dfrac{625}{527} = \dfrac{336}{527}$

(b) Since θ is in quadrant III, $\theta = 360° - \cos^{-1}\left(-\dfrac{24}{25}\right)$. Numerical support is shown in Figures 23b & 23c.

Figure 23a   Figure 23b   Figure 23c

25. Let $\theta = \cos^{-1} 1$. It follows that $\cos \theta = 1$ and $\sin \theta = 0$ (since $\theta = 0°$).

$\sin \left(2 \cos^{-1} 1\right) = \sin (2\theta) = 2 \sin \theta \cos \theta = 2(0)(1) = 0$

27. Let $\theta = \sin^{-1} \dfrac{7}{25}$. It follows that $\sin \theta = \dfrac{7}{25}$.

$\cos \left(2 \sin^{-1} \dfrac{7}{25}\right) = \cos (2\theta) = 1 - 2\sin^2 \theta = 1 - 2\left(\dfrac{7}{25}\right)^2 = 1 - \dfrac{98}{625} = \dfrac{527}{625}$

29. Let $\theta = \sin^{-1}\dfrac{5}{13}$. It follows that $\sin\theta = \dfrac{5}{13}$ and $\cos\theta = \dfrac{12}{13}$. See Figure 29.

$$\cos\left(3\,\sin^{-1}\dfrac{5}{13}\right) = \cos(3\theta) = 4\cos^3\theta - 3\cos\theta = 4\left(\dfrac{12}{13}\right)^3 - 3\left(\dfrac{12}{13}\right) = \dfrac{828}{2197}$$

31. Let $\theta = \tan^{-1} x$, where $x < 0$. It follows that $\sin\theta = \dfrac{x}{\sqrt{x^2+1}}$ and $\cos\theta = \dfrac{1}{\sqrt{x^2+1}}$. See Figure 31.

$$\sin(2\,\tan^{-1} x) = \sin(2\theta) = 2\sin\theta\cos\theta = 2\left(\dfrac{x}{\sqrt{x^2+1}}\right)\left(\dfrac{1}{\sqrt{x^2+1}}\right) = \dfrac{2x}{x^2+1}$$

Figure 29   Figure 31   Figure 33a   Figure 33b

33. Let $\alpha = \sin^{-1}\dfrac{3}{5}$. It follows that $\sin\alpha = \dfrac{3}{5}$ and $\cos\alpha = \dfrac{4}{5}$. See Figure 33a.

Let $\beta = \sin^{-1}\dfrac{4}{5}$. It follows that $\sin\beta = \dfrac{4}{5}$ and $\cos\beta = \dfrac{3}{5}$. See Figure 33b.

$$\cos\left(\sin^{-1}\dfrac{3}{5} - \sin^{-1}\dfrac{4}{5}\right) = \cos(\alpha - \beta) = \cos\alpha\cos\beta + \sin\alpha\sin\beta = \left(\dfrac{4}{5}\right)\left(\dfrac{3}{5}\right) + \left(\dfrac{3}{5}\right)\left(\dfrac{4}{5}\right) = \dfrac{24}{25}$$

35. $2\cos\theta\sin\theta = 2\sin\theta\cos\theta = \sin 2\theta$

37. $\sin\theta\cos\theta = \dfrac{1}{2}(2\sin\theta\cos\theta) = \dfrac{1}{2}\sin 2\theta$

39. $2\cos^2 2\theta - 1 = \cos 4\theta$ (let $t = 2\theta$, then $2\cos^2 t - 1 = \cos 2t$)

41. $\sin^2 3\theta + \cos^2 3\theta = 1$ (let $t = 3\theta$, then $\sin^2 t + \cos^2 t = 1$)

43. $\csc^2 5x - 1 = \cot^2 5x$ (let $t = 5x$, then $\csc^2 t - 1 = \cot^2 t$)

45. $\cos^2(22.5°) = \dfrac{1 + \cos(2 \cdot 22.5°)}{2} = \dfrac{1 + \cos 45°}{2} = \dfrac{1 + \dfrac{\sqrt{2}}{2}}{2} = \dfrac{\dfrac{2 + \sqrt{2}}{2}}{2} = \dfrac{2 + \sqrt{2}}{4}$

Numerical support is shown in Figure 45.

```
(cos(22.5))²
        .8535533906
(2+√(2))/4
        .8535533906
```

```
(tan(75))²
        13.92820323
(2+√(3))/(2-√(3))
)
        13.92820323
```

Figure 45   Figure 47

47. $\tan^2(75°) = \dfrac{1 - \cos(2 \cdot 75°)}{1 + \cos(2 \cdot 75°)} = \dfrac{1 - \cos 150°}{1 + \cos 150°} = \dfrac{1 - \left(-\dfrac{\sqrt{3}}{2}\right)}{1 + \left(-\dfrac{\sqrt{3}}{2}\right)} = \dfrac{\dfrac{2 + \sqrt{3}}{2}}{\dfrac{2 - \sqrt{3}}{2}} = \dfrac{2 + \sqrt{3}}{2 - \sqrt{3}}$

Numerical support is shown in Figure 47.

49. (a) Since the angle is in quadrant I, we use the positive value of the half-angle formula.

$$\cos 15° = \sqrt{\frac{1 + \cos 30°}{2}} = \sqrt{\frac{1 + \frac{\sqrt{3}}{2}}{2}} = \sqrt{\frac{\frac{2 + \sqrt{3}}{2}}{2}} = \sqrt{\frac{2 + \sqrt{3}}{4}} = \frac{\sqrt{2 + \sqrt{3}}}{2}$$

(b) Since the angle is in quadrant IV, we use the negative value of the half-angle formula.

$$\tan(-15°) = -\sqrt{\frac{1 - \cos(-30°)}{1 + \cos(-30°)}} = -\sqrt{\frac{1 - \frac{\sqrt{3}}{2}}{1 + \frac{\sqrt{3}}{2}}} = -\sqrt{\frac{\frac{2 - \sqrt{3}}{2}}{\frac{2 + \sqrt{3}}{2}}} = -\sqrt{\frac{2 - \sqrt{3}}{2 + \sqrt{3}}} \text{ or}$$

$-\dfrac{1}{2 + \sqrt{3}}$ or $\sqrt{3} - 2$. Numerical support is shown in Figure 49a & 49b.

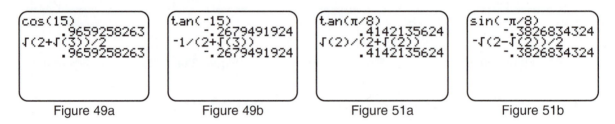

Figure 49a   Figure 49b   Figure 51a   Figure 51b

51. (a) Since the angle is in quadrant I, we use the positive value of the half-angle formula.

$$\tan\left(\frac{\pi}{8}\right) = \sqrt{\frac{1 - \cos \frac{\pi}{4}}{1 + \cos \frac{\pi}{4}}} = \sqrt{\frac{1 - \frac{\sqrt{2}}{2}}{1 + \frac{\sqrt{2}}{2}}} = \sqrt{\frac{\frac{2 - \sqrt{2}}{2}}{\frac{2 + \sqrt{2}}{2}}} = \sqrt{\frac{2 - \sqrt{2}}{2 + \sqrt{2}}} \text{ or } \frac{\sqrt{2}}{2 + \sqrt{2}} \text{ or } \sqrt{2} - 1$$

(b) Since the angle is in quadrant IV, we use the negative value of the half-angle formula.

$$\sin\left(-\frac{\pi}{8}\right) = -\sqrt{\frac{1 - \cos\left(-\frac{\pi}{4}\right)}{2}} = -\sqrt{\frac{1 - \left(\frac{\sqrt{2}}{2}\right)}{2}} = -\sqrt{\frac{\frac{2 - \sqrt{2}}{2}}{2}} = -\sqrt{\frac{2 - \sqrt{2}}{4}} = -\frac{\sqrt{2 - \sqrt{2}}}{2}$$

Numerical support is shown in Figure 51a & 51b.

53. $\sqrt{\dfrac{1 - \cos 60°}{2}} = \sin\left(\dfrac{60°}{2}\right) = \sin 30°$. Numerical support is shown in Figure 53.

55. $\sqrt{\dfrac{1 + \cos 50°}{2}} = \cos\left(\dfrac{50°}{2}\right) = \cos 25°$. Numerical support is shown in Figure 55.

Figure 53   Figure 55

57. $\sqrt{\dfrac{1 - \cos 40°}{1 + \cos 40°}} = \tan\left(\dfrac{40°}{2}\right) = \tan 20°$. Numerical support is shown in Figure 57.

59. Since $0° < \dfrac{\theta}{2} < 45°$, we know that $\dfrac{\theta}{2}$ is in quadrant I.

$$\sin\frac{\theta}{2} = \sqrt{\frac{1 - \cos\theta}{2}} = \sqrt{\frac{1 - \frac{4}{5}}{2}} = \sqrt{\frac{\frac{5-4}{5}}{2}} = \sqrt{\frac{1}{10}} = \frac{1}{\sqrt{10}}$$

$$\cos\frac{\theta}{2} = \sqrt{\frac{1 + \cos\theta}{2}} = \sqrt{\frac{1 + \frac{4}{5}}{2}} = \sqrt{\frac{\frac{5+4}{5}}{2}} = \sqrt{\frac{9}{10}} = \frac{3}{\sqrt{10}}$$

$$\tan\frac{\theta}{2} = \sqrt{\frac{1 - \cos\theta}{1 + \cos\theta}} = \sqrt{\frac{1 - \frac{4}{5}}{1 + \frac{4}{5}}} = \sqrt{\frac{\frac{5-4}{5}}{\frac{5+4}{5}}} = \sqrt{\frac{\frac{1}{5}}{\frac{9}{5}}} = \sqrt{\frac{1}{9}} = \frac{1}{3}$$

```
√((1-cos(40))/(1
+cos(40)))
          .3639702343
tan(20)
          .3639702343
```

Figure 57

61. Since $-45° < \dfrac{\theta}{2} < 0°$, we know that $\dfrac{\theta}{2}$ is in quadrant IV.

By sketching an appropriate triangle in quadrant IV, we see that $\cos\theta = \dfrac{12}{13}$.

$$\sin\frac{\theta}{2} = -\sqrt{\frac{1 - \cos\theta}{2}} = -\sqrt{\frac{1 - \frac{12}{13}}{2}} = -\sqrt{\frac{\frac{13-12}{13}}{2}} = -\sqrt{\frac{1}{26}} = -\frac{1}{\sqrt{26}}$$

$$\cos\frac{\theta}{2} = \sqrt{\frac{1 + \cos\theta}{2}} = \sqrt{\frac{1 + \frac{12}{13}}{2}} = \sqrt{\frac{\frac{13+12}{13}}{2}} = \sqrt{\frac{25}{26}} = \frac{5}{\sqrt{26}}$$

$$\tan\frac{\theta}{2} = -\sqrt{\frac{1 - \cos\theta}{1 + \cos\theta}} = -\sqrt{\frac{1 - \frac{12}{13}}{1 + \frac{12}{13}}} = -\sqrt{\frac{\frac{13-12}{13}}{\frac{13+12}{13}}} = -\sqrt{\frac{\frac{1}{13}}{\frac{25}{13}}} = -\sqrt{\frac{1}{25}} = -\frac{1}{5}$$

63. Since $45° < \dfrac{\theta}{2} < 90°$, we know that $\dfrac{\theta}{2}$ is in quadrant I.

By sketching an appropriate triangle in quadrant II, we see that $\cos\theta = -\dfrac{7}{25}$.

$$\sin\frac{\theta}{2} = \sqrt{\frac{1 - \cos\theta}{2}} = \sqrt{\frac{1 - \left(-\frac{7}{25}\right)}{2}} = \sqrt{\frac{\frac{25+7}{25}}{2}} = \sqrt{\frac{32}{50}} = \sqrt{\frac{16}{25}} = \frac{4}{5}$$

$$\cos\frac{\theta}{2} = \sqrt{\frac{1 + \cos\theta}{2}} = \sqrt{\frac{1 + \left(-\frac{7}{25}\right)}{2}} = \sqrt{\frac{\frac{25-7}{25}}{2}} = \sqrt{\frac{18}{50}} = \sqrt{\frac{9}{25}} = \frac{3}{5}$$

$$\tan\frac{\theta}{2} = \sqrt{\frac{1 - \cos\theta}{1 + \cos\theta}} = \sqrt{\frac{1 - \left(-\frac{7}{25}\right)}{1 + \left(-\frac{7}{25}\right)}} = \sqrt{\frac{\frac{25+7}{25}}{\frac{25-7}{25}}} = \sqrt{\frac{\frac{32}{25}}{\frac{18}{25}}} = \sqrt{\frac{32}{18}} = \sqrt{\frac{16}{9}} = \frac{4}{3}$$

65. $4 \sin 2x = 4(2 \sin x \cos x) = 8 \sin x \cos x$

Graph $Y_1 = 4 \sin (2X)$ and $Y_2 = 8 \sin (X) \cos (X)$ in $[-2\pi, 2\pi, \pi/2]$ by $[-6, 6, 1]$.

Graph Y_1 is shown in Figure 65a. Graph Y_2 is shown in Figure 65b. The graphs are the same.

Table $Y_1 = 4 \sin (2X)$ and $Y_2 = 8 \sin (X) \cos (X)$ starting at $x = 0$, incrementing by $\dfrac{\pi}{6}$. See Figure 65c.

The tables are the same.

Figure 65a   Figure 65b   Figure 65c

67. $\dfrac{2 - \sec^2 x}{\sec^2 x} = \dfrac{2}{\sec^2 x} - \dfrac{\sec^2 x}{\sec^2 x} = 2 \cos^2 x - 1 = \cos 2x$

Graph $Y_1 = (2 - 1/(\cos (X))^2)/(1/(\cos (X))^2)$ and $Y_2 = \cos (2X)$ in $[-2\pi, 2\pi, \pi/2]$ by $[-2, 2, 1]$.

Graph Y_1 is shown in Figure 67a. Graph Y_2 is shown in Figure 67b. The graphs are the same.

Table $Y_1 = (2 - 1/(\cos (X))^2)/(1/(\cos (X))^2)$ and $Y_2 = \cos (2X)$ starting at $x = 0$, incrementing by $\dfrac{\pi}{6}$.

See Figure 67c. The tables are the same.

Figure 67a   Figure 67b   Figure 67c

69. $\sec 2x = \dfrac{1}{\cos 2x} = \dfrac{1}{1 - 2 \sin^2 x}$

Graph $Y_1 = 1/\cos (2X)$ and $Y_2 = 1/(1 - 2 (\sin (X))^2)$ in $[-2\pi, 2\pi, \pi/2]$ by $[-4, 4, 1]$.

Graph Y_1 is shown in Figure 69a. Graph Y_2 is shown in Figure 69b. The graphs are the same.

Table $Y_1 = 1/\cos (2X)$ and $Y_2 = 1/(1 - 2 (\sin (X))^2)$ starting at $x = 0$, incrementing by $\dfrac{\pi}{6}$.

See Figure 69c. The tables are the same.

Figure 69a   Figure 69b   Figure 69c

71. $\sin 3\theta = \sin (2\theta + \theta) = \sin 2\theta \cos \theta + \cos 2\theta \sin \theta = (2 \sin \theta \cos \theta) \cos \theta + (1 - 2 \sin^2 \theta) \sin \theta =$

$2 \sin \theta \cos^2 \theta + \sin \theta - 2 \sin^3 \theta = 2 \sin \theta (1 - \sin^2 \theta) + \sin \theta - 2 \sin^3 \theta =$

$2 \sin \theta - 2 \sin^3 \theta + \sin \theta - 2 \sin^3 \theta = 3 \sin \theta - 4 \sin^3 \theta$

Graph $Y_1 = \sin (3X)$ and $Y_2 = 3 \sin (X) - 4 (\sin (X))\text{^}3$ in $[-2\pi, 2\pi, \pi/2]$ by $[-2, 2, 1]$.

Graph Y_1 is shown in Figure 71a. Graph Y_2 is shown in Figure 71b. The graphs are the same.

Table $Y_1 = \sin (3X)$ and $Y_2 = 3 \sin (X) - 4 (\sin (X))\text{^}3$ starting at $x = 0$, incrementing by $\dfrac{\pi}{6}$.

See Figure 71c. The tables are the same.

$[-2\pi, 2\pi, \pi/2]$ by $[-2, 2, 1]$ $[-2\pi, 2\pi, \pi/2]$ by $[-2, 2, 1]$

X	Y1	Y2
0	0	0
.5236	1	1
1.0472	0	0
1.5708	-1	-1
2.0944	0	0
2.618	1	1
3.1416	0	0

Y1**⯁**sin(3X)

Figure 71a   Figure 71b   Figure 71c

73. $\sin 4\theta = \sin (2 \cdot 2\theta) = 2 \sin 2\theta \cos 2\theta = 2(2 \sin \theta \cos \theta) \cos 2\theta = 4 \sin \theta \cos \theta \cos 2\theta$

Graph $Y_1 = \sin (4X)$ and $Y_2 = 4 \sin (X) \cos (X) \cos (2X)$ in $[-2\pi, 2\pi, \pi/2]$ by $[-2, 2, 1]$.

Graph Y_1 is shown in Figure 73a. Graph Y_2 is shown in Figure 73b. The graphs are the same.

Table $Y_1 = \sin (4X)$ and $Y_2 = 4 \sin (X) \cos (X) \cos (2X)$ starting at $x = 0$, incrementing by $\dfrac{\pi}{6}$.

See Figure 73c. The tables are the same.

$[-2\pi, 2\pi, \pi/2]$ by $[-2, 2, 1]$ $[-2\pi, 2\pi, \pi/2]$ by $[-2, 2, 1]$

X	Y1	Y2
0	0	0
.5236	.86603	.86603
1.0472	-.866	-.866
1.5708	0	0
2.0944	.86603	.86603
2.618	-.866	-.866
3.1416	-2E-13	0

Y1**⯁**sin(4X)

Figure 73a   Figure 73b   Figure 73c

75. $\dfrac{\sin 2\theta}{\sin \theta} = \dfrac{2 \sin \theta \cos \theta}{\sin \theta} = 2 \cos \theta$

77. $2 \cos^2 \dfrac{\theta}{2} = 2\left(\dfrac{1 + \cos \theta}{2}\right) = 1 + \cos \theta$

79. $\cos^4 \theta - \sin^4 \theta = (\cos^2 \theta - \sin^2 \theta)(\cos^2 \theta + \sin^2 \theta) = (\cos 2\theta)(1) = \cos 2\theta$

81. $\csc 2t = \dfrac{1}{\sin 2t} = \dfrac{1}{2 \sin t \cos t} = \dfrac{\csc t}{2 \cos t}$

83. $\tan \dfrac{x}{2} = \dfrac{\sin \frac{x}{2}}{\cos \frac{x}{2}} = \dfrac{2 \sin \frac{x}{2}}{2 \cos \frac{x}{2}} = \dfrac{2 \sin \frac{x}{2} \cos \frac{x}{2}}{2 \cos^2 \frac{x}{2}} = \dfrac{\sin x}{1 + (2 \cos^2 \frac{x}{2} - 1)} = \dfrac{\sin x}{1 + \cos x}$

85. (a) $\cos 50° \sin 20° = \dfrac{1}{2}(\sin (50° + 20°) - \sin (50° - 20°)) = \dfrac{1}{2}(\sin 70° - \sin 30°)$

(b) $\cos 2x \cos x = \dfrac{1}{2}(\cos (2x + x) + \cos (2x - x)) = \dfrac{1}{2}(\cos (3x) + \cos (x))$

87. (a) $\sin 7\theta \cos 3\theta = \dfrac{1}{2}(\sin(7\theta + 3\theta) + \sin(7\theta - 3\theta)) = \dfrac{1}{2}(\sin 10\theta + \sin 4\theta)$

(b) $\sin 8x \sin 4x = \dfrac{1}{2}(\cos(8x - 4x) - \cos(8x + 4x)) = \dfrac{1}{2}(\cos 4x - \cos 12x)$

89. (a) $\sin 40° + \sin 30° = 2 \sin\left(\dfrac{40° + 30°}{2}\right)\cos\left(\dfrac{40° - 30°}{2}\right) = 2 \sin 35° \cos 5°$

(b) $\cos 45° + \cos 35° = 2 \cos\left(\dfrac{45° + 35°}{2}\right)\cos\left(\dfrac{45° - 35°}{2}\right) = 2 \cos 40° \cos 5°$

91. (a) $\cos 6\theta + \cos 4\theta = 2 \cos\left(\dfrac{6\theta + 4\theta}{2}\right)\cos\left(\dfrac{6\theta - 4\theta}{2}\right) = 2 \cos 5\theta \cos \theta$

(b) $\sin 7x + \sin 4x = 2 \sin\left(\dfrac{7x + 4x}{2}\right)\cos\left(\dfrac{7x - 4x}{2}\right) = 2 \sin \dfrac{11x}{2} \cos \dfrac{3x}{2}$

93. (a) Symbolic: $\cos 2\theta = 1 \Rightarrow 2\theta = \cos^{-1} 1 \Rightarrow 2\theta = 0°, 360° \Rightarrow \theta = 0°, 180°$

(b) Graphical: Graph $Y_1 = \cos(2X)$ and $Y_2 = 1$ in degree mode in $[0°, 360°, 30°]$ by $[-2, 2, 1]$.

In $[0°, 360°)$, the x-coordinates of the intersection points are $x = 0°, 180°$. See Figure 93b.

(c) Numerical: Table $Y_1 = \cos(2X)$ and $Y_2 = 1$ in degree mode starting at $x = 0$, incrementing by 60.

In the interval $[0°, 360°)$, the tables match when $x = 0°, 180°$. See Figure 93c.

$[0°, 360°, 30°]$ by $[-2, 2, 1]$ $[0°, 360°, 30°]$ by $[-2, 2, 1]$

Figure 93b   Figure 93c   Figure 95b   Figure 95c

95. (a) Symbolic: $\sin 2\theta + \cos \theta = 0 \Rightarrow 2\sin\theta\cos\theta + \cos\theta = 0 \Rightarrow \cos\theta(2\sin\theta + 1) = 0 \Rightarrow \cos\theta = 0,$

$2\sin\theta + 1 = 0.$ $\cos\theta = 0 \Rightarrow \theta = 90°, 270°.$ $2\sin\theta + 1 = 0 \Rightarrow \sin\theta = -\dfrac{1}{2} \Rightarrow \theta = 210°, 330°.$

(b) Graphical: Graph $Y_1 = \sin(2X) + \cos(X)$ and $Y_2 = 0$ in degree mode in $[0°, 360°, 30°]$ by $[-2, 2, 1]$.

In $[0°, 360°)$, the x-coordinates of the intersection points are $x = 90°, 210°, 270°, 330°$. See Figure 95b.

(c) Numerical: Table $Y_1 = \sin(2X) + \cos(X)$ and $Y_2 = 0$ in degree mode starting at $x = 30$, incrementing

by 60. In the interval $[0°, 360°)$, the tables match when $x = 90°, 210°, 270°, 330°$. See Figure 95c.

97. (a) Symbolic: $\sin \dfrac{\theta}{2} = 1 \Rightarrow \dfrac{\theta}{2} = \sin^{-1} 1 \Rightarrow \dfrac{\theta}{2} = 90° \Rightarrow \theta = 180°$

(b) Graphical: Graph $Y_1 = \sin(X/2)$ and $Y_2 = 1$ in degree mode in $[0°, 360°, 30°]$ by $[-2, 2, 1]$.

In $[0°, 360°)$, the x-coordinate of the intersection point is $x = 180°$. See Figure 97b.

(c) Numerical: Table $Y_1 = \sin(X/2)$ and $Y_2 = 1$ in degree mode starting at $x = 0$, incrementing by 60.

In the interval $[0°, 360°)$, the tables match when $x = 180°$. See Figure 97c.

Figure 97b

Figure 97c

Figure 99b

Figure 99c

99. (a) Symbolic: $\sqrt{2} \sin \dfrac{\theta}{2} - 1 = 0 \Rightarrow \sqrt{2} \sin \dfrac{\theta}{2} = 1 \Rightarrow \sin \dfrac{\theta}{2} = \dfrac{\sqrt{2}}{2} \Rightarrow \dfrac{\theta}{2} = \sin^{-1} \dfrac{\sqrt{2}}{2} \Rightarrow$

$\dfrac{\theta}{2} = 45°, 135° \Rightarrow \theta = 90°, 270°$

(b) Graphical: graph $Y_1 = \sqrt{2} \sin (X/2) - 1$ and $Y_2 = 0$ in degree mode in $[0°, 360°, 30°]$ by $[-2, 2, 1]$. In

$[0°, 360°)$, the *x*-coordinates of the intersection points are $x = 90°, 270°$. See Figure 99b.

(c) Numeric: table $Y_1 = \sqrt{2} \sin (X/2) - 1$ and $Y_2 = 0$ in degree mode starting at $x = 0$, incrementing by 45.

In the interval $[0, 360)$, the tables match when $x = 90°, 270°$. See Figure 99c.

101. $2 \cos 2t = \sqrt{3} \Rightarrow \cos 2t = \dfrac{\sqrt{3}}{2}$. $\cos 2t = \dfrac{\sqrt{3}}{2} \Rightarrow 2t = \dfrac{\pi}{6} + 2\pi n, \dfrac{11\pi}{6} + 2\pi n \Rightarrow$

$t = \dfrac{\pi}{12} + \pi n, \dfrac{11\pi}{12} + \pi n.$

103. $\sin 2t + \sin t = 0 \Rightarrow 2 \sin t \cos t + \sin t = 0 \Rightarrow \sin t (2 \cos t + 1) = 0$

$\Rightarrow \sin t = 0$ or $2 \cos t + 1 = 0$

$\Rightarrow \sin t = 0$ or $\cos t = -\dfrac{1}{2}$

$\Rightarrow t = \sin^{-1} 0$ or $t = \cos^{-1} \left(-\dfrac{1}{2}\right)$

$\Rightarrow t = \pi n, \dfrac{2\pi}{3} + 2\pi n, \dfrac{4\pi}{3} + 2\pi n$

105. $2 \sin \dfrac{t}{2} - 1 = 0 \Rightarrow \sin \dfrac{t}{2} = \dfrac{1}{2} \Rightarrow \dfrac{t}{2} = \sin^{-1} \dfrac{1}{2}$

$\Rightarrow \dfrac{t}{2} = \dfrac{\pi}{6} + 2\pi n, \dfrac{5\pi}{6} + 2\pi n$

$\Rightarrow t = \dfrac{\pi}{3} + 4\pi n, \dfrac{5\pi}{3} + 4\pi n$

107. $\cos 2t = \sin t \Rightarrow 1 - 2 \sin^2 t = \sin t \Rightarrow 2 \sin^2 t + \sin t - 1 = 0 \Rightarrow (2 \sin t - 1)(\sin t + 1) = 0$

$\Rightarrow 2 \sin t - 1 = 0$ or $\sin t + 1 = 0$

$\Rightarrow \sin t = \dfrac{1}{2}$ or $\sin t = -1$

$\Rightarrow t = \sin^{-1} \dfrac{1}{2}$ or $t = \sin^{-1} (-1)$

$\Rightarrow t = \dfrac{\pi}{6} + 2\pi n, \dfrac{5\pi}{6} + 2\pi n, \dfrac{3\pi}{2} + 2\pi n$

109. $\tan 2t = 1 \Rightarrow 2t = \tan^{-1} 1 \Rightarrow 2t = \dfrac{\pi}{4} + 2\pi n$ or $2t = \dfrac{5\pi}{4} + 2\pi n \Rightarrow t = \dfrac{\pi}{8} + \pi n$ or $t = \dfrac{5\pi}{8} + \pi n$

or equivalently, $t = \dfrac{\pi}{8} + \dfrac{\pi n}{2}$

111. $2\cos \dfrac{t}{2} = 1 \Rightarrow \cos \dfrac{t}{2} = \dfrac{1}{2} \Rightarrow \dfrac{t}{2} = \cos^{-1} \dfrac{1}{2} \Rightarrow \dfrac{t}{2} = \dfrac{\pi}{3} + 2\pi n$ or $\dfrac{t}{2} = \dfrac{5\pi}{3} + 2\pi n \Rightarrow$

$t = \dfrac{2\pi}{3} + 4\pi n$ or $t = \dfrac{10\pi}{3} + 4\pi n$

113. $\cos 2t = 2\sin t \cos t$, let $2\sin t \cos t = \sin 2t$ and let $\theta = 2t$. So then the equation becomes $\cos \theta = \sin \theta$,

divide both sides by $\cos \theta \Rightarrow \dfrac{\cos \theta}{\cos \theta} = \dfrac{\sin \theta}{\cos \theta} \Rightarrow 1 = \tan \theta \Rightarrow \theta = \dfrac{\pi}{4}$ and $\dfrac{5\pi}{4}$. For all real number

solutions $\theta = \dfrac{\pi}{4} + 2\pi n$ or $\theta = \dfrac{5\pi}{4} + 2\pi n$. Because $\theta = 2t$, we can determine t by substituting $2t$ for $\theta \Rightarrow$

$2t = \dfrac{\pi}{4} + 2\pi n$ or $2t = \dfrac{5\pi}{4} + 2\pi n$. Therefore $t = \dfrac{\pi}{8} + \pi n$ or $t = \dfrac{5\pi}{8} + \pi n$ or equivalently, $\dfrac{\pi}{8} + \dfrac{\pi n}{2}$.

115. $2\sin^2 2t + \sin 2t - 1 = 0 \Rightarrow (2\sin 2t - 1)(\sin 2t + 1) = 0$

$\Rightarrow 2\sin 2t - 1 = 0 \quad$ or $\quad \sin 2t + 1 = 0$

$\Rightarrow \sin 2t = \dfrac{1}{2} \quad$ or $\quad \sin 2t = -1$

Let $\theta = 2t$, then

$\Rightarrow \sin \theta = \dfrac{1}{2} \quad$ or $\quad \sin \theta = -1$

$\Rightarrow \theta = \sin^{-1}\left(\dfrac{1}{2}\right) \quad$ or $\quad \theta = \sin^{-1}(-1)$

$\Rightarrow \theta = \dfrac{\pi}{6}$ or $\dfrac{5\pi}{6} \quad$ or $\quad \theta = \dfrac{3\pi}{2}$

For all real number solutions $\theta = \dfrac{\pi}{6} + 2\pi n$ or $\theta = \dfrac{5\pi}{6} + 2\pi n$ or $\theta = \dfrac{3\pi}{2} + 2\pi n$. Because $\theta = 2t$, we can

determine t by substituting $2t$ for $\theta \Rightarrow 2t = \dfrac{\pi}{6} + 2\pi n$ or $2t = \dfrac{5\pi}{6} + 2\pi n$ or $2t = \dfrac{3\pi}{2} + 2\pi n$.

Therefore $t = \dfrac{\pi}{12} + \pi n$ or $t = \dfrac{5\pi}{12} + \pi n$ or $t = \dfrac{3\pi}{4} + \pi n$.

117. $\sin t + \sin 2t = \cos t$. Graph $Y_1 = \sin(X) + \sin(2X)$ and $Y_2 = \cos(X)$ in radian mode in $[0, 2\pi, \pi/2]$ by

$[-3, 3, 1]$. See Figure 117. The x-coordinates of the intersection points are approximately 0.333 and 4.379.

[$0, 2\pi, \pi/2$] by [$-3, 3, 1$]

Figure 117

119. (a) Graph $Y_1 = (310 \sin (120\pi X))(7 \sin (120\pi X))$ in $[0, 0.04, 0.01]$ by $[-500, 2500, 500]$.

     See Figure 119.

   (b) $W = (310 \sin (120\pi x))(7 \sin (120\pi x)) = 2170 \sin^2 (120\pi x) = 2170 \cdot \dfrac{1 - \cos (240\pi x)}{2} =$

     $1085(1 - \cos (240\pi x)) = 1085 - 1085 \cos (240\pi x) = -1085 \cos (240\pi x) + 1085 \Rightarrow$

     $a = -1085, k = 240, d = 1085$

[0, 0.04, 0.01] by [–500, 2500, 500]

Figure 119   Figure 123

121. We must solve the equation $V(t) = 160$ for t: $320 \sin (120\pi t) = 160 \Rightarrow \sin (120\pi t) = \dfrac{1}{2}$.

   Let $\theta = 120\pi t$ then $\sin \theta = \dfrac{1}{2} \Rightarrow \theta = \dfrac{\pi}{6} + 2\pi n$ or $\theta = \dfrac{5\pi}{6} + 2\pi n$. Substituting $120\pi t$ for θ yields

   $\Rightarrow 120\pi t = \dfrac{\pi}{6} + 2\pi n$ or $120\pi t = \dfrac{5\pi}{6} + 2\pi n$

   $\Rightarrow t = \dfrac{1}{720} + \dfrac{n}{60}$ or $t = \dfrac{5}{720} + \dfrac{n}{60}$ sec

123. (a) $d = 600\left(1 - \cos \dfrac{80°}{2}\right) = 600(1 - \cos 40°) \approx 140.4$ ft

   (b) By bisecting β, we create a diagram as shown in Figure 123.

     Here $r = d + a$ and from right triangle trigonometry, $\cos \dfrac{\beta}{2} = \dfrac{a}{r} \Rightarrow a = r \cos \dfrac{\beta}{2}$.

     Thus, $r = d + r \cos \dfrac{\beta}{2} \Rightarrow d = r - r \cos \dfrac{\beta}{2} \Rightarrow d = r\left(1 - \cos \dfrac{\beta}{2}\right)$

   (c) No, $\cos \dfrac{\beta}{2} = \pm\sqrt{\dfrac{1 + \cos \beta}{2}} \neq \dfrac{1}{2} \cos \beta$

125. Let $f(t)$ model the tone for number 3 and let $g(t)$ model the tone for number 4.

   (a) $f(t) = \cos (2(697)\pi t) + \cos (2(1477)\pi t) = \cos (1394\pi t) + \cos (2954\pi t)$

     $g(t) = \cos (2(770)\pi t) + \cos (2(1209)\pi t) = \cos (1540\pi t) + \cos (2418\pi t)$

   (b) $f(t) = 2 \cos \left(\dfrac{1394\pi t + 2954\pi t}{2}\right) \cos \left(\dfrac{1394\pi t - 2954\pi t}{2}\right) = 2 \cos (2174\pi t) \cos (-780\pi t) =$

     $2 \cos (2174\pi t) \cos (780\pi t)$

     $g(t) = 2 \cos \left(\dfrac{1540\pi t + 2418\pi t}{2}\right) \cos \left(\dfrac{1540\pi t - 2418\pi t}{2}\right) = 2 \cos (1979\pi t) \cos (-439\pi t) =$

     $2 \cos (1979\pi t) \cos (439\pi t)$

Chapter 7 Review Exercises

1. If $\sec \theta < 0$ and $\sin \theta > 0$ then any point (x, y) on the terminal side of θ must satisfy

   $y > 0$ and $\dfrac{r}{x} < 0 \Rightarrow x < 0$. Thus, θ is contained in quadrant II.

3. $\tan \theta = \dfrac{\sin \theta}{\cos \theta} = \dfrac{\frac{3}{5}}{-\frac{4}{5}} = -\dfrac{3}{4}$ $\qquad$ $\csc \theta = \dfrac{1}{\sin \theta} = \dfrac{1}{\frac{3}{5}} = \dfrac{5}{3}$

   $\sec \theta = \dfrac{1}{\cos \theta} = \dfrac{1}{-\frac{4}{5}} = -\dfrac{5}{4}$ $\qquad$ $\cot \theta = \dfrac{1}{\tan \theta} = \dfrac{1}{-\frac{3}{4}} = -\dfrac{4}{3}$

5. $\sin \theta = \cos \theta \tan \theta = \dfrac{24}{25}\left(-\dfrac{7}{24}\right) = -\dfrac{7}{25}$ $\qquad$ $\csc \theta = \dfrac{1}{\sin \theta} = \dfrac{1}{-\frac{7}{25}} = -\dfrac{25}{7}$

   $\sec \theta = \dfrac{1}{\cos \theta} = \dfrac{1}{\frac{24}{25}} = \dfrac{25}{24}$ $\qquad$ $\cot \theta = \dfrac{1}{\tan \theta} = \dfrac{1}{-\frac{7}{24}} = -\dfrac{24}{7}$

7. Since the sine function is an odd function $\sin(-13°) = -\sin 13°$.

9. Since the secant function is an even function $\sec\left(-\dfrac{3\pi}{7}\right) = \sec \dfrac{3\pi}{7}$.

11. $\sec \theta \cot \theta \sin \theta = \dfrac{1}{\cos \theta} \cdot \dfrac{\cos \theta}{\sin \theta} \cdot \dfrac{\sin \theta}{1} = 1$

13. $(\sec^2 t - 1)(\csc^2 t - 1) = (\tan^2 t)(\cot^2 t) = 1$

15. $\dfrac{\csc \theta \sin \theta}{\sec \theta} = \dfrac{1}{\sec \theta} = \cos \theta$

17. $\tan \theta = 1.2367 \Rightarrow \theta = \tan^{-1} 1.2367$ $\qquad$ $\sin \theta = \sin(\tan^{-1} 1.2367) \approx 0.7776$

    $\cos \theta = \cos(\tan^{-1} 1.2367) \approx 0.6288$ $\qquad$ $\csc \theta = \dfrac{1}{\sin(\tan^{-1} 1.2367)} \approx 1.2860$

    $\sec \theta = \dfrac{1}{\cos(\tan^{-1} 1.2367)} \approx 1.5904$ $\qquad$ $\cot \theta = \dfrac{1}{\tan \theta} \approx 0.8086$

19. $\cos \theta = -0.4544 \Rightarrow \theta = \cos^{-1}(-0.4544)$

    $\sin \theta = \sin(\cos^{-1}(-0.4544)) \approx 0.8908$ $\qquad$ $\tan \theta = \tan(\cos^{-1}(-0.4544)) \approx -1.9604$

    $\csc \theta = \dfrac{1}{\sin \theta} \approx 1.1226$ $\qquad$ $\sec \theta = \dfrac{1}{\cos \theta} \approx -2.2007$ $\qquad$ $\cot \theta = \dfrac{1}{\tan \theta} \approx -0.5101$

21. $\sin^2 \theta + 2 \sin \theta + 1 = (\sin \theta + 1)(\sin \theta + 1)$

23. $\tan^2 \theta - 9 = (\tan \theta + 3)(\tan \theta - 3)$

25. $(\sec \theta - 1)(\sec \theta + 1) = \sec^2 \theta - 1 = \tan^2 \theta$

27. $(1 + \tan t)^2 = 1 + 2 \tan t + \tan^2 t = \sec^2 t + 2 \tan t$

29. $\sin(x - \pi) = \sin x \cos \pi - \cos x \sin \pi = \sin x \,(-1) - \cos x \,(0) = -\sin x$

31. $\sin 8x = \sin(2 \cdot 4x) = 2 \sin 4x \cos 4x$

33. $\sec 2x = \dfrac{1}{\cos 2x} = \dfrac{1}{2 \cos^2 x - 1}$

35. $\cos^4 x \sin^3 x = \cos^4 x (\sin^2 x) \sin x = \cos^4 x (1 - \cos^2 x) \sin x = (\cos^4 x - \cos^6 x) \sin x$

37. $\sec^4 \theta - \tan^4 \theta = (\sec^2 \theta - \tan^2 \theta)(\sec^2 \theta + \tan^2 \theta) = (1)((1 + \tan^2 \theta) + \tan^2 \theta) = 1 + 2 \tan^2 \theta$

39. $\theta = 240°$ is in quadrant III, $\theta_R = 240° - 180° = 60°$

41. $\theta = \dfrac{9\pi}{7}$ is in quadrant III, $\theta_R = \dfrac{9\pi}{7} - \pi = \dfrac{2\pi}{7}$

43. Estimate: $\theta = \dfrac{\pi}{2}, \dfrac{3\pi}{2}$

Symbolic: $\cos^2 \theta - 2 \cos \theta = 0 \Rightarrow \cos \theta (\cos \theta - 2) = 0$

$\Rightarrow \cos \theta = 0 \qquad$ or $\qquad \cos \theta - 2 = 0$

$\Rightarrow \cos \theta = 0 \qquad$ or $\qquad \cos \theta = 2$ (not possible)

$\Rightarrow \theta = \cos^{-1} 0 \Rightarrow \theta = \dfrac{\pi}{2}, \dfrac{3\pi}{2}$

45. (a) $\tan \theta = \sqrt{3} \Rightarrow \theta = \tan^{-1} \sqrt{3} = 60°, 240°$

(b) $\cot \theta = -\sqrt{3} \Rightarrow \tan \theta = -\dfrac{1}{\sqrt{3}} \Rightarrow \theta = \tan^{-1}\left(-\dfrac{1}{\sqrt{3}}\right) = 150°, 330°$

47. $2 \cos \theta - 1 = 0 \Rightarrow \cos \theta = \dfrac{1}{2} \Rightarrow \theta = \cos^{-1} \dfrac{1}{2} = 60°, 300°$

49. $2 \sin^2 \theta + \sin \theta - 3 = 0 \Rightarrow (\sin \theta - 1)(2 \sin \theta + 3)$

$\Rightarrow \sin \theta - 1 = 0 \qquad$ or $\qquad 2 \sin \theta + 3 = 0$

$\Rightarrow \sin \theta = 1 \qquad$ or $\qquad \sin \theta = -\dfrac{3}{2}$ (not possible)

$\Rightarrow \theta = \sin^{-1} 1 = 90°$

51. $\tan^2 t - 2 \tan t + 1 = 0 \Rightarrow (\tan t - 1)(\tan t - 1) = 0 \Rightarrow \tan t = 1 \Rightarrow t = \tan^{-1} 1 \Rightarrow t = \dfrac{\pi}{4}, \dfrac{5\pi}{4}$

53. $3 \tan^2 t - 1 = 0 \Rightarrow \tan^2 t = \dfrac{1}{3} \Rightarrow \tan t = \pm \dfrac{1}{\sqrt{3}} \Rightarrow t = \tan^{-1}\left(\pm \dfrac{1}{\sqrt{3}}\right)$

$t = \dfrac{\pi}{6} + \pi n, -\dfrac{\pi}{6} + \pi n$ or $t = 30° + 180°n, -30° + 180°n$

55. $\sin 2t + 3 \cos t = 0 \Rightarrow 2 \sin t \cos t + 3 \cos t = 0 \Rightarrow \cos t (2 \sin t + 3) = 0$

$\Rightarrow \cos t = 0 \qquad$ or $\qquad 2 \sin t + 3 = 0$

$\Rightarrow \cos t = 0 \qquad$ or $\qquad \sin t = -\dfrac{3}{2}$ (not possible)

$\Rightarrow t = \cos^{-1} 0$

$t = \dfrac{\pi}{2} + \pi n$ or $t = 90° + 180°n$

57. $\cos 105° = \cos \dfrac{210°}{2} = -\sqrt{\dfrac{1 + \cos 210°}{2}} = -\sqrt{\dfrac{1 - \left(\frac{\sqrt{3}}{2}\right)}{2}} = -\sqrt{\dfrac{\frac{2 - \sqrt{3}}{2}}{2}} = -\sqrt{\dfrac{2 - \sqrt{3}}{4}} =$

$-\sqrt{\dfrac{2 - \sqrt{3}}{2}}$. See Figure 57.

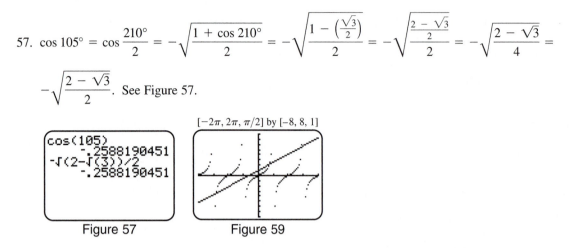

[$-2\pi, 2\pi, \pi/2$] by [$-8, 8, 1$]

Figure 57   Figure 59

59. Graph $Y_1 = \tan (X)$ and $Y_2 = X + 1$ using dot mode in $[-2\pi, 2\pi, \pi/2]$ by $[-8, 8, 1]$.

See Figure 59. The *x*-coordinates of the intersection points are approximately -4.43, 1.13 and 4.53.

61. If $\cos \alpha = \dfrac{3}{5}$ and α is in quadrant I, then $\sin \alpha = \dfrac{4}{5}$. If $\cos \beta = \dfrac{12}{13}$ and β is in quadrant I, then $\sin \beta = \dfrac{5}{13}$.

(a) $\sin (\alpha + \beta) = \sin \alpha \cos \beta + \cos \alpha \sin \beta = \left(\dfrac{4}{5}\right)\left(\dfrac{12}{13}\right) + \left(\dfrac{3}{5}\right)\left(\dfrac{5}{13}\right) = \dfrac{48 + 15}{65} = \dfrac{63}{65}$

(b) $\cos (\alpha + \beta) = \cos \alpha \cos \beta - \sin \alpha \sin \beta = \left(\dfrac{3}{5}\right)\left(\dfrac{12}{13}\right) - \left(\dfrac{4}{5}\right)\left(\dfrac{5}{13}\right) = \dfrac{36 - 20}{65} = \dfrac{16}{65}$

(c) $\tan (\alpha + \beta) = \dfrac{\sin (\alpha + \beta)}{\cos (\alpha + \beta)} = \dfrac{\frac{63}{65}}{\frac{16}{65}} = \dfrac{63}{65} \cdot \dfrac{65}{16} = \dfrac{63}{16}$

(d) Since $\sin (\alpha + \beta)$ and $\cos (\alpha + \beta)$ are positive, $\alpha + \beta$ is in quadrant I.

63. Since $\tan \theta = \dfrac{\sin \theta}{\cos \theta}$, $\cos \theta = \dfrac{\sin \theta}{\tan \theta} = \dfrac{\frac{4}{5}}{-\frac{4}{3}} = \dfrac{4}{5} \cdot \left(-\dfrac{3}{4}\right) = -\dfrac{3}{5}$

(a) $\sin 2\theta = 2 \sin \theta \cos \theta = 2\left(\dfrac{4}{5}\right)\left(-\dfrac{3}{5}\right) = -\dfrac{24}{25}$

$\cos 2\theta = 1 - 2 \sin^2 \theta = 1 - 2\left(\dfrac{4}{5}\right)^2 = 1 - \dfrac{32}{25} = -\dfrac{7}{25}$

$\tan 2\theta = \dfrac{\sin 2\theta}{\cos 2\theta} = \dfrac{-\frac{24}{25}}{-\frac{7}{25}} = -\dfrac{24}{25} \cdot \left(-\dfrac{25}{7}\right) = \dfrac{24}{7}$

(b) Since θ is in quadrant II, $\theta = \cos^{-1}\left(-\dfrac{3}{5}\right)$. Numerical support is shown in Figures 63a & 63b.

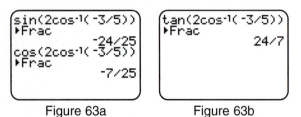

Figure 63a   Figure 63b

65. Since $\cos \theta = \dfrac{1}{4}$ and θ is in quadrant I, $\sin \theta = \sqrt{1 - \cos^2 \theta} = \sqrt{1 - \left(\dfrac{1}{4}\right)^2} = \sqrt{\dfrac{15}{16}} = \dfrac{\sqrt{15}}{4}$.

(a) $\sin \dfrac{\theta}{2} = \sqrt{\dfrac{1 - \cos \theta}{2}} = \sqrt{\dfrac{1 - \frac{1}{4}}{2}} = \sqrt{\dfrac{\frac{4-1}{4}}{2}} = \sqrt{\dfrac{3}{8}}$

$\cos \dfrac{\theta}{2} = \sqrt{\dfrac{1 + \cos \theta}{2}} = \sqrt{\dfrac{1 + \frac{1}{4}}{2}} = \sqrt{\dfrac{\frac{4+1}{4}}{2}} = \sqrt{\dfrac{5}{8}}$

$\tan \dfrac{\theta}{2} = \dfrac{\sin \frac{\theta}{2}}{\cos \frac{\theta}{2}} = \dfrac{\sqrt{\frac{3}{8}}}{\sqrt{\frac{5}{8}}} = \sqrt{\dfrac{3}{8}} \cdot \sqrt{\dfrac{8}{5}} = \sqrt{\dfrac{3}{5}}$

(b) Since θ is in quadrant I, $\theta = \cos^{-1} \dfrac{1}{4}$. Numerical support is shown in Figures 65a, 65b and 65c.

Figure 65a   Figure 65b   Figure 65c   Figure 67

67. Let $\theta = \tan^{-1} \dfrac{11}{60}$. It follows that $\sin \theta = \dfrac{11}{61}$. See Figure 67.

$\cos \left(2 \tan^{-1} \dfrac{11}{60}\right) = \cos (2\theta) = 1 - 2 \sin^2 \theta = 1 - 2\left(\dfrac{11}{61}\right)^2 = 1 - \dfrac{242}{3721} = \dfrac{3479}{3721}$

69. Graph $Y_1 = 1.2 \cos (\pi/6(X - 0.7)) + 12.1$ and $Y_2 = 11.5$ in $[0, 13, 1]$ by $[10, 15, 1]$.

See Figures 69a and 69b. The x-coordinates of the intersection points are approximately 4.7 and 8.7. The number of daylight hours equals 11.5 on about April 21 and again on about August 22.

[0, 13, 1] by [10, 15, 1]   [0, 13, 1] by [10, 15, 1]   [0, 0.06, 0.01] by [−0.012, 0.012, 0.002]

Figure 69a   Figure 69b   Figure 71

71. (a) Graph $Y_1 = 0.006 \cos (100\pi X) + 0.008 \sin (100\pi X)$ in $[0, 0.06, 0.01]$ by $[-0.012, 0.012, 0.002]$.

See Figure 71.

(b) The peaks of the graph are located at $y = 0.01$, thus $a = 0.01$. The x-value at the first peak is

$x \approx 0.00295089$. $100\pi(0.00295089) + k = \dfrac{\pi}{2} \Rightarrow k \approx 0.6435$.

(c) $P = 0.01 \sin (100\pi t + 0.6435) = 0.01[\sin (100\pi t) \cos (0.6435) + \cos (100\pi t) \sin (0.6435)] \approx$

$0.01[\sin (100\pi t)(0.8) + \cos (100\pi t)(0.6)] \approx 0.008 \sin (100\pi t) + 0.006 \cos (100\pi t)$

73. $W(t) = 7\cos^2(240\pi t) \implies W(t) = 7(1 - \sin^2(240\pi t)) \implies W(t) = 7 - 7\sin^2(240\pi t)$

When V is maximum or minimum, $W = 0$.

75. Let $F = 250 \implies 250 = 289\cos\theta \implies \cos\theta = \dfrac{250}{289}$ or $\theta = \cos^{-1}\left(\dfrac{250}{289}\right) \implies \theta \approx 30.11°$ or 0.53 radians.

Chapter 8: Further Topics in Trigonometry

8.1: Law of Sines

1. No. This situation is AAS.

3. No. This situation is SSS.

5. Yes. This situation is SSA.

7. Yes. This situation is SSA.

9. This triangle is of the form AAS. There is only one solution.

   $\beta = 180° - 45° - 75° = 60°$

   $\dfrac{a}{\sin \alpha} = \dfrac{c}{\sin \gamma} \Rightarrow a = \dfrac{c \sin \alpha}{\sin \gamma} = \dfrac{4 \sin 75°}{\sin 45°} \approx 5.46$ $\dfrac{b}{\sin \beta} = \dfrac{c}{\sin \gamma} \Rightarrow b = \dfrac{c \sin \beta}{\sin \gamma} = \dfrac{4 \sin 60°}{\sin 45°} \approx 4.90$

   $\beta = 60°, \ a \approx 5.5, \ b \approx 4.9$

11. This triangle is of the form ASA. There is only one solution.

    $\beta = 180° - 120° - 25° = 35°$

    $\dfrac{a}{\sin \alpha} = \dfrac{b}{\sin \beta} \Rightarrow a = \dfrac{b \sin \alpha}{\sin \beta} = \dfrac{10 \sin 120°}{\sin 35°} \approx 15.1$ $\dfrac{c}{\sin \gamma} = \dfrac{b}{\sin \beta} \Rightarrow c = \dfrac{b \sin \gamma}{\sin \beta} = \dfrac{10 \sin 25°}{\sin 35°} \approx 7.37$

    $\beta = 35°, \ a \approx 15.1, \ c \approx 7.4$

13. This triangle is of the form SSA. It is ambiguous.

    $\dfrac{\sin \beta}{b} = \dfrac{\sin \alpha}{a} \Rightarrow \sin \beta = \dfrac{b \sin \alpha}{a} = \dfrac{7 \sin 46°}{6} \Rightarrow \beta_R = \sin^{-1}\left(\dfrac{7 \sin 46°}{6}\right) \approx 57.1°$

    Thus, $\beta \approx 57.1°$ or $\beta \approx 180° - 57.1° \approx 122.9°$.

    Solution 1: Let $\beta \approx 57.1°$. Then $\gamma \approx 180° - 46° - 57.1° \approx 76.9°$.

    $\dfrac{c}{\sin \gamma} = \dfrac{a}{\sin \alpha} \Rightarrow c = \dfrac{a \sin \gamma}{\sin \alpha} = \dfrac{6 \sin 76.9°}{\sin 46°} \approx 8.12$ $\beta \approx 57.1°, \ \gamma \approx 76.9°, \ c \approx 8.1$

    Solution 2: Let $\beta \approx 122.9°$. Then $\gamma \approx 180° - 46° - 122.9° \approx 11.1°$.

    $\dfrac{c}{\sin \gamma} = \dfrac{a}{\sin \alpha} \Rightarrow c = \dfrac{a \sin \gamma}{\sin \alpha} = \dfrac{6 \sin 11.1°}{\sin 46°} \approx 1.61$ $\beta \approx 122.9°, \ \gamma \approx 11.1°, \ c \approx 1.6$

15. This triangle is of the form SSA. It is ambiguous.

    $\dfrac{\sin \alpha}{a} = \dfrac{\sin \beta}{b} \Rightarrow \sin \alpha = \dfrac{a \sin \beta}{b} = \dfrac{5 \sin 52°}{3} \approx 1.31$ This is not possible. No solution.

17. This triangle is of the form SSA. It is ambiguous.

    $\dfrac{\sin \beta}{b} = \dfrac{\sin \alpha}{a} \Rightarrow \sin \beta = \dfrac{b \sin \alpha}{a} = \dfrac{12 \sin 50.2°}{10} \approx 0.9219 \Rightarrow$

    $\beta_R \approx \sin^{-1}(0.9219) \approx 67.2°$ or $\beta_1 \approx 67° \ 13'$

    Then $\gamma_1 = 180° - 50.2° - 67.2° \approx 62.6°$ or $\gamma_1 \approx 62° \ 35'$

    $\dfrac{c_1}{\sin \gamma} = \dfrac{a}{\sin \alpha} \Rightarrow c_1 = \dfrac{a \sin \gamma}{\sin \alpha} = \dfrac{10 \sin 62.6°}{\sin 50.2°} \approx 11.6$ or $c_1 \approx 11.6$ $\beta_1 \approx 67°13', \gamma_1 \approx 62°35', c_1 \approx 11.6$

    However if $\beta_2 = 180° - 67.2° = 112.8°$ or $112° \ 48'$ then $\gamma_2 = 180° - 50.2° - 112.8° = 17°$ or $\gamma_2 \approx 17° \ 0'$

    $\dfrac{c_2}{\sin \gamma_2} = \dfrac{a}{\sin \alpha} \Rightarrow c_2 = \dfrac{a \sin \gamma_2}{\sin \alpha} = \dfrac{10 \sin 17°}{\sin 50.2°} \approx 3.8$ or $c_2 \approx 3.8$ $\beta_2 \approx 112°48', \gamma_2 \approx 17°0', c_2 \approx 3.8$

19. This triangle is of the form AAS. There is only one solution.

$\gamma = 180° - 32° - 55° = 93°$

$\dfrac{a}{\sin \alpha} = \dfrac{b}{\sin \beta} \Rightarrow a = \dfrac{b \sin \alpha}{\sin \beta} = \dfrac{12 \sin 32°}{\sin 55°} \approx 7.76$ $\dfrac{c}{\sin \gamma} = \dfrac{b}{\sin \beta} \Rightarrow c = \dfrac{b \sin \gamma}{\sin \beta} = \dfrac{12 \sin 93°}{\sin 55°} \approx 14.6$

$\gamma = 93°,\ a \approx 7.8,\ c \approx 14.6$

21. This triangle is of the form SSA. It is ambiguous.

$\dfrac{\sin \beta}{b} = \dfrac{\sin \alpha}{a} \Rightarrow \sin \beta = \dfrac{b \sin \alpha}{a} = \dfrac{9 \sin 20°}{7} \Rightarrow \beta_R = \sin^{-1}\left(\dfrac{9 \sin 20°}{7}\right) \approx 26.1°$

Thus, $\beta \approx 26.1°$ or $\beta \approx 180° - 26.1° \approx 153.9°$.

Solution 1: Let $\beta \approx 26.1°$. Then $\gamma \approx 180° - 20° - 26.1° \approx 133.9°$.

$\dfrac{c}{\sin \gamma} = \dfrac{a}{\sin \alpha} \Rightarrow c = \dfrac{a \sin \gamma}{\sin \alpha} = \dfrac{7 \sin 133.9°}{\sin 20°} \approx 14.7$ $\beta \approx 26.1°,\ \gamma \approx 133.9°,\ c \approx 14.7$

Solution 2: Let $\beta \approx 153.9°$. Then $\gamma \approx 180° - 20° - 153.9° \approx 6.1°$.

$\dfrac{c}{\sin \gamma} = \dfrac{a}{\sin \alpha} \Rightarrow c = \dfrac{a \sin \gamma}{\sin \alpha} = \dfrac{7 \sin 6.1°}{\sin 20°} \approx 2.17$ $\beta \approx 153.9°,\ \gamma \approx 6.1°,\ c \approx 2.2$

23. This triangle is of the form SSA. It is ambiguous.

$\dfrac{\sin \gamma}{c} = \dfrac{\sin \beta}{b} \Rightarrow \sin \gamma = \dfrac{c \sin \beta}{b} = \dfrac{20 \sin 30°}{10} = 1 \Rightarrow \gamma_R = \sin^{-1} 1 = 90°$

Thus, $\gamma = 90°$. Then $\alpha = 180° - 30° - 90° = 60°$.

$\dfrac{a}{\sin \alpha} = \dfrac{b}{\sin \beta} \Rightarrow a = \dfrac{b \sin \alpha}{\sin \beta} = \dfrac{10 \sin 60°}{\sin 30°} = 10\sqrt{3} \approx 17.3$ $\gamma = 90°,\ \alpha = 60°,\ a = 10\sqrt{3} \approx 17.3$

25. This triangle is of the form SSA. It is ambiguous.

$\dfrac{\sin \alpha}{a} = \dfrac{\sin \gamma}{c} \Rightarrow \sin \alpha = \dfrac{a \sin \gamma}{c} = \dfrac{42.1 \sin 102°}{51.6} \Rightarrow \alpha_R = \sin^{-1}\left(\dfrac{42.1 \sin 102°}{51.6}\right) \approx 52.9°$

Thus, $\alpha \approx 52.9°$ or $\alpha \approx 180° - 52.9° \approx 127.1°$ (which is not possible).

So $\alpha \approx 52.9°$. Then $\beta \approx 180° - 102° - 52.9° \approx 25.1°$.

$\dfrac{b}{\sin \beta} = \dfrac{c}{\sin \gamma} \Rightarrow b = \dfrac{c \sin \beta}{\sin \gamma} = \dfrac{51.6 \sin 25.1°}{\sin 102°} \approx 22.4$ $\alpha \approx 52.9°,\ \beta \approx 25.1°,\ b \approx 22.4$

27. This triangle is of the form ASA. There is only one solution.

$\beta = 180° - 55.2° - 114.8° = 10°$

$\dfrac{a}{\sin \alpha} = \dfrac{b}{\sin \beta} \Rightarrow a = \dfrac{b \sin \alpha}{\sin \beta} = \dfrac{19.5 \sin 55.2°}{\sin 10°} \approx 92.2$

$\dfrac{c}{\sin \gamma} = \dfrac{b}{\sin \beta} \Rightarrow c = \dfrac{b \sin \gamma}{\sin \beta} = \dfrac{19.5 \sin 114.8°}{\sin 10°} \approx 101.9$

$\beta = 10°,\ a \approx 92.2,\ c \approx 101.9$

29. This triangle is of the form SSA. It is ambiguous.

$\dfrac{\sin \gamma}{c} = \dfrac{\sin \beta}{b} \Rightarrow \sin \gamma = \dfrac{c \sin \beta}{b} = \dfrac{7.4 \sin 73°}{6.2} \approx 1.14$ This is not possible. No solution.

31. This triangle is of the form SSA. It is ambiguous.

$\dfrac{\sin \beta}{b} = \dfrac{\sin \alpha}{a} \Rightarrow \sin \beta = \dfrac{b \sin \alpha}{a} = \dfrac{12 \sin 35° 15'}{5} \approx 1.39$ This is not possible. No solution.

33. This triangle is of the form SSA. It is ambiguous.

$$\frac{\sin \alpha}{a} = \frac{\sin \beta}{b} \Rightarrow \sin \alpha = \frac{a \sin \beta}{b} = \frac{6 \sin 46° 45'}{5} \approx 0.8740 \Rightarrow \alpha_R \approx \sin^{-1}(0.8740) \approx 60.93°$$

or $\alpha_1 \approx 60° 56'$ then $\gamma_1 \approx 180° - 46.75° - 60.93° \approx 72.32°$ or $\gamma_1 \approx 72° 19'$

$$\frac{c_1}{\sin \gamma_1} = \frac{b}{\sin \beta} \Rightarrow c_1 = \frac{b \sin \gamma_1}{\sin \beta} = \frac{5 \sin 72.32°}{\sin 46.75°} \approx 6.54 \text{ or } c_1 \approx 6.5$$

However, if $\alpha_2 \approx 180° - 60.93° \approx 119.07°$ or $119° 4'$ then $\gamma_2 \approx 180° - 46.75° - 119.07° \approx 14.18°$ or

$\gamma_2 = 14° 11'$. $\dfrac{c_2}{\sin \gamma_2} = \dfrac{b}{\sin \beta} \Rightarrow c_2 = \dfrac{b \sin \gamma_2}{\sin \beta} = \dfrac{5 \sin 14.18°}{\sin 46.75°} \approx 1.68$ or $c_2 \approx 1.7$

35. This triangle is of the form ASA. There is only one solution. Note that $\alpha = 56° 30'$ or $56.5°$ and

$\beta = 23° 45'$ or $23.75°$. $\gamma = 180° - 56.5° - 23.75° = 99.75°$ or $99° 45'$

$$\frac{a}{\sin \alpha} = \frac{c}{\sin \gamma} \Rightarrow a = \frac{c \sin \alpha}{\sin \gamma} = \frac{100 \sin 56.5°}{\sin 99.75°} \approx 84.6$$

$$\frac{b}{\sin \beta} = \frac{c}{\sin \gamma} \Rightarrow b = \frac{c \sin \beta}{\sin \gamma} = \frac{100 \sin 23.75°}{\sin 99.75°} \approx 40.9$$

$\gamma = 99° 45'$, $a \approx 84.6$, $b \approx 40.9$

37. Using standard labels we have the following: $\beta = 52° - 7° = 45°$ and $\alpha = 180° - 70° - 45° = 65°$.

$$\frac{c}{\sin 70°} = \frac{3500}{\sin 65°} \Rightarrow c = \frac{3500 \sin 70°}{\sin 65°} \approx 3629 \text{ ft.}$$ The ground distance is about 3629 feet.

39. $\gamma = \theta - \alpha = 52.7430° - 52.6901° = 0.0529°$

$$\frac{a}{\sin \alpha} = \frac{c}{\sin \gamma} \Rightarrow a = \frac{c \sin \alpha}{\sin \gamma} = \frac{398.02 \sin 52.6901°}{\sin 0.0529°} \approx 342,878.7$$

The calculated distance to the moon changes to about 343,000 km, a difference of about 76,000 km. A small

error in measuring lunar angle could result in large errors when calculating the distance to the moon.

41. Using standard labels we have the following:

$\alpha = 90° - 54.3° = 35.7°$, $\beta = 325.2° - 270° = 55.2°$ and $\gamma = 180° - 35.7° - 55.2° = 89.1°$.

$$\frac{a}{\sin \alpha} = \frac{c}{\sin \gamma} \Rightarrow a = \frac{c \sin \alpha}{\sin \gamma} = \frac{15 \sin 35.7°}{\sin 89.1°} \approx 8.75$$

Then the perpendicular distance d from the ship to the shore can be found as follows:

$$\sin 55.2° = \frac{d}{8.75} \Rightarrow d = 8.75 \sin 55.2° \approx 7.2 \text{ mi.}$$ The ship is about 7.2 miles from the shore.

43. Angle DAB has measure $180° - 118° - 28° = 34°$. $\dfrac{BD}{\sin 34°} = \dfrac{24.2}{\sin 28°} \Rightarrow BD = \dfrac{24.2 \sin 34°}{\sin 28°} \approx 28.8$ ft.

45. Note that $\beta = 112° 10'$ or $112.17°$ and $\gamma = 15° 20'$ or $15.33°$.

Therefore $\alpha = 180° - 112.17° - 15.33° = 52.5°$ or $52° 30'$

$$\frac{c}{\sin \gamma} = \frac{a}{\sin \alpha} \Rightarrow c = \frac{a \sin \gamma}{\sin \alpha} = \frac{354 \sin 15.33°}{\sin 52.5°} \approx 118 \text{ meters}$$ Therefore $c \approx 118$ meters from point A to B.

47. Let A, B, and C be the positions of the observer on the left, the helicopter, and observer on the right, respectively. Triangle ABC has the following measurements: $A = 20.5°$, $C = 180° - 27.8° = 152.2°$, $B = 180° - 20.5° - 152.2° = 7.3°$, and $b = 3$. Using the law of sines,

$$\frac{a}{\sin 20.5°} = \frac{3}{\sin 7.3°} \Rightarrow a = \frac{3 \sin 20.5°}{\sin 7.3°} \Rightarrow a \approx 8.27 \text{ miles.}$$ If the perpendicular line segment joining B to the ground intersects at D, we can use right triangle CBD to find the distance d to the helicopter:

$$\sin 27.8° = \frac{d}{8.27} \Rightarrow d = 8.27 \sin 27.8° \approx 3.86 \text{ miles.}$$

49. Let A, B, and C be the positions of the town on the left, the balloon, and town on the right, respectively. Triangle ABC has the following measurements: $A = 35°$, $B = 180° - 35° - 31° = 114°$, $C = 31°$, and $b = 1.5$. Using the law of sines, $\dfrac{c}{\sin 31°} = \dfrac{1.5}{\sin 114°} \Rightarrow c = \dfrac{1.5 \sin 31°}{\sin 114°} \Rightarrow c \approx 0.85$ mile. If the perpendicular line segment joining B to the ground intersects at D, we can use right triangle ABD to find the distance d to the balloon: $\sin 35° = \dfrac{d}{0.85} \Rightarrow d = 0.85 \sin 35° \approx 0.49$ mile.

51. Using standard labels we have the distance $AB = c$, $\gamma = 55.1°$, $\alpha = 75.7°$ and $b = 97.3$.

$$\beta = 180° - 75.7° - 55.1° = 49.2° \qquad \frac{c}{\sin \gamma} = \frac{b}{\sin \beta} \Rightarrow c = \frac{b \sin \gamma}{\sin \beta} = \frac{97.3 \sin 55.1°}{\sin 49.2°} \approx 105.4$$

The distance AB is approximately 105.4 feet.

53. (a) Using standard labels we calculate the distance as follows:

$$\gamma = 180° - 28° - 60° = 92° \qquad \frac{b}{\sin \beta} = \frac{c}{\sin \gamma} \Rightarrow b = \frac{c \sin \beta}{\sin \gamma} = \frac{4.12 \sin 60°}{\sin 92°} \approx 3.57 \text{ mi.}$$

(b) Let D represent the point $(4, 0)$ and let θ represent the angle DAB.

$$\tan \theta = \frac{1}{4} \Rightarrow \theta = \tan^{-1} \frac{1}{4} \approx 14.0° \qquad \text{Thus, the bearing is } 90° - 28° - 14.0° \approx 48°.$$

55. Represent the top of the arch with point C and use standard labels.

Note that $\alpha = 180° - 64.91° = 115.09°$. Then $\gamma = 180° - 115.09° - 60.81° = 4.1°$

First we find side $b = AC$: $\dfrac{b}{\sin \beta} = \dfrac{c}{\sin \gamma} \Rightarrow b = \dfrac{c \sin \beta}{\sin \gamma} = \dfrac{57 \sin 60.81°}{\sin 4.1°} \approx 696 \text{ ft.}$

Label a point D on the ground directly below the top of the arch and let h represent the height of the arch.

From triangle ADC: $\sin 64.91° \approx \dfrac{h}{696} \Rightarrow h \approx 696 \sin 64.91° \approx 630 \text{ ft.}$

8.2: Law of Cosines

1. (a) This triangle is SAS.

    (b) Law of cosines should be used.

3. (a) This triangle is SSA.

    (b) This case is ambiguous. Law of sines should be used.

5. (a) This triangle is ASA.

    (b) Law of sines should be used.

7. (a) This triangle is ASA.

(b) Law of sines should be used.

9. $c^2 = 3^2 + 8^2 - 2(3)(8) \cos 60° \Rightarrow c^2 = 73 - (48)(0.5) \Rightarrow c^2 = 73 - 24 \Rightarrow c^2 = 49 \Rightarrow c = 7$

11. $1^2 = 1^2 + (\sqrt{3})^2 - 2(1)\sqrt{3} \cos \theta \Rightarrow 1 = 4 - 2\sqrt{3} \cos \theta \Rightarrow \cos \theta = \dfrac{-3}{-2\sqrt{3}} \Rightarrow$

$\cos \theta = \dfrac{\sqrt{3}}{2} \Rightarrow \theta = 30°$

13. $a^2 = b^2 + c^2 - 2bc \cos \alpha = 4^2 + 6^2 - 2(4)(6) \cos 61° \Rightarrow a = \sqrt{52 - 48 \cos 61°} \approx 5.35996$

$c^2 = a^2 + b^2 - 2ab \cos \gamma \Rightarrow \cos \gamma = \dfrac{c^2 - a^2 - b^2}{-2ab} \Rightarrow \gamma = \cos^{-1}\left(\dfrac{6^2 - 5.35996^2 - 4^2}{-2(5.35996)(4)}\right) \approx 78.25399°$

$\beta = 180° - 61° - 78.25399° \approx 40.74601° \qquad a \approx 5.4, \ \beta \approx 40.7°, \ \gamma \approx 78.3°$

15. $a^2 = b^2 + c^2 - 2bc \cos \alpha \Rightarrow \cos \alpha = \dfrac{a^2 - b^2 - c^2}{-2bc} \Rightarrow \alpha = \cos^{-1}\left(\dfrac{4^2 - 10^2 - 8^2}{-2(10)(8)}\right) \approx 22.33165°$

$b^2 = a^2 + c^2 - 2ac \cos \beta \Rightarrow \cos \beta = \dfrac{b^2 - a^2 - c^2}{-2ac} \Rightarrow \beta = \cos^{-1}\left(\dfrac{10^2 - 4^2 - 8^2}{-2(4)(8)}\right) \approx 108.20996°$

$\gamma = 180° - 22.33165° - 108.20996° \approx 49.45839° \qquad \alpha \approx 22.3°, \ \beta \approx 108.2°, \ \gamma \approx 49.5°$

17. $a^2 = b^2 + c^2 - 2bc \cos \alpha \Rightarrow \cos \alpha = \dfrac{a^2 - b^2 - c^2}{-2bc} \Rightarrow \alpha = \cos^{-1}\left(\dfrac{5^2 - 7^2 - 9^2}{-2(7)(9)}\right) \approx 33.55731°$

$b^2 = a^2 + c^2 - 2ac \cos \beta \Rightarrow \cos \beta = \dfrac{b^2 - a^2 - c^2}{-2ac} \Rightarrow \beta = \cos^{-1}\left(\dfrac{7^2 - 5^2 - 9^2}{-2(5)(9)}\right) \approx 50.70352°$

$\gamma = 180° - 33.55731° - 50.70352° \approx 95.73917° \qquad \alpha \approx 33.6°, \ \beta \approx 50.7°, \ \gamma \approx 95.7°$

19. $c^2 = a^2 + b^2 - 2ab \cos \gamma = 45^2 + 24^2 - 2(45)(24) \cos 35° \Rightarrow c = \sqrt{2601 - 2160 \cos 35°} \approx 28.83802$

$a^2 = b^2 + c^2 - 2bc \cos \alpha \Rightarrow \cos \alpha = \dfrac{a^2 - b^2 - c^2}{-2bc} \Rightarrow$

$\alpha = \cos^{-1}\left(\dfrac{45^2 - 24^2 - 28.83802^2}{-2(24)(28.83802)}\right) \approx 116.48753°$

$\beta = 180° - 35° - 116.48753° \approx 28.51247° \qquad c \approx 28.8, \ \alpha \approx 116.5°, \ \beta \approx 28.5°$

21. $a^2 = b^2 + c^2 - 2bc \cos \alpha \Rightarrow \cos \alpha = \dfrac{a^2 - b^2 - c^2}{-2bc} \Rightarrow \alpha = \cos^{-1}\left(\dfrac{2.4^2 - 1.7^2 - 1.4^2}{-2(1.7)(1.4)}\right) \approx 101.02145°$

$b^2 = a^2 + c^2 - 2ac \cos \beta \Rightarrow \cos \beta = \dfrac{b^2 - a^2 - c^2}{-2ac} \Rightarrow \beta = \cos^{-1}\left(\dfrac{1.7^2 - 2.4^2 - 1.4^2}{-2(2.4)(1.4)}\right) \approx 44.04863°$

$\gamma = 180° - 101.02145° - 44.04863° \approx 34.92992° \qquad \alpha \approx 101.0°, \ \beta \approx 44.0°, \ \gamma \approx 34.9°$

(The angles do not sum to 180° due to rounding)

23. $a^2 = b^2 + c^2 - 2bc \cos \alpha = 24.1^2 + 15.8^2 - 2(24.1)(15.8) \cos 10°30' \Rightarrow$

$a = \sqrt{830.45 - 761.56 \cos 10°30'} \approx 9.03562$

$b^2 = a^2 + c^2 - 2ac \cos \beta \Rightarrow \cos \beta = \dfrac{b^2 - a^2 - c^2}{-2ac} \Rightarrow$

$\beta = \cos^{-1}\left(\dfrac{24.1^2 - 9.03562^2 - 15.8^2}{-2(9.03562)(15.8)}\right) \approx 150.91785°$

$\gamma = 180° - 10.5° - 150.91785° \approx 18.58215° \qquad a \approx 9.0, \ \beta \approx 150.9°, \ \gamma \approx 18.6°$

25. $a^2 = b^2 + c^2 - 2bc \cos \alpha \Rightarrow \cos \alpha = \dfrac{a^2 - b^2 - c^2}{-2bc} \Rightarrow$

$\alpha = \cos^{-1}\left(\dfrac{10.6^2 - 25.8^2 - 20.6^2}{-2(25.8)(20.6)}\right) \approx 23.11284°$

$b^2 = a^2 + c^2 - 2ac \cos \beta \Rightarrow \cos \beta = \dfrac{b^2 - a^2 - c^2}{-2ac} \Rightarrow \beta = \cos^{-1}\left(\dfrac{25.8^2 - 10.6^2 - 20.6^2}{-2(10.6)(20.6)}\right) \approx$

$107.16957°$

$\gamma = 180° - 23.11284° - 107.16957° \approx 49.71759°$ $\alpha \approx 23.1°,\ \beta \approx 107.2°,\ \gamma \approx 49.7°$

27. $b^2 = a^2 + c^2 - 2ac \cos \beta = 20^2 + 15^2 - 2(20)(15) \cos 122°\ 10' \Rightarrow$

$b = \sqrt{625 - 600 \cos 122°\ 10'} \approx 30.73207$

$a^2 = b^2 + c^2 - 2bc \cos \alpha \Rightarrow \cos \alpha = \dfrac{a^2 - b^2 - c^2}{-2bc} \Rightarrow$

$\alpha = \cos^{-1}\left(\dfrac{20^2 - 30.73207^2 - 15^2}{-2(30.73207)(15)}\right) \approx 33.42685°$ or $33°\ 26'$

$\gamma \approx 180° - 122°\ 10' - 33°\ 26' \approx 24°\ 24'$ $b \approx 30.7,\ \alpha \approx 33°\ 26',\ \gamma \approx 24°\ 24'$

29. $a^2 = b^2 + c^2 - 2bc \cos \alpha \Rightarrow \cos \alpha = \dfrac{a^2 - b^2 - c^2}{-2bc} \Rightarrow \alpha = \cos^{-1}\left(\dfrac{5.3^2 - 6.7^2 - 7.1^2}{-2(6.7)(7.1)}\right) \approx 45.05460°$

$b^2 = a^2 + c^2 - 2ac \cos \beta \Rightarrow \cos \beta = \dfrac{b^2 - a^2 - c^2}{-2ac} \Rightarrow \beta = \cos^{-1}\left(\dfrac{6.7^2 - 5.3^2 - 7.1^2}{-2(5.3)(7.1)}\right) \approx 63.47520°$

$\gamma = 180° - 45.05460° - 63.47520° \approx 71.4702°$ $\alpha \approx 45.1,\ \beta = 63.5°,\ \gamma \approx 71.5°$

(The angles do not sum to 180° due to rounding)

31. No, since $a + b < c$.

33. No, since $89° + 112° > 180°$.

35. Yes, since we are given ASA and $\alpha + \gamma < 180°$.

37. Area $= K = \dfrac{1}{2}(18)(15) \sin 40° \approx 86.8$

39. $s = \dfrac{1}{2}(3 + 4 + 6) = 6.5 \Rightarrow$ Area $= K = \sqrt{6.5(6.5 - 3)(6.5 - 4)(6.5 - 6)} \approx 5.3$

41. Area $= K = \dfrac{1}{2}(10)(12) \sin 58° \approx 50.9$

43. Area $= K = \dfrac{1}{2}(5.5)(6.8) \sin 78° \approx 18.3$

45. Note that $\gamma = 180° - 31° - 54° = 95°$. Then $\dfrac{b}{\sin \beta} = \dfrac{a}{\sin \alpha} \Rightarrow b = \dfrac{a \sin \beta}{\sin \alpha} = \dfrac{2.6 \sin 31°}{\sin 54°} \approx 1.65522$.

Area $= K \approx \dfrac{1}{2}(1.65522)(2.6) \sin 95° \approx 2.1$

47. $s = \dfrac{1}{2}(5.5 + 6.7 + 9.2) = 10.7 \Rightarrow$ Area $= K = \sqrt{10.7(10.7 - 5.5)(10.7 - 6.7)(10.7 - 9.2)} \approx 18.3$

49. $s = \dfrac{1}{2}(11 + 13 + 20) = 22 \Rightarrow$ Area $= K = \sqrt{22(22 - 11)(22 - 13)(22 - 20)} = 66$

51. To apply the formula

    $K = \dfrac{1}{2}ac \sin \beta$, we need to find β, or to use the formula $K = \dfrac{1}{2}bc \sin \alpha$, we need to find b.

    This triangle is in the form of SSA. It is ambiguous.

    $\dfrac{\sin \gamma}{c} = \dfrac{\sin \alpha}{a} \Rightarrow \sin \gamma = \dfrac{c \sin \alpha}{a} = \dfrac{16 \sin 42°}{21} \approx 0.50981 \Rightarrow \gamma_R = \sin^{-1}(0.50981) \approx 30.65°$

    $\gamma_1 = 30.65°$ then $\beta = 180° - 42° - 30.65° = 107.35°$

    Since $\beta = 107.35°$, then Area $= K = \dfrac{1}{2}(21)(16) \sin 107.35° \approx 160.4°$

53. Using standard labels, $a = BC, b = AC, c = AB$ and $\gamma =$ angle ACB.

    $c^2 = a^2 + b^2 - 2ab \cos \gamma \Rightarrow c = \sqrt{123^2 + 143^2 - 2(123)(143) \cos 78° \, 35'} \approx 169$ ft.

55. After 1.5 hours, the first ship has gone 30 miles and the second ship has gone 21 miles. See Figure 55.

    $d^2 = 30^2 + 21^2 - 2(30)(21) \cos 69° \Rightarrow d = \sqrt{1341 - 1260 \cos 69°} \approx 29.82$

    After 1.5 hours, the ships are about 29.8 miles apart.

Figure 55

57. The shorter diagonal and the two given sides form a triangle of the form SAS. Using the law of cosines

    we will be able to find the length of the shorter diagonal. We will use $\alpha = 56°, b = 5.2$ and $c = 3.5 \Rightarrow$

    $a^2 = b^2 + c^2 - 2bc \cos \alpha = 5.2^2 + 3.5^2 - 2(5.2)(3.5) \cos 56° \Rightarrow$

    $a = \sqrt{39.29 - 36.4 \cos 56°} \approx 4.35148$ or $a = 4.4$ feet. The longer diagonal and the two given sides form a

    triangle of the form SAS. Since adjacent angles of a parallelogram are supplementary, $\alpha = 180° - 56° = 124°$.

    Using the law of cosines we will be able to find the length of the longer diagonal. We will use

    $\alpha = 124°, b = 5.2$ and $c = 3.5 \Rightarrow a^2 = b^2 + c^2 - 2bc \cos \alpha = 5.2^2 + 3.5^2 - 2(5.2)(3.5) \cos 124° \Rightarrow$

    $a = \sqrt{39.29 - 36.4 \cos 124°} \approx 7.72299$ or $a \approx 7.7$ feet.

59. Using standard labels, $a = 500, b = 410, c = 400, \alpha =$ angle $CAB, \beta =$ angle ABC and $\gamma =$ angle ACB.

    To find the bearing when traveling from A to B we first need to find α.

    $a^2 = b^2 + c^2 - 2bc \cos \alpha \Rightarrow \cos \alpha = \dfrac{a^2 - b^2 - c^2}{-2bc} \Rightarrow \alpha = \cos^{-1}\left(\dfrac{500^2 - 410^2 - 400^2}{-2(410)(400)}\right) \approx 76.2°$

    Thus, the bearing is approximately 76.2°.

    To find the bearing when traveling from B to C we first need to find β.

    $b^2 = a^2 + c^2 - 2ac \cos \beta \Rightarrow \cos \beta = \dfrac{b^2 - a^2 - c^2}{-2ac} \Rightarrow \beta = \cos^{-1}\left(\dfrac{410^2 - 500^2 - 400^2}{-2(500)(400)}\right) \approx 52.8°$

    Thus, the bearing is approximately $180° + 76.2° + 52.8° \approx 309°$.

61. (a) $\cos\alpha = \dfrac{a^2 - b^2 - c^2}{-2bc} \Rightarrow \alpha = \cos^{-1}\left(\dfrac{145.2^2 - 136.8^2 - 95.3^2}{-2(136.8)(95.3)}\right) \approx 75.08001°$

    $\cos\beta = \dfrac{b^2 - a^2 - c^2}{-2ac} \Rightarrow \beta = \cos^{-1}\left(\dfrac{136.8^2 - 145.2^2 - 95.3^2}{-2(145.2)(95.3)}\right) \approx 65.55867°$

    $\gamma = 180° - 75.08001° - 65.55867° \approx 39.36132°$ $\alpha \approx 75.1°,\ \beta \approx 65.6°,\ \gamma \approx 39.4°$

    (The angles do not sum to 180° due to rounding)

    (b) $s = \dfrac{1}{2}(145.2 + 136.8 + 95.3) = 188.65 \Rightarrow$

    Area $= K = \sqrt{188.65(188.65 - 145.2)(188.65 - 136.8)(188.65 - 95.3)} \approx 6298.76 \approx 6299\ \text{ft}^2.$

63. $13^2 = 16^2 + 20^2 - 2(16)(20)\cos\theta \Rightarrow \theta = \cos^{-1}\left(\dfrac{13^2 - 16^2 - 20^2}{-2(16)(20)}\right) \approx 40.5°$

65. Refer to Figure 65. Mark point D directly south of C on the segment joining A and B. Then triangle CDB is a

    right triangle. Since the bearing from C to B is 125°, angle $DCB = 180° - 125° = 55°$.

    Thus $\beta = 90° - 55° = 35°$. We will use $a = 357$, $c = 515$ and $\beta = 35°$.

    $b^2 = a^2 + c^2 - 2ac\cos\beta = 357^2 + 515^2 - 2(357)(515)\cos 35° \Rightarrow b = \sqrt{392{,}674 - 367{,}710\cos 35°} \approx$

    302 miles. Thus, the distance between A and C is approximately 302 miles.

Figure 65

67. We will use the labels A = Airplane, B = Battleship, and C = Submarine, with corresponding angles

    α, β, and γ, respectively. Since the angle of elevation from B to A equals the angle of depression from A to B,

    we have $\beta = 17°\ 30'$. Also $\alpha = 24°\ 10' - 17°\ 30' = 6°\ 40'$ and $\gamma = 180° - 17°\ 30' - 6°\ 40' = 155°\ 50'$.

    Using the law of sines $\dfrac{a}{\sin 6°\ 40'} = \dfrac{5120}{\sin 155°\ 50'} \Rightarrow a = \dfrac{5120\sin 6°\ 40'}{\sin 155°\ 50'} \approx 1451.9.$

    The distance between the battleship and the submarine is about 1452 ft.

69. Use the law of cosines to find the distance d. $d^2 = 402^2 + 402^2 - 2(402)(402)\cos 135°\ 40' \Rightarrow$

    $d = \sqrt{323{,}208 - 323{,}208\cos 135°\ 40'} \Rightarrow d \approx 745\ \text{mi}.$

71. The painter's triangular region is in the form of SAS where we will use $c = 25$, $b = 15$ and $\alpha = 128°$.

    If $K = \dfrac{1}{2}bc\sin\alpha$, then $K = \dfrac{1}{2}(15)(25)\sin 128° \approx 147.8$. Thus, the area of the region is 147.8 ft^2.

73. (a) $L = 6$ and $n = 3 \Rightarrow A = \dfrac{3(6)^2}{4}\cot\dfrac{\pi}{3} = 27\left(\dfrac{1}{\sqrt{3}}\right) = 9\sqrt{3} \approx 15.6\ \text{in}^2.$

    (b) $s = \dfrac{1}{2}(6 + 6 + 6) = 9 \Rightarrow$ Area $= K = \sqrt{9(9 - 6)(9 - 6)(9 - 6)} \approx 15.6\ \text{in}^2.$

    The results are equal.

75. Area $= K = \dfrac{1}{2}(500)(600)\sin 85° \approx 149{,}429\ \text{ft}^2.$

77. In triangle ADE,

$$s = \frac{1}{2}(105 + 190 + 125) = 210 \Rightarrow K_1 = \sqrt{210(210 - 105)(210 - 190)(210 - 125)} \approx 6122.5 \text{ ft}^2.$$

In triangle ABD, $s = \frac{1}{2}(190 + 110 + 195) = 247.5 \Rightarrow$

$$K_2 = \sqrt{247.5(247.5 - 190)(247.5 - 110)(247.5 - 195)} \approx 10{,}135.7 \text{ ft}^2.$$

In triangle BCD,

$$s = \frac{1}{2}(195 + 75 + 150) = 210 \Rightarrow K_3 = \sqrt{210(210 - 195)(210 - 75)(210 - 150)} \approx 5051.2 \text{ ft}^2.$$

The total area is $K_1 + K_2 + K_3 = 6122.5 + 10{,}135.7 + 5051.2 \approx 21{,}309 \text{ ft}^2.$

8.3: Vectors

1. (a) $a_1 \approx 3$, $a_2 \approx 4$

   (b) $\|\mathbf{v}\| = \sqrt{3^2 + 4^2} = \sqrt{9 + 16} = \sqrt{25} = 5$

3. (a) $a_1 \approx -5$, $a_2 \approx -12$

   (b) $\|\mathbf{v}\| = \sqrt{(-5)^2 + (-12)^2} = \sqrt{25 + 144} = \sqrt{169} = 13$

5. (a) See Figure 5.

   (b) $\mathbf{v} = \langle 0, -20 \rangle$

   (c) $2\mathbf{v} = 2\langle 0, -20 \rangle = \langle 0, -40 \rangle$; this represents a 40 mph north wind.

   $-\frac{1}{2}\mathbf{v} = -\frac{1}{2}\langle 0, -20 \rangle = \langle 0, 10 \rangle$; this represents a 10 mph south wind.

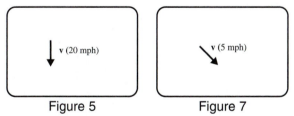

Figure 5   Figure 7

7. (a) See Figure 7.

   (b) We may think of this vector as a hypotenuse of an isosceles right triangle. Since the legs l of such a triangle have the same length, we have the following:

   $$l^2 + l^2 = 5^2 \Rightarrow 2l^2 = 25 \Rightarrow l^2 = \frac{25}{2} \Rightarrow l = \sqrt{\frac{25}{2}} \Rightarrow l = \frac{5}{\sqrt{2}}.$$

   The vector can be represented by $\mathbf{v} = \left\langle \frac{5}{\sqrt{2}}, -\frac{5}{\sqrt{2}} \right\rangle$ or $\left\langle \frac{5}{2}\sqrt{2}, -\frac{5}{2}\sqrt{2} \right\rangle.$

   (c) $2\mathbf{v} = 2\left\langle \frac{5}{\sqrt{2}}, -\frac{5}{\sqrt{2}} \right\rangle = \left\langle \frac{10}{\sqrt{2}}, -\frac{10}{\sqrt{2}} \right\rangle$ or $\langle 5\sqrt{2}, -5\sqrt{2} \rangle$; this represents a 10 mph northwest wind.

   $-\frac{1}{2}\mathbf{v} = -\frac{1}{2}\left\langle \frac{5}{2}\sqrt{2}, -\frac{5}{2}\sqrt{2} \right\rangle = \left\langle -\frac{5}{4}\sqrt{2}, \frac{5}{4}\sqrt{2} \right\rangle$; this represents a 2.5 mph southeast wind.

9. (a) See Figure 9.

(b) $\mathbf{v} = \langle 0, 30 \rangle$

(c) $2\mathbf{v} = 2\langle 0, 30 \rangle = \langle 0, 60 \rangle$; this represents a 60 lb force upward.

$-\dfrac{1}{2}\mathbf{v} = -\dfrac{1}{2}\langle 0, 30 \rangle = \langle 0, -15 \rangle$; this represents a 15 lb force downward.

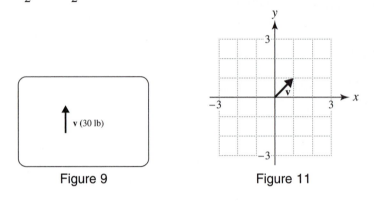

Figure 9   Figure 11

11. (a) Horizontal = 1, vertical = 1

(b) $\|\mathbf{v}\| = \sqrt{1^2 + 1^2} = \sqrt{1 + 1} = \sqrt{2}$; $\mathbf{v}$ is not a unit vector.

(c) $\|\mathbf{v}\|$ represents the length of $\mathbf{v}$. See Figure 11.

13. (a) Horizontal = 3, vertical = -4

(b) $\|\mathbf{v}\| = \sqrt{3^2 + (-4)^2} = \sqrt{9 + 16} = \sqrt{25} = 5$; $\mathbf{v}$ is not a unit vector.

(c) $\|\mathbf{v}\|$ represents the length of $\mathbf{v}$. See Figure 13.

15. (a) Horizontal = 1, vertical = 0

(b) $\|\mathbf{v}\| = \sqrt{1^2 + 0^2} = \sqrt{1 + 0} = \sqrt{1} = 1$; $\mathbf{v}$ is a unit vector.

(c) $\|\mathbf{v}\|$ represents the length of $\mathbf{v}$. See Figure 15.

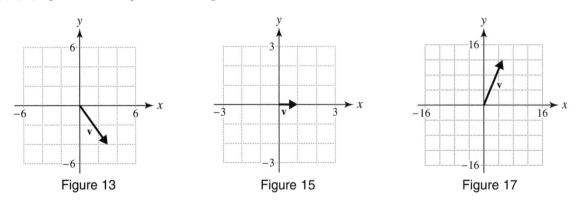

Figure 13   Figure 15   Figure 17

17. (a) Horizontal = 5, vertical = 12

(b) $\|\mathbf{v}\| = \sqrt{5^2 + 12^2} = \sqrt{25 + 144} = \sqrt{169} = 13$; $\mathbf{v}$ is not a unit vector.

(c) $\|\mathbf{v}\|$ represents the length of $\mathbf{v}$. See Figure 17.

19. $\|\mathbf{v}\| = \sqrt{3^2 + 0^2} = 3$; $\mathbf{v} = \langle 3, 0 \rangle$ lies on the positive x-axis and $\tan \theta = \dfrac{0}{3} \Rightarrow \theta = 0$.

Therefore, the direction angle is $0°$.

21. $\|\mathbf{v}\| = \sqrt{(-1)^2 + 1^2} = \sqrt{2}$; $\mathbf{v} = \langle -1, 1 \rangle$ is in quadrant II and $\tan \theta = -1 \Rightarrow \theta = 135°$.

     Therefore, the direction angle is $135°$.

23. $\|\mathbf{v}\| = \sqrt{(\sqrt{3})^2 + (-1)^2} = 2$; $\mathbf{v} = \langle \sqrt{3}, -1 \rangle$ is in quadrant IV and $\tan \theta = \dfrac{-1}{\sqrt{3}} \Rightarrow \theta = 330°$.

     Therefore, the direction angle is $330°$.

25. $\|\mathbf{v}\| = \sqrt{(-5)^2 + (-12)^2} = 13$; $\mathbf{v} = \langle -5, -12 \rangle$ is in quadrant III and $\tan \theta = \dfrac{-12}{-5} \Rightarrow \theta = 247.4°$.

     Therefore, the direction angle is $247.4°$.

27. $\|\mathbf{v}\| = \sqrt{13^2 + (-84)^2} = 85$; $\mathbf{v} = \langle 13, -84 \rangle$ is in quadrant IV and $\tan \theta = \dfrac{-84}{13} \Rightarrow \theta = 278.8°$.

     Therefore, the direction angle is $278.8°$.

29. $\|\mathbf{v}\| = \sqrt{(-20)^2 + (21)^2} = 29$; $\mathbf{v} = \langle -20, 21 \rangle$ is in quadrant II and $\tan \theta = \dfrac{21}{-20} \Rightarrow \theta = 133.6°$.

     Therefore, the direction angle is $133.6°$.

31. $a_1 = 4 \cos 180° = -4$; $a_2 = 4 \sin 180° = 0$. Horizontal component $= -4$, vertical component $= 0$.

33. $a_1 = \sqrt{2} \cos 135° = -1$; $a_2 = \sqrt{2} \sin 135° = 1$. Horizontal component $= -1$, vertical component $= 1$.

35. $a_1 = 23 \cos 54° = 13.5$; $a_2 = 23 \sin 54° = 18.6$. Horizontal component $= 13.5$, vertical component $= 18.6$.

37. $a_1 = 34 \cos 312° = 22.8$; $a_2 = 34 \sin 312° = -25.3$. Horizontal component $= 22.8$,

     vertical component $= -25.3$.

39. $\mathbf{v} = \langle 6 \cos 30°, 6 \sin 30° \rangle = \langle 3\sqrt{3}, 3 \rangle$

41. $\mathbf{v} = \langle 9 \cos 225°, 9 \sin 225° \rangle = \left\langle -\dfrac{9\sqrt{2}}{2}, -\dfrac{9\sqrt{2}}{2} \right\rangle$

43. horizontal: $\cos 40° = \dfrac{v_1}{4} \Rightarrow v_1 = 4 \cos 40° \approx 3.06$

     vertical: $\sin 40° = \dfrac{v_2}{4} \Rightarrow v_2 = 4 \sin 40° \approx 2.57$

45. horizontal: $\cos (-35°) = \dfrac{v_1}{5} \Rightarrow v_1 = 5 \cos (-35°) \approx 4.10$

     vertical: $\sin (-35°) = \dfrac{v_2}{5} \Rightarrow v_2 = 5 \sin (-35°) \approx -2.87$

47. (a) See Figure 47.

     (b) $\mathbf{v} = \langle -1 - 0, \; 2 - 0 \rangle = \langle -1, 2 \rangle$

     (c) $\|\overrightarrow{PQ}\| = \sqrt{(-1)^2 + (2)^2} = \sqrt{5}$

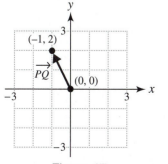

Figure 47

49. (a) See Figure 49.

   (b) $\mathbf{v} = \langle 3 - 1, \ 6 - 2 \rangle = \langle 2, \ 4 \rangle$

   (c) $\| \overrightarrow{PQ} \| = \sqrt{(2)^2 + (4)^2} = \sqrt{20}$

51. (a) See Figure 51.

   (b) $\mathbf{v} = \langle 3 - (-2), \ -2 - 4 \rangle = \langle 5, \ -6 \rangle$

   (c) $\| \overrightarrow{PQ} \| = \sqrt{(5)^2 + (-6)^2} = \sqrt{61}$

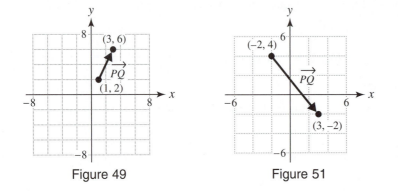

Figure 49   Figure 51

53. Complete the parallelogram and find the unknown angle: $180° - 40° = 140°$.

   Find the magnitude of the resultant force by using the law of cosines to find the length of the diagonal d.

   $d^2 = 40^2 + 60^2 - 2(40)(60) \cos 140° \Rightarrow d = \sqrt{5200 - 4800 \cos 140°} \approx 94.2$ lb

55. Complete the parallelogram and find the unknown angle: $180° - 110° = 70°$.

   Find the magnitude of the resultant force by using the law of cosines to find the length of the diagonal d.

   $d^2 = 15^2 + 25^2 - 2(15)(25) \cos 70° \Rightarrow d = \sqrt{850 - 750 \cos 70°} \approx 24.4$ lb

57. $F_1 + F_2 = (3\mathbf{i} - 4\mathbf{j}) + (-8\mathbf{i} + 16\mathbf{j}) = (3 - 8)\mathbf{i} + (-4 + 16)\mathbf{j} = -5\mathbf{i} + 12\mathbf{j}$,

   $\| \mathbf{F} \| = \sqrt{(-5)^2 + 12^2} = \sqrt{169} = 13$

59. $F_1 + F_2 = \langle 8, 9 \rangle + \langle 1, -31 \rangle = \langle 8 + 1, 9 + (-31) \rangle = \langle 9, -22 \rangle, \| \mathbf{F} \| = \sqrt{9^2 + (-22)^2} = \sqrt{565}$

61. $F_1 + F_2 = (0.5\mathbf{i} + 0.7\mathbf{j}) + (-1.5\mathbf{i} - 5.7\mathbf{j}) = (0.5 - 1.5)\mathbf{i} + (0.7 - 5.7)\mathbf{j} = -\mathbf{i} - 5\mathbf{j}$,

   $\| \mathbf{F} \| = \sqrt{(-1)^2 + (-5)^2} = \sqrt{26}$

63. (a) $\mathbf{a} + \mathbf{b} = \langle 0, 2 \rangle + \langle 3, 0 \rangle = \langle 0 + 3, 2 + 0 \rangle = \langle 3, 2 \rangle$

   (b) $\mathbf{a} - \mathbf{b} = \langle 0, 2 \rangle - \langle 3, 0 \rangle = \langle 0 - 3, 2 - 0 \rangle = \langle -3, 2 \rangle$

65. (a) $\mathbf{a} + \mathbf{b} = (2\mathbf{i} + \mathbf{j}) + (\mathbf{i} - 2\mathbf{j}) = 3\mathbf{i} - \mathbf{j}$

   (b) $\mathbf{a} - \mathbf{b} = (2\mathbf{i} + \mathbf{j}) - (\mathbf{i} - 2\mathbf{j}) = \mathbf{i} + 3\mathbf{j}$

67. (a) $\mathbf{a} + \mathbf{b} = \left\langle -\sqrt{2}, \frac{1}{2} \right\rangle + \left\langle \sqrt{2}, -\frac{3}{4} \right\rangle = \left\langle -\sqrt{2} + \sqrt{2}, \frac{1}{2} + \left(-\frac{3}{4} \right) \right\rangle = \left\langle 0, -\frac{1}{4} \right\rangle$

   (b) $\mathbf{a} - \mathbf{b} = \left\langle -\sqrt{2}, \frac{1}{2} \right\rangle - \left\langle \sqrt{2}, -\frac{3}{4} \right\rangle = \left\langle -\sqrt{2} - \sqrt{2}, \frac{1}{2} - \left(-\frac{3}{4} \right) \right\rangle = \left\langle -2\sqrt{2}, \frac{5}{4} \right\rangle$

69. (a) $\mathbf{a} + \mathbf{b} = \left[\left(\cos\dfrac{\pi}{4}\right)\mathbf{i} + \left(\sin\dfrac{\pi}{4}\right)\mathbf{j}\right] + \left[\left(\cos\dfrac{\pi}{2}\right)\mathbf{i} + \left(\sin\dfrac{\pi}{2}\right)\mathbf{j}\right] =$

$\left[\left(\cos\dfrac{\pi}{4}\right)\mathbf{i} + \left(\cos\dfrac{\pi}{2}\right)\mathbf{i} + \left(\sin\dfrac{\pi}{4}\right)\mathbf{j} + \left(\sin\dfrac{\pi}{2}\right)\mathbf{j}\right] = \left(\dfrac{\sqrt{2}}{2}\mathbf{i} + 0\mathbf{i}\right) + \left(\dfrac{\sqrt{2}}{2}\mathbf{j} + 1\mathbf{j}\right) =$

$\dfrac{\sqrt{2}}{2}\mathbf{i} + \dfrac{\sqrt{2} + 2}{2}\mathbf{j}$

(b) $\mathbf{a} - \mathbf{b} = \left[\left(\cos\dfrac{\pi}{4}\right)\mathbf{i} + \left(\sin\dfrac{\pi}{4}\right)\mathbf{j}\right] - \left[\left(\cos\dfrac{\pi}{2}\right)\mathbf{i} + \left(\sin\dfrac{\pi}{2}\right)\mathbf{j}\right] =$

$\left[\left(\cos\dfrac{\pi}{4}\right)\mathbf{i} - \left(\cos\dfrac{\pi}{2}\right)\mathbf{i} + \left(\sin\dfrac{\pi}{4}\right)\mathbf{j} - \left(\sin\dfrac{\pi}{2}\right)\mathbf{j}\right] = \left(\dfrac{\sqrt{2}}{2}\mathbf{i} - 0\mathbf{i}\right) + \left(\dfrac{\sqrt{2}}{2}\mathbf{j} - 1\mathbf{j}\right) =$

$\dfrac{\sqrt{2}}{2}\mathbf{i} + \dfrac{\sqrt{2} - 2}{2}\mathbf{j}$

71. (a) $\mathbf{a} + \mathbf{b} = \langle -8, 8\rangle + \langle 4, 8\rangle = \langle -8 + 4, 8 + 8\rangle = \langle -4, 16\rangle$

(b) $\mathbf{a} - \mathbf{b} = \langle -8, 8\rangle - \langle 4, 8\rangle = \langle -8 - 4, 8 - 8\rangle = \langle -12, 0\rangle$

(c) $-\mathbf{a} = -\langle -8, 8\rangle = \langle 8, -8\rangle$

73. (a) $\mathbf{a} + \mathbf{b} = \langle 4, 8\rangle + \langle 4, -8\rangle = \langle 4 + 4, 8 + (-8)\rangle = \langle 8, 0\rangle$

(b) $\mathbf{a} - \mathbf{b} = \langle 4, 8\rangle - \langle 4, -8\rangle = \langle 4 - 4, 8 - (-8)\rangle = \langle 0, 16\rangle$

(c) $-\mathbf{a} = -\langle 4, 8\rangle = \langle -4, -8\rangle$

75. (a) $\mathbf{a} + \mathbf{b} = \langle -8, 4\rangle + \langle 8, 8\rangle = \langle -8 + 8, 4 + 8\rangle = \langle 0, 12\rangle$

(b) $\mathbf{a} - \mathbf{b} = \langle -8, 4\rangle - \langle 8, 8\rangle = \langle -8 - 8, 4 - 8\rangle = \langle -16, -4\rangle$

(c) $-\mathbf{a} = -\langle -8, 4\rangle = \langle 8, -4\rangle$

77. (a) Graphical: See Figure 77a. The length appears to be about 2.

Symbolic: $\|\mathbf{a}\| = \|2\mathbf{i}\| = \sqrt{2^2} = \sqrt{4} = 2$

(b) Graphical: $2\mathbf{a} = 4\mathbf{i}$. See Figure 77b.

Symbolic: $2\mathbf{a} = 2(2\mathbf{i}) = 4\mathbf{i}$

(c) Graphical: $2\mathbf{a} + 3\mathbf{b} = 7\mathbf{i} + 3\mathbf{j}$. See Figure 77c.

Symbolic: $2\mathbf{a} + 3\mathbf{b} = 2(2\mathbf{i}) + 3(\mathbf{i} + \mathbf{j}) = 4\mathbf{i} + 3\mathbf{i} + 3\mathbf{j} = 7\mathbf{i} + 3\mathbf{j}$

Figure 77a

Figure 77b

Figure 77c

79. (a) Graphical: Graphical: See Figure 79a. The length appears to be about 2.25.

Symbolic: $\|\mathbf{a}\| = \sqrt{(-1)^2 + 2^2} = \sqrt{5}$

(b) Graphical: $2\mathbf{a} = \langle -2, 4 \rangle$. See Figure 79b.

Symbolic: $2\mathbf{a} = 2\langle -1, 2 \rangle = \langle -2, 4 \rangle$

(c) Graphical: $2\mathbf{a} + 3\mathbf{b} = \langle 7, 4 \rangle$. See Figure 79c.

Symbolic: $2\mathbf{a} + 3\mathbf{b} = 2\langle -1, 2 \rangle + 3\langle 3, 0 \rangle = \langle -2, 4 \rangle + \langle 9, 0 \rangle = \langle -2 + 9, 4 + 0 \rangle = \langle 7, 4 \rangle$

Figure 79a   Figure 79b   Figure 79c

81. (a) $-a + 4b = -(\mathbf{i} - 2\mathbf{j}) + 4(-5\mathbf{i} + 2\mathbf{j}) = (-\mathbf{i} + 2\mathbf{j}) + (-20\mathbf{i} + 8\mathbf{j}) = (-1 - 20)\mathbf{i} + (2 + 8)\mathbf{j} =$
$-21\mathbf{i} + 10\mathbf{j}$

(b) $2a - 3b = 2(\mathbf{i} - 2\mathbf{j}) - 3(-5\mathbf{i} + 2\mathbf{j}) = (2\mathbf{i} - 4\mathbf{j}) + (15\mathbf{i} - 6\mathbf{j}) = (2 + 15)\mathbf{i} + (-4 - 6)\mathbf{j} =$
$17\mathbf{i} - 10\mathbf{j}$

83. (a) $-a + 4b = -\langle 1, -4 \rangle + 4\langle -3, 5 \rangle = \langle -1, 4 \rangle + \langle -12, 20 \rangle = \langle -1 - 12, 4 + 20 \rangle = \langle -13, 24 \rangle =$
$-13\mathbf{i} + 24\mathbf{j}$

(b) $2a - 3b = 2\langle 1, -4 \rangle - 3\langle -3, 5 \rangle = \langle 2, -8 \rangle + \langle 9, -15 \rangle = \langle 2 + 9, -8 - 15 \rangle = \langle 11, -23 \rangle = 11\mathbf{i} - 23\mathbf{j}$

85. (a) $-a + 4b = -\langle -7, -2 \rangle + 4\langle 9, -1 \rangle = \langle 7, 2 \rangle + \langle 36, -4 \rangle = \langle 7 + 36, 2 - 4 \rangle = \langle 43, -2 \rangle = 43\mathbf{i} - 2\mathbf{j}$

(b) $2a - 3b = 2\langle -7, -2 \rangle - 3\langle 9, -1 \rangle = \langle -14, -4 \rangle + \langle -27, 3 \rangle = \langle -14 - 27, -4 + 3 \rangle = \langle -41, -1 \rangle =$
$-41\mathbf{i} - \mathbf{j}$

87. (a) $\mathbf{a} \cdot \mathbf{b} = (1)(3) + (-2)(1) = 3 - 2 = 1$

(b) $\|\mathbf{a}\| = \sqrt{1^2 + (-2)^2} = \sqrt{1 + 4} = \sqrt{5}$, $\|\mathbf{b}\| = \sqrt{3^2 + 1^2} = \sqrt{9 + 1} = \sqrt{10}$

$\theta = \cos^{-1}\left(\dfrac{\mathbf{a} \cdot \mathbf{b}}{\|\mathbf{a}\|\|\mathbf{b}\|}\right) = \cos^{-1}\left(\dfrac{1}{\sqrt{5}\sqrt{10}}\right) = \cos^{-1}\left(\dfrac{1}{\sqrt{50}}\right) \approx 81.9°$

(c) The vectors are neither parallel nor perpendicular.

89. (a) $\mathbf{a} \cdot \mathbf{b} = (6)(-4) + (8)(3) = -24 + 24 = 0$

(b) $\|\mathbf{a}\| = \sqrt{6^2 + 8^2} = \sqrt{36 + 64} = \sqrt{100} = 10$, $\|\mathbf{b}\| = \sqrt{(-4)^2 + 3^2} = \sqrt{16 + 9} = \sqrt{25} = 5$

$\theta = \cos^{-1}\left(\dfrac{\mathbf{a} \cdot \mathbf{b}}{\|\mathbf{a}\|\|\mathbf{b}\|}\right) = \cos^{-1}\left(\dfrac{0}{10 \cdot 5}\right) = \cos^{-1} 0 = 90°$

(c) The vectors are perpendicular.

91. (a) $\mathbf{a} \cdot \mathbf{b} = (5)(10) + (6)(12) = 50 + 72 = 122$

(b) $\|\mathbf{a}\| = \sqrt{5^2 + 6^2} = \sqrt{25 + 36} = \sqrt{61}, \ \|\mathbf{b}\| = \sqrt{10^2 + 12^2} = \sqrt{100 + 144} = \sqrt{244}$

$$\theta = \cos^{-1}\left(\frac{\mathbf{a} \cdot \mathbf{b}}{\|\mathbf{a}\|\|\mathbf{b}\|}\right) = \cos^{-1}\left(\frac{122}{\sqrt{61}\sqrt{244}}\right) = \cos^{-1}\left(\frac{122}{\sqrt{14,884}}\right) = \cos^{-1}\left(\frac{122}{122}\right) = \cos^{-1} 1 = 0°$$

(c) The vectors are parallel and they point in the same direction.

93. (a) $\mathbf{a} \cdot \mathbf{b} = (1)(0.5) + (3)(-1.5) = 0.5 - 4.5 = -4$

(b) $\|\mathbf{a}\| = \sqrt{1^2 + 3^2} = \sqrt{1 + 9} = \sqrt{10}, \ \|\mathbf{b}\| = \sqrt{0.5^2 + (-1.5)^2} = \sqrt{0.25 + 2.25} = \sqrt{2.5}$

$$\theta = \cos^{-1}\left(\frac{\mathbf{a} \cdot \mathbf{b}}{\|\mathbf{a}\|\|\mathbf{b}\|}\right) = \cos^{-1}\left(\frac{-4}{\sqrt{10}\sqrt{2.5}}\right) = \cos^{-1}\left(\frac{-4}{\sqrt{25}}\right) = \cos^{-1}\left(\frac{-4}{5}\right) \approx 143.1°$$

(c) The vectors are neither parallel nor perpendicular.

95. $W = 30 \cdot 5 = 150$ ft-lb

97. $W = 100 \cdot 1000 = 100,000$ ft-lb

99. $W = \mathbf{F} \cdot \mathbf{D} = \langle 10, 20 \rangle \cdot \langle 15, 22 \rangle = (10)(15) + (20)(22) = 590$ ft-lb

$\|\mathbf{F}\| = \sqrt{10^2 + 20^2} = \sqrt{100 + 400} = \sqrt{500} = 10\sqrt{5} \approx 22.4$ lb

101. $W = \mathbf{F} \cdot \mathbf{D} = (5\mathbf{i} - 3\mathbf{j}) \cdot (3\mathbf{i} - 4\mathbf{j}) = (5)(3) + (-3)(-4) = 27$ ft-lb

$\|\mathbf{F}\| = \sqrt{5^2 + (-3)^2} = \sqrt{25 + 9} = \sqrt{34} \approx 5.8$ lb

103. $\overrightarrow{PQ} = \langle 1 - (-2), 6 - 3 \rangle = \langle 3, 3 \rangle = 3\mathbf{i} + 3\mathbf{j}$

$W = \mathbf{F} \cdot \mathbf{D} = (5\mathbf{i} + 3\mathbf{j}) \cdot (3\mathbf{i} + 3\mathbf{j}) = (5)(3) + (3)(3) = 24$

105. $\overrightarrow{PQ} = \langle 4 - 2, -5 - (-3) \rangle = \langle 2, -2 \rangle = 2\mathbf{i} - 2\mathbf{j}$

$W = \mathbf{F} \cdot \mathbf{D} = (5\mathbf{i} + 3\mathbf{j}) \cdot (2\mathbf{i} - 2\mathbf{j}) = (5)(2) + (3)(-2) = 4$

107. The swimmer's velocity is modeled by $\mathbf{a} = \langle 0, 3 \rangle$. The velocity of the current is modeled by $\mathbf{b} = \langle 2, 0 \rangle$.

The resultant velocity of the swimmer is modeled by $\mathbf{a} + \mathbf{b} = \langle 0, 3 \rangle + \langle 2, 0 \rangle = \langle 0 + 2, 3 + 0 \rangle = \langle 2, 3 \rangle$.

The speed of the swimmer is $\|\mathbf{a} + \mathbf{b}\| = \sqrt{2^2 + 3^2} = \sqrt{4 + 9} = \sqrt{13} \approx 3.6$ mph.

109. The velocity of the airplane is modeled by $\mathbf{a} = \langle -400, 0 \rangle$. Refer to the referenced example when finding the appropriate vector to represent the wind. The velocity of the wind can be modeled by

$\mathbf{b} = \left\langle \dfrac{50}{\sqrt{2}}, -\dfrac{50}{\sqrt{2}} \right\rangle$. The resultant velocity of the airplane can be found by vector addition and is modeled by

$\mathbf{a} + \mathbf{b} = \langle -400, 0 \rangle + \left\langle \dfrac{50}{\sqrt{2}}, -\dfrac{50}{\sqrt{2}} \right\rangle = \left\langle -400 + \dfrac{50}{\sqrt{2}}, 0 + \left(-\dfrac{50}{\sqrt{2}}\right) \right\rangle \approx \langle -364.6, -35.4 \rangle$.

The ground speed of the airplane is $\|\mathbf{a} + \mathbf{b}\| = \sqrt{(-364.6)^2 + (-35.4)^2} \approx 366.3$ mph.

The bearing of the airplane is $270° - \tan^{-1}\left(\dfrac{-35.4}{-364.6}\right) \approx 264.5°$.

111. Since the bearing is 160°, the plane is headed for a location which makes a $-70°$ angle with the horizontal. Thus the vector location is modeled by $\langle 450\cos(-70°), 450\sin(-70°) \rangle \approx \langle 153.9, -422.9 \rangle$. The wind can be modeled by the vector $\langle 0, 20 \rangle$. The vector $\mathbf{a}$ that the plane should take can be found as follows: $\mathbf{a} + \langle 0, 20 \rangle \approx \langle 153.9, -422.9 \rangle \Rightarrow \mathbf{a} \approx \langle 153.9, -402.9 \rangle$. The airplane's speed is the magnitude of this vector: $\|\mathbf{a}\| = \sqrt{153.9^2 + (-402.9)^2} \approx 431.3$ mph = groundspeed. The angle from the horizontal of vector $\mathbf{a}$ is $\tan^{-1}\left(\dfrac{-402.9}{153.9}\right) \approx -69.1°$. Thus the final bearing should be about $90° + 69.1° = 159.1°$.

113. Since the bearing is 75°, the plane is headed for a location which makes a 15° angle with the horizontal. The airspeed is found by the use of tangent $\Rightarrow \tan 15° = \dfrac{40}{x} \Rightarrow 0.2679x = 40 \Rightarrow x = 149.3$ mph. The groundspeed is found by the use of sine $\Rightarrow \sin 15° = \dfrac{40}{x} \Rightarrow 0.2588x = 40 \Rightarrow x = 154.6$ mph.

115. (a) $\|\mathbf{R}\| = \sqrt{1^2 + (-2)^2} = \sqrt{1 + 4} = \sqrt{5} \approx 2.2$, $\|\mathbf{A}\| = \sqrt{0.5^2 + 1^2} = \sqrt{0.25 + 1} = \sqrt{1.25} \approx 1.1$

   About 2.2 inches of rain fell. The area of the opening of the rain gauge was about 1.1 in^2.

   (b) $V = |\mathbf{R} \cdot \mathbf{A}| = |(1)(0.5) + (-2)(1)| = |-1.5| = 1.5$. The volume of the rain collected was 1.5 in^3.

   (c) $\mathbf{R}$ and $\mathbf{A}$ should be parallel and point in opposite directions.

117. (a) $\mathbf{c} = \mathbf{a} + \mathbf{b} = \langle 3, 2 \rangle + \langle -2, 2 \rangle = \langle 1, 4 \rangle$

   (b) $\|\mathbf{c}\| = \sqrt{1^2 + 4^2} = \sqrt{1 + 16} = \sqrt{17} \approx 4.1$ ft.

   (c) $3\mathbf{a} + 0.5\mathbf{b} = 3\langle 3, 2 \rangle + 0.5\langle -2, 2 \rangle = \langle 9, 6 \rangle + \langle -1, 1 \rangle = \langle 8, 7 \rangle$

119. (a) Represent the point $(-2, 4)$ by the vector $\mathbf{a} = \langle -2, 4 \rangle$. The new location is given by

   $$\mathbf{b} = \mathbf{a} + \mathbf{v} = \langle -2, 4 \rangle + \langle 4, -2 \rangle = \langle -2 + 4, 4 + (-2) \rangle = \langle 2, 2 \rangle \Rightarrow (2, 2)$$

   (b) See Figure 119.

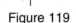

Figure 119

121. Note that 1.5 miles = 7920 feet. The x-component of the displacement vector is $d_1 = 7920 \cos 15°$.

   The y-component of the displacement vector is $d_2 = 7920 \sin 15°$.

   $W = \mathbf{F} \cdot \mathbf{D} = \langle 0, 145 \rangle \cdot \langle 7920 \cos 15°, 7920 \sin 15° \rangle =$

   $(0)(7920 \cos 15°) + (145)(7920 \sin 15°) \approx 297{,}228$ ft-lb

123. Since the bearing is 135°, the city is located on a vector which makes a $-45°$ angle with the horizontal. Thus the vector to the city is modeled by $\langle 200 \cos (-45°), 200 \sin (-45°) \rangle \approx \langle 141.4, -141.4 \rangle$. The wind can be modeled by the vector $\langle 0, -30 \rangle$. The vector $\mathbf{a}$ that the pilot should take can be found as follows:

   $\mathbf{a} + \langle 0, -30 \rangle \approx \langle 141.4, -141.4 \rangle \Rightarrow \mathbf{a} \approx \langle 141.4, -111.4 \rangle$. The airplane's speed is the magnitude of this vector: $\|\mathbf{a}\| = \sqrt{141.4^2 + (-111.4)^2} \approx 180$ mph. The angle from horizontal of vector $\mathbf{a}$ is

   $\tan^{-1}\left(\dfrac{-111.4}{141.4}\right) \approx -38.2°$. Thus, the bearing should be about $90° + 38.2° = 128.2°$. The pilot should fly approximately 180 mph on a bearing of about 128.2°.

8.4: Parametric Equations

1. (a)

t	0	1	2	3
x	−1	0	1	2
y	0	2	4	6

(b) See Figure 1.

(c) This curve is a line segment.

3. (a)

t	0	1	2	3
x	2	3	4	5
y	4	1	0	1

(b) See Figure 3.

(c) This curve is the lower portion of a parabola.

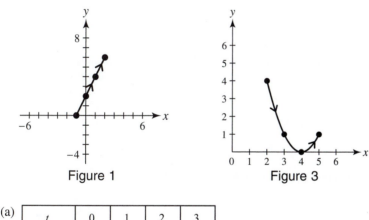

Figure 1   Figure 3

5. (a)

t	0	1	2	3
x	3	$\sqrt{8}$	$\sqrt{5}$	0
y	0	1	2	3

(b) See Figure 5.

(c) This curve is a portion of a circle with radius 3.

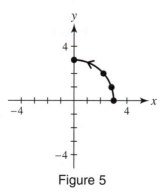

Figure 5

7. (a)

t	0	1	2	3
x	0	1	2	3
y	3	$\sqrt{8}$	$\sqrt{5}$	0

(b) See Figure 7.

(c) This curve is a portion of a circle with radius 3.

9. $x = 3t \Rightarrow t = \frac{1}{3}x$, substituting for t in $y = t - 1$ results in $y = \frac{1}{3}x - 1$. This curve is a line segment.
See Figure 9.

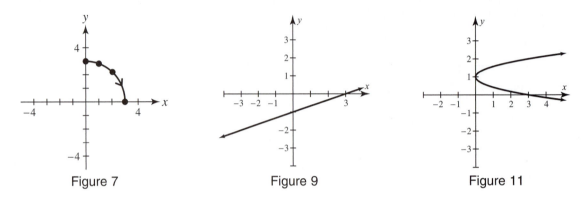

Figure 7   Figure 9   Figure 11

11. $y = t + 1 \Rightarrow t = y - 1$, substituting for t in $x = 3t^2$ results in $x = 3(y - 1)^2$. This curve is a parabola.
See Figure 11.

13. Since $y = t$, substituting y for t in $x = \sqrt{1 - t^2}$ yields $x = \sqrt{1 - y^2}$. This curve is a portion of a circle with
radius 1. See Figure 13.

15. $x = t \Rightarrow y = \frac{1}{2}x^2$. This curve is a portion of a parabola. See Figure 15.

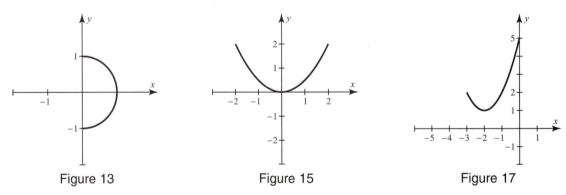

Figure 13   Figure 15   Figure 17

17. $x = t - 2 \Rightarrow x + 2 = t \Rightarrow y = (x + 2)^2 + 1 \Rightarrow y = x^2 + 4x + 5$
This curve is a portion of a parabola. See Figure 17.

19. $x^2 + y^2 = (3 \sin t)^2 + (3 \cos t)^2 = 9 \sin^2 t + 9 \cos^2 t = 9(\sin^2 t + \cos^2 t) = 9 \Rightarrow x^2 + y^2 = 9$
This curve is a circle with radius 3. See Figure 19.

21. $x^2 + y^2 = (2 \sin t)^2 + (-2 \cos t)^2 = 4 \sin^2 t + 4 \cos^2 t = 4(\sin^2 t + \cos^2 t) = 4 \Rightarrow x^2 + y^2 = 4$
This curve is a circle with radius 2. See Figure 21.

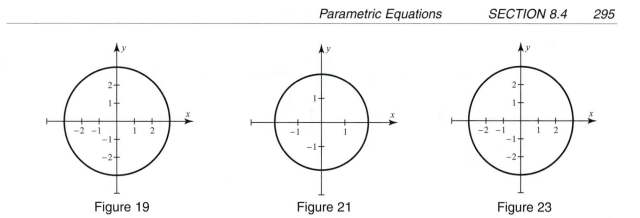

Figure 19   Figure 21   Figure 23

23. $x^2 + y^2 = (3 \cos 2t)^2 + (3 \sin 2t)^2 = 9 \cos^2 2t + 9 \sin^2 2t = 9(\cos^2 2t + \sin^2 2t) = 9 \Rightarrow x^2 + y^2 = 9$

This curve is a circle with radius 3. See Figure 23.

25. See Figure 25.

27. See Figure 27.

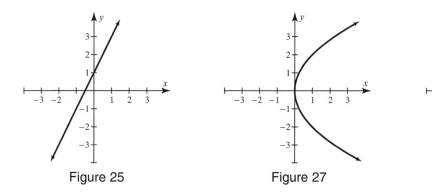

Figure 25   Figure 27   Figure 29

29. See Figure 29.

31. See Figure 31.

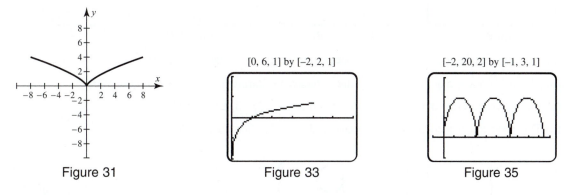

Figure 31   Figure 33   Figure 35

33. See Figure 33.

35. See Figure 35.

37. See Figure 37.

39. See Figure 39.

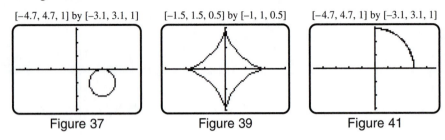

[−4.7, 4.7, 1] by [−3.1, 3.1, 1]   [−1.5, 1.5, 0.5] by [−1, 1, 0.5]   [−4.7, 4.7, 1] by [−3.1, 3.1, 1]

Figure 37   Figure 39   Figure 41

41. See Figure 41.

43. Let $x = t$. Then $2x + y = 4 \Rightarrow 2t + y = 4 \Rightarrow y = 4 - 2t$.

One representation in parametric form is $x = t, y = 4 - 2t$. *Answers may vary.*

45. Let $x = t$. Then $y = 4 - x^2 \Rightarrow y = 4 - t^2$.

One representation in parametric form is $x = t, y = 4 - t^2$. *Answers may vary.*

47. Let $y = t$. Then $x = y^2 + y - 3 \Rightarrow x = t^2 + t - 3$.

One representation in parametric form is $x = t^2 + t - 3, y = t$. *Answers may vary.*

49. Since $\cos^2 t + \sin^2 t = 1$ for all t, $4 \cos^2 t + 4 \sin^2 t = 4 \Rightarrow (2 \cos t)^2 + (2 \sin t)^2 = 4$.

One representation in parametric form is $x = 2 \cos t, y = 2 \sin t, 0 \le t \le 2\pi$. *Answers may vary.*

51. Let $x = t$. Then $\ln y = 0.1x^2 \Rightarrow \ln y = 0.1t^2 \Rightarrow y = e^{0.1t^2}$.

One representation in parametric form is $x = t, y = e^{0.1t^2}$. *Answers may vary.*

53. Let $y = t$. Then $x = y^2 - 2y + 1 \Rightarrow x = t^2 - 2t + 1$. One representation in parametric form is

$x = t^2 - 2t + 1, y = t$. *Answers may vary.*

55. (a) The curve traces a circle of radius 3 once counterclockwise, starting at $(3, 0)$. See Figure 55a.

(b) The curve traces a circle of radius 3 twice counterclockwise, starting at $(3, 0)$. See Figure 55b.

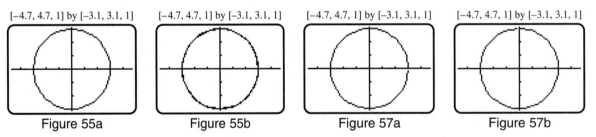

[−4.7, 4.7, 1] by [−3.1, 3.1, 1]   [−4.7, 4.7, 1] by [−3.1, 3.1, 1]   [−4.7, 4.7, 1] by [−3.1, 3.1, 1]   [−4.7, 4.7, 1] by [−3.1, 3.1, 1]

Figure 55a   Figure 55b   Figure 57a   Figure 57b

57. (a) The curve traces a circle of radius 3 once counterclockwise, starting at $(3, 0)$. See Figure 57a.

(b) The curve traces a circle of radius 3 once clockwise, starting at $(0, 3)$. See Figure 57b.

59. (a) The curve traces a circle of radius 1 centered at $(-1, 2)$. See Figure 59a.

(b) The curve traces a circle of radius 1 centered at $(1, 2)$. See Figure 59b.

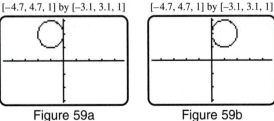

[−4.7, 4.7, 1] by [−3.1, 3.1, 1]   [−4.7, 4.7, 1] by [−3.1, 3.1, 1]

Figure 59a   Figure 59b

61. This is the letter F. See Figure 61.

63. This is the letter D. See Figure 63.

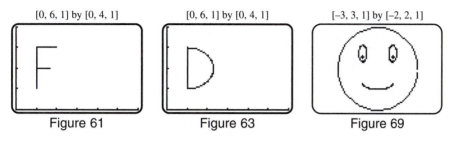

[0, 6, 1] by [0, 4, 1]   [0, 6, 1] by [0, 4, 1]   [−3, 3, 1] by [−2, 2, 1]

Figure 61   Figure 63   Figure 69

65. One possible solution is $x_1 = 0$, $y_1 = 2t$; $x_2 = t$, $y_2 = 0$ for $0 \le t \le 1$. *Answers may vary.*

67. One possible solution is $x_1 = \sin t$, $y_1 = \cos t$; $x_2 = 0$, $y_2 = t - 2$ for $0 \le t \le \pi$. *Answers may vary.*

69. One possible solution is $x_1 = 2 \cos t$, $y_1 = 2 \sin t$; $x_2 = 0.75 + 0.2 \cos t$, $y_2 = 0.75 + 0.4 \sin t$;

    $x_3 = -0.75 + 0.2 \cos t$, $y_3 = 0.75 + 0.4 \sin t$; $x_4 = 0.75 \cos 0.5t$, $y_4 = -0.7 - 0.3 \sin 0.5t$ for $0 \le t \le 2\pi$.

    Add pupils by plotting the points $(0.75, 0.6)$ and $(-0.75, 0.6)$. See Figure 69. *Answers may vary.*

71. The parametric equations which model the flight of the golf ball hit at a 35° angle are given by

    $x_1 = 66 \cos 35°(t)$, $y_1 = 66 \sin 35°(t) - 16t^2$ for $0 \le t \le 4$. This ball travels about 128 feet. See Figure 71a.

    The parametric equations which model the flight of the golf ball hit at a 50° angle are given by

    $x_2 = 66 \cos 50°(t)$, $y_2 = 66 \sin 50°(t) - 16t^2$ for $0 \le t \le 4$. This ball travels about 134 feet. See Figure 71b.

[0, 150, 10] by [−10, 50, 10]   [0, 150, 10] by [−10, 50, 10]   [0, 250, 25] by [0, 65, 5]

Figure 71a   Figure 71b   Figure 73

73. The parametric equations $x_1 = 88 \cos 45°(t)$, $y_1 = 88 \sin 45°(t) - 16t^2$ for $0 \le t \le 5$ model the flight of this

    ball. The fence can be modeled by the parametric equations $x_2 = 200$, $y_2 = 2t$ for $0 \le t \le 5$. The ball will

    go over the fence. See Figure 73.

75. Let $v = 88$, $\theta = 45°$, and $h = 50$. The parametric equations are initially

    $x = (88 \cos 45°)t$ and $y = (88 \sin 45°)t - 16t^2 + 50$.

    Since $\cos 45° = \sin 45° = \dfrac{1}{\sqrt{2}}$, these equations become $x = \left(\dfrac{88}{\sqrt{2}}\right)t$ and $y = \left(\dfrac{88}{\sqrt{2}}\right)t - 16t^2 + 50$.

    Since the ball hits the ground when $y = 0$, we can determine how long the ball was in flight by solving the

    following equation: $-16t^2 + \left(\dfrac{88}{\sqrt{2}}\right)t + 50 = 0$. By the quadratic formula, the solution is $t \approx 4.573$ seconds.

    After 4.573 seconds the ball has traveled horizontally $x = \left(\dfrac{88}{\sqrt{2}}\right)(4.573) \approx 285$ feet.

77. See Figure 77.

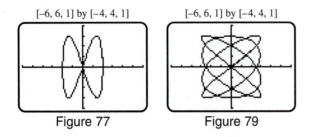

[–6, 6, 1] by [–4, 4, 1]   [–6, 6, 1] by [–4, 4, 1]

Figure 77   Figure 79

79. See Figure 79.

8.5: Polar Equations

1. See Figure 1.

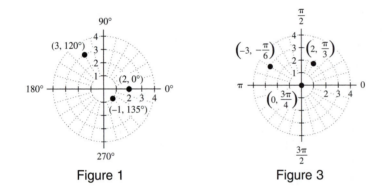

Figure 1   Figure 3

3. See Figure 3.

5. Yes. The angles $180°$ and $-180°$ are coterminal.

7. No. The angles $45°$ and $-45°$ are not coterminal.

9. Yes. The points are both located at the pole.

11. $x = r \cos \theta = 3 \cos 45° = \dfrac{3}{\sqrt{2}}, y = r \sin \theta = 3 \sin 45° = \dfrac{3}{\sqrt{2}}; \left(\dfrac{3}{\sqrt{2}}, \dfrac{3}{\sqrt{2}} \right)$ or $\left(\dfrac{3\sqrt{2}}{2}, \dfrac{3\sqrt{2}}{2} \right)$

13. $x = r \cos \theta = 10 \cos 90° = 0, y = r \sin \theta = 10 \sin 90° = 10; \ (0, 10)$

15. $x = r \cos \theta = 5 \cos 2\pi = 5, y = r \sin \theta = 5 \sin 2\pi = 0; \ (5, 0)$

17. $x = r \cos \theta = -3 \cos 60° = -3\left(\dfrac{1}{2} \right) = -\dfrac{3}{2}, y = r \sin \theta = -3 \sin 60° = -3\left(\dfrac{\sqrt{3}}{2} \right) = -\dfrac{3\sqrt{3}}{2};$

$\left(-\dfrac{3}{2}, -\dfrac{3\sqrt{3}}{2} \right)$

19. $x = r \cos \theta = -2 \cos \left(-\dfrac{3\pi}{2} \right) = -2(0) = 0, y = r \sin \theta = -2 \sin \left(-\dfrac{3\pi}{2} \right) = -2(1) = -2; \ (0, -2)$

21. (a) $r^2 = x^2 + y^2 \Rightarrow r^2 = (0)^2 + (3)^2 \Rightarrow r^2 = 9 \Rightarrow r = \pm 3 \Rightarrow r = 3$ (since $r > 0$)

$\tan \theta = \dfrac{y}{x} \Rightarrow \tan \theta = \dfrac{3}{0} \Rightarrow \tan \theta$ is undefined $\Rightarrow \theta = 90°$ or $270°$ (since $0° \le \theta < 360°$)

Since $(0, 3)$ is located on the positive y-axis, $\theta = 90°$. The polar coordinates are $(3, 90°)$.

(b) $r^2 = x^2 + y^2 \Rightarrow r^2 = (0)^2 + (3)^2 \Rightarrow r^2 = 9 \Rightarrow r = \pm 3 \Rightarrow r = -3$ (since $r < 0$)

$\tan \theta = \dfrac{y}{x} \Rightarrow \tan \theta = \dfrac{3}{0} \Rightarrow \tan \theta$ is undefined $\Rightarrow \theta = -90°$ or $-270°$ (since $-180° < \theta \le 180°$)

Since $(0, 3)$ is located on the positive y-axis, $\theta = -90°$. The polar coordinates are $(-3, -90°)$.

23. (a) $r^2 = x^2 + y^2 \Rightarrow r^2 = (-1)^2 + (-\sqrt{3})^2 \Rightarrow r^2 = 4 \Rightarrow r = \pm 2 \Rightarrow r = 2$ (since $r > 0$)

$\tan \theta = \dfrac{y}{x} \Rightarrow \tan \theta = \dfrac{-\sqrt{3}}{-1} \Rightarrow \theta = \tan^{-1} \sqrt{3} \Rightarrow \theta = 60°$ or $240°$ (since $0° \le \theta < 360°$)

Since $(-1, -\sqrt{3}\,)$ is located in quadrant III, $\theta = 240°$. The polar coordinates are $(2, 240°)$.

(b) $r^2 = x^2 + y^2 \Rightarrow r^2 = (-1)^2 + (-\sqrt{3})^2 \Rightarrow r^2 = 4 \Rightarrow r = \pm 2 \Rightarrow r = -2$ (since $r < 0$)

$\tan \theta = \dfrac{y}{x} \Rightarrow \tan \theta = \dfrac{-\sqrt{3}}{-1} \Rightarrow \theta = \tan^{-1} \sqrt{3} \Rightarrow \theta = -120°$ or $60°$ (since $-180° < \theta \le 180°$)

Since $(-1, -\sqrt{3}\,)$ is located in quadrant III, $\theta = 60°$. The polar coordinates are $(-2, 60°)$.

25. (a) $r^2 = x^2 + y^2 \Rightarrow r^2 = (3)^2 + (-3)^2 \Rightarrow r^2 = 18 \Rightarrow r = \pm \sqrt{18} \Rightarrow r = \sqrt{18}$ (since $r > 0$)

$\tan \theta = \dfrac{y}{x} \Rightarrow \tan \theta = \dfrac{-3}{3} \Rightarrow \theta = \tan^{-1}(-1) \Rightarrow \theta = 135°$ or $315°$ (since $0° \le \theta < 360°$)

Since $(3, -3)$ is located in quadrant IV, $\theta = 315°$. The polar coordinates are $(\sqrt{18}, 315°)$.

(b) $r^2 = x^2 + y^2 \Rightarrow r^2 = (3)^2 + (-3)^2 \Rightarrow r^2 = 18 \Rightarrow r = \pm \sqrt{18} \Rightarrow r = -\sqrt{18}$ (since $r < 0$)

$\tan \theta = \dfrac{y}{x} \Rightarrow \tan \theta = \dfrac{-3}{3} \Rightarrow \theta = \tan^{-1}(-1) \Rightarrow \theta = -45°$ or $135°$ (since $-180° < \theta \le 180°$)

Since $(3, -3)$ is located in quadrant IV, $\theta = 135°$. The polar coordinates are $(-\sqrt{18}, 135°)$.

27. $r^2 = x^2 + y^2 \Rightarrow r^2 = (7)^2 + (24)^2 \Rightarrow r^2 = 625 \Rightarrow r = \pm 25 \Rightarrow r = 25$ (since $r > 0$)

$\tan \theta = \dfrac{y}{x} \Rightarrow \tan \theta = \dfrac{24}{7} \Rightarrow \theta = \tan^{-1} \dfrac{24}{7} \Rightarrow \theta \approx 1.29$ or $\pi + 1.29 \approx 4.43$ (since $0 \le \theta < 2\pi$)

Since $(7, 24)$ is located in quadrant I, $\theta = 1.29$. The polar coordinates are $(25, 1.29)$.

29. $r^2 = x^2 + y^2 \Rightarrow r^2 = (-5)^2 + (12)^2 \Rightarrow r^2 = 169 \Rightarrow r = \pm 13 \Rightarrow r = 13$ (since $r > 0$)

$\tan \theta = \dfrac{y}{x} \Rightarrow \tan \theta = \dfrac{12}{-5} \Rightarrow \theta = \tan^{-1}\left(-\dfrac{12}{5}\right) \Rightarrow \theta \approx 1.97$ or $\pi + 1.97 \approx 5.11$ (since $0 \le \theta < 2\pi$)

Since $(-5, 12)$ is located in quadrant II, $\theta = 1.97$. The polar coordinates are $(13, 1.97)$.

31. See Figure 31.

Figure 31

33. (a)

θ	0°	90°	180°	270°
r	2	2	2	2

(b) See Figure 33.

35. (a)

θ	0°	90°	180°	270°
r	0	3	0	−3

(b) See Figure 35.

Figure 33

Figure 35

Figure 37

37. (a)

θ	0°	90°	180°	270°
r	2	4	2	0

(b) See Figure 37.

39. (a)

θ	0°	90°	180°	270°
r	1	2	3	2

(b) See Figure 39.

41. See Figure 41.

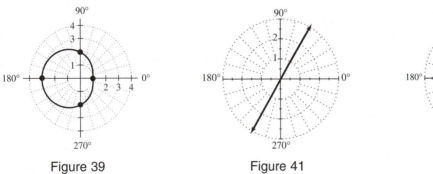

Figure 39

Figure 41

Figure 43

43. See Figure 43.

45. See Figure 45.

47. See Figure 47.

49. See Figure 49.

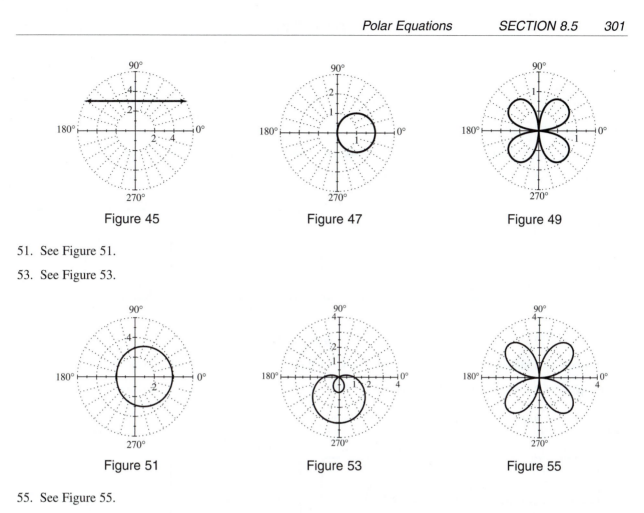

Figure 45   Figure 47   Figure 49

51. See Figure 51.

53. See Figure 53.

Figure 51   Figure 53   Figure 55

55. See Figure 55.

57. See Figure 57.

Figure 57

59. $y = 3 \Rightarrow r \sin \theta = 3$ or $r = 3 \csc \theta$

61. $y = x \Rightarrow r \sin \theta = r \cos \theta \Rightarrow \dfrac{r \sin \theta}{r \cos \theta} = 1 \Rightarrow \tan \theta = 1$ or $\theta = \dfrac{\pi}{4}$

63. $x^2 + y^2 = 9 \Rightarrow (r \cos \theta)^2 + (r \sin \theta)^2 = 9 \Rightarrow r^2(\cos^2 \theta + \sin^2 \theta) = 9 \Rightarrow r^2 = 9$ or $r = 3$

65. $x^2 + y^2 = 2x \Rightarrow (r \cos \theta)^2 + (r \sin \theta)^2 = 2(r \cos \theta) \Rightarrow r^2(\cos^2 \theta + \sin^2 \theta) = 2(r \cos \theta) \Rightarrow$
    $r^2 = 2(r \cos \theta) \Rightarrow r = 2 \cos \theta$

67. $r = 3 \Rightarrow r^2 = 9 \Rightarrow x^2 + y^2 = 9$

69. $r = 2 \sec \theta \Rightarrow r = \dfrac{2}{\cos \theta} \Rightarrow r \cos \theta = 2 \Rightarrow x = 2$

71. $r = \dfrac{3}{2 \cos \theta + 4 \sin \theta} \Rightarrow 2r \cos \theta + 4r \sin \theta = 3 \Rightarrow 2x + 4y = 3$

73. $r = \cos \theta \Rightarrow r^2 = r \cos \theta \Rightarrow x^2 + y^2 = x$

75. Graph $r_1 = 3 + 3 \cos (\theta)$ for $0° \leq \theta \leq 360°$ in $[-9.4, 9.4, 1]$ by $[-6.2, 6.2, 1]$. See Figure 75.

77. Graph $r_1 = 3 - 2 \sin (\theta)$ for $0° \leq \theta \leq 360°$ in $[-9.4, 9.4, 1]$ by $[-6.2, 6.2, 1]$. See Figure 77.

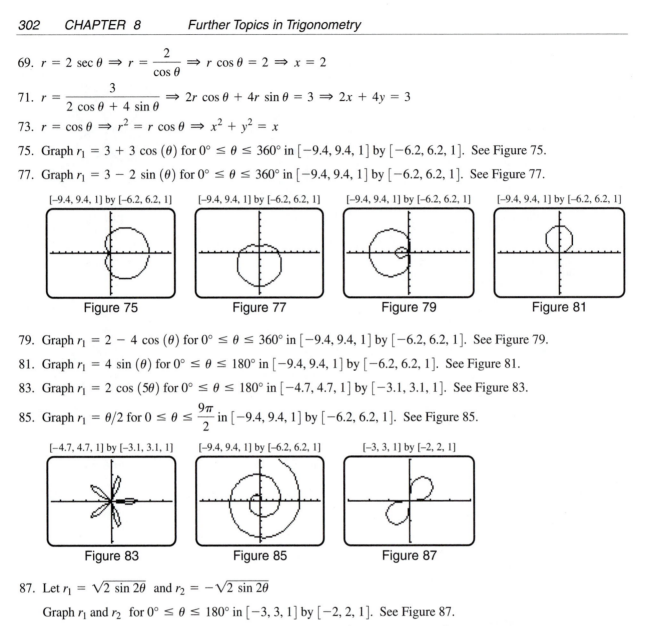

[−9.4, 9.4, 1] by [−6.2, 6.2, 1]   [−9.4, 9.4, 1] by [−6.2, 6.2, 1]   [−9.4, 9.4, 1] by [−6.2, 6.2, 1]   [−9.4, 9.4, 1] by [−6.2, 6.2, 1]

Figure 75   Figure 77   Figure 79   Figure 81

79. Graph $r_1 = 2 - 4 \cos (\theta)$ for $0° \leq \theta \leq 360°$ in $[-9.4, 9.4, 1]$ by $[-6.2, 6.2, 1]$. See Figure 79.

81. Graph $r_1 = 4 \sin (\theta)$ for $0° \leq \theta \leq 180°$ in $[-9.4, 9.4, 1]$ by $[-6.2, 6.2, 1]$. See Figure 81.

83. Graph $r_1 = 2 \cos (5\theta)$ for $0° \leq \theta \leq 180°$ in $[-4.7, 4.7, 1]$ by $[-3.1, 3.1, 1]$. See Figure 83.

85. Graph $r_1 = \theta/2$ for $0 \leq \theta \leq \dfrac{9\pi}{2}$ in $[-9.4, 9.4, 1]$ by $[-6.2, 6.2, 1]$. See Figure 85.

[−4.7, 4.7, 1] by [−3.1, 3.1, 1]   [−9.4, 9.4, 1] by [−6.2, 6.2, 1]   [−3, 3, 1] by [−2, 2, 1]

Figure 83   Figure 85   Figure 87

87. Let $r_1 = \sqrt{2} \sin 2\theta$ and $r_2 = -\sqrt{2} \sin 2\theta$

Graph r_1 and r_2 for $0° \leq \theta \leq 180°$ in $[-3, 3, 1]$ by $[-2, 2, 1]$. See Figure 87.

89. $r_1 = r_2 \Rightarrow 3 = 2 + 2 \sin \theta \Rightarrow 1 = 2 \sin \theta \Rightarrow \sin \theta = \dfrac{1}{2} \Rightarrow \theta_R = \sin^{-1} \dfrac{1}{2} = 30° \Rightarrow \theta = 30°, 150°$

91. (a) $r_1 = r_2 \Rightarrow 3 = 2 - 2 \sin \theta \Rightarrow 2 \sin \theta = -1 \Rightarrow \sin \theta = -\dfrac{1}{2} \Rightarrow \theta_R = \sin^{-1} \dfrac{1}{2} = 30° \Rightarrow$

   $\theta = 210°, 330°$

   (b) Graph $r_1 = 3$ and $r_2 = 2 - 2 \sin (\theta)$ for $0° \leq \theta \leq 360°$ in $[-9.4, 9.4, 1]$ by $[-6.2, 6.2, 1]$.

   Use the calculator's trace feature to find the solutions: $\theta = 210°, 330°$. See Figures 91a and 91b.

   (c) Table $r_1 = 3$ and $r_2 = 2 - 2 \sin (\theta)$ starting at $\theta = 180$, incrementing by 30.

   The solutions are $\theta = 210°, 330°$. See Figure 91c.

[−9.4, 9.4, 1] by [−6.2, 6.2, 1]

Figure 91a

[−9.4, 9.4, 1] by [−6.2, 6.2, 1

Figure 91b

Figure 91c

93. (a) $r_1 = r_2 \Rightarrow 1 = 2 \sin \theta \Rightarrow \sin \theta = \dfrac{1}{2} \Rightarrow \theta_R = \sin^{-1} \dfrac{1}{2} = 30° \Rightarrow \theta = 30°, 150°$

(b) Graph $r_1 = 1$ and $r_2 = 2 \sin(\theta)$ for $0° \le \theta \le 360°$ in $[-4.7, 4.7, 1]$ by $[-3.1, 3.1, 1]$.

Use the calculator's trace feature to find the solutions: $\theta = 30°, 150°$. See Figures 93a and 93b.

(c) Table $r_1 = 1$ and $r_2 = 2 \sin(\theta)$ starting at $\theta = 0$, incrementing by 30.

The solutions are $\theta = 30°, 150°$. See Figure 93c.

[−4.7, 4.7, 1] by [−3.1, 3.1, 1]

Figure 93a

[−4.7, 4.7, 1] by [−3.1, 3.1, 1]

Figure 93b

Figure 93c

95. For Saturn $r_1 = \dfrac{9.54(1 - 0.056^2)}{1 + 0.056 \cos \theta} = (9.54(1 - 0.056^2))/(1 + 0.056 \cos(\theta))$

For Uranus $r_2 = \dfrac{19.2(1 - 0.047^2)}{1 + 0.047 \cos \theta} = (19.2(1 - 0.047^2))/(1 + 0.047 \cos(\theta))$

[−30, 30, 10] by [−20, 20, 10]

Figure 95

[−3, 3, 1] by [−2, 2, 1]

Figure 97

97. For Mercury $r_1 = \dfrac{0.39(1 - 0.206^2)}{1 + 0.206 \cos \theta} = (0.39(1 - 0.206^2))/(1 + 0.206 \cos(\theta))$

For Venus $r_2 = \dfrac{0.78(1 - 0.007^2)}{1 + 0.007 \cos \theta} = (0.78(1 - 0.007^2))/(1 + 0.007 \cos(\theta))$

For Earth $r_3 = \dfrac{1.00(1 - 0.017^2)}{1 + 0.017 \cos \theta} = (1.00(1 - 0.017^2))/(1 + 0.017 \cos(\theta))$

For Mars $r_4 = \dfrac{1.52(1 - 0.093^2)}{1 + 0.093 \cos \theta} = (1.52(1 - 0.093^2))/(1 + 0.093 \cos(\theta))$

Graph r_1, r_2, r_3 and r_4 for $0° \le \theta \le 360°$ in $[-3, 3, 1]$ by $[-2, 2, 1]$. See Figure 97.

99. Let $r_1 = \sqrt{40,000\ \cos 2\theta}$ and $r_2 = -\sqrt{40,000\ \cos 2\theta}$.

Graph r_1 and r_2 for $0° \leq \theta \leq 180°$ in $[-300, 300, 100]$ by $[-200, 200, 100]$. See Figure 99.

The radio signal is received inside the "figure eight". This region is generally in an east-west direction from the two towers with a maximum distance of 200 miles.

[–300, 300, 100] by [–200, 200, 100]

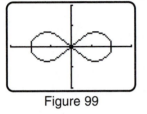

Figure 99

8.6: Trigonometric Form and Roots of Complex Numbers

1.   See Figure 1.

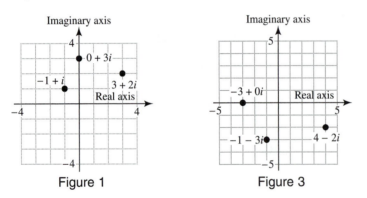

Figure 1   Figure 3

3.   See Figure 3.

5. $|1 + i| = \sqrt{1^2 + 1^2} = \sqrt{1 + 1} = \sqrt{2}$

7. $|12 - 5i| = \sqrt{12^2 + (-5)^2} = \sqrt{144 + 25} = \sqrt{169} = 13$

9. $|-6 + 0i| = \sqrt{(-6)^2 + 0^2} = \sqrt{36} = 6$

11. $|2 - 3i| = \sqrt{2^2 + (-3)^2} = \sqrt{4 + 9} = \sqrt{13}$

13. $r = \sqrt{1^2 + 1^2} = \sqrt{1 + 1} = \sqrt{2}$, $\tan \theta = \dfrac{1}{1} \Rightarrow \theta_R = \tan^{-1} 1 = 45°$

Since $-1 + i$ is located in quadrant III of the complex plane, $\theta = 225°$; $\sqrt{2}(\cos 225° + i \sin 225°)$

15. $r = \sqrt{5^2 + 0^2} = \sqrt{25 + 0} = \sqrt{25} = 5$, $\tan \theta = \dfrac{0}{5} \Rightarrow \theta_R = \tan^{-1} 0 = 0°$

Since 5 is positive and quadrantal in the complex plane, $\theta = 0°$; $5(\cos 0° + i \sin 0°)$

17. $r = \sqrt{0^2 + 4^2} = \sqrt{0 + 16} = \sqrt{16} = 4$, $\tan \theta = \dfrac{4}{0}$ is undefined $\Rightarrow \theta_R = 90°$

Since $4i$ is quadrantal in the complex plane, $\theta = 90°$; $4(\cos 90° + i \sin 90°)$

19. $r = \sqrt{(-1)^2 + (\sqrt{3})^2} = \sqrt{1 + 3} = \sqrt{4} = 2,\, \tan\theta = \dfrac{\sqrt{3}}{-1} \Rightarrow \theta_R = \tan^{-1}\sqrt{3} = 60°$

Since $-1 + i\sqrt{3}$ is located in quadrant II of the complex plane, $\theta = 120°;\ 2(\cos 120° + i\sin 120°)$

21. $r = \sqrt{(\sqrt{3})^2 + 1^2} = \sqrt{3 + 1} = \sqrt{4} = 2,\, \tan\theta = \dfrac{1}{\sqrt{3}} \Rightarrow \theta_R = \tan^{-1}\dfrac{1}{\sqrt{3}} = 30°$

Since $\sqrt{3} + i$ is located in quadrant I of the complex plane, $\theta = 30°;\ 2(\cos 30° + i\sin 30°)$

23. $r = \sqrt{(-2)^2 + 0^2} = \sqrt{4 + 0} = \sqrt{4} = 2,\, \tan\theta = \dfrac{0}{-2} \Rightarrow \theta_R = \tan^{-1} 0 = 0°$

Since -2 is negative and quadrantal in the complex plane, $\theta = \pi;\ 2(\cos\pi + i\sin\pi)$

25. $r = \sqrt{(-2)^2 + 2^2} = \sqrt{4 + 4} = \sqrt{8},\, \tan\theta = \dfrac{2}{-2} \Rightarrow \theta_R = \tan^{-1} 1 = \dfrac{\pi}{4}$

Since $-2 + 2i$ is located in quadrant II of the complex plane, $\theta = \dfrac{3\pi}{4};\ \sqrt{8}\left(\cos\dfrac{3\pi}{4} + i\sin\dfrac{3\pi}{4}\right)$

27. $5(\cos 180° + i\sin 180°) = 5(-1 + i(0)) = -5$

29. $2(\cos 45° + i\sin 45°) = 2\left(\dfrac{1}{\sqrt{2}} + i\left(\dfrac{1}{\sqrt{2}}\right)\right) = \sqrt{2} + i\sqrt{2}$

31. $2\left(\cos\dfrac{\pi}{6} + i\sin\dfrac{\pi}{6}\right) = 2\left(\dfrac{\sqrt{3}}{2} + i\left(\dfrac{1}{2}\right)\right) = \sqrt{3} + i$

33. $3(\cos 2\pi + i\sin 2\pi) = 3(1 + i(0)) = 3$

35. $z_1 z_2 = (9\cdot 3)(\cos(45° + 15°) + i\sin(45° + 15°)) = 27(\cos 60° + i\sin 60°) =$

$27\left(\dfrac{1}{2} + i\left(\dfrac{\sqrt{3}}{2}\right)\right) = \dfrac{27}{2} + \dfrac{27\sqrt{3}}{2}i$

$\dfrac{z_1}{z_2} = \left(\dfrac{9}{3}\right)(\cos(45° - 15°) + i\sin(45° - 15°)) = 3(\cos 30° + i\sin 30°) =$

$3\left(\dfrac{\sqrt{3}}{2} + i\left(\dfrac{1}{2}\right)\right) = \dfrac{3\sqrt{3}}{2} + \dfrac{3}{2}i$

37. $z_1 z_2 = (6\cdot 1)\left(\cos\left(\dfrac{3\pi}{4} + \dfrac{\pi}{4}\right) + i\sin\left(\dfrac{3\pi}{4} + \dfrac{\pi}{4}\right)\right) = 6(\cos\pi + i\sin\pi) = 6(-1 + i(0)) = -6$

$\dfrac{z_1}{z_2} = \left(\dfrac{6}{1}\right)\left(\cos\left(\dfrac{3\pi}{4} - \dfrac{\pi}{4}\right) + i\sin\left(\dfrac{3\pi}{4} - \dfrac{\pi}{4}\right)\right) = 6\left(\cos\dfrac{\pi}{2} + i\sin\dfrac{\pi}{2}\right) = 6(0 + i(1)) = 6i$

39. $z_1 z_2 = (1\cdot 1)(\cos(15° + (-45°)) + i\sin(15° + (-45°))) = 1(\cos(-30°) + i\sin(-30°)) = \dfrac{\sqrt{3}}{2} - \dfrac{1}{2}i$

$\dfrac{z_1}{z_2} = \left(\dfrac{1}{1}\right)(\cos(15° - (-45°)) + i\sin(15° - (-45°))) = 1(\cos 60° + i\sin 60°) = \dfrac{1}{2} + \dfrac{\sqrt{3}}{2}i$

41. $(2(\cos 30° + i\sin 30°))^3 = 2^3(\cos(3\cdot 30°) + i\sin(3\cdot 30°)) = 8(\cos 90° + i\sin 90°) = 8(0 + i(1)) = 8i$

43. $(\cos 10° + i\sin 10°)^{36} = 1^{36}(\cos(36\cdot 10°) + i\sin(36\cdot 10°)) = 1(\cos 360° + i\sin 360°) = 1 + i(0) = 1$

45. $(5(\cos 60° + i\sin 60°))^2 = 5^2(\cos(2\cdot 60°) + i\sin(2\cdot 60°)) = 25(\cos 120° + i\sin 120°) =$

$-\dfrac{25}{2} + \dfrac{25\sqrt{3}}{2}i$

47. $(1 + i)^3 = (\sqrt{2}(\cos 45° + i \sin 45°))^3 = (\sqrt{2})^3(\cos (3 \cdot 45°) + i \sin (3 \cdot 45°)) =$

$2\sqrt{2}(\cos 135° + i \sin 135°) = -2 + 2i$ This result is verified with a calculator in Figure 47.

49. $(\sqrt{3} + i)^5 = (2(\cos 30° + i \sin 30°))^5 = 2^5(\cos (5 \cdot 30°) + i \sin (5 \cdot 30°)) =$

$32(\cos 150° + i \sin 150°) = -16\sqrt{3} + 16i$ This result is verified with a calculator in Figure 49.

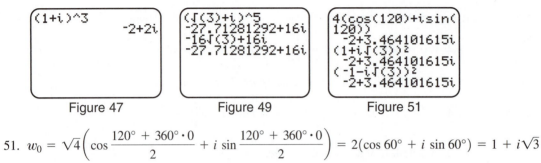

Figure 47   Figure 49   Figure 51

51. $w_0 = \sqrt{4}\left(\cos \dfrac{120° + 360° \cdot 0}{2} + i \sin \dfrac{120° + 360° \cdot 0}{2}\right) = 2(\cos 60° + i \sin 60°) = 1 + i\sqrt{3}$

$w_1 = \sqrt{4}\left(\cos \dfrac{120° + 360° \cdot 1}{2} + i \sin \dfrac{120° + 360° \cdot 1}{2}\right) = 2(\cos 240° + i \sin 240°) = -1 - i\sqrt{3}$

This result is verified with a calculator in Figure 51.

53. $w_0 = \sqrt[3]{1}\left(\cos \dfrac{180° + 360° \cdot 0}{3} + i \sin \dfrac{180° + 360° \cdot 0}{3}\right) = 1(\cos 60° + i \sin 60°) = \dfrac{1}{2} + \dfrac{\sqrt{3}}{2}i$

$w_1 = \sqrt[3]{1}\left(\cos \dfrac{180° + 360° \cdot 1}{3} + i \sin \dfrac{180° + 360° \cdot 1}{3}\right) = 1(\cos 180° + i \sin 180°) = -1$

$w_2 = \sqrt[3]{1}\left(\cos \dfrac{180° + 360° \cdot 2}{3} + i \sin \dfrac{180° + 360° \cdot 2}{3}\right) = 1(\cos 300° + i \sin 300°) = \dfrac{1}{2} - \dfrac{\sqrt{3}}{2}i$

These results are verified with a calculator in Figure 53a and 53b.

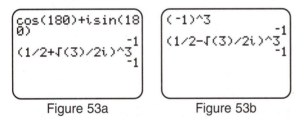

Figure 53a   Figure 53b

55. $i = \cos 90° + i \sin 90°$

$w_0 = \sqrt{1}\left(\cos \dfrac{90° + 360° \cdot 0}{2} + i \sin \dfrac{90° + 360° \cdot 0}{2}\right) = 1(\cos 45° + i \sin 45°) = \dfrac{\sqrt{2}}{2} + \dfrac{\sqrt{2}}{2}i$

$w_1 = \sqrt{1}\left(\cos \dfrac{90° + 360° \cdot 1}{2} + i \sin \dfrac{90° + 360° \cdot 1}{2}\right) = 1(\cos 225° + i \sin 225°) = -\dfrac{\sqrt{2}}{2} - \dfrac{\sqrt{2}}{2}i$

These results are verified with a calculator in Figure 55.

Figure 55

57. $-8 = 8(\cos 180° + i \sin 180°)$

$$w_0 = \sqrt[3]{8}\left(\cos \frac{180° + 360° \cdot 0}{3} + i \sin \frac{180° + 360° \cdot 0}{3}\right) = 2(\cos 60° + i \sin 60°) = 1 + i\sqrt{3}$$

$$w_1 = \sqrt[3]{8}\left(\cos \frac{180° + 360° \cdot 1}{3} + i \sin \frac{180° + 360° \cdot 1}{3}\right) = 2(\cos 180° + i \sin 180°) = -2$$

$$w_2 = \sqrt[3]{8}\left(\cos \frac{180° + 360° \cdot 2}{3} + i \sin \frac{180° + 360° \cdot 2}{3}\right) = 2(\cos 300° + i \sin 300°) = 1 - i\sqrt{3}$$

These results are verified with a calculator in Figure 57.

59. $64i = 64(\cos 90° + i \sin 90°)$

$$w_0 = \sqrt[3]{64}\left(\cos \frac{90° + 360° \cdot 0}{3} + i \sin \frac{90° + 360° \cdot 0}{3}\right) = 4(\cos 30° + i \sin 30°) = 2\sqrt{3} + 2i$$

$$w_1 = \sqrt[3]{64}\left(\cos \frac{90° + 360° \cdot 1}{3} + i \sin \frac{90° + 360° \cdot 1}{3}\right) = 4(\cos 150° + i \sin 150°) = -2\sqrt{3} + 2i$$

$$w_2 = \sqrt[3]{64}\left(\cos \frac{90° + 360° \cdot 2}{3} + i \sin \frac{90° + 360° \cdot 2}{3}\right) = 4(\cos 270° + i \sin 270°) = -4i$$

These results are verified with a calculator in Figure 59.

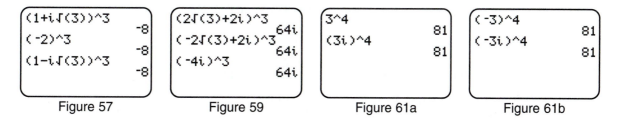

Figure 57　　　　Figure 59　　　　Figure 61a　　　　Figure 61b

61. $81 = 81(\cos 0° + i \sin 0°)$

$$w_0 = \sqrt[4]{81}\left(\cos \frac{0° + 360° \cdot 0}{4} + i \sin \frac{0° + 360° \cdot 0}{4}\right) = 3(\cos 0° + i \sin 0°) = 3$$

$$w_1 = \sqrt[4]{81}\left(\cos \frac{0° + 360° \cdot 1}{4} + i \sin \frac{0° + 360° \cdot 1}{4}\right) = 3(\cos 90° + i \sin 90°) = 3i$$

$$w_2 = \sqrt[4]{81}\left(\cos \frac{0° + 360° \cdot 2}{4} + i \sin \frac{0° + 360° \cdot 2}{4}\right) = 3(\cos 180° + i \sin 180°) = -3$$

$$w_3 = \sqrt[4]{81}\left(\cos \frac{0° + 360° \cdot 3}{4} + i \sin \frac{0° + 360° \cdot 3}{4}\right) = 3(\cos 270° + i \sin 270°) = -3i$$

These results are verified with a calculator in Figure 61a and 61b.

63. $z_0 = -0.4i$

$z_1 = (-0.4i)^2 + (-0.4i) = -0.16 - 0.4i \Rightarrow |z_1| \approx 0.4308$

$z_2 = (-0.16 - 0.4i)^2 + (-0.4i) = -0.1344 - 0.272i \Rightarrow |z_2| \approx 0.3034$

$z_3 = (-0.1344 - 0.272i)^2 + (-0.4i) \approx -0.0559 - 0.3269i \Rightarrow |z_3| \approx 0.3316$

The modulus of each consecutive z_k never exceeds 2. Thus, $-0.4i$ is in the Mandelbrot set.

65. $z_0 = 1 + i$

$z_1 = (1 + i)^2 + (1 + i) = 1 + 3i \Rightarrow |z_1| = \sqrt{10} \approx 3.162 > 2$

The modulus of z_1 exceeds 2. Thus, $1 + i$ is not in the Mandelbrot set.

67. (a) $Z = 50 + 60 + 15i + 17i = 110 + 32i$

(b) $|110 + 32i| = \sqrt{110^2 + 32^2} = \sqrt{12,100 + 1024} = \sqrt{13,124} \approx 114.6$ ohms

Chapter 8 Review Exercises

1. This triangle is of the form ASA. There is only one solution.

$\gamma = 180° - 70° - 40° = 70°$

$\dfrac{a}{\sin \alpha} = \dfrac{c}{\sin \gamma} \Rightarrow a = \dfrac{c \sin \alpha}{\sin \gamma} = \dfrac{10.1 \sin 70°}{\sin 70°} = 10.1$

$\dfrac{b}{\sin \beta} = \dfrac{c}{\sin \gamma} \Rightarrow b = \dfrac{c \sin \beta}{\sin \gamma} = \dfrac{10.1 \sin 40°}{\sin 70°} \approx 6.91$

$\gamma = 70°, a = 10.1, b \approx 6.9$

3. This triangle is of the form SAS. There is only one solution.

$a^2 = b^2 + c^2 - 2bc \cos \alpha = 7^2 + 8^2 - 2(7)(8) \cos 42° \Rightarrow a = \sqrt{113 - 112 \cos 42°} \approx 5.45599$

$b^2 = a^2 + c^2 - 2ac \cos \beta \Rightarrow \cos \beta = \dfrac{b^2 - a^2 - c^2}{-2ac} \Rightarrow \beta = \cos^{-1}\left(\dfrac{7^2 - 5.45599^2 - 8^2}{-2(5.45599)(8)}\right) \approx 59.14756°$

$\gamma = 180° - 42° - 59.14756° \approx 78.85244°$ $\quad a \approx 5.5, \beta \approx 59.1°, \gamma \approx 78.9°$

5. This triangle is of the form AAS. There is only one solution.

$\gamma = 180° - 19° - 46° = 115°$

$\dfrac{a}{\sin \alpha} = \dfrac{b}{\sin \beta} \Rightarrow a = \dfrac{b \sin \alpha}{\sin \beta} = \dfrac{13 \sin 19°}{\sin 46°} \approx 5.9 \qquad \dfrac{c}{\sin \gamma} = \dfrac{b}{\sin \beta} \Rightarrow c = \dfrac{b \sin \gamma}{\sin \beta} = \dfrac{13 \sin 115°}{\sin 46°} \approx 16.4$

$\gamma = 115°, a \approx 5.9, c \approx 16.4$

7. This triangle is of the form SSA. It is ambiguous.

$\dfrac{\sin \beta}{b} = \dfrac{\sin \gamma}{c} \Rightarrow \sin \beta = \dfrac{b \sin \gamma}{c} = \dfrac{8 \sin 20°}{11} \Rightarrow \beta_R = \sin^{-1}\left(\dfrac{8 \sin 20°}{11}\right) \approx 14.4°$

Thus, $\beta \approx 14.4°$ or $\beta \approx 180° - 14.4° \approx 165.6°$ (which is not possible).

So $\beta \approx 14.4°$. Then $\alpha \approx 180° - 20° - 14.4° \approx 145.6°$.

$\dfrac{a}{\sin \alpha} = \dfrac{c}{\sin \gamma} \Rightarrow a = \dfrac{c \sin \alpha}{\sin \gamma} = \dfrac{11 \sin 145.6°}{\sin 20°} \approx 18.2 \qquad \alpha \approx 145.6°, \beta \approx 14.4°, a \approx 18.2$

9. This triangle is of the form SAS. There is only one solution.

$c^2 = a^2 + b^2 - 2ab \cos \gamma = 18^2 + 23^2 - 2(18)(23) \cos 35° \Rightarrow c = \sqrt{853 - 828 \cos 35°} \approx 13.21901$

$a^2 = b^2 + c^2 - 2bc \cos \alpha \Rightarrow \cos \alpha = \dfrac{a^2 - b^2 - c^2}{-2bc} \Rightarrow \alpha = \cos^{-1}\left(\dfrac{18^2 - 23^2 - 13.21901^2}{-2(23)(13.21901)}\right) \approx$

$51.35453°$

$\beta = 180° - 35° - 51.35453° \approx 93.64547° \qquad c \approx 13.2, \alpha \approx 51.4°, \beta \approx 93.6°$

11. Area $= K = \dfrac{1}{2}(12.3)(13.7)\sin 39° \approx 53.0$

13. $s = \dfrac{1}{2}(34 + 67 + 53) = 77 \Rightarrow$ Area $= K = \sqrt{77(77-34)(77-67)(77-53)} \approx 891.4$

15. (a) Horizontal $= 3$, vertical $= 4$

   (b) $\|\mathbf{v}\| = \sqrt{3^2 + 4^2} = \sqrt{9 + 16} = \sqrt{25} = 5$

   (c) $\|\mathbf{v}\|$ represents the length of **v**. See Figure 15.

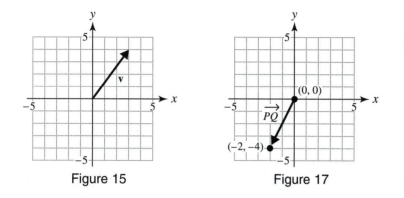

Figure 15   Figure 17

17. (a) See Figure 17.

   (b) $\mathbf{v} = \langle -2 - 0, -4 - 0 \rangle = \langle -2, -4 \rangle = -2\mathbf{i} - 4\mathbf{j}$

   (c) $\|\overrightarrow{PQ}\| = \sqrt{(-2)^2 + (-4)^2} = \sqrt{4 + 16} = \sqrt{20}$

19. (a) $2\mathbf{a} = 2\langle 3, -2 \rangle = \langle 6, -4 \rangle$

   (b) $\mathbf{a} - 3\mathbf{b} = \langle 3, -2 \rangle - 3\langle 1, 1 \rangle = \langle 3, -2 \rangle - \langle 3, 3 \rangle = \langle 3 - 3, -2 - 3 \rangle = \langle 0, -5 \rangle$

   (c) $\mathbf{a} \cdot \mathbf{b} = \langle 3, -2 \rangle \cdot \langle 1, 1 \rangle = (3)(1) + (-2)(1) = 3 - 2 = 1$

   (d) $\|\mathbf{a}\| = \sqrt{3^2 + (-2)^2} = \sqrt{9 + 4} = \sqrt{13}$, $\|\mathbf{b}\| = \sqrt{1^2 + 1^2} = \sqrt{1 + 1} = \sqrt{2}$

   $\theta = \cos^{-1}\left(\dfrac{\mathbf{a} \cdot \mathbf{b}}{\|\mathbf{a}\|\|\mathbf{b}\|}\right) = \cos^{-1}\left(\dfrac{1}{\sqrt{13}\sqrt{2}}\right) = \cos^{-1}\left(\dfrac{1}{\sqrt{26}}\right) \approx 78.7°$

21. (a) $2\mathbf{a} = 2(2\mathbf{i} + 2\mathbf{j}) = 4\mathbf{i} + 4\mathbf{j}$

   (b) $\mathbf{a} - 3\mathbf{b} = (2\mathbf{i} + 2\mathbf{j}) - 3(\mathbf{i} + \mathbf{j}) = 2\mathbf{i} + 2\mathbf{j} - 3\mathbf{i} - 3\mathbf{j} = -\mathbf{i} - \mathbf{j}$

   (c) $\mathbf{a} \cdot \mathbf{b} = (2\mathbf{i} + 2\mathbf{j}) \cdot (\mathbf{i} + \mathbf{j}) = (2)(1) + (2)(1) = 2 + 2 = 4$

   (d) $\|\mathbf{a}\| = \sqrt{2^2 + 2^2} = \sqrt{4 + 4} = \sqrt{8}$, $\|\mathbf{b}\| = \sqrt{1^2 + 1^2} = \sqrt{1 + 1} = \sqrt{2}$

   $\theta = \cos^{-1}\left(\dfrac{\mathbf{a} \cdot \mathbf{b}}{\|\mathbf{a}\|\|\mathbf{b}\|}\right) = \cos^{-1}\left(\dfrac{4}{\sqrt{8}\sqrt{2}}\right) = \cos^{-1}\left(\dfrac{4}{\sqrt{16}}\right) = \cos^{-1} 1 = 0°$

23. Complete the parallelogram and find the unknown angle: $180° - 52° = 128°$.

   Find the magnitude of the resultant force by using the law of cosines to find the length of the diagonal d.

   $d^2 = 100^2 + 130^2 - 2(100)(130)\cos 128° \Rightarrow d = \sqrt{26{,}900 - 26{,}000 \cos 128°} \approx 207.1 \text{ lb}$

25. See Figure 25.

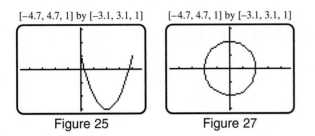

[−4.7, 4.7, 1] by [−3.1, 3.1, 1]   [−4.7, 4.7, 1] by [−3.1, 3.1, 1]

Figure 25   Figure 27

27. See Figure 27.

29. See Figure 29.

31. The graph of $r = 3 \sin (3\theta)$ for $0 \le \theta \le \pi$ is shown in Figure 31.

Figure 29   Figure 31   Figure 33

33. The graph of $r = 3 + 3 \sin (\theta)$ for $0 \le \theta \le 2\pi$ is shown in Figure 33.

35. See Figure 35.

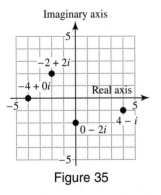

Figure 35

37. $z_1 z_2 = (4 \cdot 2)(\cos (150° + 30°) + i \sin (150° + 30°)) = 8(\cos 180° + i \sin 180°) = 8(-1 + i(0)) = -8$

$\dfrac{z_1}{z_2} = \left(\dfrac{4}{2}\right)(\cos (150° - 30°) + i \sin (150° - 30°)) = 2(\cos 120° + i \sin 120°) =$

$2\left(-\dfrac{1}{2} + i\left(\dfrac{\sqrt{3}}{2}\right)\right) = -1 + i\sqrt{3}$

39. $w_0 = \sqrt{4}\left(\cos\dfrac{60° + 360° \cdot 0}{2} + i\sin\dfrac{60° + 360° \cdot 0}{2}\right) = 2(\cos 30° + i\sin 30°) = \sqrt{3} + i$

     $w_1 = \sqrt{4}\left(\cos\dfrac{60° + 360° \cdot 1}{2} + i\sin\dfrac{60° + 360° \cdot 1}{2}\right) = 2(\cos 210° + i\sin 210°) = -\sqrt{3} - i$

This result is verified with a calculator in Figure 39.

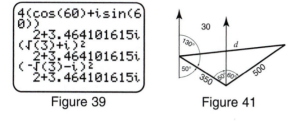

Figure 39   Figure 41

41. Refer to Figure 41.

$d^2 = 350^2 + 500^2 - 2(350)(500)\cos 110° \Rightarrow d = \sqrt{372{,}500 - 350{,}000\cos 110°} \approx 701.6$

The plane is about 701.6 miles from its takeoff point.

43. Let the airplane be located at position C and use standard labels.

Note that $\beta = 180° - 57° = 123°$. Then $\gamma = 180° - 123° - 52° = 5°$.

First we find side $a = BC$: $\dfrac{a}{\sin\alpha} = \dfrac{c}{\sin\gamma} \Rightarrow a = \dfrac{c\sin\alpha}{\sin\gamma} = \dfrac{950\sin 52°}{\sin 5°} \approx 8589.34$ ft.

Label a point D on the ground directly below the airplane and let h represent the height of the airplane.

From triangle BDC: $\sin 57° = \dfrac{h}{8589.34} \Rightarrow h = 8589.34\sin 57° \approx 7204$ ft.

45. In triangle ABC, $s = \dfrac{1}{2}(150 + 140 + 200) = 245 \Rightarrow$

     $K_1 = \sqrt{245(245 - 150)(245 - 140)(245 - 200)} \approx 10{,}486.9$ ft^2.

In triangle ACD, $s = \dfrac{1}{2}(100 + 125 + 200) = 212.5 \Rightarrow$

     $K_2 = \sqrt{212.5(212.5 - 100)(212.5 - 125)(212.5 - 200)} \approx 5113.5$ ft^2.

The total area is $K_1 + K_2 = 10{,}486.9 + 5113.5 \approx 15{,}600$ ft^2.

47. (a) $\mathbf{c} = \mathbf{a} + \mathbf{b} = (40\mathbf{i} - 20\mathbf{j}) + (20\mathbf{i} + 30\mathbf{j}) = 60\mathbf{i} + 10\mathbf{j}$

    (b) $\|\mathbf{c}\| = \sqrt{60^2 + 10^2} = \sqrt{3600 + 100} = \sqrt{3700} \approx 60.8$ cm

    (c) $\mathbf{a} + 2\mathbf{b} = (40\mathbf{i} - 20\mathbf{j}) + 2(20\mathbf{i} + 30\mathbf{j}) = (40\mathbf{i} - 20\mathbf{j}) + (40\mathbf{i} + 60\mathbf{j}) = 80\mathbf{i} + 40\mathbf{j}$

49. The parametric equations which model the flight of the ball are given by:

$x_1 = 50\cos 45°(t)$, $y_1 = 50\sin 45°(t) - 16t^2$ for $0 \le t \le 3$. The ball travels about 78.1 feet. See Figure 49.

$[0, 90, 10]$ by $[-10, 30, 10]$

Figure 49

Chapters 1-8 Cumulative Review Exercises

1. Move the decimal piont four places to the left; $91{,}200 \Rightarrow 9.12 \times 10^4$

   Move the decimal piont three places to the left; $6.734 \times 10^{-3} \Rightarrow 0.006734$

3. $D = \sqrt{[7 - 3]^2 + [(-9) - (-2)]^2} = \sqrt{16 + 49} = \sqrt{65}$

5. The domain of f includes all real numbers less than or equal to 4

   $\Rightarrow D = \{x \,|\, x \le 4\};\ f(-5) = \sqrt{4 - (-5)} = \sqrt{9} = 3.$

7. $f(x) = 4x^2;\ f(x + h) = 4(x + h)^2 = 4x^2 + 8xh + 4h^2,$ the difference quatient $= \dfrac{f(x + h) - f(x)}{h} \Rightarrow$

   $\dfrac{4x^2 + 8xh + 4h^2 - 4x^2}{h} = \dfrac{8xh + 4h^2}{h} = 8x + 4h$

9. For $4x + 3y = -12$: x-intercept, then $y = 0 \Rightarrow 4x + 3(0) = -12 \Rightarrow 4x = -12 \Rightarrow x = -3,$

   y-intercept, then $x = 0 \Rightarrow 4(0) + 3y = -12 \Rightarrow 3y = -12 \Rightarrow y = -4.$ See Figure 9.

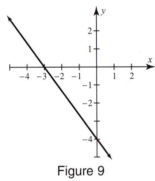

Figure 9

11. (a) $-2(1 - 2x) = -(5 - x) + 1 \Rightarrow -2 + 4x = -5 + x + 1 \Rightarrow 3x = -2 \Rightarrow x = -\dfrac{2}{3}$

    (b) $|2x - 5| = 6$ then $2x - 5 = 6$ or $2x - 5 = -6.$ When $2x - 5 = -6 \Rightarrow 2x = -1 \Rightarrow x = -\dfrac{1}{2}.$

    When $2x - 5 = 6 \Rightarrow 2x = 11 \Rightarrow x = \dfrac{11}{2}.$ Thus $x = -\dfrac{1}{2}$ or $\dfrac{11}{2}.$

    (c) $6x^2 + 22x = 8 \Rightarrow 6x^2 + 22x - 8 = 0 \Rightarrow 2(3x^2 + 11x - 4) = 0 \Rightarrow 2(3x - 1)(x + 4) = 0 \Rightarrow$

    $3x - 1 = 0 \Rightarrow x = \dfrac{1}{3}$ or $x + 4 = 0 \Rightarrow x = -4.$ Thus $x = \dfrac{1}{3}$ or -4

    (d) $x^3 = x \Rightarrow x^3 - x = 0 \Rightarrow x(x^2 - 1) = 0 \Rightarrow x(x + 1)(x - 1) = 0 \Rightarrow x = 0, -1$ or 1

    $x = -1$ or $x - 1 = 0 \Rightarrow x = 1 \Rightarrow x = 0, -1$ or 1

    (e) $x^4 - 2x^2 - 3 = 0 \Rightarrow (x^2 - 3)(x^2 + 1) = 0 \Rightarrow x^2 - 3 = 0 \Rightarrow x^2 = 3 \Rightarrow$

    $x = \pm\sqrt{3}$ or $x^2 + 1 = 0 \Rightarrow x^2 = -1 \Rightarrow x = \pm\sqrt{-1}.$ Thus $x = \pm\sqrt{3}$ or $\pm i.$

(f) $\dfrac{1}{2x} - \dfrac{2}{x^2} = -7 \Rightarrow 2x^2\left(\dfrac{1}{2x} - \dfrac{2}{x^2}\right) = 2x^2(-7) \Rightarrow x - 4 = -14x^2 \Rightarrow 14x^2 + x - 4 = 0 \Rightarrow$

$(7x + 4)(2x - 1) = 0 \Rightarrow 7x + 4 = 0 \Rightarrow x = -\dfrac{4}{7}$ or $2x - 1 = 0 \Rightarrow x = \dfrac{1}{2}$. Thus $x = -\dfrac{4}{7}$ or $\dfrac{1}{2}$.

(g) $3x^{2/3} = 12 \Rightarrow x^{2/3} = 4 \Rightarrow (x^{2/3})^3 = 4^3 \Rightarrow x^2 = 64 \Rightarrow x = \pm 8$

(h) $2e^{3x} - 1 = 50 \Rightarrow 2e^{3x} = 51 \Rightarrow e^{3x} = \dfrac{51}{2} \Rightarrow 3x = \ln\dfrac{51}{2} \Rightarrow x = \dfrac{\ln\left(\frac{51}{2}\right)}{3} \approx 1.08$

(i) $\ln x - \ln(x - 1) = \ln 2 \Rightarrow \ln\left(\dfrac{x}{x - 1}\right) = \ln 2 \Rightarrow \dfrac{x}{x - 1} = 2 \Rightarrow x = 2(x - 1) \Rightarrow x = 2$

(j) $\sin t = \dfrac{1}{2}, t = \sin^{-1}\left(\dfrac{1}{2}\right) \Rightarrow t = \dfrac{\pi}{6}$ or $\dfrac{5\pi}{6}$

(k) $\tan 2t = -\sqrt{3} \Rightarrow$ let $\theta = 2t$ so then the equation becomes $\tan\theta = -\sqrt{3}$ and in radian measure these

solutions are $\theta = \dfrac{2\pi}{3}$ and $\dfrac{5\pi}{3}$. For all real number solutions $\theta = \dfrac{2\pi}{3} + \pi n$.

Because $\theta = 2t$, we can determine t by substituting $2t$ for $\theta \Rightarrow 2t = \dfrac{2\pi}{3} + \pi n$.

Therefore $t = \dfrac{\pi}{3} + \dfrac{\pi}{2}n$.

(l) $2\cos^2 t + \cos t = 1 \Rightarrow 2\cos^2 + \cos t - 1 = 0 \Rightarrow (2\cos t - 1)(\cos t + 1) = 0$. Either

$2\cos t - 1 = 0 \Rightarrow \cos t = \dfrac{1}{2} \Rightarrow t = \dfrac{\pi}{3} + 2\pi n$ or $t = \dfrac{5\pi}{3} + 2\pi n$

or $\cos t + 1 = 0 \Rightarrow \cos t = -1 \Rightarrow t = \pi + 2\pi n$

13. (a) $4(x - 3) > 1 - x \Rightarrow 4x - 12 > 1 - x \Rightarrow 5x > 13 \Rightarrow x > \dfrac{13}{5} \Rightarrow \left(\dfrac{13}{5}, \infty\right)$

(b) The solutions to $|2x - 1| \le 3$ are $c \le x \le d$ where c and d are the solutions to $|2x - 1| = 3$.

$|2x - 1| = 3 \Rightarrow 2x - 1 = 3 \Rightarrow 2x = 4 \Rightarrow x = 2$ or $2x - 1 = -3 \Rightarrow 2x = -2 \Rightarrow x = -1$

Here $c = -1$ and $d = 2$. The solution interval is $[-1, 2]$.

(c) First solve $x^2 - 2x - 3 = 0 \Rightarrow (x + 1)(x - 3) = 0 \Rightarrow x = -1$ or $x = 3$.

Since the graph of $y = x^2 - 2x - 3$ is a parabola opening upward, the graph is above the x-axis for

$x < -1$ or $x > 3$. The solution interval is $(-\infty, -1) \cup (3, \infty)$.

(d) First solve $x^3 - 4x = 0 \Rightarrow x(x^2 - 4) = 0 \Rightarrow x(x + 2)(x - 2) = 0 \Rightarrow x = -2, 0,$ or 2.

The graph of $y = x^3 - 4x$ is above the x-axis for $-2 < x < 0$ or $x > 2$.

The solution interval is $(-2, 0) \cup (2, \infty)$.

(e) First solve $\dfrac{x}{x - 1} = 0 \Rightarrow x = 0$. The graph of $y = \dfrac{x}{x - 1}$ intersects or is on or below the x-axis for

$0 \le x < 1$. The solution interval is $[0, 1)$.

(f) $-4 \le 4 - 3x \le 12 \Rightarrow -8 \le -3x \le 8 \Rightarrow \dfrac{8}{3} \ge x \ge -\dfrac{8}{3} \Rightarrow -\dfrac{8}{3} \le x \le \dfrac{8}{3}$.

The solution interval is $\left[-\dfrac{8}{3}, \dfrac{8}{3}\right]$.

15. $f(x) = \frac{1}{2}x^2 + 2x - 1 = \frac{1}{2}(x^2 + 4x) - 1 = \frac{1}{2}(x^2 + 4x + 4) - 1 - 2 = \frac{1}{2}(x + 2)^2 - 3 \Rightarrow$

$f(x) = \frac{1}{2}(x + 2)^2 - 3$

17. To shift the graph of $y = f(x)$ 2 units to the left and 3 units down, find $g(x) = f(x + 2) - 3$.

$g(x) = (x + 2)^2 + 2(x + 2) - 3 - 3 = x^2 + 4x + 4 + 2x + 4 - 6 = x^2 + 6x + 2$

That is, $g(x) = x^2 + 6x + 2$.

19. $f(-x) = (-x)^3 - 2(-x) + 1 = -x^3 + 2x + 1$. Thus $f(-x) \neq f(x)$ and $f(-x) \neq -f(x)$. Neither.

21. (a)
$$\begin{array}{r} 3x^2 - 6x + 11 \\ x + 2 \overline{\smash{\big)}\ 3x^3 + 0x^2 - x + 2} \\ \underline{3x^3 + 6x^2} \\ -6x^2 - x + 2 \\ \underline{-6x^2 - 12x} \\ 11x + 2 \\ \underline{11x + 22} \\ -20 \end{array}$$
The solution is $3x^2 - 6x + 11 + \dfrac{-20}{x + 2}$.

(b)
$$\begin{array}{r} x^2 - x \\ 2x - 1 \overline{\smash{\big)}\ 2x^3 - 3x^2 + x - 1} \\ \underline{2x^3 - x^2} \\ -2x^2 + x - 1 \\ \underline{-2x^2 + x} \\ -1 \end{array}$$
The solution is $x^2 - x + \dfrac{-1}{2x - 1}$.

23. $f(x) = 6(x - 1)(x + 1)(x - i)(x + i);\ f(x) = 6x^4 - 6$

25. The denominator cannot equal zero $\Rightarrow 2 - 3x \neq 0 \Rightarrow x \neq \frac{2}{3}$. Then $D = \left\{ x \mid x \neq \frac{2}{3} \right\}$. The vertical

asymptotes are values when the function is undefined. The vertical asymptote is $x = \frac{2}{3}$. The horizontal

asymptote is $y = \dfrac{3x}{-3x} = -1 \Rightarrow y = -1$.

27. If $f(x) = 3x - 2$ then $y = 3x - 2 \Rightarrow y + 2 = 3x \Rightarrow \dfrac{y + 2}{3} = x \Rightarrow f^{-1}(x) = \dfrac{x + 2}{3}$

29. If the initial bacteria count is 5000 then $C = 5000$. Since it doubles every 1.5 hours, $10{,}000 = 5000a^{3/2} \Rightarrow$

$2 = a^{3/2} \Rightarrow a = 2^{2/3}$.

31. (a) $\log_3 \dfrac{1}{27} = \log_3 \dfrac{1}{3^3} = \log_3 3^{-3} = -3$

    (b) $\ln \dfrac{1}{e^3} = \ln e^{-3} = -3$

    (c) $\log \sqrt[3]{10} = \log 10^{1/3} = \dfrac{1}{3}$

    (d) $\log_4 32 - \log_4 \dfrac{1}{2} = \log_4 \dfrac{32}{\frac{1}{2}} = \log_4 64 = \log_4 4^3 = 3$

33. $2 \log_2 x + 3 \log_2 y + \dfrac{1}{2} \log_2 z = \log_2 x^2 + \log_2 y^3 + \log_2 z^{1/2} = \log_2 x^2 y^3 \sqrt{z}$

35. $225° \left(\dfrac{\pi}{180} \right) = \dfrac{5\pi}{4}$

37. $c = \sqrt{5^2 + 12^2} = \sqrt{169} = 13$

$$\sin \theta = \dfrac{5}{13} \qquad \cos \theta = \dfrac{12}{13} \qquad \tan \theta = \dfrac{5}{12}$$

$$\csc \theta = \dfrac{13}{5} \qquad \sec \theta = \dfrac{13}{12} \qquad \cot \theta = \dfrac{12}{5}$$

39. $c = \sqrt{11^2 + 60^2} = \sqrt{3721} = 61$

$$\sin \theta = -\dfrac{60}{61} \qquad \cos \theta = \dfrac{11}{61} \qquad \tan \theta = -\dfrac{60}{11}$$

$$\csc \theta = -\dfrac{61}{60} \qquad \sec \theta = \dfrac{61}{11} \qquad \cot \theta = -\dfrac{11}{60}$$

41. Since $\sec \theta < 0$, $\cos \theta = -\sqrt{1 - \sin^2 \theta} = -\sqrt{1 - (-\frac{7}{25})^2} = -\sqrt{1 - \frac{49}{625}} = -\sqrt{\frac{576}{625}} = -\dfrac{24}{25}$

$$\sin \theta = -\dfrac{7}{25} \qquad \cos \theta = -\dfrac{24}{25} \qquad \tan \theta = -\dfrac{7}{24}$$

$$\csc \theta = -\dfrac{25}{7} \qquad \sec \theta = -\dfrac{25}{24} \qquad \cot \theta = \dfrac{24}{7}$$

43. $c = \sqrt{5^2 + 12^2} = \sqrt{169} = 13$. $\sin \theta = \dfrac{5}{13} \Rightarrow \sin^{-1} \left(\dfrac{5}{13} \right) \approx 22.6° \Rightarrow \theta \approx 22.6°$

    $180° - 90° - 22.6° = 67.4° \Rightarrow \beta = 67.4°$

45. $\cot^2 \theta - 2 \cot \theta + 1 = (\cot \theta - 1)(\cot \theta - 1) = (\cot \theta - 1)^2$

47. If $\sin \alpha = \dfrac{5}{13}$ and α is in QII, $\cos \alpha = -\sqrt{1 - (\frac{5}{13})^2} = -\sqrt{1 - \frac{25}{169}} = -\sqrt{\frac{144}{169}} = -\dfrac{12}{13}$.

    If $\cos \beta = -\dfrac{4}{5}$ and β is in QII, $\sin \beta = \sqrt{1 - (-\frac{4}{5})^2} = \sqrt{1 - \frac{16}{25}} = \sqrt{\frac{9}{25}} = \dfrac{3}{5}$.

    Note: $\sin \alpha = \dfrac{5}{13}$, $\cos \alpha = -\dfrac{12}{13}$, $\sin \beta = \dfrac{3}{5}$, $\cos \beta = -\dfrac{4}{5}$

    $$\cos (\alpha - \beta) = \cos \alpha \cos \beta + \sin \alpha \sin \beta = \left(-\dfrac{12}{13} \right) \left(-\dfrac{4}{5} \right) + \left(\dfrac{5}{13} \right) \left(\dfrac{3}{5} \right) = \dfrac{48}{65} + \dfrac{15}{65} = \dfrac{63}{65}$$

49. (a) This triangle is in the form ASA. There is only one solution. $\beta = 180° - 31° - 53° = 96°$

$$\frac{a}{\sin \alpha} = \frac{b}{\sin \beta} \Rightarrow a = \frac{b \sin \alpha}{\sin \beta} = \frac{15 \sin 31°}{\sin 96°} \approx 7.8$$

$$\frac{c}{\sin \gamma} = \frac{b}{\sin \beta} \Rightarrow c = \frac{b \sin \gamma}{\sin \beta} = \frac{15 \sin 53°}{\sin 96°} \approx 12.0 \quad \beta = 96°, a \approx 7.8, c \approx 12.0$$

(b) This triangle is in the form SSA. It is ambiguous.

$$\frac{\sin \beta}{b} = \frac{\sin \alpha}{a} \Rightarrow \sin \beta = \frac{b \sin \alpha}{a} = \frac{5 \sin 31°}{6} = 0.42920 \Rightarrow \beta_R \approx \sin^{-1}(0.42920) \approx 25.4°$$

Thus $\beta = 25.4°$. Then $\gamma = 180° - 31° - 25.4° \approx 123.6°$

$$\frac{c}{\sin \gamma} = \frac{a}{\sin \alpha} \Rightarrow c = \frac{a \sin \gamma}{\sin \alpha} = \frac{6 \sin 123.6°}{\sin 31°} \approx 9.7 \quad \beta \approx 25.4°, \gamma \approx 123.6°, c \approx 9.7$$

(c) $b^2 = a^2 + c^2 - 2ac \cos \beta = 6^2 + 8^2 - 2(6)(8) \cos 56° \Rightarrow b = \sqrt{100 - 96 \cos 56°} \approx 6.8$

$$a^2 = b^2 + c^2 - 2bc \cos \alpha \Rightarrow \cos \alpha = \frac{a^2 - b^2 - c^2}{-2bc} \Rightarrow \alpha = \cos^{-1}\left(\frac{6^2 - (6.8)^2 - 8^2}{-2(6.8)(8)}\right) \approx 47.0°$$

$\gamma = 180° - 56° - 47.0° \approx 77.0° \quad b \approx 6.8, \alpha \approx 47.0°, \gamma \approx 77.0°$

(d) $a^2 = b^2 + c^2 - 2bc \cos \alpha \Rightarrow \cos \alpha = \dfrac{a^2 - b^2 - c^2}{-2bc} \Rightarrow \cos^{-1}\left(\dfrac{6^2 - 7^2 - 8^2}{-2(7)(8)}\right) \approx 46.6°$

$$b^2 = a^2 + c^2 - 2ac \cos \beta \Rightarrow \cos \beta = \frac{b^2 - a^2 - c^2}{-2ac} \Rightarrow \beta = \cos^{-1}\left(\frac{7^2 - 6^2 - 8^2}{-2(6)(8)}\right) \approx 57.9°$$

$\gamma = 180° - 46.6° - 57.9° \approx 75.5° \quad \alpha \approx 46.6°, \beta \approx 57.9°, \gamma \approx 75.5°$

51. (a) $\|\mathbf{b}\| = \sqrt{7^2 + (-24)^2} = \sqrt{49 + 576} = \sqrt{625} = 25$

(b) $2\mathbf{a} - 3\mathbf{b} = 2\langle -5, 12 \rangle - 3\langle 7, -24 \rangle = \langle -10, 24 \rangle - \langle 21, -72 \rangle \Rightarrow \langle -10 - 21, 24 - (-72) \rangle \Rightarrow$

$\langle -31, 96 \rangle$

(c) $\mathbf{a} \cdot \mathbf{b} = (-5)(7) + (12)(-24) = -35 + (-288) = -323$

(d) $\|\mathbf{a}\| = \sqrt{(-5)^2 + 12^2} = \sqrt{25 + 144} = \sqrt{169} = 13,$

$\|\mathbf{b}\| = \sqrt{7^2 + (-24)^2} = \sqrt{49 + 576} = \sqrt{625} = 25$

$$\theta = \cos^{-1}\left(\frac{\mathbf{a} \cdot \mathbf{b}}{\|\mathbf{a}\|\|\mathbf{b}\|}\right) = \cos^{-1}\left(\frac{-323}{(13)(25)}\right) = \cos^{-1}\left(\frac{-323}{325}\right) \Rightarrow \theta \approx 173.6°$$

53. See Figure 53.

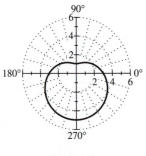

Figure 53

55. (a) $D(t) = 12t^2$, when $t = 1$, then $D(1) = 12(1)^2 = 12 \Rightarrow (1, 12)$, when $t = 3$,

    then $D(3) = 12(3)^2 = 108 \Rightarrow (3, 108)$. Rate of change from 1 to 3 $= \left(\dfrac{108 - 12}{3 - 1}\right) = 48$ ft/sec.

    When $t = 4$, then $D(4) = 12(4)^2 = 192 \Rightarrow (4, 192)$, when $t = 6$, then $D(6) = 12(6)^2 = 432 \Rightarrow$

    $(6, 432)$.

    Rate of change from 4 to 6 $= \left(\dfrac{432 - 192}{6 - 4}\right) = 120$ ft/sec.

    (b) Between the 1st and 3rd seconds, the car travels at an average rate of 48 ft/sec. Between the

    4th and 6th seconds, the car travels at an average rate of 120 ft/sec.

57. Let x = width and $x + 4$ = length. The new width of the box is $x - 6$, because two square corners, with

    sides of 3 inches have been removed. Similarily, the length of the bottom of the box is $x - 2$ inches. Because

    height times the width times the length must equal the volume, or 351 cubic inches, it follows that:

    $3(x - 6)(x - 2) = 351$ or $(x - 6)(x - 2) = 117 \Rightarrow x^2 - 2x - 6x + 12 = 117 \Rightarrow$

    $x^2 - 8x + 12 = 117 \Rightarrow x^2 - 8x - 105 = 0 \Rightarrow (x + 7)(x - 15) = 0 \Rightarrow x = -7$ or $x = 15$. Since the

    dimensions cannot be negative, the width of the cardboard is 15 inches and the length is 4 inches more or 19

    inches.

59. Let x = length, $\dfrac{1}{2}x$ = width and h = height. Then the surface area is given by

    $A = 2\left(\dfrac{1}{2}x^2\right) + 2(xh) + 2\left(\dfrac{1}{2}xh\right) = x^2 + 2xh + xh = x^2 + 3xh$. Since the surface area is 288, we have

    $x^2 + 3xh = 288$. Also, the volume is given by $V = x\left(\dfrac{1}{2}x\right)h = \dfrac{1}{2}x^2 h$. Since the volume is 288, we have

    $\dfrac{1}{2}x^2 h = 288 \Rightarrow x^2 h = 576 \Rightarrow h = \dfrac{576}{x^2}$. Substitute this value for h in the surface area equation.

    $x^2 + 3x\left(\dfrac{576}{x^2}\right) = 288 \Rightarrow x^2 + \dfrac{1728}{x} = 288 \Rightarrow x^3 + 1728 = 288x \Rightarrow x^3 - 288x + 1728 = 0$

    The solutions can be found by graphing the left side of this equation in $[0, 20, 1]$ by $[-250, 250, 50]$. (Not shown.)

    $x \approx 7.4 \Rightarrow$ length ≈ 7.4 in., width $\approx \dfrac{1}{2}(7.4) = 3.7$ in., and height $\approx \dfrac{576}{7.4^2} = 10.5$ in.

    $x = 12 \Rightarrow$ length $= 12$ in., width $= \dfrac{1}{2}(12) = 6$ in., and height $\approx \dfrac{576}{12^2} = 4$ in.

61. (a) If $(1980, 3600)$ and $(2000, 16{,}200)$ then the average rate of change is: $\dfrac{16{,}200 - 3600}{2000 - 1980} = \dfrac{12{,}600}{20} = 630$.

    Then $c(t) = 630(t - 1980) + 3600$.

    (b) $11{,}000 = 630(t - 1980) + 3600 \Rightarrow 11{,}000 = 630t - 1{,}247{,}400 + 3600 \Rightarrow$

    $11{,}000 = 630t - 1{,}243{,}800 \Rightarrow 1{,}254{,}800 = 630t \Rightarrow t = 1991.7$ or close to the year 1992. This is only

    one year off from the actual year of 1993.

63. The length of the shadow is found by the use of tangent. $\tan 63° = \dfrac{5}{x} \Rightarrow 5 = x \tan 63° \Rightarrow x = \dfrac{5}{\tan 63°} \Rightarrow$

    $x \approx 2.5$ feet.

65. The angle of elevation can be found by the use of tangent. $\tan^{-1}\dfrac{85}{57} = x \Rightarrow x = 56.2°$.

67. (a) $c^2 = a^2 + b^2 - 2ab\cos\gamma = 242^2 + 165^2 - 2(242)(165)\cos 72° \Rightarrow$

$\quad\quad c = \sqrt{85{,}789 - 79{,}860\cos 72°} \approx 247°$

(b) Area $= K = \dfrac{1}{2}(242)(165)\sin 72° \approx 18{,}988\ \text{ft}^2$.

69. Graph the parametic equations

$\quad x_1 = 96\cos 60°\,(t),\ y_1 = 96\sin 60°\,(t) - 16t^2$ for $0 \le t \le 9$ in $[0, 300, 25]$ by

$\quad [-20, 200, 50]$. See Figure 69. About 249.4 feet of horizontal distance was traveled when it strikes the ground.

Figure 69

Chapter 9: Systems of Equations and Inequalities

9.1: Functions and Systems of Equations in Two Variables

1. $A(5, 8) = \dfrac{1}{2}(5)(8) = 20.$ The area of a triangle with a base of 5 and height of 8 is 20 square units.

3. $f(2, -3)$ if $f(x, y) = x^2 + y^2 \Rightarrow f(2, -3) = 2^2 + (-3)^2 = 4 + 9 = 13$

5. $f(-2, 3)$ if $f(x, y) = 3x - 4y \Rightarrow f(-2, 3) = 3(-2) - 4(3) = -6 - 12 = -18$

7. $f\left(\dfrac{1}{2}, -\dfrac{7}{4}\right)$ if $f(x, y) = \dfrac{2x}{y + 3} \Rightarrow f\left(\dfrac{1}{2}, -\dfrac{7}{4}\right) = \dfrac{2(\frac{1}{2})}{(-\frac{7}{4}) + 3} = \dfrac{1}{\frac{5}{4}} = \dfrac{4}{5}$

9. The sum of y and twice x is computed by $f(x, y) = y + 2x.$

11. The product of x and y divided by $1 + x$ is computed by $f(x, y) = \dfrac{xy}{1 + x}.$

13. $3x - 4y = 7 \Rightarrow 3x = 4y + 7 \Rightarrow x = \dfrac{4y + 7}{3};\ 3x - 4y = 7 \Rightarrow -4y = -3x + 7 \Rightarrow y = \dfrac{3x - 7}{4}$

15. $x - y^2 = 5 \Rightarrow x = y^2 + 5;\ x - y^2 = 5 \Rightarrow -y^2 = -x + 5 \Rightarrow y^2 = x - 5 \Rightarrow y = \pm\sqrt{x - 5}$

17. $\dfrac{2x - y}{3y} = 1 \Rightarrow 2x - y = 3y \Rightarrow 2x = 4y \Rightarrow x = 2y;\ \dfrac{2x - y}{3y} = 1 \Rightarrow 2x - y = 3y \Rightarrow 2x = 4y \Rightarrow$

$y = \dfrac{x}{2}$

19. The only ordered pair that satisfies both equations is $(2, 1)$. The system is linear.

$2(2) + 1 = 5 \bigstar \qquad 2(-2) + 1 = -3 \qquad 2(1) + 0 = 2$

$2 + 1 = 3 \bigstar \qquad\quad -2 + 1 = -1 \qquad\quad 1 + 0 = 1$

21. The only ordered pair that satisfies both equations is $(4, -3)$. The system is non-linear.

$4^2 + (-3)^2 = 25 \bigstar \qquad 0^2 + 5^2 = 25 \bigstar \qquad 4^2 + 3^2 = 25 \bigstar$

$2(4) + 3(-3) = -1 \bigstar \quad 2(0) + 3(5) = 15 \qquad 2(4) + 3(3) = 17$

23. From the graph the solution is $(2, 2)$. The solution satisfies both $x - y = 0$ and $x + y = 4.$

25. From the graph the solution is $\left(\dfrac{1}{2}, -2\right)$. The solution satisfies both $6x + 4y = -5$ and $2x - 3y = 7.$

27. $x - y = 2$ and $2x + 2y = 38$. Multiply the second equation by $\dfrac{1}{2}$ and add to eliminate the y-variable.

$\begin{aligned} x - y &= 2 \\ x + y &= 19 \\ \hline 2x &= 21 \end{aligned} \Rightarrow x = 10.5$

Since, $x - y = 2, y = 8.5$. The solution is $(10.5, 8.5).$

29. $x + y = 75$ and $4x + 7y = 456$. Multiply the first equation by 7 and subtract to eliminate the y-variable.

$\begin{aligned} 7x + 7y &= 525 \\ 4x + 7y &= 456 \\ \hline 3x &= 69 \end{aligned} \Rightarrow x = 23$

Since, $x + y = 75, y = 52$. The solution is $(23, 52).$

31. Since the lines intersect, the system is consistent with a unique solution at $(2, 2)$.

    $x + y = 4 \Rightarrow y = 4 - x$; We substitute $4 - x$ for y in the other equation.

    $2x - (4 - x) = 2 \Rightarrow 3x = 6 \Rightarrow x = 2$, then $2 + y = 4 \Rightarrow y = 2$. The solution is $(2, 2)$.

33. Since the lines are parallel, the system is inconsistent with no solutions.

    $2x + y = -4 \Rightarrow y = -4 - 2x$; We substitute $-4 - 2x$ for y in the other equation.

    $-x + 2(-4 - 2x) = 2 \Rightarrow x = -2$, then $2(-2) + y = -4 \Rightarrow y = 0$. The solution is $(-2, 0)$.

35. Parallel lines $\Rightarrow$ inconsistent system. See Figure 35.

37. The lines intersect at $(2, -1)$. The system is consistent and independent. See Figure 37.

Figure 35   Figure 37   Figure 39

39. The lines intersect at $(-2, 2)$. The system is consistent and independent. See Figure 39.

41. The system has an infinite number of solutions. $\{(x, y) \mid 2x - y = -4\}$. The system is consistent and

    dependent. See Figure 41.

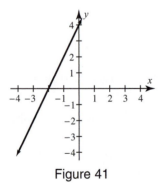

Figure 41

43. Solve the first equation for x: $x + 2y = 0 \Rightarrow x = -2y$. Substitute this into the second equation.

    $3x + 7y = 1 \Rightarrow 3(-2y) + 7y = 1 \Rightarrow -6y + 7y = 1 \Rightarrow y = 1$.

    If $y = 1$, then $x = -2(1) = -2$. The solution is $(-2, 1)$.

    Check: $x + 2y = 0 \Rightarrow (-2) + 2(1) = 0 \Rightarrow 0 = 0$; $x = -2y \Rightarrow -2 = -2(1) \Rightarrow -2 = -2$

45. Solve the first equation for x: $2x - 9y = -17 \Rightarrow 2x = 9y - 17 \Rightarrow x = \dfrac{9y - 17}{2}$. Substitute this into the

 second equation. $8x + 5y = 14 \Rightarrow 8\left(\dfrac{9y - 17}{2}\right) + 5y = 14 \Rightarrow 4(9y - 17) + 5y = 14 \Rightarrow$

 $36y - 68 + 5y = 14 \Rightarrow 41y = 82 \Rightarrow y = 2$. If $y = 2$, then $x = \dfrac{9(2) - 17}{2} = \dfrac{1}{2}$. The solution is $(0.5, 2)$.

 Check: $2x - 9y = -17 \Rightarrow 2(0.5) - 9(2) = -17 \Rightarrow 1 - 18 = -17 \Rightarrow -17 = -17$;

 $8x + 5y = 14 \Rightarrow 8(0.5) + 5(2) = 14 \Rightarrow 4 + 10 = 14 \Rightarrow 14 = 14$

47. Solve the second equation for x: $x + \dfrac{1}{2}y = 10 \Rightarrow x = -\dfrac{1}{2}y + 10$. Substitute this into the first equation.

 $\dfrac{1}{2}x - y = -5 \Rightarrow \dfrac{1}{2}\left(-\dfrac{1}{2}y + 10\right) - y = -5 \Rightarrow -\dfrac{1}{4}y + 5 - y = -5 \Rightarrow -\dfrac{5}{4}y = -10 \Rightarrow y = 8$.

 If $y = 8$, then $x = -\dfrac{1}{2}(8) + 10 = 6$. The solution is $(6, 8)$.

 Check: $x + \dfrac{1}{2}y = 10 \Rightarrow 6 + \dfrac{1}{2}(8) = 10 \Rightarrow 6 + 4 = 10 \Rightarrow 10 = 10$

 $\dfrac{1}{2}x - y = -5 \Rightarrow \dfrac{1}{2}(6) - 8 = -5 \Rightarrow 3 - 8 = -5 \Rightarrow -5 = -5$

49. Solve the first equation for y: $3x - 2y = 5 \Rightarrow -2y = -3x + 5 \Rightarrow y = \dfrac{3}{2}x - \dfrac{5}{2}$. Substitute this into the

 second equation. $-6x + 4\left(\dfrac{3}{2}x - \dfrac{5}{2}\right) = -10 \Rightarrow -6x + 6x - 10 = -10 \Rightarrow -10 = -10 \Rightarrow$ there are

 infinitely many solutions. $\{(x, y) \mid 3x - 2y = 5\}$.

51. Solve the first equation for x: $2x - 7y = 8 \Rightarrow 2x = 7y + 8 \Rightarrow x = \dfrac{7}{2}y + 4$. Substitute this into the second

 equation. $-3\left(\dfrac{7}{2}y + 4\right) + \dfrac{21}{2}y = 5 \Rightarrow -\dfrac{21}{2}y - 12 + \dfrac{21}{2}y = 5 \Rightarrow -12 = 5 \Rightarrow$ there are no real

 solutions.

53. $0.2x - 0.1y = 0.5 \Rightarrow 2x - y = 5$ and $0.4x + 0.3y = 2.5 \Rightarrow 4x + 3y = 25$

 Solve the first equation for y: $2x - y = 5 \Rightarrow y = 2x - 5$. Substitute this into the second equation.

 $4x + 3y = 25 \Rightarrow 4x + 3(2x - 5) = 25 \Rightarrow 4x + 6x - 15 = 25 \Rightarrow 10x = 40 \Rightarrow x = 4$.

 If $x = 4$, then $y = 2(4) - 5 \Rightarrow y = 3$. The solution is $(4, 3)$.

55. Solve the second equation for y: $2x + y = 0 \Rightarrow y = -2x$. Substitute this into the first equation.

 $x^2 - y = 0 \Rightarrow x^2 - (-2x) = 0 \Rightarrow x^2 + 2x = 0 \Rightarrow x(x + 2) = 0 \Rightarrow x = 0$ or $x = -2$.

 When $x = 0, y = -2(0) = 0$ and when $x = -2, y = -2(-2) = 4$. The solutions are $(0, 0)$ and $(-2, 4)$.

57. Solve the second equation for y: $x + y = 6 \Rightarrow y = 6 - x$. Substitute this into the first equation.

 $xy = 8 \Rightarrow x(6 - x) = 8 \Rightarrow 6x - x^2 = 8 \Rightarrow x^2 - 6x + 8 = 0 \Rightarrow (x - 4)(x - 2) = 0 \Rightarrow$

 $x = 4$ or $x = 2$.

 When $x = 4, y = 6 - 4 = 2$ and when $x = 2, y = 6 - 2 = 4$. The solutions are $(4, 2)$ and $(2, 4)$.

59. Substitute the second equation, $y = 2x$, into the first equation.

 $x^2 + y^2 = 20 \Rightarrow x^2 + (2x)^2 = 20 \Rightarrow x^2 + 4x^2 = 20 \Rightarrow 5x^2 = 20 \Rightarrow x^2 = 4 \Rightarrow x = \pm 2$.

 When $x = -2, y = 2(-2) = -4$ and when $x = 2, y = 2(2) = 4$. The solutions are $(-2, -4)$ and $(2, 4)$.

61. Solve the second equation for x: $x - y = -2 \Rightarrow x = y - 2$. Substitute this into the first equation.

$\sqrt{y - 2} - 2y = 0 \Rightarrow \sqrt{y - 2} = 2y \Rightarrow y - 2 = 4y^2 \Rightarrow 4y^2 - y + 2 = 0$. Using the quadratic formula

to solve we get: $\dfrac{1 \pm \sqrt{1 - 4(4)(2)}}{2(4)} = \dfrac{1 \pm \sqrt{-31}}{8} \Rightarrow$ no real solutions.

63. Solve the first equation for y: $2x^2 - y = 5 \Rightarrow -y = -2x^2 + 5 \Rightarrow y = 2x^2 - 5$. Substitute this into the

second equation. $-4x^2 + 2(2x^2 - 5) = -10 \Rightarrow -4x^2 + 4x^2 - 10 = -10 \Rightarrow -10 = -10 \Rightarrow$ there are

infinitely many solutions, $\{(x, y) \mid 2x^2 - y = 5\}$.

65. Solve the second equation for y: $x^2 + y = 4 \Rightarrow y = 4 - x^2$. Substitute this into the first equation.

$x^2 - y = 4 \Rightarrow x^2 - (4 - x^2) = 4 \Rightarrow 2x^2 = 8 \Rightarrow x^2 = 4 \Rightarrow x = \pm 2$.

When $x = -2$, $y = 4 - (-2)^2 = 0$ and when $x = 2$, $y = 4 - (2)^2 = 0$.

The solutions are $(-2, 0)$ and $(2, 0)$.

67. Solve the second equation for y: $x - y = 0 \Rightarrow y = x$. Substitute this into the first equation.

$x^3 - x = 3y \Rightarrow x^3 - x = 3x \Rightarrow x^3 - 4x = 0 \Rightarrow x(x + 2)(x - 2) = 0 \Rightarrow x = 0, x = -2, \text{ or } x = 2.$

When $x = 0$, $y = 0$, when $x = -2$, $y = -2$, and when $x = 2$, $y = 2$.

The solutions are $(-2, -2)$, $(0, 0)$, and $(2, 2)$.

69. The given equations result in the following nonlinear system of equations.

$A(l, w) = 35 \Rightarrow lw = 35$ and $P(l, w) = 24 \Rightarrow 2l + 2w = 24$

Begin solving the second equation for l. $2l + 2w = 24 \Rightarrow l + w = 12 \Rightarrow l = 12 - w$. Substitute this into

the first equation. $lw = 35 \Rightarrow (12 - w)w = 35 \Rightarrow 12w - w^2 = 35 \Rightarrow w^2 - 12w + 35 = 0$. This is a

quadratic equation that can be solved by factoring, graphing, or the quadratic formula. The solutions to this

quadratic are found using factoring. $w^2 - 12w + 35 = 0 \Rightarrow (w - 5)(w - 7) = 0 \Rightarrow w = 5 \text{ or } 7$.

Since $l = 12 - w$, if $w = 5$, then $l = 7$, and if $w = 7$, then $l = 5$. If the length is greater than the width, the

solution is $l = 7$ and $w = 5$. A rectangle with length 7 and width 5 has an area of 35 and a perimeter of 24.

71. Add the two equations together to eliminate the y-variable.

$$
\begin{array}{r}
x + y = 20 \\
x - y = 8 \\
\hline
2x \phantom{{}+ y} = 28
\end{array} \Rightarrow x = 14
$$

Since $x + y = 20$, it follows that $y = 6$. The unique solution is $(14, 6)$. The system is consistent and

independent. Graphical and numerical support are shown in Figure 71a & 71b, where

$Y_1 = 20 - X$ and $Y_2 = X - 8$.

[0, 24, 4] by [0, 16, 4]

Figure 71a　　　　Figure 71b

73. Subtract the two equations to eliminate the *x*-variable.

$$\begin{array}{rcl} x + 3y &=& 10 \\ x - 2y &=& -5 \\ \hline 5y &=& 15 \Rightarrow y = 3 \end{array}$$

Since $x + 3y = 10$, it follows that $x = 1$. The solution is $(1, 3)$. The system is consistent and independent.

75. Multiply the second equation by -1 and subtract to eliminate both variables.

$$\begin{array}{rcl} x + y &=& 500 \\ x + y &=& 500 \\ \hline 0 &=& 0 \Rightarrow \text{infinite number of solutions} \end{array}$$

The solution is $\{(x, y) \mid x + y = 500\}$. The system is consistent but dependent.

77. Multiply the second equation by 2 and add. This eliminates both variables.

$$\begin{array}{rcl} 2x + 4y &=& 7 \\ -2x - 4y &=& 10 \\ \hline 0 &=& 17 \Rightarrow \text{no solution} \end{array}$$

Since $0 \neq 17$, the system is inconsistent.

79. Multiply the second equation by 2 and subtract to eliminate the *x*-variable.

$$\begin{array}{rcl} 2x + 3y &=& 2 \\ 2x - 4y &=& -10 \\ \hline 7y &=& 12 \Rightarrow y = \dfrac{12}{7} \end{array}$$

Since $2x + 3y = 2$, it follows that $x = \dfrac{2 - 3y}{2} \Rightarrow x = -\dfrac{11}{7}$. The solution is $\left(-\dfrac{11}{7}, \dfrac{12}{7}\right)$. The system is consistent and independent.

81. Multiply the second equation by $-\dfrac{1}{2}$ and add to eliminate the *x*-variable.

$$\begin{array}{rcl} \dfrac{1}{2}x - y &=& 5 \\ -\dfrac{1}{2}x + \dfrac{1}{4}y &=& -2 \\ \hline -\dfrac{3}{4}y &=& 3 \Rightarrow y = -4 \end{array}$$

Then, $x - \dfrac{1}{2}y = 4 \Rightarrow x = 4 + \dfrac{1}{2}y \Rightarrow x = 2$. The solution is $(2, -4)$.

83. Multiply the first equation by 3 and add to eliminate both variables.

$$\begin{array}{rcl} 21x - 9y &=& -51 \\ -21x + 9y &=& 51 \\ \hline 0 &=& 0 \Rightarrow \text{infinite number of solutions} \end{array}$$

There are infinitely many solutions of the form $\{(x, y) \mid 7x - 3y = -17\}$.

85. Multiply the first equation by 3 and add to eliminate both variables.

$$\begin{array}{rcl} 2x + 4y &=& 1 \\ -2x - 4y &=& 5 \\ \hline 0 &=& 6 \Rightarrow \text{no solutions} \end{array}$$

87. Clear decimals: $0.2x + 0.3y = 8 \Rightarrow 2x + 3y = 80$ and $-0.4x + 0.2y = 0 \Rightarrow -4x + 2y = 0$.

    Multiply the first equation by 2 and add to eliminate the x-variable.

$$\begin{array}{rcl} 4x + 6y &=& 160 \\ -4x + 2y &=& 0 \\ \hline 8y &=& 160 \end{array} \Rightarrow y = 20$$

    Then, $-4x + 2y = 0 \Rightarrow 4x = 2y \Rightarrow x = \dfrac{1}{2}y \Rightarrow x = \dfrac{1}{2}(20) = 10$. The solution is $(10, 20)$.

89. Multiply the first equation by 3 and the second equation by 2. Add to eliminate the x-variable.

$$\begin{array}{rcl} 6x + 9y &=& 21 \\ -6x + 4y &=& -8 \\ \hline 13y &=& 13 \end{array} \Rightarrow y = 1$$

    Then, $2x + 3y = 7 \Rightarrow 2x + 3 = 7 \Rightarrow 2x = 4 \Rightarrow x = 2$. The solution is $(2, 1)$.

91. Multiply the first equation by 3, and the second equation by 5. Add to elimate the y-variable.

$$\begin{array}{rcl} 21x - 15y &=& -45 \\ -10x + 15y &=& -10 \\ \hline 11x &=& -55 \end{array} \Rightarrow x = -5$$

    Then, $-10x + 15y = -10 \Rightarrow 50 + 15y = -10 \Rightarrow 15y = -60 \Rightarrow y = -4$. The solution is $(-5, -4)$.

93. Add the two equations:

$$\begin{array}{rcl} x^2 + y &=& 12 \\ x^2 - y &=& 6 \\ \hline 2x^2 &=& 18 \end{array} \Rightarrow x^2 = 9 \Rightarrow x = \pm 3.$$

    If $x = 3$, then $3^2 + y = 12 \Rightarrow y = 3$, and if $x = -3$ then $(-3)^2 + y = 12 \Rightarrow y = 3$. Therefore the solutions are: $(3, 3)$ and $(-3, 3)$.

95. Subtract the two equations:

$$\begin{array}{rcl} x^2 + y^2 &=& 25 \\ x^2 + 7y &=& 37 \\ \hline y^2 - 7y &=& -12 \end{array} \Rightarrow y^2 - 7y + 12 = 0 \Rightarrow (y - 3)(y - 4) = 0 \Rightarrow y = 3, 4.$$

    If $y = 3$, then $x^2 + 7(3) = 37 x^2 = 16 \Rightarrow x = \pm 4$, and if $y = 4$ then $x^2 + 7(4) = 37 \Rightarrow x^2 = 9 \Rightarrow$

    $x = \pm 3$. Therefore the solutions are: $(4, 3), (-4, 3), (3, 4)$, and $(-3, 4)$.

97. Subtract the two equations:

$$\begin{array}{rcl} x^2 + y^2 &=& 4 \\ 2x^2 + y^2 &=& 8 \\ \hline -x^2 &=& -4 \end{array} \Rightarrow x^2 = 4 \Rightarrow x = \pm 2.$$

    If $x = -2$, then $(-2)^2 + y^2 = 4 \Rightarrow 4 + y^2 = 4 \Rightarrow y^2 = 0 \Rightarrow y = 0$,

    and if $x = 2$ then $(2)^2 + y^2 = 4 \Rightarrow y^2 = 0 \Rightarrow y = 0$. Therefore the solutions are: $(-2, 0)$ and $(2, 0)$.

99. $x^2 + y^2 = 16 \Rightarrow y = \pm\sqrt{16 - x^2}$ and $x - y = 0 \Rightarrow y = x$

    Graph $Y_1 = \sqrt{(16 - X^2)}$, $Y_2 = -\sqrt{(16 - X^2)}$, and $Y_3 = X$. Their graphs intersect near the points

    $(-2.828, -2.828)$ and $(2.828, 2.828)$. See Figures 99a & 99b.

    Substituting $y = x$ into the first equation gives $x^2 + x^2 = 16 \Rightarrow 2x^2 = 16 \Rightarrow x^2 = 8 \Rightarrow x = \pm\sqrt{8}$

    Since $y = x$, the solutions are $(-\sqrt{8}, -\sqrt{8})$ and $(\sqrt{8}, \sqrt{8})$

Figure 99a   Figure 99b   Figure 101a   Figure 101b

101. $xy = 12 \Rightarrow y = \dfrac{12}{x}$ and $x - y = 4 \Rightarrow y = x - 4$. Graph $Y_1 = 12/X$, $Y_2 = X - 4$.

Their graphs intersect near the points $(6, 2)$ and $(-2, -6)$. See Figures 101a & 101b.

Substituting $y = x - 4$ into the first equation gives $x(x - 4) = 12 \Rightarrow x^2 - 4x - 12 = 0 \Rightarrow$

$(x + 2)(x - 6) = 0 \Rightarrow x = -2$ or 6. Since $y = x - 4$, the solutions are $(-2, -6)$ and $(6, 2)$.

103. (a) $2x + y = 1 \Rightarrow y = 1 - 2x$ and $x - 2y = 3 \Rightarrow y = \dfrac{1}{2}(x - 3)$

Graph $Y_1 = 1 - 2X$ and $Y_2 = 0.5(X - 3)$. Their graphs intersect at the point $(1, -1)$, which is the

solution. See Figure 103a.

(b) Table $Y_1 = 1 - 2X$ and $Y_2 = 0.5(X - 3)$ starting at 0 and incrementing by 0.5. See Figure 103b.

Here $Y_1 = Y_2 = -1$ when $x = 1$. The solution is $(1, -1)$.

(c) Substituting $y = 1 - 2x$ into the equation $x - 2y = 3$ gives $x - 2(1 - 2x) = 3 \Rightarrow$

$x - 2 + 4x = 3 \Rightarrow 5x = 5 \Rightarrow x = 1$. If $x = 1$, then $y = 1 - 2(1) = -1$. The solution is $(1, -1)$.

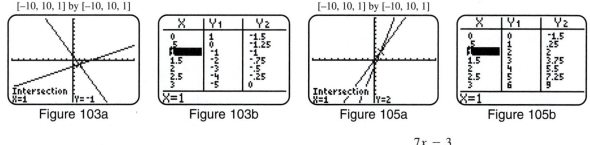

Figure 103a   Figure 103b   Figure 105a   Figure 105b

105. (a) $-2x + y = 0 \Rightarrow y = 2x$ and $7x - 2y = 3 \Rightarrow 7x - 3 = 2y \Rightarrow y = \dfrac{7x - 3}{2}$

Graph $Y_1 = 2X$ and $Y_2 = (7X - 3)/2$. Their graphs intersect at the point $(1, 2)$, which is the

solution. See Figure 105a.

(b) Table $Y_1 = 2X$ and $Y_2 = (7X - 3)/2$ starting at 0 and incrementing by 0.5. See Figure 105b.

Here $Y_1 = Y_2 = 2$ when $x = 1$. The solution is $(1, 2)$.

(c) Substituting $y = 2x$ into the second equation gives $7x - 2(2x) = 3 \Rightarrow 3x = 3 \Rightarrow x = 1$.

If $x = 1$, then $y = 2(1) = 2$. The solution is $(1, 2)$.

107. $x^3 - 3x + y = 1 \Rightarrow y = 1 + 3x - x^3$ and $x^2 + 2y = 3 \Rightarrow y = \dfrac{3 - x^2}{2}$. Graph $Y_1 = 1 + 3X - X^3$

and $Y_2 = (3 - X^2)/2$. See Figure 107. There are three points of intersection. The coordinates of these

points are near $(-1.588, 0.239)$, $(0.164, 1.487)$, and $(1.924, -0.351)$.

109. $2x^3 - x^2 = 5y \Rightarrow y = \dfrac{2x^3 - x^2}{5}$ and $2^{-x} - y = 0 \Rightarrow y = 2^{-x}$. Graph $Y_1 = (2X^3 - X^2)/5$

and $Y_2 = 2^{\wedge}(-X)$. See Figure 109. There is one point of intersection. The coordinates of this point are near

$(1.220, 0.429)$.

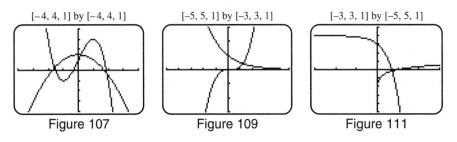

[-4, 4, 1] by [-4, 4, 1]	[-5, 5, 1] by [-3, 3, 1]	[-3, 3, 1] by [-5, 5, 1]
Figure 107	Figure 109	Figure 111

111. $e^{2x} + y = 4 \Rightarrow y = 4 - e^{2x}$ and $\ln x - 2y = 0 \Rightarrow y = \dfrac{\ln x}{2}$. Graph $Y_1 = 4 - e^{\wedge}(2X)$

and $Y_2 = \ln (X)/2$. See Figure 111. There is one point of intersection. The coordinates of this point are

near $(0.714, -0.169)$.

113. (a) Let $x =$ population of Minneapolis and $y =$ population of St. Paul. Then $x + y = 670$ and $x - y = 96$.

(b) $x + y = 670$

$\underline{\quad x - y = \quad 96 \quad}$

$\quad 2x \quad\quad = 766 \Rightarrow x = 383.$

Then, $x - y = 96 \Rightarrow 383 - y = 96 \Rightarrow y = 287$. The solution is $(383, 287)$.

(c) The system is consistent and independent.

115. $W_1 + \sqrt{2}W_2 = 300 \Rightarrow W_2 = \dfrac{300 - W_1}{\sqrt{2}}$ and $\sqrt{3}W_1 - \sqrt{2}W_2 = 0 \Rightarrow W_2 = \dfrac{\sqrt{3}W_1}{\sqrt{2}}$

Graph $Y_1 = (300 - X)/\sqrt{}(2)$ and $Y_2 = \sqrt{}(3)X/\sqrt{}(2)$. Their graphs intersect near the point

$(109.81, 134.49)$ as shown in Figure 115a. The forces on the rafters are approximately 110 and 134 pounds.

To find the solution numerically, table Y_1 and Y_2 starting at 107 and incrementing by 1. We find that

$Y_1 = Y_2 \approx 134$ when $x \approx 110$. See Figure 115b.

We can find the solution symbolically by using the substitution method. Since $W_1 + \sqrt{2}W_2 = 300 \Rightarrow$

$W_1 = 300 - \sqrt{2}W_2$, we will substitute into the other equation. $\sqrt{3}(300 - \sqrt{2}W_2) - \sqrt{2}W_2 = 0 \Rightarrow$

$300\sqrt{3} - \sqrt{6}W_2 - \sqrt{2}W_2 = 0 \Rightarrow 300\sqrt{3} = (\sqrt{6} + \sqrt{2})W_2 \Rightarrow W_2 = \dfrac{300\sqrt{3}}{\sqrt{6} + \sqrt{2}}$ and

$W_1 = 300 - \sqrt{2}\left[\dfrac{300\sqrt{3}}{\sqrt{6} + \sqrt{2}}\right] \Rightarrow W_1 = 300 - \dfrac{300\sqrt{3}}{1 + \sqrt{3}} \Rightarrow W_1 = \dfrac{300}{1 + \sqrt{3}}$

$W_1 = \dfrac{300}{1 + \sqrt{3}} \approx 109.8 \text{ lbs}, \ W_2 = \dfrac{300\sqrt{3}}{\sqrt{6} + \sqrt{2}} \approx 134.5 \text{ lbs}$

[0, 200, 50] by [0, 200, 50]   [0, 4, 1] by [0, 20, 1]

Figure 115a   Figure 115b   Figure 117

117. We must solve the system of nonlinear equations: $\pi r^2 h = 50$ and $2\pi r h = 65$.

Solving each equation for h results in the following: $\pi r^2 h = 50 \Rightarrow h = \dfrac{50}{\pi r^2}$ and $2\pi r h = 65 \Rightarrow h = \dfrac{65}{2\pi r}$.

Graph $Y_1 = 50/(\pi X\text{^}2)$ and $Y_2 = 65/(2\pi X)$. Their graphs intersect near $(1.538, 6.724)$. See Figure 117.

A cylinder with approximate measurements of $r \approx 1.538$ inches and $h \approx 6.724$ inches has a volume of 50

cubic inches and a lateral surface area of 65 square inches.

119. Let x = the length of each side of the base and let y = the height of the box. Since the volume = 576 in^3,

$x^2 y = 576$, and so $y = \dfrac{576}{x^2}$. Since the surface are is 336 in^2, $x^2 + 4xy = 336$. Substituting for y in this

equation yields $x^2 + 4x \cdot \dfrac{576}{x^2} = 336$. Simplifying we get: $x^3 - 336x + 2304 = 0$. By graphing the left side

of the equation, the x-intercepts are approximately 9.1 and 12. See Figure 119. When $x \approx 9.1$, the

dimensions are: 9.1 by 9.1 by $\dfrac{576}{(9.1)^2} \approx 6.96$ inches. When

$x = 12$, the dimensions are: 12 by 12 by $\dfrac{576}{(12)^2} = 4$ inches.

[0, 20, 2] by [−500, 1000, 100]   [0, 800,000, 100,000] by [0, 800,000, 100,000]

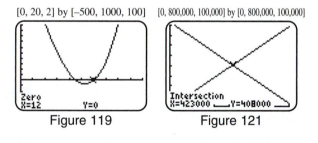

Figure 119   Figure 121

121. (a) Let x represent the number of robberies in 2000 and let y represent the number of robberies in 2001.

The required system of equations is $x + y = 831,000$ and $x - y = 15,000$.

(b) Adding the two equations results in $2x = 846,000 \Rightarrow x = 423,000$. From the first equation,

$423,000 + y = 831,000 \Rightarrow y = 408,000$. There were 423,000 robberies in 2000 and 408,000 in 2001.

(c) Solve each equation for y and graph $Y_1 = 831,000 - X$ and $Y_2 = X - 15,000$ as shown in Figure 121.

The solution is the intersection point $(423,000, 408,000)$.

123. (a) To solve this problem start by letting x represent the amount of the 8% loan and y the 10% loan. Since the total of both loans is $\$3000$, the equation $x + y = 3000$ must be satisfied. The annual interest rate for the 8% loan is given by $0.08x$, while the annual interest rate for the 10% loan is expressed by $0.10y$. the total interest for both loans is their sum $0.08x + 0.10y = 264$.

   (b) Thus, to determine a solution, the following linear system of equations could be solved.

   $x + y = 3000$ and $0.08x + 0.10y = 264$. Start by solving each for y.

   $x + y = 3000 \Rightarrow y = 3000 - x$

   and $0.08x + 0.10y = 264 \Rightarrow 0.10y = 264 - 0.08x \Rightarrow y = \dfrac{264 - 0.08x}{0.10} \Rightarrow y = 2640 - 0.8x$. The

   two equations can be solved using the intersection-of-graphs method. Let $Y_1 = 3000 - X$ and

   $Y_2 = 2640 - 0.8X$. Since the system of equations is linear, each graph is a line. The lines are not

   parallel and intersect at the point $(1800, 1200)$, as shown in Figure 123. Thus, the 8% loan is for $\$1800$

   and the 10% loan is for $\$1200$.

[0, 3000, 1000] by [0, 3000, 1000]   [0, 3000, 1000] by [0, 3000, 1000]   [0, 25, 5] by [0, 150, 25]

Figure 123   Figure 125   Figure 127

125. With these conditions the system of equations becomes $x + y = 3000$ and $0.10x + 0.10y = 264$.

   Solving each equation for y provides the following results, $y = 3000 - x$ and $y = 2640 - x$.

   Graphs of $Y_1 = 3000 - X$ and $Y_2 = 2640 - X$ are shown in Figure 125. Notice that their graphs are parallel

   lines with slope -1 that do not intersect. There is no solution. This means that there is no way to have two

   loans totaling $\$3000$, both with an interest rate of 10%, and only pay $\$264$ in interest each year. The interest

   must be 10% of $\$3000$ or $\$300$. This system of equations is inconsistent - there is no solution. Graphs of

   inconsistent systems in two variables consist of parallel lines.

127. (a) The perimeter is 40, thus $2l + 2w = 40 \Rightarrow l + w = 20 \Rightarrow l = 20 - w$. Since the area is 91, we have

   that $lw = 91$. Use the substitution method to solve this system.

   $(20 - w)w = 91 \Rightarrow 20w - w^2 = 91 \Rightarrow w^2 - 20w + 91 = 0 \Rightarrow (w - 13)(w - 7) = 0 \Rightarrow$

   $w = 7$ or $w = 13$. When $w = 7, l = 13$ and when $w = 13, l = 7$. Since length is longer than width, the

   solution is $l = 13$ feet and $w = 7$ feet.

   (b) $P = 2l + 2w = 40 \Rightarrow l + w = 20 \Rightarrow l = 20 - w$. Then, $A = lw = (20 - w)w = 20w - w^2$.

   Graph $Y_1 = 20X - X^2$ in $[0, 25, 5]$ by $[0, 150, 25]$.

   (c) The area can be any positive number less than or equal to 100 square feet. The maximum area of 100

   square feet occurs when $w = 10$ and $l = 10$. See Figure 127. Since all sides are equal to 10, the shape of

   the rectangle is a square. That is, a square pen will provide the largest area.

129. Let x represent the air speed of the plane and y the wind speed. Traveling with the wind, the average ground speed of the plane is $\dfrac{1680}{3} = 560$ mph, while its ground speed against the wind was $\dfrac{1680}{3.5} = 480$ mph. Thus,

$$
\begin{aligned}
x + y &= 560 \\
\underline{x - y} &= \underline{480} \\
2x &= 1040 \Rightarrow x = 520
\end{aligned}
$$

Thus, $y = 560 - x = 40$. The air speed of the plane is 520 mph and the wind speed is 40 mph.

131. Let x represent the number of deaths in WWI and y the number of deaths in WWII, then

$x + y = 345{,}000$ and $y = 5.5x$. Substituting the second equation into the first we get:

$x + 5.5x = 345{,}000 \Rightarrow 6.5x = 345{,}000 \Rightarrow x \approx 53{,}077$, then $y \approx 291{,}923$. Thus there were 53,077 deaths in WWI and 291,923 deaths in WWII.

133. (a) A 6' 11" person is 83 inches tall.

$w = 7.46(83) - 374 = 245.18 \approx 245$ lbs; $\quad w = 7.93(83) - 405 = 253.19 \approx 253$ lbs

(b) Graph $Y_1 = 7.46X - 374$ and $Y_2 = 7.93X - 405$. Their graphs intersect near $(65.96, 118.04)$.

See Figure 133. The models agree when $h \approx 65.96$ inches and $w \approx 118$ pounds.

(c) The first model's coefficient for h is 7.46. Thus, for each increase in height of 1 inch, the weight increases by 7.46 lbs. Similarly, for the second equation the increase is 7.93 lbs.

[65, 70, 1] by [100, 150, 10]

Figure 133

135. Substitute 165.1 for h and 70 for w: $S(70, 165.1) = 0.007184(70)^{0.425}(165.1)^{0.725} \approx 1.77 \text{m}^2$

137. $w = 132 \text{ lb} \approx \dfrac{132}{2.2} \text{ kg} = 60 \text{ kg}; \quad h = 62 \text{ inches} \approx 62 \cdot 2.54 \text{ cm} = 157.48 \text{ cm}$

$S(w, h) = 0.007184(w^{0.425})(h^{0.725}) \Rightarrow S(60, 157.48) = 0.007184(60^{0.425})(157.48^{0.725}) \approx 1.6 \text{ m}^2$

139. Since $z = kx^2y^3$ and $z = 31.9$ when $x = 2$ and $y = 2.5$.

$31.9 = k(2)^2(2.5)^3 \Rightarrow 31.9 = 62.5k \Rightarrow k = \dfrac{31.9}{62.5} \approx 0.51$

141. $z = k\sqrt{x} \cdot \sqrt[3]{y} \Rightarrow 10.8 = k\sqrt{4} \cdot \sqrt[3]{8} \Rightarrow 10.8 = k \cdot 2 \cdot 2 \Rightarrow 10.8 = 4k \Rightarrow k = 2.7$

Therefore: $z = 2.7\sqrt{x} \cdot \sqrt[3]{y}$ and now $z = 2.7\sqrt{16} \cdot \sqrt[3]{27} \Rightarrow z = 2.7(4)(3) \Rightarrow z = 32.4$.

143. Let d represent the diameter of the blades, v represent the wind velocity, and w represent the watts of power generated by the windmill. Then $w = kd^2v^3$ and $w = 2405$ when $d = 8$ and $v = 10$.

$2405 = k(8)^2(10)^3 \Rightarrow 2405 = 64{,}000k \Rightarrow k = \dfrac{481}{12{,}800}$. The variation equation becomes

$w = \dfrac{481}{12{,}800}d^2v^3$. Thus, when $d = 6$ and $v = 20$; $w = \dfrac{481}{12{,}800}(6)^2(20)^3 = 10{,}822.5$.

With six-foot blades and a 20 mile-per-hour wind, the windmill will produce about 10,823 watts.

145. From the example $V = 0.00132h^{1.12}d^{1.98}$. When $h = 105$ and $d = 38$ we have

$V = 0.00132(105)^{1.12}(38)^{1.98} \approx 325.295$. A tree which is 105 feet tall with a diameter of 38 inches contains

approximately 325.295 cubic feet of wood. To find the number of cords, divide this result by 128:

$\frac{325.295}{128} \approx 2.54$ cords.

147. From exercise 135, $S(w, h) = 0.007184(w^{0.425})(h^{0.725})$ where w is weight in kilograms and h is height in

centimeters. Let $S = 1.77$, $w = 154$, and $h = 65$. Then, $1.77 = k(154^{0.425})(65^{0.725})$, which results in

$k \approx 0.0101$. Thus, $S(w, h) = 0.0101(w^{0.425})(h^{0.725})$, where w is in pounds, h is in inches, and S is in square

meters.

9.2: Systems of Inequalities in Two Variables

1. Graph the boundary line $y = x$. Choose $(1, 0)$ as a test point. Since $1 \geq 0$, the region containing

   $(1, 0)$ is part of the solution set. The inequality $x \geq y$ includes this region and the line $y = x$. See Figure 1.

3. Graph the boundary line $x = 1$. Choose $(0, 0)$ as a test point. Since $0 < 1$, the region containing $(0, 0)$ is the

   solution set. The boundary line $x = 1$ is not part of the solution set. See Figure 3.

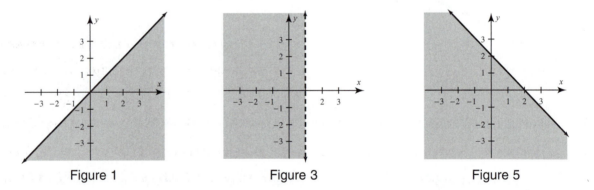

Figure 1   Figure 3   Figure 5

5. Graph the boundary line $x + y = 2 \Rightarrow y = 2 - x$. Choose $(0, 0)$ as a test point. Since $0 + 0 \leq 2$, the

   region containing $(0, 0)$ is part of the solution set. The boundary line $y = 2 - x$ is also part of the solution set.

   See Figure 5.

7. Graph the boundary line $2x + y = 4 \Rightarrow y = 4 - 2x$. Choose $(3, 0)$ as a test point. Since $0 > 4 - 2(3)$, the

   region containing $(3, 0)$ is the solution set. The boundary line $y = 4 - 2x$ is not part of the solution set.

   See Figure 7.

9. Graph the circle determined by $x^2 + y^2 = 4$. Choose $(0, 0)$ as a test point. Since $0 > 4$, the region inside the

   circle is not part of the solution $\Rightarrow$ shade the region outside the circle. Note, the circle is not part of the

   solution. See Figure 9.

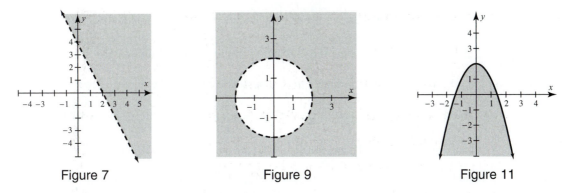

Figure 7   Figure 9   Figure 11

11. Graph the boundary parabola determined by $x^2 + y = 2 \Rightarrow y = -x^2 + 2$. Choose $(0, 0)$ as a test point.

    Since $0 \le 2$, the region inside the parabola is part of the solution $\Rightarrow$ shade the region inside of the parabola.

    Note, the parabola is part of the solution. See Figure 11.

13. $x + y \ge 2 \Rightarrow y \ge 2 - x$ is above the line $y = 2 - x$. It includes the line.

    $x - y \le 1 \Rightarrow y \ge x - 1$ is above the line $y = x - 1$. It includes the line.

    The solution is above two lines, which matches Figure c. One solution is $(2, 3)$. *Answers may vary.*

15. $\frac{1}{2}x^3 - y > 0 \Rightarrow y < \frac{1}{2}x^3$ is below the curve $y = \frac{1}{2}x^3$. It does not include the curve.

    $2x - y \le 1 \Rightarrow y \ge 2x - 1$ is above the line $y = 2x - 1$. It includes the line.

    The solution is below a dotted curve and above a solid line, which matches Figure d. One solution is $(-1, -1)$.

    Answers may vary.

17. The solution region is above the parabola $y = x^2$ and below the line $y = 6 - x$. It includes the boundary. See

    Figure 17. One solution is $(0, 2)$. *Answers may vary.*

19. The solution region lies between the parallel lines $y = -\frac{1}{2}x - 1$ and $y = -\frac{1}{2}x + \frac{5}{2}$. It does not include the

    boundary. See Figure 19. One solution is $(0, 0)$. *Answers may vary.*

Figure 17   Figure 19   Figure 21

21. The solution region lies inside the circle centered at the origin with radius 4 and below the line $y = -x + 2$. It

    does not include the boundary determined by the line. See Figure 21. One solution is $(-1, 1)$.

    Answers may vary.

23. $x + 2y \leq 4 \Rightarrow y \leq -\dfrac{1}{2}x + 2$ and $2x - y \geq 6 \Rightarrow y \leq 2x - 6$. Graph the boundary lines

$y = -\dfrac{1}{2}x + 2$ and $y = 2x - 6$. The region satisfying the system is below the line $x + 2y = 4$ and below the

line $2x - y = 6$. Because equality is included, the boundaries are part of the region. See Figure 23.

25. $3x + 2y < 6 \Rightarrow y < -\dfrac{3}{2}x + 3$ and $x + 3y \leq 6 \Rightarrow y \leq -\dfrac{1}{3}x + 2$. Graph the boundary lines

$y = -\dfrac{3}{2}x + 3$ and $y = -\dfrac{1}{3}x + 2$. The region satisfying the system is below the line $3x + 2y = 6$, not

including the boundary and below the line $x + 3y = 6$, including the boundary. See Figure 25.

Figure 23   Figure 25   Figure 27

27. $x - 2y \geq 0 \Rightarrow y \leq \dfrac{1}{2}x$ and $x - 3y \leq 3 \Rightarrow y \geq \dfrac{1}{3}x - 1$. Graph the boundary lines $y = \dfrac{1}{2}x$ and

$y = \dfrac{1}{3}x - 1$. The region satisfying the system is below the line $y = \dfrac{1}{2}x$ and above the line $y = \dfrac{1}{3}x - 1$.

Because equality is included, the boundaries are part of the region. See Figure 27.

29. $x^2 + y^2 \leq 4$ and $y \geq 1$. Graph the boundary circle $x^2 + y^2 = 4$ and the boundary line $y = 1$. The region

satisfying the system is inside the circle $x^2 + y^2 = 4$ and above the line $y = 1$. Because equality is included,

the boundaries are part of the region. See Figure 29.

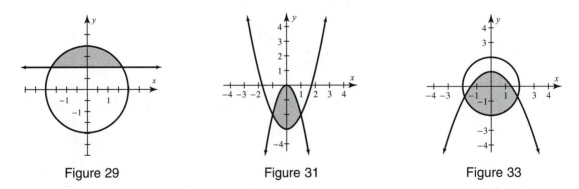

Figure 29   Figure 31   Figure 33

31. $2x^2 + y \leq 0 \Rightarrow y \leq -2x^2$ and $x^2 - y \leq 3 \Rightarrow y \geq x^2 - 3$. Graph the boundary parabolas $y = -2x^2$ and

$y = x^2 - 3$. Because equality is included, the boundaries are part of the region. See Figure 31.

33. $x^2 + 2y \leq 2 \Rightarrow y \leq 1 - \dfrac{x^2}{2}$ and $x^2 + y^2 \leq 4$. Graph the boundary parabola $y = 1 - \dfrac{x^2}{2}$ and the boundary

circle $x^2 + y^2 = 4$. Because equality is included, the boundaries are part of the region. See Figure 33.

35. The total number of vehicles entering intersection A is $500 + 150 = 650$ vehicles per hour. The expression $x + y$ represents the number of vehicles leaving intersection A each hour. Therefore, we have $x + y = 650$. The total number of vehicles leaving intersection B is $50 + 400 = 450$. There are 100 vehicles entering intersection B from the south and y vehicles entering intersection B from the west. Thus, $y + 100 = 450$. We must solve the system;

$$\begin{array}{rcl} x + y & = & 650 \\ \underline{y + 100} & = & \underline{450} \\ x \qquad - 100 & = & 200 \quad \Rightarrow x = 300 \end{array}$$

Thus, $y = 350$ and $x = 300$. At intersection A, a stoplight should allow for 300 vehicles per hour to travel south and 350 vehicles per hour to continue traveling east.

37. This region corresponds to weights that are less and heights that are greater than recommended. This individual has weight that is less than recommended for his or her height.

39. The upper left boundary is given by $25h - 7w = 800$ or $h = \dfrac{7w + 800}{25}$. The region is below and includes this line, which is described by $h \le \dfrac{7w + 800}{25}$ or $25h - 7w \le 800$. The lower right boundary of this region is given by $5h - w = 170$. The region is above and includes this line, which is described by $5h - w \ge 170$. Thus, the region can be described by the system of inequalities: $25h - 7w \le 800, 5h - w \ge 170$.

41. See Figure 41.

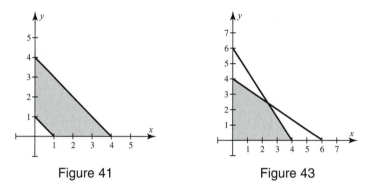

Figure 41   Figure 43

43. See Figure 43.

45. To find the maximum and minimum values of P in the region, we must evaluate P at each of the vertices. These values are shown in the table. See Figure 45. From the table the maximum is 65 and the minimum is 8.

47. To find the maximum and minimum values of C in the region, we must evaluate C at each of the vertices. These values are shown in the table. See Figure 47. From the table the maximum is 66 and the minimum is 3.

Vertex	$P = 3x + 5y$
(1, 1)	$3(1) + 5(1) = 8$
(6, 3)	$3(6) + 5(3) = 33$
(5, 10)	$3(5) + 5(10) = 65$
(2, 7)	$3(2) + 5(7) = 41$

Figure 45

Vertex	$C = 3x + 5y$
(1, 0)	$3(1) + 5(0) = 3$
(7, 6)	$3(7) + 5(6) = 51$
(7, 9)	$3(7) + 5(9) = 66$
(1, 10)	$3(1) + 5(10) = 53$

Figure 47

49. To find the maximum and minimum values of C in the region, we must evaluate C at each of the vertices.

These values are shown in the table. See Figure 49. From the table the maximum is 100 and the minimum is 0.

Vertex	$C = 10y$
(1, 0)	$10(0) = 0$
(7, 6)	$10(6) = 60$
(7, 9)	$10(9) = 90$
(1, 10)	$10(10) = 100$

Figure 49

51. The line that goes through the points (0, 4) and (4, 0) has the equation $x + y = 4$. The shaded region is also bounded by the line $x = 0$ and the line $y = 0$. Thus, the shaded region is described by the system:

$x + y \leq 4$, $x \geq 0$, and $y \geq 0$.

53. The region of feasible solutions is shown in Figure 53a. The vertices of this region are (3, 0), (6, 0), (0, 4), and (0, 3). To find the minimum value of C in the region, we must evaluate C at each of the vertices. These values are shown in Figure 53b. From the table the minimum is 6 at the point (0, 3).

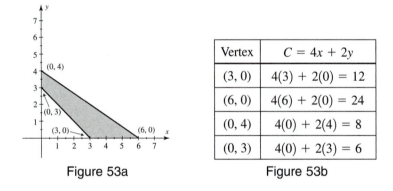

Vertex	$C = 4x + 2y$
(3, 0)	$4(3) + 2(0) = 12$
(6, 0)	$4(6) + 2(0) = 24$
(0, 4)	$4(0) + 2(4) = 8$
(0, 3)	$4(0) + 2(3) = 6$

Figure 53a   Figure 53b

55. The region of feasible solutions is shown in Figure 55a. The vertices of this region are (0, 4), (0, 8), (4, 0), and (8, 0). To find the maximum and minimum values of $z = 7x + 6y$ in the region, we must evaluate z at each of the vertices. These values are shown in Figure 55b. From the table the maximum is 56 and the minimum is 24.

Vertex	$z = 7x + 6y$
(0, 4)	$7(0) + 6(4) = 24$
(0, 8)	$7(0) + 6(8) = 48$
(4, 0)	$7(4) + 6(0) = 28$
(8, 0)	$7(8) + 6(0) = 56$

Figure 55a   Figure 55b

57. The new objective equation is $P = 20x + 15y$. To find the maximum profit we must evaluate P at each of the vertices. These values are shown in Figure 57. From the table the maximum is 950 at the vertex (25, 30). The maximum profit will be $950 when 25 radios and 30 CD players are manufactured.

Vertex	$P = 20x + 15y$
(5, 5)	$20(5) + 15(5) = 175$
(25, 25)	$20(25) + 15(25) = 875$
(25, 30)	$20(25) + 15(30) = 950$
(5, 30)	$20(5) + 15(30) = 550$

Figure 57

59. Make a table to list the information given. See Figure 59a. Using the table, we can write the linear programming problem as follows;

Cost: $C = 80x + 50y$, Protein: $15x + 20y \geq 60$, Fat: $10x + 5y \geq 30$, $x \geq 0$, and $y \geq 0$.

Brand	Units	Protein	Fat	Cost
A	x	15	10	80¢
B	y	20	5	50¢
Minimum		60	30	

Figure 59a

The region of feasible solutions is shown in Figure 59b. The vertices of this region are (0, 6), (2.4, 1.2), and (4, 0). To find the minimum value of C in the region, we must evaluate C at each of the vertices. These values are shown in Figure 59c. The minimum cost occurs when 2.4 units of Brand A and 1.2 units of Brand B are mixed, to give a cost of $2.52 per serving.

Figure 59b

Vertex	$C = 80x + 50y$
(0, 6)	$80(0) + 50(6) = 300$
(2.4, 1.2)	$80(2.4) + 50(1.2) = 252$
(4, 0)	$80(4) + 50(0) = 320$

Figure 59c

61. Let x and y represent the number of hamsters and mice respectively. Since the total number of animals cannot exceed 50, $x + y \leq 50$. Because no more than 20 hamsters can be raised, $x \leq 20$. Here the revenue function is $R = 15x + 10y$. From the graph of the region of feasible solutions (not shown), the vertices are (0, 0), (0, 50), (20, 30), and (20, 0). To find (20, 30) solve the equations $x + y = 50$ and $x = 20$. The maximum value of R occurs at one of the vertices. For (0, 50), $R = 15(0) + 10(50) = 500$. For (20, 30), $R = 15(20) + 10(30) = 600$. For (20, 0), $R = 15(20) + 10(0) = 300$. The maximum revenue is $600.

63. The number of hours on machine A needed to manufacture x units of part X is $4x$ while the number of hours on machine A needed to manufacture y units of part Y is $1y$. Since machine A is only available for 40 hours each week we have the constraint $4x + y \leq 40$. Similarly, the number of hours on machine B needed to manufacture x units of part X is $2x$ while the number of hours on machine B to make y units of part Y is $3y$. Since machine B is only available for 30 hours each week we have the constraint $2x + 3y \leq 30$. It should be noted that the number of parts of each type cannot be negative. This gives the constraint $x \geq 0$ and $y \geq 0$. The profit earned on x units of part X is $500x$ while the profit earned on y units of part Y is $600y$. Thus, the total weekly profit is $P = 500x + 600y$. This is our objective function. Graph the constraints and shade the region. The region of feasible solutions is shown in Figure 63a. The vertices of this region are $(0, 0)$, $(10, 0)$, $(9, 4)$, and $(0, 10)$. To find the maximum value of P in the region, we must evaluate P at each of the vertices. These values are shown in Figure 63b. From the table the maximum is 6900 at the point $(9, 4)$. The maximum profit is \$6900 when there are 9 parts of type X and 4 parts of type Y manufactured.

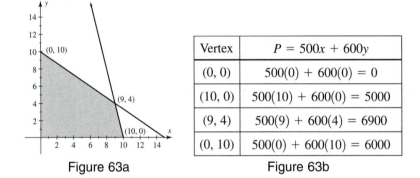

Vertex	$P = 500x + 600y$
$(0, 0)$	$500(0) + 600(0) = 0$
$(10, 0)$	$500(10) + 600(0) = 5000$
$(9, 4)$	$500(9) + 600(4) = 6900$
$(0, 10)$	$500(0) + 600(10) = 6000$

Figure 63a   Figure 63b

9.3 Systems of Linear Equations in Three Variables

1. No, systems of linear equations can have zero, one, or infinitely many solutions.

3. 2, the same as the number of variables.

5. Testing $(0, 2, -2)$: $0 + 2 - (-2) = 4$, true; $-0 + 2 + (-2) = 2$, false; $0 + 2 + (-2) = 0$, true; since the second equation is false, $(0, 2, -2)$ is not a solution for the system.

   Testing $(-1, 3, -2)$: $-1 + 3 - (-2) = 4$, true; $-(-1) + 3 + (-2) = 2$, true; $-1 + 3 + (-2) = 0$, true; since all equation are true, $(-1, 3, -2)$ is a solution for the system.

7. Testing $\left(-\dfrac{5}{11}, \dfrac{20}{11}, -2\right)$: $-\dfrac{5}{11} + 3\left(\dfrac{20}{11}\right) - 2(-2) = 9 \Rightarrow -\dfrac{5}{11} + \dfrac{60}{11} + \dfrac{44}{11} = 9 \Rightarrow \dfrac{99}{11} = 9$, true;

$-3\left(-\dfrac{5}{11}\right) + 2\left(\dfrac{20}{11}\right) + 4(-2) = -3 \Rightarrow \dfrac{15}{11} + \dfrac{40}{11} - \dfrac{88}{11} = -3 \Rightarrow -\dfrac{33}{11} = -3$, true;

$-2\left(-\dfrac{5}{11}\right) + 5\left(\dfrac{20}{11}\right) + 2(-2) = 6 \Rightarrow \dfrac{10}{11} + \dfrac{100}{11} - \dfrac{44}{11} = 6 \Rightarrow \dfrac{66}{11} = 6$, true; since all of the equations

are true, $\left(-\dfrac{5}{11}, \dfrac{20}{11}, -2\right)$ is a solution for the system.

Testing $(1, 2, -1)$: $1 + 3(2) - 2(-1) = 9 \Rightarrow 1 + 6 + 2 = 9 \Rightarrow 9 = 9$, true;

$-3(1) + 2(2) + 4(-1) = -3 \Rightarrow -3 + 4 + (-4) = -3 \Rightarrow -3 = -3$, true;

$-2(1) + 5(2) + 2(-1) = 6 \Rightarrow (-2) + 10 + (-2) = 6 \Rightarrow 6 = 6$, true; since all equations are true,

$(1, 2, -1)$ is a solution for the system.   Therefore they are both solutions.

9. Add The first two equations:

$$\begin{array}{r} x + y + z = 6 \\ -x + 2y + z = 6 \\ \hline 3y + 2z = 12. \end{array}$$

Subtract 2 times the third equation from this equation:

$$\begin{array}{r} 3y + 2z = 12 \\ 2y + 2z = 10 \\ \hline y \quad\quad = 2. \end{array}$$

Substitute $y = 2$ into the equation: $3y + 2z = 12 \Rightarrow 3(2) + 2z = 12 \Rightarrow 6 + 2z = 12 \Rightarrow 2z = 6 \Rightarrow$

$z = 3$. Finally, substitute $y = 2$ and $z = 3$ into one of the original equations: $x + 2 + 3 = 6 \Rightarrow x = 1$.

The solution is $(1, 2, 3)$.

11. Multiply the first equation by 2 and subtract the second equation:

$$\begin{array}{r} 2x + 4y + 6z = 8 \\ 2x + y + 3z = 5 \\ \hline 3y + 3z = 3. \end{array}$$

Subtract the third equation from the first:

$$\begin{array}{r} x + 2y + 3z = 4 \\ x - y + z = 2 \\ \hline 3y + 2z = 2. \end{array}$$

Now subtract these two equations:

$$\begin{array}{r} 3y + 3z = 3 \\ 3y + 2z = 2 \\ \hline z = 1. \end{array}$$

Substitute $z = 1$ into $3y + 2z = 2$: $3y + 2(1) = 2 \Rightarrow 3y = 0 \Rightarrow y = 0$.

Finally substitute $y = 0$ and $z = 1$ into the original equation $x - y + z = 2$: $x - 0 + 1 = 2 \Rightarrow x = 1$.

The solution is: $(1, 0, 1)$.

13. Add the second and third equations:

$$4x + 2y + z = 1$$
$$2x - 2y - z = 2$$
$$6x \qquad\quad = 3 \Rightarrow x = \frac{1}{2}.$$

Subtract the first two equations:

$$3x + y + z = 0$$
$$4x + 2y + z = 1$$
$$-x - y \quad\;\; = -1.$$

Substitute $x = \frac{1}{2}$ into $-x - y = -1$: $-\frac{1}{2} - y = -1 \Rightarrow -y = -\frac{1}{2} \Rightarrow y = \frac{1}{2}.$

Substitute $x = \frac{1}{2}$ and $y = \frac{1}{2}$ into $3x + y + z = 0$: $3\left(\frac{1}{2}\right) + \frac{1}{2} + z = 0 \Rightarrow \frac{3}{2} + \frac{1}{2} + z = 0 \Rightarrow$

$2 + z = 0 \Rightarrow z = -2.$

The solution is: $\left(\dfrac{1}{2}, \dfrac{1}{2}, -2\right).$

15. Subtract the third equation from the first:

$$x + 3y + z = 6$$
$$x - y - z = 0$$
$$\qquad 4y + 2z = 6.$$

Multiply the third equation by 3 and subtract it from the second equation:

$$3x + y - z = 6$$
$$3x - 3y - 3z = 0$$
$$\qquad 4y + 2z = 6.$$

Subtracting these two equations we get:

$$4y + 2z = 6$$
$$4y + 2z = 6$$
$$\qquad 0 = 0.$$

Therefore infinitely many solutions and $4y + 2z = 6 \Rightarrow 4y = -2z + 6 \Rightarrow y = \dfrac{-z + 3}{2}.$

Adding the last two original equations we get:

$$3x + y - z = 6$$
$$x - y - z = 0$$
$$4x \qquad - 2z = 6 \Rightarrow 4x = 2z + 6 \Rightarrow x = \dfrac{z + 3}{2}.$$

We have infinitely many solutions: $\left(\dfrac{z + 3}{2}, \dfrac{-z + 3}{2}, z\right).$

17. Add the first two equations:

$$x - 4y + 2z = -2$$
$$\underline{x + 2y - 2z = -3}$$
$$2x - 2y \qquad = -5.$$

Multiply the third equation by 2 and subtract from $2x - 2y = -5$:

$$2x - 2y = \quad -5$$
$$\underline{2x - 2y = \quad 8}$$
$$0 = -13. \quad \text{Therefore we have no solutions.}$$

19. Add the last two equations:

$$2a + \quad b - c = -11$$
$$\underline{2a - 2b + c = \quad 3}$$
$$4a - \quad b \qquad = \quad -8.$$

Multiply the second equation by 2 and add to the first equation:

$$4a - \quad b + 2c = \quad 0$$
$$\underline{4a + 2b - 2c = -22}$$
$$8a + \quad b \qquad = -22.$$

Now add the equation $4a - b = -8$ to $8a + b = -8$:

$$4a - b = \quad -8$$
$$\underline{8a + b = -22}$$
$$12a \qquad = -30 \Rightarrow a = \frac{-30}{12} \Rightarrow a = \frac{-5}{2}.$$

Substitute $a = \frac{-5}{2}$: $8\left(\frac{-5}{2}\right) + b = -22 \Rightarrow -20 + b = -22 \Rightarrow b = -2.$

Substitute $a = \frac{-5}{2}$ and $b = -2$ into $4\left(\frac{-5}{2}\right) - (-2) + 2c = 0 \Rightarrow -10 + 2 + 2c = 0 \Rightarrow 2c = 8 \Rightarrow$

$c = 4$. The solution is: $\left(\frac{-5}{2}, -2, 4\right)$.

21. Subtract the first and third equations:

$$a + \quad b + \quad c = \quad 0$$
$$\underline{a + 3b + 3c = \quad 5}$$
$$-2b - 2c = -5.$$

Subtract the second and third equations:

$$a - \quad b - \quad c = \quad 3$$
$$\underline{a + 3b + 3c = \quad 5}$$
$$-4b - 4c = -2.$$

Multiply $-2b - 2c = -5$ by 2 and subtract $-4b - 4c = -2$:

$$-4b - 4c = -10$$
$$\underline{-4b - 4c = \quad -2}$$
$$0 = \quad -8.$$

Therefore, we have no solution.

23. Add the first two equations:

$$3x + 2y + z = -1$$
$$\underline{3x + 4y - z = 1}$$
$$6x + 6y = 0.$$

Add the last two equations:

$$3x + 4y - z = 1$$
$$\underline{x + 2y + z = 0}$$
$$4x + 6y = 1.$$

Now, subtract $6x + 6y = 0$ and $4x + 6y = 1$:

$$6x + 6y = 0$$
$$\underline{4x + 6y = 1}$$
$$2x = -1 \Rightarrow x = -\frac{1}{2}.$$

Substitute $x = -\dfrac{1}{2}$ into $6x + 6y = 0$: $6\left(-\dfrac{1}{2}\right) + 6y = 0 \Rightarrow 6y = 3 \Rightarrow y = \dfrac{1}{2}$.

Substitute $x = -\dfrac{1}{2}$ and $y = \dfrac{1}{2}$ into $x + 2y + x = 0 \Rightarrow -\dfrac{1}{2} + 2\left(\dfrac{1}{2}\right) + z = 0 \Rightarrow z = -\dfrac{1}{2}$.

The solution is: $\left(-\dfrac{1}{2}, \dfrac{1}{2}, -\dfrac{1}{2}\right)$.

25. Multiply the first equation by 2 and add the second equation:

$$-2x + 6y + 2z = 6$$
$$\underline{2x + 7y + 4z = 13}$$
$$13y + 6z = 19.$$

Multiply the second equation by 2 and subtract the third equation:

$$4x + 14y + 8z = 26$$
$$\underline{4x + y + 2z = 7}$$
$$13y + 6z = 19.$$

Subtracting the two new equations we get:

$$13y + 6z = 19$$
$$\underline{13y + 6z = 19}$$
$$ 0 = 0.$$

Therefore, we have infinitely many solutions.

$$13y + 6z = 19 \Rightarrow 13y = -6z + 19 \Rightarrow y = \frac{-6z + 19}{13}.$$

Multiply the third equation by 3 and subtract from the first equation:

$$-x + 3y + z = 3$$
$$\underline{12x + 3y + 6z = 21}$$
$$-13x - 5z = -18 \Rightarrow x = \frac{-5z + 18}{13}.$$

We have infinitely many solutions: $\left(\dfrac{-5z + 18}{13}, \dfrac{-6z + 19}{13}, z\right)$.

27. Subtract the second and third equations:

$$\begin{aligned} y + 4z &= -13 \\ \underline{3x + y } &= 13 \\ -3x + 4z &= -26. \end{aligned}$$

Multiply the first equation by 3 and subtract $-3x + 4z = -26$ from it:

$$\begin{aligned} -3x + 6z &= -27 \\ \underline{-3x + 4z} &= -26 \\ 2z &= -1 \Rightarrow z = -\frac{1}{2}. \end{aligned}$$

Substitute $z = -\frac{1}{2}$ into $y + 4z = -13$: $y + 4\left(-\frac{1}{2}\right) = -13 \Rightarrow y = -11.$

Substitute $z = -\frac{1}{2}$ into $-x + 2z = -9 \Rightarrow -x + 2\left(-\frac{1}{2}\right) = -9 \Rightarrow -x = -8 \Rightarrow x = 8.$

The solution is: $\left(8, -11, -\frac{1}{2}\right).$

29. Multiply the first equation by 2 and subtract the second equation:

$$\begin{aligned} x - 2y + z &= -8 \\ \underline{x + 2y - 3z} &= 20 \\ -4y + 4z &= -28. \end{aligned}$$

Add the first and third equations:

$$\begin{aligned} \tfrac{1}{2}x - y + \tfrac{1}{2}z &= -4 \\ \underline{-\tfrac{1}{2}x + 3y + 2z} &= 0 \\ 2y + \tfrac{5}{2}z &= -4. \end{aligned}$$

Multiply $2y + \frac{5}{2}z = -4$ by 2 and add $-4y + 4z = -28$:

$$\begin{aligned} 4y + 5z &= -8 \\ \underline{-4y + 4z} &= -28 \\ 9z &= -36 \Rightarrow z = -4. \end{aligned}$$

Substitute $z = -4$ into $-4y + 4z = -28$: $-4y + 4(-4) = -28 \Rightarrow -4y - 16 = -28 \Rightarrow$

$-4y = -12 \Rightarrow y = 3.$

Substitute $y = 3$ and $z = -4$ into $x + 2y - 3z = 20$: $x + 2(3) - 3(-4) = 20 \Rightarrow x + 18 = 20 \Rightarrow x = 2$

The solution is: $(2, 3, -4).$

31. Let x = children tickets sold, y = student tickets sold, and z = adult tickets sold. Then: $x + y + z = 500$,

    $5x + 7y + 10z = 3560$, and $y = z + 180$ or $y - z = 180$.

    Multiply the first equation by 5 and subtract the second equation:

    $$\begin{array}{r} 5x + 5y + 5z = 2500 \\ 5x + 7y + 10z = 3560 \\ \hline -2y - 5z = -1060. \end{array}$$

    Now, multiply $y - z = 180$ by 2 and add $-2y - 5z = -1060$:

    $$\begin{array}{r} 2y - 2z = 360 \\ -2y - 5z = -1060 \\ \hline -7z = -700 \Rightarrow z = 100. \end{array}$$

    Substitute $z = 100$ into $y - z = 180$: $y - 100 = 180 \Rightarrow y = 280$.

    Substitute $y = 280$ and $z = 100$ into $x + y + z = 500$: $x + 280 + 100 = 500 \Rightarrow x = 120$.

    There were 120 children tickets, 280 student tickets, and 100 adult tickets sold.

33. Let x = cost of a hamburger, y = cost of fries, and z = cost of a soda. Then $2x + 2y + z = 9$;

    $x + y + z = 5$; and $x + y = 5$.

    Multiply the equation $x + y + z = 5$ by 2 and subtract from the equation $2x + 2y + z = 9$:

    $$\begin{array}{r} 2x + 2y + z = 9 \\ 2x + 2y + 2z = 10 \\ \hline -z = -1 \Rightarrow z = 1. \end{array}$$

    Subtract $x + y + z = 5$ and $x + y = 5$:

    $$\begin{array}{r} x + y + z = 5 \\ x + y = 5 \\ \hline z = 0. \end{array}$$

    z cannot equal both 0 and 1 therefore there is no solution, at least one student was charged incorrectly.

35. (a) $x + y + z = 180$; $x = z + 25 \Rightarrow x - z = 25$; and $y + z = x + 30 \Rightarrow -x + y + z = 30$.

    (b) Add $x - z = 25$ to $-x + y + z = 30$:

    $$\begin{array}{r} x - z = 25 \\ -x + y + z = 30 \\ \hline y = 55. \end{array}$$

    Add $x + y + z = 180$ to $x - z = 25$:

    $$\begin{array}{r} x + y + z = 180 \\ x - z = 25 \\ \hline 2x + y = 205. \end{array}$$

    Substitute $y = 55$ into $2x + y = 205$: $2x + 55 = 205 \Rightarrow 2x = 150 \Rightarrow x = 75$.

    Substitute $x = 75$ and $y = 55$ into $x + y + z = 180$: $75 + 55 + z = 180 \Rightarrow z = 50$.

    The angles are: $75°$, $55°$, and $50°$.

    Check: $75 + 55 + 50 = 180 \Rightarrow 180 = 180$; $75 - 50 = 25 \Rightarrow 25 = 25$; $-75 + 55 + 50 = 30 \Rightarrow$

    $30 = 30$.

37. Let x, y, and z equal the amounts invested in the three mutual funds.

    Then $x + y + z = 20,000$; $0.05x + 0.07y + 0.10z = 1650$; and $4x = z \Rightarrow 4x - z = 0$.

    Multiply $x + y + z = 20,000$ by 0.07 and subtract $0.05x + 0.07y + 0.10z = 1650$:

    $$\begin{array}{r} 0.07x + 0.07y + 0.07z = 1400 \\ 0.05x + 0.07y + 0.10z = 1650 \\ \hline 0.02x - 0.03z = -250. \end{array}$$

    Now multiply $0.02x - 0.03z = -250$ by 200 and subtract $4x - z = 0$:

    $$\begin{array}{r} 4x - 6z = -50,000 \\ 4x - z = 0 \\ \hline -5z = -50,000 \Rightarrow z = 10,000. \end{array}$$

    Substitute $z = 10,000$ into $4x - z = 0$: $4x - 10,000 = 0 \Rightarrow 4x = 10,000 \Rightarrow x = 2500$.

    Substitute $x = 2500$ and $z = 10,000$ into $x + y + z = 20,000$: $2500 + y + 10,000 = 20,000 \Rightarrow y = 7500$.

    The fund amounts are: $2500 at 5%, $7500 at 7%, and $10,000 at 10%.

39. (a) $N + P + K = 80$
    $N + P - K = 8$
    $9P - K = 0$

    (b) Using technology to solve the system, the solution is $(40, 4, 36)$.

    The sample contains 40 pounds of nitrogen, 4 pounds of phosphorus and 36 pounds of potassium.

9.4 Solutions to Linear Systems Using Matrices

1. (a) Since there are three rows and one column, its dimension is 3×1.

   (b) Since there are two rows and three columns, its dimension is 2×3.

   (c) Since there are two rows and two columns, its dimension is 2×2.

3. This system can be written using a 2×3 matrix:

$$\left[\begin{array}{rr|r} 5 & -2 & 3 \\ -1 & 3 & -1 \end{array}\right]$$

5. This system can be written using a 3×4 matrix:

$$\left[\begin{array}{rrr|r} -3 & 2 & 1 & -4 \\ 5 & 0 & -1 & 9 \\ 1 & -3 & -6 & -9 \end{array}\right]$$

7. $3x + 2y = 4$ and $y = 5$

9. $3x + y + 4z = 0, 5y + 8z = -1$, and $-7z = 1$

11. (a) Yes

    (b) No, since $a_{22} = -1$ and $a_{32} \neq 0$. The diagonal is not all 1's and there are not all 0's below the diagonal.

    (c) Yes

13. The system can be written as $x + 2y = 3$ and $y = -1$. Substituting $y = -1$ into the first equation gives

    $x + 2(-1) = 3 \Rightarrow x = 5$. The solution is $(5, -1)$.

15. The system can be written as $x - y = 2$ and $y = 0$. Substituting $y = 0$ into the first equation gives

$x - 0 = 2 \Rightarrow x = 2$. The solution is $(2, 0)$.

17. The system can be written as $x + y - z = 4$, $y - z = 2$, and $z = 1$. Substituting $z = 1$ into the second

equation gives $y - (1) = 2 \Rightarrow y = 3$. Substituting $y = 3$ and $z = 1$ into the first equation gives

$x + (3) - (1) = 4 \Rightarrow x = 2$. The solution is $(2, 3, 1)$.

19. The system can be written as $x + 2y - z = 5$, $y - 2z = 1$, and $0 = 0$. Since $0 = 0$, there are an infinite

number of solutions. The second equation gives $y = 1 + 2z$. Substituting this into the first equation gives

$x + 2(1 + 2z) - z = 5 \Rightarrow x = 3 - 3z$. The solution can be written as

$\{(3 - 3z, 1 + 2z, z) \,|\, z \text{ is a real number}\}$.

21. The system can be written as $x + 2y + z = -3$, $y - 3z = \dfrac{1}{2}$, and $0 = 4$. Since $0 = 4$ is false, there are no

solutions.

23. $\begin{array}{l} (1/2)R_1 \to \\ \\ (1/4)R_3 \to \end{array} \left[\begin{array}{ccc|c} 1 & -2 & 3 & 5 \\ -3 & 5 & 3 & 2 \\ 1 & 2 & 1 & -2 \end{array} \right]$

25. $\begin{array}{l} \\ R_2 + R_1 \to \\ R_3 - R_1 \to \end{array} \left[\begin{array}{ccc|c} 1 & -1 & 1 & 2 \\ 0 & 1 & -1 & 2 \\ 0 & 8 & -1 & 3 \end{array} \right]$

27. The system can be written as follows:

$\left[\begin{array}{cc|c} 1 & 2 & 3 \\ -1 & -1 & 7 \end{array} \right] R_2 + R_1 \to \left[\begin{array}{cc|c} 1 & 2 & 3 \\ 0 & 1 & 10 \end{array} \right]$

The solution is $y = 10$ and $x + 2y = 3 \Rightarrow x + 2(10) = 3 \Rightarrow x = -17$. The solution is $(-17, 10)$.

29. The system can be written as follows:

$\left[\begin{array}{ccc|c} 1 & 2 & 1 & 3 \\ 1 & 1 & -1 & 3 \\ -1 & -2 & 1 & -5 \end{array} \right] \begin{array}{l} \\ R_2 - R_1 \to \\ R_3 + R_1 \to \end{array} \left[\begin{array}{ccc|c} 1 & 2 & 1 & 3 \\ 0 & -1 & -2 & 0 \\ 0 & 0 & 2 & -2 \end{array} \right] \begin{array}{l} \\ (-1)R_2 \to \\ (1/2)R_3 \to \end{array} \left[\begin{array}{ccc|c} 1 & 2 & 1 & 3 \\ 0 & 1 & 2 & 0 \\ 0 & 0 & 1 & -1 \end{array} \right]$

Back substitution produces $z = -1$; $y + 2z = 0 \Rightarrow y = 2$; $x + 2y + z = 3 \Rightarrow x = 0$.

The solution is $(0, 2, -1)$.

31. The system can be written as follows:

$\left[\begin{array}{ccc|c} 1 & 2 & -1 & -1 \\ 2 & -1 & 1 & 0 \\ -1 & -1 & 2 & 7 \end{array} \right] \begin{array}{l} \\ -2R_1 + R_2 \to \\ R_1 + R_3 \to \end{array} \left[\begin{array}{ccc|c} 1 & 2 & -1 & -1 \\ 0 & -5 & 3 & 2 \\ 0 & 1 & 1 & 6 \end{array} \right] R_3 \Leftrightarrow R_2 \to \left[\begin{array}{ccc|c} 1 & 2 & -1 & -1 \\ 0 & 1 & 1 & 6 \\ 0 & -5 & 3 & 2 \end{array} \right]$

$5R_2 + R_3 \to \left[\begin{array}{ccc|c} 1 & 2 & -1 & -1 \\ 0 & 1 & 1 & 6 \\ 0 & 0 & 8 & 32 \end{array} \right] \dfrac{1}{8}R_3 \to \left[\begin{array}{ccc|c} 1 & 2 & -1 & -1 \\ 0 & 1 & 1 & 6 \\ 0 & 0 & 1 & 4 \end{array} \right]$

Back substitution produces $z = 4$; $y + z = 6 \Rightarrow y = 2$; $x + 2y - z = -1 \Rightarrow x = -1$.

The solution is $(-1, 2, 4)$.

33. The system can be written as follows:

$$\begin{bmatrix} 3 & 1 & 3 & | & 14 \\ 1 & 1 & 1 & | & 6 \\ -2 & -2 & 3 & | & -7 \end{bmatrix} \begin{matrix} R_2 \to \\ R_1 \to \\ \end{matrix} \begin{bmatrix} 1 & 1 & 1 & | & 6 \\ 3 & 1 & 3 & | & 14 \\ -2 & -2 & 3 & | & -7 \end{bmatrix} \begin{matrix} \\ R_2 - 3R_1 \to \\ R_3 + 2R_1 \to \end{matrix} \begin{bmatrix} 1 & 1 & 1 & | & 6 \\ 0 & -2 & 0 & | & -4 \\ 0 & 0 & 5 & | & 5 \end{bmatrix}$$

$$\begin{matrix} \\ (-1/2)R_2 \to \\ (1/5)R_3 \to \end{matrix} \begin{bmatrix} 1 & 1 & 1 & | & 6 \\ 0 & 1 & 0 & | & 2 \\ 0 & 0 & 1 & | & 1 \end{bmatrix}$$

Back substitution produces $z = 1$; $y = 2$; $x + y + z = 6 \Rightarrow x = 3$. The solution is $(3, 2, 1)$.

35. The system can be written as follows:

$$\begin{bmatrix} 1 & 2 & -1 & | & 2 \\ 2 & 5 & 1 & | & 8 \\ 3 & 7 & 0 & | & 5 \end{bmatrix} \begin{matrix} \\ R_2 - 2R_1 \to \\ R_3 - 3R_1 \to \end{matrix} \begin{bmatrix} 1 & 2 & -1 & | & 2 \\ 0 & 1 & 3 & | & 4 \\ 0 & 1 & 3 & | & -1 \end{bmatrix} \begin{matrix} \\ \\ R_3 - R_2 \to \end{matrix} \begin{bmatrix} 1 & 2 & -1 & | & 2 \\ 0 & 1 & 3 & | & 4 \\ 0 & 0 & 0 & | & -5 \end{bmatrix}$$

The last equation indicates that $0 = -5$, which is false. Therefore, there are no solutions.

37. The system can be written as follows:

$$\begin{bmatrix} -1 & 2 & 4 & | & 10 \\ 3 & -2 & -2 & | & -12 \\ 1 & 2 & 6 & | & 8 \end{bmatrix} \begin{matrix} (-1)R_1 \to \\ R_2 + 3R_1 \to \\ R_1 + R_3 \to \end{matrix} \begin{bmatrix} 1 & -2 & -4 & | & -10 \\ 0 & 4 & 10 & | & 18 \\ 0 & 4 & 10 & | & 18 \end{bmatrix} \begin{matrix} R_1 + (1/2)R_2 \to \\ (1/4)R_2 \to \\ R_2 - R_3 \to \end{matrix} \begin{bmatrix} 1 & 0 & 1 & | & -1 \\ 0 & 1 & \frac{5}{2} & | & \frac{9}{2} \\ 0 & 0 & 0 & | & 0 \end{bmatrix}$$

The last equation indicates that $0 = 0$, which is true. Therefore, there is an infinite number of solutions.

The second equation gives: $y + \dfrac{5}{2}z = \dfrac{9}{2} \Rightarrow y = \dfrac{-5z + 9}{2}$. The first equation gives: $x + z = -1 \Rightarrow$

$x = -1 - z$. Therefore there are infinitely many solutions which can be written as $\left(-1 - z, \dfrac{-5z + 9}{2}, z \right)$.

39. $\begin{bmatrix} 1 & -1 & 1 & | & 1 \\ 1 & 2 & -1 & | & 2 \\ 0 & 1 & -1 & | & 0 \end{bmatrix} R_2 - R_1 \to \begin{bmatrix} 1 & -1 & 1 & | & 1 \\ 0 & 3 & -2 & | & 1 \\ 0 & 1 & -1 & | & 0 \end{bmatrix} (1/3)R_2 \to \begin{bmatrix} 1 & -1 & 1 & | & 1 \\ 0 & 1 & -\frac{2}{3} & | & \frac{1}{3} \\ 0 & 1 & -1 & | & 0 \end{bmatrix}$

$$R_3 - R_2 \to \begin{bmatrix} 1 & -1 & 1 & | & 1 \\ 0 & 1 & -\frac{2}{3} & | & \frac{1}{3} \\ 0 & 0 & -\frac{1}{3} & | & -\frac{1}{3} \end{bmatrix} -3R_3 \to \begin{bmatrix} 1 & -1 & 1 & | & 1 \\ 0 & 1 & -\frac{2}{3} & | & \frac{1}{3} \\ 0 & 0 & 1 & | & 1 \end{bmatrix}$$

The matrix is now in row-echelon form. We see that $z = 1$. Thus, $y - \dfrac{2}{3}z = \dfrac{1}{3} \Rightarrow$

$y - \dfrac{2}{3}(1) = \dfrac{1}{3} \Rightarrow y = 1$ and $x - y + z = 1 \Rightarrow x - 1 + 1 = 1 \Rightarrow x = 1$. The solution is $(1, 1, 1)$.

41. $\begin{bmatrix} 2 & -4 & 2 & | & 11 \\ 1 & 3 & -2 & | & -9 \\ 4 & -2 & 1 & | & 7 \end{bmatrix}$ $(1/2)R_1 \rightarrow$ $\begin{bmatrix} 1 & -2 & 1 & | & \frac{11}{2} \\ 1 & 3 & -2 & | & -9 \\ 4 & -2 & 1 & | & 7 \end{bmatrix}$ $\begin{matrix} \\ R_2 - R_1 \rightarrow \\ R_3 - 4R_1 \rightarrow \end{matrix}$ $\begin{bmatrix} 1 & -2 & 1 & | & \frac{11}{2} \\ 0 & 5 & -3 & | & -\frac{29}{2} \\ 0 & 6 & -3 & | & -15 \end{bmatrix}$

$(1/5)R_2 \rightarrow$ $\begin{bmatrix} 1 & -2 & 1 & | & \frac{11}{2} \\ 0 & 1 & -\frac{3}{5} & | & -\frac{29}{10} \\ 0 & 6 & -3 & | & -15 \end{bmatrix}$ $\begin{matrix} \\ \\ R_3 - 6R_2 \rightarrow \end{matrix}$ $\begin{bmatrix} 1 & -2 & 1 & | & \frac{11}{2} \\ 0 & 1 & -\frac{3}{5} & | & -\frac{29}{10} \\ 0 & 0 & \frac{3}{5} & | & \frac{24}{10} \end{bmatrix}$ $(5/3)R_3 \rightarrow$ $\begin{bmatrix} 1 & -2 & 1 & | & \frac{11}{2} \\ 0 & 1 & -\frac{3}{5} & | & -\frac{29}{10} \\ 0 & 0 & 1 & | & 4 \end{bmatrix}$

The matrix is now in row-echelon form. We see that $z = 4$. Thus, $y - \frac{3}{5}z = -\frac{29}{10} \Rightarrow$

$y - \frac{3}{5}(4) = -\frac{29}{10} \Rightarrow y = -\frac{1}{2}$ and $x - 2y + z = \frac{11}{2} \Rightarrow x - 2\left(-\frac{1}{2}\right) + 4 = \frac{11}{2} \Rightarrow x = \frac{1}{2}$.

The solution is $\left(\frac{1}{2}, -\frac{1}{2}, 4\right)$.

43. $\begin{bmatrix} 3 & -2 & 2 & | & -18 \\ -1 & 2 & -4 & | & 16 \\ 4 & -3 & -2 & | & -21 \end{bmatrix}$ $(1/3)R_1 \rightarrow$ $\begin{bmatrix} 1 & -\frac{2}{3} & \frac{2}{3} & | & -6 \\ -1 & 2 & -4 & | & 16 \\ 4 & -3 & -2 & | & -21 \end{bmatrix}$ $\begin{matrix} \\ R_2 + R_1 \rightarrow \\ R_3 - 4R_1 \rightarrow \end{matrix}$ $\begin{bmatrix} 1 & -\frac{2}{3} & \frac{2}{3} & | & -6 \\ 0 & \frac{4}{3} & -\frac{10}{3} & | & 10 \\ 0 & -\frac{1}{3} & -\frac{14}{3} & | & 3 \end{bmatrix}$

$\begin{matrix} (3/4)R_2 \rightarrow \\ 4R_3 + R_2 \rightarrow \end{matrix}$ $\begin{bmatrix} 1 & -\frac{2}{3} & \frac{2}{3} & | & -6 \\ 0 & 1 & -\frac{5}{2} & | & \frac{15}{2} \\ 0 & 0 & -22 & | & 22 \end{bmatrix}$ $(-1/22)R_3 \rightarrow$ $\begin{bmatrix} 1 & -\frac{2}{3} & \frac{2}{3} & | & -6 \\ 0 & 1 & -\frac{5}{2} & | & \frac{15}{2} \\ 0 & 0 & 1 & | & -1 \end{bmatrix}$

The matrix is now in row-echelon form. We see that $z = -1$. Thus, $y - \frac{5}{2}z = \frac{15}{2} \Rightarrow$

$y - \frac{5}{2}(-1) = \frac{15}{2} \Rightarrow y = 5$ and $x - \frac{2}{3}y + \frac{2}{3}z = -6 \Rightarrow x - \frac{2}{3}(5) + \frac{2}{3}(-1) = -6 \Rightarrow x = -2$.

The solution is $(-2, 5, -1)$.

45. $\begin{bmatrix} 1 & -4 & 3 & | & 26 \\ -1 & 3 & -2 & | & -19 \\ 0 & -1 & 1 & | & 10 \end{bmatrix}$ $R_2 + R_1 \rightarrow$ $\begin{bmatrix} 1 & -4 & 3 & | & 26 \\ 0 & -1 & 1 & | & 7 \\ 0 & -1 & 1 & | & 10 \end{bmatrix}$ $\begin{matrix} -R_2 \rightarrow \\ R_3 - R_2 \rightarrow \end{matrix}$ $\begin{bmatrix} 1 & -4 & 3 & | & 26 \\ 0 & 1 & -1 & | & -7 \\ 0 & 0 & 0 & | & 3 \end{bmatrix}$

The last equation indicates that $0 = 3$, which is always false. Therefore, the system has no solutions.

47. $\begin{bmatrix} 5 & 0 & 4 & | & 7 \\ 2 & -4 & 0 & | & 6 \\ 0 & 3 & 3 & | & 3 \end{bmatrix}$ $\begin{matrix} (1/5)R_1 \rightarrow \\ 2R_1 - 5R_2 \rightarrow \\ (1/3)R_3 \rightarrow \end{matrix}$ $\begin{bmatrix} 1 & 0 & \frac{4}{5} & | & \frac{7}{5} \\ 0 & 20 & 8 & | & -16 \\ 0 & 1 & 1 & | & 1 \end{bmatrix}$ $\begin{matrix} \\ (1/20)R_2 \rightarrow \\ R_2 - 20R_3 \rightarrow \end{matrix}$ $\begin{bmatrix} 1 & 0 & \frac{4}{5} & | & \frac{7}{5} \\ 0 & 1 & \frac{2}{5} & | & -\frac{4}{5} \\ 0 & 0 & -12 & | & -36 \end{bmatrix}$

$(-1/12)R_3 \rightarrow$ $\begin{bmatrix} 1 & 0 & \frac{4}{5} & | & \frac{7}{5} \\ 0 & 1 & \frac{2}{5} & | & -\frac{4}{5} \\ 0 & 0 & 1 & | & 3 \end{bmatrix}$ $\begin{matrix} (-4/5)R_3 + R_1 \rightarrow \\ (-2/5)R_3 + R_2 \rightarrow \end{matrix}$ $\begin{bmatrix} 1 & 0 & 0 & | & -1 \\ 0 & 1 & 0 & | & -2 \\ 0 & 0 & 1 & | & 3 \end{bmatrix}$ The solution is $(-1, -2, 3)$.

49. $\begin{bmatrix} 5 & -2 & 1 & | & 5 \\ 1 & 1 & -2 & | & -2 \\ 4 & -3 & 3 & | & 7 \end{bmatrix}$ $\begin{matrix} R_2 \rightarrow \\ 5R_2 - R_1 \rightarrow \\ 4R_2 - R_3 \rightarrow \end{matrix}$ $\begin{bmatrix} 1 & 1 & -2 & | & -2 \\ 0 & 7 & -11 & | & -15 \\ 0 & 7 & -11 & | & -15 \end{bmatrix}$ $\begin{matrix} (1/7)R_2 \rightarrow \\ \\ R_2 - R_3 \rightarrow \end{matrix}$ $\begin{bmatrix} 1 & 1 & -2 & | & -2 \\ 0 & 1 & -\frac{11}{7} & | & -\frac{15}{7} \\ 0 & 0 & 0 & | & 0 \end{bmatrix}$

$(-1)R_2 + R_1 \rightarrow \begin{bmatrix} 1 & 0 & -\frac{3}{7} & | & \frac{1}{7} \\ 0 & 1 & -\frac{11}{7} & | & -\frac{15}{7} \\ 0 & 0 & 0 & | & 0 \end{bmatrix}$

The last equation indicates that $0 = 0$, which is true. Therefore, there is an infinite number of solutions.

The second equation gives: $y - \dfrac{11}{7}z = -\dfrac{15}{7} \Rightarrow y = \dfrac{11z - 15}{7}$. The first equation gives: $x - \dfrac{3}{7}z = \dfrac{1}{7} \Rightarrow$

$x = \dfrac{3z + 1}{7}$. Therefore, there are infinitely many solutions which can be written as: $\left(\dfrac{3z + 1}{7}, \dfrac{11z - 15}{7}, z \right)$.

51. The equations are: $x = 12$ and $y = 3 \Rightarrow (12, 3)$.

53. The equations are: $x = -2$, $y = 4$, and $z = \dfrac{1}{2} \Rightarrow \left(-2, 4, \dfrac{1}{2} \right)$.

55. The last equation indicates that $0 = 0$, which is true. Therefore, there is an infinite number of solutions.

The first equation gives: $x + 2z = 4 \Rightarrow x = -2z + 4$. The second equation gives: $y - z = -3 \Rightarrow$

$y = z - 3$. The system has infinitely many solutions which can be written as: $(-2z + 4, z - 3, z)$.

57. The last equation indicates that $0 = \dfrac{2}{3}$, which is always false. Therefore, there are no solutions.

59. $\begin{bmatrix} 1 & -1 & | & 1 \\ 1 & 1 & | & 5 \end{bmatrix}$ $\begin{matrix} \\ R_2 - R_1 \rightarrow \end{matrix}$ $\begin{bmatrix} 1 & -1 & | & 1 \\ 0 & 2 & | & 4 \end{bmatrix}$ $(1/2)R_2 \rightarrow \begin{bmatrix} 1 & -1 & | & 1 \\ 0 & 1 & | & 2 \end{bmatrix}$ $\begin{matrix} R_1 + R_2 \rightarrow \\ \end{matrix}$ $\begin{bmatrix} 1 & 0 & | & 3 \\ 0 & 1 & | & 2 \end{bmatrix}$

The solution is $(3, 2)$

61. $\begin{bmatrix} 1 & 2 & 1 & | & 3 \\ 0 & 1 & -1 & | & -2 \\ -1 & -2 & 2 & | & 6 \end{bmatrix}$ $\begin{matrix} \\ \\ R_1 + R_3 \rightarrow \end{matrix}$ $\begin{bmatrix} 1 & 2 & 1 & | & 3 \\ 0 & 1 & -1 & | & -2 \\ 0 & 0 & 3 & | & 9 \end{bmatrix}$ $\begin{matrix} R_1 - 2R_2 \rightarrow \\ \\ (1/3)R_3 \rightarrow \end{matrix}$ $\begin{bmatrix} 1 & 0 & 3 & | & 7 \\ 0 & 1 & -1 & | & -2 \\ 0 & 0 & 1 & | & 3 \end{bmatrix}$

$\begin{matrix} R_1 - 3R_3 \rightarrow \\ R_2 + R_3 \rightarrow \\ \end{matrix}$ $\begin{bmatrix} 1 & 0 & 0 & | & -2 \\ 0 & 1 & 0 & | & 1 \\ 0 & 0 & 1 & | & 3 \end{bmatrix}$ The solution is $(-2, 1, 3)$.

63. $\begin{bmatrix} 1 & -1 & 2 & | & 7 \\ 2 & 1 & -4 & | & -27 \\ -1 & 1 & -1 & | & 0 \end{bmatrix}$ $\begin{matrix} \\ R_2 - 2R_1 \rightarrow \\ R_1 + R_3 \rightarrow \end{matrix}$ $\begin{bmatrix} 1 & -1 & 2 & | & 7 \\ 0 & 3 & -8 & | & -41 \\ 0 & 0 & 1 & | & 7 \end{bmatrix}$ $(1/3)R_2 \rightarrow \begin{bmatrix} 1 & -1 & 2 & | & 7 \\ 0 & 1 & -\frac{8}{3} & | & -\frac{41}{3} \\ 0 & 0 & 1 & | & 7 \end{bmatrix}$

$R_2 + R_1 \rightarrow \begin{bmatrix} 1 & 0 & -\frac{2}{3} & | & -\frac{20}{3} \\ 0 & 1 & -\frac{8}{3} & | & -\frac{41}{3} \\ 0 & 0 & 1 & | & 7 \end{bmatrix}$ $\begin{matrix} (2/3)R_3 + R_1 \rightarrow \\ (8/3)R_3 + R_2 \rightarrow \\ \end{matrix}$ $\begin{bmatrix} 1 & 0 & 0 & | & -2 \\ 0 & 1 & 0 & | & 5 \\ 0 & 0 & 1 & | & 7 \end{bmatrix}$ The solution is $(-2, 5, 7)$.

65. $\begin{bmatrix} 2 & 1 & -1 & | & 2 \\ 1 & -2 & 1 & | & 0 \\ 1 & 3 & -2 & | & 4 \end{bmatrix}$ $\begin{matrix} R_2 \rightarrow \\ 2R_2 - R_1 \rightarrow \\ R_2 - R_3 \rightarrow \end{matrix}$ $\begin{bmatrix} 1 & -2 & 1 & | & 0 \\ 0 & -5 & 3 & | & -2 \\ 0 & -5 & 3 & | & -4 \end{bmatrix}$ $\begin{matrix} \\ (-1/5)R_2 \rightarrow \\ R_2 - R_3 \rightarrow \end{matrix}$ $\begin{bmatrix} 1 & -2 & 1 & | & 0 \\ 0 & 1 & -\frac{3}{5} & | & \frac{2}{5} \\ 0 & 0 & 0 & | & 2 \end{bmatrix}$

The last equation indicates that $0 = 2$, which is always false. Therefore, there are no solutions.

67. Enter the coefficients of the linear system into a 3 × 4 matrix, as shown in Figure 67a. Reduce the matrix to reduced row-echelon form, as shown in Figure 67b. The solution is the ordered triple (−9.266, −9.167, 2.440).

| Figure 67a | Figure 67b | Figure 69a | Figure 69b |

69. Enter the coefficients of the linear system into a 3 × 4 matrix, as shown in Figure 69a. Reduce the matrix to reduced row-echelon form, as shown in Figure 69b. The solution is the ordered triple (5.211, 3.739, −4.655).

71. Enter the coefficients of the linear system into a 3 × 4 matrix, as shown in Figure 71a. Reduce the matrix to reduced row-echelon form, as shown in Figure 71b. The solution is the ordered triple (7.993, 1.609, −0.401).

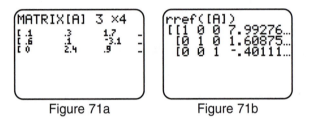

| Figure 71a | Figure 71b |

73. (a) The constants a, b, and c should satisfy the following three equations:

$1300 = a(1800) + b(5000) + c, 5300 = a(3200) + b(12,000) + c,$

and $6500 = a(4500) + b(13,000) + c$

This can be written:

$1800a + 5000b + c = 1300, 3200a + 12,000b + c = 5300,$ and $4500a + 13,000b + c = 6500$

The associated augmented matrix is: $\begin{bmatrix} 1800 & 5000 & 1 & 1300 \\ 3200 & 12,000 & 1 & 5300 \\ 4500 & 13,000 & 1 & 6500 \end{bmatrix}$

Using technology, the solution is $a \approx 0.5714$, $b \approx 0.4571$, and $c \approx -2014$. See Figure 73. Thus, the equation modeling the data can be expressed (approximately) as $F = 0.5714N + 0.4571R - 2014$.

(b) To predict the food costs for a shelter that serves 3500 people and receives charitable receipts of $12,500, let $N = 3500$ and $R = 12,500$ and evaluate the equation.

$F = 0.5714(3500) + 0.4571(12,500) - 2014 = 5699.65$. This model predicts monthly food costs of approximately $5700.

Figure 73

75. Let x represent the fraction of the pool that the first pump can empty each hour, and y the fraction of the pool that the second pump can empty each hour, and z the fraction for the third pump. Since the first pump is twice as fast we have $x = 2y$ or $x - 2y = 0$. Since the first two pumps can empty the pool in 8 hours, it follows that they can empty $\frac{1}{8}$ of the pool in an hour. Thus, $x + y = \frac{1}{8}$. Similarly, all three pumps can empty the pool in 6 hours, so $x + y + z = \frac{1}{6}$. Putting these equations in an augmented matrix results in the following:

$$\begin{bmatrix} 1 & -2 & 0 & | & 0 \\ 1 & 1 & 0 & | & \frac{1}{8} \\ 1 & 1 & 1 & | & \frac{1}{6} \end{bmatrix} \begin{bmatrix} 1 & -2 & 0 & | & 0 \\ 0 & 3 & 0 & | & \frac{1}{8} \\ 0 & 3 & 1 & | & \frac{1}{6} \end{bmatrix} \begin{bmatrix} 1 & -2 & 0 & | & 0 \\ 0 & 1 & 0 & | & \frac{1}{24} \\ 0 & 0 & 1 & | & \frac{1}{24} \end{bmatrix} \begin{bmatrix} 1 & 0 & 0 & | & \frac{1}{12} \\ 0 & 1 & 0 & | & \frac{1}{24} \\ 0 & 0 & 1 & | & \frac{1}{24} \end{bmatrix}$$

Thus, $x = \frac{1}{12}, y = z = \frac{1}{24}$. The first pump could empty $\frac{1}{12}$ of the pool in one hour or the entire pool in 12 hours, while the second and third pumps individually could empty the pool in 24 hours. Using technology, this solution can be obtained as shown in Figure 75.

Figure 75   Figure 79

77. $I_1 = I_2 + I_3 \Rightarrow I_1 - I_2 - I_3 = 0, 15 + 4I_3 = 14I_2 \Rightarrow -14I_2 + 4I_3 = -15$, and

$10 + 4I_3 = 5I_1 \Rightarrow -5I_1 + 4I_3 = -10$

Therefore the matrix is:

$$\begin{bmatrix} 1 & -1 & -1 & | & 0 \\ 0 & -14 & 4 & | & -15 \\ -5 & 0 & 4 & | & -10 \end{bmatrix} \begin{matrix} \\ (-1/14)R_2 \rightarrow \\ 5R_1 + R_3 \rightarrow \end{matrix} \begin{bmatrix} 1 & -1 & -1 & | & 0 \\ 0 & 1 & -\frac{2}{7} & | & \frac{15}{14} \\ 0 & -5 & -1 & | & -10 \end{bmatrix} 5R_2 + R_3 \rightarrow \begin{bmatrix} 1 & -1 & -1 & | & 0 \\ 0 & 1 & -\frac{2}{7} & | & \frac{15}{14} \\ 0 & 0 & -\frac{17}{7} & | & -\frac{65}{14} \end{bmatrix}$$

$$(-7/17)R_3 \rightarrow \begin{bmatrix} 1 & -1 & -1 & | & 0 \\ 0 & 1 & -\frac{2}{7} & | & \frac{15}{14} \\ 0 & 0 & 1 & | & \frac{65}{34} \end{bmatrix} \begin{matrix} R_1 + R_3 \rightarrow \\ (2/7)R_3 + R_2 \rightarrow \end{matrix} \begin{bmatrix} 1 & -1 & 0 & | & \frac{65}{34} \\ 0 & 1 & 0 & | & \frac{385}{238} \\ 0 & 0 & 1 & | & \frac{65}{34} \end{bmatrix} R_2 + R_1 \rightarrow \begin{bmatrix} 1 & 0 & 0 & | & \frac{840}{238} \\ 0 & 1 & 0 & | & \frac{385}{238} \\ 0 & 0 & 1 & | & \frac{65}{34} \end{bmatrix}$$

The solution is: $(3.53, 1.62, 1.91)$.

79. Let x equal the amount invested at 8%, y equal the amount invested at 11% and z equal the amount invested at 14%.

(a) $x + y + z = 5000, x + y - z = 0$, and $0.08x + 0.11y + 0.14z = 595$

(b) Enter the coefficients of the following 3×4 augmented matrix into a calculator.

$$\begin{bmatrix} 1 & 1 & 1 & | & 5000 \\ 1 & 1 & -1 & | & 0 \\ 0.08 & 0.11 & 0.14 & | & 595 \end{bmatrix}$$

Using technology, the solution is $x = 1000, y = 1500$, and $z = 2500$. See Figure 79. Thus, $1000 needs to be invested at 8%, $1500 at 11%, and $2500 at 14% in order to earn the total amount interest of $595.

81. (a) At intersection A incoming traffic is equal to $x + 5$. The outgoing traffic is given by $y + 7$. Therefore,

$x + 5 = y + 7$, which is the first equation. The incoming traffic at intersection B is $z + 6$ and the

outgoing traffic is $x + 3$, so $z + 6 = x + 3$. Finally at intersection C, the incoming flow is $y + 3$ and the

outgoing flow is $z + 4$, so $y + 3 = z + 4$.

(b) These three equations can be written as: $x - y = 2$, $x - z = 3$, and $y - z = 1$.

The system of linear equations can be represented by the following augmented matrix:

$$\left[\begin{array}{ccc|c} 1 & -1 & 0 & 2 \\ 1 & 0 & -1 & 3 \\ 0 & 1 & -1 & 1 \end{array}\right]$$

Begin by subtracting the first row from the second, followed by subtracting the second row from the third.

Gaussian elimination results in the following augmented matrix:

$$\left[\begin{array}{ccc|c} 1 & -1 & 0 & 2 \\ 0 & 1 & -1 & 1 \\ 0 & 0 & 0 & 0 \end{array}\right]$$

The last row of zeros indicates that the linear system is dependent and has an infinite number of solutions.

Back solving produces $y - z = 1 \Rightarrow y = z + 1$. Substituting into the first equation gives

$x - (z + 1) = 2 \Rightarrow x = z + 3$. Thus, the solution can be written

$\{(z + 3, z + 1, z) \mid z \text{ is any nonnegative real number}\}$.

(c) There are an infinite number of solutions to the system. However, solutions such as $z = 1000$, $x = 1003$,

and $y = 1001$ are likely, unless a large number of people are simply driving around the block. In reality there is

an average traffic flow rate for z that could be measured. From this, values for both x and y could be determined.

83. (a) The three equations can be written as: $3 = 0^2 + 0 + c$, $55 = 2^2a + 2b + c$, and

$150 = 4^2a + 4b + c \Rightarrow 16a + 4b + c = 150$, $4a + 2b + c = 55$, and $c = 3$.

Therefore, the matrix is:

$$\left[\begin{array}{ccc|c} 16 & 4 & 1 & 150 \\ 4 & 2 & 1 & 55 \\ 0 & 0 & 1 & 3 \end{array}\right]$$

(b) Using technology, $a = 5.375$, $b = 15.25$, and $c = 3$. $f(x) = 5.375x^2 + 15.25x + 3$.

(c) See Figure 83.

(d) For example, in 2012 the predicted sales is $f(8) = 5.375(8)^2 + 15.25(8) + 3 = 469$. *Answers may vary.*

[−0.5, 5, 1] by [−10, 175, 25]

Figure 83

85. (a) Let $f(x) = ax^2 + bx + c$. The constants a, b, and c must satisfy the following equations:

$f(1990) = a(1990)^2 + b(1990) + c = 11$, $f(2010) = a(2010)^2 + b(2010) + c = 10$, and

$f(2030) = a(2030)^2 + b(2030) + c = 6$. This system of equations can be represented by the following

augmented matrix:

$$\begin{bmatrix} 1990^2 & 1990 & 1 & 11 \\ 2010^2 & 2010 & 1 & 10 \\ 2030^2 & 2030 & 1 & 6 \end{bmatrix}$$

(b) Using technology to solve the system results in the reduced row-echelon form shown in Figure 85b. The

solution is $a = -0.00375$, $b = 14.95$, and $c = -14{,}889.125$. Therefore, the symbolic representation of

f is $f(x) = -0.00375x^2 + 14.95x - 14{,}889.125$.

(c) The data and f are graphed in Figure 85c. Notice that the graph of f passes through each data point.

(d) *Answers may vary.* For example, in 2015 the ratio could be $f(2015) \approx 9.3$.

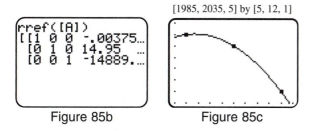

[1985, 2035, 5] by [5, 12, 1]

Figure 85b   Figure 85c

9.5 Properties and Applications of Matrices

1. (a) $a_{12} = 2$ and $a_{21} = 4$. The element a_{32} is undefined since there is no row three.

    (b) $a_{11}a_{22} + 3a_{23} = (1)(5) + 3(6) = 23$

3. (a) $a_{12} = -1$, $a_{21} = 3$ and $a_{32} = 0$

    (b) $a_{11}a_{22} + 3a_{23} = (1)(-2) + 3(5) = 13$

5. $x = 1$ and $y = 1$

7. The matrices are not the same size, therefore not possible.

9. (a) $a_{12} = 3$, $b_{32} = 1$, $b_{22} = 0$

    (b) $a_{11} = 1$, $b_{11} = 1 \Rightarrow a_{11}b_{11} = 1$; $a_{12} = 3$, $b_{21} = 3 \Rightarrow a_{12}b_{21} = 9$;

    $a_{13} = -4$, $b_{31} = 3 \Rightarrow a_{13}b_{31} = -12 \Rightarrow a_{11}b_{11} + a_{12}b_{21} + a_{13}b_{31} = -2$

    (c) If $A = B$, then all elements of A must be equal to all elements of B. Then $x = 3$.

11. (a) $\begin{bmatrix} 4 & -1 \\ -1 & 4 \end{bmatrix} + \begin{bmatrix} -1 & 4 \\ 4 & -1 \end{bmatrix} = \begin{bmatrix} 3 & 3 \\ 3 & 3 \end{bmatrix}$

    (b) $\begin{bmatrix} -1 & 4 \\ 4 & -1 \end{bmatrix} + \begin{bmatrix} 4 & -1 \\ -1 & 4 \end{bmatrix} = \begin{bmatrix} 3 & 3 \\ 3 & 3 \end{bmatrix}$

    (c) $\begin{bmatrix} 4 & -1 \\ -1 & 4 \end{bmatrix} - \begin{bmatrix} -1 & 4 \\ 4 & -1 \end{bmatrix} = \begin{bmatrix} 5 & -5 \\ -5 & 5 \end{bmatrix}$

13. (a) $\begin{bmatrix} 3 & 4 & -1 \\ 0 & -3 & 2 \\ -2 & 5 & 10 \end{bmatrix} + \begin{bmatrix} 11 & 5 & -2 \\ 4 & -7 & 12 \\ 6 & 6 & 6 \end{bmatrix} = \begin{bmatrix} 14 & 9 & -3 \\ 4 & -10 & 14 \\ 4 & 11 & 16 \end{bmatrix}$

(b) $\begin{bmatrix} 11 & 5 & -2 \\ 4 & -7 & 12 \\ 6 & 6 & 6 \end{bmatrix} + \begin{bmatrix} 3 & 4 & -1 \\ 0 & -3 & 2 \\ -2 & 5 & 10 \end{bmatrix} = \begin{bmatrix} 14 & 9 & -3 \\ 4 & -10 & 14 \\ 4 & 11 & 16 \end{bmatrix}$

(c) $\begin{bmatrix} 3 & 4 & -1 \\ 0 & -3 & 2 \\ -2 & 5 & 10 \end{bmatrix} - \begin{bmatrix} 11 & 5 & -2 \\ 4 & -7 & 12 \\ 6 & 6 & 6 \end{bmatrix} = \begin{bmatrix} -8 & -1 & 1 \\ -4 & 4 & -10 \\ -8 & -1 & 4 \end{bmatrix}$

15. (a) $A + B = \begin{bmatrix} 2 & -6 \\ 3 & 1 \end{bmatrix} + \begin{bmatrix} -1 & 0 \\ -2 & 3 \end{bmatrix} = \begin{bmatrix} 1 & -6 \\ 1 & 4 \end{bmatrix}$

(b) $3A = 3\begin{bmatrix} 2 & -6 \\ 3 & 1 \end{bmatrix} = \begin{bmatrix} 6 & -18 \\ 9 & 3 \end{bmatrix}$

(c) $2A - 3B = 2\begin{bmatrix} 2 & -6 \\ 3 & 1 \end{bmatrix} - 3\begin{bmatrix} -1 & 0 \\ -2 & 3 \end{bmatrix} = \begin{bmatrix} 7 & -12 \\ 12 & -7 \end{bmatrix}$

17. (a) $A + B$ is undefined since A is 3×3 and B is 2×3. They do not have the same dimension.

(b) $3A = 3\begin{bmatrix} 1 & -1 & 0 \\ 1 & 5 & 9 \\ -4 & 8 & -5 \end{bmatrix} = \begin{bmatrix} 3 & -3 & 0 \\ 3 & 15 & 27 \\ -12 & 24 & -15 \end{bmatrix}$

(c) $2A - 3B$ is undefined since A is 3×3 and B is 2×3. They do not have the same dimension.

19. (a) $A + B = \begin{bmatrix} -2 & -1 \\ -5 & 1 \\ 2 & -3 \end{bmatrix} + \begin{bmatrix} 2 & -1 \\ 3 & 1 \\ 7 & -5 \end{bmatrix} = \begin{bmatrix} 0 & -2 \\ -2 & 2 \\ 9 & -8 \end{bmatrix}$

(b) $3A = 3\begin{bmatrix} -2 & -1 \\ -5 & 1 \\ 2 & -3 \end{bmatrix} = \begin{bmatrix} -6 & -3 \\ -15 & 3 \\ 6 & -9 \end{bmatrix}$

(c) $2A - 3B = 2\begin{bmatrix} -2 & -1 \\ -5 & 1 \\ 2 & -3 \end{bmatrix} - 3\begin{bmatrix} 2 & -1 \\ 3 & 1 \\ 7 & -5 \end{bmatrix} = \begin{bmatrix} -10 & 1 \\ -19 & -1 \\ -17 & 9 \end{bmatrix}$

21. $2\begin{bmatrix} 2 & -1 \\ 5 & 1 \\ 0 & 3 \end{bmatrix} + \begin{bmatrix} 5 & 0 \\ 7 & -3 \\ 1 & 1 \end{bmatrix} - \begin{bmatrix} 9 & -4 \\ 4 & 4 \\ 1 & 6 \end{bmatrix} = \begin{bmatrix} 0 & 2 \\ 13 & -5 \\ 0 & 1 \end{bmatrix}$

23. $\begin{bmatrix} 4 & 6 \\ 3 & -7 \end{bmatrix} - 2\begin{bmatrix} 1 & 0 \\ -4 & 1 \end{bmatrix} = \begin{bmatrix} 2 & 6 \\ 11 & -9 \end{bmatrix}$

25. $2\begin{bmatrix} 2 & -1 & -1 \\ -1 & 2 & -1 \\ -1 & -1 & 2 \end{bmatrix} + 3\begin{bmatrix} 1 & 2 & 3 \\ 2 & 1 & 3 \\ 2 & 3 & 1 \end{bmatrix} = \begin{bmatrix} 7 & 4 & 7 \\ 4 & 7 & 7 \\ 4 & 7 & 7 \end{bmatrix}$

27. The "1" is dark gray and the background is light gray.

$$A = \begin{bmatrix} 1 & 2 & 1 \\ 1 & 2 & 1 \\ 1 & 2 & 1 \end{bmatrix}$$

29. To enhance the contrast, change light gray to white and change dark gray to black. This could be accomplished by adding the 3×3 matrix B to A.

$$B = \begin{bmatrix} -1 & 1 & -1 \\ -1 & 1 & -1 \\ -1 & 1 & -1 \end{bmatrix}; \; A + B = \begin{bmatrix} 1 & 2 & 1 \\ 1 & 2 & 1 \\ 1 & 2 & 1 \end{bmatrix} + \begin{bmatrix} -1 & 1 & -1 \\ -1 & 1 & -1 \\ -1 & 1 & -1 \end{bmatrix} = \begin{bmatrix} 0 & 3 & 0 \\ 0 & 3 & 0 \\ 0 & 3 & 0 \end{bmatrix}$$

31. A and B are both 2×2 so AB and BA are also both 2×2.

$$AB = \begin{bmatrix} 1 & -1 \\ 2 & 0 \end{bmatrix}\begin{bmatrix} -2 & 3 \\ 1 & 2 \end{bmatrix} = \begin{bmatrix} -3 & 1 \\ -4 & 6 \end{bmatrix}; \; BA = \begin{bmatrix} -2 & 3 \\ 1 & 2 \end{bmatrix}\begin{bmatrix} 1 & -1 \\ 2 & 0 \end{bmatrix} = \begin{bmatrix} 4 & 2 \\ 5 & -1 \end{bmatrix}$$

33. Since both A and B are 2×3, the number of rows in B is not equal to the number of columns in A, so AB is undefined. Also, the number of rows in A is not equal to the number of columns in B so BA is undefined.

35. $AB = \begin{bmatrix} 3 & -1 \\ 1 & 0 \\ -2 & -4 \end{bmatrix}\begin{bmatrix} -2 & 5 & -3 \\ 9 & -7 & 0 \end{bmatrix} =$

$$\begin{bmatrix} 3(-2) + (-1)(9) & 3(5) + (-1)(-7) & 3(-3) + (-1)(0) \\ 1(-2) + 0(9) & 1(5) + 0(-7) & 1(-3) + 0(0) \\ -2(-2) + (-4)(9) & -2(5) + (-4)(-7) & -2(-3) + (-4)(0) \end{bmatrix} \Rightarrow AB = \begin{bmatrix} -15 & 22 & -9 \\ -2 & 5 & -3 \\ -32 & 18 & 6 \end{bmatrix}$$

$$BA = \begin{bmatrix} -2 & 5 & -3 \\ 9 & -7 & 0 \end{bmatrix}\begin{bmatrix} 3 & -1 \\ 1 & 0 \\ -2 & -4 \end{bmatrix} =$$

$$\begin{bmatrix} -2(3) + 5(1) + (-3)(-2) & -2(-1) + (5)(0) + (-3)(-4) \\ 9(3) + (-7)(1) + 0(-2) & 9(-1) + (-7)(0) + 0(-4) \end{bmatrix} \Rightarrow BA = \begin{bmatrix} 5 & 14 \\ 20 & -9 \end{bmatrix}$$

37. AB is undefined, we cannot multiply a 3×3 by a 2×3.

$$BA = \begin{bmatrix} -1 & 3 & -1 \\ 7 & -7 & 1 \end{bmatrix}\begin{bmatrix} 1 & -1 & 0 \\ 2 & -1 & 5 \\ 6 & 1 & -4 \end{bmatrix} =$$

$$\begin{bmatrix} -1(1) + 3(2) + (-1)(6) & -1(-1) + 3(-1) + (-1)(1) & -1(0) + 3(5) + (-1)(-4) \\ 7(1) + (-7)(2) + 1(6) & 7(-1) + (-7)(-1) + 1(1) & 7(0) + (-7)(5) + 1(-4) \end{bmatrix} \Rightarrow$$

$$BA = \begin{bmatrix} -1 & -3 & 19 \\ -1 & 1 & -39 \end{bmatrix}$$

39. Since A is 2×2 and B is 3×1, the number of rows in B is not equal to the number of columns in A, so AB is undefined. Also, the number of rows in A is not equal to the number of columns in B so BA is undefined.

41. A and B are both 3×3 so AB and BA are also both 3×3.

$$AB = \begin{bmatrix} 2 & -1 & 3 \\ 0 & 1 & 0 \\ 2 & -2 & 3 \end{bmatrix} \begin{bmatrix} 1 & 5 & -1 \\ 0 & 1 & 3 \\ -1 & 2 & 1 \end{bmatrix} = \begin{bmatrix} -1 & 15 & -2 \\ 0 & 1 & 3 \\ -1 & 14 & -5 \end{bmatrix};$$

$$BA = \begin{bmatrix} 1 & 5 & -1 \\ 0 & 1 & 3 \\ -1 & 2 & 1 \end{bmatrix} \begin{bmatrix} 2 & -1 & 3 \\ 0 & 1 & 0 \\ 2 & -2 & 3 \end{bmatrix} = \begin{bmatrix} 0 & 6 & 0 \\ 6 & -5 & 9 \\ 0 & 1 & 0 \end{bmatrix}$$

43. A is 2×2 and B is 2×1 so AB is 2×1. However BA is undefined since the number of rows in A is not equal to the number of columns in B.

$$AB = \begin{bmatrix} 2 & -1 \\ 3 & 1 \end{bmatrix} \begin{bmatrix} 1 \\ 3 \end{bmatrix} = \begin{bmatrix} -1 \\ 6 \end{bmatrix}$$

45. A is 2×2 and B is 2×3 so AB is 2×3. However BA is undefined since the number of rows in A is not equal to the number of columns in B.

$$AB = \begin{bmatrix} -3 & 1 \\ 2 & -4 \end{bmatrix} \begin{bmatrix} 1 & 0 & -2 \\ -4 & 8 & 1 \end{bmatrix} = \begin{bmatrix} -7 & 8 & 7 \\ 18 & -32 & -8 \end{bmatrix}$$

47. A is 3×3 and B is 3×1 so AB is 3×1. However BA is undefined since the number of rows in A is not equal to the number of columns in B.

$$AB = \begin{bmatrix} 1 & 0 & -2 \\ 3 & -4 & 1 \\ 2 & 0 & 5 \end{bmatrix} \begin{bmatrix} 1 \\ -1 \\ 3 \end{bmatrix} = \begin{bmatrix} -5 \\ 10 \\ 17 \end{bmatrix}$$

49. Using technology; $BA = \begin{bmatrix} 3 & -2 & 4 \\ 5 & 2 & 3 \\ 7 & 5 & 4 \end{bmatrix} \begin{bmatrix} 1 & 1 & -5 \\ -1 & 0 & -7 \\ -6 & 4 & 3 \end{bmatrix} = \begin{bmatrix} -19 & 19 & 11 \\ 21 & -7 & -48 \\ -22 & 23 & -58 \end{bmatrix}$

51. Using technology; $3A^2 + 2B = 3 \begin{bmatrix} 3 & -2 & 4 \\ 5 & 2 & 3 \\ 7 & 5 & 4 \end{bmatrix}^2 + 2 \begin{bmatrix} 1 & 1 & -5 \\ -1 & 0 & -7 \\ -6 & 4 & 3 \end{bmatrix} = \begin{bmatrix} 83 & 32 & 92 \\ 10 & -63 & -8 \\ 210 & 56 & 93 \end{bmatrix}$

53. (a) $B + C = \begin{bmatrix} 6 & 2 & 7 \\ 3 & -4 & -5 \\ 7 & 1 & 0 \end{bmatrix} + \begin{bmatrix} 1 & 4 & -3 \\ 8 & 1 & -1 \\ 4 & 6 & -2 \end{bmatrix} = \begin{bmatrix} 7 & 6 & 4 \\ 11 & -3 & -6 \\ 11 & 7 & -2 \end{bmatrix}$

$$A(B + C) = \begin{bmatrix} 2 & -1 & 3 \\ 1 & 3 & -5 \\ 0 & -2 & 1 \end{bmatrix} \begin{bmatrix} 7 & 6 & 4 \\ 11 & -3 & -6 \\ 11 & 7 & -2 \end{bmatrix} = \begin{bmatrix} 36 & 36 & 8 \\ -15 & -38 & -4 \\ -11 & 13 & 10 \end{bmatrix}$$

(b) $AB = \begin{bmatrix} 2 & -1 & 3 \\ 1 & 3 & -5 \\ 0 & -2 & 1 \end{bmatrix} \begin{bmatrix} 6 & 2 & 7 \\ 3 & -4 & -5 \\ 7 & 1 & 0 \end{bmatrix} = \begin{bmatrix} 30 & 11 & 19 \\ -20 & -15 & -8 \\ 1 & 9 & 10 \end{bmatrix}$

$$AC = \begin{bmatrix} 2 & -1 & 3 \\ 1 & 3 & -5 \\ 0 & -2 & 1 \end{bmatrix} \begin{bmatrix} 1 & 4 & -3 \\ 8 & 1 & -1 \\ 4 & 6 & -2 \end{bmatrix} = \begin{bmatrix} 6 & 25 & -11 \\ 5 & -23 & 4 \\ -12 & 4 & 0 \end{bmatrix}; \quad AB + AC = \begin{bmatrix} 36 & 36 & 8 \\ -15 & -38 & -4 \\ -11 & 13 & 10 \end{bmatrix}$$

$A(B + C) = AB + AC$, which indicates that the distributive property holds for matrices.

55. (a) $(A - B)^2 = \begin{bmatrix} -4 & -3 & -4 \\ -2 & 7 & 0 \\ -7 & -3 & 1 \end{bmatrix} \begin{bmatrix} -4 & -3 & -4 \\ -2 & 7 & 0 \\ -7 & -3 & 1 \end{bmatrix} = \begin{bmatrix} 50 & 3 & 12 \\ -6 & 55 & 8 \\ 27 & -3 & 29 \end{bmatrix}$

(b) $A^2 - AB - BA + B^2 =$

$\begin{bmatrix} 3 & -11 & 14 \\ 5 & 18 & -17 \\ -2 & -8 & 11 \end{bmatrix} - \begin{bmatrix} 30 & 11 & 19 \\ -20 & -15 & -8 \\ 1 & 9 & 10 \end{bmatrix} - \begin{bmatrix} 14 & -14 & 15 \\ 2 & -5 & 24 \\ 15 & -4 & 16 \end{bmatrix} + \begin{bmatrix} 91 & 11 & 32 \\ -29 & 17 & 41 \\ 45 & 10 & 44 \end{bmatrix} =$

$\begin{bmatrix} 50 & 3 & 12 \\ -6 & 55 & 8 \\ 27 & -3 & 29 \end{bmatrix}$

$(A - B)^2 = A^2 - AB - BA + B^2$, which indicates that matrices seem to conform to common rules of algebra except for the commutative property since $AB \neq BA$, in general.

57. To make a negative image, subtract the matrix A from a completely black image matrix B.

$B = \begin{bmatrix} 3 & 3 & 3 \\ 3 & 3 & 3 \\ 3 & 3 & 3 \end{bmatrix}$; $B - A = \begin{bmatrix} 3 & 3 & 3 \\ 3 & 3 & 3 \\ 3 & 3 & 3 \end{bmatrix} - \begin{bmatrix} 0 & 3 & 0 \\ 0 & 3 & 0 \\ 0 & 3 & 0 \end{bmatrix} = \begin{bmatrix} 3 & 0 & 3 \\ 3 & 0 & 3 \\ 3 & 0 & 3 \end{bmatrix}$

59. $A = \begin{bmatrix} 3 & 3 & 3 & 3 \\ 3 & 0 & 0 & 0 \\ 3 & 3 & 3 & 0 \\ 3 & 0 & 0 & 0 \\ 3 & 0 & 0 & 0 \end{bmatrix}$

61. (a) One possible solution for a "Z" is $A = \begin{bmatrix} 3 & 3 & 3 & 3 \\ 0 & 0 & 3 & 0 \\ 0 & 3 & 0 & 0 \\ 3 & 3 & 3 & 3 \end{bmatrix}$

(b) If A is the matrix in part (a) then

$B = \begin{bmatrix} 3 & 3 & 3 & 3 \\ 3 & 3 & 3 & 3 \\ 3 & 3 & 3 & 3 \\ 3 & 3 & 3 & 3 \end{bmatrix}$ and $B - A = \begin{bmatrix} 3 & 3 & 3 & 3 \\ 3 & 3 & 3 & 3 \\ 3 & 3 & 3 & 3 \\ 3 & 3 & 3 & 3 \end{bmatrix} - \begin{bmatrix} 3 & 3 & 3 & 3 \\ 0 & 0 & 3 & 0 \\ 0 & 3 & 0 & 0 \\ 3 & 3 & 3 & 3 \end{bmatrix} = \begin{bmatrix} 0 & 0 & 0 & 0 \\ 3 & 3 & 0 & 3 \\ 3 & 0 & 3 & 3 \\ 0 & 0 & 0 & 0 \end{bmatrix}$

63. (a) One possible solution for a "L" is $A = \begin{bmatrix} 3 & 0 & 0 & 0 \\ 3 & 0 & 0 & 0 \\ 3 & 0 & 0 & 0 \\ 3 & 3 & 3 & 3 \end{bmatrix}$

(b) If A is the matrix in part (a) then

$B = \begin{bmatrix} 3 & 3 & 3 & 3 \\ 3 & 3 & 3 & 3 \\ 3 & 3 & 3 & 3 \\ 3 & 3 & 3 & 3 \end{bmatrix}$ and $B - A = \begin{bmatrix} 3 & 3 & 3 & 3 \\ 3 & 3 & 3 & 3 \\ 3 & 3 & 3 & 3 \\ 3 & 3 & 3 & 3 \end{bmatrix} - \begin{bmatrix} 3 & 0 & 0 & 0 \\ 3 & 0 & 0 & 0 \\ 3 & 0 & 0 & 0 \\ 3 & 3 & 3 & 3 \end{bmatrix} = \begin{bmatrix} 0 & 3 & 3 & 3 \\ 0 & 3 & 3 & 3 \\ 0 & 3 & 3 & 3 \\ 0 & 0 & 0 & 0 \end{bmatrix}$

65. (a) These tables can be represented by the matrices A and B where $A = \begin{bmatrix} 12 & 4 \\ 8 & 7 \end{bmatrix}$ and $B = \begin{bmatrix} 55 \\ 70 \end{bmatrix}$.

(b) The product AB of these matrices calculates tuition cost for each student.

$$AB = \begin{bmatrix} 12 & 4 \\ 8 & 7 \end{bmatrix}\begin{bmatrix} 55 \\ 70 \end{bmatrix} = \begin{bmatrix} 12(55) + 4(70) \\ 8(55) + 7(70) \end{bmatrix} = \begin{bmatrix} 940 \\ 930 \end{bmatrix}$$

Student 1 is taking 12 credits at $55 each and 4 credits at $70 each. The total tuition for student 1 is

$12(\$55) + 4(\$70) = \$940$. Similarly, the tuition for student 2 is $930.

67. (a) These tables can be represented by the matrices A and B where $A = \begin{bmatrix} 10 & 5 \\ 9 & 8 \\ 11 & 3 \end{bmatrix}$ and $B = \begin{bmatrix} 60 \\ 70 \end{bmatrix}$.

(b) The product AB of these matrices calculates tuition cost for each student.

$$AB = \begin{bmatrix} 10 & 5 \\ 9 & 8 \\ 11 & 3 \end{bmatrix}\begin{bmatrix} 60 \\ 70 \end{bmatrix} = \begin{bmatrix} 10(60) + 5(70) \\ 9(60) + 8(70) \\ 11(60) + 3(70) \end{bmatrix} = \begin{bmatrix} 950 \\ 1100 \\ 870 \end{bmatrix}$$

The total tuition for student 1 is $10(\$60) + 5(\$70) = \$950$. Similarly, the tuition for student 2 is $1100

and the tuition for student 3 is $870.

69. $AB = \begin{bmatrix} 3 & 4 & 8 \\ 5 & 6 & 2 \end{bmatrix}\begin{bmatrix} 10 \\ 20 \\ 30 \end{bmatrix} = \begin{bmatrix} 350 \\ 230 \end{bmatrix}$; The total cost of Order 1 is $350, and the total cost of Order 2 is $230.

70. $AB = \begin{bmatrix} 1 & 3 & 8 & 4 \\ 3 & 5 & 7 & 0 \end{bmatrix}\begin{bmatrix} 15 \\ 21 \\ 28 \\ 38 \end{bmatrix} = \begin{bmatrix} 454 \\ 346 \end{bmatrix}$

Dealer 1 spent a total of $454,000 on the cars sold, and Dealer 2 spent a total of $346,000 on the cars sold.

9.6 Inverses of Matrices

1. B is the inverse of A.

$$AB = \begin{bmatrix} 4 & 3 \\ 5 & 4 \end{bmatrix}\begin{bmatrix} 4 & -3 \\ -5 & 4 \end{bmatrix} = \begin{bmatrix} 1 & 0 \\ 0 & 1 \end{bmatrix} \text{ and } BA = \begin{bmatrix} 4 & -3 \\ -5 & 4 \end{bmatrix}\begin{bmatrix} 4 & 3 \\ 5 & 4 \end{bmatrix} = \begin{bmatrix} 1 & 0 \\ 0 & 1 \end{bmatrix}$$

3. B is the inverse of A.

$$AB = \begin{bmatrix} 1 & -1 & 2 \\ 0 & 1 & -1 \\ 1 & 0 & 2 \end{bmatrix}\begin{bmatrix} 2 & 2 & -1 \\ -1 & 0 & 1 \\ -1 & -1 & 1 \end{bmatrix} = \begin{bmatrix} 1 & 0 & 0 \\ 0 & 1 & 0 \\ 0 & 0 & 1 \end{bmatrix};$$

$$BA = \begin{bmatrix} 2 & 2 & -1 \\ -1 & 0 & 1 \\ -1 & -1 & 1 \end{bmatrix}\begin{bmatrix} 1 & -1 & 2 \\ 0 & 1 & -1 \\ 1 & 0 & 2 \end{bmatrix} = \begin{bmatrix} 1 & 0 & 0 \\ 0 & 1 & 0 \\ 0 & 0 & 1 \end{bmatrix}$$

5.   *B* is not the inverse of *A*.

$$AB = \begin{bmatrix} 2 & 1 & -1 \\ 3 & 0 & 2 \\ -1 & 0 & 1 \end{bmatrix} \begin{bmatrix} 0 & 1 & -2 \\ 1 & -3 & 7 \\ 0 & -1 & 3 \end{bmatrix} = \begin{bmatrix} 1 & 0 & 0 \\ 0 & 1 & 0 \\ 0 & -2 & 5 \end{bmatrix};$$

$$BA = \begin{bmatrix} 0 & 1 & -2 \\ 1 & -3 & 7 \\ 0 & -1 & 3 \end{bmatrix} \begin{bmatrix} 2 & 1 & -1 \\ 3 & 0 & 2 \\ -1 & 0 & 1 \end{bmatrix} = \begin{bmatrix} 5 & 0 & 0 \\ -14 & 1 & 0 \\ -6 & 0 & 1 \end{bmatrix}$$

7. $AA^{-1} = \begin{bmatrix} 1 & 1 \\ 1 & 2 \end{bmatrix} \begin{bmatrix} 2 & -1 \\ -1 & k \end{bmatrix} = \begin{bmatrix} 1 & -1 + k \\ 0 & 2k - 1 \end{bmatrix}$

We must have $-1 + k = 0$ and $2k - 1 = 1$. The solution to both equations is $k = 1$.

9. $AA^{-1} = \begin{bmatrix} 1 & 3 \\ -1 & -5 \end{bmatrix} \begin{bmatrix} k & 1.5 \\ -0.5 & -0.5 \end{bmatrix} = \begin{bmatrix} k - 1.5 & 0 \\ -k + 2.5 & 1 \end{bmatrix}$

We must have $k - 1.5 = 1$ and $-k + 2.5 = 0$. The solution to both equations is $k = 2.5$.

11. I_2 multiplied by any 2×2 matrix *A* is equal to *A*, that is, $I_2A = AI_2 = A$.

13. I_3 multiplied by any 3×3 matrix *A* is equal to *A*, that is, $I_3A = AI_3 = A$.

15. $A \,|\, I_2 = \begin{bmatrix} 1 & 2 & | & 1 & 0 \\ 1 & 3 & | & 0 & 1 \end{bmatrix} \begin{matrix} \\ R_2 - R_1 \rightarrow \end{matrix} \begin{bmatrix} 1 & 2 & | & 1 & 0 \\ 0 & 1 & | & -1 & 1 \end{bmatrix} \begin{matrix} R_1 - 2R_2 \rightarrow \\ \end{matrix} \begin{bmatrix} 1 & 0 & | & 3 & -2 \\ 0 & 1 & | & -1 & 1 \end{bmatrix}; \; A^{-1} = \begin{bmatrix} 3 & -2 \\ -1 & 1 \end{bmatrix}$

17. $A \,|\, I_2 = \begin{bmatrix} -1 & 2 & | & 1 & 0 \\ 3 & -5 & | & 0 & 1 \end{bmatrix} \begin{matrix} -1R_1 \rightarrow \\ \end{matrix} \begin{bmatrix} 1 & -2 & | & -1 & 0 \\ 3 & -5 & | & 0 & 1 \end{bmatrix} \begin{matrix} \\ R_2 - 3R_1 \rightarrow \end{matrix} \begin{bmatrix} 1 & -2 & | & -1 & 0 \\ 0 & 1 & | & 3 & 1 \end{bmatrix}$

$R_1 + 2R_2 \rightarrow \begin{bmatrix} 1 & 0 & | & 5 & 2 \\ 0 & 1 & | & 3 & 1 \end{bmatrix}; \; A^{-1} = \begin{bmatrix} 5 & 2 \\ 3 & 1 \end{bmatrix}$

19. $A \,|\, I_2 = \begin{bmatrix} 8 & 5 & | & 1 & 0 \\ 2 & 1 & | & 0 & 1 \end{bmatrix} \begin{matrix} (1/8)R_1 \rightarrow \\ R_1 - 4R_2 \rightarrow \end{matrix} \begin{bmatrix} 1 & \frac{5}{8} & | & \frac{1}{8} & 0 \\ 0 & 1 & | & 1 & -4 \end{bmatrix} \begin{matrix} (-5/8)R_2 + R_1 \rightarrow \\ \end{matrix} \begin{bmatrix} 1 & 0 & | & -\frac{1}{2} & \frac{5}{2} \\ 0 & 1 & | & 1 & -4 \end{bmatrix};$

$A^{-1} = \begin{bmatrix} -\frac{1}{2} & \frac{5}{2} \\ 1 & -4 \end{bmatrix}$

21. $A \,|\, I_3 = \begin{bmatrix} 0 & 0 & 1 & | & 1 & 0 & 0 \\ 1 & 0 & 0 & | & 0 & 1 & 0 \\ 0 & 1 & 0 & | & 0 & 0 & 1 \end{bmatrix} \begin{matrix} R_2 \rightarrow \\ R_3 \rightarrow \\ R_1 \rightarrow \end{matrix} \begin{bmatrix} 1 & 0 & 0 & | & 0 & 1 & 0 \\ 0 & 1 & 0 & | & 0 & 0 & 1 \\ 0 & 0 & 1 & | & 1 & 0 & 0 \end{bmatrix}; \; A^{-1} = \begin{bmatrix} 0 & 1 & 0 \\ 0 & 0 & 1 \\ 1 & 0 & 0 \end{bmatrix}$

23. $A \,|\, I_3 = \begin{bmatrix} 1 & 0 & 1 & | & 1 & 0 & 0 \\ 2 & 1 & 3 & | & 0 & 1 & 0 \\ -1 & 1 & 1 & | & 0 & 0 & 1 \end{bmatrix} \begin{matrix} \\ R_2 - 2R_1 \rightarrow \\ R_3 + R_1 \rightarrow \end{matrix} \begin{bmatrix} 1 & 0 & 1 & | & 1 & 0 & 0 \\ 0 & 1 & 1 & | & -2 & 1 & 0 \\ 0 & 1 & 2 & | & 1 & 0 & 1 \end{bmatrix} \begin{matrix} \\ \\ R_3 - R_2 \rightarrow \end{matrix}$

$\begin{bmatrix} 1 & 0 & 1 & | & 1 & 0 & 0 \\ 0 & 1 & 1 & | & -2 & 1 & 0 \\ 0 & 0 & 1 & | & 3 & -1 & 1 \end{bmatrix} \begin{matrix} R_1 - R_3 \rightarrow \\ R_2 - R_3 \rightarrow \\ \end{matrix} \begin{bmatrix} 1 & 0 & 0 & | & -2 & 1 & -1 \\ 0 & 1 & 0 & | & -5 & 2 & -1 \\ 0 & 0 & 1 & | & 3 & -1 & 1 \end{bmatrix}; \; A^{-1} = \begin{bmatrix} -2 & 1 & -1 \\ -5 & 2 & -1 \\ 3 & -1 & 1 \end{bmatrix}$

25. $\begin{bmatrix} 1 & 2 & -1 & | & 1 & 0 & 0 \\ 2 & 5 & 0 & | & 0 & 1 & 0 \\ -1 & -1 & 2 & | & 0 & 0 & 1 \end{bmatrix} \begin{array}{l} 2R_1 - R_2 \to \\ R_3 + R_1 \to \end{array} \begin{bmatrix} 1 & 2 & -1 & | & 1 & 0 & 0 \\ 0 & -1 & -2 & | & 2 & -1 & 0 \\ 0 & 1 & 1 & | & 1 & 0 & 1 \end{bmatrix}$

$\begin{array}{l} (-1)R_2 \to \\ R_2 + R_3 \to \end{array} \begin{bmatrix} 1 & 2 & -1 & | & 1 & 0 & 0 \\ 0 & 1 & 2 & | & -2 & 1 & 0 \\ 0 & 0 & -1 & | & 3 & -1 & 1 \end{bmatrix} \begin{array}{l} R_1 - R_3 \to \\ 2R_3 + R_2 \to \\ (-1)R_3 \to \end{array} \begin{bmatrix} 1 & 2 & 0 & | & -2 & 1 & -1 \\ 0 & 1 & 0 & | & 4 & -1 & 2 \\ 0 & 0 & 1 & | & -3 & 1 & -1 \end{bmatrix}$

$R_1 - 2R_2 \to \begin{bmatrix} 1 & 0 & 0 & | & -10 & 3 & -5 \\ 0 & 1 & 0 & | & 4 & -1 & 2 \\ 0 & 0 & 1 & | & -3 & 1 & -1 \end{bmatrix}; \quad A^{-1} = \begin{bmatrix} -10 & 3 & -5 \\ 4 & -1 & 2 \\ -3 & 1 & -1 \end{bmatrix}$

27. $\begin{bmatrix} -2 & 1 & -3 & | & 1 & 0 & 0 \\ 0 & 1 & 2 & | & 0 & 1 & 0 \\ 1 & -2 & 1 & | & 0 & 0 & 1 \end{bmatrix} \begin{array}{l} (-1/2)R_1 \to \\ \\ 2R_3 + R_1 \to \end{array} \begin{bmatrix} 1 & -\frac{1}{2} & \frac{3}{2} & | & -\frac{1}{2} & 0 & 0 \\ 0 & 1 & 2 & | & 0 & 1 & 0 \\ 0 & -3 & -1 & | & 1 & 0 & 2 \end{bmatrix}$

$3R_2 + R_3 \to \begin{bmatrix} 1 & -\frac{1}{2} & \frac{3}{2} & | & -\frac{1}{2} & 0 & 0 \\ 0 & 1 & 2 & | & 0 & 1 & 0 \\ 0 & 0 & 5 & | & 1 & 3 & 2 \end{bmatrix} (1/5)R_3 \to \begin{bmatrix} 1 & -\frac{1}{2} & \frac{3}{2} & | & -\frac{1}{2} & 0 & 0 \\ 0 & 1 & 2 & | & 0 & 1 & 0 \\ 0 & 0 & 1 & | & \frac{1}{5} & \frac{3}{5} & \frac{2}{5} \end{bmatrix}$

$\begin{array}{l} R_1 - (3/2)R_3 \to \\ (-2)R_3 + R_2 \to \end{array} \begin{bmatrix} 1 & -\frac{1}{2} & 0 & | & -\frac{4}{5} & -\frac{9}{10} & -\frac{3}{5} \\ 0 & 1 & 0 & | & -\frac{2}{5} & -\frac{1}{5} & -\frac{4}{5} \\ 0 & 0 & 1 & | & \frac{1}{5} & \frac{3}{5} & \frac{2}{5} \end{bmatrix} R_1 + (1/2)R_2 \to \begin{bmatrix} 1 & 0 & 0 & | & -1 & -1 & -1 \\ 0 & 1 & 0 & | & -\frac{2}{5} & -\frac{1}{5} & -\frac{4}{5} \\ 0 & 0 & 1 & | & \frac{1}{5} & \frac{3}{5} & \frac{2}{5} \end{bmatrix};$

$A^{-1} = \begin{bmatrix} -1 & -1 & -1 \\ -\frac{2}{5} & -\frac{1}{5} & -\frac{4}{5} \\ \frac{1}{5} & \frac{3}{5} & \frac{2}{5} \end{bmatrix}$

29. $A = \begin{bmatrix} 0.5 & -1.5 \\ 0.2 & -0.5 \end{bmatrix} \Rightarrow A^{-1} = \begin{bmatrix} -10 & 30 \\ -4 & 10 \end{bmatrix}$ as shown in Figure 29.

31. $A = \begin{bmatrix} 1 & 2 & 0 \\ -1 & 4 & -1 \\ 2 & -1 & 0 \end{bmatrix} \Rightarrow A^{-1} = \begin{bmatrix} 0.2 & 0 & 0.4 \\ 0.4 & 0 & -0.2 \\ 1.4 & -1 & -1.2 \end{bmatrix}$ as shown in Figure 31.

Figure 29

Figure 31

33. $A = \begin{bmatrix} 2 & -2 & 1 \\ 0 & 5 & 8 \\ 0 & 0 & -1 \end{bmatrix} \Rightarrow A^{-1} = \begin{bmatrix} 0.5 & 0.2 & 2.1 \\ 0 & 0.2 & 1.6 \\ 0 & 0 & -1 \end{bmatrix}$ as shown in Figure 33.

```
[A]⁻¹
 [[.5   .25 .25]
  [.25  .5  .25]
  [.25  .25 .5 ]]
```

Figure 35

35. $A = \begin{bmatrix} 3 & -1 & -1 \\ -1 & 3 & -1 \\ -1 & -1 & 3 \end{bmatrix} \Rightarrow A^{-1} = \begin{bmatrix} 0.5 & 0.25 & 0.25 \\ 0.25 & 0.5 & 0.25 \\ 0.25 & 0.25 & 0.5 \end{bmatrix}$ as shown in Figure 35.

```
[A]⁻¹
 [[.5   .25 .25]
  [.25  .5  .25]
  [.25  .25 .5 ]]
```

Figure 35

37. $A = \begin{bmatrix} 1 & -1 & 0 & 0 \\ -1 & 5 & -1 & 0 \\ 0 & -1 & 5 & -1 \\ 0 & 0 & -1 & 1 \end{bmatrix} \Rightarrow A^{-1} = \begin{bmatrix} 1.2\overline{6} & 0.2\overline{6} & 0.0\overline{6} & 0.0\overline{6} \\ 0.2\overline{6} & 0.2\overline{6} & 0.0\overline{6} & 0.0\overline{6} \\ 0.0\overline{6} & 0.0\overline{6} & 0.2\overline{6} & 0.2\overline{6} \\ 0.0\overline{6} & 0.0\overline{6} & 0.2\overline{6} & 1.2\overline{6} \end{bmatrix}$ as shown in Figure 37.

```
[A]⁻¹
 [[1.266666667  ...
  [.266666667   ...
  [.0666666667  ...
  [.0666666667  ...
```

Figure 37

39. $\begin{array}{l} 2x - 3y = 7 \\ -3x - 4y = 9 \end{array} \Rightarrow AX = \begin{bmatrix} 2 & -3 \\ -3 & -4 \end{bmatrix}\begin{bmatrix} x \\ y \end{bmatrix} = \begin{bmatrix} 7 \\ 9 \end{bmatrix} = B$

41. $\begin{array}{l} \frac{1}{2}x - \frac{3}{2}y = \frac{1}{4} \\ -x + 2y = 5 \end{array} \Rightarrow AX = \begin{bmatrix} \frac{1}{2} & -\frac{3}{2} \\ -1 & 2 \end{bmatrix}\begin{bmatrix} x \\ y \end{bmatrix} = \begin{bmatrix} \frac{1}{4} \\ 5 \end{bmatrix} = B$

43. $\begin{array}{l} x - 2y + z = 5 \\ 3y - z = 6 \\ 5x - 4y - 7z = 0 \end{array} \Rightarrow AX = \begin{bmatrix} 1 & -2 & 1 \\ 0 & 3 & -1 \\ 5 & -4 & -7 \end{bmatrix}\begin{bmatrix} x \\ y \\ z \end{bmatrix} = \begin{bmatrix} 5 \\ 6 \\ 0 \end{bmatrix} = B$

45. $\begin{array}{l} 4x - y + 3z = -2 \\ x + 2y + 5z = 11 \\ 2x - 3y = -1 \end{array} \Rightarrow AX = \begin{bmatrix} 4 & -1 & 3 \\ 1 & 2 & 5 \\ 2 & -3 & 0 \end{bmatrix}\begin{bmatrix} x \\ y \\ z \end{bmatrix} = \begin{bmatrix} -2 \\ 11 \\ -1 \end{bmatrix} = B$

47. (a) $\begin{matrix} x + 2y = 3 \\ x + 3y = 6 \end{matrix} \Rightarrow AX = \begin{bmatrix} 1 & 2 \\ 1 & 3 \end{bmatrix}\begin{bmatrix} x \\ y \end{bmatrix} = \begin{bmatrix} 3 \\ 6 \end{bmatrix} = B$

(b) If $AX = B \Rightarrow \begin{bmatrix} 1 & 2 \\ 1 & 3 \end{bmatrix}\begin{bmatrix} x \\ y \end{bmatrix} = \begin{bmatrix} 3 \\ 6 \end{bmatrix} \Rightarrow X = A^{-1}B \Rightarrow \begin{bmatrix} x \\ y \end{bmatrix} = \begin{bmatrix} 3 & -2 \\ -1 & 1 \end{bmatrix}\begin{bmatrix} 3 \\ 6 \end{bmatrix} = \begin{bmatrix} -3 \\ 3 \end{bmatrix}$

The solution to the system is $(-3, 3)$.

49. (a) $\begin{matrix} -x + 2y = 5 \\ 3x - 5y = -2 \end{matrix} \Rightarrow AX = \begin{bmatrix} -1 & 2 \\ 3 & -5 \end{bmatrix}\begin{bmatrix} x \\ y \end{bmatrix} = \begin{bmatrix} 5 \\ -2 \end{bmatrix} = B$

(b) If $AX = B \Rightarrow \begin{bmatrix} -1 & 2 \\ 3 & -5 \end{bmatrix}\begin{bmatrix} x \\ y \end{bmatrix} = \begin{bmatrix} 5 \\ -2 \end{bmatrix} \Rightarrow X = A^{-1}B \Rightarrow \begin{bmatrix} x \\ y \end{bmatrix} = \begin{bmatrix} 5 & 2 \\ 3 & 1 \end{bmatrix}\begin{bmatrix} 5 \\ -2 \end{bmatrix} = \begin{bmatrix} 21 \\ 13 \end{bmatrix}$

The solution to the system is $(21, 13)$.

51. (a) $\begin{matrix} x + z = -7 \\ 2x + y + 3z = -13 \\ -x + y + z = -4 \end{matrix} \Rightarrow AX = \begin{bmatrix} 1 & 0 & 1 \\ 2 & 1 & 3 \\ -1 & 1 & 1 \end{bmatrix}\begin{bmatrix} x \\ y \\ z \end{bmatrix} = \begin{bmatrix} -7 \\ -13 \\ -4 \end{bmatrix} = B$

(b) If $AX = B \Rightarrow \begin{bmatrix} 1 & 0 & 1 \\ 2 & 1 & 3 \\ -1 & 1 & 1 \end{bmatrix}\begin{bmatrix} x \\ y \\ z \end{bmatrix} = \begin{bmatrix} -7 \\ -13 \\ -4 \end{bmatrix} \Rightarrow X = A^{-1}B = \begin{bmatrix} x \\ y \\ z \end{bmatrix} = \begin{bmatrix} -2 & 1 & -1 \\ -5 & 2 & -1 \\ 3 & -1 & 1 \end{bmatrix}\begin{bmatrix} -7 \\ -13 \\ -4 \end{bmatrix} =$

$\begin{bmatrix} 5 \\ 13 \\ -12 \end{bmatrix}$ The solution to the system is $(5, 13, -12)$.

53. (a) $\begin{matrix} x + 2y - z = 2 \\ 2x + 5y = -1 \\ -x - y + 2z = 0 \end{matrix} \Rightarrow AX = \begin{bmatrix} 1 & 2 & -1 \\ 2 & 5 & 0 \\ -1 & -1 & 2 \end{bmatrix}\begin{bmatrix} x \\ y \\ z \end{bmatrix} = \begin{bmatrix} 2 \\ -1 \\ 0 \end{bmatrix} = B$

(b) If $AX = B \Rightarrow \begin{bmatrix} 1 & 2 & -1 \\ 2 & 5 & 0 \\ -1 & -1 & 2 \end{bmatrix}\begin{bmatrix} x \\ y \\ z \end{bmatrix} = \begin{bmatrix} 2 \\ -1 \\ 0 \end{bmatrix} \Rightarrow X = A^{-1}B = \begin{bmatrix} x \\ y \\ z \end{bmatrix} = \begin{bmatrix} -10 & 3 & -5 \\ 4 & -1 & 2 \\ -3 & 1 & -1 \end{bmatrix}\begin{bmatrix} 2 \\ -1 \\ 0 \end{bmatrix} =$

$\begin{bmatrix} -23 \\ 9 \\ -7 \end{bmatrix}$ The solution to the system is $(-23, 9, -7)$.

55. (a) $AX = B \Rightarrow \begin{bmatrix} 1.5 & 3.7 \\ -0.4 & -2.1 \end{bmatrix}\begin{bmatrix} x \\ y \end{bmatrix} = \begin{bmatrix} 0.32 \\ 0.36 \end{bmatrix}$

(b) See Figure 55. $X = A^{-1}B \Rightarrow X = \begin{bmatrix} 1.2 \\ -0.4 \end{bmatrix}$

```
[A]⁻¹*[B]
          [[1.2]
           [-.4]]
```

Figure 55

57. (a) $AX = B \Rightarrow \begin{bmatrix} 0.08 & -0.7 \\ 1.1 & -0.05 \end{bmatrix} \begin{bmatrix} x \\ y \end{bmatrix} = \begin{bmatrix} -0.504 \\ 0.73 \end{bmatrix}$

(b) See Figure 57. $X = A^{-1}B \Rightarrow X = \begin{bmatrix} 0.7 \\ 0.8 \end{bmatrix}$

59. (a) $AX = B \Rightarrow \begin{bmatrix} 3.1 & 1.9 & -1 \\ 6.3 & 0 & -9.9 \\ -1 & 1.5 & 7 \end{bmatrix} \begin{bmatrix} x \\ y \\ z \end{bmatrix} = \begin{bmatrix} 1.99 \\ -3.78 \\ 5.3 \end{bmatrix}$

(b) See Figure 59. $X = A^{-1}B \Rightarrow X = \begin{bmatrix} 0.5 \\ 0.6 \\ 0.7 \end{bmatrix}$

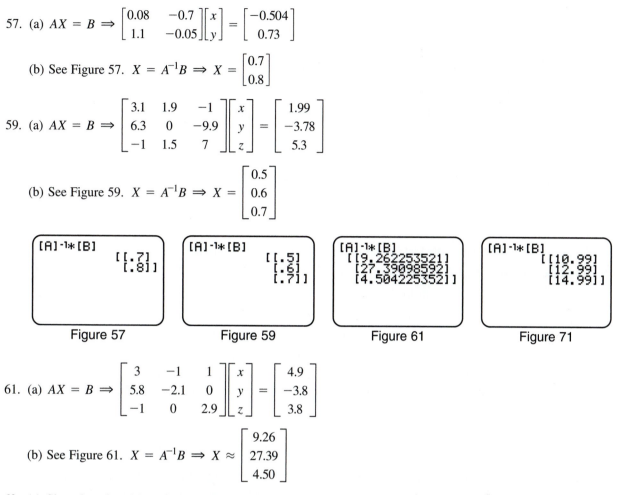

Figure 57   Figure 59   Figure 61   Figure 71

61. (a) $AX = B \Rightarrow \begin{bmatrix} 3 & -1 & 1 \\ 5.8 & -2.1 & 0 \\ -1 & 0 & 2.9 \end{bmatrix} \begin{bmatrix} x \\ y \\ z \end{bmatrix} = \begin{bmatrix} 4.9 \\ -3.8 \\ 3.8 \end{bmatrix}$

(b) See Figure 61. $X = A^{-1}B \Rightarrow X \approx \begin{bmatrix} 9.26 \\ 27.39 \\ 4.50 \end{bmatrix}$

63. (a) Since $h = 2$ and $k = 3$, the matrix A will translate the point $(0, 1)$ to the right 2 units and up 3 units. Its new location will be $(0 + 2, 1 + 3) = (2, 4)$. This is verified by the following computation:

$$AX = \begin{bmatrix} 1 & 0 & 2 \\ 0 & 1 & 3 \\ 0 & 0 & 1 \end{bmatrix} \begin{bmatrix} 0 \\ 1 \\ 1 \end{bmatrix} = \begin{bmatrix} 2 \\ 4 \\ 1 \end{bmatrix}$$

(b) $A^{-1}Y = X$. That is, A^{-1} will translate $(2, 4)$ back to $(0, 1)$ by moving it left 2 units and down 3 units. Thus $h = -2$ and $k = -3$ in A^{-1}.

$$A^{-1}Y = \begin{bmatrix} 1 & 0 & -2 \\ 0 & 1 & -3 \\ 0 & 0 & 1 \end{bmatrix} \begin{bmatrix} 2 \\ 4 \\ 1 \end{bmatrix} = \begin{bmatrix} 0 \\ 1 \\ 1 \end{bmatrix}$$

(c) The product $AA^{-1} = I_3 = A^{-1}A$, since they are 3×3 inverse matrices.

65. Three units left implies that $h = -3$ and five units down implies that $k = -5$.

$$A = \begin{bmatrix} 1 & 0 & -3 \\ 0 & 1 & -5 \\ 0 & 0 & 1 \end{bmatrix} \text{ and } A^{-1} = \begin{bmatrix} 1 & 0 & 3 \\ 0 & 1 & 5 \\ 0 & 0 & 1 \end{bmatrix}$$

A^{-1} will translate a point 3 units to the right and 5 units up.

67. (a) $BX = \begin{bmatrix} \dfrac{1}{\sqrt{2}} & \dfrac{1}{\sqrt{2}} & 0 \\[2ex] -\dfrac{1}{\sqrt{2}} & \dfrac{1}{\sqrt{2}} & 0 \\[2ex] 0 & 0 & 1 \end{bmatrix} \begin{bmatrix} -\sqrt{2} \\ -\sqrt{2} \\ 1 \end{bmatrix} = \begin{bmatrix} -2 \\ 0 \\ 1 \end{bmatrix} = Y$

(b) $B^{-1}Y = \begin{bmatrix} \dfrac{1}{\sqrt{2}} & -\dfrac{1}{\sqrt{2}} & 0 \\[2ex] \dfrac{1}{\sqrt{2}} & \dfrac{1}{\sqrt{2}} & 0 \\[2ex] 0 & 0 & 1 \end{bmatrix} \begin{bmatrix} -2 \\ 0 \\ 1 \end{bmatrix} = \begin{bmatrix} -\sqrt{2} \\ -\sqrt{2} \\ 1 \end{bmatrix} = X$

B^{-1} rotates the point represented by Y counterclockwise $45°$ about the origin.

69. (a) In the computation ABX, B translates $(1, 1)$ left 3 units and up 3 units to $(-2, 4)$. Then, A translates $(-2, 4)$ right 4 units and down 2 units to $(2, 2)$.

$ABX = \begin{bmatrix} 1 & 0 & 4 \\ 0 & 1 & -2 \\ 0 & 0 & 1 \end{bmatrix} \begin{bmatrix} 1 & 0 & -3 \\ 0 & 1 & 3 \\ 0 & 0 & 1 \end{bmatrix} \begin{bmatrix} 1 \\ 1 \\ 1 \end{bmatrix} = \begin{bmatrix} 2 \\ 2 \\ 1 \end{bmatrix} = Y;$ This represents the point $(2, 2)$ as expected.

(b) The net result of A and B is to translate a point 1 unit right and 1 unit up. Therefore, it is reasonable to expect that $h = 1$ and $k = 1$ in the matrix of the form AB.

$AB = \begin{bmatrix} 1 & 0 & 4 \\ 0 & 1 & -2 \\ 0 & 0 & 1 \end{bmatrix} \begin{bmatrix} 1 & 0 & -3 \\ 0 & 1 & 3 \\ 0 & 0 & 1 \end{bmatrix} = \begin{bmatrix} 1 & 0 & 1 \\ 0 & 1 & 1 \\ 0 & 0 & 1 \end{bmatrix}$

(c) Yes. If a point is translated left 3 units and up 3 units followed by right 4 units and down 2 units, the final result will be the same as the translation obtained when the point is first translated right 4 units and down 2 units followed by left 3 units and up 3. Therefore, we might expect that $AB = BA$.

$BA = \begin{bmatrix} 1 & 0 & -3 \\ 0 & 1 & 3 \\ 0 & 0 & 1 \end{bmatrix} \begin{bmatrix} 1 & 0 & 4 \\ 0 & 1 & -2 \\ 0 & 0 & 1 \end{bmatrix} = \begin{bmatrix} 1 & 0 & 1 \\ 0 & 1 & 1 \\ 0 & 0 & 1 \end{bmatrix} = AB$

(d) Since AB translates a point 1 unit right and 1 unit up, the inverse of AB would translate a point 1 unit left and 1 unit down. So $h = -1$ and $k = -1$ in $(AB)^{-1}$.

$(AB)^{-1} = \begin{bmatrix} 1 & 0 & -1 \\ 0 & 1 & -1 \\ 0 & 0 & 1 \end{bmatrix};$ Notice that $(AB)(AB)^{-1} = I_3$ as expected.

71. Let x be the number of CDs of type A purchased, let y be the CDs of type B and let z be the CDs of type C. The first row in the table implies that $2x + 3y + 4z = 120.91$. The other rows can be interpreted similarly. The system in matrix form is shown below.

$AX = B \Rightarrow \begin{bmatrix} 2 & 3 & 4 \\ 1 & 4 & 0 \\ 2 & 1 & 3 \end{bmatrix} \begin{bmatrix} x \\ y \\ z \end{bmatrix} = \begin{bmatrix} 120.91 \\ 62.95 \\ 79.94 \end{bmatrix}$

Use a graphing calculator to find the solution as shown in Figure 71. Type A CDs cost $10.99, type B cost $12.99, and type C cost $14.99.

73. (a) The following three equations must be solved, using the equation $P = a + bS + cC$.

$$a + b(1500) + c(\ 8) = 122$$
$$a + b(2000) + c(\ 5) = 130$$
$$a + b(2200) + c(10) = 158$$

These equations can be written in matrix form as follows:

$$AX = B \Rightarrow \begin{bmatrix} 1 & 1500 & 8 \\ 1 & 2000 & 5 \\ 1 & 2200 & 10 \end{bmatrix} \begin{bmatrix} a \\ b \\ c \end{bmatrix} = \begin{bmatrix} 122 \\ 130 \\ 158 \end{bmatrix}$$

The solution is given by $X = A^{-1}B$ as shown in Figure 73. It is $a = 30$, $b = 0.04$, and $c = 4$.

That is, P is given by $P = 30 + 0.04S + 4C$

(b) $P = 30 + 0.04(1800) + 4(7) = 130$, or $130,000

Figure 73

75. (a) Although this system of equations can be solved using matrices, we will use elimination to avoid performing row operations on variables. Subtracting the second equation from the third equations gives

$$\begin{array}{r} 0.4S + 0.4E + 0.1T = T \\ - \ 0.4S + 0.2E + 0.1T = E \\ \hline 0.2E \qquad\qquad = T - E \Rightarrow E = \dfrac{5}{6}T. \end{array}$$

Substituting $E = \dfrac{5}{6}T$ in the first equation gives $0.2S + 0.4\left(\dfrac{5}{6}T\right) + 0.8T = S \Rightarrow$

$\dfrac{17}{15}T = \dfrac{4}{5}S \Rightarrow S = \dfrac{17}{12}T$. The solution is $\left(\dfrac{17}{12}T, \dfrac{5}{6}T, T\right)$.

(b) When $T = 60$ units, $S = \dfrac{17}{12}(60) = 85$ units and $E = \dfrac{5}{6}(60) = 50$ units. That is, service should produce 85 units and electrical should produce 50 units.

9.7 Determinants

1. $\det A = \det \begin{bmatrix} 4 & 3 \\ 5 & 4 \end{bmatrix} = (4)(4) - (5)(3) = 1 \neq 0$; A is invertible.

3. $\det A = \det \begin{bmatrix} -4 & 6 \\ -8 & 12 \end{bmatrix} = (-4)(12) - (-8)(6) = 0$; A is not invertible.

5. Deleting the first row and second column gives $M_{12} = \det A = \det \begin{bmatrix} 2 & -2 \\ 0 & 5 \end{bmatrix} = (2)(5) - (0)(-2) = 10.$

The cofactor is $A_{12} = (-1)^{1+2}M_{12} = -1(10) = -10.$

7. Deleting the second row and second column gives $M_{22} = \det A = \det\begin{bmatrix} 7 & 1 \\ 1 & -2 \end{bmatrix} = (7)(-2) - (1)(1) = -15.$

The cofactor is $A_{22} = (-1)^{2+2}M_{22} = 1(-15) = -15.$

9. $\det A = a_{11}A_{11} + a_{21}A_{21} + a_{31}A_{31} = a_{11}M_{11} - a_{21}M_{21} + a_{31}M_{31} \Rightarrow$

$\det A = (1)\det\begin{bmatrix} 2 & -3 \\ -1 & 3 \end{bmatrix} - (0)\det\begin{bmatrix} 4 & -7 \\ -1 & 3 \end{bmatrix} + (0)\det\begin{bmatrix} 4 & -7 \\ 2 & -3 \end{bmatrix} = (1)(6 - 3) = 3$

Since $\det A = 3 \neq 0$, A^{-1} exists.

11. $\det A = a_{11}A_{11} + a_{21}A_{21} + a_{31}A_{31} = a_{11}M_{11} - a_{21}M_{21} + a_{31}M_{31} \Rightarrow$

$\det A = (5)\det\begin{bmatrix} -2 & 0 \\ 4 & 0 \end{bmatrix} - (0)\det\begin{bmatrix} 1 & 6 \\ 4 & 0 \end{bmatrix} + (0)\det\begin{bmatrix} 1 & 6 \\ -2 & 0 \end{bmatrix} = (5)(0 - 0) = 0$

Since $\det A = 0$, A^{-1} does not exist.

13. Expanding about the first column results in $\det A = 2\det\begin{bmatrix} 3 & 0 \\ 0 & 5 \end{bmatrix} - 0 + 0 = (2)(15 - 0) = 30.$

15. Expanding about the first row results in $\det A = 0$.

17. Expanding about the first column results in

$\det A = 3\det\begin{bmatrix} 5 & 7 \\ 0 & -1 \end{bmatrix} - 0 + 1\det\begin{bmatrix} -1 & 2 \\ 5 & 7 \end{bmatrix} = (3)(-5 - 0) + (1)(-7 - 10) = -32.$

19. Expanding about the first column results in

$\det A = 1\det\begin{bmatrix} 1 & 3 \\ 4 & -2 \end{bmatrix} - (-7)\det\begin{bmatrix} -5 & 2 \\ 4 & -2 \end{bmatrix} + 0 = (1)(-2 - 12) - (-7)(10 - 8) = 0.$

21. $\det A = \det\begin{bmatrix} 11 & -32 \\ 1.2 & 55 \end{bmatrix} = 643.4$

23. $\det A = \det\begin{bmatrix} 2.3 & 5.1 & 2.8 \\ 1.2 & 4.5 & 8.8 \\ -0.4 & -0.8 & -1.2 \end{bmatrix} = -4.484$

25. By Cramer's rule, the solution can be found as follows:

$E = \det\begin{bmatrix} 5 & 2 \\ 1 & 3 \end{bmatrix} = 13;\ F = \det\begin{bmatrix} -1 & 5 \\ 3 & 1 \end{bmatrix} = -16;\ D = \det\begin{bmatrix} -1 & 2 \\ 3 & 3 \end{bmatrix} = -9$

Thus $x = \dfrac{E}{D} = \dfrac{13}{-9}$ and $y = \dfrac{F}{D} = \dfrac{-16}{-9} = \dfrac{16}{9}$. The solution is $\left(-\dfrac{13}{9}, \dfrac{16}{9}\right)$.

27. By Cramer's rule, the solution can be found as follows:

$E = \det\begin{bmatrix} 8 & 3 \\ 3 & -5 \end{bmatrix} = -49;\ F = \det\begin{bmatrix} -2 & 8 \\ 4 & 3 \end{bmatrix} = -38;\ D = \det\begin{bmatrix} -2 & 3 \\ 4 & -5 \end{bmatrix} = -2$

Thus $x = \dfrac{E}{D} = \dfrac{-49}{-2} = \dfrac{49}{2}$ and $y = \dfrac{F}{D} = \dfrac{-38}{-2} = 19$. The solution is $\left(\dfrac{49}{2}, 19\right)$.

29. By Cramer's rule, the solution can be found as follows:

$$E = \det\begin{bmatrix} 23 & 4 \\ 70 & -5 \end{bmatrix} = -395; \quad F = \det\begin{bmatrix} 7 & 23 \\ 11 & 70 \end{bmatrix} = 237; \quad D = \det\begin{bmatrix} 7 & 4 \\ 11 & -5 \end{bmatrix} = -79$$

Thus $x = \dfrac{E}{D} = \dfrac{-395}{-79} = 5$ and $y = \dfrac{F}{D} = \dfrac{237}{-79} = -3$. The solution is $(5, -3)$.

31. By Cramer's rule, the solution can be found as follows:

$$E = \det\begin{bmatrix} -0.91 & -2.5 \\ 0.423 & 0.9 \end{bmatrix} = 0.2385; \quad F = \det\begin{bmatrix} 1.7 & -0.91 \\ -0.4 & 0.423 \end{bmatrix} = 0.3551; \quad D = \det\begin{bmatrix} 1.7 & -2.5 \\ -0.4 & 0.9 \end{bmatrix} = 0.53$$

Thus $x = \dfrac{E}{D} = \dfrac{0.2385}{0.53} = 0.45$ and $y = \dfrac{F}{D} = \dfrac{0.3551}{0.53} = 0.67$. The solution is $(0.45, 0.67)$.

33. Enter the vertices as columns in a counterclockwise direction.

$$D = \frac{1}{2}\det\begin{bmatrix} 0 & 4 & 1 \\ 0 & 2 & 4 \\ 1 & 1 & 1 \end{bmatrix} = 7; \quad \text{The area of the triangle is 7 square units.}$$

35. A line segment between $(1, 3)$ and $(3, 2)$ divides the quadrangle into two triangles whose areas can be found using determinants. Enter the vertices of each triangle as columns in a counterclockwise direction.

$$D = \frac{1}{2}\det\begin{bmatrix} 0 & 3 & 1 \\ 0 & 2 & 3 \\ 1 & 1 & 1 \end{bmatrix} + \frac{1}{2}\det\begin{bmatrix} 1 & 3 & 5 \\ 3 & 2 & 4 \\ 1 & 1 & 1 \end{bmatrix} = 6.5; \quad \text{The area of the quadrangle is 6.5 square units.}$$

37. If the three points form a triangle with no area ($D = 0$ using determinants), then the points must be collinear.

$$D = \frac{1}{2}\det\begin{bmatrix} 1 & -3 & 2 \\ 3 & 11 & 1 \\ 1 & 1 & 1 \end{bmatrix} = 0; \quad \text{The points are collinear.}$$

39. If the three points form a triangle with no area ($D = 0$ using determinants), then the points must be collinear.

$$D = \frac{1}{2}\det\begin{bmatrix} -2 & 4 & 2 \\ -5 & 4 & 3 \\ 1 & 1 & 1 \end{bmatrix} = 6 \neq 0; \quad \text{The points are not collinear.}$$

41. Use cofactors to expand about row 1 of $\begin{bmatrix} x & y & 1 \\ 2 & 1 & 1 \\ -1 & 4 & 1 \end{bmatrix} = 0 \Rightarrow x(-3) - y(3) + 9 = 0 \Rightarrow$

$-3x - 3y = -9 \Rightarrow x + y = 3$.

43. Use cofactors to expand about row 1 of $\begin{bmatrix} x & y & 1 \\ 6 & -7 & 1 \\ 4 & -3 & 1 \end{bmatrix} = 0 \Rightarrow x(-4) - y(2) + (10) = 0 \Rightarrow 2x + y = 5$.

Chapter 9 Review Exercises

1. $A(b, h) = \dfrac{1}{2}bh \implies A(3, 6) = \dfrac{1}{2}(3)(6) = 9$

3. (a) $3x + y = 1 \implies y = 1 - 3x$ and $2x - 3y = 8 \implies y = \dfrac{2x - 8}{3}$

    Graph $Y_1 = 1 - 3X$ and $Y_2 = (2X - 8)/3$. The graphs intersect at the point $(1, -2)$. See Figure 3.

    (b) Substituting $y = 1 - 3x$ into the second equation gives $2x - 3(1 - 3x) = 8 \implies 2x - 3 + 9x = 8 \implies$

    $11x = 11 \implies x = 1$. If $x = 1$, then $y = 1 - 3(1) = -2$. The solution is $(1, -2)$.

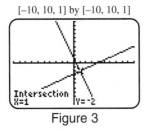

[-10, 10, 1] by [-10, 10, 1]

Figure 3

5. Multiply the first equation by 2 and add to eliminate the y-variable.

    $$\begin{array}{rcl} 4x + 2y &=& 14 \\ x - 2y &=& -4 \\ \hline 5x &=& 10 \implies x = 2 \end{array}$$

    Since $2x + y = 7$, it follows that $y = 3$. The solution is $(2, 3)$. The system is consistent.

7. Multiply the first equation by 2, the second equation by 3, and add to eliminate the y-variable.

    $$\begin{array}{rcl} 12x - 30y &=& 24 \\ -12x + 30y &=& -24 \\ \hline 0 &=& 0 \implies \text{infinitely many solutions of the form } \{(x, y) \mid 2x - 5y = 4\} \end{array}$$

    Since $0 = 0$, the system is consistent and there are infinitely many solutions.

9. Subtract the equations to eliminate the x-variable.

    $$\begin{array}{rcl} x^2 - 3y &=& 3 \\ x^2 + 2y^2 &=& 5 \\ \hline \end{array}$$

    $-3y - 2y^2 = -2 \implies 0 = 2y^2 + 3y - 2 \implies (2y - 1)(y + 2) = 0 \implies y = -2, \dfrac{1}{2}$.

    If $y = -2$ then $x^2 - 3(-2) = 3 \implies x^2 + 6 = 3 \implies x^2 = -3$. There is no real solution to this $\implies y \neq -2$.

    If $y = \dfrac{1}{2}$ then $x^2 - 3\left(\dfrac{1}{2}\right) = 3 \implies x^2 = \dfrac{9}{2} \implies x = \pm\dfrac{3}{\sqrt{2}} \implies x = \pm\dfrac{3\sqrt{2}}{2}$.

    The solutions are $\left(\dfrac{3\sqrt{2}}{2}, \dfrac{1}{2}\right)$ and $\left(\dfrac{-3\sqrt{2}}{2}, \dfrac{1}{2}\right)$.

11. See Figure 11.

Figure 11

13. The solution region is inside the circle or radius 3 and above the line $y = 3 - x$. Their graphs intersect at the points $(0, 3)$ and $(3, 0)$. It does not include the boundary. See Figure 13. One solution to the system is $(2, 2)$.

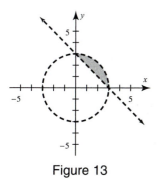

Figure 13

15. First, subtract the first two equations to eliminate the x-variable.

$$
\begin{array}{rcr}
x - y + z &=& -2 \\
x + 2y - z &=& 2 \\
\hline
-3y + 2z &=& -4
\end{array}
$$

Multiply this equation by 2 and multiply the last equation by 3 to eliminate the y-variable.

$$
\begin{array}{rcr}
-6y + 4z &=& -8 \\
6y + 9z &=& 21 \\
\hline
13z &=& 13 \Rightarrow z = 1
\end{array}
$$

Substituting $z = 1$ into $2y + 3z = 7$ gives $2y + 3 = 7 \Rightarrow 2y = 4 \Rightarrow y = 2$. Substituting $y = 2$ and $z = 1$ into the original first equation gives $x - 2 + 1 = -2 \Rightarrow x - 1 = -2 \Rightarrow x = -1$.

The solution to the system is $(-1, 2, 1)$.

17. Add the first two equations to eliminate the *x*-variable.

$$-x + 2y + 2z = 9$$
$$\underline{x + y - 3z = 6}$$
$$\phantom{-x + {}}3y - z = 15$$

Subtract this new equation from the third equation.

$$3y - z = 8$$
$$\underline{3y - z = 15}$$
$$\phantom{3y - z = {}}0 = -7$$

Since $0 \neq -7$, there is no solution.

19. The system is $x + 5y = 6$ and $y = 3$.

Substituting $y = 3$ into the first equation gives $x + 5(3) = 6 \Rightarrow x = -9$. The solution is $(-9, 3)$.

21. The augmented matrix is in Row-Echelon Form $\Rightarrow x = -2, y = 3, z = 0 \Rightarrow (-2, 3, 0)$.

23.
$$\begin{bmatrix} 2 & -1 & 2 & | & 10 \\ 1 & -2 & 1 & | & 8 \\ 3 & -1 & 2 & | & 11 \end{bmatrix} \begin{matrix} (1/2)R_1 \to \\ R_1 - 2R_2 \to \\ 3R_2 - R_3 \to \end{matrix} \begin{bmatrix} 1 & -\frac{1}{2} & 1 & | & 5 \\ 0 & 3 & 0 & | & -6 \\ 0 & -5 & 1 & | & 13 \end{bmatrix} (1/3)R_2 \to \begin{bmatrix} 1 & -\frac{1}{2} & 1 & | & 5 \\ 0 & 1 & 0 & | & -2 \\ 0 & -5 & 1 & | & 13 \end{bmatrix}$$

$$5R_2 + R_3 \to \begin{bmatrix} 1 & -\frac{1}{2} & 1 & | & 5 \\ 0 & 1 & 0 & | & -2 \\ 0 & 0 & 1 & | & 3 \end{bmatrix}$$

Backward substitution produces $z = 3; y = -2; x - \dfrac{1}{2}(-2) + 3 = 5 \Rightarrow x + 1 + 3 = 5 \Rightarrow x = 1$.

The solution is $(1, -2, 3)$.

25. (a) $a_{12} = 3$ and $a_{22} = 2 \Rightarrow a_{12} + a_{22} = 3 + 2 \Rightarrow a_{12} + a_{22} = 5$

(b) $a_{11} = -2$ and $a_{23} = 4 \Rightarrow a_{11} - 2a_{23} = -2 - 2(4) \Rightarrow a_{11} - 2a_{23} = -10$

27. (a) $A + 2B = \begin{bmatrix} 1 & -3 \\ 2 & -1 \end{bmatrix} + 2\begin{bmatrix} 3 & 2 \\ -5 & 1 \end{bmatrix} = \begin{bmatrix} 7 & 1 \\ -8 & 1 \end{bmatrix}$

(b) $A - B = \begin{bmatrix} 1 & -3 \\ 2 & -1 \end{bmatrix} - \begin{bmatrix} 3 & 2 \\ -5 & 1 \end{bmatrix} = \begin{bmatrix} -2 & -5 \\ 7 & -2 \end{bmatrix}$

(c) $-4A = -4\begin{bmatrix} 1 & -3 \\ 2 & -1 \end{bmatrix} = \begin{bmatrix} -4 & 12 \\ -8 & 4 \end{bmatrix}$

29. *A* and *B* are both 2×2 so *AB* and *BA* are also both 2×2.

$$AB = \begin{bmatrix} 2 & 0 \\ -5 & 3 \end{bmatrix}\begin{bmatrix} -1 & -2 \\ 4 & 7 \end{bmatrix} = \begin{bmatrix} -2 & -4 \\ 17 & 31 \end{bmatrix}; BA = \begin{bmatrix} -1 & -2 \\ 4 & 7 \end{bmatrix}\begin{bmatrix} 2 & 0 \\ -5 & 3 \end{bmatrix} = \begin{bmatrix} 8 & -6 \\ -27 & 21 \end{bmatrix}$$

31. *A* is 2×3 and *B* is 3×2 so *AB* is 2×2 and *BA* is 3×3.

$$AB = \begin{bmatrix} 2 & -1 & 3 \\ 2 & 4 & 0 \end{bmatrix}\begin{bmatrix} 1 & 0 \\ -1 & 2 \\ 0 & 3 \end{bmatrix} = \begin{bmatrix} 3 & 7 \\ -2 & 8 \end{bmatrix}; BA = \begin{bmatrix} 1 & 0 \\ -1 & 2 \\ 0 & 3 \end{bmatrix}\begin{bmatrix} 2 & -1 & 3 \\ 2 & 4 & 0 \end{bmatrix} = \begin{bmatrix} 2 & -1 & 3 \\ 2 & 9 & -3 \\ 6 & 12 & 0 \end{bmatrix}$$

33. *B* is the inverse of *A*.

$$AB = \begin{bmatrix} 8 & 5 \\ 6 & 4 \end{bmatrix}\begin{bmatrix} 2 & -2.5 \\ -3 & 4 \end{bmatrix} = \begin{bmatrix} 1 & 0 \\ 0 & 1 \end{bmatrix} \text{ and } BA = \begin{bmatrix} 2 & -2.5 \\ -3 & 4 \end{bmatrix}\begin{bmatrix} 8 & 5 \\ 6 & 4 \end{bmatrix} = \begin{bmatrix} 1 & 0 \\ 0 & 1 \end{bmatrix}$$

35. (a) $\begin{array}{c} x - 3y = 4 \\ 2x - y = 3 \end{array} \Rightarrow AX = \begin{bmatrix} 1 & -3 \\ 2 & -1 \end{bmatrix} \begin{bmatrix} x \\ y \end{bmatrix} = \begin{bmatrix} 4 \\ 3 \end{bmatrix} = B$

(b) $X = A^{-1}B \Rightarrow \begin{bmatrix} x \\ y \end{bmatrix} = \begin{bmatrix} -\frac{1}{5} & \frac{3}{5} \\ -\frac{2}{5} & \frac{1}{5} \end{bmatrix} \begin{bmatrix} 4 \\ 3 \end{bmatrix} = \begin{bmatrix} 1 \\ -1 \end{bmatrix}$

37. (a) $AX = B \Rightarrow \begin{bmatrix} 12 & 7 & -3 \\ 8 & -11 & 13 \\ -23 & 0 & 9 \end{bmatrix} \begin{bmatrix} x \\ y \\ z \end{bmatrix} = \begin{bmatrix} 14.6 \\ -60.4 \\ -14.6 \end{bmatrix}$

(b) See Figure 37. $X = A^{-1}B \Rightarrow X = \begin{bmatrix} -0.5 \\ 1.7 \\ -2.9 \end{bmatrix}$

Figure 37

39. $A | I_2 = \begin{bmatrix} 1 & -2 & | & 1 & 0 \\ -1 & 1 & | & 0 & 1 \end{bmatrix} R_2 + R_1 \rightarrow \begin{bmatrix} 1 & -2 & | & 1 & 0 \\ 0 & -1 & | & 1 & 1 \end{bmatrix} (-1)R_2 \rightarrow \begin{bmatrix} 1 & -2 & | & 1 & 0 \\ 0 & 1 & | & -1 & -1 \end{bmatrix}$

$R_1 + 2R_2 \rightarrow \begin{bmatrix} 1 & 0 & | & -1 & -2 \\ 0 & 1 & | & -1 & -1 \end{bmatrix}; \; A^{-1} = \begin{bmatrix} -1 & -2 \\ -1 & -1 \end{bmatrix}$

41. Expanding about the first column results in

$\det A = 2 \det \begin{bmatrix} 3 & 4 \\ 0 & 5 \end{bmatrix} - 0 + 1 \det \begin{bmatrix} 1 & 3 \\ 3 & 4 \end{bmatrix} = (2)(15 - 0) + (1)(4 - 9) = 25.$

43. $\det A = \det \begin{bmatrix} 13 & 22 \\ 55 & -57 \end{bmatrix} = (13)(-57) - (55)(22) = -1951 \neq 0; \; A$ is invertible.

45. The given equations result in the following nonlinear system of equations: $A(l, w) = 77 \Rightarrow lw = 77$ and

$P(l, w) = 36 \Rightarrow 2l + 2w = 36.$ Begin by solving the second equation for l.

$2l + 2w = 36 \Rightarrow 2l = 36 - 2w \Rightarrow l = 18 - w.$ Substitute this into the first equation.

$lw = 77 \Rightarrow (18 - w)w = 77 \Rightarrow 18w - w^2 = 77 \Rightarrow w^2 - 18w + 77 = 0.$ This is a quadratic equation

that can be solved by factoring. $w^2 - 18w + 77 = 0 \Rightarrow (w - 7)(w - 11) = 0 \Rightarrow w = 7$ or $11.$

Since $l = 18 - w,$ if $w = 7,$ then $l = 11$ and if $w = 11$ then $l = 7.$ For this rectangle $l = 11$ and $w = 7.$

47. (a) Let x represent the amount of the 7% loan and let y represent the amount of the 9% loan. Then the system

of equations is $x + y = 2000$ and $0.07x + 0.09y = 156$. Multiply the first equation by 0.09 and subtract.

$$0.09x + 0.09y = 180$$
$$\underline{0.07x + 0.09y = 156}$$
$$0.02x \qquad\;\; = 24 \;\Rightarrow\; x = 1200 \text{ and } y = 2000 - 1200 = 800$$

The loan amounts are $1200 at 7% and $800 at 9%.

(b) Graph $Y_1 = 2000 - X$ and $Y_2 = (156 - 0.07X)/0.09$. The graph intersect at (1200, 800). See Figure 47.

The loan amounts are $1200 at 7% and $800 at 9%.

[0, 2000, 100] by [0, 2000, 100]

Figure 47

49. Let x be the number of CDs of type A purchased and let y be the number of CDs of type B. Then from the

table we see that $1x + 2y = 37.47$ and $2x + 3y = 61.95$.

$$AX = B \;\Rightarrow\; \begin{bmatrix} 1 & 2 \\ 2 & 3 \end{bmatrix}\begin{bmatrix} x \\ y \end{bmatrix} = \begin{bmatrix} 37.47 \\ 61.95 \end{bmatrix}$$

Using a graphing calculator to solve the system yeilds that type A CDs cost $11.49 and type B CD's cost $12.99.

51. Enter the vertices as columns in a counterclockwise direction.

$$D = \frac{1}{2}\det\begin{bmatrix} 0 & 5 & 2 \\ 0 & 2 & 5 \\ 1 & 1 & 1 \end{bmatrix} = 10.5; \text{ The area of the triangle is } 10.5 \text{ square units.}$$

53. Since P varies jointly as the square of x and the cube of y, the variation equation $P = kx^2y^3$ must hold.

If $P = 432$ when $x = 2$ and $y = 3$, then $432 = k(2)^2(3)^3 \Rightarrow 432 = 108k \Rightarrow k = 4$.

Our variation equation becomes $P = 4x^2y^3$. Thus, when $x = 3$ and $y = 5$, $P = 4(3)^2(5)^3 = 4500$.

Chapter 10: Conic Sections

10.1: Parabolas

1. See Figure 1.

3. See Figure 3.

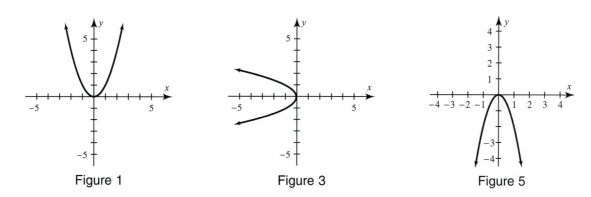

| Figure 1 | Figure 3 | Figure 5 |

5. See Figure 5.

7. See Figure 7.

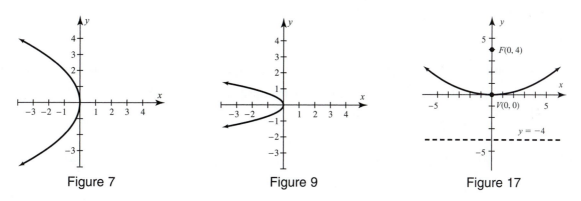

| Figure 7 | Figure 9 | Figure 17 |

9. See Figure 9.

11. Opens upward; e

13. Opens to the left; a

15. Opens to the right, passes through (2, 2); d

17. The equation $16y = x^2$ is in the form $x^2 = 4py$, and the vertex is $V(0, 0)$. Thus, $16 = 4p$ or $p = 4$. The focus

   is $F(0, 4)$, the equation of the directrix is $y = -4$, and the parabola opens upward. The graph of

   $y = \dfrac{1}{16}x^2$ is shown in Figure 17.

19. The equation $x = \dfrac{1}{8}y^2$ can be written as $y^2 = 8x$, which is in the form $y^2 = 4px$. The vertex is $V(0, 0)$. Thus, $8 = 4p$ or $p = 2$. The focus is $F(2, 0)$, the equation of the directrix is $x = -2$, and the parabola opens to the right. The graph of $x = \dfrac{1}{8}y^2$ is shown in Figure 19.

21. The equation $-4x = y^2$ can be written as $y^2 = -4x$, which is in the form $y^2 = 4px$. The vertex is $V(0, 0)$. Thus, $-4 = 4p$ or $p = -1$. The focus is $F(-1, 0)$, the equation of the directrix is $x = 1$, and the parabola opens to the left. The graph of $x = -\dfrac{1}{4}y^2$ is shown in Figure 21.

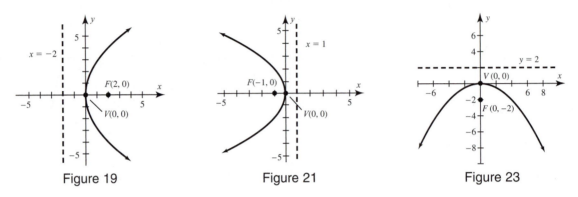

Figure 19   Figure 21   Figure 23

23. The equation $x^2 = -8y$ is in the form $x^2 = 4py$, and the vertex is $V(0, 0)$. Thus, $-8 = 4p$ or $p = -2$. The focus is $F(0, -2)$, the equation of the directrix is $y = 2$, and the parabola opens downward. The graph of $y = -\dfrac{1}{8}x^2$ is shown in Figure 23.

25. The equation $2y^2 = -8x$ can be written as $y^2 = -4x$, which is in the form $y^2 = 4px$. The vertex is $V(0, 0)$. Thus, $-4 = 4p$ or $p = -1$. The focus is $F(-1, 0)$, the equation of the directrix is $x = 1$, and the parabola opens to the left. The graph of $x = -\dfrac{1}{4}y^2$ is shown in Figure 25.

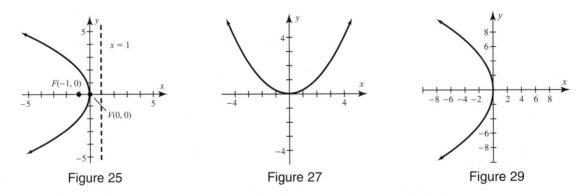

Figure 25   Figure 27   Figure 29

27. The focus is $F(0, 1)$ and the vertex is $V(0, 0)$. The distance between these points is 1. Since the focus is above the directrix, the parabola opens upward, so $p = 1$. Since the line passing through F and V is vertical, the parabola has a vertical axis. Its equation is given by $x^2 = 4py$ or $x^2 = 4y$. The graph of $y = \dfrac{1}{4}x^2$ is shown in Figure 27.

29. The focus is $F(-3, 0)$ and the vertex is $V(0, 0)$. The distance between these points is 3. Since the focus is left of the directrix, the parabola opens to the left, so $p = -3$. Since the line passing through F and V is horizontal, the parabola has a horizontal axis. Its equation is given by $y^2 = 4px$ or $y^2 = -12x$. The graph of $x = -\dfrac{1}{12}y^2$ is shown in Figure 29.

31. If the vertex is $V(0, 0)$ and the focus is $F\left(0, \dfrac{3}{4}\right)$, then the parabola opens upward and $p = \dfrac{3}{4}$. Thus, $x^2 = 4py \Rightarrow x^2 = 3y$. Its graph is shown in Figure 31.

33. If the vertex is $V(0, 0)$ and the directrix is $x = 2$, then the parabola opens to the left and $p = -2$. Thus, $y^2 = 4px \Rightarrow y^2 = -8x$. Its graph is shown in Figure 33.

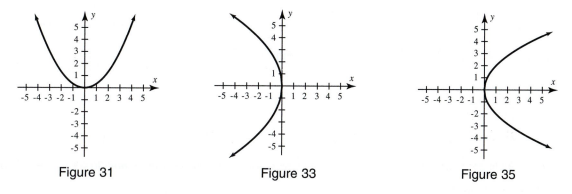

Figure 31   Figure 33   Figure 35

35. If the vertex is $V(0, 0)$ and the focus is $F(1, 0)$, then the parabola opens to the right and $p = 1$. Thus, $y^2 = 4px \Rightarrow y^2 = 4x$. Its graph is shown in Figure 35.

37. If the vertex is $V(0, 0)$ and the directrix is $x = \dfrac{1}{4}$, then the parabola opens to the left and $p = -\dfrac{1}{4}$. Thus, $y^2 = 4px \Rightarrow y^2 = -x$. Its graph is shown in Figure 37.

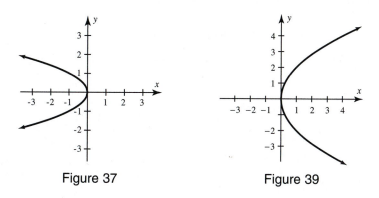

Figure 37   Figure 39

39. If the vertex is $V(0, 0)$ and the parabola has a horizontal axis, the equation is in the form $y^2 = 4px$. Find the value of p by using the fact that the parabola passes through $(1, -2)$. Thus, $(-2)^2 = 4p(1) \Rightarrow p = 1$. The equation is $y^2 = 4x$. Its graph is shown in Figure 39.

41. If the focus is $F(0, -3)$ and the equation of the directrix is $y = 3$, the vertex is $V(0, 0)$, the parabola opens downward, and $p = -3$. Thus, $x^2 = 4py \Rightarrow x^2 = -12y$.

43. If the focus is $F(-1, 0)$ and the equation of the directrix is $x = 1$, the vertex is $V(0, 0)$, the parabola opens to the left, and $p = -1$. Thus, $y^2 = 4px \Rightarrow y^2 = -4x$.

45. See Figure 45.

47. See Figure 47.

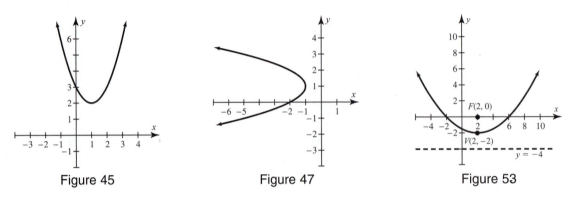

Figure 45   Figure 47   Figure 53

49. Vertex at $(1, 1)$; c

51. Vertex at $(0, 2)$; a

53. The equation $(x - 2)^2 = 8(y + 2)$ is in the form $(x - h)^2 = 4p(y - k)$, with $(h, k) = (2, -2)$ and $p = 2$. The parabola opens upward, with vertex at $(2, -2)$, focus at $(2, 0)$, and equation of directrix $y = -4$. The graph is shown in Figure 53.

55. The equation $x = -\dfrac{1}{4}(y + 3)^2 + 2$ can be written as $(y + 3)^2 = -4(x - 2)$, which is in the form $(y - k)^2 = 4p(x - h)$, with $(h, k) = (2, -3)$ and $p = -1$. The parabola opens to the left, with vertex at $(2, -3)$, focus at $(1, -3)$, and equation of directrix $x = 3$. The graph is shown in Figure 55.

57. The equation $y = -\dfrac{1}{4}(x + 2)^2$ can be written as $(x + 2)^2 = -4y$, which is in the form $(x - h)^2 = 4p(y - k)$, with $(h, k) = (-2, 0)$ and $p = -1$. The parabola opens downward, with vertex at $(-2, 0)$, focus at $(-2, -1)$, and equation of directrix $y = 1$. The graph is shown in Figure 57.

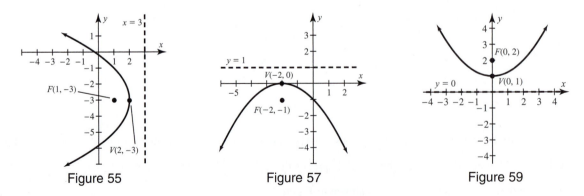

Figure 55   Figure 57   Figure 59

59. If the focus is at $(0, 2)$ and the vertex at $(0, 1)$, the parabola opens upward and $p = 1$. Substituting in $(x - h)^2 = 4p(y - k)$, we get $(x - 0)^2 = 4(1)(y - 1)$ or $x^2 = 4(y - 1)$. See Figure 59.

61. If the focus is at $(0, 0)$ and the directrix has equation $x = -2$, the vertex is at $(-1, 0)$, $p = 1$, and the parabola

    opens to the right. Substituting in $(y - k)^2 = 4p(x - h)$, we get

    $(y - 0)^2 = 4(1)(x - (-1))$ or $y^2 = 4(x + 1)$. See Figure 61.

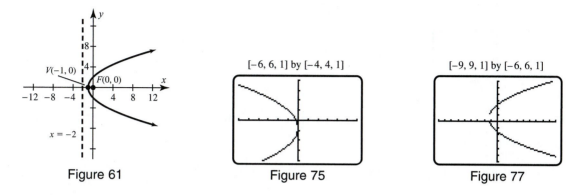

[−6, 6, 1] by [−4, 4, 1]   [−9, 9, 1] by [−6, 6, 1]

Figure 61   Figure 75   Figure 77

63. If the focus is at $(-1, 3)$ and the directrix has equation $y = 7$, the vertex is at $(-1, 5)$, $p = -2$, and the

    parabola opens downward. Substituting in $(x - h)^2 = 4p(y - k)$, we get $(x + 1)^2 = 4(-2)(y - 5)$ or

    $(x + 1)^2 = -8(y - 5)$.

65. Since the parabola has a horizontal axis, the equation is in the form $(y - k)^2 = a(x - h)$. Find the value of a

    by using the fact that the parabola passes through $(-4, 0)$ and the vertex is $V(-2, 3)$.

    Substituting $x = -4, y = 0, h = -2$ and $k = 3$ yields $(0 - 3)^2 = a(-4 - (-2)) \Rightarrow a = -\dfrac{9}{2}$.

    The equation is $(y - 3)^2 = -\dfrac{9}{2}(x + 2)$.

67. $-2x = y^2 + 6x + 10 \Rightarrow y^2 = -8x - 10 \Rightarrow (y - 0)^2 = -8\left(x + \dfrac{5}{4}\right)$

69. $x = 2y^2 + 4y - 1 \Rightarrow 2y^2 + 4y = x + 1 \Rightarrow y^2 + 2y = \dfrac{1}{2}(x + 1) \Rightarrow$

    $y^2 + 2y + 1 = \dfrac{1}{2}(x + 1) + 1 \Rightarrow (y + 1)^2 = \dfrac{1}{2}(x + 1 + 2) \Rightarrow (y + 1)^2 = \dfrac{1}{2}(x + 3)$

71. $x^2 - 3x + 4 = 2y \Rightarrow x^2 - 3x = 2y - 4 \Rightarrow x^2 - 3x + \dfrac{9}{4} = 2y - 4 + \dfrac{9}{4} \Rightarrow$

    $\left(x - \dfrac{3}{2}\right)^2 = 2y - \dfrac{7}{4} \Rightarrow \left(x - \dfrac{3}{2}\right)^2 = 2\left(y - \dfrac{7}{8}\right)$

73. $4y^2 + 4y - 5 = 5x \Rightarrow 4y^2 + 4y = 5x + 5 \Rightarrow y^2 + y = \dfrac{5}{4}(x + 1) \Rightarrow$

    $y^2 + y + \dfrac{1}{4} = \dfrac{5}{4}(x + 1) + \dfrac{1}{4} \Rightarrow \left(y + \dfrac{1}{2}\right)^2 = \dfrac{5}{4}\left(x + 1 + \dfrac{1}{5}\right) \Rightarrow \left(y + \dfrac{1}{2}\right)^2 = \dfrac{5}{4}\left(x + \dfrac{6}{5}\right)$

75. $y = -0.75 \pm \sqrt{-3x}$; See Figure 75.

77. $y = 0.5 \pm \sqrt{3.1(x + 1.3)}$; See Figure 77. Note: If a break in the graph appears near the vertex, it should

    not be there. It is a result of the low resolution of the graphing calculator screen.

79. $y = -1 \pm \sqrt{\dfrac{x}{2.3}}$; See Figure 79.

[-6, 6, 1] by [-4, 4, 1]

Figure 79

81. $x^2 = 2y$ and $x^2 = y + 1 \Rightarrow 2y = y + 1 \Rightarrow y = 1$; $x^2 = 2y$ when $y = 1 \Rightarrow x^2 = 2 \Rightarrow x = \pm\sqrt{2}$; the

solution is $(\pm\sqrt{2}, 1)$.

83. $\dfrac{1}{3}y^2 = -3x$ and $y^2 = x + 1 \Rightarrow -9x = x + 1 \Rightarrow x = -\dfrac{1}{10}$; $y^2 = x + 1$ when $x = -\dfrac{1}{10} \Rightarrow$

$y^2 = -\dfrac{1}{10} + 1 \Rightarrow y = \pm\sqrt{0.9}$; the solution is $\left(-\dfrac{1}{10}, \pm\sqrt{0.9}\right)$.

85. $(y - 1)^2 = x + 1$ and $(y + 2)^2 = -x + 4 \Rightarrow (y - 1)^2 + (y + 2)^2 = 5 \Rightarrow$

$y^2 - 2y + 1 + y^2 + 4y + 4 = 5 \Rightarrow 2y^2 + 2y = 0 \Rightarrow 2y(y + 1) = 0 \Rightarrow y = 0$ or

$y = -1$; $(y - 1)^2 = x + 1$ when $y = 0 \Rightarrow 1 = x + 1 \Rightarrow x = 0$, when $y = -1 \Rightarrow 4 = x + 1 \Rightarrow$

$x = 3$, the solution is $(0, 0)$, $(3, -1)$.

87. Substitute the point $(3, 0.75)$ into $x^2 = 4py$ and solve for p; $9 = 4p(0.75) \Rightarrow 9 = 3p \Rightarrow p = 3$.

The receiver should be 3 feet from the vertex.

89. (a) Substitute the point $(105, 32)$ into $y = ax^2$ and solve for a; $32 = a(105)^2 \Rightarrow a = \dfrac{32}{11{,}025}$.

The equation is $y = \dfrac{32}{11{,}025}x^2$.

(b) Rewriting the answer in (a) we have $x^2 = \dfrac{11{,}025}{32}y$, so $4p = \dfrac{11{,}025}{32}$ and $p = \dfrac{11{,}025}{128} \approx 86.1$.

The receiver should be located about 86.1 feet from the vertex.

91. (a) Since $y^2 = 100x$, $4p = 100$ and $p = 25$. Thus, the coordinates of the sun are $(25, 0)$.

(b) The minimum distance occurs when the comet is at the vertex of the parabola, so the minimum distance is

25 million miles.

93. The pipe should be at the focus, so $p = 18$, $k = 4p = 72$ inches or 6 ft.

10.2: Ellipses

1. $\dfrac{x^2}{4} + \dfrac{y^2}{9} = 1 \Rightarrow a = 3$ and $b = 2$. $a^2 - b^2 = 3^2 - 2^2 = 5 = c^2 \Rightarrow c = \sqrt{5}$. The foci are $(0, \pm\sqrt{5})$, the

endpoints of the major axis (vertices) are $(0, \pm3)$, while the endpoints of the minor axis are $(\pm2, 0)$. The

ellipse is graphed in Figure 1.

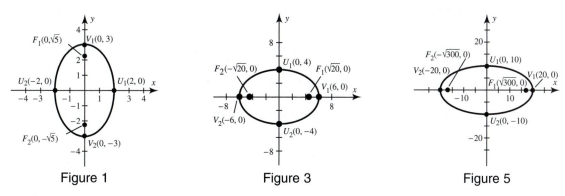

Figure 1   Figure 3   Figure 5

3. $\dfrac{x^2}{36} + \dfrac{y^2}{16} = 1 \Rightarrow a = 6$ and $b = 4$. $a^2 - b^2 = 6^2 - 4^2 = 20 = c^2 \Rightarrow c = \sqrt{20}$. The foci are $(\pm\sqrt{20}, 0)$,

the endpoints of the major axis (vertices) are $(\pm 6, 0)$, while the endpoints of the minor axis are $(0, \pm 4)$. The

ellipse is graphed in Figure 3.

5. $x^2 + 4y^2 = 400 \Rightarrow \dfrac{x^2}{400} + \dfrac{y^2}{100} = 1 \Rightarrow a = 20$ and $b = 10$.

$a^2 - b^2 = 400 - 100 = 300 = c^2 \Rightarrow c = \sqrt{300}$. The foci are $(\pm\sqrt{300}, 0)$, the endpoints of the major

axis (vertices) are $(\pm 20, 0)$, while the endpoints of the minor axis are $(0, \pm 10)$. The ellipse is graphed in

Figure 5.

7. $25x^2 + 9y^2 = 225 \Rightarrow \dfrac{x^2}{9} + \dfrac{y^2}{25} = 1 \Rightarrow a = 5$ and $b = 3$. $a^2 - b^2 = 25 - 9 = 16 = c^2 \Rightarrow c = 4$. The

foci are $(0, \pm 4)$, the endpoints of the major axis (vertices) are $(0, \pm 5)$, while the endpoints of the minor axis

are $(\pm 3, 0)$. The ellipse is graphed in Figure 7.

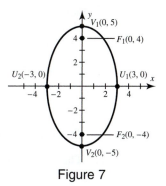

Figure 7

9. Vertices $(0, \pm 6)$; b

11. Vertices $(\pm 4, 0)$; c

13. The ellipse is centered at $(0, 0)$ and has a horizontal major axis. Its standard equation has the form

$\dfrac{x^2}{a^2} + \dfrac{y^2}{b^2} = 1$. The endpoints of the major axis are $(\pm 6, 0)$ and the endpoints of the minor axis are $(0, \pm 4)$.

It follows that $a = 6$ and $b = 4$, and the standard equation is $\dfrac{x^2}{36} + \dfrac{y^2}{16} = 1$. The foci lie on the horizontal

major axis and can be determined as follows. $c^2 = a^2 - b^2 = 36 - 16 = 20$. Thus, $c = \sqrt{20}$, and the

coordinates of the foci are $(\pm\sqrt{20}, 0)$.

15. The ellipse is centered at $(0, 0)$ and has a vertical major axis. Its standard equation has the form

$\dfrac{x^2}{b^2} + \dfrac{y^2}{a^2} = 1$. The endpoints of the major axis are $(0, \pm 4)$ and the endpoints of the minor axis are $(\pm 2, 0)$.

It follows that $a = 4$ and $b = 2$, and the standard equation is $\dfrac{x^2}{4} + \dfrac{y^2}{16} = 1$. The foci lie on the vertical

major axis and can be determined as follows. $c^2 = a^2 - b^2 = 16 - 4 = 12$. Thus, $c = \sqrt{12}$, and the

coordinates of the foci are $(0, \pm\sqrt{12})$.

17. To sketch a graph of an ellipse centered at the origin, it is helpful to plot the vertices and the endpoints of the

minor axis. The vertices are $V(\pm 5, 0)$ so $a = 5$, the foci are $F(\pm 4, 0)$ so $c = 4$, and the endpoints of the minor

axis are $U(0, \pm 3)$ so $b = 3$. A graph of the ellipse is shown in Figure 17. Its equation is $\dfrac{x^2}{25} + \dfrac{y^2}{9} = 1$.

19. The vertices are $V(0, \pm 3)$ so $a = 3$, the foci are $F(0, \pm 2)$ so $c = 2$, and the endpoints of the minor axis are

$U(\pm\sqrt{5}, 0)$ so $b = \sqrt{5}$. A graph of the ellipse is shown in Figure 19. Its equation is $\dfrac{x^2}{5} + \dfrac{y^2}{9} = 1$.

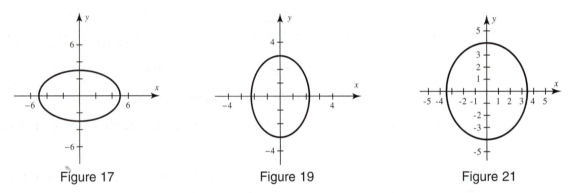

Figure 17   Figure 19   Figure 21

21. Foci of $F(0, \pm 2) \Rightarrow c = 2$ and $V(0, \pm 4) \Rightarrow a = 4$. The major axis lies on the y-axis. The value of b is as

follows: $a^2 - b^2 = c^2 \Rightarrow a^2 - c^2 = b^2 \Rightarrow 4^2 - 2^2 = b^2 \Rightarrow b^2 = 12 \Rightarrow b = \sqrt{12}$. The equation of the

ellipse is $\dfrac{x^2}{12} + \dfrac{y^2}{16} = 1$. Its graph is shown in Figure 21.

23. Foci of $F(\pm 5, 0) \Rightarrow c = 5$ and $V(\pm 6, 0) \Rightarrow a = 6$. The major axis lies on the x-axis. The value of b is as

follows: $a^2 - b^2 = c^2 \Rightarrow a^2 - c^2 = b^2 \Rightarrow 6^2 - 5^2 = b^2 \Rightarrow b^2 = 11 \Rightarrow b = \sqrt{11}$. The equation of the

ellipse is $\dfrac{x^2}{36} + \dfrac{y^2}{11} = 1$. Its graph is shown in Figure 23.

25. Horizontal major axis of length $8 \Rightarrow a = 4$. Minor axis of length $6 \Rightarrow b = 3$. The major axis lies on the

x-axis. The equation of the ellipse is $\dfrac{x^2}{16} + \dfrac{y^2}{9} = 1$. Its graph is shown in Figure 25.

27. $e = \dfrac{c}{a} = \dfrac{2}{3} \Rightarrow 3c = 2a \Rightarrow c = \dfrac{2}{3}a$. Since the major axis is length $6 \Rightarrow a = 3$. Thus, $c = \dfrac{2}{3}(3) = 2$.

Then the value of b is given by the following: $a^2 - b^2 = c^2 \Rightarrow a^2 - c^2 = b^2 \Rightarrow 3^2 - 2^2 = b^2 \Rightarrow$

$b^2 = 5 \Rightarrow b = \sqrt{5}$. The equation of the ellipse is $\dfrac{x^2}{9} + \dfrac{y^2}{5} = 1$. Its graph is shown in Figure 27.

Figure 23   Figure 25   Figure 27

29. To translate the center from $(0, 0)$ to $(2, -1)$ replace x with $(x - 2)$ and y with $(y + 1)$. This new equation is
$\dfrac{(x - 2)^2}{4} + \dfrac{(y + 1)^2}{3} = 1.$ See Figure 29.

31. To translate the center from $(0, 0)$ to $(-3, -4)$ replace x with $(x + 3)$ and y with $(y + 4)$. This new equation
is $\dfrac{(x + 3)^2}{2} + \dfrac{(y + 4)^2}{9} = 1.$ See Figure 31.

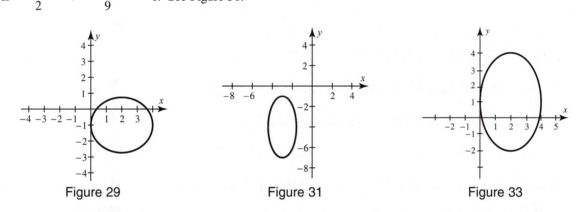

Figure 29   Figure 31   Figure 33

33. The ellipse is centered at $(2, 1)$. The major axis has length $2a = 6$ and the length of the minor axis is $2b = 4$. The major axis is parallel to the y-axis. The graph is shown in Figure 33.

35. The ellipse is centered at $(-1, -2)$. The major axis has length $2a = 10$ and the length of the minor axis is $2b = 8$. The major axis is parallel to the y-axis. The graph is shown in Figure 35.

37. The ellipse is centered at $(-2, 0)$. The horizontal major axis has length $2a = 4$ and the vertical minor axis has length $2b = 2$. The graph is shown in Figure 37.

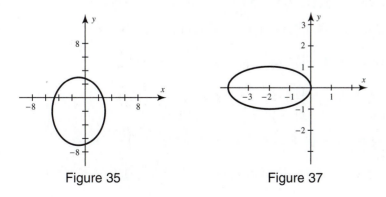

Figure 35   Figure 37

39. Center at $(2, -4)$; d

41. Center at $(-1, 1)$; c

43. Center at $(1, 1)$, $a = 5$, $b = 3$, major axis vertical. $c^2 = a^2 - b^2 = 25 - 9 = 16 \Rightarrow c = 4$.

Foci: $(1, 1 \pm 4)$; veritices: $(1, 1 \pm 5)$; the graph is shown in Figure 43.

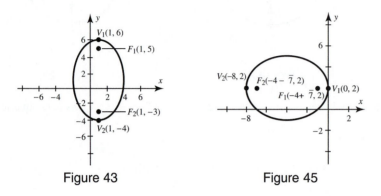

Figure 43   Figure 45

45. Center at $(-4, 2)$, $a = 4$, $b = 3$, major axis horizontal. $c^2 = a^2 - b^2 = 16 - 9 = 7 \Rightarrow c = \sqrt{7}$.

Foci: $(-4 \pm \sqrt{7}, 2)$; veritices: $(-4 \pm 4, 2)$; the graph is shown in Figure 45.

47. Since the center is $(2, 1)$ and a focus is $(2, 3)$, $c = 2$. Since the vertex is $(2, 4)$, $a = 3$;

$b^2 = a^2 - c^2 = 9 - 4 = 5$; the major axis is vertical. The equation is $\dfrac{(x - 2)^2}{5} + \dfrac{(y - 1)^2}{9} = 1$.

49. The center is halfway between the vertices at $(0, 2)$; $a = 3$ and $c = 2$; $b^2 = a^2 - c^2 = 9 - 4 = 5$; the

major axis is horizontal. The equation is $\dfrac{x^2}{9} + \dfrac{(y - 2)^2}{5} = 1$.

51. Center at $(2, 4)$, $a = 4$ and $b = 2$, major axis parallel to the x-axis; the equation is

$\dfrac{(x - 2)^2}{16} + \dfrac{(y - 4)^2}{4} = 1$.

53. $9x^2 + 18x + 4y^2 - 8y - 23 = 0 \Rightarrow 9(x^2 + 2x) + 4(y^2 - 2y) = 23 \Rightarrow$

$9(x^2 + 2x + 1) + 4(y^2 - 2y + 1) = 23 + 9 + 4 \Rightarrow 9(x + 1)^2 + 4(y - 1)^2 = 36 \Rightarrow$

$\dfrac{(x + 1)^2}{4} + \dfrac{(y - 1)^2}{9} = 1$; The center is $(-1, 1)$. The vertices are

$(-1, 1 - 3), (-1, 1 + 3)$ or $(-1, -2), (-1, 4)$.

55. $4x^2 + 8x + y^2 + 2y + 1 = 0 \Rightarrow 4(x^2 + 2x) + (y^2 + 2y) = -1 \Rightarrow$

$4(x^2 + 2x + 1) + (y^2 + 2y + 1) = -1 + 4 + 1 \Rightarrow 4(x + 1)^2 + (y + 1)^2 = 4 \Rightarrow$

$\dfrac{(x + 1)^2}{1} + \dfrac{(y + 1)^2}{4} = 1$; The center is $(-1, -1)$. The vertices are

$(-1, -1 - 2), (-1, -1 + 2)$ or $(-1, -3), (-1, 1)$.

57. $4x^2 + 16x + 5y^2 - 10y + 1 = 0 \Rightarrow 4(x^2 + 4x) + 5(y^2 - 2y) = -1 \Rightarrow$

$4(x^2 + 4x + 4) + 5(y^2 - 2y + 1) = -1 + 16 + 5 \Rightarrow 4(x + 2)^2 + 5(y - 1)^2 = 20 \Rightarrow$

$\dfrac{(x + 2)^2}{5} + \dfrac{(y - 1)^2}{4} = 1$; The center is $(-2, 1)$. The vertices are $(-2 - \sqrt{5}, 1), (-2 + \sqrt{5}, 1)$.

59. $16x^2 - 16x + 4y^2 + 12y = 51 \Rightarrow 16(x^2 - x) + 4(y^2 + 3y) = 51 \Rightarrow$

$16\left(x^2 - x + \dfrac{1}{4}\right) + 4\left(y^2 + 3y + \dfrac{9}{4}\right) = 51 + 4 + 9 \Rightarrow 16\left(x - \dfrac{1}{2}\right)^2 + 4\left(y + \dfrac{3}{2}\right)^2 = 64 \Rightarrow$

$\dfrac{(x - \frac{1}{2})^2}{4} + \dfrac{(y + \frac{3}{2})^2}{16} = 1$

The center is $\left(\dfrac{1}{2}, -\dfrac{3}{2}\right)$. The vertices are $\left(\dfrac{1}{2}, -\dfrac{3}{2} - 4\right), \left(\dfrac{1}{2}, -\dfrac{3}{2} + 4\right)$ or $\left(\dfrac{1}{2}, -\dfrac{11}{2}\right), \left(\dfrac{1}{2}, \dfrac{5}{2}\right)$.

61. $y = \pm\sqrt{10\left(1 - \dfrac{x^2}{15}\right)}$; See Figure 61.

63. $y = \pm\sqrt{\dfrac{25 - 4.1x^2}{6.3}}$; See Figure 63.

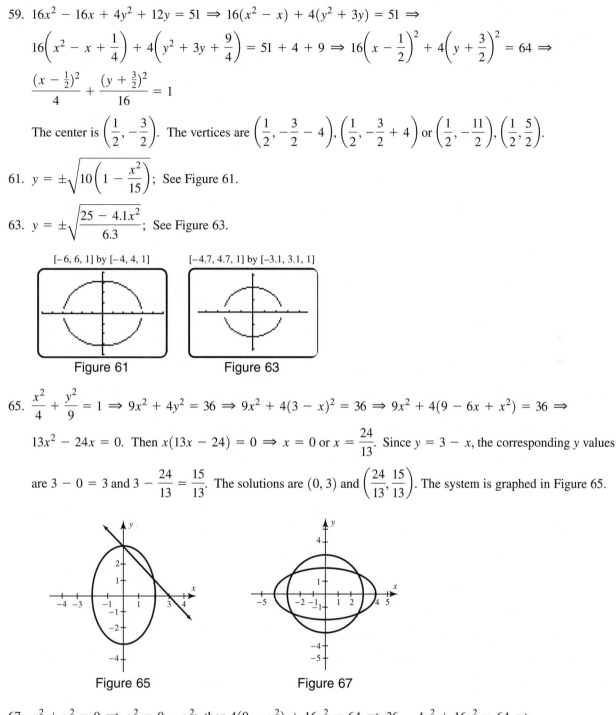

$[-6, 6, 1]$ by $[-4, 4, 1]$ $[-4.7, 4.7, 1]$ by $[-3.1, 3.1, 1]$

Figure 61   Figure 63

65. $\dfrac{x^2}{4} + \dfrac{y^2}{9} = 1 \Rightarrow 9x^2 + 4y^2 = 36 \Rightarrow 9x^2 + 4(3 - x)^2 = 36 \Rightarrow 9x^2 + 4(9 - 6x + x^2) = 36 \Rightarrow$

$13x^2 - 24x = 0$. Then $x(13x - 24) = 0 \Rightarrow x = 0$ or $x = \dfrac{24}{13}$. Since $y = 3 - x$, the corresponding y values

are $3 - 0 = 3$ and $3 - \dfrac{24}{13} = \dfrac{15}{13}$. The solutions are $(0, 3)$ and $\left(\dfrac{24}{13}, \dfrac{15}{13}\right)$. The system is graphed in Figure 65.

Figure 65   Figure 67

67. $x^2 + y^2 = 9 \Rightarrow x^2 = 9 - y^2$; then $4(9 - y^2) + 16y^2 = 64 \Rightarrow 36 - 4y^2 + 16y^2 = 64 \Rightarrow$

$y^2 = \dfrac{7}{3} \Rightarrow y = \pm\sqrt{\dfrac{7}{3}}$. Substituting in

$x^2 + y^2 = 9$ we find $x^2 + \dfrac{7}{3} = 9$, $x^2 = \dfrac{27}{3} - \dfrac{7}{3} = \dfrac{20}{3}$, so $x = \pm\sqrt{\dfrac{20}{3}}$.

There are four solutions: $\left(\pm\sqrt{\dfrac{20}{3}}, \pm\sqrt{\dfrac{7}{3}}\right)$. The system is graphed in Figure 67.

69. $x^2 + y^2 = 9 \Rightarrow y^2 = 9 - x^2$; then $2x^2 + 3(9 - x^2) = 18 \Rightarrow 2x^2 + 27 - 3x^2 = 18 \Rightarrow$

$x^2 = 9 \Rightarrow x = \pm 3$;

Substituting in $x^2 + y^2 = 9$ we find $9 + y^2 = 9$, so $y = 0$. There are two solutions: $(\pm 3, 0)$. The system is graphed in Figure 69.

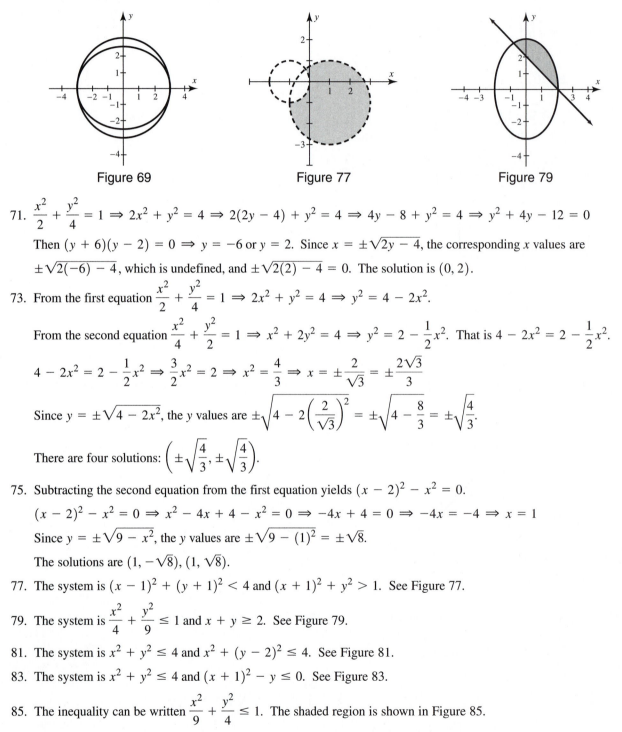

Figure 69   Figure 77   Figure 79

71. $\dfrac{x^2}{2} + \dfrac{y^2}{4} = 1 \Rightarrow 2x^2 + y^2 = 4 \Rightarrow 2(2y - 4) + y^2 = 4 \Rightarrow 4y - 8 + y^2 = 4 \Rightarrow y^2 + 4y - 12 = 0$

Then $(y + 6)(y - 2) = 0 \Rightarrow y = -6$ or $y = 2$. Since $x = \pm\sqrt{2y - 4}$, the corresponding x values are

$\pm\sqrt{2(-6) - 4}$, which is undefined, and $\pm\sqrt{2(2) - 4} = 0$. The solution is $(0, 2)$.

73. From the first equation $\dfrac{x^2}{2} + \dfrac{y^2}{4} = 1 \Rightarrow 2x^2 + y^2 = 4 \Rightarrow y^2 = 4 - 2x^2$.

From the second equation $\dfrac{x^2}{4} + \dfrac{y^2}{2} = 1 \Rightarrow x^2 + 2y^2 = 4 \Rightarrow y^2 = 2 - \dfrac{1}{2}x^2$. That is $4 - 2x^2 = 2 - \dfrac{1}{2}x^2$.

$4 - 2x^2 = 2 - \dfrac{1}{2}x^2 \Rightarrow \dfrac{3}{2}x^2 = 2 \Rightarrow x^2 = \dfrac{4}{3} \Rightarrow x = \pm\dfrac{2}{\sqrt{3}} = \pm\dfrac{2\sqrt{3}}{3}$

Since $y = \pm\sqrt{4 - 2x^2}$, the y values are $\pm\sqrt{4 - 2\left(\dfrac{2}{\sqrt{3}}\right)^2} = \pm\sqrt{4 - \dfrac{8}{3}} = \pm\sqrt{\dfrac{4}{3}}$.

There are four solutions: $\left(\pm\sqrt{\dfrac{4}{3}}, \pm\sqrt{\dfrac{4}{3}}\right)$.

75. Subtracting the second equation from the first equation yields $(x - 2)^2 - x^2 = 0$.

$(x - 2)^2 - x^2 = 0 \Rightarrow x^2 - 4x + 4 - x^2 = 0 \Rightarrow -4x + 4 = 0 \Rightarrow -4x = -4 \Rightarrow x = 1$

Since $y = \pm\sqrt{9 - x^2}$, the y values are $\pm\sqrt{9 - (1)^2} = \pm\sqrt{8}$.

The solutions are $(1, -\sqrt{8}), (1, \sqrt{8})$.

77. The system is $(x - 1)^2 + (y + 1)^2 < 4$ and $(x + 1)^2 + y^2 > 1$. See Figure 77.

79. The system is $\dfrac{x^2}{4} + \dfrac{y^2}{9} \leq 1$ and $x + y \geq 2$. See Figure 79.

81. The system is $x^2 + y^2 \leq 4$ and $x^2 + (y - 2)^2 \leq 4$. See Figure 81.

83. The system is $x^2 + y^2 \leq 4$ and $(x + 1)^2 - y \leq 0$. See Figure 83.

85. The inequality can be written $\dfrac{x^2}{9} + \dfrac{y^2}{4} \leq 1$. The shaded region is shown in Figure 85.

Here $a = 3$ and $b = 2$. The area is $A = \pi ab = \pi(3)(2) = 6\pi \approx 18.85$ ft^2.

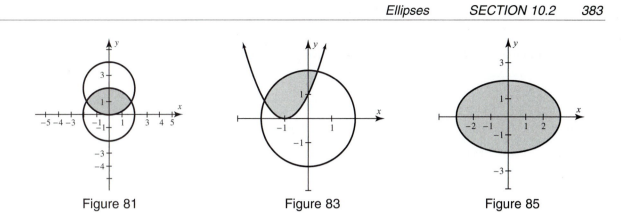

Figure 81   Figure 83   Figure 85

87. The shaded region is shown in Figure 87.

Here $a = 5$ and $b = 4$. The area is $A = \pi ab = \pi(5)(4) = 20\pi \approx 62.83$ ft^2.

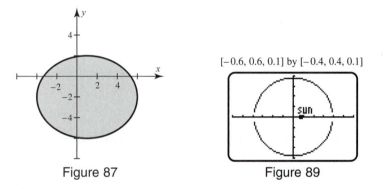

$[-0.6, 0.6, 0.1]$ by $[-0.4, 0.4, 0.1]$

Figure 87   Figure 89

89. $e = \dfrac{c}{a} = 0.206 \Rightarrow c = 0.206a$. Since $a = 0.387$, it follows that $c = 0.206(0.387) \approx 0.0797$. Then, the value

of b is given by the following: $a^2 - b^2 = c^2 \Rightarrow a^2 - c^2 = b^2 \Rightarrow 0.387^2 - 0.0797^2 = b^2 \Rightarrow$

$b^2 \approx 0.1434 \Rightarrow b \approx 0.379$. The major axis could be located on either the x- or y-axis. We will choose the

x-axis. Thus, the equation of the orbit is $\dfrac{x^2}{0.387^2} + \dfrac{y^2}{0.379^2} = 1$. The sun can be located on either of the foci.

We will locate the sun at $(0.0797, 0)$. To graph the orbit with a graphing calculator, solve the equation for the

ellipse for the variable y: $\dfrac{x^2}{0.387^2} + \dfrac{y^2}{0.379^2} = 1 \Rightarrow y = \pm 0.379\sqrt{1 - \dfrac{x^2}{0.387^2}}$. Graph each of the equations

and plot the sun at $(0.0797, 0)$. $Y_1 = 0.379\sqrt{((1 - X^2)/0.387^2)}$, $Y_2 = -Y_1$. See Figure 89.

91. The source and the stone are at the two foci of the ellipse, so the distance between them is $2c$.

$c^2 = a^2 - b^2 = 4^2 - (2.5)^2 = 16 - 6.25 = 9.75 \Rightarrow c = \sqrt{9.75}$. Thus, $2c = 2\sqrt{9.75} \approx 6.245$. The stone

should be 6.245 inches from the source.

93. $c = 50$ and $b = 40$ for the elliptical floor. $a = \sqrt{50^2 + 40^2} = 10 \cdot \sqrt{41} \approx 64.03$. The area of the ellipse is

πab, so the area of the floor is $\pi(64.03)(40) \approx 8046.25$ square feet.

95. The ellipse with a mjor axis of 620 feet and minor axis of 513 feet inplies vertices of

$(\pm 310, 0)$ and $(0, \pm 256.5)$. $c^2 = a^2 - b^2 \Rightarrow c^2 = (310)^2 - (256.5)^2 \Rightarrow c \approx 174.1$. The distance between

the foci is given as $2c = 2(174.1) = 348.2$ feet.

97. The equation of the ellipse is $\dfrac{x^2}{30^2} + \dfrac{y^2}{25^2} = 1$. Solving for y we get $y = 25\sqrt{1 - \dfrac{x^2}{900}}$. When $x = 15$,

$y = 25\sqrt{1 - \dfrac{225}{900}} \approx 21.65$ feet.

99. The minimum height is $4464 - (3960 + 164) = 340$ miles; the maximum height is

$4464 - (3960 - 164) = 668$ miles.

Section 10.3: Hyperbolas

1. The transverse axis is horizontal with $a = 3$ and $b = 7$. The vertices are $(\pm 3, 0)$. The asymptotes are

$y = \pm\dfrac{7}{3}x$. See Figure 1.

Since $c^2 = a^2 + b^2 \Rightarrow c = \pm\sqrt{9 + 49} \Rightarrow \sqrt{58}$, the foci are $(\pm\sqrt{58}, 0)$.

3. The transverse axis is vertical with $a = 6$ and $b = 4$. The vertices are $(0, \pm 6)$. The asymptotes are

$y = \pm\dfrac{3}{2}x$. See Figure 3.

Since $c^2 = a^2 + b^2 \Rightarrow c = \pm\sqrt{36 + 16} \Rightarrow \sqrt{52}$, the foci are $(0, \pm\sqrt{52})$.

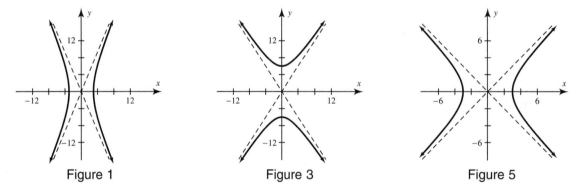

Figure 1   Figure 3   Figure 5

5. $x^2 - y^2 = 9 \Rightarrow \dfrac{x^2}{9} - \dfrac{y^2}{9} = 1$. The transverse axis is horizontal with $a = 3$ and $b = 3$. The vertices are

$(\pm 3, 0)$. The asymptotes are $y = \pm x$. See Figure 5.

Since $c^2 = a^2 + b^2 \Rightarrow c = \pm\sqrt{9 + 9} \Rightarrow \sqrt{18}$, the foci are $(\pm\sqrt{18}, 0)$.

7. $9y^2 - 16x^2 = 144 \Rightarrow \dfrac{y^2}{16} - \dfrac{x^2}{9} = 1$. The transverse axis is vertical with $a = 4$ and $b = 3$. The vertices are

$(0, \pm 4)$. The asymptotes are $y = \pm\dfrac{4}{3}x$. See Figure 7.

Since $c^2 = a^2 + b^2 \Rightarrow c = \pm\sqrt{16 + 9} \Rightarrow \pm 5$, the foci are $(0, \pm 5)$.

9. Horizontal transverse axis, vertices $(\pm 2, 0)$; d

11. Vertical transverse axis, vertices $(0, \pm 3)$; a

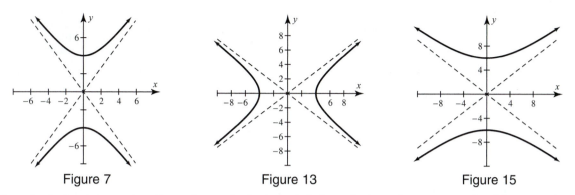

Figure 7   Figure 13   Figure 15

13. Since the foci and the vertices lie on the *x*-axis, the hyperbola has a horizontal transverse axis with an equation

of the form $\dfrac{x^2}{a^2} - \dfrac{y^2}{b^2} = 1$. $F(\pm 5, 0) \Rightarrow c = 5$ and $V(\pm 4, 0) \Rightarrow a = 4$. $c^2 = a^2 + b^2 \Rightarrow$

$b^2 = c^2 - a^2 = 25 - 16 = 9 \Rightarrow b = 3$. The equation of the hyperbola is $\dfrac{x^2}{16} - \dfrac{y^2}{9} = 1$, and the asymptotes

have the equation $y = \pm\dfrac{b}{a}x$ or $y = \pm\dfrac{3}{4}x$. The hyperbola is graphed in Figure 13.

15. Since the foci and the vertices lie on the *y*-axis, the hyperbola has a vertical transverse axis with an equation

of the form $\dfrac{y^2}{a^2} - \dfrac{x^2}{b^2} = 1$. $F(0, \pm 10) \Rightarrow c = 10$ and $V(0, \pm 6) \Rightarrow a = 6$. $c^2 = a^2 + b^2 \Rightarrow$

$b^2 = c^2 - a^2 = 100 - 36 = 64 \Rightarrow b = 8$. The equation of the hyperbola is $\dfrac{y^2}{36} - \dfrac{x^2}{64} = 1$, and the

asymptotes have the equation $y = \pm\dfrac{a}{b}x$ or $y = \pm\dfrac{3}{4}x$. The hyperbola is graphed in Figure 15.

17. Since the foci and the vertices lie on the *y*-axis, the hyperbola has a vertical transverse axis with an equation

of the form $\dfrac{y^2}{a^2} - \dfrac{x^2}{b^2} = 1$. $F(0, \pm 13) \Rightarrow c = 13$ and $V(0, \pm 12) \Rightarrow a = 12$. $c^2 = a^2 + b^2 \Rightarrow$

$b^2 = c^2 - a^2 = 169 - 144 = 25 \Rightarrow b = 5$. The equation of the hyperbola is $\dfrac{y^2}{144} - \dfrac{x^2}{25} = 1$, and the

asymptotes have the equation $y = \pm\dfrac{a}{b}x$ or $y = \pm\dfrac{12}{5}x$. The hyperbola is graphed in Figure 17.

19. Since the foci are $(0, \pm 5)$, $c = 5$. Since the transverse axis is vertical of length 4, $a = 2$. The equation has the

form $\dfrac{y^2}{a^2} - \dfrac{x^2}{b^2} = 1$. $c^2 = a^2 + b^2 \Rightarrow b^2 = c^2 - a^2 = 25 - 4 = 21 \Rightarrow b = \sqrt{21} \approx 4.58$. The equation of

the hyperbola is $\dfrac{y^2}{4} - \dfrac{x^2}{21} = 1$, and the asymptotes have the equation $y = \pm\dfrac{a}{b}x$ or $y = \pm\dfrac{2}{\sqrt{21}}x$. The

hyperbola is graphed in Figure 19.

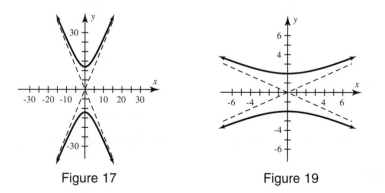

Figure 17   Figure 19

21. Since the vertices lie on the x-axis and are $(\pm 3, 0)$, $a = 3$. The equation has the form $\dfrac{x^2}{a^2} - \dfrac{y^2}{b^2} = 1$. Since

$y = \pm\dfrac{b}{a}x = \pm\dfrac{2}{3}x$ and $a = 3$, it follows that $b = 2$. The equation of the hyperbola is $\dfrac{x^2}{9} - \dfrac{y^2}{4} = 1$. The

hyperbola is graphed in Figure 21.

Figure 21   Figure 23   Figure 25

23. Since the endpoints of the conjugate axis are $(0, \pm 3)$, $b = 3$. The vertices $(\pm 4, 0)$ lie on the x-axis so $a = 4$

and the equation of the hyperbola is $\dfrac{x^2}{a^2} - \dfrac{y^2}{b^2} = 1$ or $\dfrac{x^2}{16} - \dfrac{y^2}{9} = 1$. The asymptotes have the equation

$y = \pm\dfrac{b}{a}x$ or $y = \pm\dfrac{3}{4}x$. The hyperbola is graphed in Figure 23.

25. Since the vertices lie on the x-axis and are $(\pm\sqrt{10}, 0)$, $a^2 = 10$. The equation has the form $\dfrac{x^2}{a^2} - \dfrac{y^2}{b^2} = 1$.

The value of b^2 can be found by substituting $a^2 = 10$, $x = 10$, and $y = 9$ in this equation.

$\dfrac{(10)^2}{10} - \dfrac{(9)^2}{b^2} = 1 \Rightarrow \dfrac{100}{10} - \dfrac{81}{b^2} = 1 \Rightarrow 10 - \dfrac{81}{b^2} = 1 \Rightarrow 9 = \dfrac{81}{b^2} \Rightarrow b^2 = \dfrac{81}{9} \Rightarrow b^2 = 9$

The equation of the hyperbola is $\dfrac{x^2}{10} - \dfrac{y^2}{9} = 1$. The asymptotes have the equation $y = \pm\dfrac{b}{a}x$ or $y = \pm\dfrac{3}{\sqrt{10}}x$.

The hyperbola is graphed in Figure 25.

27. See Figure 27. The hyperbola has a horizontal transverse axis and its center is $(1, 2)$. Since

$a^2 = 16$ and $b^2 = 4$, it follows that $c^2 = a^2 + b^2 = 16 + 4 = 20$. Thus, $a = 4, b = 2$, and $c = \sqrt{20}$. The

vertices are located 4 units to the left and right of center and the foci are located $\sqrt{20}$ units to the left and right

of center. That is, the vertices are $(1 \pm 4, 2)$ and the foci are $(1 \pm \sqrt{20}, 2)$. The asymptotes are given by

$y = \pm\dfrac{b}{a}(x - h) + k \Rightarrow y = \pm\dfrac{1}{2}(x - 1) + 2.$

29. See Figure 29. The hyperbola has a vertical transverse axis and its center is $(-2, 2)$. Since

$a^2 = 36$ and $b^2 = 4$, it follows that $c^2 = a^2 + b^2 = 36 + 4 = 40$. Thus, $a = 6, b = 2$, and $c = \sqrt{40}$. The

vertices are located 6 units above and below center and the foci are located $\sqrt{40}$ units above and below center.

That is, the vertices are $(-2, 2 \pm 6)$ and the foci are $(-2, 2 \pm \sqrt{40})$. The asymptotes are given by

$y = \pm\dfrac{b}{a}(x - h) + k \Rightarrow y = \pm 3(x + 2) + 2.$

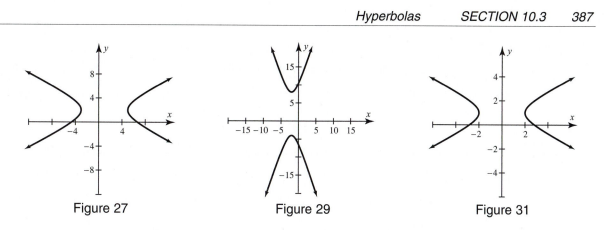

Figure 27   Figure 29   Figure 31

31. See Figure 31. The hyperbola has a horizontal transverse axis and its center is $(0, 1)$. Since $a^2 = 4$ and $b^2 = 1$, it follows that $c^2 = a^2 + b^2 = 4 + 1 = 5$. Thus, $a = 2$, $b = 1$, and $c = \sqrt{5}$. The vertices are located 2 units to the left and right of center and the foci are located $\sqrt{5}$ units to the left and right of center. That is, the vertices are $(\pm 2, 1)$ and the foci are $(\pm \sqrt{5}, 1)$. The asymptotes are given by

$$y = \pm \frac{b}{a}(x - h) + k \Rightarrow y = \pm \frac{1}{2}x + 1.$$

33. Center at $(2, -4)$; b

35. Center at $(2, -1)$; c

37. The hyperbola is centered at $(4, -4)$. The vertical transverse axis has length 8 so $a = 4$. The conjugate axis has length 4, so $b = 2$. Its equation is $\dfrac{(y + 4)^2}{16} - \dfrac{(x - 4)^2}{4} = 1$ with vertices $(4, -4 \pm 4)$, foci $(4, -4 \pm \sqrt{20})$. The asymptotes are given by $y = \pm \dfrac{b}{a}(x - h) + k \Rightarrow y = \pm 2(x - 4) - 4$.

39. The hyperbola has a horizontal transverse axis, and its center is $(1, 1)$. Since $a^2 = 4$, $b^2 = 4$, and $c^2 = a^2 + b^2 = 8$, $a = 2$, $b = 2$, and $c = \sqrt{8}$. Thus, the vertices are $(1 \pm 2, 1)$ and the foci are $(1 \pm \sqrt{8}, 1)$. The asymptotes are the lines $y = \pm(x - 1) + 1$. See Figure 39.

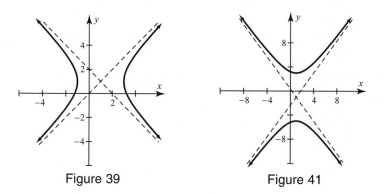

Figure 39   Figure 41

41. The hyperbola has a vertical transverse axis, and its center is $(1, -1)$. Since $a^2 = 16$, $b^2 = 9$, and $c^2 = a^2 + b^2 = 25$, $a = 4$, $b = 3$, and $c = 5$. Thus, the vertices are $(1, -1 \pm 4)$ and the foci are $(1, -1 \pm 5)$. The asymptotes are the lines $y = \pm \dfrac{4}{3}(x - 1) - 1$. See Figure 41.

43. Center $(2, -2) \Rightarrow h = 2$ and $k = -2$. Rewrite the coordinates for the given vertex:

$(3, -2) \Rightarrow (1 + 2, -2)$. Rewrite the coordinates for the given focus: $(4, -2) \Rightarrow (2 + 2, -2)$. Thus, the

transverse axis is horizontal with $a = 1$ and $c = 2$. $b^2 = c^2 - a^2 = 3$. Using the standard equation form, we

get the equation: $\dfrac{(x - h)^2}{a^2} - \dfrac{(y - k)^2}{b^2} = 1 \Rightarrow (x - 2)^2 - \dfrac{(y + 2)^2}{3} = 1$.

45. Since the vertices $(-1, \pm 1)$ and the foci $(-1, \pm 3)$ have the same x-coordinate, $h = -1$ and the transverse axis

is vertical. Rewrite the vertices: $(-1, \pm 1) \Rightarrow (-1, 0 \pm 1)$, so $k = 0$ and $a = 1 \Rightarrow a^2 = 1$. Rewrite the foci:

$(-1, \pm 3) \Rightarrow (-1, 0 \pm 3)$, so $k = 0$ and $c = 3$. So, $b^2 = c^2 - a^2 = 8$. Using the standard equation form, we

get the equation: $\dfrac{(y - k)^2}{a^2} - \dfrac{(x - h)^2}{b^2} = 1 \Rightarrow y^2 - \dfrac{(x + 1)^2}{8} = 1$.

47. $x^2 - 2x - y^2 + 2y = 4 \Rightarrow (x^2 - 2x + 1) - (y^2 - 2y + 1) = 4 + 1 - 1 \Rightarrow$

$(x - 1)^2 - (y - 1)^2 = 4 \Rightarrow \dfrac{(x - 1)^2}{4} - \dfrac{(y - 1)^2}{4} = 1$. The center is $(1, 1)$. The vertices are

$(1 - 2, 1), (1 + 2, 1)$ or $(-1, 1), (3, 1)$.

49. $3y^2 + 24y - 2x^2 + 12x + 24 = 0 \Rightarrow 3(y^2 + 8y) - 2(x^2 - 6x) = -24 \Rightarrow$

$3(y^2 + 8y + 16) - 2(x^2 - 6x + 9) = -24 + 48 - 18 \Rightarrow 3(y + 4)^2 - 2(x - 3)^2 = 6 \Rightarrow$

$\dfrac{(y + 4)^2}{2} - \dfrac{(x - 3)^2}{3} = 1$. The center is $(3, -4)$. The vertices are $(3, -4 - \sqrt{2}), (3, -4 + \sqrt{2})$.

51. $x^2 - 6x - 2y^2 + 7 = 0 \Rightarrow (x^2 - 6x + 9) - 2y^2 = -7 + 9 \Rightarrow (x - 3)^2 - 2(y - 0)^2 = 2 \Rightarrow$

$\dfrac{(x - 3)^2}{2} - \dfrac{(y - 0)^2}{1} = 1$. The center is $(3, 0)$. The vertices are $(3 - \sqrt{2}, 0), (3 + \sqrt{2}, 0)$.

53. $4y^2 + 32y - 5x^2 - 10x + 39 = 0 \Rightarrow 4(y^2 + 8y) - 5(x^2 + 2x) = -39 \Rightarrow$

$4(y^2 + 8y + 16) - 5(x^2 + 2x + 1) = -39 + 64 - 5 \Rightarrow 4(y + 4)^2 - 5(x + 1)^2 = 20 \Rightarrow$

$\dfrac{(y + 4)^2}{5} - \dfrac{(x + 1)^2}{4} = 1$. The center is $(-1, -4)$. The vertices are $(-1, -4 - \sqrt{5}), (-1, -4 + \sqrt{5})$.

55. Solve for y: $\dfrac{(y - 1)^2}{11} - \dfrac{x^2}{5.9} = 1 \Rightarrow \dfrac{(y - 1)^2}{11} = 1 + \dfrac{x^2}{5.9} \Rightarrow (y - 1)^2 = 11\left(1 + \dfrac{x^2}{5.9}\right) \Rightarrow$

$y - 1 = \pm\sqrt{11\left(1 + \dfrac{x^2}{5.9}\right)} \Rightarrow y = 1 \pm \sqrt{11\left(1 + \dfrac{x^2}{5.9}\right)}$. Graph $Y_1 = 1 + \sqrt{(11 \cdot (1 + (X^2/5.9)))}$ and

$Y_2 = 1 - \sqrt{(11 \cdot (1 + (X^2/5.9)))}$. See Figure 55.

[−15, 15, 5] by [−10, 10, 5]   [−9, 9, 1] by [−6, 6, 1]

Figure 55   Figure 57

57. Solve for y: $3y^2 - 4x^2 = 15 \Rightarrow 3y^2 = 4x^2 + 15 \Rightarrow y^2 = \dfrac{4x^2 + 15}{3} \Rightarrow y = \pm\sqrt{\dfrac{4x^2 + 15}{3}}$.

Graph $Y_1 = \sqrt{((4X^2 + 15)/3)}$ and $Y_2 = -\sqrt{((4X^2 + 15)/3)}$. See Figure 57.

59. Add both equations together to eliminate the y^2-term:

$$\begin{array}{rcl} x^2 - y^2 &=& 4 \\ x^2 + y^2 &=& 9 \\ \hline 2x^2 &=& 13 \end{array} \Rightarrow x^2 = \frac{13}{2} \Rightarrow x = \pm\sqrt{\frac{13}{2}}$$

Substitute $x^2 = \frac{13}{2}$ into the first equation and solve for y: $x^2 - y^2 = 4 \Rightarrow \frac{13}{2} - y^2 = 4 \Rightarrow y^2 = \frac{5}{2} \Rightarrow$

$y = \pm\sqrt{\frac{5}{2}}$ There are four solutions to the system:

$\left(\sqrt{\frac{13}{2}}, \sqrt{\frac{5}{2}}\right), \left(-\sqrt{\frac{13}{2}}, \sqrt{\frac{5}{2}}\right), \left(\sqrt{\frac{13}{2}}, -\sqrt{\frac{5}{2}}\right),$ and $\left(-\sqrt{\frac{13}{2}}, -\sqrt{\frac{5}{2}}\right)$. See Figure 59.

Figure 59

61. Solve the second equation for y: $x + y = 2 \Rightarrow y = 2 - x$. Substitute this result into the first equation and

solve for x: $\frac{x^2}{4} - \frac{y^2}{9} = 1 \Rightarrow \frac{x^2}{4} - \frac{(2-x)^2}{9} = 1 \Rightarrow 9x^2 - 4(4 - 4x + x^2) = 36 \Rightarrow$

$9x^2 - 16 + 16x - 4x^2 = 36 \Rightarrow 5x^2 + 16x - 52 = 0 \Rightarrow (5x + 26)(x - 2) = 0 \Rightarrow$

$x = -\frac{26}{5} = -5.2$ or $x = 2$. Substitute for x to find y: $y = 2 - x \Rightarrow y = 2 - (-5.2) = 7.2$ and

$y = 2 - x \Rightarrow y = 2 - 2 = 0$. There are two solutions to the system: (2, 0) and (–5.2, 7.2). See Figure 61.

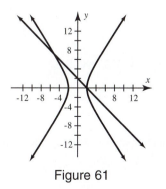

Figure 61

63. Multiply the second equation by 2 and add the equations to eliminate the y^2-term:

$$8x^2 - 6y^2 = 24$$
$$10x^2 + 6y^2 = 48$$
$$\overline{18x^2 \qquad = 72} \Rightarrow x^2 = 4 \Rightarrow x = \pm 2$$

Substitute $x^2 = 4$ into the first equation and solve for y:

$$8x^2 - 6y^2 = 24 \Rightarrow 8(4) - 6y^2 = 24 \Rightarrow 6y^2 = 8 \Rightarrow y^2 = \frac{4}{3} \Rightarrow y = \pm\sqrt{\frac{4}{3}} \Rightarrow y = \pm\frac{2}{\sqrt{3}}.$$

There are four solutions to the system: $\left(-2, -\frac{2}{\sqrt{3}}\right), \left(-2, \frac{2}{\sqrt{3}}\right), \left(2, -\frac{2}{\sqrt{3}}\right), \left(2, \frac{2}{\sqrt{3}}\right)$. See Figure 63.

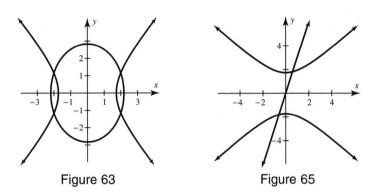

Figure 63   Figure 65

65. Solve the second equation for y: $3x - y = 0 \Rightarrow y = 3x$. Substitute this result into the first equation and

solve for x: $\frac{y^2}{3} - \frac{x^2}{4} = 1 \Rightarrow \frac{(3x)^2}{3} - \frac{x^2}{4} = 1 \Rightarrow \frac{9x^2}{3} - \frac{x^2}{4} = 1 \Rightarrow 3x^2 - \frac{x^2}{4} = 1 \Rightarrow 12x^2 - x^2 = 4 \Rightarrow$

$11x^2 = 4 \Rightarrow x^2 = \frac{4}{11} \Rightarrow x = \pm\frac{2}{\sqrt{11}}$.

Substitute for x to find y: When $x = \frac{2}{\sqrt{11}}, y = 3x \Rightarrow y = 3\left(\frac{2}{\sqrt{11}}\right) \Rightarrow y = \frac{6}{\sqrt{11}}$.

When $-\frac{2}{\sqrt{11}}, y = 3x \Rightarrow y = 3\left(-\frac{2}{\sqrt{11}}\right) \Rightarrow y = -\frac{6}{\sqrt{11}}$.

The two solutions are: $\left(\frac{2}{\sqrt{11}}, \frac{6}{\sqrt{11}}\right), \left(-\frac{2}{\sqrt{11}}, -\frac{6}{\sqrt{11}}\right)$. See Figure 65.

67. (a) $k = 2.82 \times 10^7$ and $D = 42.5 \times 10^6 \Rightarrow \frac{k}{\sqrt{D}} = \frac{2.82 \times 10^7}{\sqrt{42.5 \times 10^6}} \Rightarrow \frac{k}{\sqrt{D}} \approx 4325.68$. Since

$V = 2090$ and $V < \frac{k}{\sqrt{D}}$, the trajectory is elliptic.

(b) For $V > \frac{k}{\sqrt{D}}, V > 4326$. the speed of Explorer IV should be 4326 meters per second or greater so its

trajectory is hyperbolic.

(c) If D is larger, then $\frac{k}{\sqrt{D}}$ is smaller, so smaller values for V satisfy $V > \frac{k}{\sqrt{D}}$.

Chapter 10 Review Exercises

1. The equation is $-x^2 = y$. See Figure 1.

3. The equation is $\dfrac{x^2}{25} + \dfrac{y^2}{49} = 1$. See Figure 3.

5. The equation is $\dfrac{y^2}{4} - \dfrac{x^2}{9} = 1$. See Figure 5.

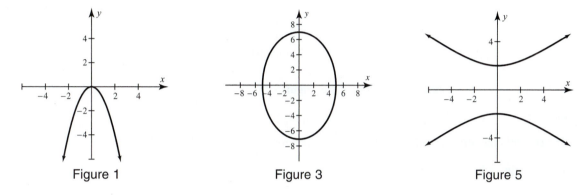

Figure 1   Figure 3   Figure 5

7. A parabola opening upwards; d

9. A circle; a

11. A hyperbola with horizontal transverse axis; e

13. The parabola opens to the right and $p = 2$. $y^2 = 4px$, so $y^2 = 8x$. See Figure 13.

15. The major axis is horizontal, $a = 5$ and $c = 4$; $b = \sqrt{a^2 - c^2} = \sqrt{25 - 16} = 3$.

    The equation is $\dfrac{x^2}{25} + \dfrac{y^2}{9} = 1$. See Figure 15.

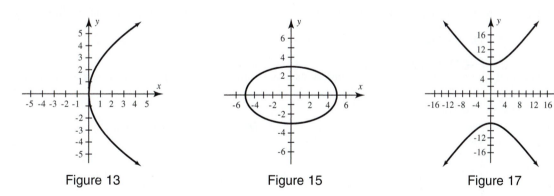

Figure 13   Figure 15   Figure 17

17. The transverse axis is vertical, $c = 10$ and $b = 6$; $a = \sqrt{100 - 36} = 8$.

    The equation is $\dfrac{y^2}{64} - \dfrac{x^2}{36} = 1$. See Figure 17.

19. $x^2 = 4py \implies 4p = -4$, so $p = -1$. The vertex is at the origin and the focus is $(0, -1)$. See Figure 19.

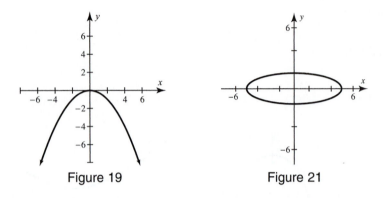

Figure 19   Figure 21

21. The major axis is horizontal, the center is at the origin, $a = 5$ and $b = 2$. $c^2 = 25 - 4 = 21$, so $c = \sqrt{21}$; the foci are $(\pm\sqrt{21}, 0)$. See Figure 21.

23. The center is the origin and the transverse axis is horizontal. $c^2 = a^2 + b^2 = 16 + 9 = 25$, so $c = 5$. The foci are $(\pm 5, 0)$. See Figure 23.

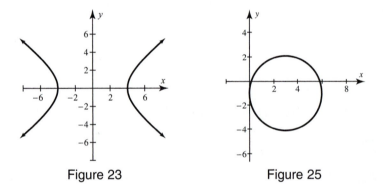

Figure 23   Figure 25

25. The equation represents a circle of radius 3, centered at $(3, -1)$. If we think of the circle as an ellipse, both foci are at $(3, -1)$. See Figure 25.

27. The equation $\dfrac{(x-1)^2}{4} - \dfrac{(y+1)^2}{4} = 1$ represents a hyperbola with horizontal transverse axis and center $(1, -1)$. See Figure 27.

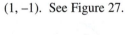
Figure 27

29. The equation $(y - 4)^2 = -8(x - 8)$ has the form $(y - k)^2 = 4p(x - h)$ with vertex

(h, k) at $(8, 4)$, and $p = -2$.

The focus is $(6, 4)$ and the directrix is $x = 10$. See Figure 29.

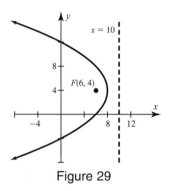

Figure 29

31. $y = \pm\sqrt{\dfrac{1}{8.2}(60 - 7.1x^2)}$; See Figure 31.

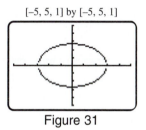

$[-5, 5, 1]$ by $[-5, 5, 1]$

Figure 31

33. $-2x = y^2 + 8x + 14 \Rightarrow y^2 = -10x - 14 \Rightarrow (y - 0)^2 = -10\left(x + \dfrac{7}{5}\right)$

35. $4x^2 + 8x + 25y^2 - 250y = -529 \Rightarrow 4(x^2 + 2x) + 25(y^2 - 10y) = -529 \Rightarrow$

$4(x^2 + 2x + 1) + 25(y^2 - 10y + 25) = -529 + 4 + 625 \Rightarrow 4(x + 1)^2 + 25(y - 5)^2 = 100 \Rightarrow$

$\dfrac{(x + 1)^2}{25} + \dfrac{(y - 5)^2}{4} = 1$; The center is $(-1, 5)$. The vertices are

$(-1 - 5, 5), (-1 + 5, 5)$ or $(-6, 5), (4, 5)$.

37. $x^2 + 4x - 4y^2 + 24y = 36 \Rightarrow (x^2 + 4x + 4) - 4(y^2 - 6y + 9) = 36 + 4 - 36 \Rightarrow$

$(x + 2)^2 - 4(y - 3)^2 = 4 \Rightarrow \dfrac{(x + 2)^2}{4} - \dfrac{(y - 3)^2}{1} = 1$.

The center is $(-2, 3)$. The vertices are $(-2 - 2, 3), (-2 + 2, 3)$ or $(-4, 3), (0, 3)$.

39. Clear fractions: $x^2 + y^2 = 4 \Rightarrow y^2 = 4 - x^2$. Substituting in $x^2 + 4y^2 = 8$ gives $x^2 + 4(4 - x^2) = 8$ or

$3x^2 = 8$, so $x = \pm\sqrt{\dfrac{8}{3}}$. Substituting in $x^2 + y^2 = 4$ we find $y^2 = 4 - \dfrac{8}{3} = \dfrac{4}{3}$, so $y = \pm\sqrt{\dfrac{4}{3}}$. There are four

solutions: $\left(\pm\sqrt{\dfrac{8}{3}}, \pm\sqrt{\dfrac{4}{3}}\right)$.

41. The system is $\dfrac{x^2}{9} + \dfrac{y^2}{4} \le 1$ and $x + y \le 3$. See Figure 41.

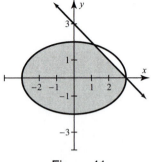

Figure 41

43. (a) $c = \sqrt{500^2 - 70^2} \approx 495.08$, and $a = 500$. The minimum distance is $a - c \approx 4.92$ million miles. The maximum distance is $a + c \approx 995.08$ million miles.

(b) $2\pi\sqrt{\dfrac{500^2 + 70^2}{2}} \approx 2243$ million miles or 2.243 billion miles.

45. The equation of the ellipse is $\dfrac{x^2}{40^2} + \dfrac{y^2}{30^2} = 1$. Solving for y gives $y = 30\sqrt{1 - \dfrac{x^2}{40^2}}$. When $x = 10$,

$y = 30\sqrt{1 - \left(\dfrac{10}{40}\right)^2} \approx 29.05$ feet.

Chapter 11: Further Topics in Algebra

11.1: Sequences

1. $a_1 = 2(1) + 1 = 3$; $a_2 = 2(2) + 1 = 5$; $a_3 = 2(3) + 1 = 7$; $a_4 = 2(4) + 1 = 9$.

   The first four terms are 3, 5, 7, and 9.

3. $a_1 = 4(-2)^{1-1} = 4$; $a_2 = 4(-2)^{2-1} = -8$; $a_3 = 4(-2)^{3-1} = 16$; $a_4 = 4(-2)^{4-1} = -32$.

   The first four terms are 4, -8, 16, and -32.

5. $a_1 = \dfrac{1}{1^2 + 1} = \dfrac{1}{2}$; $a_2 = \dfrac{2}{2^2 + 1} = \dfrac{2}{5}$; $a_3 = \dfrac{3}{3^2 + 1} = \dfrac{3}{10}$; $a_4 = \dfrac{4}{4^2 + 1} = \dfrac{4}{17}$.

   The first four terms are $\dfrac{1}{2}, \dfrac{2}{5}, \dfrac{3}{10}$, and $\dfrac{4}{17}$.

7. $a_1 = (-1)^1 \left(\dfrac{1}{2}\right)^1 = -\dfrac{1}{2}$; $a_2 = (-1)^2 \left(\dfrac{1}{2}\right)^2 = \dfrac{1}{4}$; $a_3 = (-1)^3 \left(\dfrac{1}{2}\right)^3 = -\dfrac{1}{8}$; $a_4 = (-1)^4 \left(\dfrac{1}{2}\right)^4 = \dfrac{1}{16}$.

   The first four terms are $-\dfrac{1}{2}, \dfrac{1}{4}, -\dfrac{1}{8}$, and $\dfrac{1}{16}$.

9. $a_1 = (-1)^0 \left(\dfrac{2}{1 + 2}\right) = \dfrac{2}{3}$; $a_2 = (-1)^1 \left(\dfrac{4}{1 + 4}\right) = -\dfrac{4}{5}$; $a_3 = (-1)^2 \left(\dfrac{8}{1 + 8}\right) = \dfrac{8}{9}$;

   $a_4 = (-1)^3 \left(\dfrac{16}{1 + 16}\right) = -\dfrac{16}{17}$. The first four terms are $\dfrac{2}{3}, -\dfrac{4}{5}, \dfrac{8}{9}$, and $-\dfrac{16}{17}$.

11. $a_1 = 2 + 1^2 = 3$; $a_2 = 4 + 2^2 = 8$; $a_3 = 8 + 3^2 = 17$; $a_4 = 16 + 4^2 = 32$.

    The first four terms are 3, 8, 17, and 32.

13. The points (1, 2), (2, 4), (3, 3), (4, 5), (5, 3), (6, 6), (7, 4) lie on the graph. Therefore, the terms of the sequence are 2, 4, 3, 5, 3, 6, 4.

15. (a) $a_1 = 1$; $a_2 = 2a_1 = 2(1) = 2$; $a_3 = 2a_2 = 2(2) = 4$; $a_4 = 2a_3 = 2(4) = 8$.

    The first four terms are 1, 2, 4, and 8.

    (b) The graph of the points (1, 1), (2, 2), (3, 4), and (4, 8) is shown in Figure 15.

17. (a) $a_1 = -3$; $a_2 = a_1 + 3 = -3 + 3 = 0$; $a_3 = a_2 + 3 = 0 + 3 = 3$; $a_4 = a_3 + 3 = 3 + 3 = 6$.

    The first four terms are -3, 0, 3, and 6.

    (b) See Figure 17.

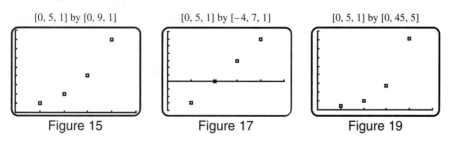

| [0, 5, 1] by [0, 9, 1] | [0, 5, 1] by [−4, 7, 1] | [0, 5, 1] by [0, 45, 5] |

Figure 15   Figure 17   Figure 19

19. (a) $a_1 = 2$; $a_2 = 3a_1 - 1 = 3(2) - 1 = 5$; $a_3 = 3a_2 - 1 = 3(5) - 1 = 14$;

    $a_4 = 3a_3 - 1 = 3(14) - 1 = 41$. The first four terms are 2, 5, 14, and 41.

    (b) See Figure 19.

21. (a) $a_1 = 2$; $a_2 = 5$; $a_3 = a_2 - a_1 = 5 - 2 = 3$; $a_4 = a_3 - a_2 = 3 - 5 = -2$.

   The first four terms are 2, 5, 3, and –2.

   (b) The graph of the points (1, 2), (2, 5), (3, 3), and (4, –2) is shown in Figure 21.

23. (a) $a_1 = 2$; $a_2 = a_1^2 = 2^2 = 4$; $a_3 = a_2^2 = 4^2 = 16$; $a_4 = a_3^2 = 16^2 = 256$.

   The first four terms are 2, 4, 16, and 256.

   (b) The graph of the points (1, 2), (2, 4), (3, 16), and (4, 256) is shown in Figure 23.

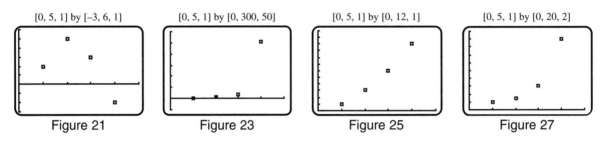

[0, 5, 1] by [–3, 6, 1]   [0, 5, 1] by [0, 300, 50]   [0, 5, 1] by [0, 12, 1]   [0, 5, 1] by [0, 20, 2]

Figure 21   Figure 23   Figure 25   Figure 27

25. (a) $a_1 = 1$; $a_2 = a_1 + 2 = 1 + 2 = 3$; $a_3 = a_2 + 3 = 3 + 3 = 6$; $a_4 = a_3 + 4 = 6 + 4 = 10$.

   The first four terms are 1, 3, 6, and 10.

   (b) The graph of the points (1, 1), (2, 3), (3, 6), and (4, 10) is shown in Figure 25.

27. (a) $a_1 = 2$; $a_2 = 3$; $a_3 = a_2 \cdot a_1 = 2 \cdot 3 = 6$; $a_4 = a_3 \cdot a_2 = 6 \cdot 3 = 18$

   The first four terms are 2, 3, 6, and 18.

   (b) The graph of the points (1, 2), (2, 3), (3, 6), and (4, 18) is shown in Figure 27.

29. (a) Each term can be found by adding 2 to the previous term. A numerical representation for the first eight

   terms is shown in Figure 29a.

[0, 10, 1] by [0, 16, 1]

n	1	2	3	4	5	6	7	8
a_n	1	3	5	7	9	11	13	15

Figure 29a

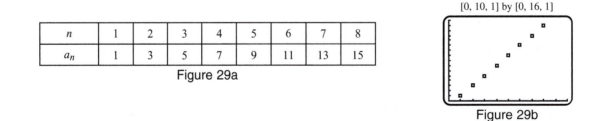

Figure 29b

   (b) A graphical representation is shown in Figure 29b and includes the points in Figure 29a. Notice that the

   points lie on a line with slope 2 because the sequence is arithmetic.

   (c) To find the symbolic representation, we will use the formula $a_n = a_1 + (n - 1)d$, where $a_n = f(n)$. The

   common difference of this sequence is $d = 2$ and the first term is $a_1 = 1$. Therefore, a symbolic

   representation of the sequence is given by $a_n = 1 + (n - 1)2$ or $a_n = 2n - 1$.

31. (a) Each term can be found by subtracting 1.5 to the previous term. A numerical representation for the first

eight terms is shown in Figure 31a.

n	1	2	3	4	5	6	7	8
a_n	7.5	6	4.5	3	1.5	0	-1.5	-3

Figure 31a

[0, 12, 1] by [−4, 8, 1]

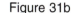

Figure 31b

(b) A graphical representation is shown in Figure 31b and includes the points in Figure 31a. Notice that the

points lie on a line with slope –1.5 because the sequence is arithmetic.

(c) To find the symbolic representation, we will use the formula $a_n = a_1 + (n - 1)d$, where $a_n = f(n)$. The

common difference of this sequence is $d = -1.5$ and the first term is $a_1 = 7.5$. Therefore, a symbolic

representation of the sequence is given by $a_n = 7.5 + (n - 1)(-1.5)$ or $a_n = -1.5n + 9$.

33. (a) Each term can be found by adding $\dfrac{3}{2}$ to the previous term. A numerical representation for the first eight

terms is shown in Figure 33a.

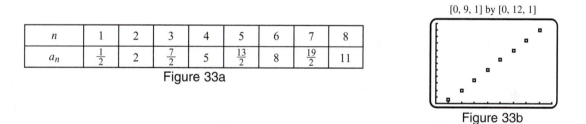

[0, 9, 1] by [0, 12, 1]

n	1	2	3	4	5	6	7	8
a_n	$\dfrac{1}{2}$	2	$\dfrac{7}{2}$	5	$\dfrac{13}{2}$	8	$\dfrac{19}{2}$	11

Figure 33a

Figure 33b

(b) A graphical representation is shown in Figure 33b and includes the points in Figure 33a. Notice that the

points lie on a line with slope $\dfrac{3}{2}$ because the sequence is arithmetic.

(c) To find the symbolic representation, we will use the formula $a_n = a_1 + (n - \!1)d$, where $a_n = f(n)$. The

common difference of this sequence is $d = \dfrac{3}{2}$ and the first term is $a_1 = \dfrac{1}{2}$. Therefore, a symbolic

representation of the sequence is given by $a_n = \dfrac{1}{2} + (n - 1)\left(\dfrac{3}{2}\right)$ or $a_n = \dfrac{3}{2}n - 1$.

35. (a) Each term can be found by multiplying the previous term by $\frac{1}{2}$. A numerical representation for the first

eight terms is shown in Figure 35a.

[0, 10, 1] by [−1, 9, 1]

n	1	2	3	4	5	6	7	8
a_n	8	4	2	1	$\frac{1}{2}$	$\frac{1}{4}$	$\frac{1}{8}$	$\frac{1}{16}$

Figure 35a

Figure 35b

(b) A graphical representation is shown in Figure 35b and includes the points in Figure 35a. Notice that the

points lie on a curve that is decaying exponentially, because the sequence is geometric and the common

ratio is less than one in absolute value.

(c) To find the symbolic representation, we will use the formula $a_n = a_1 r^{n-1}$, where $a_n = f(n)$. The common

ratio of this sequence is $r = \frac{1}{2}$ and the first term is $a_1 = 8$. Therefore, a symbolic representation of the

sequence is given by $a_n = 8\left(\frac{1}{2}\right)^{n-1}$.

37. (a) Each term can be found by multiplying the previous term by 2. A numerical representation for the first

eight terms is shown in Figure 37a.

[0, 10, 1] by [−10, 110, 10]

n	1	2	3	4	5	6	7	8
a_n	$\frac{3}{4}$	$\frac{3}{2}$	3	6	12	24	48	96

Figure 37a

Figure 37b

(b) A graphical representation is shown in Figure 37b and includes the points in Figure 37a. Notice that the

points lie on a curve that is increasing exponentially, because the sequence is geometric and the common

ratio is greater than one in absolute value.

(c) To find the symbolic representation, we will use the formula $a_n = a_1 r^{n-1}$, where $a_n = f(n)$. The common

ratio of this sequence is $r = 2$ and the first term is $a_1 = \frac{3}{4}$. Therefore, a symbolic representation of the

sequence is given by $a_n = \frac{3}{4}(2)^{n-1}$.

39. (a) Each term can be found by multiplying the previous term by 2. A numerical representation for the first eight terms is shown in Figure 39a.

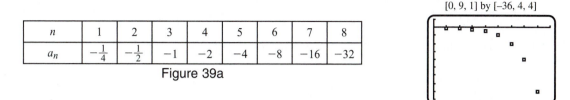

n	1	2	3	4	5	6	7	8
a_n	$-\frac{1}{4}$	$-\frac{1}{2}$	-1	-2	-4	-8	-16	-32

Figure 39a

$[0, 9, 1]$ by $[-36, 4, 4]$

Figure 39b

(b) A graphical representation is shown in Figure 39b and includes the points in Figure 39a. Notice that the points lie on a curve that is increasing exponentially, because the sequence is geometric and the common ratio is greater than one in absolute value.

(c) To find the symbolic representation, we will use the formula $a_n = a_1 r^{n-1}$, where $a_n = f(n)$. The common ratio of this sequence is $r = 2$ and the first term is $a_1 = -\frac{1}{4}$. Therefore, a symbolic representation of the sequence is given by $a_n = -\frac{1}{4}(2)^{n-1}$.

41. Let $a_n = f(n)$, where $f(n) = dn + c$. Then, $f(n) = -2n + c$. Since $f(1) = -2(1) + c = 5 \Rightarrow c = 7$. Thus, $a_n = f(n) = -2n + 7$. An alternate solution is to use the formula:
$a_n = a_1 + (n - 1)d = 5 + (n - 1)(-2) \Rightarrow a_n = -2n + 7$.

43. Let $a_n = f(n)$, where $f(n) = dn + c$. Then, $f(n) = 3n + c$. Since $f(3) = 3(3) + c = 1 \Rightarrow c = -8$. Thus, $a_n = f(n) = 3n - 8$.

45. Let $a_n = f(n)$, where $f(n) = dn + c$. $a_2 = 5 \Rightarrow f(2) = 5$ and $a_6 = 13 \Rightarrow f(6) = 13$.
Thus, $d = \dfrac{f(6) - f(2)}{6 - 2} = \dfrac{13 - 5}{4} = 2$. Then, $f(n) = 2n + c$. $f(2) = 2(2) + c = 5 \Rightarrow c = 1$.
Thus, $a_n = f(n) = 2n + 1$.

47. Let $a_n = f(n)$, where $f(n) = dn + c$. $a_1 = 8 \Rightarrow f(1) = 8$ and $a_4 = 17 \Rightarrow f(4) = 17$.
Thus, $d = \dfrac{f(4) - f(1)}{4 - 1} = \dfrac{17 - 8}{3} = 3$. Then, $f(n) = 3n + c$. $f(1) = 3(1) + c = 8 \Rightarrow c = 5$.
Thus, $a_n = f(n) = 3n + 5$.

49. Let $a_n = f(n)$, where $f(n) = dn + c$. $a_5 = -4 \Rightarrow f(5) = -4$ and $a_8 = -2.5 \Rightarrow f(8) = -2.5$.
Thus, $d = \dfrac{f(8) - f(5)}{8 - 5} = \dfrac{-2.5 - (-4)}{3} = 0.5$.
Then, $f(n) = 0.5n + c$. $f(5) = 0.5(5) + c = -4 \Rightarrow c = -6.5$. Thus, $a_n = f(n) = 0.5n - 6.5$.

51. Let $a_n = f(n)$, where $f(n) = a_1 r^{n-1}$. Then, $a_n = f(n) = 2\left(\dfrac{1}{2}\right)^{n-1}$.

53. Let $a_n = f(n)$, where $f(n) = a_1 r^{n-1}$. Then, $a_n = f(n) = a_1\left(-\dfrac{1}{4}\right)^{n-1}$.

Since $f(3) = a_1\left(-\dfrac{1}{4}\right)^{3-1} = \dfrac{1}{32} \Rightarrow a_1 = \dfrac{16}{32} = \dfrac{1}{2}$. Thus, $a_n = f(n) = \dfrac{1}{2}\left(-\dfrac{1}{4}\right)^{n-1}$.

55. Let $a_n = f(n)$, where $f(n) = a_1 r^{n-1}$. $a_3 = 2$ and $a_6 = \dfrac{1}{4} \Rightarrow \dfrac{1}{8} = \dfrac{\frac{1}{4}}{2} = \dfrac{a_6}{a_3} = \dfrac{a_1 r^{6-1}}{a_1 r^{3-1}} = \dfrac{r^5}{r^2} = r^3 \Rightarrow$

$r = \dfrac{1}{2}$. Then, $f(n) = a_1 \left(\dfrac{1}{2}\right)^{n-1}$. $f(3) = a_1 \left(\dfrac{1}{2}\right)^{3-1} = 2 \Rightarrow a_1 = 8$. Thus, $a_n = f(n) = 8\left(\dfrac{1}{2}\right)^{n-1}$.

57. Let $a_n = f(n)$, where $f(n) = a_1 r^{n-1}$. $a_1 = -5$ and $a_3 = -125 \Rightarrow$

$25 = \dfrac{-125}{-5} = \dfrac{a_3}{a_1} = \dfrac{a_1 r^{3-1}}{a_1 r^{1-1}} = \dfrac{r^2}{r^0} = r^2 \Rightarrow r = -5$. Thus, $a_n = f(n) = -5(-5)^{n-1}$.

59. Let $a_n = f(n)$, where $f(n) = a_1 r^{n-1}$. $a_2 = -1$ and $a_7 = -32 \Rightarrow$

$32 = \dfrac{-32}{-1} = \dfrac{a_7}{a_2} = \dfrac{a_1 r^{7-1}}{a_1 r^{2-1}} = \dfrac{r^6}{r^1} = r^5 \Rightarrow r = 2$.

Then, $f(n) = a_1(2)^{n-1}$. $f(2) = a_1(2)^{2-1} = -1 \Rightarrow a_1 = -\dfrac{1}{2}$. Thus, $a_n = f(n) = -\dfrac{1}{2}(2)^{n-1}$.

61. Since $f(n) = 4 - 3n^3$ is not a linear function, it does not represent an arithmetic sequence.

63. Since $f(n) = 4n - (3 - n) = 5n - 3$ is a linear function, it represents an arithmetic sequence.

65. Since the plotted points appear to be collinear, the graph represents an arithmetic sequence.

67. Since the common difference is -2, the table represents an arithmetic sequence.

69. Since $f(n) = 4(2)^{n-1}$ is written in the form $f(n) = cr^{n-1}$, it represents a geometric sequence.

71. Since $f(n) = -3(n)^2$ cannot be written in the form $f(n) = cr^{n-1}$, it does not represents a geometric sequence.

73. Since the plotted points appear to be collinear, the graph does not represent a geometric sequence.

75. Since there is no common ratio, the table does not represent a geometric sequence.

77. These terms represent an arithmetic sequence. Each term can be obtained by adding 7 to the previous term.

79. These terms represent a geometric sequence. Each term can be obtained by multiplying the previous term by 4.

81. These terms represent neither an arithmetic nor a geometric sequence. There is no common ratio or difference.

83. This sequence is arithmetic since the points lie on a line (and are evenly spaced). Since the sequence is decreasing the common difference must be negative. The slope of the line passing through these points is -1, so the common difference is $d = -1$.

85. The sequence is either geometric or arithmetic. This sequence must be geometric, since the points do not lie on a line. Since the terms alternate sign, the common ratio r is negative. The absolute value of the terms are dampening to a value of 0. Therefore, $|r| < 1$.

87. The insect population increases rapidly and then levels off at 5000 per acre.

89. (a) The initial density is 500. The population density each successive year is 0.8 of the previous year. Therefore $a_1 = 500$ and $a_n = 0.8a_{n-1}$.

    (b) $a_1 = 500$, $a_2 = 0.8(500) = 400$, $a_3 = 0.8(400) = 320$, $a_4 = 0.8(320) = 256$, $a_5 = 0.8(256) = 204.8$ and $a_6 = 0.8(204.8) = 163.84$. The population density is decreasing each year by 20%

    (c) The terms of the sequence 500, 400, 320, 256, ... are a geometric sequence with $a_1 = 500$ and $r = 0.8$. Therefore, the nth term is given by $a_n = 500(0.8)^{n-1}$.

91. (a) $a_1 = 8, a_2 = 2.9a_1 - 0.2a_1^2 = 2.9(8) - 0.2(8)^2 = 10.4$,

    $a_3 = 2.9a_2 - 0.2a_2^2 = 2.9(10.4) - 0.2(10.4)^2 = 8.528$.

    (b) Figures 91a & 91b show how to enter the sequence and a graph of the first twenty terms. The population density oscillates above and below 9.5 (approximately).

$[0, 21, 1]$ by $[0, 14, 1]$ $[0, 30, 10]$ by $[0, 150{,}000, 50{,}000]$

Figure 91a   Figure 91b   Figure 95

93. (a) The terms in this sequence can be found by adding the previous two terms. $a_1 = 1, a_2 = 1, a_3 = 2$, $a_4 = 3, a_5 = 5, a_6 = 8, a_7 = 13, a_8 = 21, a_9 = 34, a_{10} = 55, a_{11} = 89$, and $a_{12} = 144$.

(b) $\dfrac{a_2}{a_1} = \dfrac{1}{1} = 1, \dfrac{a_3}{a_2} = \dfrac{2}{1} = 2, \dfrac{a_4}{a_3} = \dfrac{3}{2} = 1.5, \dfrac{a_5}{a_4} = \dfrac{5}{3} \approx 1.6667, \dfrac{a_6}{a_5} = \dfrac{8}{5} = 1.6, \dfrac{a_7}{a_6} = \dfrac{13}{8} = 1.625$,

$\dfrac{a_8}{a_7} = \dfrac{21}{13} \approx 1.6154, \dfrac{a_9}{a_8} = \dfrac{34}{21} = 1.6190, \dfrac{a_{10}}{a_9} = \dfrac{55}{34} \approx 1.6176, \dfrac{a_{11}}{a_{10}} = \dfrac{89}{55} = 1.6182$, and

$\dfrac{a_{12}}{a_{11}} = \dfrac{144}{89} \approx 1.6180$. These ratios seem to be approaching a number near 1.618. This number is called the golden ratio.

(c) $n = 2$: $a_1 \cdot a_3 - a_2^2 = (1)(2) - (1)^2 = 1 = (-1)^2$;

$n = 3$: $a_2 \cdot a_4 - a_3^2 = (1)(3) - (2)^2 = -1 = (-1)^3$;

$n = 4$: $a_3 \cdot a_5 - a_4^2 = (2)(5) - (3)^2 = 1 = (-1)^4$

95. (a) The salary of the first employee at the beginning of the nth year is given by:

$a_n = 30{,}000 + (n - 1)(2000) = 2000n + 28{,}000$.

This is a arithmetic sequence with $a_1 = 30{,}000$ and $d = 2000$.

(b) The salary of the second employee at the beginning of the nth year is given by: $b_n = 30{,}000(1.05)^{n-1}$.

This is a geometric sequence with $b_1 = 30{,}000$ and $d = 1.05$.

(c) At the beginning of the 10th year each salary is:

$a_{10} = 2000(10) + 28{,}000 = \$48{,}000, b_{10} = 30{,}000(1.05)^{10-1} \approx \$46{,}540$.

At the beginning of the 20th year each salary is:

$a_{20} = 2000(20) + 28{,}000 = \$68{,}000, b_{20} = 30{,}000(1.05)^{20-1} \approx \$75{,}809$.

(d) The graph of each sequence is shown in Figure 95. With time the geometric sequence overtakes the arithmetic sequence since $r > 1$.

97. Let $a_1 = 2$. Then, $a_2 = \dfrac{1}{2}\left(a_1 + \dfrac{k}{a_1}\right) = \dfrac{1}{2}\left(2 + \dfrac{2}{2}\right) = 1.5, \ a_3 = \dfrac{1}{2}\left(a_2 + \dfrac{k}{a_2}\right) = \dfrac{1}{2}\left(1.5 + \dfrac{2}{1.5}\right) = 1.41\overline{6}$

In a similar manner, $a_4 \approx 1.414215686, a_5 \approx 1.414213562$, and $a_6 \approx 1.414213562$. Since $\sqrt{2} \approx 1.414213562$ this is a very accurate approximation.

99. Let $a_1 = 21$. Then, $a_2 = \frac{1}{2}\left(a_1 + \frac{k}{a_1}\right) = \frac{1}{2}\left(21 + \frac{21}{21}\right) = 11$, $a_3 = \frac{1}{2}\left(a_2 + \frac{k}{a_2}\right) = \frac{1}{2}\left(11 + \frac{21}{11}\right) = 6.\overline{45}$

In a similar manner, $a_4 \approx 4.854033291$, $a_5 \approx 4.59016621$, and $a_6 \approx 4.582581971$. Since

$\sqrt{21} \approx 4.582575695$ this is an accurate approximation.

101. By definition $a_n = a_1 + (n-1)d_1$ and $b_n = b_1 + (n-1)d_2$. Then

$c_n = a_n + b_n = [a_1 + (n-1)d_1] + [b_1 + (n-1)d_2] = (a_1 + b_1) + [(n-1)d_1 + (n-1)d_2] =$

$(a_1 + b_1) + (n-1)(d_1 + d_2) = c_1 + (n-1)d$ where $c_1 = a_1 + b_1$ and $d = d_1 + d_2$.

11.2: Series

1. The first six positive even integers are 2, 4, 6, 8, 10, 12.

3. The sum of the first six positive even integers is shown by $2 + 4 + 6 + 8 + 10 + 12$. The sum is 42.

5. Since A_n represents the number of AIDS deaths after 2000, the sum from 2005 to 2009 is given by

$A_5 + A_6 + A_7 + A_8 + A_9$.

7. $S_5 = 3(1) + 3(2) + 3(3) + 3(4) + 3(5) = 3 + 6 + 9 + 12 + 15 = 45$

9. $S_5 = (2(1) - 1) + (2(2) - 1) + (2(3) - 1) + (2(4) - 1) + (2(5) - 1) = 1 + 3 + 5 + 7 + 9 = 25$

11. $S_5 = (1^2 + 1) + (2^2 + 1) + (3^2 + 1) + (4^2 + 1) + (5^2 + 1) = 2 + 5 + 10 + 17 + 26 = 60$

13. $S_5 = \frac{1}{1+1} + \frac{2}{2+1} + \frac{3}{3+1} + \frac{4}{4+1} + \frac{5}{5+1} = \frac{1}{2} + \frac{2}{3} + \frac{3}{4} + \frac{4}{5} + \frac{5}{6} = \frac{71}{20}$

15. The first term is $a_1 = 3$ and the last term is $a_8 = 17$. To find the sum use:

$S_n = n\left(\frac{a_1 + a_n}{2}\right) \Rightarrow S_8 = 8\left(\frac{3+17}{2}\right) = 80$. The sum is 80.

17. The first term is $a_1 = 1$ and the last term is $a_{50} = 50$. To find the sum use:

$S_n = n\left(\frac{a_1 + a_n}{2}\right) \Rightarrow S_{50} = 50\left(\frac{1+50}{2}\right) = 1275$. The sum is 1275.

19. The first term is $a_1 = -7$ and $d = 3$. The last term is 101, but we must determine n. The term of 101 is

$\frac{101 - (-7)}{3} = 36$ terms after a_1. Thus, the last term is $a_{37} = 101$. To find the sum use the following:

$S_n = n\left(\frac{a_1 + a_n}{2}\right) \Rightarrow S_{37} = 37\left(\frac{-7+101}{2}\right) = 1739$. The sum is 1739.

21. The number of terms to be added is 40, so $n = 40$. The first term is $a_1 = 5(1) = 5$. The last term is

$a_{40} = 5(40) = 200$. $S_n = n\left(\frac{a_1 + a_n}{2}\right) \Rightarrow S_{40} = 40\left(\frac{5+200}{2}\right) = 4100$. The sum is 4100.

23. $S_{15} = 15\left(\frac{a_1 + a_{15}}{2}\right) \Rightarrow 255 = 15\left(\frac{3 + a_{15}}{2}\right) \Rightarrow 17 = \frac{3 + a_{15}}{2} \Rightarrow 34 = 3 + a_{15} \Rightarrow a_{15} = 31$

25. Since $a_1 = 4$ and $d = 2$, use the formula $S_n = \frac{n}{2}(2a_1 + (n-1)d)$.

$S_{20} = \frac{20}{2}(2(4) + (20-1)2) = 10(8 + 19(2)) = 10(8 + 38) = 10(46) = 460$

27. Since $a_1 = 10$ and $d = -\dfrac{1}{2}$, use the formula $S_n = \dfrac{n}{2}(2a_1 + (n-1)d)$.

$$S_{20} = \frac{20}{2}\left(2(10) + (20-1)\left(-\frac{1}{2}\right)\right) = 10\left(20 + 19\left(-\frac{1}{2}\right)\right) = 10\left(20 - \frac{19}{2}\right) = 10\left(\frac{21}{2}\right) = 105$$

29. Since $a_1 = 4$ and $a_{20} = 190.2$, use the formula $S_n = n\left(\dfrac{a_1 + a_n}{2}\right)$.

$$S_{20} = 20\left(\frac{4 + 190.2}{2}\right) = 20\left(\frac{194.2}{2}\right) = 20(97.1) = 1942$$

31. Since $a_1 = -2$ and $a_{11} = 50$, the common difference is $d = \dfrac{50 - (-2)}{11 - 1} = \dfrac{52}{10} = 5.2$.

Use the formula $S_n = \dfrac{n}{2}(2a_1 + (n-1)d)$.

$$S_{20} = \frac{20}{2}(2(-2) + (20-1)5.2) = 10(-4 + 19(5.2)) = 10(-4 + 98.8) = 10(94.8) = 948$$

33. Since $a_2 = 6$ and $a_{12} = 31$, the common difference is $d = \dfrac{31 - 6}{12 - 2} = \dfrac{25}{10} = 2.5$ and so $a_1 = 6 - 2.5 = 3.5$.

Use the formula $S_n = \dfrac{n}{2}(2a_1 + (n-1)d)$.

$$S_{20} = \frac{20}{2}(2(3.5) + (20-1)2.5) = 10(7 + 19(2.5)) = 10(7 + 47.5) = 10(54.5) = 545$$

35. The first term is $a_1 = 1$ and the common ratio is $r = 2$. Since there are 8 terms, the sum is

$$S_n = a_1\left(\frac{1 - r^n}{1 - r}\right) \Rightarrow S_8 = 1\left(\frac{1 - 2^8}{1 - 2}\right) = 255.$$

This sum can be verified by adding $1 + 2 + 4 + 8 + 16 + 32 + 64 + 128 = 255$.

37. The first term is $a_1 = 0.5$ and the common ratio is $r = 3$. Since there are 7 terms, the sum is

$$S_n = a_1\left(\frac{1 - r^n}{1 - r}\right) \Rightarrow S_7 = 0.5\left(\frac{1 - (3)^7}{1 - 3}\right) = 546.5.$$

This sum can be verified by adding $0.5 + 1.5 + 4.5 + 13.5 + 40.5 + 121.5 + 364.5 = 546.5$.

39. The first term is $a_1 = 3(2)^0 = 3$ and the common ratio is $r = 2$. Since there are 20 terms, the sum is

$$S_n = a_1\left(\frac{1 - r^n}{1 - r}\right) \Rightarrow S_{20} = 3\left(\frac{1 - 2^{20}}{1 - 2}\right) = 3{,}145{,}725.$$

41. Using the formula $S_n = a_1\left(\dfrac{1 - r^n}{1 - r}\right)$ with $a_1 = 1$ and $r = -\dfrac{1}{2}$,

$$S_4 = 1\left(\frac{1 - \left(-\frac{1}{2}\right)^4}{1 - \left(-\frac{1}{2}\right)}\right) = 0.625; \quad S_7 = 1\left(\frac{1 - \left(-\frac{1}{2}\right)^7}{1 - \left(-\frac{1}{2}\right)}\right) = 0.671875; \quad S_{10} = 1\left(\frac{1 - \left(-\frac{1}{2}\right)^{10}}{1 - \left(-\frac{1}{2}\right)}\right) = 0.666015625$$

43. Using the formula $S_n = a_1\left(\dfrac{1 - r^n}{1 - r}\right)$ with $a_1 = \dfrac{1}{3}$ and $r = 2$,

$$S_4 = \frac{1}{3}\left(\frac{1 - 2^4}{1 - 2}\right) = 5; \quad S_7 = \frac{1}{3}\left(\frac{1 - 2^7}{1 - 2}\right) = 42.\overline{3}; \quad S_{10} = \frac{1}{3}\left(\frac{1 - 2^{10}}{1 - 2}\right) = 341$$

45. The first term is $a_1 = 1$ and the common ratio is $r = \dfrac{1}{3}$. The sum is $S = a_1\left(\dfrac{1}{1 - r}\right) \Rightarrow S = 1\left(\dfrac{1}{1 - \frac{1}{3}}\right) = \dfrac{3}{2}$.

47. The first term is $a_1 = 6$ and the common ratio is $r = -\dfrac{2}{3}$. The sum is

$$S = a_1\left(\frac{1}{1 - r}\right) \Rightarrow S = 6\left(\frac{1}{1 - (-\frac{2}{3})}\right) = \frac{18}{5}.$$

49. The first term is $a_1 = 1$ and the common ratio is $r = -\dfrac{1}{10}$ or -0.1. The sum is

$$S = a_1\left(\frac{1}{1 - r}\right) \Rightarrow S = 1\left(\frac{1}{1 - (-0.1)}\right) = \frac{1}{1.1} = \frac{10}{11}.$$

51. $\dfrac{2}{3} = 0.6666666\ldots = 0.6 + 0.06 + 0.006 + 0.0006 + 0.00006 + \cdots$

53. $\dfrac{9}{11} = 0.81818181\ldots = 0.81 + 0.0081 + 0.000081 + 0.00000081 + \cdots$

55. $\dfrac{1}{7} = 0.142857142857\ldots = 0.142857 + 0.000000142857 + 0.000000000000142857 + \cdots$

57. The series $0.8 + 0.08 + 0.008 + 0.0008 + \cdots$ is an infinite geometric series with $a_1 = 0.8$ and $r = 0.1$.

$$S = \frac{a_1}{1 - r} = \frac{0.8}{1 - 0.1} = \frac{8}{9}.$$

59. The series $0.45 + 0.0045 + 0.000045 + \cdots$ is an infinite geometric series with $a_1 = 0.45$ and $r = 0.01$.

$$S = \frac{a_1}{1 - r} = \frac{0.45}{1 - 0.01} = \frac{45}{99} = \frac{5}{11}.$$

61. $\displaystyle\sum_{k=1}^{4}(k + 1) = (1 + 1) + (2 + 1) + (3 + 1) + (4 + 1) = 2 + 3 + 4 + 5 = 14$

63. $\displaystyle\sum_{k=1}^{8}4 = 4 + 4 + 4 + 4 + 4 + 4 + 4 + 4 = 32$

65. $\displaystyle\sum_{k=1}^{7}k^3 = 1^3 + 2^3 + 3^3 + 4^3 + 5^3 + 6^3 + 7^3 = 1 + 8 + 27 + 64 + 125 + 216 + 343 = 784$

67. $\displaystyle\sum_{k=4}^{5}(k^2 - k) = (4^2 - 4) + (5^2 - 5) = 12 + 20 = 32$

69. $1^4 + 2^4 + 3^4 + 4^4 + 5^4 + 6^4 = \displaystyle\sum_{k=1}^{6}k^4$

71. $1 + \dfrac{4}{3} + \dfrac{6}{4} + \dfrac{8}{5} + \dfrac{10}{6} + \dfrac{12}{7} + \dfrac{14}{8} = \displaystyle\sum_{k=1}^{7}\left(\frac{2k}{k + 1}\right)$

73. $1 + \dfrac{1}{2^2} + \dfrac{1}{3^2} + \dfrac{1}{4^2} + \dfrac{1}{5^2} + \cdots = \displaystyle\sum_{k=1}^{\infty}\left(\frac{1}{k^2}\right)$

75. Because $\displaystyle\sum_{k=6}^{9}k^3 = 6^3 + 7^3 + 8^3 + 9^3$, the summation has 4 terms. We will write $k = 6, 7, 8, 9$ as

$n = 1, 2, 3, 4$. It follows that $n + 5 = k$, and $\displaystyle\sum_{n=1}^{4}(n + 5)^3$.

77. Because $\displaystyle\sum_{k=9}^{32}(3k - 2) = (3(9) - 2) + (3(10) - 2) + \cdots + (3(32) - 2)$, the summation has 24 terms. We

will write $k = 9, 10, 11, \ldots 32$ as $n = 1, 2, 3, 4, \ldots 24$. It follows that $n + 8 = k$, and

$$\sum_{n=1}^{24}(3(n + 8) - 2) \Rightarrow \sum_{n=1}^{24}(3n + 24 - 2) \Rightarrow \sum_{n=1}^{24}(3n + 22).$$

79. Because $\displaystyle\sum_{k=16}^{52} (k^2 - 3k) = (16^2 - 3(16)) + (17^2 - 3(17)) + \cdots + (52^2 - 3(52))$, the summation has 37

terms. We will write $k = 16, 17, 18, \ldots 52$ as $n = 1, 2, 3, \ldots 37$. It follows that $n + 15 = k$, and

$$\sum_{n=1}^{37} ((n + 15)^2 - 3(n + 15)) \implies \sum_{n=1}^{37} (n^2 + 30n + 225 - 3n - 45) \implies \sum_{n=1}^{37} (n^2 + 27n + 180).$$

81. $\displaystyle\sum_{k=1}^{60} 9 = 60(9) = 540$

83. $\displaystyle\sum_{k=1}^{15} 5k = 5 \sum_{k=1}^{15} k = 5 \left[\frac{15(16)}{2}\right] = 600$

85. $\displaystyle\sum_{k=1}^{31} (3k - 3) = 3 \sum_{k=1}^{31} k - \sum_{k=1}^{31} 3 = 3\left[\frac{31(32)}{2}\right] - 31(3) = 1488 - 93 = 1395$

87. $\displaystyle\sum_{k=1}^{25} k^2 = \frac{(25)(26)(51)}{6} = 5525$

89. $\displaystyle\sum_{k=1}^{16} (k^2 - k) = \sum_{k=1}^{16} k^2 - \sum_{k=1}^{16} k = \frac{(16)(17)(33)}{6} - \frac{(16)(17)}{2} = 1496 - 136 = 1360$

91. $\displaystyle\sum_{k=5}^{24} k = \sum_{k=1}^{24} k - \sum_{k=1}^{4} k = \frac{(24)(25)}{2} - \frac{(4)(5)}{2} = 300 - 10 = 290$

93. $\displaystyle\sum_{k=1}^{n} k = 1 + 2 + 3 + 4 + \cdots + n$. This is an arithmetic series with $a_1 = 1$ and $a_n = n$. The sum is given by

$$S_n = n\left(\frac{a_1 + a_n}{2}\right) = n\left(\frac{1 + n}{2}\right) = \frac{n(n + 1)}{2}.$$

95. (a) The arithmetic sequence describing the salary during year n is computed by $a_n = 42{,}000 + 1800(n - 1)$.

   The 1st and 15th years salaries are: $a_1 = 42{,}000 + 1800(1 - 1) = 42{,}000$ and

   $a_{15} = 42{,}000 + 1800(15 - 1) = 67{,}200$. The total amount earned during the 15 year period is

   $$S_{15} = \frac{15(42{,}000 + 67{,}200)}{2} = 819{,}000.$$

   (b) Verify with a calculator, compute the sum $a_1 + a_2 + a_3 + \cdots + a_{15}$, where

   $a_n = 42{,}000 + 1800(n - 1)$. See Figure 95.

Figure 95

97. Future value is given by $S_n = A_0 \left(\dfrac{(1 + i)^n - 1}{i}\right)$, where $A_0 = a_1$.

$$S_{20} = 2000\left(\frac{(1 + 0.08)^{20} - 1}{0.08}\right) \approx 91{,}523.93.$$ If $\$2000$ is deposited in an account at the end of each year for

20 years and the account pays 8% interest, the future value of this annuity will be $\$91{,}523.93$.

99. Future value is given by $S_n = A_0\left(\dfrac{(1+i)^n - 1}{i}\right)$, where $A_0 = a_1$.

$S_5 = 10{,}000\left(\dfrac{(1+0.11)^5 - 1}{0.11}\right) \approx 62{,}278.01$. If \$10,000 is deposited in an account at the end of each year

for 5 years and the account pays 11% interest, the future value of this annuity will be \$62,278.01.

101. The number of logs in the stack is given by $7 + 8 + 9 + 10 + 11 + 12 + 13 + 14 + 15$. This series is

arithmetic with first term $a_1 = 7$ and the last term $a_9 = 15$ so its sum is given by $S_9 = 9\left(\dfrac{7+15}{2}\right) = 99$. So,

there are 99 logs in the stack.

103. (a) Since each filter lets half of the impurities through, the series is can be written as

$0.5(1) + 0.5(0.5) + 0.5(0.25) + \cdots = \displaystyle\sum_{k=1}^{n} 0.5(0.5)^{k-1}$.

(b) The series sums to 1 only if an infinite number of filters are used.

105. The area of the largest square is 1, the area of the next largest square is $\dfrac{1}{2}$, and each successive square has an

area that is $\dfrac{1}{2}$ the area of the previous square. If there are an infinite number of squares, then the area is

represented by the geometric series $1 + \dfrac{1}{2} + \dfrac{1}{4} + \dfrac{1}{8} + \dfrac{1}{16} + \cdots$. This is an infinite geometric series with

$a_1 = 1$ and $r = \dfrac{1}{2}$, whose sum is $S = \dfrac{1}{1-r} = \dfrac{1}{1-\frac{1}{2}} = 2$. The area is 2.

107. The value of e^a is approximated by $e^a = 1 + a + \dfrac{a^2}{2!} + \dfrac{a^3}{3!} + \cdots + \dfrac{a^n}{n!}$. Apply the series by letting $a = 1$,

since $e^1 = e$. The first 8 terms sum to $e^1 \approx 1 + 1 + \dfrac{1}{2!} + \dfrac{1}{3!} + \dfrac{1}{4!} + \dfrac{1}{5!} + \dfrac{1}{6!} + \dfrac{1}{7!} =$

$1 + 1 + \dfrac{1}{2} + \dfrac{1}{6} + \dfrac{1}{24} + \dfrac{1}{120} + \dfrac{1}{720} + \dfrac{1}{5040} \approx 2.718254$. The actual value is $e \approx 2.718282$, so only 8

terms of the series provides an approximation for e that is accurate to 4 decimal places.

109. One can either add the terms directly or apply the formula $S_n = a_1\left(\dfrac{1-r^n}{1-r}\right)$. We will apply the formula. In

this sequence $a_1 = 1$ and $r = \dfrac{1}{3}$. $S_2 = 1\left(\dfrac{1-\left(\frac{1}{3}\right)^2}{1-\left(\frac{1}{3}\right)}\right) = \dfrac{4}{3} \approx 1.3333$; $S_4 = 1\left(\dfrac{1-\left(\frac{1}{3}\right)^4}{1-\left(\frac{1}{3}\right)}\right) = \dfrac{40}{27} \approx 1.4815$;

$S_8 = 1\left(\dfrac{1-\left(\frac{1}{3}\right)^8}{1-\left(\frac{1}{3}\right)}\right) \approx 1.49977$; $S_{16} = 1\left(\dfrac{1-\left(\frac{1}{3}\right)^{16}}{1-\left(\frac{1}{3}\right)}\right) \approx 1.49999997$. The sum of the infinite series

$\displaystyle\sum_{k=1}^{\infty}\left(\dfrac{1}{3}\right)^{k-1}$ is given by $S = \dfrac{a_1}{1-r} = \dfrac{1}{1-\frac{1}{3}} = \dfrac{3}{2} = 1.5$. As n increases, the partial sums $S_2, S_4, S_8,$ and S_{16}

become closer and closer to the value of 1.5.

111. One can either add the terms directly or apply the formula $S_n = a_1\left(\dfrac{1 - r^n}{1 - r}\right)$. We will apply the formula. In

this sequence $a_1 = 4$ and $r = -\dfrac{1}{10}$. $S_1 = 4\left(\dfrac{1 - (-\frac{1}{10})^1}{1 - (-\frac{1}{10})}\right) = 4$; $S_2 = 4\left(\dfrac{1 - (-\frac{1}{10})^2}{1 - (-\frac{1}{10})}\right) = 3.6$;

$S_3 = 4\left(\dfrac{1 - (-\frac{1}{10})^3}{1 - (-\frac{1}{10})}\right) = 3.64$; $S_4 = 4\left(\dfrac{1 - (-\frac{1}{10})^4}{1 - (-\frac{1}{10})}\right) = 3.636$; $S_5 = 4\left(\dfrac{1 - (-\frac{1}{10})^5}{1 - (-\frac{1}{10})}\right) = 3.6364$;

$S_6 = 4\left(\dfrac{1 - (-\frac{1}{10})^6}{1 - (-\frac{1}{10})}\right) = 3.63636$; The sum of the infinite series $\displaystyle\sum_{k=1}^{\infty} 4\left(-\dfrac{1}{10}\right)^{k-1}$ is given by

$S = \dfrac{a_1}{1 - r} = \dfrac{4}{1 - (-0.1)} = \dfrac{40}{11} = 3.\overline{63}$. As n increases, the partial sums become closer and closer to the

value of $3.\overline{63}$.

11.3: Counting

1. There are $2 \cdot 2 \cdot 2 \cdot 2 \cdot 2 \cdot 2 \cdot 2 \cdot 2 \cdot 2 \cdot 2 = 2^{10} = 1024$ different ways to answer the exam.

3. There are $2 \cdot 2 \cdot 2 \cdot 2 \cdot 2 \cdot 4 \cdot 4 \cdot 4 \cdot 4 \cdot 4 \cdot 4 \cdot 4 \cdot 4 \cdot 4 \cdot 4 = 2^5 \cdot 4^{10} = 33{,}554{,}432$ different ways.

5. There are 10 digits and 26 letters. There are $10 \cdot 10 \cdot 10 \cdot 26 \cdot 26 \cdot 26 = 17{,}576{,}000$ different license plates.

7. There are 36 digits and letters. There are $26 \cdot 26 \cdot 26 \cdot 36 \cdot 36 \cdot 36 = 820{,}025{,}856$ different license plates.

9. There are 3 choices for each of the 5 letters. There are $3 \cdot 3 \cdot 3 \cdot 3 \cdot 3 = 3^5 = 243$ different strings.

11. There are 5 choices for each of the 5 letters. There are $5 \cdot 5 \cdot 5 \cdot 5 \cdot 5 = 5^5 = 3125$ different strings.

13. Since there are 2 letters that can only be used once, each string must be 2 letters long. For the first position there are 2 choices and for the second position there is only 1 choice. There are $2 \cdot 1 = 2$ possible strings.

15. There are $4 \cdot 3 \cdot 2 \cdot 1 = 24$ possible strings.

17. Each combination can vary between 000 and 999. Thus, there are 1000 possibilities for each lock. That is, there are $1000 \cdot 1000 = 1{,}000{,}000$ different combinations in all.

19. There are 2 settings (on or off) for each of the 12 switches.

There are $2 \cdot 2 \cdot 2 \cdot 2 \cdot 2 \cdot 2 \cdot 2 \cdot 2 \cdot 2 \cdot 2 \cdot 2 \cdot 2 = 2^{12} = 4096$ different codes for the garage door opener.

21. For the first letter there are 2 possibilities, whereas there are 26 possibilities for each of the last 3 positions. There are $2 \cdot 26 \cdot 26 \cdot 26 = 35{,}152$ different call letters possible. There is no shortage of call letters.

23. There are $2 \cdot 3 \cdot 4 = 24$ different packages that can be purchased.

25. There are $8 \cdot 10 \cdot 10 \cdot 10 \cdot 10 \cdot 10 \cdot 10 = 8{,}000{,}000$ such phone numbers.

27. $6! = 6 \cdot 5 \cdot 4 \cdot 3 \cdot 2 \cdot 1 = 720$

29. $10! = 10 \cdot 9 \cdot 8 \cdot 7 \cdot 6 \cdot 5 \cdot 4 \cdot 3 \cdot 2 \cdot 1 = 3,628,800$

31. $P(n, r) = \dfrac{n!}{(n - r)!} \Rightarrow P(5, 3) = \dfrac{5!}{(5 - 3)!} = \dfrac{5!}{2!} = \dfrac{120}{2} = 60$

33. $P(n, r) = \dfrac{n!}{(n - r)!} \Rightarrow P(8, 1) = \dfrac{8!}{(8 - 1)!} = \dfrac{8!}{7!} = \dfrac{8 \cdot 7!}{7!} = 8$

35. $P(n, r) = \dfrac{n!}{(n - r)!} \Rightarrow P(7, 3) = \dfrac{7!}{(7 - 3)!} = \dfrac{7!}{4!} = \dfrac{7 \cdot 6 \cdot 5 \cdot 4!}{4!} = 7 \cdot 6 \cdot 5 = 210$

37. $P(n, r) = \dfrac{n!}{(n - r)!} \Rightarrow P(25, 2) = \dfrac{25!}{(25 - 2)!} = \dfrac{25!}{23!} = \dfrac{25 \cdot 24 \cdot 23!}{23!} = 25 \cdot 24 = 600$

39. $P(n, r) = \dfrac{n!}{(n - r)!} \Rightarrow P(10, 4) = \dfrac{10!}{(10 - 4)!} = \dfrac{10!}{6!} = \dfrac{10 \cdot 9 \cdot 8 \cdot 7 \cdot 6!}{6!} = 10 \cdot 9 \cdot 8 \cdot 7 = 5040$

41. $P(4, 4) = \dfrac{4!}{(4 - 4)!} = \dfrac{4!}{0!} = 4! = 24$

43. There are $15 \cdot 14 \cdot 13 = 2730$ different arrangements of 3 students from a class of 15.

45. There are $7 \cdot 6 \cdot 5 = 210$ different routes.

47. The remaining 4 digits for the 7 digit number are independent events. Since the last 4 digits can be any number from 0 to 9. The total is given by $10 \cdot 10 \cdot 10 \cdot 10 = 10,000$. The first 3 numbers are restricted to 3 different possibilities and there will be $(3) \cdot (10,000)$ or 30,000 phone numbers.

49. Initially any person can sit at the table. Then the remaining 6 people can sit in $6! = 720$ different ways. Since there is no difference between the people sitting clockwise or counterclockwise around the table, we must divide this result by 2 for the final answer. There are 360 different ways to seat the 7 people.

51. $P(9, 9) = \dfrac{9!}{(9 - 9)!} = \dfrac{9!}{0!} = 9! = 362,880$

53. Counting February 29th, $P(366, 5) = \dfrac{366!}{(366 - 5)!} = \dfrac{366!}{361!} = \dfrac{366 \cdot 365 \cdot 364 \cdot 363 \cdot 362 \cdot 361!}{361!} \approx 6.39 \times 10^{12}$.

55. There are 8 choices for the first and fourth digits and 10 choices for each of the other digits.

$8 \cdot 10 \cdot 10 \cdot 8 \cdot 10 \cdot 10 \cdot 10 \cdot 10 \cdot 10 \cdot 10 = 6,400,000,000$

57. $C(n, r) = \dfrac{n!}{(n - r)! \, r!} \Rightarrow C(3, 1) = \dfrac{3!}{(3 - 1)! \, 1!} = \dfrac{3!}{2! \, 1!} = \dfrac{6}{2} = 3$

59. $C(n, r) = \dfrac{n!}{(n - r)! \, r!} \Rightarrow C(6, 3) = \dfrac{6!}{(6 - 3)! \, 3!} = \dfrac{6!}{3! \, 3!} = \dfrac{720}{36} = 20$

61. $C(n, r) = \dfrac{n!}{(n - r)! \, r!} \Rightarrow C(5, 0) = \dfrac{5!}{(5 - 0)! \, 0!} = \dfrac{5!}{5! \, 0!} = \dfrac{5!}{5!} = 1$

63. $\dbinom{n}{r} = \dfrac{n!}{(n - r)! \, r!} \Rightarrow \dbinom{8}{2} = \dfrac{8!}{(8 - 2)! \, 2!} = \dfrac{8!}{6! \, 2!} = \dfrac{8 \cdot 7 \cdot 6!}{6! \, 2!} = \dfrac{56}{2} = 28$

65. $\binom{n}{r} = \frac{n!}{(n-r)!\, r!} \Rightarrow \binom{20}{18} = \frac{20!}{(20-18)!\, 18!} = \frac{20!}{2!\, 18!} = \frac{20 \cdot 19 \cdot 18!}{2!\, 18!} = \frac{20 \cdot 19}{2} = 190$

67. From 39 numbers a player picks 5 numbers. Since order is unimportant, there are $C(39, 5) = 575{,}757$ different ways of doing this.

69. Two women can be selected from 5 women in $C(5, 2)$ different ways. Two men can be selected from 3 men in $C(3, 2)$ different ways. The total number of committees is $C(5, 2) \cdot C(3, 2) = 10 \cdot 3 = 30$.

71. Three questions can be selected from 5 questions in $C(5, 3)$ different ways. Four questions can be selected from 5 questions in $C(5, 4)$ different ways. The total number of possibilities is $C(5, 3) \cdot C(5, 4) = 10 \cdot 5 = 50$.

73. Three red marbles can be drawn from 10 red marbles in $C(10, 3)$ different ways. Two blue marbles can be drawn from 12 blue marbles in $C(12, 2)$ different ways. There are $C(10, 3) \cdot C(12, 2) = 120 \cdot 66 = 7920$ ways.

75. Since order is not important, there are $C(24, 3) = 2024$ ways to do this.

77. $P(n, n-1) = \frac{n!}{(n-(n-1))!} = \frac{n!}{1} = n!$ and $P(n, n) = \frac{n!}{(n-n)!} = \frac{n!}{0!} = \frac{n!}{1} = n!$

For example $P(7, 6) = 5040 = P(7, 7)$.

11.4: The Binomial Theorem

1. $\binom{5}{4} = \frac{5!}{1!\, 4!} = \frac{5 \cdot 4!}{1 \cdot 4!} = 5$

3. $\binom{4}{0} = \frac{4!}{4!\, 0!} = \frac{4!}{4! \cdot 1} = 1$

5. $\binom{6}{5} = \frac{6!}{1!\, 5!} = \frac{6 \cdot 5!}{1 \cdot 5!} = \frac{6}{1} = 6$

7. $\binom{3}{3} = \frac{3!}{0!\, 3!} = \frac{3!}{1 \cdot 3!} = 1$

9. Three a's, two b's $\Rightarrow C(5, 2) = 10$ different strings.

11. Four a's, four b's $\Rightarrow C(8, 4) = 70$ different strings.

13. Five a's, zero b's $\Rightarrow C(5, 0) = 1$ string.

15. Four a's, one $b \Rightarrow C(5, 1) = 5$ different strings.

17. $(x + y)^2 = \binom{2}{0}x^2y^0 + \binom{2}{1}x^1y^1 + \binom{2}{2}x^0y^2 = x^2 + 2xy + y^2$

19. $(m + 2)^3 = \binom{3}{0}m^3(2)^0 + \binom{3}{1}m^2(2)^1 + \binom{3}{2}m^1(2)^2 + \binom{3}{3}m^0(2)^3 = m^3 + 6m^2 + 12m + 8$

21. $(2x - 3)^3 = \binom{3}{0}(2x)^3(-3)^0 + \binom{3}{1}(2x)^2(-3)^1 + \binom{3}{2}(2x)^1(-3)^2 + \binom{3}{3}(2x)^0(-3)^3 =$

$8x^3 - 36x^2 + 54x - 27$

23. $(p - q)^6 = \binom{6}{0}p^6(-q)^0 + \binom{6}{1}p^5(-q)^1 + \binom{6}{2}p^4(-q)^2 + \binom{6}{3}p^3(-q)^3 + \binom{6}{4}p^2(-q)^4 +$

$\binom{6}{5}p^1(-q)^5 + \binom{6}{6}p^0(-q)^6 = p^6 - 6p^5q + 15p^4q^2 - 20p^3q^3 + 15p^2q^4 - 6pq^5 + q^6$

25. $(2m + 3n)^3 = \binom{3}{0}(2m)^3(3n)^0 + \binom{3}{1}(2m)^2(3n)^1 + \binom{3}{2}(2m)^1(3n)^2 + \binom{3}{3}(2m)^0(3n)^3 =$

$8m^3 + 36m^2n + 54mn^2 + 27n^3$

27. $(1 - x^2)^4 = \binom{4}{0}(1)^4(-x^2)^0 + \binom{4}{1}(1)^3(-x^2)^1 + \binom{4}{2}(1)^2(-x^2)^2 + \binom{4}{3}(1)^1(-x^2)^3 + \binom{4}{4}(1)^0(-x^2)^4$

$= 1 - 4x^2 + 6x^4 - 4x^6 + x^8$

29. $(2p^3 - 3)^3 = \binom{3}{0}(2p^3)^3(-3)^0 + \binom{3}{1}(2p^3)^2(-3)^1 + \binom{3}{2}(2p^3)^1(-3)^2 + \binom{3}{3}(2p^3)^0(-3)^3 =$

$8p^9 - 36p^6 + 54p^3 - 27$

31. Using Pascal's triangle, the coefficients are 1, 2, and 1.

$(x + y)^2 = 1x^2y^0 + 2x^1y^1 + 1x^0y^2 = x^2 + 2xy + y^2$

33. Using Pascal's triangle, the coefficients are 1, 4, 6, 4, and 1.

$(3x + 1)^4 = 1(3x)^4(1)^0 + 4(3x)^3(1)^1 + 6(3x)^2(1)^2 + 4(3x)^1(1)^3 + 1(3x)^0(1)^4 =$

$81x^4 + 108x^3 + 54x^2 + 12x + 1$

35. Using Pascal's triangle, the coefficients are 1, 5, 10, 10, 5, and 1.

$(2 - x)^5 = 1(2)^5(-x)^0 + 5(2)^4(-x)^1 + 10(2)^3(-x)^2 + 10(2)^2(-x)^3 + 5(2)^1(-x)^4 + 1(2)^0(-x)^5 =$

$32 - 80x + 80x^2 - 40x^3 + 10x^4 - x^5$

37. Using Pascal's triangle, the coefficients are 1, 4, 6, 4, and 1.

$(x^2 + 2)^4 = 1(x^2)^4(2)^0 + 4(x^2)^3(2)^1 + 6(x^2)^2(2)^2 + 4(x^2)^1(2)^3 + 1(x^2)^0(2)^4 =$

$x^8 + 8x^6 + 24x^4 + 32x^2 + 16$

39. Using Pascal's triangle, the coefficients are 1, 4, 6, 4, and 1.

$(4x - 3y)^4 = 1(4x)^4(-3y)^0 + 4(4x)^3(-3y)^1 + 6(4x)^2(-3y)^2 + 4(4x)^1(-3y)^3 + 1(4x)^0(-3y)^4 =$

$256x^4 - 768x^3y + 864x^2y^2 - 432xy^3 + 81y^4$

41. Using Pascal's triangle, the coefficients are 1, 6, 15, 20, 15, 6, and 1.

$(m + n)^6 = m^6 + 6m^5n + 15m^4n^2 + 20m^3n^3 + 15m^2n^4 + 6mn^5 + n^6$

43. Using Pascal's triangle, the coefficients are 1, 3, 3, and 1.

$(2x^3 - y^2)^3 = 1(2x^3)^3(-y^2)^0 + 3(2x^3)^2(-y^2)^1 + 3(2x^3)^1(-y^2)^2 + 1(2x^3)^0(-y^2)^3 =$

$8x^9 - 12x^6y^2 + 6x^3y^4 - y^6$

45. The fourth term of $(a + b)^9$ is $\binom{9}{6}a^{9-3}b^3 = \binom{9}{6}a^6b^3 = 84a^6b^3$.

47. The fifth term of $(x + y)^8$ is $\binom{8}{4}x^{8-4}y^4 = \binom{8}{4}x^4y^4 = 70x^4y^4$.

49. The fourth term of $(2x + y)^5$ is $\binom{5}{2}(2x)^{5-3}y^3 = \binom{5}{2}(2x)^2y^3 = 40x^2y^3$.

51. The sixth term of $(3x - 2y)^6$ is $\binom{6}{1}(3x)^{6-5}(-2y)^5 = \binom{6}{1}(3x)(-2y)^5 = -576xy^5$.

11.5: Mathematical Induction

1. $3 + 6 + 9 + \cdots + 3n = \dfrac{3n(n + 1)}{2}$

   (i) Show that the statement is true for $n = 1$: $3(1) = \dfrac{3(1)(2)}{2} \Rightarrow 3 = 3$

   (ii) Assume that S_k is true: $3 + 6 + 9 + \cdots + 3k = \dfrac{3k(k + 1)}{2}$

   Show that S_{k+1} is true: $3 + 6 + \cdots + 3(k + 1) = \dfrac{3(k + 1)(k + 2)}{2}$

   Add $3(k + 1)$ to each side of S_k: $3 + 6 + 9 + \cdots + 3k + 3(k + 1) = \dfrac{3k(k + 1)}{2} + 3(k + 1) =$

   $\dfrac{3k(k + 1) + 6(k + 1)}{2} = \dfrac{(k + 1)(3k + 6)}{2} = \dfrac{3(k + 1)(k + 2)}{2}$

   Since S_k implies S_{k+1}, the statement is true for every positive integer n.

3. $5 + 10 + 15 + \cdots + 5n = \dfrac{5n(n + 1)}{2}$

   (i) Show that the statement is true for $n = 1$: $5(1) = \dfrac{5(1)(2)}{2} \Rightarrow 5 = 5$

   (ii) Assume that S_k is true: $5 + 10 + 15 + \cdots + 5k = \dfrac{5k(k + 1)}{2}$

   Show that S_{k+1} is true: $5 + 10 + \cdots + 5(k + 1) = \dfrac{5(k + 1)(k + 2)}{2}$

   Add $5(k + 1)$ to each side of S_k: $5 + 10 + 15 + \cdots + 5k + 5(k + 1) = \dfrac{5k(k + 1)}{2} + 5(k + 1) =$

   $\dfrac{5k(k + 1) + 10(k + 1)}{2} = \dfrac{(k + 1)(5k + 10)}{2} = \dfrac{5(k + 1)(k + 2)}{2}$

   Since S_k implies S_{k+1}, the statement is true for every positive integer n.

5. $3 + 3^2 + 3^3 + \cdots + 3^n = \dfrac{3(3^n - 1)}{2}$

   (i) Show that the statement is true for $n = 1$: $3^1 = \dfrac{3(3^1 - 1)}{2} \Rightarrow 3 = 3$

   (ii) Assume that S_k is true: $3 + 3^2 + 3^3 + \cdots + 3^k = \dfrac{3(3^k - 1)}{2}$

   Show that S_{k+1} is true: $3 + 3^2 + 3^3 + \cdots + 3^{k+1} = \dfrac{3(3^{k+1} - 1)}{2}$

   Add 3^{k+1} to each side of S_k: $3 + 3^2 + 3^3 + \cdots + 3^k + 3^{k+1} = \dfrac{3(3^k - 1)}{2} + 3^{k+1} =$

   $\dfrac{3(3^k - 1) + 2(3^{k+1})}{2} = \dfrac{3^{k+1} - 3 + 2(3^{k+1})}{2} = \dfrac{3(3^{k+1}) - 3}{2} = \dfrac{3(3^{k+1} - 1)}{2}$

   Since S_k implies S_{k+1}, the statement is true for every positive integer n.

7. $1^3 + 2^3 + 3^3 + \cdots + n^3 = \dfrac{n^2(n + 1)^2}{4}$

   (i) Show that the statement is true for $n = 1$: $1^3 = \dfrac{1^2(1 + 1)^2}{4} \Rightarrow 1 = 1$

   (ii) Assume that S_k is true: $1^3 + 2^3 + 3^3 + \cdots + k^3 = \dfrac{k^2(k + 1)^2}{4}$

   Show that S_{k+1} is true: $1^3 + 2^3 + \cdots + (k + 1)^3 = \dfrac{(k + 1)^2(k + 2)^2}{4}$

   Add $(k + 1)^3$ to each side of S_k: $1^3 + 2^3 + 3^3 + \cdots + k^3 + (k + 1)^3 = \dfrac{k^2(k + 1)^2}{4} + (k + 1)^3 =$

   $\dfrac{k^2(k + 1)^2 + 4(k + 1)^3}{4} = \dfrac{(k + 1)^2(k^2 + 4k + 4)}{4} = \dfrac{(k + 1)^2(k + 2)^2}{4}$

   Since S_k implies S_{k+1}, the statement is true for every positive integer n.

9. $\dfrac{1}{1 \cdot 2} + \dfrac{1}{2 \cdot 3} + \cdots + \dfrac{1}{n(n + 1)} = \dfrac{n}{n + 1}$

   (i) Show that the statement is true for $n = 1$: $\dfrac{1}{1(1 + 1)} = \dfrac{1}{1 + 1} \Rightarrow \dfrac{1}{2} = \dfrac{1}{2}$

   (ii) Assume that S_k is true: $\dfrac{1}{1 \cdot 2} + \dfrac{1}{2 \cdot 3} + \cdots + \dfrac{1}{k(k + 1)} = \dfrac{k}{k + 1}$

   Show that S_{k+1} is true: $\dfrac{1}{1 \cdot 2} + \dfrac{1}{2 \cdot 3} + \cdots + \dfrac{1}{(k + 1)(k + 2)} = \dfrac{k + 1}{k + 2}$

   Add $\dfrac{1}{(k + 1)(k + 2)}$ to each side of S_k: $\dfrac{1}{1 \cdot 2} + \dfrac{1}{2 \cdot 3} + \cdots + \dfrac{1}{(k + 1)(k + 2)} =$

   $\dfrac{k}{k + 1} + \dfrac{1}{(k + 1)(k + 2)} = \dfrac{k(k + 2) + 1}{(k + 1)(k + 2)} = \dfrac{k^2 + 2k + 1}{(k + 1)(k + 2)} = \dfrac{(k + 1)(k + 1)}{(k + 1)(k + 2)} = \dfrac{k + 1}{k + 2}$

   Since S_k implies S_{k+1}, the statement is true for every positive integer n.

11. $\dfrac{4}{5} + \dfrac{4}{5^2} + \dfrac{4}{5^3} + \cdots + \dfrac{4}{5^n} = 1 - \dfrac{1}{5^n}$

(i) Show that the statement is true for $n = 1$: $\dfrac{4}{5^1} = 1 - \dfrac{1}{5^1} \Rightarrow \dfrac{4}{5} = \dfrac{4}{5}$

(ii) Assume that S_k is true: $\dfrac{4}{5} + \dfrac{4}{5^2} + \dfrac{4}{5^3} + \cdots + \dfrac{4}{5^k} = 1 - \dfrac{1}{5^k}$

Show that S_{k+1} is true: $\dfrac{4}{5} + \dfrac{4}{5^2} + \cdots + \dfrac{4}{5^{k+1}} = 1 - \dfrac{1}{5^{k+1}}$

Add $\dfrac{4}{5^{k+1}}$ to each side of S_k: $\dfrac{4}{5} + \dfrac{4}{5^2} + \dfrac{4}{5^3} + \cdots + \dfrac{4}{5^k} + \dfrac{4}{5^{k+1}} = 1 - \dfrac{1}{5^k} + \dfrac{4}{5^{k+1}} =$

$1 - \dfrac{1}{5^k} \cdot \dfrac{5}{5} + \dfrac{4}{5^{k+1}} = 1 - \dfrac{5}{5^{k+1}} + \dfrac{4}{5^{k+1}} = 1 - \dfrac{1}{5^{k+1}}$

Since S_k implies S_{k+1}, the statement is true for every positive integer n.

13. $\dfrac{1}{1\cdot 4} + \dfrac{1}{4\cdot 7} + \cdots + \dfrac{1}{(3n-2)(3n+1)} = \dfrac{n}{3n+1}$

(i) Show that the statement is true for $n = 1$: $\dfrac{1}{1\cdot 4} = \dfrac{1}{3(1)+1} \Rightarrow \dfrac{1}{4} = \dfrac{1}{4}$

(ii) Assume that S_k is true: $\dfrac{1}{1\cdot 4} + \cdots + \dfrac{1}{(3k-2)(3k+1)} = \dfrac{k}{3k+1}$

Show that S_{k+1} is true: $\dfrac{1}{1\cdot 4} + \cdots + \dfrac{1}{[3(k+1)-2][3(k+1)+1]} = \dfrac{k+1}{3(k+1)+1}$

Add $\dfrac{1}{[3(k+1)-2][3(k+1)+1]}$ to each side of S_k: $\dfrac{1}{1\cdot 4} + \cdots + \dfrac{1}{[3(k+1)-2][3(k+1)+1]}$

$= \dfrac{k}{3k+1} + \dfrac{1}{[3(k+1)-2][3(k+1)+1]} = \dfrac{k}{3k+1} + \dfrac{1}{(3k+1)(3k+4)} = \dfrac{k(3k+4)+1}{(3k+1)(3k+4)} =$

$\dfrac{3k^2 + 4k + 1}{(3k+1)(3k+4)} = \dfrac{(3k+1)(k+1)}{(3k+1)(3k+4)} = \dfrac{k+1}{3k+4} = \dfrac{k+1}{3(k+1)+1}$

Since S_k implies S_{k+1}, the statement is true for every positive integer n.

15. When $n = 1$, $3^1 < 6(1) \Rightarrow 3 < 6$. When $n = 2$, $3^2 < 6(2) \Rightarrow 9 < 12$.

When $n = 3$, $3^3 > 6(3) \Rightarrow 27 > 18$. For all $n \geq 3$, $3^n > 6n$. The only values are 1 and 2.

17. When $n = 1$, $2^1 > 1^2 \Rightarrow 2 > 1$. When $n = 2$, $2^2 = 2^2 \Rightarrow 4 = 4$. When $n = 3$, $2^3 < 3^2 \Rightarrow 8 < 9$.

When $n = 4$, $2^4 = 4^2 \Rightarrow 16 = 16$. For all $n \geq 5$, $2^n > n^2$. The only values are 2, 3, and 4.

19. $(a^m)^n = a^{mn}$

(i) Show that the statement is true for $n = 1$: $(a^m)^1 = a^{m\cdot 1} \Rightarrow a^m = a^m$

(ii) Assume that S_k is true: $(a^m)^k = a^{mk}$

Show that S_{k+1} is true: $(a^m)^{k+1} = a^{m(k+1)}$

Multiply each side of S_k by a^m: $(a^m)^k \cdot (a^m)^1 = a^{mk} \cdot a^m \Rightarrow (a^m)^{k+1} = a^{mk+m} \Rightarrow (a^m)^{k+1} = a^{m(k+1)}$

Since S_k implies S_{k+1}, the statement is true for every positive integer n.

21. $2^n > 2n$, if $n \geq 3$

    (i) Show that the statement is true for $n = 3$: $2^3 > 2(3) \Rightarrow 8 > 6$

    (ii) Assume that S_k is true: $2^k > 2k$

    Show that S_{k+1} is true: $2^{k+1} > 2(k + 1)$

    Multiply each side of S_k by 2: $2^k \cdot 2 > 2k \cdot 2 \Rightarrow 2^{k+1} > 2(k + 1)$

    Since S_k implies S_{k+1}, the statement is true for every positive integer $n \geq 3$.

23. $a^n > 1$, if $a > 1$

    (i) Show that the statement is true for $n = 1$: $a^1 > 1 \Rightarrow a > 1$, which is true by the given restriction.

    (ii) Assume that S_k is true: $a^k > 1$

    Show that S_{k+1} is true: $a^{k+1} > 1$

    Multiply each side of S_k by a: $a^k \cdot a > 1 \cdot a \Rightarrow a^{k+1} > a$

    Because $a > 1$, we may substitute 1 for a in the expression. That is $a^{k+1} > 1$

    Since S_k implies S_{k+1}, the statement is true for every positive integer n.

25. $a^n < a^{n-1}$, if $0 < a < 1$

    (i) Show that the statement is true for $n = 1$: $a^1 < a^0 \Rightarrow a < 1$, which is true by the given restriction.

    (ii) Assume that S_k is true: $a^k < a^{k-1}$

    Show that S_{k+1} is true: $a^{k+1} < a^k$

    Multiply each side of S_k by a: $a^k \cdot a < a^{k-1} \cdot a \Rightarrow a^{k+1} < a^k$

    Since S_k implies S_{k+1}, the statement is true for every positive integer n.

27. $n! > 2^n$, if $n \geq 4$

    (i) Show that the statement is true for $n = 4$: $4! > 2^4 \Rightarrow 24 > 16$

    (ii) Assume that S_k is true: $k! > 2^k$

    Show that S_{k+1} is true: $(k + 1)! > 2^{k+1}$

    Multiply each side of S_k by $k + 1$: $(k + 1)k! > 2^k(k + 1) \Rightarrow (k + 1)! > 2^k(k + 1)$

    Because $(k + 1) > 2$ for all $k \geq 4$, we may substitute 2 for $(k + 1)$ in the expression.

    That is $(k + 1)! > 2^k(2)$ or $(k + 1)! > 2^{k+1}$.

    Since S_k implies S_{k+1}, the statement is true for every positive integer $n \geq 4$.

29. The number of handshakes is $\dfrac{n^2 - n}{2}$ if $n \geq 2$.

    (i) Show that the statement is true for $n = 2$: The number of handshakes for 2 people is $\dfrac{2^2 - 2}{2} = \dfrac{2}{2} = 1$,

    which is true.

    (ii) Assume that S_k is true: The number of handshakes for k people is $\dfrac{k^2 - k}{2}$.

    Show that S_{k+1} is true: The number of handshakes for $k + 1$ people is

    $$\frac{(k + 1)^2 - (k + 1)}{2} = \frac{k^2 + 2k + 1 - k - 1}{2} = \frac{k^2 + k}{2}.$$

    When a person joins a group of k people, each person must shake hands with the new person.

    Since there are a total of k people that will shake hands with the new person, the total number of handshakes

    for $k + 1$ people is $\dfrac{k^2 - k}{2} + k = \dfrac{k^2 - k + 2k}{2} = \dfrac{k^2 + k}{2}$.

    Since S_k implies S_{k+1}, the statement is true for every positive integer $n \geq 2$.

31. The first figure has perimeter $P = 3$. When a new figure is generated, each side if the previous figure increases in length by a factor of $\dfrac{4}{3}$. Thus, the second figure has perimeter $P = 3\left(\dfrac{4}{3}\right)$, the third figure has perimeter $P = 3\left(\dfrac{4}{3}\right)^2$, and so on. In general, the nth figure has perimeter $P = 3\left(\dfrac{4}{3}\right)^{n-1}$.

33. With 1 ring, 1 move is required. With 2 rings, 3 moves are required. Note that $3 = 2 + 1$. With 3 rings, 7 moves are required. Note that $7 = 2^2 + 2 + 1$.

With n rings $2^{n-1} + 2^{n-2} + \cdots + 2^1 + 1 = 2^n - 1$ moves are required.

(i) Show that the statement is true for $n = 1$: The number of moves for 1 ring is $2^1 - 1 = 1$, which is true.

(ii) Assume that S_k is true: The number of moves for k rings is $2^k - 1$.

Show that S_{k+1} is true: The number of moves for $k + 1$ rings is $2^{k+1} - 1$.

Assume $k + 1$ rings are on the first peg. Since S_k is true, the top k rings can be moved to the second peg in $2^k - 1$ moves. Now move the bottom ring to the third peg. Since S_k is true, move the k rings from the second peg on top of the ring on the third peg in $2^k - 1$ moves. The total number of moves is

$$(2^k - 1) + 1 + (2^k - 1) = 2 \cdot 2^k - 1 = 2^{k+1} - 1$$

Since S_k implies S_{k+1}, the statement is true for every positive integer n.

11.6: Probability

1. Yes. The number is between 0 and 1.

3. No. The number is greater 1.

5. Yes. The number is between 0 and 1, inclusive.

7. No. This number is less than 0.

9. A head when tossing a fair coin $\Rightarrow \dfrac{1}{2}$.

11. Rolling a 2 with a fair die $\Rightarrow \dfrac{1}{6}$.

13. Guessing the correct answer for a true-false question $\Rightarrow \dfrac{1}{2}$.

15. Drawing a king from a standard deck of 52 cards $\Rightarrow \dfrac{4}{52} = \dfrac{1}{13}$.

17. The access code can vary between 0000 and 9999. This is 10,000 possibilities, so the probability of guessing the ATM code at random is $\dfrac{1}{10,000}$.

19. (a) The probability that their favorite is pepperoni is 0.43, so the probability it is not pepperoni is

$$1 - 0.43 = 0.57 \text{ or } 57\%.$$

(b) $19\% + 14\% = 33\%$ or 0.33

21. The probability of one tail is $\frac{1}{2}$. The events of tossing a coin are independent so the probability of tossing two tails is $\frac{1}{2} \cdot \frac{1}{2} = \frac{1}{4}$.

23. The probability of rolling either a 5 or a 6 is $\frac{2}{6} = \frac{1}{3}$. The probability of obtaining a 5 or 6 on three consecutive roles is $\frac{1}{3} \cdot \frac{1}{3} \cdot \frac{1}{3} = \frac{1}{27}$.

25. To obtain a sum of 2, both die must show a 1. The probability of rolling a 1 with one die is $\frac{1}{6}$. Since the events are independent, the probability of obtaining a 1 on two dice is $\frac{1}{6} \cdot \frac{1}{6} = \frac{1}{36}$.

27. The probability of rolling a die and not obtaining a 6 is $\frac{5}{6}$. The probability of rolling a die four times and not obtaining a 6 is $\frac{5}{6} \cdot \frac{5}{6} \cdot \frac{5}{6} \cdot \frac{5}{6} = \frac{625}{1296} \approx 0.482$.

29. The probability of drawing the first ace is $\frac{4}{52}$, the second ace $\frac{3}{51}$, the third ace $\frac{2}{50}$, and the fourth ace $\frac{1}{49}$. The probability of drawing four aces is $\frac{4}{52} \cdot \frac{3}{51} \cdot \frac{2}{50} \cdot \frac{1}{49} = \frac{24}{6,497,400} = \frac{1}{270,725}$.

31. There are $\binom{13}{3}$ ways to draw 3 hearts, and there are $\binom{13}{2}$ ways to draw 2 diamonds. there are $\binom{52}{5}$ different poker hands. Thus, the probability of drawing 3 hearts and 2 diamonds is
$$P(E) = \frac{n(E)}{n(S)} = \frac{\binom{13}{3} \cdot \binom{13}{2}}{\binom{52}{5}} = \frac{286 \cdot 78}{2,598,960} = 0.0086, \text{ or a } 0.86\% \text{ chance.}$$

33. There are 4 strings out of 20 that are defective. Therefore, there is a $\frac{4}{20} = 0.2$ probability or 20% chance of drawing a defective string and rejecting the box.

35. (a) Figure 35 shows a Venn diagram of the data.

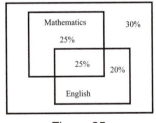

Figure 35

(b) The percentage of students needing help with mathematics, English, or both is $25 + 25 + 20 = 70\%$. The probability of needing help in mathematics, English, or both is 0.7.

(c) If M represents the event of a student needing help with mathematics, and E represents the event that a student needs help with English, then
$$P(M \text{ or } E) = P(M \cup E) = P(M) + P(E) - P(M \cap E) = 0.5 + 0.45 - 0.25 = 0.7.$$

37. (a) There were $4481 + 1614 = 6095$ books published in the areas of art and music out of a possible 119,923.

    The probability of a new book or edition being in the area of art or music is $\dfrac{6095}{119,923} \approx 0.051$.

    (b) The number of books published in science or religion was $7032 + 6659 = 13,691$. The probability of a new book or edition not being published in the areas of science or religion is

    $$1 - \dfrac{13,691}{119,923} = \dfrac{106,232}{119,923} \approx 0.886.$$

39. The probability is $\dfrac{94}{100,000} \approx 0.00094$.

41. (a) The probability is $\dfrac{158,502}{984,155} \approx 0.161$.

    (b) This is the complement of the event in part (a), which is $1 - \dfrac{158,502}{984,155} = \dfrac{825,653}{984,155} \approx 0.839$.

    (c) The probability is $\dfrac{49,913 + 29,092}{984,155} \approx 0.080$.

43. The possible outcomes for a sum of 5 or 6 are (1, 4), (2, 3), (3, 2), (4, 1), (1, 5), (2, 4), (3, 3), (4, 2), and (5, 1).

    This is a total of 9 outcomes out of 36 possible outcomes. The probability is $\dfrac{9}{36} = \dfrac{1}{4}$.

45. There are four 2's, four 3's and four 4's in a standard deck. The probability is $\dfrac{12}{52} = \dfrac{3}{13}$.

47. (a) Since the events of rolling a 4, 5, or 6 are mutually exclusive, the probability of either a 4, 5, or 6 is

    $0.2 + 0.2 + 0.3 = 0.7$.

    (b) The events of rolling a 6 followed by a second 6 are independent events. The probability of two consecutive

    6's is $(0.3)(0.3) = 0.09$.

49. (a) The only way to obtain a sum of 12 is for both dice to show a 6. Since these events are independent, the

    probability of two 6's is $(0.3)(0.3) = 0.09$ or 9%.

    (b) There are two ways to obtain a sum of 11. The red die shows a 5 and the blue die a 6, or vice versa. The

    roll of the two dice are independent. The probability of the red die showing a 5 is 0.2 and the probability

    of the blue die showing a 6 is 0.3. The probability of a sum of 11 is $(0.2)(0.3) = 0.06$. Similarly, the

    probability of the red die showing a 6 and the blue die showing a 5 is 0.06. Since these events are mutually

    exclusive, the probability of a sum of 11 is $0.06 + 0.06 = 0.12$ or 12%.

51. There are ten possibilities for each of the three digits. By the fundamental counting principle, there are

    $10 \cdot 10 \cdot 10 = 10^3 = 1000$ different ways to pick these numbers. There is only one winning number so the

    probability is $\dfrac{1}{1000}$.

53. (a) There are a total of $22 + 18 + 10 = 50$ marbles in the jar. Since 22 of them are red, there is a probability

    of $\dfrac{22}{50} = 0.44$ of drawing a red ball.

    (b) Since there is a 0.44 probability of drawing a red ball, there is a $1 - 0.44 = 0.56$ probability of not drawing

    a red ball.

    (c) Drawing a blue or green ball is equivalent to not drawing a red ball. The probability is the same as in

    part (b), which was 0.56.

55. Since one card, a queen, has been drawn, there are 3 queens left in the set of 51 cards. So the probability of drawing a queen is $\dfrac{3}{51}$.

57. Since there are 4 kings in a total of 12 face cards, the probability of drawing a king is $\dfrac{4}{12} = \dfrac{1}{3}$.

59. Since 2 red out of 10 red marbles and 4 blue out of 23 blue marbles have already been drawn, 6 marbles out of 33 marbles have been removed from the jar. There are $23 - 4 = 19$ blue marbles left out of $33 - 6 = 27$ marbles in the jar. The probability of drawing a blue marble next is $\dfrac{19}{27}$.

61. Let E_1 denote the event that it is cloudy and E_2 denote the event that it is windy. $P(E_1) = 0.30$ and $P(E_1 \text{ and } E_2) = P(E_1 \cap E_2) = 0.12$. $P(E_1 \cap E_2) = P(E_1) \cdot P(E_2, \text{given that } E_1 \text{ has occured}) \Rightarrow 0.12 = 0.30 \cdot P(E_2, \text{given that } E_1 \text{ has occured}) \Rightarrow 0.4 = P(E_2, \text{given that } E_1 \text{ has occured})$. The probability that it will be windy given that the day is cloudy is 40%.

63. The possibilities for the result that the first die is a 2 and the sum of the two dice is 7 or more, can be represented by $\{(2, 5), (2, 6)\}$. The sample space can be represented by $\{(2, 1), (2, 2), (2, 3), (2, 4), (2, 5), \text{ and } (2, 6)\}$. Out of a total of 6, 2 outcomes satisfy the conditions, so the probability is $\dfrac{2}{6} = \dfrac{1}{3}$.

65. (a) Let D represent the event that part is defective. Since 18 of the 235 parts are defective, $P(D) = \dfrac{18}{235}$.

    (b) Let A represent the event that part is type A. Since 7 of the 18 parts are type A, $P(A, \text{given } D) = \dfrac{7}{18}$.

    (c) $P(D \text{ and } A) = P(D \cap A) = P(D) \cdot P(A, \text{given } D) = \dfrac{18}{235} \cdot \dfrac{7}{18} = \dfrac{7}{235}$.

67. (a) Let O represent the event that the number is odd. Since 8 of the 15 numbers are odd, $P(O) = \dfrac{8}{15}$.

    (b) Let E represent the event that the number is even. Since 7 of the 15 numbers are even, $P(E) = \dfrac{7}{15}$.

    (c) Let M represent the event that the number is prime. Since 6 of the 15 numbers are prime,
    $$P(M) = \dfrac{6}{15} = \dfrac{2}{5}.$$

    (d) Since 5 of the 6 prime numbers are odd, $P(M \cap O) = \dfrac{5}{15} = \dfrac{1}{3}$.

    (e) Since 1 of the 6 prime numbers is even, $P(M \cap E) = \dfrac{1}{15}$.

Chapter 11 Review Exercises

1. $a_1 = -3(1) + 2 = -1;\ a_2 = -3(2) + 2 = -4;\ a_3 = -3(3) + 2 = -7;\ a_4 = -3(4) + 2 = -10$
    The first four terms are $-1, -4, -7$, and -10.

3. $a_1 = 0;\ a_2 = 2a_1 + 1 = 2(0) + 1 = 1;\ a_3 = 2a_2 + 1 = 2(1) + 1 = 3;\ a_4 = 2a_3 + 1 = 2(3) + 1 = 7$
    The first four terms are $0, 1, 3$, and 7.

5. The points $(1, 5), (2, 3), (3, 1), (4, 2), (5, 4)$, and $(6, 6)$ lie on the graph. Therefore, the terms of this sequence are $5, 3, 1, 2, 4, 6$.

7. (a) Each term can be found by adding –2 to the previous term, beginning with 3. It is an arithmetic sequence. A numerical representation for the first eight terms is shown in Figure 7a.

[0, 10, 1] by [–12, 4, 1]

Figure 7b

n	1	2	3	4	5	6	7	8
a_n	3	1	-1	-3	-5	-7	-9	-11

Figure 7a

(b) A graphical representation for the first eight terms is shown in Figure 7b. Notice that the points lie on a line with slope –2 because the sequence is arithmetic with $d = -2$.

(c) To find the symbolic representation, we will use the formula $a_n = a_1 + (n - 1)d$, where $a_n = f(n)$. The common difference of the sequence is $d = -2$ and the first term is $a_1 = 3$. Therefore a symbolic representation is given by $a_n = 3 + (n - 1)(-2)$ or $a_n = -2n + 5$.

9. Let $a_n = f(n)$, where $f(n) = dn + c$. Then, $f(n) = 4n + c$. Since $f(3) = 4(3) + c = -3 \Rightarrow$. $c = -15$. Thus $a_n = f(n) = 4n - 15$.

11. Since f is a linear function, the sequence is arithmetic.

13. Since f can be written in the form $f(n) = cr^{n-1}$, the sequence is geometric.

15. $S_5 = (4(1) + 1) + (4(2) + 1) + (4(3) + 1) + (4(4) + 1) + (4(5) + 1) =$ $5 + 9 + 13 + 17 + 21 = 65$

17. The first term is $a_1 = -2$ and the last term is $a_9 = 22$.

To find the sum use $S_n = n\left(\dfrac{a_1 + a_n}{2}\right) \Rightarrow S_9 = 9\left(\dfrac{-2 + 22}{2}\right) = 90$. The sum is 90.

19. The first term is $a_1 = 1$ and the common ratio is $r = 3$. Since there are 8 terms, the sum is

$$S_n = a_1\left(\frac{1 - r^n}{1 - r}\right) \Rightarrow S_8 = 1\left(\frac{1 - 3^8}{1 - 3}\right) = 3280.$$

21. The first term is $a_1 = 4$ and the common ratio is $r = -\dfrac{1}{3}$. The sum is

$$S_n = a_1\left(\frac{1}{1 - r}\right) \Rightarrow S = 4\left(\frac{1}{1 + \frac{1}{3}}\right) = 3.$$

23. $\displaystyle\sum_{k=1}^{5}(5k + 1) = (5 + 1) + (10 + 1) + (15 + 1) + (20 + 1) + (25 + 1) = 6 + 11 + 16 + 21 + 26$

25. $1^3 + 2^3 + 3^3 + 4^3 + 5^3 + 6^3 = \displaystyle\sum_{k=1}^{6} k^3$

27. Since $a_1 = 5$ and $d = -3$, use the formula $S_n = \dfrac{n}{2}(2a_1 + (n - 1)d)$.

$$S_{30} = \frac{30}{2}(2(5) + (30 - 1)(-3)) = 15(10 + 29(-3)) = 15(-77) = -1155$$

29. $\dfrac{2}{11} = 0.18181818\ldots = 0.18 + 0.0018 + 0.000018 + 0.00000018 + \cdots$

31. $P(n, r) = \dfrac{n!}{(n-r)!} \Rightarrow P(6, 3) = \dfrac{6!}{(6-3)!} = \dfrac{6!}{3!} = \dfrac{6 \cdot 5 \cdot 4 \cdot 3!}{3!} = 6 \cdot 5 \cdot 4 = 120$

33. $1 + 3 + 5 + \cdots + (2n - 1) = n^2$

    (i) Show that the statement is true for $n = 1$: $2(1) - 1 = 1^2 \Rightarrow 1 = 1$

    (ii) Assume that S_k is true: $1 + 3 + 5 + \cdots + (2k - 1) = k^2$

        Show that S_{k+1} is true: $1 + 3 + \cdots + (2(k + 1) - 1) = (k + 1)^2$

        Add $2k + 1$ to each side of S_k: $1 + 3 + 5 + \cdots + (2k - 1) + (2k + 1) = k^2 + 2k + 1 = (k + 1)^2$

    Since S_k implies S_{k+1}, the statement is true for every positive integer n.

35. Since 1, 2, or 3 are three outcomes out of six possible outcomes, the probability is $\dfrac{3}{6} = \dfrac{1}{2}$.

37. There are $P(5, 5) = 5! = 120$ different arrangements.

39. There are $4^{20} \approx 1.1 \times 10^{12}$ different ways to answer the exam.

41. There are 50 choices for each number in the combination. So there are $50^4 = 6{,}250{,}000$ possible combinations.

43. (a) The initial height is 4 feet. On the first rebound it reaches 90% of 4 or 3.6 feet. On the second rebound it

      attains a height of 90% of 3.6 or 3.24 feet. Each term in this sequence is found by multiplying the previous

      term by 0.9. Thus, the first five terms are 4, 3.6, 3.24, 2.916, 2.6244. This is a geometric sequence.

    (b) Plot the points (1, 4), (2, 3.6), (3, 3.24), (4, 2.916), and (5, 2.6244) as shown in Figure 43.

    (c) The first term is 4 and the common ratio is 0.9. Thus, $a_n = 4(0.9)^{n-1}$.

[0, 6, 1] by [0, 6, 1]   [0, 16, 1] by [0, 5000, 1000]

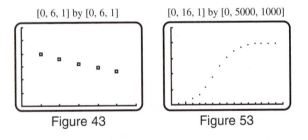

Figure 43   Figure 53

45. From a group of 6 people, we must select a set of 3 people. This can be done $C(6, 3) = 20$ different ways.

47. They can be selected in $C(10, 6) = 210$ different ways.

49. There are $C(16, 2)$ different ways to select 2 batteries from a pack of 16. In order to avoid testing a defective

    battery, we must pick 2 batteries from the 14 good ones. This can be done in $C(14, 2)$ different ways. The

    probability that the box is not rejected is $\dfrac{C(14, 2)}{C(16, 2)} = \dfrac{91}{120} \approx 0.758$.

51. (a) The total number of marbles in the jar is $13 + 27 + 20 = 60$. Since 27 of the marbles are blue, there is a

      probability of $\dfrac{27}{60} = 0.45$ of drawing a blue marble.

    (b) Since there is a 0.45 probability of drawing a blue marble, there is a $1 - 0.45 = 0.55$ probability of drawing

      a marble that is not blue.

    (c) There is a $\dfrac{13}{60} \approx 0.217$ probability of drawing a red marble.

53. The sequence is graphed in Figure 53. Initially, the population density grows slowly, then it increases rapidly.

    After some time, it levels off near 4000 thousand (4,000,000) per acre.

Chapters 1-11 Cumulative Review Exercises

1. $34{,}500 = 3.45 \times 10^4$; $1.52 \times 10^{-4} = 0.000152$

3. $d = \sqrt{(1 - (-4))^2 + (-2 - 2)^2} = \sqrt{5^2 + (-4)^2} = \sqrt{25 + 16} = \sqrt{41}$

5. (a) $f(-3) = \sqrt{1 - (-3)} = \sqrt{4} = 2$; $f(a + 1) = \sqrt{1 - (a + 1)} = \sqrt{-a}$

   (b) For f to be defined, $1 - x \geq 0$. Thus, the domain is $D = \{x \mid x \leq 1\}$.

7. $\dfrac{f(-1) - f(-2)}{-1 - (-2)} = \dfrac{((-1)^3 - 4) - ((-2)^3 - 4)}{1} = -5 - (-12) = 7$

9. A line passing through the points $(2, -4)$ and $(-3, 2)$ has slope $m = \dfrac{2 - (-4)}{-3 - 2} = \dfrac{6}{-5} = -\dfrac{6}{5}$.

   Using the point $(2, -4)$, the equation is $y = -\dfrac{6}{5}(x - 2) - 4 \Rightarrow y = -\dfrac{6}{5}x - \dfrac{8}{5}$.

11. (a) The graph passes through the points $(0, -1)$ and $(4, 2)$. The slope is $m = \dfrac{2 - (-1)}{4 - 0} = \dfrac{3}{4}$.

    The y-intercept is -1. The x-intercept can be found by noting that a 1-unit rise from the point $(0, -1)$

    would require a $\dfrac{4}{3}$-unit run to return to a line with slope $\dfrac{3}{4}$. That is, $m = \dfrac{\text{rise}}{\text{run}} = \dfrac{1}{\frac{4}{3}} = \dfrac{3}{4}$. The x-intercept is $\dfrac{4}{3}$.

    (b) Since the slope is $\dfrac{3}{4}$ and the y-intercept is -1, the formula for f is $f(x) = \dfrac{3}{4}x - 1$.

    (c) $\dfrac{3}{4}x - 1 = 0 \Rightarrow \dfrac{3}{4}x = 1 \Rightarrow x = \dfrac{4}{3}$

13. (a) $4(x - 2) + 1 = 3 - \dfrac{1}{2}(2x + 3) \Rightarrow 4x - 8 + 1 = 3 - x - \dfrac{3}{2} \Rightarrow 5x = \dfrac{17}{2} \Rightarrow x = \dfrac{17}{10}$

    (b) $6x^2 = 13x + 5 \Rightarrow 6x^2 - 13x - 5 = 0 \Rightarrow (3x + 1)(2x - 5) = 0 \Rightarrow x = -\dfrac{1}{3} \text{ or } x = \dfrac{5}{2}$

    (c) By the quadratic formula, $x = \dfrac{-b \pm \sqrt{b^2 - 4ac}}{2a} \Rightarrow x = \dfrac{-(-1) \pm \sqrt{(-1)^2 - 4(1)(-3)}}{2(1)} = \dfrac{1 \pm \sqrt{13}}{2}$

    (d) $x^3 + x^2 = 4x + 4 \Rightarrow x^3 + x^2 - 4x - 4 = 0 \Rightarrow x^2(x + 1) - 4(x + 1) = 0 \Rightarrow$

       $(x^2 - 4)(x + 1) = 0 \Rightarrow (x + 2)(x - 2)(x + 1) = 0 \Rightarrow x = -2, -1, \text{ or } 2$

    (e) $x^4 - 4x^2 + 3 = 0 \Rightarrow (x^2 - 3)(x^2 - 1) = 0 \Rightarrow x^2 = 3 \text{ or } x^2 = 1 \Rightarrow x = \pm\sqrt{3} \text{ or } \pm 1$

    (f) $\dfrac{1}{x - 3} = \dfrac{4}{x + 5} \Rightarrow (x - 3)(x + 5) \cdot \dfrac{1}{x - 3} = \dfrac{4}{x + 5} \cdot (x - 3)(x + 5) \Rightarrow x + 5 = 4(x - 3) \Rightarrow$

       $x + 5 = 4x - 12 \Rightarrow -3x = -17 \Rightarrow x = \dfrac{17}{3}$

    (g) $3e^{2x} - 5 = 23 \Rightarrow 3e^{2x} = 28 \Rightarrow e^{2x} = \dfrac{28}{3} \Rightarrow \ln e^{2x} = \ln\dfrac{28}{3} \Rightarrow 2x = \ln\dfrac{28}{3} \Rightarrow x = \dfrac{\ln\left(\frac{28}{3}\right)}{2} \approx 1.117$

    (h) $2 \log (x + 1) - 1 = 2 \Rightarrow 2 \log (x + 1) = 3 \Rightarrow \log (x + 1) = \dfrac{3}{2} \Rightarrow 10^{\log (x+1)} = 10^{3/2} \Rightarrow$

       $x + 1 = 10^{3/2} \Rightarrow x = 10^{3/2} - 1 \approx 30.623$

    (i) $\sqrt{x + 3} + 4 = x + 1 \Rightarrow \sqrt{x + 3} = x - 3 \Rightarrow (\sqrt{x + 3})^2 = (x - 3)^2 \Rightarrow$

       $x + 3 = x^2 - 6x + 9 \Rightarrow x^2 - 7x + 6 = 0 \Rightarrow (x - 1)(x - 6) = 0 \Rightarrow x = 1 \text{ or } x = 6$. Note, 1 is

       extraneous. The only solution is 6.

    (j) $|3x - 1| = 5 \Rightarrow 3x - 1 = -5 \text{ or } 3x - 1 = 5 \Rightarrow 3x = -4 \text{ or } 3x = 6 \Rightarrow x = -\dfrac{4}{3} \text{ or } 2$

15. f is continuous. See Figure 15.

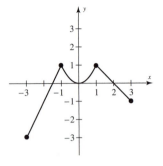

Figure 15

17. (a) The solution to $f(x) = 0$ is the x-intercepts. The solutions are $-2, -1, 1$, and 2.

   (b) The graph of f is above the x-axis on the interval $(-\infty, -2) \cup (-1, 1) \cup (2, \infty)$.

   (c) The graph of f is on or below the x-axis on the interval $[-2, -1] \cup [1, 2]$.

19. The x-coordinate of the vertex is $x = -\dfrac{b}{2a} = -\dfrac{9}{2(-3)} = \dfrac{3}{2}$. The y-coordinate of the vertex is $f\left(\dfrac{3}{2}\right)$.

   $$y = f\left(\frac{3}{2}\right) = -3\left(\frac{3}{2}\right)^2 + 9\left(\frac{3}{2}\right) + 1 = -\frac{27}{4} + \frac{27}{2} + 1 = -\frac{27}{4} + \frac{54}{4} + \frac{4}{4} = \frac{31}{4}. \text{ The vertex is } \left(\frac{3}{2}, \frac{31}{4}\right).$$

21. (a) The graph of f is increasing on $(-\infty, -2] \cup [1, \infty)$ and decreasing on $[-2, 1]$.

   (b) The zeros of f are the x-intercepts which approximately $-3.3, 0$, and 1.8.

   (c) The turning points are approximately $(-2, 2)$ and $(1, -0.7)$.

   (d) There is a local minimum of -0.7 and a local maximum of 2.

23. (a) $\dfrac{6x^4 - 2x^2 + 1}{2x^2} = \dfrac{6x^4}{2x^2} - \dfrac{2x^2}{2x^2} + \dfrac{1}{2x^2} = 3x^2 - 1 + \dfrac{1}{2x^2}$

   (b)
   $$
   \begin{array}{r}
   2x^3 - 5x^2 + 5x - 6 \\
   x + 1 \overline{)2x^4 - 3x^3 + 0x^2 - x + 2} \\
   \underline{2x^4 + 2x^3} \\
   -5x^3 + 0x^2 - x + 2 \\
   \underline{-5x^3 - 5x^2} \\
   5x^2 - x + 2 \\
   \underline{5x^2 + 5x} \\
   -6x + 2 \\
   \underline{-6x - 6} \\
   8
   \end{array}
   $$
   The solution is: $2x^3 - 5x^2 + 5x - 6 + \dfrac{8}{x + 1}$

25. Since $3i$ is a zero its conjugate, $-3i$, is also a zero. The complete factored form is

   $$f(x) = 3(x - (-1))(x - (-3i))(x - 3i) \text{ or } f(x) = 3(x + 1)(x + 3i)(x - 3i).$$

   To find the expanded form, first multiply $(x + 3i)(x - 3i) = x^2 + 9$. Continuing to multiply out gives

   $$f(x) = 3(x + 1)(x^2 + 9) = 3(x^3 + x^2 + 9x + 9) \text{ or } f(x) = 3x^3 + 3x^2 + 27x + 27.$$

27. $x = \dfrac{-b \pm \sqrt{b^2 - 4ac}}{2a} \Rightarrow x = \dfrac{-2 \pm \sqrt{2^2 - 4(1)(5)}}{2(1)} = \dfrac{-2 \pm \sqrt{-16}}{2} = \dfrac{-2 \pm 4i}{2} = -1 \pm 2i$

29. $\sqrt[5]{(x+1)^3} = (x+1)^{3/5}$. When $x = 31$, $(x+1)^{3/5} = (31+1)^{3/5} = 32^{3/5} = (\sqrt[5]{32})^3 = 2^3 = 8$

31. (a) $(f+g)(2) = f(2) + g(2) = 3 + 0 = 3$

   (b) $(fg)(0) = f(0) \cdot g(0) = -1(1) = -1$

   (c) $(g \circ f)(1) = g(f(1)) = g(0) = 1$

   (d) Note, for $g^{-1}(3)$, read the graph backward. $(g^{-1} \circ f)(-2) = g^{-1}(f(-2)) = g^{-1}(3) = -4$

33. Let $y = f(x)$: $y = \dfrac{x}{x+1}$. Interchange x and y: $x = \dfrac{y}{y+1}$. Then solve for y.

   $x = \dfrac{y}{y+1} \Rightarrow x(y+1) = y \Rightarrow xy + x = y \Rightarrow x = y - xy \Rightarrow x = y(1-x) \Rightarrow \dfrac{x}{1-x} = y$

   The inverse function is $f^{-1}(x) = \dfrac{x}{1-x}$ or $f^{-1}(x) = -\dfrac{x}{x-1}$.

35. Since each unit increase in x results in $f(x)$ increasing by a factor of 3, the function is exponential with base 3.

   Since $f(0) = 2$, the initial value is 2 and the function is $f(x) = 2(3^x)$.

37. $A = 500\left(1 + \dfrac{0.06}{4}\right)^{15 \cdot 4} \approx \$1221.61$

39. (a) A quadratic function is defined for all real inputs. The domain is $D = (-\infty, \infty)$.

   The graph of f is a parabola opening upward with vertex $(1, 0)$. The range is $R = [0, \infty)$.

   (b) An exponential function is defined for all real inputs. The domain is $D = (-\infty, \infty)$.

   The function $f(x) = 10^x$ has only positive outputs. The range is $R = (0, \infty)$.

   (c) The natural logarithm function is defined for only positive real inputs. The domain is $D = (0, \infty)$.

   The natural logarithm function can output any real value. The range is $R = (-\infty, \infty)$.

   (d) The reciprocal function is defined for all real inputs except 0. The domain is $D = (-\infty, 0) \cup (0, \infty)$.

   The reciprocal function can output any real value except 0. The range is $R = (-\infty, 0) \cup (0, \infty)$.

41. $2 \log x + 3 \log y - \dfrac{1}{3} \log z = \log x^2 + \log y^3 - \log z^{1/3} = \log \dfrac{x^2 y^3}{z^{1/3}} = \log \dfrac{x^2 y^3}{\sqrt[3]{z}}$

43. (a) See Figure 43a.

(b) See Figure 43b.

(c) See Figure 43c.

(d) See Figure 43d.

Figure 43a

Figure 43b

Figure 43c

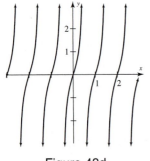

Figure 43d

45. (a) $\cos t = -\dfrac{\sqrt{3}}{2} \Rightarrow t = \cos^{-1}\left(-\dfrac{\sqrt{3}}{2}\right) = \dfrac{5\pi}{6}$ or $\dfrac{7\pi}{6}$. Therefore $t = \dfrac{5\pi}{6} + 2\pi n$ or $t = \dfrac{7\pi}{6} + 2\pi n$.

(b) $\tan^2 t - 1 = 0 \Rightarrow \tan^2 t = 1 \Rightarrow \tan t = \pm 1 \Rightarrow t_R = \tan^{-1} 1 = \dfrac{\pi}{4}$. Since tangent is positive or negative

in all quadrant., $t = \dfrac{\pi}{4} + \dfrac{\pi}{2}n$.

(c) $\sin^2 t + \dfrac{1}{2}\sin t = 0 \Rightarrow 2\left(\sin^2 t + \dfrac{1}{2}\sin t\right) = 2(0) \Rightarrow 2\sin^2 t + \sin t = 0 \Rightarrow$

$\sin t\,(2\sin t + 1) = 0 \Rightarrow \sin t = 0$ or $2\sin t + 1 = 0$. When $\sin t = 0, t = \pi n$ and when

$2\sin t + 1 = 0 \Rightarrow 2\sin t = -1 \Rightarrow \sin t = -\dfrac{1}{2}, t = \dfrac{7\pi}{6} + 2\pi n, \dfrac{11\pi}{6} + 2\pi n.$

(d) $\cos 2t = -1$, let $\theta = 2t$ so then the equation becomes $\cos\theta = -1$, in radian measure this solution is π. For

all real solutions $\theta = \pi + 2\pi n \Rightarrow$ because $\theta = 2t$, we can substitute $2t$ for $\theta \Rightarrow 2t = \pi + 2\pi n$ or

$t = \dfrac{\pi}{2} + \pi n.$

47. $5.54° = 5° + (0.54)(60) = 5° + 32.4' = 5° + 32' + (0.4)(60) = 5° + 32' + 24'' = 5° 32' 24''$

49. $\dfrac{5\pi}{4} \cdot \dfrac{180}{\pi} = 225°$

51. If the terminal side passes through the point $(-7, 24)$, θ lies in quadrant II.

$$r^2 = x^2 + y^2 \Rightarrow r^2 = 24^2 + 7^2 \Rightarrow r^2 = 625 \Rightarrow r = 25$$

$$\sin \theta = \frac{y}{r} = \frac{24}{25} \qquad \cos \theta = \frac{x}{r} = -\frac{7}{25} \qquad \tan \theta = \frac{y}{x} = -\frac{24}{7}$$

$$\csc \theta = \frac{r}{y} = \frac{25}{24} \qquad \sec \theta = \frac{r}{x} = -\frac{25}{7} \qquad \cot \theta = \frac{x}{y} = -\frac{7}{24}$$

53. If $\cos \theta < 0$ and $\csc \theta < 0$, θ lies in quadrant III, $\cos \theta = -\dfrac{11}{61}$.

$$\sin^2 \theta + \cos^2 \theta = 1 \Rightarrow \sin^2 \theta + \left(-\frac{11}{61}\right)^2 = 1 \Rightarrow \sin^2 \theta = 1 - \frac{121}{3721} \Rightarrow \sin^2 \theta = \frac{3600}{3721} \Rightarrow \sin \theta = -\frac{60}{61}$$

$$\tan \theta = \frac{\sin \theta}{\cos \theta} = \frac{-\frac{60}{61}}{-\frac{11}{61}} = \frac{60}{11} \qquad \csc \theta = \frac{1}{\sin \theta} = \frac{1}{-\frac{60}{61}} = -\frac{61}{60}$$

$$\sec \theta = \frac{1}{\cos \theta} = \frac{1}{-\frac{11}{61}} = -\frac{61}{11} \qquad \cot \theta = \frac{\cos \theta}{\sin \theta} = \frac{-\frac{11}{61}}{-\frac{60}{61}} = \frac{11}{60}$$

55. From the pythagorean theorem $a^2 + b^2 = c^2 \Rightarrow 6^2 + b^2 = 10^2 \Rightarrow 36 + b^2 = 100 \Rightarrow b^2 = 64 \Rightarrow b = 8$

$$\tan \alpha = \frac{6}{8} \Rightarrow \alpha = \tan^{-1}\left(\frac{3}{4}\right) \approx 36.9° \qquad \tan \beta = \frac{8}{6} \Rightarrow \beta = \tan^{-1}\left(\frac{4}{3}\right) \approx 53.1°.$$

57. $1 - \sin^2 \theta + \cot^2 \theta - \sin^2 \theta \cot^2 \theta = 1 - \sin^2 \theta - \sin^2 \theta \cot^2 \theta + \cot^2 \theta =$

$$1 - \sin^2 \theta (1 + \cot^2 \theta) + \cot^2 \theta = 1 - \sin^2 \theta (\csc^2 \theta) + \cot^2 \theta = 1 - \sin^2 \theta \left(\frac{1}{\sin^2 \theta}\right) + \cot^2 \theta =$$

$$1 - 1 + \cot^2 \theta = \cot^2 \theta$$

59. (a) This triangle is of the form of ASA. There is only one solution. $\alpha = 180° - 42° - 31° = 107°$

$$\frac{b}{\sin \beta} = \frac{a}{\sin \alpha} \Rightarrow b = \frac{a \sin \beta}{\sin \alpha} = \frac{22 \sin 42°}{\sin 107°} \approx 15.4$$

$$\frac{c}{\sin \gamma} = \frac{a}{\sin \alpha} \Rightarrow c = \frac{a \sin \gamma}{\sin \alpha} = \frac{22 \sin 31°}{\sin 107°} \approx 11.8 \quad \alpha = 107°, b \approx 15.4, c \approx 11.8$$

(b) This triangle is of the form of SSA. It is ambiguous.

$$\frac{\sin \beta}{b} = \frac{\sin \gamma}{c} \Rightarrow \sin \beta = \frac{b \sin \gamma}{c} = \frac{8 \sin 50°}{7} \approx 0.87548 \Rightarrow \beta_R = \sin^{-1}(0.87548) \approx 61.1°$$

Thus $\beta = 61.1°$ or $\beta = 180° - 61.1° \approx 118.9°$

Solution 1: Let $\beta = 61.1°$. Then $\alpha \approx 180° - 50° - 61.1° \approx 68.9°$.

$$\frac{a}{\sin \alpha} = \frac{c}{\sin \gamma} \Rightarrow a = \frac{c \sin \alpha}{\sin \gamma} = \frac{7 \sin 68.9°}{\sin 50°} \approx 8.5 \quad \beta \approx 61.1°, \alpha \approx 68.9°, a \approx 8.5$$

Solution 2: Let $\beta = 118.9°$. Then $\alpha \approx 180° - 50° - 118.9° \approx 11.1°$.

$$\frac{a}{\sin \alpha} = \frac{c}{\sin \gamma} \Rightarrow a = \frac{c \sin \alpha}{\sin \gamma} = \frac{7 \sin 11.1°}{\sin 50°} \approx 1.8 \quad \beta \approx 118.9°, \alpha \approx 11.1°, a \approx 1.8$$

(c) This triangle is of the form SAS.

$$a^2 = b^2 + c^2 - 2bc \cos \alpha = 7^2 + 8^2 - 2(7)(8) \cos 44° \Rightarrow a = \sqrt{113 - 112 \cos 44°} \approx 5.69508$$

$$c^2 = a^2 + b^2 - 2ab \cos \gamma \Rightarrow \cos \gamma = \frac{c^2 - a^2 - b^2}{-2ab} \Rightarrow \alpha = \cos^{-1}\left(\frac{8^2 - (5.69508)^2 - 7^2}{-2(5.69508)(7)}\right) \approx$$

$77.36971°$

$$\beta = 180° - 44° - 77.36971° \approx 58.63029° \quad a \approx 5.7, \beta = 58.6°, \gamma \approx 77.4°$$

(d) This Triangle is of the form SSS.

$$a^2 = b^2 + c^2 - 2bc \cos \alpha \Rightarrow \cos \alpha = \frac{a^2 - b^2 - c^2}{-2bc} \Rightarrow \alpha = \cos^{-1}\left(\frac{10^2 - 11^2 - 12^2}{-2(11)(12)}\right) \approx 51.31781°$$

$$b^2 = a^2 + c^2 - 2ac \cos \beta \Rightarrow \cos \beta = \frac{b^2 - a^2 - c^2}{-2ac} \Rightarrow \beta = \cos^{-1}\left(\frac{11^2 - 10^2 - 12^2}{-2(10)(12)}\right) \approx 59.16950°$$

$$\gamma = 180° - 51.31781° - 59.16950 \approx 69.51269° \quad \alpha \approx 51.3°, \beta \approx 59.2°, \gamma \approx 69.5°$$

61. (a) $\|\mathbf{a}\| = \sqrt{(3)^2 + (-4)^2} = \sqrt{25} = 5$

(b) $4\mathbf{b} - 2\mathbf{a} = 4\langle -5, 12 \rangle - 2\langle 3, -4 \rangle = \langle -20, 48 \rangle - \langle 6, -8 \rangle \Rightarrow \langle -20 - 6, 48 - (-8) \rangle = \langle -26, 56 \rangle$

(c) $a \cdot b = \langle 3, -4 \rangle \cdot \langle -5, 12 \rangle = (3)(-5) + (-4)(12) = -15 + (-48) = -63$

(d) $\theta = \cos^{-1}\left(\frac{\mathbf{a} \cdot \mathbf{b}}{\|\mathbf{a}\| \|\mathbf{b}\|}\right) \Rightarrow$ we found $\mathbf{a} \cdot \mathbf{b}$ from part (c) above to be -63.

$\|\mathbf{a}\| = \sqrt{(3)^2 + (-4)^2} = \sqrt{25} = 5$ and $\|\mathbf{b}\| = \sqrt{(-5)^2 + (12)^2} = \sqrt{169} = 13 \Rightarrow$

$\theta = \cos^{-1}\left(\frac{-63}{(5)(13)}\right) \approx 165.75°$

63. See Figure 63.

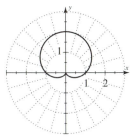

Figure 63

65. (a) Multiplying the first equation by -1 and adding the two equations will eliminate the variable x.

$$
\begin{array}{r}
-2x - 3y = -4 \\
2x - 5y = -12 \\
\hline
-8y = -16
\end{array}
$$
Thus, $y = 2$. And so $2x + 3(2) = 4 \Rightarrow x = -1$. The solution is $(-1, 2)$.

(b) Multiplying the first equation by 2 and adding the two equations will eliminate both variables.

$$
\begin{array}{r}
-4x + y = 2 \\
4x - y = -2 \\
\hline
0 = 0
\end{array}
$$
This is an identity. The system is dependent with solutions $\{(x, y)\,|\,4x - y = -2\}$.

(c) Adding the two equations will eliminate the variable y.

$$
\begin{array}{r}
x^2 + y^2 = 16 \\
2x^2 - y^2 = 11 \\
\hline
3x^2 = 27
\end{array}
$$
Thus, $x = \pm 3$. And so $(\pm 3)^2 + y^2 = 16 \Rightarrow y^2 = 7 \Rightarrow y = \pm\sqrt{7}$.

There are four solutions, $(\pm 3, \pm\sqrt{7})$.

(d) Multiply the first equation by 2 and subtract the second equations to eliminate the variable x.

$$
\begin{array}{r}
2x + 2y - 4z = -12 \\
2x - y - 3z = -18 \\
\hline
3y - z = 6
\end{array}
$$

Subtract the third equation from this *new* equation to eliminate both y and z.

$$
\begin{array}{r}
3y - z = 6 \\
3y - z = 6 \\
\hline
0 = 0
\end{array}
$$

This is an identity which means there are an infinite number of solutions to the system.

Solving the third equation for y yields $3y - z = 6 \Rightarrow 3y = z + 6 \Rightarrow y = \dfrac{z + 6}{3}$. Substituting this

result for y in the first equation yields $x + \dfrac{z + 6}{3} - 2z = -6 \Rightarrow x = 2z - \dfrac{z + 6}{3} - 6 \Rightarrow$

$x = \dfrac{5z - 24}{3}$. The solutions set is $\left\{ (x, y, z)\,|\,x = \dfrac{5z - 24}{3}, y = \dfrac{z + 6}{3}, \text{and } z = z \right\}$.

67. For the graphs of parts (a) – (d), see Figures 67a – 67b.

Figure 67a   Figure 67b

69. $\begin{bmatrix} 1 & -2 & | & 1 & 0 \\ -3 & 4 & | & 0 & 1 \end{bmatrix} R_2 + 3R_1 \rightarrow \begin{bmatrix} 1 & -2 & | & 1 & 0 \\ 0 & -2 & | & 3 & 1 \end{bmatrix} R_1 - R_2 \rightarrow \begin{bmatrix} 1 & 0 & | & -2 & -1 \\ 0 & -2 & | & 3 & 1 \end{bmatrix}$

$-\frac{1}{2}R_2 \rightarrow \begin{bmatrix} 1 & 0 & | & -2 & -1 \\ 0 & 1 & | & -\frac{3}{2} & -\frac{1}{2} \end{bmatrix}$. That is $A^{-1} = \begin{bmatrix} -2 & -1 \\ -1.5 & -0.5 \end{bmatrix}$.

71. (a) $\det\left(\begin{bmatrix} -1 & 4 \\ 2 & 3 \end{bmatrix} \right) = -1(3) - 2(4) = -11$

(b) $\det\left(\begin{bmatrix} 2 & 3 & -1 \\ 3 & -1 & 5 \\ 0 & 0 & -2 \end{bmatrix} \right) = 0(3(5) - (-1)(-1)) - 0(2(5) - 3(-1)) + -2(2(-1) - 3(3)) =$

$-2(-11) = 22$

73. If the focus is at $\left(\frac{3}{4}, 0 \right)$ and the vertex at $(0, 0)$, the parabola opens to the right and $p = \frac{3}{4}$. Substituting in

$(y - h)^2 = 4p(x - k)$, we get $(y - 0)^2 = 4\left(\frac{3}{4} \right)(x - 0)$ or $y^2 = 3x$.

75. Since the foci and the vertices lie on the y-axis, the hyperbola has a vertical transverse axis with an equation

of the form $\frac{y^2}{a^2} - \frac{x^2}{b^2} = 1$. $F(0, \pm 13) \Rightarrow c = 13$ and $V(0, \pm 5) \Rightarrow a = 5$. $c^2 = a^2 + b^2 \Rightarrow$

$b^2 = c^2 - a^2 = 169 - 25 = 144 \Rightarrow b = 12$. The equation of the hyperbola is $\frac{y^2}{25} - \frac{x^2}{144} = 1$.

77. Let $a_n = f(n)$, where $f(n) = dn + c$. $a_1 = 4 \Rightarrow f(1) = 4$ and $a_3 = 12 \Rightarrow f(3) = 12$.

Thus, $d = \frac{f(3) - f(1)}{3 - 1} = \frac{12 - 4}{2} = 4$. Then, $f(n) = 4n + c$. $f(1) = 4(1) + c = 4 \Rightarrow c = 0$.

Thus, $a_n = f(n) = 4n$.

79. (a) The first term is $a_1 = 2$ and the last term is $a_8 = 74$. There are 25 terms. To find the sum use:

$S_n = n\left(\frac{a_1 + a_n}{2} \right) \Rightarrow S_{25} = 25\left(\frac{2 + 74}{2} \right) = 950$. The sum is 950.

(b) The first term is $a_1 = 0.2$ and the common ratio is $r = \frac{1}{10}$ or 0.1. The sum is

$S = a_1\left(\frac{1}{1 - r} \right) \Rightarrow S = 0.2\left(\frac{1}{1 - (0.1)} \right) = \frac{0.2}{0.9} = \frac{2}{9}$.

81. There are $26 \cdot 26 \cdot 26 \cdot 10 \cdot 10 \cdot 10 \cdot 10 = 175,760,000$ such license plates.

83. $(2x - 1)^4 = \binom{4}{0}(2x)^4(-1)^0 + \binom{4}{1}(2x)^3(-1)^1 + \binom{4}{2}(2x)^2(-1)^2 + \binom{4}{3}(2x)^1(-1)^3 + \binom{4}{4}(2x)^0(-1)^4$

   $= 16x^4 - 32x^3 + 24x^2 - 8x + 1$

85. The card must be one of the 16 cards (out of 52) that is either a heart or an ace. The probability is $\dfrac{16}{52} = \dfrac{4}{13}$.

87. There are 8 primes from 1 to 20 (2, 3, 5, 7, 11, 13, 17, 19). The probability is $\dfrac{8}{20} = \dfrac{2}{5}$.

89. From noon to 1:45 PM car A travelled $1.75(50) = 87.5$ miles and thus would be located $87.5 + 30 = 117.5$ miles north of the original location of car B. From noon to 1:45 PM car B travelled $1.75(50) = 87.5$ miles and thus would be located 87.5 miles east of its original location. The paths of the two cars form a right triangle with legs of length 117.5 and 87.5. The length of the hypotenuse is equal to the distance between the cars. $c^2 = 117.5^2 + 87.5^2 \Rightarrow c^2 = 21,462.5 \Rightarrow c \approx 146.5$ miles.

91. Let x represent the time spent jogging at 7 miles per hour and let y represent the time spent jogging at 9 miles per hour. Then the system of equations to be solved is $x + y = 1.3$ and $7x + 9y = 10.5$. Solving the first equation for y gives $y = 1.3 - x$. Substituting this expression in the second equation and solving for x yields $7x + 9(1.3 - x) = 10.5 \Rightarrow -2x + 11.7 = 10.5 \Rightarrow -2x = -1.2 \Rightarrow x = 0.6$. From the first equation $0.6 + y = 1.3 \Rightarrow y = 0.7$. The jogger ran for 0.6 hours at 7 mph and 0.7 hours at 9 mph.

93. Let x represent the number of hours they work together. The the first person mows $\dfrac{x}{5}$ of the lawn in x hours and the second person mows $\dfrac{x}{4}$ of the lawn in x hours. Together they mow $\dfrac{x}{5} + \dfrac{x}{4}$ of the lawn in x hours. The job is complete when the fraction of the lawn mowed is 1. That is, we must solve $\dfrac{x}{5} + \dfrac{x}{4} = 1$.

   The LCD is 20. The first step is to multiply each side of the equation by the LCD.

   $20 \cdot \left(\dfrac{x}{5} + \dfrac{x}{4} \right) = 1 \cdot 20 \Rightarrow 4x + 5x = 20 \Rightarrow 9x = 20 \Rightarrow x = \dfrac{20}{9} \approx 2.22$ hours.

95. (a) The cost of 1 ticket is $1(405 - 1(5)) = 400$. The cost for 2 tickets is $2(405 - 2(5)) = \$790$. The cost for 3 tickets is $3(405 - 3(5)) = \$1170$. Following this pattern, the cost for x tickets is

   $C(x) = x(405 - 5x)$.

   (b) $x(405 - 5x) = 7000 \Rightarrow 405x - 5x^2 = 7000 \Rightarrow 5x^2 - 405x + 7000 = 0 \Rightarrow$

   $x^2 - 81x + 1400 = 0 \Rightarrow (x - 25)(x - 56) = 0 \Rightarrow x = 25$ or $x = 56$. It costs $7000 to buy either 25 or 56 tickets.

   (c) The maximum for the function occurs at the vertex of its graph - a parabola. For $C(x) = -5x^2 + 405x$,

   $x = -\dfrac{b}{2a} = -\dfrac{405}{2(-5)} = 40.5$. Since the number of tickets must be an integer, the maximum occurs at either 40 or 41 tickets being purchased. In either case, the maximum expenditure is

   $40(405 - 5(40)) = \$8200$.

97. Substitute the point $(1.5, 0.5)$ into $x^2 = 4py$ and solve for p; $1.5^2 = 4p(0.5) \Rightarrow 2.25 = 2p \Rightarrow p = 1.125$.

   The receiver should be 1.125 feet from the vertex.

99. There are 49 marbles that are not blue and there are a total of 77 marbles. The probability of drawing a

   marble that is not blue is $\dfrac{49}{77} = \dfrac{7}{11}$.

101. $\tan 42° = \dfrac{105}{x} \Rightarrow x = 116.6$ feet

103. $\tan \theta = \dfrac{6}{7} \Rightarrow \tan^{-1}\left(\dfrac{6}{7}\right) \approx 40.6°$

Chapter R: Basic Concepts from Algebra and Geometry

R.1: Formulas from Geometry

1. $A = LW = (15)(7) = 105 \text{ ft}^2$; $P = 2L + 2W = 2(15) + 2(7) = 30 + 14 = 44 \text{ ft}$

3. $A = LW = (100)(35) = 3500 \text{ m}^2$; $P = 2L + 2W = 2(100) + 2(35) = 200 + 70 = 270 \text{ m}$

5. $A = LW = (3x)(y) = 3xy$ square units; $P = 2L + 2W = 2(3x) + 2(y) = 6x + 2y$ units

7. Since the width is half of the length, $L = 2W$.

   $A = LW = (2W)(W) = 2W^2$ square units; $P = 2L + 2W = 2(2W) + 2(W) = 6W$ units

9. Since the length equals the width plus 5, $L = W + 5$.

   $A = LW = (W + 5)(W) = W(W + 5)$ square units; $P = 2L + 2W = 2(W + 5) + 2(W) = 4W + 10$ units

11. $A = \frac{1}{2}bh = \left(\frac{1}{2}\right)(8)(5) = 20 \text{ cm}^2$

13. $A = \frac{1}{2}bh = \left(\frac{1}{2}\right)(5)(8) = 20 \text{ in.}^2$

15. $A = \frac{1}{2}bh = \left(\frac{1}{2}\right)(10.1)(730) = 3686.5 \text{ m}^2$

17. $A = \frac{1}{2}bh = \left(\frac{1}{2}\right)(2x)(6x) = 6x^2$ square units

19. $A = \frac{1}{2}bh = \left(\frac{1}{2}\right)(z)(5z) = \frac{5}{2}z^2$ square units

21. $C = 2\pi r = 2\pi(4) = 8\pi \approx 25.1 \text{ m}$; $A = \pi r^2 = \pi(4)^2 = 16\pi \approx 50.3 \text{ m}^2$

23. $C = 2\pi r = 2\pi(19) = 38\pi \approx 119.4 \text{ in.}$; $A = \pi r^2 = \pi(19)^2 = 361\pi \approx 1134.1 \text{ in.}^2$

25. $C = 2\pi r = 2\pi(2x) = 4\pi x$ units; $A = \pi r^2 = \pi(2x)^2 = 4\pi x^2$ square units

27. $c^2 = a^2 + b^2 \Rightarrow c^2 = (60)^2 + (11)^2 \Rightarrow c^2 = 3721 \Rightarrow c = \sqrt{3721} \Rightarrow c = 61 \text{ ft}$

    $P = 60 + 11 + 61 = 132 \text{ ft}$

29. $c^2 = a^2 + b^2 \Rightarrow b^2 = c^2 - a^2 \Rightarrow b^2 = (13)^2 - (5)^2 \Rightarrow b^2 = 144 \Rightarrow b = \sqrt{144} \Rightarrow b = 12 \text{ cm}$

    $P = 5 + 12 + 13 = 30 \text{ cm}$

31. $c^2 = a^2 + b^2 \Rightarrow a^2 = c^2 - b^2 \Rightarrow a^2 = (10)^2 - (7)^2 \Rightarrow a^2 = 51 \Rightarrow a = \sqrt{51} \Rightarrow a \approx 7.1 \text{ mm}$

    $P \approx 7.1 + 7 + 10 = 24.1 \text{ mm}$

33. $A = \frac{1}{2}bh = \left(\frac{1}{2}\right)(6)(3) = 9 \text{ ft}^2$

35. $c^2 = a^2 + b^2 \Rightarrow b^2 = c^2 - a^2 \Rightarrow b^2 = (15)^2 - (11)^2 \Rightarrow b^2 = 104 \Rightarrow b = \sqrt{104} \Rightarrow b \approx 10.2 \text{ in.}$

    $A = \frac{1}{2}bh = \left(\frac{1}{2}\right)(11)(\sqrt{104}) \approx 56.1 \text{ in.}^2$

37. $V = LWH = (4)(3)(2) = 24 \text{ ft}^3$; $S = 2LW + 2WH + 2LH = 2(4)(3) + 2(3)(2) + 2(4)(2) = 52 \text{ ft}^2$

39. Since 1 foot is 12 inches, $V = LWH = (4.5)(4)(12) = 216 \text{ in.}^3$

    $S = 2LW + 2WH + 2LH = 2(4.5)(4) + 2(4)(12) + 2(4.5)(12) = 240 \text{ in.}^2$

115. $3x^2 - 5x - 8 = (x + 1)(3x - 8)$

117. $7a^3 + 20a^2 - 3a = a(7a^2 + 20a - 3) = a(a + 3)(7a - 1)$

119. $2x^3 - x^2 + 6x - 3 = x^2(2x - 1) + 3(2x - 1) = (x^2 + 3)(2x - 1)$

121. $2x^4 - 5x^3 - 25x^2 = x^2(2x^2 - 5x - 25) = x^2(2x + 5)(x - 5)$

123. $2x^4 + 5x^2 + 3 = (2x^2 + 3)(x^2 + 1)$

125. $x^3 + 3x^2 + x + 3 = x^2(x + 3) + (x + 3) = (x + 3)(x^2 + 1)$

127. $5x^3 - 5x^2 + 10x - 10 = 5x^2(x - 1) + 10(x - 1) = (x - 1)(5x^2 + 10) = 5(x^2 + 2)(x - 1)$

129. $ax + bx - ay - by = x(a + b) - y(a + b) = (a + b)(x - y)$

131. $18x^2 + 12x + 2 = 2(9x^2 + 6x + 1) = 2(3x + 1)^2$

133. $-4x^3 + 24x^2 - 36x = -4x(x^2 - 6x + 9) = -4x(x - 3)^2$

135. $27x^3 - 8 = (3x - 2)(9x^2 + 6x + 4)$

137. $-x^4 - 8x = -x(x^3 + 8) = -x(x + 2)(x^2 - 2x + 4)$

139. $x^4 - 2x^3 - x + 2 = x^3(x - 2) - 1(x - 2) = (x - 2)(x^3 - 1) = (x - 2)(x - 1)(x^2 + x + 1)$

141. $r^4 - 16 = (r^2 - 4)(r^2 + 4) = (r + 2)(r - 2)(r^2 + 4)$

143. $25x^2 - 4a^2 = (5x - 2a)(5x + 2a)$

145. $(2x^4 - 2y^4) = 2(x^4 - y^4) = 2(x^2 - y^2)(x^2 + y^2) = 2(x - y)(x + y)(x^2 + y^2)$

147. $9x^3 + 6x^2 - 3x = 3x(3x^2 + 2x - 1) = 3x(3x - 1)(x + 1)$

149. $(z - 2)^2 - 9 = (z - 2 - 3)(z - 2 + 3) = (z - 5)(z + 1)$

151. $3x^5 - 27x^3 + 3x^2 - 27 = 3(x^5 - 9x^3 + x^2 - 9) = 3[x^3(x^2 - 9) + (x^2 - 9)] =$

$3(x^2 - 9)(x^3 + 1) = 3(x + 3)(x - 3)(x + 1)(x^2 - x + 1)$

153. $(x + 2)^2(x + 4)^4 + (x + 2)^3(x + 4)^3 = (x + 2)^2(x + 4)^3[x + 4 + x + 2] =$

$(x + 2)^2(x + 4)^3(2x + 6) = 2(x + 2)^2(x + 4)^3(x + 3)$

155. $(6x + 1)(8x - 3)^4 - (6x + 1)^2(8x - 3)^3 = (6x + 1)(8x - 3)^3[8x - 3 - (6x + 1)] =$

$(6x + 1)(8x - 3)^2(2x - 4) = 2(6x + 1)(8x - 3)^3(x - 2)$

157. $4x^2(5x - 1)^5 + 2x(5x - 1)^6 = 2x(5x - 1)^5[2x + 5x - 1] = 2x(5x - 1)^5(7x - 1)$

R.5: Rational Expressions

1. $\dfrac{10x^3}{5x^2} = 2x$

3. $\dfrac{(x - 5)(x + 5)}{x - 5} = x + 5$

5. $\dfrac{x^2 - 16}{x - 4} = \dfrac{(x - 4)(x + 4)}{x - 4} = x + 4$

7. $\dfrac{x + 3}{2x^2 + 5x - 3} = \dfrac{x + 3}{(2x - 1)(x + 3)} = \dfrac{1}{2x - 1}$

9. $-\dfrac{z+2}{4z+8} = -\dfrac{z+2}{4(z+2)} = -\dfrac{1}{4}$

11. $\dfrac{x^2+2x}{x^2+3x+2} = \dfrac{x(x+2)}{(x+1)(x+2)} = \dfrac{x}{x+1}$

13. $\dfrac{a^3+b^3}{a+b} = \dfrac{(a+b)(a^2-ab+b^2)}{a+b} = a^2-ab+b^2$

15. $\dfrac{5}{8} \cdot \dfrac{4}{15} = \dfrac{1}{6}$

17. $\dfrac{5}{6} \cdot \dfrac{3}{10} \cdot \dfrac{8}{3} = \dfrac{2}{3}$

19. $\dfrac{4}{7} \div \dfrac{8}{7} = \dfrac{4}{7} \cdot \dfrac{7}{8} = \dfrac{1}{2}$

21. $\dfrac{1}{2} \div \dfrac{3}{4} \div \dfrac{5}{6} = \left(\dfrac{1}{2} \cdot \dfrac{4}{3}\right) \div \dfrac{5}{6} = \dfrac{2}{3} \cdot \dfrac{6}{5} = \dfrac{4}{5}$

23. $\dfrac{3}{8} + \dfrac{5}{8} = \dfrac{8}{8} = 1$

25. $\dfrac{3}{7} - \dfrac{4}{7} = -\dfrac{1}{7}$

27. $\dfrac{2}{3} + \dfrac{5}{11} = \dfrac{22}{33} + \dfrac{15}{33} = \dfrac{37}{33}$

29. $\dfrac{4}{5} - \dfrac{1}{10} = \dfrac{8}{10} - \dfrac{1}{10} = \dfrac{7}{10}$

31. $\dfrac{1}{3} + \dfrac{3}{4} - \dfrac{3}{7} = \dfrac{28}{84} + \dfrac{63}{84} - \dfrac{36}{84} = \dfrac{55}{84}$

33. $\dfrac{1}{x^2} \cdot \dfrac{3x}{2} = \dfrac{3}{2x}$

35. $\dfrac{5x}{3} \div \dfrac{10x}{6} = \dfrac{5x}{3} \cdot \dfrac{6}{10x} = \dfrac{2}{2} = 1$

37. $\dfrac{x+1}{2x-5} \cdot \dfrac{x}{x+1} = \dfrac{x}{2x-5}$

39. $\dfrac{(x-5)(x+3)}{3x-1} \cdot \dfrac{x(3x-1)}{(x-5)} = \dfrac{x+3}{1} \cdot \dfrac{x}{1} = x(x+3)$

41. $\dfrac{x^2-2x-35}{2x^3-3x^2} \cdot \dfrac{x^3-x^2}{2x-14} = \dfrac{(x-7)(x+5)}{x^2(2x-3)} \cdot \dfrac{x^2(x-1)}{2(x-7)} = \dfrac{x+5}{2x-3} \cdot \dfrac{x-1}{2} = \dfrac{(x-1)(x+5)}{2(2x-3)}$

43. $\dfrac{6b}{b+2} \div \dfrac{3b^4}{2b+4} = \dfrac{6b}{b+2} \cdot \dfrac{2(b+2)}{3b^4} = \dfrac{2}{1} \cdot \dfrac{2}{b^3} = \dfrac{4}{b^3}$

45. $\dfrac{3a+1}{a^7} \div \dfrac{a+1}{3a^8} = \dfrac{3a+1}{a^7} \cdot \dfrac{3a^8}{a+1} = \dfrac{3a+1}{1} \cdot \dfrac{3a}{a+1} = \dfrac{3a(3a+1)}{a+1}$

47. $\dfrac{x+5}{x^3-x} \div \dfrac{x^2-25}{x^3} = \dfrac{x+5}{x(x^2-1)} \cdot \dfrac{x^3}{(x-5)(x+5)} = \dfrac{1}{x^2-1} \cdot \dfrac{x^2}{x-5} = \dfrac{x^2}{(x-5)(x^2-1)}$

49. $\dfrac{x-2}{x^3-x} \div \dfrac{x^2-2x}{x^2-1} = \dfrac{x-2}{x(x+1)(x-1)} \cdot \dfrac{(x+1)(x-1)}{x(x-2)} = \dfrac{1}{x^2}$

51. $\dfrac{x^2 - 3x + 2}{x^2 + 5x + 6} \div \dfrac{x^2 + x - 2}{x^2 + 2x - 3} = \dfrac{(x-2)(x-1)}{(x+3)(x+2)} \cdot \dfrac{(x+3)(x-1)}{(x+2)(x-1)} = \dfrac{(x-2)(x-1)}{(x+2)^2}$

53. $\dfrac{x^2 - 4}{x^2 + x - 2} \div \dfrac{x-1}{x-2} = \dfrac{(x+2)(x-2)}{(x+2)(x-1)} \cdot \dfrac{x-1}{x-2} = 1$

55. $\dfrac{3y}{x^2} \div \dfrac{y^2}{x} \div \dfrac{y}{5x} = \left(\dfrac{3y}{x^2} \cdot \dfrac{x}{y^2}\right) \div \dfrac{y}{5x} = \dfrac{3}{xy} \cdot \dfrac{5x}{y} = \dfrac{15}{y^2}$

57. $\dfrac{x-3}{x-1} \div \dfrac{x^2}{x-1} \div \dfrac{x-3}{x} = \left(\dfrac{x-3}{x-1} \cdot \dfrac{x-1}{x^2}\right) \div \dfrac{x-3}{x} = \dfrac{x-3}{x^2} \cdot \dfrac{x}{x-3} = \dfrac{1}{x}$

59. Since $12 = 2 \cdot 2 \cdot 3$ and $18 = 2 \cdot 3 \cdot 3$, the LCM is $2 \cdot 2 \cdot 3 \cdot 3 = 36$.

61. Since $5a^3 = 5 \cdot a \cdot a \cdot a$ and $10a = 2 \cdot 5 \cdot a$, the LCM is $2 \cdot 5 \cdot a \cdot a \cdot a = 10a^3$.

63. Since $z^2 - 4z = z(z - 4)$ and $(z - 4)^2 = (z - 4)(z - 4)$, the LCM is $z(z - 4)(z - 4) = z(z - 4)^2$.

65. Since $x^2 - 6x + 9 = (x - 3)(x - 3)$ and $x^2 - 5x + 6 = (x - 2)(x - 3)$, the LCM is

$(x - 2)(x - 3)(x - 3) = (x - 2)(x - 3)^2$.

67. The factored denominators are $x + 1$ and 7. The LCD is $7(x + 1)$.

69. The factored denominators are $x + 4$ and $(x + 4)(x - 4)$. The LCD is $(x + 4)(x - 4)$.

71. The factored denominators are 2 and $2x + 1$ and $2(x - 2)$. The LCD is $2(2x + 1)(x - 2)$.

73. $\dfrac{4}{x+1} + \dfrac{3}{x+1} = \dfrac{7}{x+1}$

75. $\dfrac{2}{x^2 - 1} - \dfrac{x+1}{x^2 - 1} = \dfrac{2 - (x+1)}{x^2 - 1} = \dfrac{2 - x - 1}{x^2 - 1} = \dfrac{-x + 1}{(x-1)(x+1)} = \dfrac{-(x-1)}{(x+1)(x-1)} = -\dfrac{1}{x+1}$

77. $\dfrac{x}{x+4} - \dfrac{x+1}{x(x+4)} = \dfrac{x^2}{x(x+4)} - \dfrac{x+1}{x(x+4)} = \dfrac{x^2 - (x+1)}{x(x+4)} = \dfrac{x^2 - x - 1}{x(x+4)}$

79. $\dfrac{2}{x^2} - \dfrac{4x-1}{x} = \dfrac{2}{x^2} - \dfrac{x(4x-1)}{x^2} = \dfrac{2 - (4x^2 - x)}{x^2} = \dfrac{-4x^2 + x + 2}{x^2}$

81. $\dfrac{x+3}{x-5} + \dfrac{5}{x-3} = \dfrac{(x+3)(x-3)}{(x-5)(x-3)} + \dfrac{5(x-5)}{(x-5)(x-3)} = \dfrac{x^2 - 9 + 5x - 25}{(x-5)(x-3)} = \dfrac{x^2 + 5x - 34}{(x-5)(x-3)}$

83. $\dfrac{3}{x-5} - \dfrac{1}{x-3} - \dfrac{2x}{x-5} = \dfrac{3(x-3)}{(x-5)(x-3)} - \dfrac{x-5}{(x-5)(x-3)} - \dfrac{2x(x-3)}{(x-5)(x-3)} = \dfrac{-2(x^2 - 4x + 2)}{(x-5)(x-3)}$

85. $\dfrac{x}{x^2 - 9} + \dfrac{5x}{x-3} = \dfrac{x}{(x-3)(x+3)} + \dfrac{5x(x+3)}{(x-3)(x+3)} = \dfrac{x + 5x^2 + 15x}{(x-3)(x+3)} = \dfrac{x(5x + 16)}{(x-3)(x+3)}$

87. $\dfrac{b}{2b-4} - \dfrac{b-1}{b-2} = \dfrac{b}{2(b-2)} - \dfrac{2(b-1)}{2(b-2)} = \dfrac{b - (2b-2)}{2(b-2)} = \dfrac{-(b-2)}{2(b-2)} = -\dfrac{1}{2}$

89. $\dfrac{2x}{x-5} + \dfrac{2x-1}{3x^2 - 16x + 5} = \dfrac{2x(3x-1)}{(x-5)(3x-1)} + \dfrac{2x-1}{(x-5)(3x-1)} = \dfrac{6x^2 - 2x + 2x - 1}{(x-5)(3x-1)} =$

$\dfrac{6x^2 - 1}{(x-5)(3x-1)}$

91. $\dfrac{x}{(x-1)^2} - \dfrac{1}{(x-1)(x+3)} = \dfrac{x(x+3)}{(x-1)^2(x+3)} - \dfrac{x-1}{(x-1)^2(x+3)} = \dfrac{x^2+3x-x+1}{(x-1)^2(x+3)} =$

$\dfrac{x^2+2x+1}{(x-1)^2(x+3)} = \dfrac{(x+1)^2}{(x-1)^2(x+3)}$

93. $\dfrac{x}{x^2-5x+4} + \dfrac{2}{x^2-2x-8} = \dfrac{x}{(x-4)(x-1)} + \dfrac{2}{(x-4)(x+2)} =$

$\dfrac{x(x+2)}{(x-4)(x+2)(x-1)} + \dfrac{2(x-1)}{(x-4)(x+2)(x-1)} = \dfrac{x^2+2x+2x-2}{(x-4)(x+2)(x-1)} =$

$\dfrac{x^2+4x-2}{(x-4)(x+2)(x-1)}$

95. $\dfrac{x}{x^2-4} - \dfrac{1}{x^2+4x+4} = \dfrac{x}{(x-2)(x+2)} - \dfrac{1}{(x+2)(x+2)} =$

$\dfrac{x(x+2)}{(x+2)^2(x-2)} - \dfrac{x-2}{(x+2)^2(x-2)} = \dfrac{x^2+2x-x+2}{(x+2)^2(x-2)} = \dfrac{x^2+x+2}{(x+2)^2(x-2)}$

97. $\dfrac{3x}{x-y} - \dfrac{3y}{x^2-2xy+y^2} = \dfrac{3x}{x-y} - \dfrac{3y}{(x-y)(x-y)} =$

$\dfrac{3x(x-y)}{(x-y)^2} - \dfrac{3y}{(x-y)^2} = \dfrac{3x^2-3xy-3y}{(x-y)^2} = \dfrac{3(x^2-xy-y)}{(x-y)^2}$

99. $x + \dfrac{1}{x-1} - \dfrac{1}{x+1} = \dfrac{x(x-1)(x+1)}{(x-1)(x+1)} + \dfrac{x+1}{(x-1)(x+1)} - \dfrac{x-1}{(x-1)(x+1)} =$

$\dfrac{x^3-x+x+1-x+1}{(x-1)(x+1)} = \dfrac{x^3-x+2}{(x-1)(x+1)}$

101. $\dfrac{6}{t-1} + \dfrac{2}{t-2} + \dfrac{1}{t} = \dfrac{6t(t-2)}{t(t-1)(t-2)} + \dfrac{2t(t-1)}{t(t-1)(t-2)} + \dfrac{(t-1)(t-2)}{t(t-1)(t-2)} =$

$\dfrac{6t^2-12t+2t^2-2t+t^2-3t+2}{t(t-1)(t-2)} = \dfrac{9t^2-17t+2}{t(t-1)(t-2)}$

103. The factored denominators are x and x^2. The LCD is x^2.

$\dfrac{1}{x} + \dfrac{3}{x^2} = 0 \Rightarrow \dfrac{1(x^2)}{x} + \dfrac{3(x^2)}{x^2} = 0(x^2) \Rightarrow x+3=0 \Rightarrow x=-3$

105. The factored denominators are x and $2x-1$. The LCD is $x(2x-1)$.

$\dfrac{1}{x} + \dfrac{3x}{2x-1} = 0 \Rightarrow \dfrac{1(x)(2x-1)}{x} + \dfrac{3x(x)(2x-1)}{2x-1} = 0(x)(2x-1) \Rightarrow 2x-1+3x^2=0 \Rightarrow$

$3x^2+2x-1=0 \Rightarrow (3x-1)(x+1)=0 \Rightarrow x=\dfrac{1}{3}$ or $x=-1$

107. The factored denominators are $(3-x)(3+x)$ and $3-x$. The LCD is $(3-x)(3+x)$.

$\dfrac{2x}{9-x^2} + \dfrac{1}{3-x} = 0 \Rightarrow \dfrac{2x(3-x)(3+x)}{9-x^2} + \dfrac{1(3-x)(3+x)}{3-x} = 0(3-x)(3+x) \Rightarrow$

$2x+3+x=0 \Rightarrow 3x+3=0 \Rightarrow x=-1$

109. The factored denominators are $2x, 2x^2$ and x^3. The LCD is $2x^3$.

$$\frac{1}{2x} + \frac{1}{2x^2} - \frac{1}{x^3} = 0 \Rightarrow \frac{1(2x^3)}{2x} + \frac{1(2x^3)}{2x^2} - \frac{1(2x^3)}{x^3} = 0(2x^3) \Rightarrow x^2 + x - 2 = 0 \Rightarrow$$

$$(x + 2)(x - 1) = 0 \Rightarrow x = -2 \text{ or } x = 1$$

111. The factored denominators are $x, x + 5$ and $x - 5$. The LCD is $x(x + 5)(x - 5)$.

$$\frac{1}{x} - \frac{2}{x + 5} + \frac{1}{x - 5} = 0 \Rightarrow$$

$$\frac{1(x)(x + 5)(x - 5)}{x} - \frac{2(x)(x + 5)(x - 5)}{x + 5} + \frac{1(x)(x + 5)(x - 5)}{x - 5} = 0(x)(x + 5)(x - 5) \Rightarrow$$

$$(x + 5)(x - 5) - 2x(x - 5) + x(x + 5) = 0 \Rightarrow x^2 - 25 - 2x^2 + 10x + x^2 + 5x = 0 \Rightarrow$$

$$15x - 25 = 0 \Rightarrow x = \frac{25}{15} = \frac{5}{3}$$

113. $\dfrac{1 + \dfrac{1}{x}}{1 - \dfrac{1}{x}} = \dfrac{1 + \dfrac{1}{x}}{1 - \dfrac{1}{x}} \cdot \dfrac{x}{x} = \dfrac{x + 1}{x - 1}$

115. $\dfrac{\dfrac{1}{x - 5}}{\dfrac{4}{x} - \dfrac{1}{x - 5}} = \dfrac{\dfrac{1}{x - 5}}{\dfrac{4}{x} - \dfrac{1}{x - 5}} \cdot \dfrac{x(x - 5)}{x(x - 5)} = \dfrac{x}{4(x - 5) - x} = \dfrac{x}{4x - 20 - x} = \dfrac{x}{3x - 20}$

117. $\dfrac{\dfrac{1}{x} + \dfrac{2 - x}{x^2}}{\dfrac{3}{x^2} - \dfrac{1}{x}} = \dfrac{\dfrac{1}{x} + \dfrac{2 - x}{x^2}}{\dfrac{3}{x^2} - \dfrac{1}{x}} \cdot \dfrac{x^2}{x^2} = \dfrac{x + 2 - x}{3 - x} = \dfrac{2}{3 - x}$

119. $\dfrac{\dfrac{1}{x + 3} + \dfrac{2}{x - 3}}{2 - \dfrac{1}{x - 3}} = \dfrac{\dfrac{1}{x + 3} + \dfrac{2}{x - 3}}{2 - \dfrac{1}{x - 3}} \cdot \dfrac{(x - 3)(x + 3)}{(x - 3)(x + 3)} = \dfrac{x - 3 + 2(x + 3)}{2(x - 3)(x + 3) - (x + 3)} = \dfrac{3(x + 1)}{(x + 3)(2x - 7)}$

121. $\dfrac{\dfrac{4}{x - 5}}{\dfrac{1}{x + 5} + \dfrac{1}{x}} = \dfrac{\dfrac{4}{x - 5}}{\dfrac{1}{x + 5} + \dfrac{1}{x}} \cdot \dfrac{x(x - 5)(x + 5)}{x(x - 5)(x + 5)} = \dfrac{4x(x + 5)}{x(x - 5) + (x - 5)(x + 5)} = \dfrac{4x(x + 5)}{(x - 5)(2x + 5)}$

123. $\dfrac{\dfrac{1}{2a} - \dfrac{1}{2b}}{\dfrac{1}{a^2} - \dfrac{1}{b^2}} = \dfrac{\dfrac{1}{2a} - \dfrac{1}{2b}}{\dfrac{1}{a^2} - \dfrac{1}{b^2}} \cdot \dfrac{2a^2b^2}{2a^2b^2} = \dfrac{ab^2 - a^2b}{2b^2 - 2a^2} = \dfrac{ab(b - a)}{2(b + a)(b - a)} = \dfrac{ab}{2(a + b)}$

R.6: Radical Notation and Rational Exponents

1. $-\sqrt{25} = -\sqrt{5^2} = -5$ and $\sqrt{25} = \sqrt{5^2} = 5$

3. $-\sqrt{\dfrac{16}{25}} = -\sqrt{\left(\dfrac{4}{5}\right)^2} = -\dfrac{4}{5}$ and $\sqrt{\dfrac{16}{25}} = \sqrt{\left(\dfrac{4}{5}\right)^2} = \dfrac{4}{5}$

5. $-\sqrt{11} \approx -3.32$ and $\sqrt{11} \approx 3.32$

7. $\sqrt{144} = \sqrt{12^2} = 12$

9. $\sqrt{23} \approx 4.80$

11. $\sqrt{\dfrac{4}{49}} = \sqrt{\left(\dfrac{2}{7}\right)^2} = \dfrac{2}{7}$

13. Since $b < 0$, $\sqrt{b^2} = -b$.

15. $\sqrt[3]{27} = \sqrt[3]{3^3} = 3$

17. $\sqrt[3]{-8} = \sqrt[3]{(-2)^3} = -2$

19. $\sqrt[3]{\dfrac{1}{27}} = \sqrt[3]{\left(\dfrac{1}{3}\right)^3} = \dfrac{1}{3}$

21. $\sqrt[3]{b^9} = \sqrt[3]{(b^3)^3} = b^3$

23. $\sqrt{9} = 3$

25. $-\sqrt{5} \approx -2.24$

27. $\sqrt[3]{27} = 3$

29. $\sqrt[3]{-64} = -4$

31. $\sqrt[3]{5} \approx 1.71$

33. $-\sqrt[3]{x^9} = -\sqrt[3]{(x^3)^3} = -x^3$

35. $\sqrt[3]{(2x)^6} = \sqrt[3]{((2x)^2)^3} = (2x)^2 = 4x^2$

37. $\sqrt[4]{81} = 3$

39. $\sqrt[5]{-7} \approx -1.48$

41. $6^{1/2} = \sqrt{6}$

43. $(xy)^{1/2} = \sqrt{xy}$

45. $y^{-1/5} = \dfrac{1}{\sqrt[5]{y}}$

47. The expression can be written $27^{2/3} = \sqrt[3]{27^2}$ or $(\sqrt[3]{27})^2$ and evaluated as $(\sqrt[3]{27})^2 = 3^2 = 9$.

49. The expression can be written $(-1)^{4/3} = \sqrt[3]{(-1)^4}$ or $(\sqrt[3]{-1})^4$ and evaluated as $(\sqrt[3]{-1})^4 = (-1)^4 = 1$.

51. The expression can be written $8^{-1/3} = \dfrac{1}{8^{1/3}} = \dfrac{1}{\sqrt[3]{8}}$ and evaluated as $\dfrac{1}{\sqrt[3]{8}} = \dfrac{1}{2}$.

53. The expression can be written $13^{-3/5} = \dfrac{1}{13^{3/5}} = \dfrac{1}{\sqrt[5]{13^3}}$ or $\dfrac{1}{(\sqrt[5]{13})^3}$. The result is not an integer.

55. $16^{1/2} = \sqrt{16} = 4$

57. $256^{1/4} = \sqrt[4]{256} = 4$

59. $32^{1/5} = \sqrt[5]{32} = 2$

61. $(-8)^{4/3} = (\sqrt[3]{-8})^4 = (-2)^4 = 16$

63. $2^{1/2} \cdot 2^{2/3} = 2^{1/2+2/3} = 2^{7/6} \approx 2.24$

65. $\left(\dfrac{4}{9}\right)^{1/2} = \dfrac{4^{1/2}}{9^{1/2}} = \dfrac{\sqrt{4}}{\sqrt{9}} = \dfrac{2}{3}$

67. $\dfrac{4^{2/3}}{4^{1/2}} = 4^{2/3-1/2} = 4^{1/6} \approx 1.26$

69. $4^{-1/2} = \dfrac{1}{4^{1/2}} = \dfrac{1}{\sqrt{4}} = \dfrac{1}{2}$

71. $(-8)^{-1/3} = \dfrac{1}{(-8)^{1/3}} = \dfrac{1}{\sqrt[3]{-8}} = \dfrac{1}{-2} = -\dfrac{1}{2}$

73. $\left(\dfrac{1}{16}\right)^{-1/4} = 16^{1/4} = \sqrt[4]{16} = 2$

75. $(2^{1/2})^3 = 2^{1/2 \cdot 3} = 2^{3/2} \approx 2.83$

77. $(x^2)^{3/2} = x^{2 \cdot 3/2} = x^3$

79. $(x^2 y^8)^{1/2} = x^{2 \cdot 1/2} \cdot y^{8 \cdot 1/2} = xy^4$

81. $\sqrt[3]{x^3 y^6} = (x^3 y^6)^{1/3} = x^{3 \cdot 1/3} \cdot y^{6 \cdot 1/3} = xy^2$

83. $\sqrt{\dfrac{y^4}{x^2}} = \left(\dfrac{y^4}{x^2}\right)^{1/2} = \dfrac{y^{4 \cdot 1/2}}{x^{2 \cdot 1/2}} = \dfrac{y^2}{x}$

85. $\sqrt{y^3} \cdot \sqrt[3]{y^2} = (y^3)^{1/2} \cdot (y^2)^{1/3} = y^{3 \cdot 1/2} \cdot y^{2 \cdot 1/3} = y^{3/2} \cdot y^{2/3} = y^{3/2+2/3} = y^{13/6}$

87. $\left(\dfrac{x^6}{27}\right)^{2/3} = \dfrac{x^{6 \cdot 2/3}}{27^{2/3}} = \dfrac{x^4}{(\sqrt[3]{27})^2} = \dfrac{x^4}{3^2} = \dfrac{x^4}{9}$

89. $\left(\dfrac{x^2}{y^6}\right)^{-1/2} = \left(\dfrac{y^6}{x^2}\right)^{1/2} = \dfrac{y^{6 \cdot 1/2}}{x^{2 \cdot 1/2}} = \dfrac{y^3}{x}$

91. $\sqrt{\sqrt{y}} = (y^{1/2})^{1/2} = y^{1/2 \cdot 1/2} = y^{1/4}$

93. $(a^{-1/2})^{4/3} = a^{-1/2 \cdot 4/3} = a^{-2/3} = \dfrac{1}{a^{2/3}}$

95. $(a^3 b^6)^{1/3} = a^{3 \cdot 1/3} \cdot b^{6 \cdot 1/3} = ab^2$

97. $\dfrac{(k^{1/2})^{-3}}{(k^2)^{1/4}} = \dfrac{k^{-3/2}}{k^{1/2}} = k^{-3/2-1/2} = k^{-4/2} = k^{-2} = \dfrac{1}{k^2}$

99. $\sqrt{b} \cdot \sqrt[4]{b} = b^{1/2} \cdot b^{1/4} = b^{1/2+1/4} = b^{3/4}$

101. $\sqrt{z} \cdot \sqrt[3]{z^2} \cdot \sqrt[4]{z^3} = z^{1/2} \cdot z^{2/3} \cdot z^{3/4} = z^{1/2+2/3+3/4} = z^{23/12}$

103. $p^{1/2}(p^{3/2} + p^{1/2}) = p^{1/2+3/2} + p^{1/2+1/2} = p^2 + p$

105. $\sqrt[3]{x}(\sqrt{x} - \sqrt[3]{x^2}) = x^{1/3}(x^{1/2} - x^{2/3}) = x^{1/3+1/2} - x^{1/3+2/3} = x^{5/6} - x$

R.7: Radical Expressions

1. $\sqrt{3} \cdot \sqrt{3} = \sqrt{3 \cdot 3} = \sqrt{9} = 3$

3. $\sqrt{2} \cdot \sqrt{50} = \sqrt{2 \cdot 50} = \sqrt{100} = 10$

5. $\sqrt[3]{4} \cdot \sqrt[3]{16} = \sqrt[3]{4 \cdot 16} = \sqrt[3]{64} = 4$

7. $\sqrt{\dfrac{9}{25}} = \dfrac{\sqrt{9}}{\sqrt{25}} = \dfrac{3}{5}$

9. $\sqrt{\dfrac{1}{2}} \cdot \sqrt{\dfrac{1}{8}} = \sqrt{\dfrac{1 \cdot 1}{2 \cdot 8}} = \sqrt{\dfrac{1}{16}} = \dfrac{\sqrt{1}}{\sqrt{16}} = \dfrac{1}{4}$

11. $\sqrt{\dfrac{x}{2}} \cdot \sqrt{\dfrac{x}{8}} = \sqrt{\dfrac{x \cdot x}{2 \cdot 8}} = \sqrt{\dfrac{x^2}{16}} = \dfrac{\sqrt{x^2}}{\sqrt{16}} = \dfrac{x}{4}$

13. $\dfrac{\sqrt{45}}{\sqrt{5}} = \sqrt{\dfrac{45}{5}} = \sqrt{9} = 3$

15. $\sqrt[4]{9} \cdot \sqrt[4]{9} = \sqrt[4]{9 \cdot 9} = \sqrt[4]{81} = 3$

17. $\dfrac{\sqrt[5]{64}}{\sqrt[5]{-2}} = \sqrt[5]{\dfrac{64}{-2}} = \sqrt[5]{-32} = -2$

19. $\dfrac{\sqrt{a^2 b}}{\sqrt{b}} = \sqrt{\dfrac{a^2 b}{b}} = \sqrt{a^2} = a$

21. $\sqrt[3]{\dfrac{x^3}{8}} = \dfrac{\sqrt[3]{x^3}}{\sqrt[3]{8}} = \dfrac{x}{2}$

23. $\sqrt{4x^4} = \sqrt{4} \cdot \sqrt{(x^2)^2} = 2x^2$

25. $\sqrt[4]{16x^4 y} = \sqrt[4]{16} \cdot \sqrt[4]{x^4} \cdot \sqrt[4]{y} = 2x\sqrt[4]{y}$

27. $\sqrt{3x} \cdot \sqrt{12x} = \sqrt{3 \cdot 12 \cdot x \cdot x} = \sqrt{36x^2} = \sqrt{36} \cdot \sqrt{x^2} = 6x$

29. $\sqrt[3]{8x^6 y^3 z^9} = \sqrt[3]{8} \cdot \sqrt[3]{(x^2)^3} \cdot \sqrt[3]{y^3} \cdot \sqrt[3]{(z^3)^3} = 2x^2 y z^3$

31. $\sqrt[4]{\dfrac{3}{4}} \cdot \sqrt[4]{\dfrac{27}{4}} = \sqrt[4]{\dfrac{3}{4} \cdot \dfrac{27}{4}} = \sqrt[4]{\dfrac{81}{16}} = \dfrac{\sqrt[4]{81}}{\sqrt[4]{16}} = \dfrac{3}{2}$

33. $\sqrt[4]{25z} \cdot \sqrt[4]{25z} = \sqrt[4]{625z^2} = \sqrt[4]{625} \cdot \sqrt[4]{z^2} = 5\sqrt{z}$

35. $\sqrt[5]{\dfrac{7a}{b^2}} \cdot \sqrt[5]{\dfrac{b^2}{7a^6}} = \sqrt[5]{\dfrac{7ab^2}{7a^6 b^2}} \sqrt[5]{\dfrac{1}{a^5}} = \dfrac{1}{a}$

37. $\sqrt{200} = \sqrt{100 \cdot 2} = \sqrt{100} \cdot \sqrt{2} = 10\sqrt{2}$

39. $\sqrt[3]{81} = \sqrt[3]{27 \cdot 3} = \sqrt[3]{27} \cdot \sqrt[3]{3} = 3\sqrt[3]{3}$

41. $\sqrt[4]{64} = \sqrt[4]{16 \cdot 4} = \sqrt[4]{16} \cdot \sqrt[4]{4} = 2\sqrt[4]{4} = 2\sqrt[4]{2^2} = 2\sqrt{2}$

43. $\sqrt[5]{-64} = \sqrt[5]{-2^6} = \sqrt[5]{-2^5 \cdot 2} = \sqrt[5]{-2^5} \cdot \sqrt[5]{2} = -2\sqrt[5]{2}$

45. $\sqrt{8n^3} = \sqrt{(2n^2 \cdot 2n)} = \sqrt{(2n)^2} \cdot \sqrt{2n} = 2n\sqrt{2n}$

47. $\sqrt{12a^2 b^5} = \sqrt{(2ab^2)^2 \cdot 3b} = \sqrt{(2ab^2)^2} \cdot \sqrt{3b} = 2ab^2\sqrt{3b}$

49. $\sqrt[3]{-125x^4 y^5} = \sqrt[3]{(-5xy)^3 \cdot xy^2} = \sqrt[3]{(-5xy)^3} \cdot \sqrt[3]{xy^2} = -5xy\sqrt[3]{xy^2}$

51. $\sqrt[3]{5t} \cdot \sqrt[3]{125t} = \sqrt[3]{625t^2} = \sqrt[3]{5^4 t^2} = \sqrt[3]{5^3 \cdot 5t^2} = \sqrt[3]{5^3} \cdot \sqrt[3]{5t^2} = 5\sqrt[3]{5t^2}$

53. $\sqrt[4]{\dfrac{9t^5}{r^8}} \cdot \sqrt[4]{\dfrac{9r}{5t}} = \sqrt[4]{\dfrac{81rt^5}{5r^8 t}} = \sqrt[4]{\dfrac{81t^4}{5r^7}} = \dfrac{\sqrt[4]{(3t)^4}}{\sqrt[4]{r^4 \cdot 5r^3}} = \dfrac{3t}{r\sqrt[4]{5r^3}}$

55. $\sqrt{3} \cdot \sqrt[3]{3} = 3^{1/2} \cdot 3^{1/3} = 3^{1/2 + 1/3} = 3^{5/6} = \sqrt[6]{3^5}$

57. $\sqrt[4]{8} \cdot \sqrt[3]{4} = \sqrt[4]{2^3} \cdot \sqrt[3]{2^2} = 2^{3/4} \cdot 2^{2/3} = 2^{3/4 + 2/3} = 2^{17/12} = 2^{12/12 + 5/12} = 2 \cdot 2^{5/12} = 2\sqrt[12]{2^5}$

59. $\sqrt[4]{x^3} \cdot \sqrt[3]{x} = x^{3/4} \cdot x^{1/3} = x^{3/4 + 1/3} = x^{13/12} = x^{12/12} \cdot x^{1/12} = x\sqrt[12]{x}$

61. $\sqrt[4]{rt} \cdot \sqrt[3]{r^2 t} = (rt)^{1/4} \cdot (r^2 t)^{1/3} = r^{1/4} t^{1/4} \cdot r^{2/3} t^{1/3} = r^{1/4 + 2/3} t^{1/4 + 1/3} = r^{11/12} t^{7/12} = \sqrt[12]{r^{11} t^7}$

63. $2\sqrt{3} + 7\sqrt{3} = 9\sqrt{3}$

65. $\sqrt{x} + \sqrt{x} - \sqrt{y} = 2\sqrt{x} - \sqrt{y}$

67. $2\sqrt[3]{6} - 7\sqrt[3]{6} = -5\sqrt[3]{6}$

69. $3\sqrt{28} + 3\sqrt{7} = 3\sqrt{4\cdot 7} + 3\sqrt{7} = 3\cdot 2\sqrt{7} + 3\sqrt{7} = 9\sqrt{7}$

71. $\sqrt{44} - 4\sqrt{11} = \sqrt{4\cdot 11} - 4\sqrt{11} = 2\sqrt{11} - 4\sqrt{11} = -2\sqrt{11}$

73. $2\sqrt[3]{16} + \sqrt[3]{2} - \sqrt{2} = 2\sqrt[3]{8\cdot 2} + \sqrt[3]{2} - \sqrt{2} = 2\cdot 2\sqrt[3]{2} + \sqrt[3]{2} - \sqrt{2} = 5\sqrt[3]{2} - \sqrt{2}$

75. $\sqrt[3]{xy} - 2\sqrt[3]{xy} = -\sqrt[3]{xy}$

77. $\sqrt{4x+8} + \sqrt{x+2} = \sqrt{4(x+2)} + \sqrt{x+2} = 2\sqrt{x+2} + \sqrt{x+2} = 3\sqrt{x+2}$

79. $\dfrac{15\sqrt{8}}{4} - \dfrac{2\sqrt{2}}{5} = \dfrac{15\cdot 2\sqrt{2}}{4}\cdot\dfrac{5}{5} - \dfrac{2\sqrt{2}}{5}\cdot\dfrac{4}{4} = \dfrac{150\sqrt{2}}{20} - \dfrac{8\sqrt{2}}{20} = \dfrac{150\sqrt{2} - 8\sqrt{2}}{20} = \dfrac{142\sqrt{2}}{20} = \dfrac{71\sqrt{2}}{10}$

81. $2\sqrt[4]{64} - \sqrt[4]{324} + \sqrt[4]{4} = 2\sqrt[4]{16\cdot 4} - \sqrt[4]{81\cdot 4} + \sqrt[4]{4} = 4\sqrt[4]{4} - 3\sqrt[4]{4} + \sqrt[4]{4} = 2\sqrt[4]{4} = 2\sqrt{2}$

83. $2\sqrt{3z} + 3\sqrt{12z} + 3\sqrt{48z} = 2\sqrt{3z} + 3\sqrt{4\cdot 3z} + 3\sqrt{16\cdot 3z} = 2\sqrt{3z} + 6\sqrt{3z} + 12\sqrt{3z} = 20\sqrt{3z}$

85. $\sqrt[4]{81a^5b^5} - \sqrt[4]{ab} = \sqrt[4]{(3ab)^4\cdot ab} - \sqrt[4]{ab} = 3ab\sqrt[4]{ab} - \sqrt[4]{ab} = (3ab - 1)\sqrt[4]{ab}$

87. $5\sqrt[3]{\dfrac{n^4}{125}} - 2\sqrt[3]{n} = 5\sqrt[3]{\dfrac{n^3}{125}\cdot n} - 2\sqrt[3]{n} = 5\cdot\dfrac{n}{5}\sqrt[3]{n} - 2\sqrt[3]{n} = n\sqrt[3]{n} - 2\sqrt[3]{n} = (n - 2)\sqrt[3]{n}$

89. $(3 + \sqrt{7})(3 - \sqrt{7}) = 3^2 - (\sqrt{7})^2 = 9 - 7 = 2$

91. $(\sqrt{x} + 8)(\sqrt{x} - 8) = (\sqrt{x})^2 - 8^2 = x - 64$

93. $(\sqrt{ab} - \sqrt{c})(\sqrt{ab} + \sqrt{c}) = (\sqrt{ab})^2 - (\sqrt{c})^2 = ab - c$

95. $(\sqrt{x} - 7)(\sqrt{x} + 8) = (\sqrt{x})^2 + 8\sqrt{x} - 7\sqrt{x} - 56 = x + \sqrt{x} - 56$

97. $\dfrac{4}{\sqrt{3}} = \dfrac{4}{\sqrt{3}}\cdot\dfrac{\sqrt{3}}{\sqrt{3}} = \dfrac{4\sqrt{3}}{3}$

99. $\dfrac{5}{3\sqrt{5}} = \dfrac{5}{3\sqrt{5}}\cdot\dfrac{\sqrt{5}}{\sqrt{5}} = \dfrac{5\sqrt{5}}{3\cdot 5} = \dfrac{5\sqrt{5}}{15} = \dfrac{\sqrt{5}}{3}$

101. $\sqrt{\dfrac{b}{12}} = \dfrac{\sqrt{b}}{\sqrt{12}} = \dfrac{\sqrt{b}}{\sqrt{12}}\cdot\dfrac{\sqrt{12}}{\sqrt{12}} = \dfrac{\sqrt{12b}}{12} = \dfrac{\sqrt{4\cdot 3b}}{12} = \dfrac{2\sqrt{3b}}{12} = \dfrac{\sqrt{3b}}{6}$

103. $\dfrac{1}{3 - \sqrt{2}} = \dfrac{1}{3 - \sqrt{2}}\cdot\dfrac{3 + \sqrt{2}}{3 + \sqrt{2}} = \dfrac{3 + \sqrt{2}}{9 - 2} = \dfrac{3 + \sqrt{2}}{7}$

105. $\dfrac{\sqrt{2}}{\sqrt{5} + 2} = \dfrac{\sqrt{2}}{\sqrt{5} + 2}\cdot\dfrac{\sqrt{5} - 2}{\sqrt{5} - 2} = \dfrac{\sqrt{10} - 2\sqrt{2}}{5 - 4} = \dfrac{\sqrt{10} - 2\sqrt{2}}{1} = \sqrt{10} - 2\sqrt{2}$

107. $\dfrac{1}{\sqrt{7} - \sqrt{6}} = \dfrac{1}{\sqrt{7} - \sqrt{6}}\cdot\dfrac{\sqrt{7} + \sqrt{6}}{\sqrt{7} + \sqrt{6}} = \dfrac{\sqrt{7} + \sqrt{6}}{7 - 6} = \dfrac{\sqrt{7} + \sqrt{6}}{1} = \sqrt{7} + \sqrt{6}$

109. $\dfrac{\sqrt{z}}{\sqrt{z} - 3} = \dfrac{\sqrt{z}}{\sqrt{z} - 3}\cdot\dfrac{\sqrt{z} + 3}{\sqrt{z} + 3} = \dfrac{z + 3\sqrt{z}}{z - 9}$

111. $\dfrac{\sqrt{a} + \sqrt{b}}{\sqrt{a} - \sqrt{b}} = \dfrac{\sqrt{a} + \sqrt{b}}{\sqrt{a} - \sqrt{b}}\cdot\dfrac{\sqrt{a} + \sqrt{b}}{\sqrt{a} + \sqrt{b}} = \dfrac{a + 2\sqrt{ab} + b}{a - b}$

Appendix C: Partial Fractions

1. Multiply $\dfrac{5}{3x(2x + 1)} = \dfrac{A}{3x} + \dfrac{B}{2x + 1}$ by $3x(2x + 1) \Rightarrow 5 = A(2x + 1) + B(3x)$.

   Let $x = 0 \Rightarrow 5 = A(1) \Rightarrow A = 5$. Let $x = -\dfrac{1}{2} \Rightarrow 5 = B\left(-\dfrac{3}{2}\right) \Rightarrow B = -\dfrac{10}{3}$.

   The expression can be written $\dfrac{5}{3x} + \dfrac{-10}{3(2x + 1)}$.

3. Multiply $\dfrac{4x + 2}{(x + 2)(2x - 1)} = \dfrac{A}{x + 2} + \dfrac{B}{2x - 1}$ by $(x + 2)(2x - 1) \Rightarrow 4x + 2 = A(2x - 1) + B(x + 2)$.

   Let $x = -2 \Rightarrow -6 = A(-5) \Rightarrow A = \dfrac{6}{5}$. Let $x = \dfrac{1}{2} \Rightarrow 4 = B\left(\dfrac{5}{2}\right) \Rightarrow B = \dfrac{8}{5}$.

   The expression can be written $\dfrac{6}{5(x + 2)} + \dfrac{8}{5(2x - 1)}$.

5. Factoring $\dfrac{x}{x^2 + 4x - 5}$ results in $\dfrac{x}{(x + 5)(x - 1)}$.

   Multiply $\dfrac{x}{(x + 5)(x - 1)} = \dfrac{A}{x + 5} + \dfrac{B}{x - 1}$ by $(x + 5)(x - 1) \Rightarrow x = A(x - 1) + B(x + 5)$.

   Let $x = -5 \Rightarrow -5 = A(-6) \Rightarrow A = \dfrac{5}{6}$. Let $x = 1 \Rightarrow 1 = B(6) \Rightarrow B = \dfrac{1}{6}$.

   The expression can be written $\dfrac{5}{6(x + 5)} + \dfrac{1}{6(x - 1)}$.

7. Multiply $\dfrac{2x}{(x + 1)(x + 2)^2} = \dfrac{A}{x + 1} + \dfrac{B}{x + 2} + \dfrac{C}{(x + 2)^2}$ by $(x + 1)(x + 2)^2 \Rightarrow$

   $2x = A(x + 2)^2 + B(x + 1)(x + 2) + C(x + 1)$. Let $x = -1 \Rightarrow -2 = A(1) \Rightarrow A = -2$.

   Let $x = -2 \Rightarrow -4 = C(-1) \Rightarrow C = 4$. Let $x = 0$ with $A = -2$ and $C = 4 \Rightarrow$

   $0 = -2(4) + B(2) + 4(1) \Rightarrow 4 = 2B \Rightarrow B = 2$. The expression can be written $\dfrac{-2}{x + 1} + \dfrac{2}{x + 2} + \dfrac{4}{(x + 2)^2}$.

9. Multiply $\dfrac{4}{x(1 - x)} = \dfrac{A}{x} + \dfrac{B}{1 - x}$ by $x(1 - x) \Rightarrow 4 = A(1 - x) + B(x)$.

   Let $x = 0 \Rightarrow 4 = A(1) \Rightarrow A = 4$. Let $x = 1 \Rightarrow 4 = B(1) \Rightarrow B = 4$.

   The expression can be written $\dfrac{4}{x} + \dfrac{4}{1 - x}$.

11. Multiply $\dfrac{4x^2 - x - 15}{x(x + 1)(x - 1)} = \dfrac{A}{x} + \dfrac{B}{x + 1} + \dfrac{C}{x - 1}$ by $x(x + 1)(x - 1) \Rightarrow$

    $4x^2 - x - 15 = A(x - 1)(x + 1) + B(x)(x - 1) + C(x)(x + 1)$. Let $x = 0 \Rightarrow -15 = A(-1) \Rightarrow$

    $A = 15$. Let $x = 1 \Rightarrow -12 = C(1)(2) \Rightarrow C = -6$. Let $x = -1 \Rightarrow -10 = B(-1)(-2) \Rightarrow B = -5$.

    The expression can be written $\dfrac{15}{x} + \dfrac{-5}{x + 1} + \dfrac{-6}{x - 1}$.

13. By long division $\dfrac{x^2}{x^2 + 2x + 1} = 1 + \dfrac{-2x - 1}{(x + 1)^2}$.

Multiply $\dfrac{-2x - 1}{(x + 1)^2} = \dfrac{A}{x + 1} + \dfrac{B}{(x + 1)^2}$ by $(x + 1)^2 \Rightarrow -2x - 1 = A(x + 1) + B$.

Let $x = -1 \Rightarrow 1 = B$. Let $x = 0$ with $B = 1 \Rightarrow -1 = A + 1 \Rightarrow A = -2$.

The expression can be written $1 + \dfrac{-2}{x + 1} + \dfrac{1}{(x + 1)^2}$.

15. By long division $\dfrac{2x^5 + 3x^4 - 3x^3 - 2x^2 + x}{2x^2 + 5x + 2} = x^3 - x^2 + \dfrac{x}{2x^2 + 5x + 2} = x^3 - x^2 + \dfrac{x}{(2x + 1)(x + 2)}$.

Multiply $\dfrac{x}{(2x + 1)(x + 2)} = \dfrac{A}{2x + 1} + \dfrac{B}{x + 2}$ by $(2x + 1)(x + 2) \Rightarrow x = A(x + 2) + B(2x + 1)$.

Let $x = -\dfrac{1}{2} \Rightarrow -\dfrac{1}{2} = A\left(\dfrac{3}{2}\right) \Rightarrow A = -\dfrac{1}{3}$. Let $x = -2 \Rightarrow -2 = B(-3) \Rightarrow B = \dfrac{2}{3}$.

The expression can be written $x^3 - x^2 + \dfrac{-1}{3(2x + 1)} + \dfrac{2}{3(x + 2)}$.

17. By long division $\dfrac{x^3 + 4}{9x^3 - 4x} = \dfrac{1}{9} + \dfrac{\frac{4}{9}x + 4}{9x^3 - 4x} = \dfrac{1}{9} + \dfrac{\frac{4}{9}x + 4}{x(3x + 2)(3x - 2)}$.

Multiply $\dfrac{\frac{4}{9}x + 4}{x(3x + 2)(3x - 2)} = \dfrac{A}{x} + \dfrac{B}{3x + 2} + \dfrac{C}{3x - 2}$ by $x(3x + 2)(3x - 2) \Rightarrow$

$\dfrac{4}{9}x + 4 = A(3x + 2)(3x - 2) + B(x)(3x - 2) + C(x)(3x + 2)$. Let $x = 0 \Rightarrow 4 = A(-4) \Rightarrow A = -1$.

Let $x = -\dfrac{2}{3} \Rightarrow -\dfrac{8}{27} + 4 = B\left(-\dfrac{2}{3}\right)(-4) \Rightarrow \dfrac{100}{27} = \dfrac{8}{3}B \Rightarrow B = \dfrac{25}{18}$.

Let $x = \dfrac{2}{3} \Rightarrow \dfrac{8}{27} + 4 = C\left(\dfrac{2}{3}\right)(4) \Rightarrow \dfrac{116}{27} = \dfrac{8}{3}C \Rightarrow C = \dfrac{29}{18}$.

The expression can be written $\dfrac{1}{9} + \dfrac{-1}{x} + \dfrac{25}{18(3x + 2)} + \dfrac{29}{18(3x - 2)}$.

19. Multiply $\dfrac{-3}{x^2(x^2 + 5)} = \dfrac{A}{x} + \dfrac{B}{x^2} + \dfrac{Cx + D}{x^2 + 5}$ by $x^2(x^2 + 5) \Rightarrow$

$-3 = A(x)(x^2 + 5) + B(x^2 + 5) + (Cx + D)(x^2) \Rightarrow -3 = Ax^3 + 5Ax + Bx^2 + 5B + Cx^3 + Dx^2$.

Equate coefficients. For x^3: $0 = A + C$.

For x^2: $0 = B + D$. For x: $0 = 5A \Rightarrow A = 0$. For the constants: $-3 = 5B \Rightarrow B = -\dfrac{3}{5}$.

Substitute $A = 0$ in the first equation. $C = 0$. Substitute $B = -\dfrac{3}{5}$ in the second equation. $D = \dfrac{3}{5}$.

The expression can be written $\dfrac{-3}{5x^2} + \dfrac{3}{5(x^2 + 5)}$.

21. Multiply $\dfrac{3x - 2}{(x + 4)(3x^2 + 1)} = \dfrac{A}{x + 4} + \dfrac{Bx + C}{3x^2 + 1}$ by $(x + 4)(3x^2 + 1) \Rightarrow$

$3x - 2 = A(3x^2 + 1) + (Bx + C)(x + 4) \Rightarrow 3x - 2 = 3Ax^2 + A + Bx^2 + 4Bx + Cx + 4C.$

Let $x = -4 \Rightarrow -14 = 49A \Rightarrow A = -\dfrac{2}{7}$. Equate coefficients.

For x^2: $0 = 3A + B \Rightarrow 0 = -\dfrac{6}{7} + B \Rightarrow B = \dfrac{6}{7}$. For x: $3 = 4B + C \Rightarrow 3 = \dfrac{24}{7} + C \Rightarrow C = -\dfrac{3}{7}$.

The expression can be written $\dfrac{-2}{7(x + 4)} + \dfrac{6x - 3}{7(3x^2 + 1)}$.

23. Multiply $\dfrac{1}{x(2x + 1)(3x^2 + 4)} = \dfrac{A}{x} + \dfrac{B}{2x + 1} + \dfrac{Cx + D}{3x^2 + 4}$ by $x(2x + 1)(3x^2 + 4) \Rightarrow$

$1 = A(2x + 1)(3x^2 + 4) + B(x)(3x^2 + 4) + (Cx + D)(x)(2x + 1).$

Let $x = 0 \Rightarrow 1 = A(1)(4) \Rightarrow A = \dfrac{1}{4}$. Let $x = -\dfrac{1}{2} \Rightarrow 1 = B\left(-\dfrac{1}{2}\right)\left(\dfrac{19}{4}\right) \Rightarrow B = -\dfrac{8}{19}$.

Multiply the right side out. $1 = A(6x^3 + 3x^2 + 8x + 4) + 3Bx^3 + 4Bx + 2Cx^3 + Cx^2 + 2Dx^2 + Dx \Rightarrow$

$1 = 6Ax^3 + 3Ax^2 + 8Ax + 4A + 3Bx^3 + 4Bx + 2Cx^3 + Cx^2 + 2Dx^2 + Dx.$ Equate coefficients.

For x^3: $0 = 6A + 3B + 2C \Rightarrow 0 = 6\left(\dfrac{1}{4}\right) + 3\left(-\dfrac{8}{19}\right) + 2C \Rightarrow 0 = \dfrac{9}{38} + 2C \Rightarrow C = -\dfrac{9}{76}$.

For x^2: $0 = 3A + C + 2D \Rightarrow 0 = \dfrac{3}{4} - \dfrac{9}{76} + 2D \Rightarrow 0 = \dfrac{48}{76} + 2D \Rightarrow D = -\dfrac{24}{76}$.

The expression can be written $\dfrac{1}{4x} + \dfrac{-8}{19(2x + 1)} + \dfrac{-9x - 24}{76(3x^2 + 4)}$.

25. Multiply $\dfrac{3x - 1}{x(2x^2 + 1)^2} = \dfrac{A}{x} + \dfrac{Bx + C}{2x^2 + 1} + \dfrac{Dx + E}{(2x^2 + 1)^2}$ by $x(2x^2 + 1)^2 \Rightarrow$

$3x - 1 = A(2x^2 + 1)^2 + (Bx + C)(x)(2x^2 + 1) + (Dx + E)(x).$

Let $x = 0 \Rightarrow -1 = A(1) \Rightarrow A = -1$. Multiply the right side out.

$3x - 1 = A(4x^4 + 4x^2 + 1) + 2Bx^4 + Bx^2 + Cx + 2Cx^3 + Dx^2 + Ex \Rightarrow$

$3x - 1 = 4Ax^4 + 4Ax^2 + A + 2Bx^4 + Bx^2 + Cx + 2Cx^3 + Dx^2 + Ex.$

Equate coefficients. For x^4: $0 = 4A + 2B \Rightarrow 0 = -4 + 2B \Rightarrow B = 2$. For x^3: $0 = 2C \Rightarrow C = 0$.

For x^2: $0 = 4A + B + D \Rightarrow 0 = -4 + 2 + D \Rightarrow D = 2$. For x: $3 = C + E \Rightarrow 3 = 0 + E \Rightarrow$

$E = 3$. The expression can be written $\dfrac{-1}{x} + \dfrac{2x}{2x^2 + 1} + \dfrac{2x + 3}{(2x^2 + 1)^2}$.

27. Multiply $\dfrac{-x^4 - 8x^2 + 3x - 10}{(x+2)(x^2+4)^2} = \dfrac{A}{x+2} + \dfrac{Bx+C}{x^2+4} + \dfrac{Dx+E}{(x^2+4)^2}$ by $(x+2)(x^2+4)^2 \Rightarrow$

$-x^4 - 8x^2 + 3x - 10 = A(x^2+4)^2 + (Bx+C)(x+2)(x^2+4) + (Dx+E)(x+2)$.

Let $x = -2 \Rightarrow -64 = A(64) \Rightarrow A = -1$. Multiply the right side out.

$-x^4 - 8x^2 + 3x - 10 = Ax^4 + 8Ax^2 + 16A + Bx^4 + 2Bx^3 + 4Bx^2 + 8Bx + Cx^3 + 2Cx^2 +$.

$4Cx + 8C + Dx^2 + 2Dx + Ex + 2E$.

Equate coefficients.

For x^4: $-1 = A + B \Rightarrow -1 = -1 + B \Rightarrow B = 0$. For x^3: $0 = 2B + C \Rightarrow 0 = 0 + C \Rightarrow C = 0$.

For x^2: $-8 = 8A + 4B + 2C + D \Rightarrow -8 = -8 + 0 + 0 + D \Rightarrow D = 0$.

For x: $3 = 8B + 4C + 2D + E \Rightarrow 3 = 0 + 0 + 0 + E \Rightarrow E = 3$.

The expression can be written $\dfrac{-1}{x+2} + \dfrac{3}{(x^2+4)^2}$.

29. By long division $\dfrac{5x^5 + 10x^4 - 15x^3 + 4x^2 + 13x - 9}{x^3 + 2x^2 - 3x} = 5x^2 + \dfrac{4x^2 + 13x - 9}{x^3 + 2x^2 - 3x} = 5x^2 + \dfrac{4x^2 + 13x - 9}{x(x+3)(x-1)}$.

Multiply $\dfrac{4x^2 + 13x - 9}{x(x+3)(x-1)} = \dfrac{A}{x} + \dfrac{B}{x+3} + \dfrac{C}{x-1}$ by $x(x+3)(x-1) \Rightarrow$

$4x^2 + 13x - 9 = A(x+3)(x-1) + B(x)(x-1) + C(x)(x+3)$ Let $x = 0 \Rightarrow -9 = A(-3) \Rightarrow$

$A = 3$. Let $x = -3 \Rightarrow -12 = B(-3)(-4) \Rightarrow B = -1$. Let $x = 1 \Rightarrow 8 = C(4) \Rightarrow C = 2$.

The expression can be written $5x^2 + \dfrac{3}{x} + \dfrac{-1}{x+3} + \dfrac{2}{x-1}$.

Appendix D: Rotation of Axes

1. $4x^2 + 3y^2 + 2xy - 5x = 8 \Rightarrow B^2 - 4AC = 2^2 - 4(4)(3) = 4 - 48 < 0.$ Circle, ellipse, or a point.

3. $2x^2 + 3xy - 4y^2 = 0 \Rightarrow B^2 - 4AC = 3^2 - 4(2)(-4) = 9 + 32 > 0.$ Hyperbola or 2 intersecting lines.

5. $4x^2 + 4xy + y^2 + 15 = 0 \Rightarrow B^2 - 4AC = 4^2 - 4(4)(1) = 16 - 16 = 0.$ Parabola, one line, or 2 parallel lines.

7. $2x^2 + \sqrt{3}xy + y^2 + x = 5 \Rightarrow \cot 2\theta = \dfrac{A - C}{B} = \dfrac{2 - 1}{\sqrt{3}} \Rightarrow \cot 2\theta = \dfrac{1}{\sqrt{3}} \Rightarrow 2\theta = 60° \Rightarrow \theta = 30°$

9. $3x^2 + \sqrt{3}xy + 4y^2 + 2x - 3y = 12 \Rightarrow \cot 2\theta = \dfrac{A - C}{B} = \dfrac{3 - 4}{\sqrt{3}} \Rightarrow \cot 2\theta = -\dfrac{1}{\sqrt{3}} \Rightarrow$
   $2\theta = 120° \Rightarrow \theta = 60°$

11. $x^2 - 4xy + 5y^2 = 18 \Rightarrow \cot 2\theta = \dfrac{A - C}{B} = \dfrac{1 - 5}{-4} = 1 \Rightarrow \cot 2\theta = 1 \Rightarrow 2\theta = 45° \Rightarrow \theta = 22.5°$

13. $x^2 - xy + y^2 = 6$ [1]; $\theta = 45°$; $x = x' \cos\theta - y' \sin\theta = \dfrac{\sqrt{2}}{2}x' - \dfrac{\sqrt{2}}{2}y'$ [2];

    $y = x' \sin\theta + y' \cos\theta = \dfrac{\sqrt{2}}{2}x' + \dfrac{\sqrt{2}}{2}y'$ [3]. Substitute [2] and [3] in [1].

    $\left(\dfrac{\sqrt{2}}{2}x' - \dfrac{\sqrt{2}}{2}y'\right)^2 - \left(\dfrac{\sqrt{2}}{2}x' - \dfrac{\sqrt{2}}{2}y'\right)\left(\dfrac{\sqrt{2}}{2}x' + \dfrac{\sqrt{2}}{2}y'\right) + \left(\dfrac{\sqrt{2}}{2}x' + \dfrac{\sqrt{2}}{2}y'\right)^2 = 6 \Rightarrow$

    $\dfrac{1}{2}x'^2 - x'y' + \dfrac{1}{2}y'^2 - \dfrac{1}{2}x'^2 + \dfrac{1}{2}y'^2 + \dfrac{1}{2}x'^2 + x'y' + \dfrac{1}{2}y'^2 = 6 \Rightarrow \dfrac{1}{2}x'^2 + \dfrac{3}{2}y'^2 = 6 \Rightarrow$

    $\dfrac{x'^2}{12} + \dfrac{y'^2}{4} = 1.$ See Figure 13.

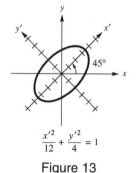

$\dfrac{x'^2}{12} + \dfrac{y'^2}{4} = 1$

Figure 13

15. $8x^2 - 4xy + 5y^2 = 36$ [1]; $\sin\theta = \dfrac{2}{\sqrt{5}}, y = 2, r = \sqrt{5}, x = \sqrt{5-4} = 1 \Rightarrow \cos\theta = \dfrac{1}{\sqrt{5}}$;

$x = x'\cos\theta - y'\sin\theta = \dfrac{1}{\sqrt{5}}x' - \dfrac{2}{\sqrt{5}}y'$ [2]; $y = x'\sin\theta + y'\cos\theta = \dfrac{2}{\sqrt{5}}x' + \dfrac{1}{\sqrt{5}}y'$ [3].

Substitute [2] and [3] in [1].

$8\left(\dfrac{1}{\sqrt{5}}x' - \dfrac{2}{\sqrt{5}}y'\right)^2 - 4\left(\dfrac{1}{\sqrt{5}}x' - \dfrac{2}{\sqrt{5}}y'\right)\left(\dfrac{2}{\sqrt{5}}x' + \dfrac{1}{\sqrt{5}}y'\right) + 5\left(\dfrac{2}{\sqrt{5}}x' + \dfrac{1}{\sqrt{5}}y'\right)^2 = 36 \Rightarrow$

$8\left(\dfrac{1}{5}x'^2 - \dfrac{4}{5}x'y' + \dfrac{4}{5}y'^2\right) - 4\left(\dfrac{2}{5}x'^2 - \dfrac{3}{5}x'y' - \dfrac{2}{5}y'^2\right) + 5\left(\dfrac{4}{5}x'^2 + \dfrac{4}{5}x'y' + \dfrac{1}{5}y'^2\right) = 36 \Rightarrow$

$\dfrac{8}{5}x'^2 - \dfrac{32}{5}x'y' + \dfrac{32}{5}y'^2 - \dfrac{8}{5}x'^2 + \dfrac{12}{5}x'y' + \dfrac{8}{5}y'^2 + 4x'^2 + 4x'y' + y'^2 = 36 \Rightarrow$

$4x'^2 + 9y'^2 = 36 \Rightarrow \dfrac{x'^2}{9} + \dfrac{y'^2}{4} = 1$. See Figure 15.

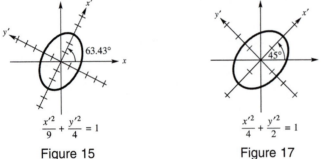

$\dfrac{x'^2}{9} + \dfrac{y'^2}{4} = 1$ $\dfrac{x'^2}{4} + \dfrac{y'^2}{2} = 1$

Figure 15   Figure 17

17. $3x^2 - 2xy + 3y^2 = 8$ [1]; $\cot 2\theta = \dfrac{A-C}{B} = \dfrac{3-3}{-2} = 0 \Rightarrow 2\theta = 90° \Rightarrow \theta = 45°$;

$x = x'\cos\theta - y'\sin\theta = \dfrac{\sqrt{2}}{2}x' - \dfrac{\sqrt{2}}{2}y'$ [2]; $y = x'\sin\theta + y'\cos\theta = \dfrac{\sqrt{2}}{2}x' + \dfrac{\sqrt{2}}{2}y'$ [3].

Substitute [2] and [3] in [1].

$3\left(\dfrac{\sqrt{2}}{2}x' - \dfrac{\sqrt{2}}{2}y'\right)^2 - 2\left(\dfrac{\sqrt{2}}{2}x' - \dfrac{\sqrt{2}}{2}y'\right)\left(\dfrac{\sqrt{2}}{2}x' + \dfrac{\sqrt{2}}{2}y'\right) + 3\left(\dfrac{\sqrt{2}}{2}x' + \dfrac{\sqrt{2}}{2}y'\right)^2 = 8 \Rightarrow$

$3\left(\dfrac{1}{2}x'^2 - x'y' + \dfrac{1}{2}y'^2\right) - 2\left(\dfrac{1}{2}x'^2 - \dfrac{1}{2}y'^2\right) + 3\left(\dfrac{1}{2}x'^2 + x'y' + \dfrac{1}{2}y'^2\right) = 8 \Rightarrow$

$\dfrac{3}{2}x'^2 - 3x'y' + \dfrac{3}{2}y'^2 - x'^2 + y'^2 + \dfrac{3}{2}x'^2 + 3x'y' + \dfrac{3}{2}y'^2 = 8 \Rightarrow 2x'^2 + 4y'^2 = 8 \Rightarrow \dfrac{x'^2}{4} + \dfrac{y'^2}{2} = 1$

See Figure 17.

19. $x^2 - 4xy + y^2 = -5$ [1]; $\cot 2\theta = \dfrac{A - C}{B} = \dfrac{1 - 1}{-4} = 0 \Rightarrow 2\theta = 90° \Rightarrow \theta = 45°$;

$x = x' \cos \theta - y' \sin \theta = \dfrac{\sqrt{2}}{2} x' - \dfrac{\sqrt{2}}{2} y'$ [2]; $y = x' \sin \theta + y' \cos \theta = \dfrac{\sqrt{2}}{2} x' + \dfrac{\sqrt{2}}{2} y'$ [3].

Substitute [2] and [3] in [1].

$\left(\dfrac{\sqrt{2}}{2} x' - \dfrac{\sqrt{2}}{2} y'\right)^2 - 4\left(\dfrac{\sqrt{2}}{2} x' - \dfrac{\sqrt{2}}{2} y'\right)\left(\dfrac{\sqrt{2}}{2} x' + \dfrac{\sqrt{2}}{2} y'\right) + \left(\dfrac{\sqrt{2}}{2} x' + \dfrac{\sqrt{2}}{2} y'\right)^2 = -5 \Rightarrow$

$\dfrac{1}{2} x'^2 - x'y' + \dfrac{1}{2} y'^2 - 4\left(\dfrac{1}{2} x'^2 - \dfrac{1}{2} y'^2\right) + \dfrac{1}{2} x'^2 + x'y' + \dfrac{1}{2} y'^2 = -5 \Rightarrow$

$\dfrac{1}{2} x'^2 - x'y' + \dfrac{1}{2} y'^2 - 2x'^2 + 2y'^2 + \dfrac{1}{2} x'^2 + x'y' + \dfrac{1}{2} y'^2 = -5 \Rightarrow -x'^2 + 3y'^2 = -5 \Rightarrow$

$\dfrac{x'^2}{5} - \dfrac{3y'^2}{5} = 1.$ See Figure 19.

$\dfrac{x'^2}{5} - \dfrac{3y'^2}{5} = 1$

Figure 19

$\dfrac{x'^2}{4} + \dfrac{y'^2}{16} = 1$

Figure 21

21. $7x^2 + 6\sqrt{3}xy + 13y^2 = 64$ [1]; $\cot 2\theta = \dfrac{A - C}{B} = \dfrac{7 - 13}{6\sqrt{3}} = \dfrac{-6}{6\sqrt{3}} = -\dfrac{1}{\sqrt{3}} \Rightarrow 2\theta = 120° \Rightarrow \theta = 60°$;

$x = x' \cos \theta - y' \sin \theta = \dfrac{1}{2} x' - \dfrac{\sqrt{3}}{2} y'$ [2]; $y = x' \sin \theta + y' \cos \theta = \dfrac{\sqrt{3}}{2} x' + \dfrac{1}{2} y'$ [3].

Substitute [2] and [3] in [1].

$7\left(\dfrac{1}{2} x' - \dfrac{\sqrt{3}}{2} y'\right)^2 + 6\sqrt{3}\left(\dfrac{1}{2} x' - \dfrac{\sqrt{3}}{2} y'\right)\left(\dfrac{\sqrt{3}}{2} x' + \dfrac{1}{2} y'\right) + 13\left(\dfrac{\sqrt{3}}{2} x' + \dfrac{1}{2} y'\right)^2 = 64 \Rightarrow$

$7\left(\dfrac{1}{4} x'^2 - \dfrac{\sqrt{3}}{2} x'y' + \dfrac{3}{4} y'^2\right) + 6\sqrt{3}\left(\dfrac{\sqrt{3}}{4} x'^2 - \dfrac{1}{2} x'y' - \dfrac{\sqrt{3}}{4} y'^2\right) +$

$13\left(\dfrac{3}{4} x'^2 + \dfrac{\sqrt{3}}{2} x'y' + \dfrac{1}{4} y'^2\right) = 64 \Rightarrow$

$\dfrac{7}{4} x'^2 - \dfrac{7\sqrt{3}}{2} x'y' + \dfrac{21}{4} y'^2 + \dfrac{18}{4} x'^2 - \dfrac{6\sqrt{3}}{2} x'y' - \dfrac{18}{4} y'^2 + \dfrac{39}{4} x'^2 + \dfrac{13\sqrt{3}}{2} x'y' + \dfrac{13}{4} y'^2 = 64 \Rightarrow$

$16x'^2 + 4y'^2 = 64 \Rightarrow \dfrac{x'^2}{4} + \dfrac{y'^2}{16} = 1.$ See Figure 21.

23. $3x^2 - 2\sqrt{3}xy + y^2 - 2x - 2\sqrt{3}y = 0$ [1]; $\cot 2\theta = \dfrac{A - C}{B} = \dfrac{3 - 1}{-2\sqrt{3}} = -\dfrac{1}{\sqrt{3}} \Rightarrow$

$2\theta = 120° \Rightarrow \theta = 60°$;

$x = x' \cos \theta - y' \sin \theta = \dfrac{1}{2}x' - \dfrac{\sqrt{3}}{2}y'$ [2]; $y = x' \sin \theta + y' \cos \theta = \dfrac{\sqrt{3}}{2}x' + \dfrac{1}{2}y'$ [3].

Substitute [2] and [3] in [1]. $3\left(\dfrac{1}{2}x' - \dfrac{\sqrt{3}}{2}y'\right)^2 - 2\sqrt{3}\left(\dfrac{1}{2}x' - \dfrac{\sqrt{3}}{2}y'\right)\left(\dfrac{\sqrt{3}}{2}x' + \dfrac{1}{2}y'\right) +$

$\left(\dfrac{\sqrt{3}}{2}x' + \dfrac{1}{2}y'\right)^2 - 2\left(\dfrac{1}{2}x' - \dfrac{\sqrt{3}}{2}y'\right) - 2\sqrt{3}\left(\dfrac{\sqrt{3}}{2}x' + \dfrac{1}{2}y'\right) = 0 \Rightarrow 3\left(\dfrac{1}{4}x'^2 - \dfrac{\sqrt{3}}{2}x'y' + \dfrac{3}{4}y'^2\right) -$

$2\sqrt{3}\left(\dfrac{\sqrt{3}}{4}x'^2 - \dfrac{1}{2}x'y' - \dfrac{\sqrt{3}}{4}y'^2\right) + \left(\dfrac{3}{4}x'^2 + \dfrac{\sqrt{3}}{2}x'y' + \dfrac{1}{4}y'^2\right) - x' + \sqrt{3}y' - 3x' - \sqrt{3}y' = 0 \Rightarrow$

$\dfrac{3}{4}x'^2 - \dfrac{3\sqrt{3}}{2}x'y' + \dfrac{9}{4}y'^2 - \dfrac{3}{2}x'^2 + \sqrt{3}x'y' + \dfrac{3}{2}y'^2 + \dfrac{3}{4}x'^2 + \dfrac{\sqrt{3}}{2}x'y' + \dfrac{1}{4}y'^2 - 4x' = 0 \Rightarrow$

$4y'^2 - 4x' = 0 \Rightarrow 4y'^2 = 4x' \Rightarrow y'^2 = x'$. See Figure 23.

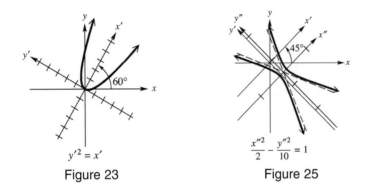

Figure 23   Figure 25

25. $x^2 + 3xy + y^2 - 5\sqrt{2}y = 15$ [1]; $\cot 2\theta = \dfrac{A - C}{B} = \dfrac{1 - 1}{3} = 0 \Rightarrow 2\theta = 90° \Rightarrow \theta = 45°$;

$x = x' \cos \theta - y' \sin \theta = \dfrac{\sqrt{2}}{2}x' - \dfrac{\sqrt{2}}{2}y'$ [2]; $y = x' \sin \theta + y' \cos \theta = \dfrac{\sqrt{2}}{2}x' + \dfrac{\sqrt{2}}{2}y'$ [3].

Substitute [2] and [3] in [1]. $\left(\dfrac{\sqrt{2}}{2}x' - \dfrac{\sqrt{2}}{2}y'\right)^2 + 3\left(\dfrac{\sqrt{2}}{2}x' - \dfrac{\sqrt{2}}{2}y'\right)\left(\dfrac{\sqrt{2}}{2}x' + \dfrac{\sqrt{2}}{2}y'\right) +$

$\left(\dfrac{\sqrt{2}}{2}x' + \dfrac{\sqrt{2}}{2}y'\right)^2 - 5\sqrt{2}\left(\dfrac{1}{2}x' - \dfrac{\sqrt{2}}{2}y'\right) = 15 \Rightarrow$

$\dfrac{1}{2}x'^2 - x'y' + \dfrac{1}{2}y'^2 + 3\left(\dfrac{1}{2}x'^2 - \dfrac{1}{2}y'^2\right) + \dfrac{1}{2}x'^2 + x'y' + \dfrac{1}{2}y'^2 - 5x' - 5y' = 15 \Rightarrow$

$\dfrac{1}{2}x'^2 - x'y' + \dfrac{1}{2}y'^2 + \dfrac{3}{2}x'^2 - \dfrac{3}{2}y'^2 + \dfrac{1}{2}x'^2 + x'y' + \dfrac{1}{2}y'^2 - 5x' - 5y' = 15 \Rightarrow$

$\dfrac{5}{2}x'^2 - \dfrac{1}{2}y'^2 - 5x' - 5y' = 15 \Rightarrow 5x'^2 - 10x' - y'^2 - 10y' = 30 \Rightarrow$

$5(x'^2 - 2x' + 1) - (y'^2 + 10y' + 25) = 30 + 5 - 25 \Rightarrow 5(x' - 1)^2 - (y' + 5)^2 = 10 \Rightarrow$

$\dfrac{(x' - 1)^2}{2} - \dfrac{(y' + 5)^2}{10} = 1$. See Figure 25.

The graph of the equation is a hyperbola with its center at $(1, -5)$. By translating the axes of the $x'y'$-system down 5 units and right 1 unit, we get an $x''y''$-coordinate system, in which the hyperbola is centered at the origin. Thus $\dfrac{x''^2}{2} - \dfrac{y''^2}{10} = 1$.

27. $4x^2 + 4xy + y^2 - 24x + 38y - 19 = 0$ [1]; $\cot 2\theta = \dfrac{A - C}{B} = \dfrac{4 - 1}{4} = \dfrac{3}{4} \Rightarrow$

$2\theta \approx 53.15° \Rightarrow \theta \approx 26.57°$; For 2θ; $x = 3, y = 4, r = \sqrt{3^2 + 4^2} = 5 \Rightarrow \cos 2\theta = \dfrac{3}{5}$.

$\sin \theta = \sqrt{\dfrac{1 - \cos 2\theta}{2}} = \sqrt{\dfrac{1 - \frac{3}{5}}{2}} = \sqrt{\dfrac{2}{10}} = \dfrac{\sqrt{5}}{5}$; $\cos \theta = \sqrt{\dfrac{1 + \cos 2\theta}{2}} = \sqrt{\dfrac{1 + \frac{3}{5}}{2}} = \sqrt{\dfrac{8}{10}} = \dfrac{2\sqrt{5}}{5}$;

$x = x' \cos \theta - y' \sin \theta = \dfrac{2\sqrt{5}}{5}x' - \dfrac{\sqrt{5}}{5}y'$ [2]; $y = x' \sin \theta + y' \cos \theta = \dfrac{\sqrt{5}}{5}x' + \dfrac{2\sqrt{5}}{5}y'$ [3].

Substitute [2] and [3] in [1]. $4\left(\dfrac{2\sqrt{5}}{5}x' - \dfrac{\sqrt{5}}{5}y'\right)^2 + 4\left(\dfrac{2\sqrt{5}}{5}x' - \dfrac{\sqrt{5}}{5}y'\right)\left(\dfrac{\sqrt{5}}{5}x' + \dfrac{2\sqrt{5}}{5}y'\right) -$

$\left(\dfrac{\sqrt{5}}{5}x' + \dfrac{2\sqrt{5}}{5}y'\right)^2 - 24\left(\dfrac{2\sqrt{5}}{5}x' - \dfrac{\sqrt{5}}{5}y'\right) + 38\left(\dfrac{\sqrt{5}}{5}x' + \dfrac{2\sqrt{5}}{5}y'\right) = 19 \Rightarrow$

$4\left(\dfrac{4}{5}x'^2 - \dfrac{4}{5}x'y' + \dfrac{1}{5}y'^2\right) + 4\left(\dfrac{2}{5}x'^2 + \dfrac{3}{5}x'y' - \dfrac{2}{5}y'^2\right) + \left(\dfrac{1}{5}x'^2 + \dfrac{4}{5}x'y' + \dfrac{4}{5}y'^2\right) -$

$\dfrac{48\sqrt{5}}{5}x' + \dfrac{25\sqrt{5}}{5}y' + \dfrac{38\sqrt{5}}{5}x' + \dfrac{76\sqrt{5}}{5}y' = 19 \Rightarrow$

$\dfrac{16}{5}x'^2 - \dfrac{16}{5}x'y' + \dfrac{4}{5}y'^2 + \dfrac{8}{5}x'^2 + \dfrac{12}{5}x'y' - \dfrac{8}{5}y'^2 + \dfrac{1}{5}x'^2 + \dfrac{4}{5}x'y' + \dfrac{4}{5}y'^2 - \dfrac{48\sqrt{5}}{5}x' + \dfrac{24\sqrt{5}}{5}y' +$

$\dfrac{38\sqrt{5}}{5}x' + \dfrac{76\sqrt{5}}{5}y' = 19 \Rightarrow 5x'^2 - 2\sqrt{5}x' + 20\sqrt{5}y' = 19 \Rightarrow$

$5\left(x'^2 - \dfrac{2\sqrt{5}}{5}x' + \dfrac{1}{5}\right) + 20\sqrt{5}y' = 19 + 1 \Rightarrow 5\left(x' - \dfrac{\sqrt{5}}{5}\right)^2 = 20 - 20\sqrt{5}y \Rightarrow$

$\left(x' - \dfrac{\sqrt{5}}{5}\right)^2 = 4 - 4\sqrt{5}y' \Rightarrow \left(x' - \dfrac{\sqrt{5}}{5}\right)^2 = -4\sqrt{5}\left(y' - \dfrac{\sqrt{5}}{5}\right)$. See Figure 27.

$x''^2 \approx -8.94y''$

Figure 27

29. $16x^2 + 24xy + 9y^2 - 130x + 90y = 0$ [1]; $\cot 2\theta = \dfrac{A - C}{B} = \dfrac{16 - 9}{24} = \dfrac{7}{24} \Rightarrow$

$2\theta \approx 73.74° \Rightarrow \theta \approx 36.87°$; For 2θ; $x = 7, y = 24, r = \sqrt{7^2 + 24^2} = 25 \Rightarrow \cos 2\theta = \dfrac{7}{25}.$

$\sin \theta = \sqrt{\dfrac{1 - \cos 2\theta}{2}} = \sqrt{\dfrac{1 - \frac{7}{25}}{2}} = \sqrt{\dfrac{18}{50}} = \dfrac{3}{5}; \cos \theta = \sqrt{\dfrac{1 + \cos 2\theta}{2}} = \sqrt{\dfrac{1 + \frac{7}{25}}{2}} = \sqrt{\dfrac{32}{50}} = \dfrac{4}{5};$

$x = x' \cos \theta - y' \sin \theta = \dfrac{4}{5}x' - \dfrac{3}{5}y'$ [2]; $y = x' \sin \theta + y' \cos \theta = \dfrac{3}{5}x' + \dfrac{4}{5}y'$ [3].

Substitute [2] and [3] in [1]. $16\left(\dfrac{4}{5}x' - \dfrac{3}{5}y'\right)^2 + 24\left(\dfrac{4}{5}x' - \dfrac{3}{5}y'\right)\left(\dfrac{3}{5}x' + \dfrac{4}{5}y'\right) +$

$9\left(\dfrac{3}{5}x' + \dfrac{4}{5}y'\right)^2 - 130\left(\dfrac{4}{5}x' - \dfrac{3}{5}y'\right) + 90\left(\dfrac{3}{5}x' + \dfrac{4}{5}y'\right) = 0 \Rightarrow$

$16\left(\dfrac{16}{25}x'^2 - \dfrac{24}{25}x'y' + \dfrac{9}{25}y'^2\right) + 24\left(\dfrac{12}{25}x'^2 + \dfrac{7}{25}x'y' - \dfrac{12}{25}y'^2\right) + 9\left(\dfrac{9}{25}x'^2 + \dfrac{24}{25}x'y' + \dfrac{16}{25}y'^2\right) -$

$104x' + 78y' + 54x' + 72y' = 0 \Rightarrow \dfrac{256}{25}x'^2 - \dfrac{384}{25}x'y' + \dfrac{144}{25}y'^2 + \dfrac{288}{25}x'^2 + \dfrac{168}{25}x'y' - \dfrac{288}{25}y'^2 +$

$\dfrac{81}{25}x'^2 + \dfrac{216}{25}x'y' + \dfrac{144}{25}y'^2 - 50x' + 150y' = 0 \Rightarrow 25x'^2 - 50x' + 150y' = 0 \Rightarrow$

$25(x'^2 - 2x' + 1) = -150y' + 25 \Rightarrow 25(x' - 1)^2 = -150\left(y' - \dfrac{1}{6}\right) \Rightarrow (x' - 1)^2 = -6\left(y' - \dfrac{1}{6}\right).$

See Figure 29. The graph of the equation is a parabola with its vertex at $\left(1, \dfrac{1}{6}\right)$. By translating the axes of the

$x'y'$-system up $\dfrac{1}{6}$ units and right 1 unit, we get an $x''y''$-coordinate system, in which the parabola is centered at

the origin. Thus $x''^2 = -6y''$.

$x''^2 = -6y''$

Figure 29